Lecture Notes in Computer Science 1723

Edited by G. Goos, J. Hartmanis and J. van Leeuwen

Springer

Berlin
Heidelberg
New York
Barcelona
Hong Kong
London
Milan
Paris
Singapore
Tokyo

Robert France   Bernhard Rumpe (Eds.)

«UML»'99 –
The Unified
Modeling Language

Beyond the Standard

Second International Conference
Fort Collins, CO, USA, October 28-30, 1999
Proceedings

Springer

Series Editors

Gerhard Goos, Karlsruhe University, Germany
Juris Hartmanis, Cornell University, NY, USA
Jan van Leeuwen, Utrecht University, The Netherlands

Volume Editors

Robert France
Department of Computer Science
Colorado State University
Fort Collins, CO 80523-1873, USA
E-mail: france@cs.colostate.edu

Bernhard Rumpe
Institut für Informatik, Technische Universität München
D-80290 München, Germany
E-mail: rumpe@in.tum.de

Cataloging-in-Publication data applied for
Die Deutsche Bibliothek - CIP-Einheitsaufnahme

The unified modeling language : second international conference ; proceedings /
UML '99: Beyond the Standard, Fort Collins, CO, USA, October 28 - 30, 1999.
Robert France ; Bernhard Rumpe (ed.). - Berlin ; Heidelberg ; New York ;
Barcelona ; Hong Kong ; London ; Milan ; Paris ; Singapore ; Tokyo : Springer,
1999
 (Lecture notes in computer science ; Vol. 1723)
ISBN 3-540-66712-1

CR Subject Classification (1998): D.2, D.3

ISSN 0302-9743
ISBN 3-540-66712-1 Springer-Verlag Berlin Heidelberg New York

© Springer-Verlag Berlin Heidelberg 1999
Printed in Germany

Typesetting: Camera-ready by author
SPIN: 10705440   06/3142 – 5 4 3 2 1 0   Printed on acid-free paper

≪UML≫'99 Preface

"While in geometry attempts to square the circle never succeeded, the UML has achieved it: states can be implemented as classes." – "We have made much progress from the time clouds were used."

The Unified Modeling Language is described as a language for "specifying, visualizing, constructing, and documenting the artifacts of software systems" and for business modeling (OMG UML V1.x documents). The UML reflects some of the best experiences in object-oriented modeling, thus it has the potential to become a widely-used standard object-oriented modeling language.

As a generally-applicable standard the UML has to be both flexible (extensible, adaptable, modifiable) and precise. Flexibility is needed if the UML is to be used in a variety of application domains. Tailoring of UML syntax and adaptation of UML semantics to system domains is highly desirable. Incorporating domain-specific concepts into the language will yield modeling languages that more effectively support system development in these domains. Tailoring may involve determining a subset of the UML that is applicable to the domain, extending or modifying existing language elements, or defining new language elements. One can envisage UML variants that are tailored to specific domains, for example, UML for real-time systems, multimedia systems, and for internet-based systems. Furthermore, one can also define UML variants that determine levels of sophistication in the use of the UML. For example, one can define a "UML-Light" that utilizes basic UML concepts, a "UML-Advanced" that utilizes more advanced concepts, and a "UML-Expert" that uses concepts that require substantial experiences in the use of the UML. In this respect, one can consider the UML to be a family of languages rather than a single, coherent language.

As in the case of natural languages, one does not need to understand the full language before one can express oneself. Consequently, lightweight versions for different purposes are needed, but extensions of the UML beyond stereotypes and tagged-values wherever necessary should be considered in the future. In the fields of business modeling, timed and analogous systems, as well as architectural descriptions, enhancements will surely come, perhaps bringing new specialized kinds of diagrams into the UML.

Precision is needed if the UML is to effectively serve as a standard. A precise language supports effective communication of intent and enables the development of rigorous analysis tools. Work on developing precise semantics for the UML is the main thrust of UML research in academia. The development of a pragmatic and precise semantics for the UML requires both technical and social processes. It is imperative that the semantics support a common-sense usage of the UML in practice. It is not good enough to propose a precise semantics in a formal notation. One must also demonstrate that the proposed semantics supports commonly held views of how the UML is to be applied and that the semantics is consistent with widely-perceived successful industrial applications of the language. Furthermore, the semantics should give tool-developers useful insight to support the development of semantic analysis tools.

The flexibility and precision qualities may seem at odds with each other. Regarding UML as a family of languages suggests that there cannot be a single precise UML semantics. On the other hand, the multiple languages must have a common language core if they are to be considered UML variants and not new languages. Work on defining a precise semantics for the UML should focus on (1) identifying this core, (2) developing precise characterizations of the core concepts, and (3) developing mechanisms that can be used to extend and modify the core semantics to support the tailoring of the UML to different usages and domains.

Balancing the demands for UML extensions and adaptations with the need to consolidate and unify concepts to create a coherent standard will be a major challenge as the UML evolves. Both forces can contribute significantly to the development of the UML only if appropriately balanced. Demands for extensions and adaptations can be analyzed together to identify common concepts that can be usefully and consistently added to a UML core, but identifying common concepts and determining the consistency of new concepts with existing standard UML concepts are challenging activities.

The evolution of the UML can benefit significantly from the best experiences in other computer science communities. Experiences that can be exploited in the development of the UML include work on conceptual modeling and knowledge engineering in the Artificial Intelligence community, work on rigorous/formal software development in the Software Engineering community, work on data modeling in the Database community, and work on denotational and operational semantics, type theories, and higher-level programming languages in the Programming Language community. For example, it is conceivable that one can use a sub-language of the UML as a higher-level programming language, thus paving the way for the use of the UML as a wide-spectrum development language.

Closely linked to UML issues are questions related to how and where to use and apply it. Current interest in methodical issues and the definition of development processes reflects this awareness. Methods-in-the-Large and project management issues are rather well elaborated, and the "methods in the small" will receive far more attention in the future. We need more techniques that allow composing or refining of the various kinds of diagram types, translate between them, and trace information across diagrams. Proprietary solutions for some techniques are coded in the tools, and need scientific examination to allow further improvement.

We are waiting for the day when the (core) UML will be regarded as a semantically sound and precise language.

The objective of the ≪UML≫'99 conference is to bring together researchers and developers from academia and industry, and from a variety of computer science communities, to present and discuss works that can potentially contribute to the evolution of the UML. In particular, the ≪UML≫'99 conference aims to foster closer working relationships between researchers and developers in industry and researchers in academia. As indicated above, the successful evolution of the UML will require theoretical and industry-driven contributions. Past work on the UML provides ample evidence that concepts developed in academia can be effectively interwoven with practical experiences. The intent of the UML conferences is to enhance such interactions by providing an open forum for discussing and analyzing theoretical and practical challenges facing the development of the UML.

In keeping with the scientific orientation of ≪UML≫'99, the conference is primarily structured around paper presentations and discussion panels. The presentations and panels are targeted to an audience that is at least familiar with the basic elements of the UML, and has a significant interest in the development of the UML as a well-founded standard. In total 166 papers were submitted to the ≪UML≫'99 conference, of which 44 were selected by the programme committee for presentation. The selected papers touch upon a variety of issues and reflect numerous perspectives on how the UML should evolve. The concerns and issues mentioned above, and more, are addressed in varying degrees in the selected papers.

We would like to express our deepest appreciation to the authors of submitted papers, the programme committee members, those committee members who also acted as shepherds for some of the papers, the external referees, Ljiljana Döhring for handling the paper printing process, Adrian Bunk for setting up and handling the electronic submission process, and Matthias Rahlf for setting up the Web page for the electronic programme committee meeting. We would also like to thank the numerous people who have been involved in the organisation of ≪UML≫'99 and, in particular, the organisers of last year's conference in Mulhouse, Jean Bézivin and Pierre-Alain Muller for their helpful advice, the publicity chairs, in particular, Jean-Michel Bruel for maintaining the mailing list, the poster chair, Jim Bieman, and the conference coordinator, Kathy Krell, who kept all the pieces together and made the organisation a much smoother process. We would also like to thank the IEEE-CS conference support staff for their invaluable help.

September 1999   Robert France, Bernhard Rumpe

Organisation

≪UML≫'99 was organised by Robert France from the Department of Computer Science at Colorado State University, and by Bernhard Rumpe from the Computer Science Department at the Technische Universität München, under the auspices of IEEE Computer Society Technical Committee on Complexity in Computing, and in cooperation with ACM SIGSOFT and SIGPLAN (Association for Computing Machinery, Special Interest Group for Software Engineering, Special Interest Group on Programming Languages).

Executive Committee

Conference Chair:	Robert France	(Colorado State University, USA)
Programme Chair:	Bernhard Rumpe	(Technische Universität München,
		Munich, Germany)

Organising Team

Conference Coordinator:	Kathy Krell
Poster Chair:	Jim Bieman
Panel Chair:	Bernhard Rumpe
Publicity Chair (Europe, Africa):	Jean-Michel Bruel
Publicity Chair (Americas):	Jim Bieman
Publicity Chair (Asia, Pacific):	Junichi Suzuki

Adrian Bunk, Ljiljana Döhring, Emanuel Grant, Matthias Rahlf, and all our on-site student volunteers.

Programme Committee

Programme Committee (continued)

Ed Seidewitz (DHR Technologies, USA)
Bran Selic (ObjecTime Limited, Canada)
Richard Mark Soley (OMG, USA)
Jos Warmer (Klasse Objecten, The Netherlands)
Anthony Wasserman (Software Methods and Tools, USA)
Alan Wills (TriReme International, UK)
Rebecca Wirfs-Brock (Wirfs-Brock Associates, USA)

Additional Referees

Daniel Amyot	Joaquim Aparicio
João Araujo	Michael Breu
Jean-Michel Bruel	Christian Bunse
Luis Caires	S. Jeromy Carriere
John Cheesman	Robert G Clark
Birgit Demuth	Ralph Depke
Jawad Drissi	Mike Fischer
Falk Fünfstück	Emanuel Grant
Reiko Heckel	Martin Hitz
James Ivers	Erik Kamsties
Elisabeth Kapsammer	Ismail Khriss
Thomas Khüne	Frank-Ulrich Kumichel
Annig Lacayrelle	Oliver Laitenberger
Katharina Mehner	Paul Mukherjee
John O'Hara	Gianna Reggio
Aziz Salah	Stefan Sauer
Stephane Some	Jean Vaucher
Annika Wagner	Jörg Zettel

Sponsoring Association

 IEEE Computer Society Technical Committee on Complexity in Computing, http://www.computer.org/

Cooperating and Supporting Associations

 ACM SIGSOFT (Association for Computing Machinery, Special Interest Group for Software Engineering). http://www.acm.org/sigsoft/

 ACM SIGPLAN (Association for Computing Machinery, Special Interest Group for Programming Languages). http://www.acm.org/sigplan/

 OMG (The Object Management Group), http://www.omg.org/. UML is a trademark of OMG.

Sponsoring Company

 Rational Software Corporation, http://www.rational.com/

Sponsoring Association

 IEEE Computer Society Technical Committee on Complexity in Computing, http://www.computer.org/

Cooperating and Supporting Associations

 ACM-SIGSOFT (Association for Computing Machinery, Special Interest Group for Software Engineering, http://www.acm.org/sigsoft)

 ACM-SIGPLAN (Association for Computing Machinery, Special Interest Group for Programming Languages) http://www.acm.org/sigplan/

 OMG (The Object Management Group), http://www.omg.org/. UML is a trademark of OMG.

Sponsoring Company

RATIONAL Rational Software Corporation, http://www.rational.com/

Table of Contents

Invited Talk 1 (Abstract)

Architecting Web-Based Systems with the Unified Modeling Language 1
  Grady Booch

Software Architecture

Extending Architectural Representation in UML with View Integration ... 2
  Alexander Egyed, Nenad Medvidovic

Enabling the Refinement of a Software Architecture into a Design 17
  Marwan Abi-Antoun, Nenad Medvidovic

Using the UML for Architectural Description 32
  Rich Hilliard

UML and Other Notations

Viewing the OML as a Variant of the UML 49
  Brian Henderson-Sellers, Colin Atkinson, Don Firesmith

A Comparison of the Business Object Notation and the Unified Modeling
Language 67
  Richard F. Paige, Jonathan S. Ostroff

Formalizing the UML Class Diagram Using Object-Z 83
  Soon-Kyeong Kim, David Carrington

Formalizing Interactions

A Formal Approach to Collaborations in the Unified Modeling Language .. 99
  Gunnar Övergaard

A Formal Semantics for UML Interactions 116
  Alexander Knapp

Panel 1

UML 2.0 Architectural Crossroads: Sculpting or Mudpacking? 131
  Moderator: Chris Kobryn
  Michael Jesse Chonoles, Steve Cook, Desmond D'Souza,
  Sridhar Iyengar, Guus Ramackers

Meta-Modeling

Core Meta-Modelling Semantics of UML: The pUML Approach 140
  Andy Evans, Stuart Kent

A Metamodel for OCL 156
  Mark Richters, Martin Gogolla

Tools

Tool-Supported Compressing of UML Class Diagrams 172
  Ferenc Dósa Rácz, Kai Koskimies

A Pragmatic Approach for Building a User-Friendly and Flexible UML
Model Repository 188
  Mariano Belaunde

Components

Modeling Dynamic Software Components in UML 204
  Axel Wienberg, Florian Matthes, Marko Boger

Extending UML for Modeling Reflective Software Components 220
  Junichi Suzuki, Yoshikazu Yamamoto

UML Extension Mechanisms

Nine Suggestions for Improving UML Extensibility 236
  Nathan Dykman, Martin Griss, Robert Kessler

A Classification of Stereotypes for Object-Oriented Modeling Languages . . 249
  Stefan Berner, Martin Glinz, Stefan Joos

First-Class Extensibility for UML - Packaging of Profiles, Stereotypes,
Patterns 265
  Desmond D'Souza, Aamod Sane, Alan Birchenough

Process Modeling

UML-based Fusion Analysis 278
  Shane Sendall, Alfred Strohmeier

Using UML for Modelling the Static Part of a Software Process 292
  Xavier Franch, Josep M. Ribó

Framework for Describing UML Compatible Development Processes 308
  Pavel Hruby

Invited Talk 2

On the Behavior of Complex Object-Oriented Systems................... 324
  David Harel

Real-Time Systems

UML-RT as a Candidate for Modeling Embedded Real-Time Systems in
the Telecommunication Domain..................................... 330
  Dominikus Herzberg

Modeling Hard Real Time Systems with UML – The *OOHARTS* Approach 339
  Laila Kabous, Wolfgang Nebel

UML Based Performance Modeling Framework for Object-Oriented
Distributed Systems... 356
  Pekka Kähkipuro

Constraint Languages

Defining the Context of OCL Expressions 372
  Steve Cook, Anneke Kleppe, Richard Mitchell, Jos Warmer, Alan Wills

Mixing Visual and Textual Constraint Languages 384
  Stuart Kent, John Howse

Correct Realizations of Interface Constraints with OCL 399
  Michel Bidoit, Rolf Hennicker, Françoise Tort, Martin Wirsing

Analyzing UML Models 1

Generating Tests from UML Specifications 416
  Jeff Offutt, Aynur Abdurazik

Formalising UML State Machines for Model Checking 430
  Johan Lilius, Iván Porres Paltor

Panel 2

SDL as UML: Why and What 446
  Moderator: Bran Selic
  Philippe Dhaussy, Anders Ek, Øystein Haugen, Philippe Leblanc,
  Birger Møller-Pedersen

Coding 1

UML Behavior: Inheritance and Implementation in Current Object-Oriented
Languages 457
   Jean Louis Sourrouille

UML Collaboration Diagrams and Their Transformation to Java 473
   Gregor Engels, Roland Hücking, Stefan Sauer, Annika Wagner

Analyzing UML Models 2

Towards Three-Dimensional Representation and Animation of UML
Diagrams 489
   Martin Gogolla, Oliver Radfelder, Mark Richters

Typechecking UML Static Models 503
   Tony Clark

Precise Behavioral Modeling

Analysing UML Use Cases as Contracts 518
   Ralph-Johan Back, Luigia Petre, Iván Porres Paltor

Closing the Gap Between Object-Oriented Modeling of Structure and
Behavior 534
   Holger Giese, Jörg Graf, Guido Wirtz

Static Modeling

Black and White Diamonds 550
   Brian Henderson-Sellers, Franck Barbier

Interconnecting Objects via Contracts 566
   Luís Filipe Andrade, José Luiz Fiadeiro

How Can a Subsystem be Both a Package and a Classifier? 584
   Joaquin Miller, Rebecca Wirfs-Brock

Applying the UML

Using UML/OCL Constraints for Relational Database Design 598
   Birgit Demuth, Heinrich Hussmann

Towards a UML Extension for Hypermedia Design 614
   Hubert Baumeister, Nora Koch, Luis Mandel

Why Unified is Not Universal? – UML Shortcomings for Coping with
Round-Trip Engineering 630
   Serge Demeyer, Stéphane Ducasse, Sander Tichelaar

Sequence Diagrams

Timed Sequence Diagrams and Tool-Based Analysis – A Case Study 645
  Thomas Firley, Michaela Huhn, Karsten Diethers, Thomas Gehrke, Ursula Goltz

Timing Analysis of UML Sequence Diagrams 661
  Xuandong Li, Johan Lilius

Coding 2

The Normal Object Form: Bridging the Gap from Models to Code 675
  Christian Bunse, Colin Atkinson

Modeling Exceptional Behavior 691
  Neelam Soundarajan, Stephen Fridella

Panel 3

Advanced Methods and Tools for a Precise UML 706
  Moderator: Andy Evans
  Steve Cook, Steve Mellor, Jos Warmer, Alan Wills

Author Index 723

Sequence Diagrams

Timed Sequence Diagrams and Tool-Based Analysis – A Case Study 618
Thomas Firley, Michaela Huhn, Karsten Diethers, Thomas Gehrke, Ursula Goltz

Timing Analysis of UML Sequence Diagrams 661
Xuandong Li, Johan Lilius

Coding 2

The Normal Object Form: Bridging the Gap from Models to Code 675
Christian Bunse, Colin Atkinson

Modeling Exceptional Behavior 691
Neelam Soundarajan, Stephen Fridella

Panel 3

Advanced Methods and Tools for a Precise UML 706
Moderator: Andy Evans
Steng Cook, Steve Mellor, Joe Warmer, Alan Wills

Author Index 723

Architecting Web-Based Systems with the Unified Modeling Language

Grady Booch

Rational Software Corporation
18880 Homestead Rd.
Cupertino, CA 95014, USA

Abstract. The Web and its related technologies have made possible new means of access to and visualization of information. As such, it has shaped the structure of systems in many domains, from enterprise, mission critical e-business to distributed, embedded systems. Although experience with developing and deploying quality systems for the Web, by the Web and to the Web is a fairly recent activity, it appears that there are a relatively small number of canonical architectures that work. It is possible and desirable to codify those architectural patterns in order to offer a vehicle for controlling the development of a Web system over its lifetime, and to accelerate the development of new Web systems. The UML is well suited to representing these architectural patterns.

In this presentation, we will examine the nature of Web-centric systems, and will study the common architectural patterns that apply. We will also examine the dozen or so underlying mechanisms upon which these architectures build, and show how both these architectural patterns and design patterns can be represented in the UML. We will conclude by addressing how these representations can be used in the lifecycle, supporting the notions of executable architectures and round trip engineering.

Extending Architectural Representation in UML with View Integration

Alexander Egyed and Nenad Medvidovic

Center for Software Engineering
University of Southern California
Los Angeles, CA 90089-0781, USA
{aegyed,neno}@sunset.usc.edu

Abstract. UML has established itself as the leading OO analysis and design methodology. Recently, it has also been increasingly used as a foundation for representing numerous (diagrammatic) views that are outside the standardized set of UML views. An example are architecture description languages. The main advantages of representing other types of views in UML are 1) a common data model and 2) a common set of tools that can be used to manipulate that model. However, attempts at representing additional views in UML usually fall short of their full integration with existing views. Integration extends representation by also describing interactions among multiple views, thus capturing the inter-view relationships. Those inter-view relationships are essential to enable automated identification of consistency and conformance mismatches. This work describes a view integration framework and demonstrates how an architecture description language, which was previously only represented in UML, can now be fully integrated into UML.

1 Introduction

Software systems are characterized by unprecedented complexity. One effective means of dealing with that complexity is to consider a system from a particular perspective, or view. Views enable software developers to reduce the amount of information they have to deal with at any given time. It has been recognized that "it is not the number of details, as such, that contributes to complexity, but the number of details of which we have to be aware at the same time." [1].

A major drawback of describing systems as collections of views is that the software development process tends to become rather view centric (seeing views instead of the big picture). Such a view centric approach exhibits a fair amount of redundancy across different views as a side effect. That redundancy is the cause for inter-view mismatches, such as inconsistencies or incompletenesses. On top of that, views are used independently, concurrently, are subjected to different audiences (interpretations) and the manner in which model information is shared is extremely inconsistent. All this implies that information about a system must be captured multiple times and must be kept consistent.

To deal with this problem, a major emphasis needs to be placed on mismatch identification and reconciliation within and among views (view integration). We design

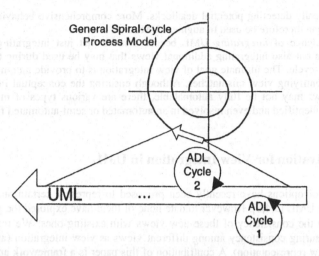

Figure 1. Substituting UML (general-purpose) with ADLs (specific)

not only because we want to build (compose) but also because we want to understand. Thus, a major focus of software development is to analyze and verify the conceptual integrity, consistency, and completeness of the model of a system.

There are numerous reasons for the lack of automated assistance in identifying view mismatches. We believe that one of the major reasons is improper integration of views on a meta-level: although both the notation and semantics of individual views may be very well defined, meta-models that integrate these different views are often inadequate or missing. The Unified Modeling Language (UML) [2] is a good example of this. UML defines a set of views (such as class diagrams, sequence diagrams, and state diagrams) and defines a meta-model for those views. However, the UML meta-model was primarily designed to deal with the issue of capturing and representing modeling elements of views in a common data model (repository).

Another problem with UML is that its views had to be designed to be generally understandable and they are therefore rather simple. The result is that UML view semantics are not well defined so as not to over-constrain the language or limit its usability. Therefore, UML becomes less suitable in domains were more precision (performance, reliability, etc.) is required. One way of addressing this deficiency is the use of additional views. Figure 1 shows an example on how a general purpose development methodology such as UML may be used together with more (domain) specific description languages such as architecture description languages (ADLs). What we mean by architecture is a course grain description of a system with high-level components, connectors, and their configuration. A design further refines the architecture by elaborating on the details of individual components/connectors as well as their interactions. In Figure 1, UML serves as a general development model, whereas more specific views can be generated by taking excursions off the main process to investigate specific concerns, e.g., deadlock detection among modeling elements. Although some UML views support behavioral design, these are inadequate in automatically detecting potential deadlocks. More comprehensive behavior models and views can therefore be used to augment UML.

The challenge of *integrating* UML becomes more that just integrating existing UML views but also integrating additional views that may be used during the development life-cycle. The ultimate goal of view integration is to provide automatic assistance in identifying view mismatches. Although ensuring the conceptual integrity of models/views may not be fully automatable, there are various types of mismatches that can be identified and even resolved in an automated or semi-automated fashion.

2   Motivation for View Integration in UML

A number of options have recently been proposed to represent certain architectural concerns in UML [3,4,5]. However to date none of them have explored the possibility of ensuring the consistency of these new views with existing ones. We refer to this issue of ensuring consistency among different views as view integration (as opposed to mere view representation). A contribution of this paper is a framework and a set of techniques for integrating existing views with newly introduced architectural views.

This section explains in a bit more detail the existing support in UML for view representation and its current limitations with regard to view integration.

2.1 Architectural Representation in UML

In UML, a number of views are captured and sufficiently represented through the use of the UML meta-model. UML views capture both structural and behavioral aspects of software development. Structural views make use of classes, packages, use cases, and so forth. Behavioral views are represented through scenarios, states, and activities.

Furthermore, UML views may be general or instantiated. Generalized views capture model information and configuration that are true during the entire life time of a model. Instantiated views, on the other hand, depict examples or possible scenarios which usually only describe a subset of the interactions a model element goes through during its life.

Although the UML meta-model could be extended to capture additional non-UML views (such as ADLs), this approach would result in a model that would become UML incompatible. UML, however, supports the need of refinement and augmentation of its specifications through three built-in extension mechanisms:

- *Constraints* place semantic restrictions on particular design elements. UML uses the Object Constraint Language (OCL) to define constraints [2].
- *Tagged values* allow new attributes to be added to particular elements of the model.
- *Stereotypes* allow groups of constraints and tagged values to be given descriptive names and applied to other model elements; the semantic effect is as if the constraints and tagged values were applied directly to those elements.

Using above mechanisms enables us to represent new concepts in UML. For instance, we could choose to incorporate Entity-Relationship models expressed via stereotyped and constrained class diagrams. We could also start representing architectural description languages (e.g. C2 [6], Wright [7], Rapide [8]) within the UML framework. The advantages of doing so are several:

- *Common Model Representation*: Modeling information of different types of views (UML and non-UML) can be physically stored in the same repository. This eliminates problems associated with distributed development information (e.g., access, loss) and their interpretation (e.g., exotic data format).
- *Reduced Toolset for Model Manipulation*: Being able to use UML elements to represent non-UML artifacts enables us to use existing UML toolsets to create those views. For instance, one diagram drawing tool can be used on different types of views.
- *Unified Way of Cross-Referencing Model Information*: Having modeling information stored at one physical location further enables us to cross-reference that information. Cross-referencing is useful for maintaining the traceability among development elements.

2.2 Deficiencies of pure View Representation

The UML extension mechanism discussed above is adequate for representing numerous diagrammatic views. The only major limitation is that existing UML modeling elements (such as classes, objects, activities, states, or packages) must be used (stereotyped and constrained) to represent new concepts. This becomes a particular problem when external concepts cannot be represented by what UML provides.

A more severe shortfall of view representation is that it comes up short in fully integrating views with each other: Although UML and its meta-model define notational and semantic aspects of individual views in detail, inter-view relationships are not captured in sufficient detail. Without these information, the (UML) model is nothing more than a collection of loosely coupled (or completely unrelated) views. Figure 2 illustrates this by showing UML views as being separate entities within a common environment (UML meta-model). Although, some views are weakly integrated (e.g. class and sequence diagrams), in general, UML views are independent.

Figure 2 shows, the lack of view integration extends beyond existing UML views to non-UML views represented in UML (e.g. ADLs). For instance, if a system is specified in some architectural fashion then its realization (in the design and implementation) must adhere to the constraints imposed by that architecture. UML view representation only limits how information can be described in UML, but does not concern itself whether that information is consistent with other parts of the model. View representation alone would allow creation of multiple views, each of which would correctly conform to its specifications; however, their combination would not build a coherent unit. We therefore speak of view integration as an extension to view representation to ensure the conceptual integrity (consistency and completeness) of the entire model *across* the boundaries of individual views.

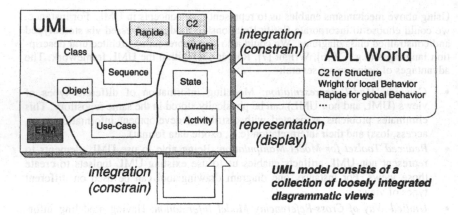

Figure 2: Views and ADLs represented in UML

2.3 Outline

The remainder of this paper will be organized as follows. First we present a simple example of the difference between view representation and view integration. We then introduce a view integration framework and its corresponding activities to describe the problem of integration. To illustrate this framework, we will describe the representation and subsequent integration of an architecture style, C2, in more detail. We conclude the paper by summarizing the key issues and solutions.

3 Example: Layered Architecture constrains UML Design

Before we demonstrate how to represent and integrate a more complex architectural style, C2, we highlight the differences between representation and integration using a simpler, well-understood, layered architectural style.

"The layered architectural [style] *helps to structure applications that can be decomposed into groups of subtasks in which each group of subtasks is at a particular level of abstraction."* [9] The layered style defines which part of a system are allowed to interact and which are not. For instance, assume that we have a trivial layered system with four layers: (1) User Interface, (2) Application Framework, (3) Network and (4) Database. The layered style defines that components within layer 1 (User Interface) may talk to other components in layer 1 as well as to components that are part of layer 2. Similarly, layer 2 components may interact among themselves, as well as with layer 1 and layer 3 components. The layered architecture, however, disallows a user interface component (layer 1) to talk directly to, say, the database in layer 4 without going through the intermediate layers 2 and 3.

In order to make use of the layered architectural style in UML, we need to represent layers in UML. An easy way of doing this is by using stereotyped UML packages. For instance, we may create a *User Interface* package with the stereotype *layer*

1, an *Application Framework* package with the stereotype *layer 2* and so forth. Next, we may use OCL to constrain the ways in which the layers (stereotyped packages) may interact. Thus, with OCL we could specify that layer 1 may depend on layer 2, layer 2 may depend on layer 3 and so forth. Note that we need not specify that layer 1 is allowed to talk to itself because that *knowledge* is already implicit in packages. Having specified how to represent layers, a UML design tool may now be used to create layered architectural diagrams.

Thus, we now have the means of *representing* an layered architectural view in UML but nothing more. Fully integrating the layered style into UML also requires that the realization of a system (i.e., its design and subsequent implementation) still conforms to the architectural style rules. Both design and implementation will make use of different types of views and, thus, we need to ensure that both are still consistent with the constraints imposed by the architecture.

For instance, if we design our system using UML class or sequence diagram(s) then the way those classes may interact is limited by both the notation of UML class diagrams and the layered style. The former constraint is usually supported by UML design tools (e.g. Rational Rose). However, the latter cannot yet be supported since we did not yet specify that relationship. To do this we need to ensure that classes are always associated with layers and that calling dependencies between those classes correspond to calling dependencies of associated layers. There are basically two types of mismatches that may happen at this stage:

- If the architecture defines some layers for which there are no associated classes in the design, then this may indicate a potential conformance mismatch.
- If the design contains a class dependency that contradicts the layer dependency in the architecture, then this may indicate a potential consistency mismatch.

4 UML Integration

To address the view mismatch problem, we have investigated ways of describing and identifying the causes of architectural mismatches across UML views. To this end, we have devised and applied a view integration framework, accompanied by a set of activities and techniques for identifying mismatches in an automated fashion; this framework is depicted in Figure 3 and described below.

The system model represents the model repository (e.g. UML model) of the designed software system. Software developers use views to add new information to the system model and to modify existing information (view synthesis). The view analysis activity interacts with the system model and the view synthesis so that new information can be validated against the existing information in the model to ensure their conceptual integrity.

This approach exploits redundancy between views. For instance, view A contains information about view B; this information can be seen as a constraint on B. The view integration framework is used to verify those constraints and, thereby, the consistency across views. Since there is more to view integration than constraints and consistency rules, our view integration framework also provides an environment where we can apply those rules in a meaningful way. Therefore, as already discussed, we see view

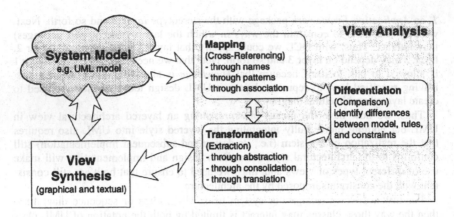

Figure 3. View Integration Framework and Activities

integration as an extension to view representation. The former extends the latter not only by rules and constraints but also by defining *what* information can be exchanged and *how* it can be exchanged. Only after the *what* and *how* have been established, can inconsistencies be identified and resolved automatically.

- Mapping: Identifies related pieces of information and thereby describes *what* information is overlapping.
- Transformation: Extracts and manipulates model elements of views in such a manner that they can be interpreted and used by other views (*how* to address information exchange).
- Differentiation: Traverses the model to identify (potential) mismatches within its elements. Mismatch identification rules can frequently be complemented by mismatch resolution rules.

It is out of the scope of this paper to deal with automated mapping and transformation in detail. Automation techniques for both are described in [10] and [11]. We will primarily focus on *Differentiation* in this paper. To illustrate the behavioral integration of an ADL later on, we will also demonstrate one *Transformation* technique.

5 C2 and UML

Section 3 highlighted the difference between view representation and integration in the case of the layered architectural style. However, this example fell short of conveying in detail how the two are actually accomplished. This section will complement that discussion by showing how the C2 architectural style [6] can be represented and then integrated into UML using the approach we propose.

5.1 Overview of C2

C2 is an architectural style intended for highly distributed software systems [6]. In a C2-style architecture, *connectors* (buses) transmit messages between components, while *components* maintain state, perform operations, and exchange messages with other components via two interfaces (named "top" and "bottom"). Each interface consists of sets of messages that may be sent and received. Inter-component messages are either *requests* for a component to perform an operation, or *notifications* that a given component has performed an operation or changed state.

A C2 component consists of two main internal parts. An *internal object* stores state and implements the operations that the component provides, while a *dialog specification* maps from messages received to operations on the internal object and from results of those operations to outgoing messages. Two components' dialogs may not directly exchange messages; they may only do so via connectors. Each component may be attached to at most one connector at the top and one at the bottom. A connector may be attached to any number of other components and connectors. Request messages may only be sent "upward" through the architecture, and notification messages may only be sent "downward."

The C2 style further demands that components communicate with each other only through message-passing, never through shared memory. Also, C2 requires that notifications sent from a component correspond to the operations of its internal object, rather than the needs of any components that receive those notifications. This constraint on notifications helps to ensure *substrate independence*, which is the ability to reuse a C2 component in architectures with differing substrate components (e.g., different window systems).

Left side of Figure 4 shows an example C2-style architecture. This system consists of four components and two connectors. One component is a *database manager*. Interacting with the *database manager* are the *database administrator*, who has direct access to the database, and the *transaction manager*, who uses the database either via

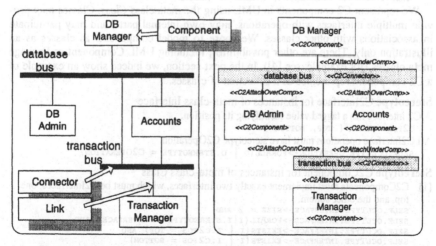

Figure 4. Simple C2 Architecture Example and its UML representation

the *Account* component (handles the transaction on a given account) or directly via the connector to connector link. This C2 diagram roughly corresponds to a layered system where the top component is the data source, the bottom components constitute the user interface, and the middle component shows the mitigating application layer. The right side of Figure 4 shows how this same C2 system can be represented in UML using a UML class diagram as a template. We chose not to modify the UML meta-model in order to stay consistent with the UML definition. Instead we adapted existing UML model elements to represent new concepts and used the UML extensibility mechanism to be able to distinguish between them. The following section will elaborate on that.

5.2 C2 Representation

In our previous work, we have begun exploring the issues in representing ADLs in UML, both using existing UML diagrams [12] and extending them via stereotypes [3]. In this section, we represent the key aspects of C2 in UML. This section is also intended to sensitize the reader to the issues inherent in representing certain (external) architectural concerns in UML.

The UML meta class *Operation* matches the C2 concept of *Message Specification*. UML Operations consist of a name and a parameter list and indicate whether they will be provided or required. To model C2 message specifications we add a tag to differentiate notifications from requests and constrain operation to have no return values. Unlike UML operations, C2 messages are all public, but that constraint is built into the UML meta-class *Interface* used below.

Stereotype C2Operation for instances of meta-class Operation
[1]   C2Operations are tagged as either notifications or requests.
        C2MSGTYPE : ENUM { NOTIFICATION, REQUEST }
[2]   C2 messages do not have return values.
        SELF.PARAMETER->FORALL(P | P.KIND <> RETURN)

We represent C2 components in UML using the meta class *Class*. Classes may provide multiple interfaces with operations, may own internal parts, and may participate in associations with other classes. We chose to model components as classes as an illustration only. There are other possibilities, including UML Component or Package meta classes (for example, see [4]). In the next section, we indeed show an example of a C2 component represented as a collection of classes.

Stereotype C2Interface for instances of meta-class Interface
A C2 interface has a tagged value identifying its position.
        c2POS : ENUM { TOP, BOTTOM }
All C2Interface operations must have stereotype C2Operation.
        SELF.OCLTYPE.OPERATION->FORALL(O | O.STEREOTYPE = C2OPERATION)

Stereotype C2Component for instances of meta-class Class
[1]   C2Components must implement exactly two interfaces, which must be C2Interfaces, one
        top, and the other bottom.
        SELF.OCLTYPE.INTERFACE->SIZE = 2 AND
        SELF.OCLTYPE.INTERFACE->FORALL(I|I.STEREOTYPE=C2INTERFACE) AND
        SELF.OCLTYPE.INTERFACE->EXISTS(I | I.c2POS = TOP) AND
        SELF.OCLTYPE.INTERFACE->EXISTS(I | I.c2POS = BOTTOM)

[2] Requests travel "upward" only, i.e., they are sent through top interfaces and received through bottom interfaces.

```
LET TOPINT = SELF.OCLTYPE.INTERFACE->SELECT(I|I.C2POS = TOP),
LET BOTINT = SELF.OCLTYPE.INTERFACE->SELECT(I|I.C2POS = BOTTOM),
TOPINT.OPERATION->FORALL(O| (O.C2MSGTYPE=REQUEST) IMPLIES O.DIR=REQUIRE)AND
BOTINT.OPERATION->FORALL(O| (O.C2MSGTYPE=REQUEST) IMPLIES O.DIR=PROVIDE)
```

[3] Notifications travel "downward" only. Similar to the constraint above.

[4] C2Components participate in at most two whole-part relationships named internalObject, and dialog.

```
LET WHOLES = SELF.OCLTYPE.ASSOCEND->SELECT(AGGREGATION = COMPOSITE),
(WHOLE->SIZE <= 4) AND
((WHOLES.ASSOCIATION.NAME->ASSET)-SET{"INTERNALOBJECT","DIALOG"})->SIZE=0
```

[5] Each operation on the internal object has a corresponding notification which is sent from the component's bottom interface.

```
LET OPS = SELF.INTERNALOBJECT.FEATURE->SELECT(F | F-
>ISKINDOF(OPERATION)),
LET BOTINT = SELF.OCLTYPE.INTERFACE->SELECT(I | I.C2POS = BOTTOM),
OPS->FORALL(OP | BOTINT->EXISTS(NOTE | (OP.NAME = NOTE.NAME AND
             OP.PARAMETER = NOTE.PARAMETER) IMPLIES
             NOT.DIR = REQUIRED AND NOTE.C2MSGTYPE=NOTIFICATION))
```

C2 connectors share many of the constraints of C2 components. One difference is that they do not have any prescribed internal structure. Components and connectors are treated differently in the architecture composition rules discussed below. Another difference is that connectors do not define their own interfaces; instead their interfaces are determined by the components that they connect. We omit the constraints specifying attachments between components and connectors in the interest of brevity.

Stereotype C2AttachOverComp for instances of meta-class Association

Stereotype C2AttachUnderComp for instances of meta-class Association.

Stereotype C2AttachConnConn for instances of meta-class Association

Stereotype C2Connector for instances of meta-class Class

[1-5] Same as constraints 1-5 on C2Component.

[6] The top interface of a connector is determined by the components and connectors attached to its bottom.

```
LET TOPINT = SELF.OCLTYPE.INTERFACE->SELECT(I | I.C2POS = TOP),
LET DOWNATTACH = SELF.OCLTYPE.ASSOCEND.ASSOCIATION->SELECT(A |
     A.ASSOCEND[2] = SELF.OCLTYPE),
LET TOPSINTSBELOW=DOWNATTACH.ASSOCEND[1].INTERFACE->SELECT(I|I.C2POS=
     TOP), TOPSINTSBELOW.OPERATION->ASSET = TOPINT.OPERATION->ASSET
```

[7] The bottom interface of a connector is determined by the components and connectors attached to its top. This is similar to the constraint above.

Finally, we specify the overall composition of components and connectors in the architecture of a system. Recall that well-formed C2 architectures consist of components and connectors, components may be attached to one connector on the top and one on the bottom, and the top (bottom) of a connector may be attached to any number of other connectors' bottoms (tops). Below, we also add two new rules that guard against degenerate cases.

Stereotype C2Architecture for instances of meta-class Model

[8] A C2 architecture is made up of only C2 model elements.

```
SELF.OCLTYPE.MODELELEMENT->FORALL(ME|ME.STEREOTYPE= C2COMPONENT OR
     ME.STEREOTYPE = C2CONNECTOR OR ME.STEREOTYPE = C2ATTACHOVERCOMP OR
ME.STEREOTYPE = C2ATTACHUNDERCOMP OR ME.STEREOTYPE = C2ATTACHCONNCONN)
```

[9] Each C2Component has at most one C2AttachOverComp.

```
LET COMPS=SELF.OCLTYPE.MODELELEMENT->SELECT(ME|ME.STEREOTYPE=C2COMPONENT),
```

```
COMPS->FORALL(C | C.ASSOCEND.ASSOCIATION->SELECT(A |
                 A.STEREOTYPE = C2ATTACHUNDERCOMP)->SIZE <= 1)
```

[10] Each C2Component has at most one C2AttachUnderComp. Similar to the constraint above.

[11] Each C2Component must be attached to some connector.
```
LET COMPS=SELF.OCLTYPE.MODELELEMENT->SELECT(ME|ME.STEREOTYPE=C2COMPONENT),
COMPS->FORALL(C | C.ASSOCEND.ASSOCIATION->SIZE > 0)
```

[12] Each C2Connector must be attached to some connector or component. Similar to the constraint above.

5.3 C2 Integration

To demonstrate the C2/UML integration we need to consider an architecture defined in the C2 style and a corresponding design in UML. Figure 5 (right side) shows a UML class diagram that realizes or refines the C2 architecture presented in Figure 4. Mapping from UML classes to C2 components/connectors is shown with dotted lines. Basically, C2 components and connectors may be seen as the interfaces for compact, self-sustaining sections of the implementation. Since C2 elements (components and connectors) are often coarse grain, it is reasonable to assume that a collection of classes is needed to implement a single C2 element.

Because of the fact that a C2 element is a black box, nothing can be said about how classes that are part of a single element are supposed to interact. However, the interaction of classes belonging to different C2 elements are constrained by the C2 style. For

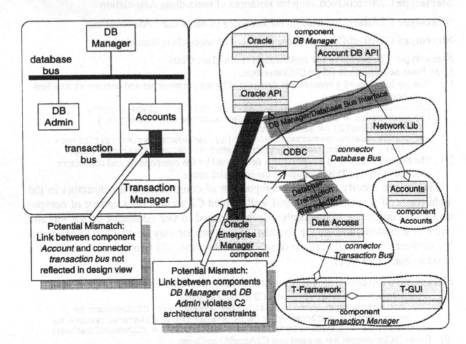

Figure 5. C2 and corresponding Class View plus Mismatches

instance, if design classes *Oracle*, *Oracle API*, and *Account DB API* correspond to the
C2 component *DB Manager*, and if the design classes *ODBC* and *Network Lib* correspond to the C2 connector *Database Bus*, then the interaction between these two
groups of classes needs to be consistent with the corresponding C2 architecture constraints. In the right half of Figure 5, the class links corresponding to the C2 compo-
nent/connector link from *DB Manager* to *Database Bus* can be seen in the shaded
area in the upper half of the diagram. The shaded area in the lower half is the class
link corresponding to the C2 connector to connector link from *Database Bus* to
Transaction Bus.

5.3.1 C2 Structural Integration

Having (manually) identified the *Mapping* from C2 to UML, we can now automatically identify mismatches between the C2 view and the UML class view in Figure 5. In this example, no transformation is needed since the structures of both views
are similar (the need for transformation for the C2/UML integration will be illustrated
later). Figure 6 shows a simplified algorithm that can be used to identify mismatches.

The first step in Figure 6 corresponds to the *Mapping* activity outlined above. In
the second step we need to traverse both the C2 diagram and the UML class diagram
and mark all links with corresponding counterpart. For instance, in case of the link
from *DB Manager* to *Database Bus* in Figure 4, the equivalent design-level class
dependencies corresponding to that link are the dependency arrows from *ODBC* to
Oracle API and from *Network Lib* to *Account DB API* (see Figure 5). Thus, we mark
all those links and repeat that process for the remaining ones.

Step 3 further marks all links between classes that are part of a single C2 component or connector. This is necessary because the C2 architecture in Figure 4 does not
specify what happens within a component/connector and thus we must assume that
class configurations corresponding to single C2 elements are consistent with the
architecture by default.

Once steps 2 and 3 are concluded, we should have ideally marked all links in both
the C2 view and the class view. If this is not the case, then we have identified potential mismatches. Figure 5 already shows this for both the C2 and UML class diagrams
with the unmarked links highlighted and pointed to by arrows:

* a C2 link between *Accounts* and *Transaction Bus* has no corresponding class
  relationship. This indicates a potential nonconformance since the design does not
  reflect everything the architecture demands.

```
1. FOR EACH C2 COMPONENT AND C2 CONNECTOR FIND CORRESPONDING UML CLASSES
2. FOR EACH C2 LINK
     FIND AND MARK C2 LINK AS WELL AS CORRESPONDING CLASS LINKS (INTERFACE)
3. FOR EACH UML CLASS LINK
     FIND AND MARK LINKS BETWEEN TWO CLASSES, WHERE BOTH CLASSES CORRESPOND TO
     ONLY ONE C2 COMPONENT OR CONNECTOR
4. FOR EACH UNMARKED C2 LINK RAISE NONCONFORMANCE MISMATCH
5. FOR EACH UNMARKED CLASS LINK RAISE INCONSISTENCY MISMATCH
6. FOR EACH C2 COMPONENT FIND AT LEAST ONE CLASS CALL DEPENDENCY BETWEEN CLASSES
     CORRESPONDING TO THAT C2 COMPONENT AND OTHER C2 COMPONENTS CONNECTED VIA
     THE SAME CONNECTOR
```

Figure 6. Differentiation Algorithm to identify Mismatches between Views

• a class dependency link from *Oracle API* to *Oracle Enterprise Manager*. These two classes belong to different C2 components and in C2 a direct link from one component to another is illegal. Thus, this unmarked link indicates a potential inconsistency since the design seems to contradict the architecture.

5.3.2 C2 Behavioral Integration

The algorithm in Figure 6 has thus far ensured the structural integration of the C2 and class diagrams. Having ensured that the proper model elements interact and none of the interaction is missing or inconsistent does not ensure that those modeling elements interact the proper way. However, the final integration step (step 6) of our algorithm addresses behavioral integration.

A C2 architecture is more than just a structure with interfaces. It also describes how components and connectors are supposed to interact. For instance, in Figure 4 the *Transaction Manager* may interact with *DB Manager* and *Accounts* and vice versa; however, the *Transaction Manager* is not allowed to interact with *DB Admin* (although they both link to the same connector). Thus, a full integration of C2 and class diagrams must also ensure that the class view dependencies adhere to C2 behavioral constraints. In order to verify that the behavior of classes corresponding to *Transaction Manager* follows the C2 architectural guidelines we need to verify the following: there must be at least one calling dependency[1] between a class corresponding to *Transaction Manager* and one corresponding to *DB Manager* and *Accounts*, and, there must be no calling dependencies between classes corresponding to *Transaction Manager* and any other classes (e.g., *DB Admin*).

In order to do that, we need a technique that allows us to abstract away intermediate helper classes. For instance, in above example of *Transaction Manager* to *DB Manager* we need to know the relationship of T-Framework (point of external access to *Transaction Manager*) and *Oracle API* or *Account DB API* (two points of external access to *DB Manager*). However, the relationship from, *T-Framework* to say *Oracle API* is obscured by the intermediate classes *Data Access* and *ODBC*. Thus, we need a technique that can eliminate both intermediate classes and leave only the pure (transitive) relationship from *T-Framework* to *Oracle API*. To this end, we use the transformation technique Rose/Architect [13].

Rose/Architect (RA) (described in detail in [13]) identifies patterns of groups of three classes and replaces them with simpler patterns using transitive relationships. Currently, the RA model consists of roughly 80 rules of abstraction. Rule 2, for instance, describes the case of a class which is generalized by a second class (opposite of inheritance), which in turn is dependent on a third class (see Figure 7). This three-class pattern can now be simplified by eliminating the middle class and creating a transitive relationship (a dependency in this case) which spans from the first class to the third one. The underlying RA model describes these rules and how they must be applied to yield an effective result.

Figure 7 shows the RA refinement steps for the case of the *Transaction Manager* to *DB Manager* relationship of our design view (Figure 5). After applying two rules

---

[1] Note that C2 prohibts the use of shared variables. Since classes do not make use of events or other triggering mechanism this leave only procedure calls as an option.

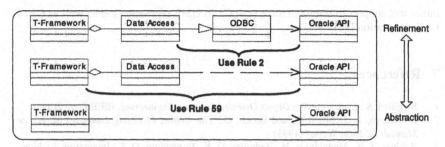

Figure 7. Using Rose/Architect to derive Behavior Dependencies

(rules 2 and 59 respectively, see [13]) we get a simplified pattern of two classes and a dependency relationship between them. If this is also done for the other classes discussed above, we can automatically verify whether the behavior of a class diagram conforms to the behavioral constraints opposed by a C2 architecture. Through this process, we find a potential mismatch between the classes corresponding to *Accounts* and *DB Manager:* no dependency relationship (mandated by the C2 architecture) could be found after abstracting away the *Network Lib* helper class.

6 Conclusion

This work outlined the differences between view representation and view integration. The former is satisfied by merely using some predefined modeling elements and adapting them so that they may be used to represent new modeling elements. The works of Robbins et al. [3] on C2 and Wright, Hofmeister et al. [4] on various conceptual architectural views, and Lyons [5] on ROOM are examples of representing architectural notations in UML. These approaches fall short of fully integrating the notations with UML, however, their view representations provide a valuable starting point for view integration. Automated integration is only possible once views are represented within a single model that supports both a common way of accessing modeling elements and cross-referencing them.

In order to integrate C2 into UML, we presented a view integration framework and demonstrated its use. For view integration, we needed to accomplish the three activities discussed in Figure 3: *Mapping, Transformation,* and *Differentiation. Differentiation* was demonstrated using the algorithm in Figure 6; Rose/Architect was presented as a *Transformation* technique to support behavioral analysis; we did not discuss *Mapping* in detail, as it was discussed elsewhere.

As mentioned previously, views are nothing more than an abstraction of relevant information from its model. Views are necessary to present that information in some meaningful way to a stakeholder (developer, architect, customer, etc.). When we talk about the need to integrate views, we are really talking about the need of having a system model integrated with its views. Although a full integration effort may seem improbable at this point, our initial experience indicates that it can still be attempted and (semi) automated for significant parts of a system. To date we have provided initial tool support to semi-automate both C2 to UML representation as well as C2 to UML integration [14].

7 References

1. Siegfried, S.: *Understanding Object-Oriented Software Engineering*. IEEE Press (1996)
2. Rumbaugh, J., Jacobson, I., and Booch, G.: *The Unified Modeling Language Reference Manual*. Addison-Wesley (1998)
3. Robbins, J. E., Medvidovic, N., Redmiles, D. F., Rosenblum, D. S.: Integrating Architecture Description Languages with a Standard Design Method. *Proceedings of the 20th International Conference on Software Engineering*, Kyoto, Japan (1998)
4. Hofmeister, C., Nord, R. L., and Soni, D.: Describing Software Architecture in UML. *Proceedings of the First Working IFIP Conference on Software Architecture (WICSA1)*, Kluwer Academic Publishers, Boston, Dordrecht, London (1999) 145-159
5. Lyons, A.: UML for Real-Time Overview. White Paper, ObjectTime (1998)
6. Taylor, R. N., Medvidovic, N., Anderson, K., Whitehead, Jr., E. J., Robbins, J. E., Nies, K. A., Oreizy, P., and Dubrow, D. L. A component and message-based architectural style for GUI software, *IEEE Trans. Software Engineering*, June, Vol.22, No.6 (1996) 390-406
7. Allen, R. and Garlan, D. A Formal Basis for Architectural Connection, *ACM Transactions on Software Engineering and Methodology*, Vol. 6, No. 3, July (1997) 213-249
8. Luckham, D. C. and Vera J. An Event-Based Architecture Definition Language, *IEEE Transactions on Software Engineering*, September (1995)
9. Buschman, F., Meunier, R., Rohnert, H., Sommerlad, P., Stal, M.: *A System of Patterns: Pattern-Oriented Software Architecture*. Wiley (1996)
10. Egyed, A.: Automating Architectural View Integration in UML. *Technical Report USC-CSE-99-511*, Center for Software Engineering, University of Southern California, Los Angeles, CA 90089-0781 (1999)
11. Egyed, A.: Using Patterns to Integrate UML Views. *Technical Report USCCSE-99-515*, Center for Software Engineering, University of Southern California, Los Angeles, CA 90089-0781 (1999)
12. N., Medvidovic, D.S. Rosenblum: Assessing the Suitability of a Standard Design Method for Modeling Software Architectures. *Proceedings of the TC2 First Working IFIP Conference on Software Architecture (WICSA1)*, Kluwer Academic Publishers (1999) 161-182
13. Egyed, A. and Kruchten, P.: Rose/Architect: a tool to visualize software architecture. *Proceedings of the 32nd Annual Hawaii Conference on Systems Sciences* (1999)
14. Center for Software Engineering, University of Southern California: Software Architecture, *http://sunset.usc.edu/software_architecture/*

Enabling the Refinement of a Software Architecture into a Design

Marwan Abi-Antoun and Nenad Medvidovic

{marwan,neno}@sunset.usc.edu
Computer Science Department
University of Southern California
Los Angeles, CA 90089-0781, USA

Abstract. Software architecture research has thus far mainly addressed formal specification and analysis of coarse-grained software models. The formality of architectural descriptions, their lack of support for downstream development activities, and their poor integration with mainstream approaches have made them unattractive to a large segment of the development community. This paper demonstrates how a mainstream design notation, the Unified Modeling Language (UML), can help address these concerns. We describe a semi-automated approach developed to assist in refining a high-level architecture specified in an architecture description language (ADL) into a design described with UML. To this end, we have integrated DRADEL, an environment for architecture modeling and analysis, with Rational Rose®, a commercial off-the-shelf (COTS) UML modeling tool. We have defined a set of rules to transform an architectural representation into an initial UML model that can then be further refined. We believe this approach to be easily adaptable to different ADLs, to the changes in our understanding of UML, and to the changes in UML itself.

1. Introduction

Architecture-based development is currently receiving a lot of attention, from both the academic and industrial communities. This is evidenced by numerous architecture-based approaches to software development that have emerged and an increasing number of conferences and symposia that focus on software architectures [1, 2, 3, 4, 5]. However, to a large degree, the claimed potential of software architectures remains unfulfilled. One reason is that most existing research is centered around highly specialized architecture description languages (ADLs), only addressing *modeling* and *analysis* of specific aspects of architectures (e.g., their structure), and rarely focusing on the broader *development* picture. A detailed survey of the area [6] shows limited overall tool support that repeatedly focuses only on well-understood problems: editing, parsing, syntactic and structural analysis, and so forth. The narrow focus and limited tool support partially account for the lack of transitioning of architecture research into the software development mainstream.

On the other hand, the industrial segment of the software development community has focused on comprehensive approaches to software development, through integrated methodologies and tools that address the entire software lifecycle. One important recent development is the Unified Modeling Language (UML) [7, 8], a general purpose modeling language that provides an expressive and extensible graphical notation. UML is emerging as a *de facto* software design standard with the potential for industry wide adoption and extensive, sophisticated tool support. However, existing work [10,11] has identified areas in architectural design where the expressiveness of standard UML alone is not entirely adequate.

We hypothesize that software architecture researchers can benefit from integrating their approaches with UML. The benefits of doing so would be manifold: easier adoption of their work, much more extensive software modeling and tool support, the ability to exploit UML in refining coarse-grained architectural elements and implementing them, and so on. This would, in turn, benefit mainstream UML users by providing them with powerful notations and analysis tools that have an explicit architectural focus and are specialized for certain development situations.

In order to validate our hypothesis, we have begun studying the relationship between UML and ADLs. In our previous work, we discussed and demonstrated the possible strategies for marrying architecture-based approaches with UML [9, 10, 11]. This paper further refines our ideas. In particular, we define a set of rules for transforming ADL-based models into their UML counterparts and provide tool support for automatically generating an initial UML model corresponding to the ADL specification. The UML model becomes a starting point for refining the architecture into a design and eventually an implementation. To this end, we have integrated two environments: DRADEL [12], which focuses on ADL-based modeling, analysis, and simulation of architectures, and Rational Rose® [13], which supports software development using UML. Although the resulting environment supports transformation from a specific ADL (C2SADEL [12]) to UML, we believe that the underlying methodology is general and flexible enough to be applicable to other ADLs.

The remainder of the paper is organized as follows. Section 2 discusses the issues in refining an architecture into a design while ensuring consistency and traceability. In Section 3, we propose our approach for refining C2SADEL into UML, while Section 4 discusses our implementation of this approach. An overview of related work and conclusions round out the paper.

2. Motivation

The problem of refining a coarse-grained architecture specified in an ADL into a design described in a modeling language comprises two separate tasks:

1. Refining the architecture into the design (forward engineering)
2. Abstracting the architecture from the design (reverse engineering)

Although the focus of our work to date has been on refinement, we briefly discuss both aspects of the problem below. In an iterative design process, the designer may need to apply both refinement and abstraction repeatedly.

2.1 Refinement

For our purposes, the starting point of refinement is an architecture composed of coarse-grained components and connectors, and their configurations. The architecture adheres to some architectural style (e.g., client-server, pipe-and-filter, layered) and its

representation may be
formalized using an ADL.
That architecture is refined
into a design, and eventually,
an implementation.

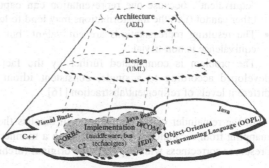

Given that architectures are
intended to describe systems
at a high-level of abstraction,
directly refining an
architectural model into a
design or implementation may
not be possible. One reason is
that the design space rapidly

Fig. 1. From Architecture to Implementation

expands with the decrease in abstraction levels, as shown in Fig. 1.

A solution to this problem is to bound the target (implementation) space by employing specific middleware technologies. A more general refinement approach includes design as an intermediate step. During design, constructs such as classes with attributes, operations, and associations, instances of objects collaborating in a scenario, and so forth, are identified. These are more effectively expressed in a notation like UML than in an ADL. However, various problems might arise. First, the design may no longer be faithful to the rules of the selected architectural style. Second, maintaining traceability is inherently difficult, because of the possible many-to-many mappings from the elements in the problem domain to the elements in the solution space (Fig. 2): a given element from one space can map to zero, one, or more elements in the "lower level" space. Third, some architectural elements, such as connectors, are given first class status in architectures [14], but may not have direct design or implementation counterparts. Typically, connectors are "designed away" into various class associations and object interactions or are "coded away" into programming language statements distributed across different components.

3. Abstraction

Since any design process is inherently
iterative, and software designers often have
to deal with legacy systems, it is also
desirable to be able to perform the reverse
step, i.e., to abstract the architecture from
the design. This involves obtaining from
the UML model a description similar to the
one represented in an ADL, using either
formal transformations [9] or various
reverse engineering heuristics (e.g., [15]).

However, converting an architectural
representation in an ADL to a
representation in UML and then back might
lead to either of the following cases:

- The resulting representations are not

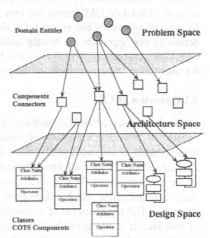

Fig. 2. Traceability during Refinement

"equivalent", because one representation can capture some information that the other cannot (i.e., the transformations may lead to loss of information).

- The resulting representations are equivalent, but not identical, and proving the equivalence is non-trivial.

The problem is complicated further by the fact that systems are not usually developed according to a single, consistent idiom and variations may occur at different levels of refinement/abstraction [16].

In the remainder of the paper, we will discuss the specifics of our approach for mapping from a candidate ADL to UML in a manner that maintains traceability and preserves correctness with respect to the original (architectural) model.

3. Mapping from an ADL to the UML

There are several possible strategies for refining ADL models into UML [11]. We briefly summarize them here, and describe how we used specific instances of those strategies.

- Strategy #1 consists of using standard UML constructs to simulate modeling architectural concerns as would be done in an ADL [10].
- Strategy #2 consists of using UML's built-in extension mechanisms (stereotypes and tagged values) [17] and the Object Constraint Language (OCL) [18] to constrain the semantics of meta-classes to those of ADL constructs.
- Strategy #3 consists of augmenting the UML meta-model to directly support architectural concerns. Although this is a potentially effective approach, it would result in a notation that is incompatible with standard UML. Since one of our goals for this work is conformance with standard UML and corresponding tools, we do not currently pursue this strategy.

We illustrate how we combined strategies #1 and #2 using C2SADEL, an ADL for C2-style architectures [19]. In our previous work [9, 10], we manually mapped several ADLs into UML using the two strategies. We use those mappings as a basis for providing automated support for our current task. Before proceeding with the details of our approach, we briefly summarize the relevant rules of the C2 style and C2SADEL constructs. We also give a brief example of a C2-style architecture used for illustration in the remainder of the paper.

3.1 Overview of C2

We have selected the C2 architectural style as a vehicle for exploring our ideas because it provides a number of useful rules for high-level system composition, demonstrated in numerous applications across several domains [19]; at the same time, the rules of the C2 style are broad enough to render it widely applicable [20,21].

An architecture in the C2 style consists of components, connectors (buses), and their configurations. Each component has two connection points, a "top" and a "bottom." The top (bottom) of a component can only be attached to the bottom (top) of one bus. It is not possible for components to be attached directly to each other: buses always have to act as intermediaries between them. Furthermore, a component cannot be attached to itself. However, buses can be attached together: in such a case, each bus considers the other as a component with regard to the publication and forwarding of events. Components communicate by exchanging two types of events: service requests to components above and notifications of completed services to components below.

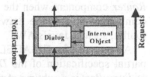

Fig. 3. Internal Architecture of a C2 Component

A C2-style component has a canonical internal architecture, consisting of an object with a defined interface and a dialog, as shown in Fig. 3. The internal object of a component can be arbitrarily complex. It is fully encapsulated inside the component, so that only the dialog can directly invoke the access routines of the object. The dialog, in turn, is in charge of interacting with the rest of the architecture via events.

A unique aspect of C2 buses is their context-reflective property [19]: a C2 bus is not defined to have a specific interface; instead, the "interface" it exports is a function of the interfaces of the components and connectors attached to it. This property of buses is a direct enabler of C2's support for dynamic adaptation [22].

C2's accompanying ADL, C2SADEL [12], specifies architectures in three parts: component types, connector types, and topology (or configuration). The topology, in turn, defines component and connector instances for a given system and their interconnections. C2SADEL specifies a component type with an invariant and sets of services a component provides and requires. A service consists of an interface and an operation. A single operation may export multiple interfaces (see the map at the bottom of Fig. 6). Invariants and operations (with their pre- and post-conditions) are specified as first-order logic expressions. A component may be subtyped from another component, using heterogeneous subtyping that preserves the supertype component's naming, interface, behavior, implementation, or a combination of them [12, 23].

We illustrate these concepts and subsequent discussion with an example. The example architecture is a variant of the logistics system for routing incoming cargo to a set of warehouses, first introduced in [22] and shown in Fig 4. The *DeliveryPort*, *Vehicle,* and *Warehouse* components keep track of the state of a port, a transportation vehicle, and a warehouse, respectively; each of them may be instantiated multiple times in a system. The *DeliveryPortArtist, VehicleArtist,* and *WarehouseArtist* components are responsible for graphically depicting the state of their respective components to the end-user. The *Layout Manager* organizes the display based on the actual number of port, vehicle, and warehouse instances. *SystemClock* provides consistent time measurement to interested components, while the *Map* component informs vehicles of routes and distances. The *Router* component determines when cargo arrives at a port, keeps track of available transport vehicles at each port, and tracks the cargo during its delivery to a warehouse. *RouterArtist* allows entry of new cargo as it arrives at a port and informs the

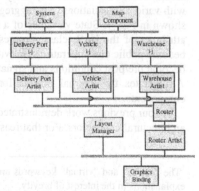

Fig. 4. C2 Architecture of the Cargo Routing System

Router component when the end-user decides to route cargo. The *GraphicsBinding* component renders the drawing information sent from the artists using a graphics toolkit, such as Java's AWT.

An extract of a C2SADEL specification of this architecture is given in Fig. 5[1]. A partial specification of the *DeliveryPort* component is shown in Fig. 6[2]. Note that *DeliveryPort* is a subtype of the more general *CargoRouteEntity* component, which is evolved by preserving both its interface and behavior.

```
architecture CargoRouteSystem is {
  component_types {
    component DeliveryPort is extern (Port.c2;)
    component GraphicsBinding is virtual ()
    ...
  }
  connector_types {
    connector FiltConn is {filter msg_filter;}
    connector RegConn is {filter no_filter;}
  }
  architectural_topology {
    component_instances {
      Runway : DeliveryPort;
      Binding : GraphicsBinding;
      ...
    }
    connector_instances {
      UtilityConn : FiltConn;
      BindingConn : RegConn;
      ...
    }
    connections {
      connector UtilityConn {
        top SimClock, DistanceCalc;
        bottom Runway, Truck;
      }
      connector BindingConn {
        top LayoutArtist, RouteArt;
        bottom Binding;
      }
    }
  }
}
```

```
component DeliveryPort is
  subtype CargoRouteEntity (int \and beh) {
    state {
      cargo          : \set Shipment;
      selected       : Integer;
      ...
    }
    invariant {
      (cap >= 0) \and (cap <= max_cap);
    }
    interface {
      prov ip_selshp: Select(sel : Integer);
      req  ir_clktck: ClockTick();
      ...
    }
    operations {
      prov op_selshp: {
        let  num : Integer;
        pre  num <= #cargo;
        post ~selected = num;
      }
      req or_clktck: {
        let  time : STATE_VARIABLE;
        post ~time = time + 1;
      }
      ...
    }
    map {
      ip_selshp -> op_selshp (sel -> num);
      ir_clktck -> or_clktck ();
    }
  }
```

Fig. 5. Cargo Routing System architecture specified in C2SADEL

Fig.6. DeliveryPort component type specified in C2SADEL

3.2 Mapping C2SADEL to UML Using Strategy #1

We initially map the internal object of each C2 component to a set of UML classes determined by the state maintained in the C2 component. The designer can then further refine the internal object by adding native UML constructs, such as classes with various associations (e.g., aggregation) and relationships (e.g., generalization), as shown in Fig. 7. State variables of a C2 component's internal object become private attributes of the UML classes representing the internal object; provided operations become public class operations; provided operation pre/post conditions and signatures become pre/post conditions and signatures on the corresponding class operations. Fig. 7 illustrates the internal objects for the *DeliveryPort* and the *DeliveryPortArtist* components.

As our previous work demonstrated [9,10], there are no direct UML counterparts to architectural connectors. For that reason, we map C2 buses to UML interfaces, where

---

[1]The "extern" and "virtual" keywords are not relevant to this discussion. We omit their explanations in the interest of brevity.

[2]"~" denotes the value of a variable after the operation has been performed, while "#" denotes set cardinality.

a UML interface is a collection of operations that are used to specify services provided by a class. To satisfy the context-reflective property of the buses, the operations provided by the interface (shown in Fig. 8) are roughly the union of the provided operations of all components attached to the bus, as discussed in [10]. In Fig. 7, the interface is graphically rendered as a circle, whereas in Fig. 8, it is graphically rendered as a stereotyped <<interface>> class in order to expose its operations.

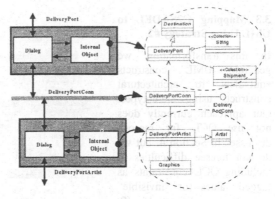

Fig.7. Designing the Internal Object. *Artist, Destination, ShipmentCollection, StringCollection* are not a part of the transformation, but were added manually by the designer.

Finally, in Fig. 9, we use a UML object diagram (i.e., instance form of a class diagram) to represent the architectural configuration: in this case, object links[3] represent instances of associations. Note that the architecture is currently based on explicit method invocations and completely bypasses the dialogs: *aDeliveryPort* is

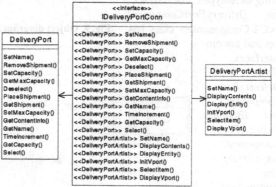

Fig. 8. Interface for the *DeliveryPortConn* connector

an instance of the *DeliveryPort* class (in Fig. 7); *DeliveryPortConn* is an instance of a class that realizes the interface *IDeliveryPortConn*, such that each method of *DeliveryPortConn* delegates the received messages to methods of the attached objects (*aDeliveryPort* and *theDeliveryPortArtist*). The object *DeliveryPortConn* performs the combined functionality of the dialogs of the C2 components *DeliveryPort* and *DeliveryPortArtist* and the C2 connector *DeliveryPortConn*, namely routing the requests with only (explicit) method invocations, instead of (implicit) event notifications coupled with dialogs invoking access routines of the internal objects. In summary, this transformation does not include certain style-specific concerns, such as event notification or implicit invocation.

---

[3]The importance of OCL constraints for preserving correctness becomes apparent in this case: although a C2 component cannot be attached to itself, links to self are legal in UML object interaction diagrams.

3.3 Mapping C2SADEL to UML Using Strategy #2

In this transformation, we generate UML constructs equivalent to architectural constructs using stereotypes. Our approach currently does not use OCL for reasons discussed in Section 4, but we do not foresee difficulties in including OCL constraints as tagged values (or "invisible" properties), as discussed in [9, 11]. Fig. 10 illustrates a fragment of the Cargo Routing System architecture, modeled using stereotypes. For instance, the *DeliveryPortComponent* <<C2-Component>> class has top and bottom interfaces, *ITop-DeliveryPort-Component* and *IBottom-DeliveryPort-Component*, respectively. The intermediate connector is mapped to a <<C2-Connector>> class, *Delivery-PortConn-Connector*. In addition, we generate UML Components,

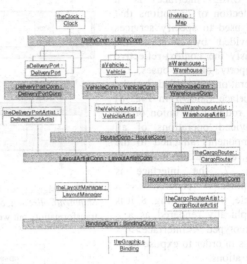

Fig. 9. Object diagram representing architecture

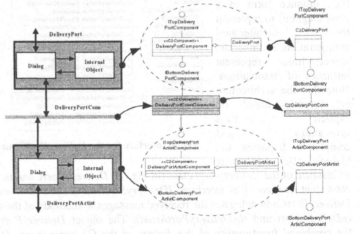

Fig. 10. Representing Architectural Constructs in UML. Although not shown in the figure, each C2 component will also have to realize the classes added by the designer to represent the internal object

which correspond to modules in UML and can realize a number of classes and interfaces (in that case, the interface of the component is represented by the interfaces it realizes). The UML Component corresponding to a C2 component realizes the <<C2-Component>> class, its top and bottom interfaces, and the classes representing the internal object discussed in Section 3.2. For example, *C2DeliveryPort* realizes

25

DeliveryPortComponent, *ITopDeliveryPortComponent,* *IBottomDeliveryPortComponent* and *DeliveryPort.* In order to satisfy the context-reflective property, the UML Component for a C2 connector realizes the <<C2-Connector>> class, as well as the bottom interfaces of Components/Connectors above, and the top interfaces of all Components/Connectors below. We also use a UML component diagram to represent the architecture: a UML dependency relationship between two UML components represents an attachment of a C2 component (or connector) with another connector.

3.4 Transformation Rules

As described above in the context of an example, the transformation from C2SADEL to UML is defined by a set of rules. The internal object is transformed using the rules shown in Table 1, while the architectural style concerns are addressed by the rules shown in Table 2. Note that, in the case of C2, it is not possible to represent in UML the architectural concerns without representing the internal objects first. The most important requirement of the transformation rules is that the generated UML design model be initially correct by construction with respect to the rules of the architectural style. However, as the design is modified, the problems discussed in Section 2 may arise. Detecting and resolving those inconsistencies is critical, but has been outside the scope of our work to date.

Table 1. Transformation Rules for the Internal Object
Internal Object → Class
    State Variable → Class Private Attribute
    Component Invariant → Tagged Value + Class Documentation
    Provided Operation → Class Operation
    Required Operation → Class Documentation
    Operation Pre/Post Condition → Pre/Post Condition on Class Operation
    Message Return Type → Return Type on Class Operation
    Message Parameter → Parameter (Name + Type) on Class Operation
Connector → Interface (<<Interface>> Class)
    Connector Interface → Union of Operations of attached Objects/Interfaces
        Message Originator → Operation <<Stereotype>>
Architecture Configuration (explicit invocation) → (Object) Collaboration Diagram
    Component Instance → Internal Object Class Instance
        Connector Instance → <<Interface>> Class Instance
        Component/Connector Binding → Object Link (instance of an association)

Table 2. Transformation Rules for Architectural Constructs
Component → <<C2-Component>> Class
    Internal Object → <<C2-Component>> Class Attribute
    Component Top Interface → <<Interface>> Class
    Component Bottom Interface → <<Interface>> Class
    Outgoing Request → <<Interface>> Class <<out>> Operation
    Incoming Notification → <<Interface>> Class <<in>> Operation
Connector → <<C2-Connector>> Class
    Connector Top Interface → Union of Bottom Interfaces of attached Components/Connectors
    Connector Bottom Interface → Union of Top Interfaces of attached Components/Connectors
Architecture Configuration (implicit invocation + event notification) → Component Diagram
    Component Instance → Component realizing...
    Connector Instance → Component realizing...

3.5 Limitations of UML to represent software architectures

Standard UML constructs and its built-in extensions are well suited for strategies #1 and #2. However, as mentioned above, UML may not be able to express all the information represented in an ADL [11]. In this particular case, UML's support for subtyping may not be able to adequately express the heterogeneous component subtyping mechanisms provided in C2SADEL, where different aspects of a component are preserved (e.g., interface, behavior, implementation, or a combination of them) [12, 23]. UML has been strongly influenced by object-oriented programming languages that typically support only a subset of possible subtyping/subclassing relationships: for example, in C++, inheritance means both interface and implementation inheritance [24]; this also appears to be the case in UML.

While using standard UML constructs for representing architectural concerns, we tried to maintain a parallel between type (instance) forms in the ADL and the type (instance) forms in UML. Thus, a class instance (i.e., an object) should correspond to a component instance. Similarly, an object type (i.e., a class) should correspond to a component type. However, the mapping is not obvious when dealing with connectors. C2 connectors defined as interfaces (equivalent to abstract classes, i.e., type form) cannot be inherently defined because they do not have interfaces of their own. Instead, we need to know the architectural configuration (instance), and specifically the instances of components and connectors attached to a given connector, to be able to define connector interfaces.

4. Integrating Rational Rose® with DRADEL

To integrate architecture-based development with mainstream approaches, one will need to bridge the gap between the tool support for architecture-based approaches and the support for modeling and design. To that end, we integrated DRADEL, a tool for modeling, analysis, and evolution of architectures described in C2SADEL, with Rational Rose®, a commercial tool that supports UML-based software modeling and development. The integrated environment[4] enables the designer to generate a UML representation of an architecture. The designer can then use the expressive power of UML to iteratively refine the resulting UML model and eventually produce an implementation using the code generation capabilities of Rose.

4.1 Architecture of the Rational Rose® Environment

Rose provides support for UML and its built-in extension mechanisms, such as stereotypes and tagged values (available as properties in Rose). Rose also provides code generation and reverse engineering capabilities for several programming languages [25, 26]. Rose itself is an extensible tool [27]: the Rose Extensibility Interface (REI) provides read/write access to the model elements (packages, classes, attributes, operations, stereotypes, etc.), their properties, the diagrams (graphical properties of model elements), and to the application itself (to execute scripts, set the visibility of the main application window, etc.).

---

[4]Note that the UML diagrams shown above as examples were generated using the environment. In some cases, Rational Rose is not fully compliant with the UML standard.

In addition, Rose also makes the REI Automation Objects (model elements, properties, diagrams, and application) available to external applications using the Microsoft Component Object Model (COM) [28]. For our purposes, we are using Rose as an Automation Server and an external application (developed in Microsoft Visual J++) as the Automation Controller.

4.2 Architecture of the DRADEL Environment

DRADEL [12] is a prototype environment for architecture modeling in C2SADEL, analysis of internal architectural consistency, topological constraint checking, type conformance checking among interacting components, and generation of application skeletons using the C2 implementation infrastructure. DRADEL itself is designed in the C2 style (see the left side of Fig. 11) and implemented using the C2 implementation infrastructure [29].

4.3 Integration Approach

Integrating Rose with DRADEL presented a challenge: DRADEL is implemented in Java, whereas Rose is COTS software, with no available source code. We solved this problem by porting DRADEL to Microsoft Visual J++ 6.0 and by generating Java wrappers for the Rose Automation objects [30]. Porting DRADEL to Visual J++ did not require modifications to the DRADEL code, since it uses standard Java syntax. The integration itself required a new C2 component in the DRADEL architecture, *UMLGenerator*, and some modifications for the user interface component, *UserPalette* (Fig. 11). The *UMLGenerator* component traverses the architectural representation maintained by DRADEL, initializes the Rose application, and creates Rose model elements and diagrams, based on the transformation rules described in Section 3.

The advantage of the approach is that DRADEL is still independently extensible: one can exploit the C2 style to add new components to DRADEL without any effect on its interaction with Rose. Similarly, Rose is still extensible independently of this application: in particular, it can still act as an Automation Controller for an external Automation Server (such as Microsoft Word), e.g., to generate a formatted report of the model.

There are several potential disadvantages of the approach we adopted. First, the *UMLGenerator* component communicates with Rose using synchronous remote procedure calls, mandated by COM, unlike the rest of the C2-style architecture, where communication is based on asynchronous events and implicit invocation. In our future work, we plan to address this problem using the technique described in [31]. Second, the integration mechanism is platform dependent, because the Java wrappers to the Rose COM objects require extensions specific to the Microsoft Java Virtual Machine. Finally, the most important limitation of our current integration of Rose and DRADEL is that Rose cannot generate notifications reflecting changes in the internal state (e.g., *ClassAdded*) in response to requests (e.g., *AddClass*). The reason is that Rose is not a Connectable Object [32]. Had that been the case, we could have packaged Rose as an internal object of the *UMLGenerator* C2 component, and enabled full two-way communication with DRADEL.

28

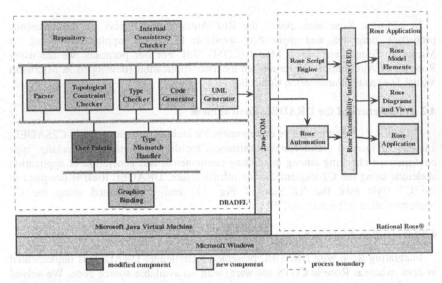

Fig. 11. Architecture of the integrated environment. *DRADEL and Rose are running in separate processes and communicating through synchronous COM remote procedure calls*

5. Related Work

Our work has been influenced by a large body of research and practical experience. In the interest of brevity, we only compare it to the most relevant approaches. Since this project builds on our previously published results, we do not focus on its relation to other work (e.g., [33, 34]), described elsewhere [9, 11].

Our approach to software architectures has emerged from the part of the research community that focuses on specifying structural and possibly behavioral aspects of a software system centered around a (formal) ADL. Another part of the community has tried to identify useful architectural perspectives, or *views*. Two representative examples are provided by Kruchten [35] and Hofmeister et al. [17, 36]. Although these approaches are in certain ways more comprehensive than ADL-centered approaches, our technique for refining architectures into UML models is applicable to them as well. Indeed, Hofmeister et al. demonstrate how UML can be constrained to model their four architectural views (conceptual, module, execution, and code). However, their technique is currently entirely manual.

As already discussed, architecture researchers have largely ignored the problem of refining architectures into designs and/or implementations. One exception is the approach proposed by Moriconi et al. [37], which incrementally transforms an architecture across levels of abstraction using a series of refinement maps. The maps must satisfy a correctness-preserving criterion, which mandates that all decisions made at a given level be maintained at all subsequent levels and that no new decisions be introduced. We believe this to be overly stringent. It sacrifices design flexibility to a notion of (absolute) correctness. The role of the human designer is virtually eliminated. Finally, formally proving the relative correctness of architectures at different refinement levels may prove impractical for large architectures and large numbers of levels. Such an approach can be of value, however, if only applied to the most critical parts of a system, and complemented by a more pragmatic technique, such as the one proposed in this paper.

6. Conclusions and Future Research Directions

Software architectures provide a promising basis for improving the state-of-the-art in software development. However, no improvement can be achieved simply by focusing solely on architectures, just like a new programming language cannot by itself solve the problems of software engineering. A programming language is only a tool that allows (but does not force) developers to put sound software engineering techniques into practice. Similarly, one can think of software architectures and ADLs as tools that also must be supported with specific techniques to achieve desired properties. Additionally, ensuring system properties at the level of architecture is of little value unless it can also be ensured that those properties will be preserved in the resulting implementation.

In this paper, we have presented a practical approach for transferring architecture-level decisions to the design and, subsequently, the implementation. The specific details of the approach, outlined in the transformation rules in Tables 1 and 2, are likely to change as our understanding of the relationship between UML and software architectures evolves, as support for refining additional ADLs into UML is added, and as UML itself evolves. However, we believe that the approach can be easily adapted to accommodate any such changes.

Our future work will exploit the integrated environment we have produced to gain additional insights along several dimensions, including:

- transformation of architectural elements that do not have direct design counterparts (e.g., connectors) into UML;
- analysis of a design represented in UML for conformance to a given architectural style. To this end, we plan to abstract architectural information from Rose and analyze it in DRADEL. We are currently planning to use Rose Architect [38] to abstract the architecture from the design, and Rose's reverse engineering capabilities to abstract the design from the implementation;
- enforcement of style rules, by attaching OCL constraints to the relevant UML model elements and integrating tool support to enforce the constraints [39]; and
- expansion of the current support for modeling architectural behavior to include dynamic behavior, e.g., by using UML statechart diagrams.

References

1 Garlan, D., Paulisch, F.N., Tichy, W.F., editors: *Summary of the Dagstuhl Workshop on Software Architecture*, February 1995
2 Garlan, D., editor: *Proceedings of the First International Workshop on Architectures for Software Systems (ISAW-1)*, Seattle, WA, April 1995
3 Wolf, A.L., editor: *Proceedings of the Second International Software Architecture Workshop (ISAW-2)*, San Francisco, CA, October 1996
4 Magee, J., and Perry, D.E., editors: *Proceedings of the Third International Software Architecture Workshop (ISAW-3)*, Orlando, FL, November 1998

5 Donohoe, P., editor: *Proceedings of the First Working IFIP Conference on Software Architecture (WICSA1)*, San Antonio, TX, February 1999
6 Medvidovic, N., Taylor, R.N.: A Classification and Comparison Framework for Software Architecture Description Languages. In *IEEE Transactions on Software Engineering*, to appear
7 Booch, G., Jacobson, I., Rumbaugh, J.: *The Unified Modeling Language User Guide*, Addison-Wesley, 1998
8 Rumbaugh, J., Jacobson, I., Booch, G.: *The Unified Modeling Language Reference Manual*, Addison-Wesley, 1998
9 Robbins, J.E., Medvidovic, N., Redmiles, D.F., Rosenblum, D.S.: Integrating Architecture Description Languages with a Standard Design Method. In *Proceedings of the 20th International Conference on Software Engineering (ICSE'98)*, Kyoto, Japan, April 1998
10 Medvidovic, N., Rosenblum, D.S.: Assessing the Suitability of a Standard Design Method for Modeling Software Architectures. In *Proceedings of the First IFIP Working Conference on Software Architecture (WICSA1)*, San Antonio, TX, February 1999
11 Medvidovic, N., Rosenblum, D.S., Robbins, J.E., Redmiles, D.F.: Modeling Software Architectures in the Unified Modeling Language. In submission
12 Medvidovic, N., Rosenblum, D.S., Taylor, R.N.: A Language and Environment for Architecture-Based Software Development and Evolution. In *Proceedings of the 21st International Conference on Software Engineering (ICSE'99)*, Los Angeles, CA, May 16-22, 1999
13 Rational Software Corporation, *Rational Rose 98: Using Rational Rose*
14 Shaw, M.: Procedure Calls are the Assembly Language of Software Interconnections: Connectors Deserve First-Class Status. *Workshop on Studies of Software Design*, 1993
15 Harris, D.R., Reubenstein, H.B., Yeh, A.S.: Reverse Engineering to the Architectural Level, In *Proceedings of the 17th International Conference on Software Engineering (ICSE'95)*, Seattle, Washington, 1995
16 Garlan, D., Shaw, M.: An Introduction to Software Architecture, *Advances in Software Engineering*, vol. 1, World Scientific Publishing Company, 1993
17 Hofmeister, C., Nord, R.L., and Soni, D.: Describing Software Architecture with UML. In *Proceedings of the First IFIP Working Conference on Software Architecture (WICSA1)*, San Antonio, TX, February 1999
18 Warmer, J.B., Kleppe, A.G.: *The Object Constraint Language: Precise Modeling With UML*, Addison-Wesley, 1999
19 Taylor, R.N., Medvidovic, N., Anderson, K.M., Whitehead, E.J. Jr., Robbins, J.E., Nies, K.A., Oreizy, P., Dubrow, D.L.: A Component- and Message-Based Architectural Style for GUI Software. *IEEE Transactions on Software Engineering*, vol. 22, no. 6, pp. 390-406, June 1996
20 Di Nitto, E., and Rosenblum, D.S.: Exploiting ADLs to Specify Architectural Styles Induced by Middleware Infrastructures. In *Proceedings of the 21st International Conference on Software Engineering*, pp. 13-22, Los Angeles, CA, May 1999
21 Yakimovich, D., Bieman, J.M., and Basili, V.R.: Software Architecture Classification for Estimating the Cost of COTS Integration. In *Proceedings of the 21st International Conference on Software Engineering*, pp. 296-302, Los Angeles, CA, May 1999
22 Oreizy, P., Medvidovic, N., Taylor, R.N.: Architecture-Based Runtime Software Evolution. In *Proceedings of the 20th International Conference on Software Engineering (ICSE'98)*, pp. 177-186, Kyoto, Japan, April 1998
23 Medvidovic, N., Oreizy, P., Robbins, J.E., Taylor, R.N.: Using Object-Oriented Typing to Support Architectural Design in the C2 Style. In *Proceedings of the Fourth ACM SIGSOFT Symposium on the Foundations of Software Engineering (FSE4)*, pp. 24-32, San Francisco, CA, October 16-18, 1996
24 Gamma, E., Helm, R., Johnson,R., Vlissides, J.: *Design Patterns: Elements of Reusable Object-Oriented Software*, Addison-Wesley, 1994
25 Rational Software Corporation: *Rational Rose 98: Roundtrip Engineering with Java*, 1998
26 Rational Software Corporation: *Rational Rose 98: Roundtrip Engineering with C++*, 1998
27 Rational Software Corporation: *Rational Rose 98 Extensibility User Guide*, 1998
28 Williams, S., Kindel, C.: The Component Object Model, *Dr. Dobb's Journal*, December 1994

31

29 Medvidovic, N., Oreizy, P., Taylor, R.N.: Reuse of Off-the-Shelf Components in C2-Style Architectures. In *Proceedings of the 1997 International Conference on Software Engineering (ICSE'97)*, Boston, MA, May 1997

30 Verbowski, C.: Integrating Java and COM, Microsoft Corporation, January 1999. http://www.microsoft.com/java/resource/java_com.htm

31 Dashofy, E.M., Medvidovic, N., Taylor, R.N.: Using Off-The-Shelf Middleware to Implement Connectors in Distributed Software Architectures, In *Proceedings of the 21st International Conference on Software Engineering (ICSE'99)*, Los Angeles, CA, May 1999

32 Brockschmidt, K.: *Inside OLE*, Second Edition, Microsoft Press, 1995

33 Garlan, D., Monroe, R., Wile, D.: ACME: An Architecture Description Interchange Language. *CASCON'97*, November 1997

34 Wang, E.Y., Richter, H.A., Cheng, B.H.C.: Formalizing and Integrating the Dynamic Model within OMT. In *Proceedings of the 19th International Conference on Software Engineering*, Boston, MA, May 1997

35 Kruchten, P.B.: The 4+1 view model of architecture, *IEEE Software*, Nov. 1995. pp. 42-50

36 Soni, D., Nord, R.L., and Hofmeister, C.: Software Architecture in Industrial Applications. In *Proceedings of the 17th International Conference on Software Engineering (ICSE'95)*, Seattle, WA, 1995

37 Moriconi, M., Qian, X., Riemenschneider, R.A.: Correct Architecture Refinement. In *IEEE Transactions on Software Engineering*, April 1995

38 Egyed, A., Kruchten, P.: Rose/Architect: A Tool to Visualize Architecture. In *Proceedings of the 32nd Hawaii International Conference on System Sciences (HICSS-32)*, January 1999

39 IBM Corporation, The Object Constraint Language: (OCL): the expression language for the UML http://www.software.ibm.com/ad/standards/ocl.html

Using the UML for Architectural Description

Rich Hilliard

Integrated Systems and Internet Solutions, Inc.
Concord, MA USA
rh@isis2000.com

Abstract. There is much interest in using the Unified Modeling Language (UML) for architectural description – those techniques by which architects sketch, capture, model, document and analyze architectural knowledge and decisions about software-intensive systems. IEEE P1471, the *Recommended Practice for Architectural Description*, represents an emerging consensus for specifying the content of an architectural description for a software-intensive system. Like the UML, IEEE P1471 does not prescribe a particular architectural method or life cycle, but may be used within a variety of such processes. In this paper, I provide an overview of IEEE P1471, describe its conceptual framework, and investigate the issues of applying the UML to meet the requirements of IEEE P1471.
Keywords: IEEE P1471, architectural description, multiple views, viewpoints, Unified Modeling Language

1 Introduction

The Unified Modeling Language (UML) is rapidly maturing into the de facto standard for modeling of software-intensive systems. Standardized by the Object Management Group (OMG) in November 1997, it is being adopted by many organizations, and being supported by numerous tool vendors.

At present, there is much interest in using the UML for *architectural description*: the techniques by which architects sketch, capture, model, document and analyze architectural knowledge and decisions about software-intensive systems. Such techniques enable architects to record what they are doing, modify or manipulate candidate architectures, reuse portions of existing architectures, and communicate architectural information to others. These descriptions may the be used to analyze and reason about the architecture – possibly with automated support. Analyses range from assessing feasibility, *Can the system be built?* [12] to certifying its implementation, *Does the system as implemented conform to the architecture?* [20].

The *Recommended Practice for Architectural Description* (IEEE P1471) [16] represents an emerging consensus for the description of the architectures of software-intensive systems.

In this paper, I investigate the applicability of the UML within the context established by IEEE P1471. I begin with some background on the history and goals for the standard. I then introduce the conceptual framework of P1471. In the main sections of the paper, I review key requirements of IEEE P1471 which pertain to the use of the UML and examine several approaches to using the UML to meet the requirements of P1471. I then address a few issues pertaining to the use of the UML for architectural description which are not specific to P1471. I close with a summary of the key issues and a review of some related work.

Note: This paper is based on version 1.1 of the UML specification which is the currently approved version of the standard. Some changes have been made in version 1.3 of the UML specification, particularly in the area of Model Management, which might affect my analysis. However, at the time of this writing, version 1.3 is not widely available.

2 What is IEEE P1471?

IEEE P1471 is the Draft *Recommended Practice for Architectural Description*.[1] It was developed by the IEEE's Architecture Working Group, chartered and sponsored by the Software Engineering Standards Committee of the IEEE Computer Society. The draft *Recommended Practice* was produced between 1995 and 1998 by a group of approximately thirty participants, and over 140 international reviewers.

2.1 IEEE Goals for P1471

Given the widespread interest in the architecture of software-intensive systems, IEEE recognized the need for providing direction in this area, for both industry and academic application. IEEE set the following goals for the standard:

1. **To take a "wide scope" interpretation of *architecture* applicable to software-intensive systems.** This includes computer-based systems ranging from software applications, information systems, embedded systems, systems-of-systems, product lines and product families – wherever software plays a substantial role in the development, operation, or evolution of a system.
2. **To establish a conceptual framework and vocabulary for talking about architectural issues of systems.** Despite the widespread interest in architecture in both the systems and software engineering communities, there is no common frame of reference for practitioners and researchers in these communities to talk with one another. There are no agreed-upon definitions for terms such as "architecture," "architectural description," and "view."
3. **To identify and promulgate sound architectural practices.** There are already a wide range of software and systems architecture practices. It is a goal of IEEE P1471 to provide a basis on which all of these practices may be defined, contrasted and applied.

---

[1] At the time of this writing, IEEE P1471 has been balloted by the IEEE, and is expected to be approved for use by the time this paper appears. Up-to-date information about IEEE P1471 can be obtained from the IEEE Architecture Working Group (http://www.pithecanthropus.com/~awg).

4. **To allow for the evolution of those practices as relevant technologies mature.** The IEEE recognized that software systems architectural practices are rapidly evolving, both in industrial use and in the research arena, with respect to architecture description languages, architectural methods, analysis techniques, and architecting processes. It is hoped these practices can be communicated, documented and shared via the framework of P1471. For this reason, the framework should be general enough to encompass current techniques and flexible enough to evolve.

2.2 Using P1471

IEEE P1471 is a *recommended practice* – which is one type of IEEE standard.[2] The important ingredients of IEEE P1471 are:

1. a normative set of definitions for terms including *architectural description, architectural view, architectural viewpoint*;
2. a conceptual framework which establishes these terms in the context of the many of architectural descriptions for system construction, analysis and system evolution; and,
3. a set of requirements on an architectural description of a system.

P1471 applies to *architectural descriptions* (ADs) – any collection of products that purports to describe the architecture of a software-intensive system. An AD is said to *conform* to IEEE P1471 if it meets the requirements of IEEE P1471. Requirements in P1471 are signalled with *shalls*, following usual standards practice. In this way, ADs may be readily checked for conformance to the recommended practice. The requirements of IEEE P1471 are designed to be independent of any individual architectural technique, and therefore should be applicable within a variety of architectural methods and architecture frameworks. It is further hoped that having a common frame of reference will allow greater understanding and sharing between different approaches.

P1471 neither describes nor requires any kind of conformance of systems, projects, organizations, processes, methods, or tools – which are the province of individual methods, frameworks and practicing organizations.

2.3 The P1471 Conceptual Framework

Figure 1, adapted from IEEE P1471, depicts the major conceptual entities referred to by the standard. The conceptual framework is presented as a UML class diagram. I will further discuss the conceptual framework below, emphasizing elements specific to the UML discussion herein. (For a complete discussion of the framework, please refer to the standard.)

The central abstraction, and primary focus, of the standard is Architectural Description. In P1471, an Architectural Description is a collection of products

---

[2] There are three types of IEEE standard: (i) standards, (ii) recommended practices and (iii) guides.

to document the architecture of a system. P1471 does not specify the format or media for an architectural description. What P1471 *does* specify is certain minimal required content of an AD reflecting current practices and industry consensus.

A key tenet of that consensus is the notion of multiple views. In P1471, an Architectural Description is organized into one or more architectural Views. Most architectural methods and frameworks advocate the use of one or more views of the system as a part of the architectural description. However, the exact views used vary from technique to technique. Rather than require a particular set of views, P1471 leaves this selection to users of the standard.

One of P1471's contributions is to make explicit the notion of an architectural Viewpoint to embody the rules governing a view. It is anticipated that this will allow the definition and reuse of viewpoints, so that varying approaches to architecture may better be able to exchange results, and that in general the growth of the discipline will be facilitated by codifying certain useful patterns of description.

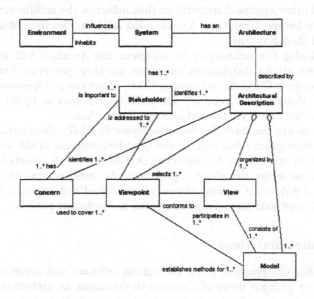

Fig. 1. The P1471 Conceptual Model

IEEE P1471 has been developed to be "notation-independent" – it does not specify any particular notations to be used in an architectural description, leaving this to individual architectural methods or practices. Thus, the question arises, *How does the UML apply in the context of IEEE P1471?* After examining the requirements of IEEE P1471 I will return to this in section 4.

3 P1471 Requirements on Architectural Descriptions

IEEE P1471 establishes requirements on what it means to be a "conforming" architectural description [16]. In this section I highlight those requirements with special implications for the use of the UML. The leading ideas in the IEEE P1471 requirements are **bolded**, and the requirements are paraphrased within the text which follows. For the full set of normative requirements on ADs, readers should refer to the standard.

3.1 Stakeholders and Concerns

As shown in figure 1, P1471 posits two key abstractions in the system environment that influence an architectural description: Stakeholders and Concerns. *Stakeholders* are any individual, class, organization or role that has an interest in the system. Those interests P1471 refers to as Concerns – the architecturally relevant areas of interest on the part of the stakeholders for the system. Concerns capture considerations including the "ilities" (e.g., security, reliability, maintainability) and other system characteristics that influence the architecture.

ADs are interest-relative: A conforming AD identifies the system's stakeholders and their concerns.

In developing the architecture for a system, the Architect will seek to understand who are its stakeholders and what are their concerns. Stakeholders typically include a Client for the system, Users, Maintainers, Operators, System Developers, Vendors, and so on [9]. The stakeholders form an upper bound on the potential audiences for the architectural description.

Concerns are the basis for completeness: In P1471, the content of an architectural description is bounded by the identified concerns of the stakeholders for the system of interest. A conforming AD addresses all stakeholders' concerns identified under the above requirement. An architectural description is incomplete if it does not at least address all such identified concerns. Individual architectural methods may specify additional completeness criteria.

3.2 Architectural Views

P1471 codifies the practice, found throughout software and systems architecture, of using multiple views of a system to document an architecture. Views are recognized as a mechanism to separate concerns, both to reduce perceived complexity and to address the needs of diverse audiences [8].

Multiple views: An AD consists of one or more views. In IEEE P1471, a *view* is a representation of a whole system from the perspective of a related set of concerns.

Views are modular: A view consists of one or more *architectural models*. Each model may use a different representational scheme. In the example below, a capability view is developed using two representational schemes: UML component diagrams and class diagrams.

Inter-view consistency: A conforming AD documents any known inconsistencies among the views it contains. IEEE P1471 does not prescribe any specific consistency-checking techniques between views, which are the realm of individual methods or organizations. It only requires that any inconsistencies, discovered by whatever means, be documented.

3.3 Architectural Viewpoints

The perspective from which a view is constructed is called a *viewpoint*.

Views are well-formed: Each view in a conforming AD is governed by exactly one viewpoint. Viewpoints define the rules for creating and using views. In the IEEE Architecture Working Group, the slogan we have used to relate views and viewpoints is this:[3]

views : viewpoint :: programs : programming language

Concerns drive viewpoint selection: In a conforming AD, each concern is addressed by one or more architectural views. Furthermore, a conforming AD identifies the selected viewpoints and provides the rationale for their selection.

No fixed set of viewpoints: Various architectural methods prescribe a fixed, or starting set of viewpoints, such as Kruchten's 4+1 view model [19], ISO's Reference Model for Open Distributed Processing [17], Siemens [26]. IEEE P1471 does not require any specific viewpoints; leaving this to individual methodological (or religious!) considerations. Instead, P1471 provides mechanisms for insuring that whatever viewpoints are used in a conforming AD, these are documented and understandable. IEEE P1471 does not take a position on where views come from. In the literature, one finds several *stances* toward views [11]. One stance treats views as projective: *a view is a partial projection of a full architectural description*; another stance is constructive: *the architectural description is constructed from one or more separate views*.

Viewpoints are first-class: Each viewpoint used in an AD is declared before use. Like a legend on a map or chart, a viewpoint provides a guide for interpreting and using a view, and appears in a conforming AD together with the view it defines.

3.4 Viewpoint Example

The remainder of this section presents a brief example of declaring and using a viewpoint conforming to IEEE P1471. The example is drawn from an architectural description for the InternetEngine—a product line architecture for electronic commerce.

---

[3] The notation $x : y :: z : w$ is read "x is to y as z is to w."

Declaring a Viewpoint. IEEE P1471 establishes the minimal information that a conforming AD must contain for each viewpoint used therein. Each viewpoint is specified by:

- the viewpoint name;
- the stakeholders addressed by the viewpoint;
- the stakeholder concerns to be addressed by the viewpoint;
- the viewpoint language, modeling techniques, or analytical methods used; and,
- the source, if any, of the viewpoint (e.g., author, literature citation).

A viewpoint definition may additionally include:

- any formal or informal consistency or completeness checks associated with the underlying method to be applied to models within the view;
- any evaluation or analysis techniques to be applied to models within the view; and
- any heuristics, patterns, or other guidelines which aid in the synthesis of an associated view or its models.

The Capability Viewpoint. This is a brief example of a viewpoint declaration and a resulting view. The Architect of a large, enterprise-wide, distributed information system needs to devise a strategy for the organization of system capabilities and the rules by which those capabilities are constructed and fielded. The "capabilities" are small to mid-sized components intended to encourage reuse across the enterprise and facilitate "plug-and-play" composition.

    Viewpoint Name: Capability

    Stakeholders: the client, producers, developers and integrators

    Concerns: How is functionality packaged? How is it fielded? What interfaces are managed?

    Viewpoint language: Components and their dependencies (≪provides≫, ≪requires≫, ≪client-server≫) (using enhanced UML component diagrams); interfaces and their attributes (using UML class diagrams).

    Sources: *Also known as:* Static, Application, Conceptual [14]

A Capability View. An application of the capability viewpoint might look like figure 2. The capability view covers all system functionality for operating on data; it is therefore intended as a reference model (template) for new construction and integration. What the cartoon does not show are the assertions associated with each element. Capabilities are constructed using a 5-tier, layered organization with interfaces at each pair of layers. Each layer is a capability. Capabilities can serve other capabilities (horizontal integration). The entire stack is a deployable capability. Rules for interaction among layers, and rules for allowed "content" of a layer are stated in terms of this diagram using OCL. All capabilities must provide certain basic operations, conforming to the **Generalized**

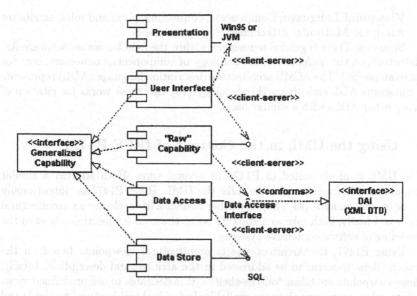

Fig. 2. A Capability View

Capability Interface, which is used for the dynamic discovery of capabilities. The **Data Access Interface** conforms to an XML document type description (DTD). Other interfaces are constrained by well-known, off-the-shelf APIs.

This capability viewpoint is defined in terms of two existing UML diagram types (class diagrams and component diagrams), stereotype extensions to the component diagram, and certain relations between the diagrams. This ensemble constitutes the *viewpoint language* part of a viewpoint declaration.

Viewpoint Reuse. Unlike a system's stakeholders and its views, viewpoints are not particular to a system. Since they are first-class, it should be possible to keep viewpoints "on-the-shelf" for (re)use. Thus, the architect may be able to reuse viewpoint descriptions; in IEEE P1471, these are referred to as *library viewpoints*.

So the Architect, instead of defining a capability viewpoint from scratch as above, ought to be able to go to the library and find a viewpoint to meet her needs. One candidate, well-known from Software Architecture, is the structural viewpoint.

Viewpoint Name: Structural

Stakeholders:[4]

Concerns: Define the computational elements of a system and their organization. What elements comprise the system? What are their interfaces? How do they interconnect? What are the mechanisms for interconnection?

---

[4] Not in library, to be filled in at use time.

Viewpoint Language: Components, connectors, ports and roles, attributes
Analytic Methods: Attachment, type consistency.

Sources: There is general agreement within the field known as Software Architecture, on the usefulness of the ontology of components, connectors, etc. See for example [25]. The ACME architecture description language (ADL) represents a consensus ADL embodying this ontology [10]. See those works for citation of many other ADLs with a similar basis.

4   Using the UML in the Context of IEEE P1471

The UML is nicely suited to P1471 in several ways. Both address a similar scope: software-intensive systems. Like the UML, IEEE P1471 is "intentionally process-independent" [23, §4.2]. It neither defines a life cycle nor an architectural process. Finally, both take as a starting point the need for multiple views in the modeling of software-intensive systems.

Using P1471, the Architect selects architectural viewpoints based on the stakeholders' concerns to be addressed in the architectural description. Ideally, these viewpoints are taken "off-the-shelf" – it is desirable to use predefined viewpoints, when such viewpoints are available. Individual architectural methods and architectural frameworks typically espouse a set of predefined architectural viewpoints (even if they are not referred to as "viewpoints"). For example, the UML *User Guide* [4] advocates a "user-case driven, architecture-centric, iterative, and incremental process" which employs five view(point)s:

> [T]he architecture of a software-intensive system can best be described by five interlocking views. Each view is a projection into the organization and structure of the system, focused on a particular aspect of that system. [4, p31]

There are a range of approaches one might take to applying the UML. In light of the requirements discussed in section 3, we now examine four approaches to using the UML within the context of IEEE P1471.

(1) "Out of the Box" The Architect can adopt the UML as a tool kit of useful notations to be applied to architectural subjects, using it "out of the box." Each of the nine predefined diagram (techniques) is potentially applicable to architectural description. The easiest way to use the UML within the P1471 context is to develop viewpoints which utilize one or more predefined diagram types as viewpoint languages. IEEE P1471 offers a way to understand, and therefore reuse, existing diagram types, as shown in table 1.

(2) "Lightweight Extension" The Architect can exploit the UML's "lightweight extension mechanisms" (tagged values, stereotypes and constraints) [4] to create new vocabularies for architectural description. Either by creating a UML extension or a variant:

Table 1. Predefined UML Diagram Types as Viewpoint Languages

Architectural Viewpoint	UML Diagram Type(s)
structural	component diagrams and class diagrams
behavioral	interaction diagrams, activity, state diagrams
user	use case diagrams, interaction diagrams
distribution	deployment diagrams, interaction diagrams

- A *UML extension* is a predefined set of stereotypes, tagged values, and constraints that extend and tailor the UML for a specific domain of application. See [23, §4.1.1]
- A *UML variant* is language built on top of the UML metamodel, specializing that metamodel, without changing any UML semantics. See [23, §4.1.1]

The key point is that such extensions will be developed on a *per-viewpoint* basis – potentially each viewpoint will necessitate its own extension. It would be useful to have a standard way to document viewpoint declarations in the UML, such that they may be notationally depicted, stored and manipulated by tools. Once viewpoints are represented in this form, more interesting uses are possible, such as specialization and combination of viewpoints. It seems such a mechanism needs to be more than what is defined by an extension, since it needs to refer to diagram types; but less than a UML profile [1] – which appears to be much too heavy a mechanism for individual architects, if it has to be applied individually to viewpoints.

(3) "UML as Integration Framework" The Architect, or more likely the architecture team, or firm, might adopt the UML (and its metamodel) as an integrating framework for architectural description. At this level of commitment, one would seek a close correspondence between the P1471 conceptual framework and the UML metamodel.

As noted, the UML and P1471 are *philosophically compatible* on the matter of multiple views; however, while views and viewpoints are "first-class" citizens in the P1471 conceptual framework (figure 1), these concepts appear only informally in the UML specification.[5] For example,

Every complex system is best approached through a small set of nearly independent views of a model; No single view is sufficient. [23, §4.1.2]

The notion of *viewpoint* appears in the definition of model:

A *model* is an abstraction of a modeled system, specifying the modeled system from a certain viewpoint and at a certain level of abstraction. A model is complete in the sense that it fully describes the whole modeled system at the chosen level of abstraction and viewpoint. [22, §12.2]

[5] The entity ViewElement does occur in the UML metamodel (vintage 1.1), but pertains to the graphical presentation (rendering) of ModelElements.

Perhaps closest in spirit to IEEE 1471's notion of a view is the UML notion of a diagram:

> In terms of the views of a model, the UML defines the following graphical diagrams: [use case diagram; class diagram; behavior diagrams: statechart diagram, activity diagram; interaction diagrams: sequence diagram, collaboration diagram; implementation diagrams: component diagram, deployment diagram.]
> These diagrams provide multiple perspectives of the system under analysis or development. ... [23, §4.1.2]

To be precise, we are less interested in individual diagrams than in the rules governing diagram instances. In the UML specification, these are referred to variously as a *diagram type*, a *diagram kind*, or a *diagram technique*. Although there are nine predefined instances of this entity in the UML specification, it has no status in the metamodel either. The "standard diagram types" are defined not in UML itself, but by external rules.

One way to fully support P1471 is to introduce representatives of diagram technique, view and viewpoint into the UML metamodel. This would support both end-user extensiblity by architects and could simplify the specification of the UML. One way to do this is shown in figure 3.

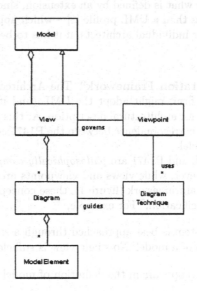

Fig. 3. Enhanced Metamodel Fragment

By reifying Diagram Technique we provide a way to document existing diagram types; provide a substrate for differential expression and extensions; and give the end user a means to define new diagram techniques.

Now we may complete the integration of IEEE P1471 and the UML. An architectural description is a model; a model composed of one or more views. Diagram techniques are candidate viewpoint languages which may be referenced in the declaration of viewpoints, and then used to guide the construction of well-formed diagrams which make up the view. A View is one kind of Model, governed by a well-defined (declared) Viewpoint. A view may be documented with one or more diagrams. Each diagram is guided by the rules of a diagram technique. Diagrams are made up of (owned) ModelElements and references to ModelElements (which may occur in other views).

A useful consequence of the reification of views and viewpoints would be the ability to map out the set of viewpoints employed by an AD and their relationships. A "big picture" notation for this would allow the Architect to sketch and document the expected content of an AD. Figure 4 suggests what a big picture might include: stakeholders (icons), concerns (arrows), viewpoints (triangles) and views (packages). Relationships between packages can be used to express traceability relations and other linkages. In some methods, there is a need to be able to articulate precedence among views – e.g., the requirements view is developed before the design view.

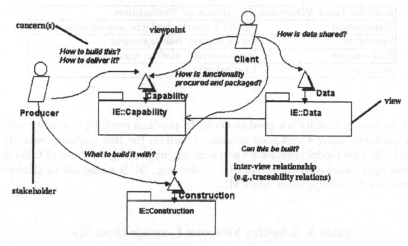

Fig. 4. Fragment of a Big Picture

The notions above are useful outside of the realm of architecture. Similar distinctions are useful in requirements and design, as well. A generalized view and viewpoint mechanism would be valuable throughout the UML.

(4) "Outside UML Ontology"

The ideal architect should be a man of letters, a skillful draftsman, a mathematician, familiar with historical studies, a diligent student of philosophy, aquainted with music, not ignorant of medicine, learned in the responses of jurisconsults, familiar with astronomy and astronomical calculations.

— Vitruvius, *De Architectura* (25 BC)

So far, I have stayed within the subject matters of viewpoints to which the UML is well-suited: the structure, organization and behaviors of a software-intensive system. But architecture – even software-intensive systems architecture – is a multi-disciplinary endeavor [24], and many diverse disciplines may be brought to bear on the Architect's challenge.

There are many existing system disciplines – each with their own existing notations, models and analytic methods. Table 2 suggests a few. For these cases, it is perhaps more efficient to use existing techniques than to recast them in the UML. This does, however, exacerbate the view-integration problem (see section 5).

Table 2. Other Viewpoints

Architectural Viewpoint	Notations or Techniques
Management	GANTT charts, budgets, organization charts
Reliability	block models, failure models, ...
Performance	discrete-event models, queueing models ...
Security	(see below)

Information security is a good example. In previous work [6], a security view was developed using a viewpoint language based on the Bell-LaPadula security model [2]. This model provides a viewpoint language along the lines of table 3. Recent approaches adopt other security models e.g., [3]. It would not be efficient to build each such approach into UML.

Table 3. A Security Viewpoint Language (from [6])

(Security) Objects	The resources being protected
(Security) Subjects	Active objects which can perform operations on objects
Policy	The set of rules which specify, for each object and each subject, what operations that subject is allowed to perform on that object
Operations	The ways in which subjects interact with objects
Info domains	Subjects and objects live in domains. Domains may have different security levels, and interconnect with other domains

5 General Issues

There are a number of other issues with the use of the UML for architectural description that are not particular to IEEE P1471. Many of these issues pertain to the generality of the UML ontology of concepts (and its metamodel) – which has been developed for object-oriented analysis and design applications – and the adequacy of that ontology for architectural concerns. For a discussion of expressiveness in architectural description, not specific to the UML, see [13]. In this section, I highlight two issues: view integration and the fixed nature of certain UML "built ins".

View Integration Multiple views necessitate means for view integration: maintaining consistency among views. At present, the UML provides only minimal mechanisms of this kind, such as the trace and refine relationships. The trace relationship is used to represent historical deriviations of an element. The refine relationship is used to represent the same element at different levels of abstraction. The assumption of identity for each of these relations is too restrictive for architectural view integration. Within the P1471 framework, one would like to be able to state such relations at:

1. the *viewpoint level*; e.g., subjects and objects in a security viewpoint are always components in a structural viewpoint;
2. the *view level*: Data access interfaces in the data view are implemented as capabilities (or, components) in the structural view;
3. the *element level*: element a in view $\mathcal{X}$ and element b in view $\mathcal{Y}$ are descriptions of the same thing.

Notations for view integration can be built as relationship stereotypes, to handle the first two cases. For the third case, element-to-element mapping, additional work would be needed, whether in UML, OCL or elsewhere.

Built-In Features Some of the built-in features of the UML, inspired by analogous programming language mechanisms, are not sufficiently general for architectural use, but cannot be overridden in the metamodel, because they are not first-class. For example, the built-in model of visibility only allows a fixed set of values (public, private, protected). Architecturally, one might like to tailor the visibility model, e.g., to support a particular security model, or for a particular distributed systems paradigm (such as the Distributed Systems Annex of Ada 95).

6 Closing

The UML is well-suited for use with IEEE P1471, providing a ready-made suite of notations to model many of the aspects of the architectures of systems. The predefined diagram techniques are all applicable within commonly used architectural viewpoints (see table 1).

The "lightweight extension mechanisms" – stereotypes, tagged values, and constraints – provide the means to develop specialized vocabularies which the Architect may use as viewpoint languages to address specific areas of concern. For example, the structural viewpoint, well-studied within academic software architecture, is readily captured in this way via component and class diagrams and a set of extensions. Most existing architecture description languages fall within this viewpoint, while other architectural concerns fall outside the current ontology of UML.

By reifying views, viewpoints and diagram techniques, the Architect has the basis to more flexibly specify and manipulate viewpoints than with the proposed profile mechanism. This generality is applicable outside of architecture, as well.

Relation to Other Work. Although first-class views and viewpoints are somewhat new to the architecture community with IEEE P1471, these concepts are found in requirements engineering [7] and aspect-oriented programming [18].

Kruchten was probably the first to use UML for architectural description (even before it was UML!) [19]. Medvidovic and Rosenblum show how to support an architectural style (the C2 style) using the UML's lightweight extension mechanisms [21]. Much of the work on architectural styles in Software Architecture takes place within the Structural Viewpoint. An interesting question is whether the UML can support a "layered" approach to extensions, such that the C2 style "extension" could be defined atop the Structural Viewpoint, as defined above. Hofmeister *et al.* at Siemens define several viewpoints (which they call "views") using UML [15]. As proposed above, each viewpoint uses more than one diagram technique for its expression: their Conceptual Viewpoint comprises UML class, state and sequence diagrams (with the ROOM extensions); their Module Viewpoint comprises class and package diagrams; their Execution Viewpoint comprises class, sequence and state diagrams; and their Code Viewpoint uses component diagrams and a supplementary table [15]. Egyed is using the UML as an integration framework for architectural views [5]. Most work on general ontologies for architecture outside of the system ontology provided by the UML metamodel is qualitative, rather than model-based.

Acknowlegments. This paper is based on a talk given in April 1999 at the *Workshop on Architecture and UML* in Denver, sponsored by Rational. Comments on that presentation by the participants have improved the presentation here. I would also like to thank the members of the IEEE Architecture Working Group for many useful discussions on views and viewpoints. Thanks to the reviewers for useful suggestions and to D. Emery (The MITRE Corporation) for comments on an earlier version of this paper.

References

[1] OMG Analysis and Design Platform Task Force. White paper on the profile mechanism (version 1.0). OMG Document ad/99-04-07, April 1999.

[2] D. Bell and L. J. LaPadula. Secure computer systems: unified exposition and Multics interpretation. Technical Report MTR-2297, The MITRE Corporation, Bedford, MA, 1976.

[3] Christophe Bidan and Valérie Issarny. Dealing with multi-policy security in large open distributed systems. In *Proceedings of the 5th European Symposium on Research in Computer Security*, number 1485 in Lecture Notes in Computer Science, pages 51-66, Belgium, September 1998. Springer-Verlag.

[4] Grady Booch, James Rumbaugh, and Ivar Jacobson. *The Unified Modeling Language User Guide*. Addison-Wesley, 1999.

[5] Alexander Egyed. Integrating architectural views in uml. Technical Report USC/CSE-99-TR-514, Center for Software Engineering, University of Southern California, Los Angeles, CA, 1999.

[6] David E. Emery, Rich Hilliard, and Timothy B. Rice. Experiences applying a practical architectural method. In Alfred Strohmeier, editor, *Reliable Software Technologies-Ada-Europe '96*, number 1088 in Lecture Notes in Computer Science. Springer, 1996.

[7] A. Finkelstein, J. Kramer, B. Nuseibeh, L. Finkelstein, and M. Goedicke. Viewpoints: a framework for integrating multiple perspectives in system development. *International Journal of Software Engineering and Knowledge Engineering*, 2(1):31-57, March 1992.

[8] A. Finkelstein and I Sommerville. The viewpoints FAQ. *Software Engineering Journal*, 11(1):2-4, 1996. Also available from ftp://cs.ucl.ac.uk/acwf/papers/viewfaq.ps.gz.

[9] Cristina Gacek, Ahmed Abd-Allah, Bradford Clark, and Barry W. Boehm. On the definition of software system architecture. In *Proceedings of the First International Workshop on Architectures for Software Systems*, Seattle, WA, 1995.

[10] David Garlan, Robert T. Monroe, and David Wile. Acme: An architecture description interchange language. In *Proceedings of CASCON '97*, pages 169-183, November 1997.

[11] Rich Hilliard. Views and viewpoints in software systems architecture. Position paper from the *First Working IFIP Conference on Software Architecture*, San Antonio, 1999.

[12] Rich Hilliard, Michael J. Kurland, and Steven D. Litvintchouk. MITRE's Architecture Quality Assessment. In *1997 MITRE Software Engineering and Economics Conference*, 1997.

[13] Rich Hilliard and Timothy B. Rice. Expressiveness in architecture description languages. In Jeff N. Magee and Dewayne E. Perry, editors, *Proceedings of the 3rd International Software Architecture Workshop*, pages 65-68. ACM Press, 1997. 1 and 2 November 1998, Orlando FL.

[14] C. Hofmeister, R. L. Nord, and D. Soni. Architectural descriptions of software systems. In D. Garlan, editor, *Proceedings of the First International Workshop on Architectures for Software Systems*, pages 127-137, Seattle, WA, 1995. Published as CMU-CS-TR-95-151.

[15] C. Hofmeister, R. L. Nord, and D. Soni. Describing software architectures with UML. In Patrick Donohoe, editor, *Proceedings of the First Working IFIP Conference on Software Architecture*, pages 145-160. Kluwer Academic Publishers, 1999.

[16] IEEE Architecture Working Group. *IEEE P1471/D5.0 Information Technology—Draft Recommended Practice for Architectural Description*, August 1999. Available by request from http://www.pithecanthropus.com/~awg/.

[17] International Organization for Standardization. *ISO/IEC 10746 1-4 Open Distributed Processing - Reference Model - Parts 1-4*, July 1995. ITU Recommendation X.901-904.

[18] Gregor Kiczales, John Lamping, Anurag Mendhekar, Chris Maeda, Cristina Lopes, Jean-Marc Loingtier, and John Irwin. Aspect-oriented programming. Xerox Palo Alto Research Center, 1997.

[19] Philippe B. Kruchten. The 4+1 view model of architecture. *IEEE Software*, 28(11):42-50, November 1995.

[20] David C. Luckham, John J. Kenney, Larry M. Augustin, James Vera, Doug Bryan, and Walter Mann. Specification and analysis of system architecture using Rapide. *IEEE Transactions on Software Engineering*, 21(4), April 1995.

[21] Nenad Medvidovic and David S. Rosenblum. Assessing the suitability of a standard design method for modeling software architectures. In Patrick Donohoe, editor, *Proceedings of the First Working IFIP Conference on Software Architecture*, pages 161-182. Kluwer Academic Publishers, 1999.

[22] Object Management Group. *Unified Modeling Language - Semantics (version 1.1)*, September 1997. OMG ad/97-08-04.

[23] Object Management Group. *Unified Modeling Language - Summary (version 1.1)*, September 1997. OMG ad/97-08-03.

[24] Eberhard Rechtin and Mark Maier. *The art of systems architecting*. CRC Press, 1996.

[25] Mary Shaw and David Garlan. *Software Architecture: Perspectives on an emerging discipline*. Prentice Hall, 1996.

[26] D. Soni, R. L. Nord, and C. Hofmeister. Software architecture in industrial applications. In *Proceedings of the 17th International Conference on Software Engineering*, Seattle, Washington, 1995.

Viewing the OML as a Variant of the UML

Brian Henderson-Sellers[1], Colin Atkinson[2], and Don Firesmith[3]

[1] University of Technology, Sydney, PO Box 123, Broadway, NSW 2007, Australia
brian@socs.uts.edu.au
[2] Fraunhofer Institute, Kaiserslautern, Germany
atkinson@iese.fhg.de
[3] Lante Corporation, Dallas, USA
FiresmithD@aol.com

Abstract. The OPEN Modelling Language, OML, was published during the standardization process which finally led to UML version 1.3.
While being contributory to this process, there are still some features
of the OML which have not been adopted in the current version of the
UML. These features offer capabilities which are complementary to those
of the UML. This paper describes how these features of the OML can be
made available to UML developers by viewing the OML as a variant of
the UML.

1 Introduction

The UML [1] and OML [2] are two object-oriented modelling languages which
were both developed in response to the unease in the software industry about
the growing divergence of object-oriented methods, and the often unnecessary
differences in object-oriented modelling notations. Both notations represent an
attempt to capture the core concepts of object-orientation and standardize upon
a set of intuitive graphical icons. As such, they share many core concepts and,
in fact, have many common roots. However, there are certain areas in which the
OML and UML do differ significantly, and where UML developers may benefit
directly from the OML features not currently supported in the OMG standard.
Alternatively, OML developers may benefit from access to UML features not
supported in OML.

Given the large overlap between the core concepts, it would seem desirable
to provide object-oriented modellers with the union of features in the OML and
UML. Fortunately, the UML provides a couple of ways to extend the features of
the UML with new concepts: one is called a UML variant and the other a UML
extension.

An extension uses special "built in" features at the M1 level (Fig. 1). These
features are stereotypes, tagged values and constraints, together with appropriate notational elements. The changes are made at the model level. The UML
documentation contains two such pre-defined extensions: one for business engineering and one for supporting the Objectory process [1, 3]. So for instance, we
might choose (as does Objectory: [1] (p4-7)) to specialize the class concept into

boundary classes, entity classes and control classes by application of an appropriate stereotype at the M1 (model) level. The resulting combination of UML and these additional user-defined stereotypes is known as a "UML extension".

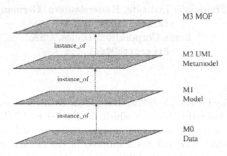

Fig. 1. UML's four layer architecture

A UML variant, on the other hand, extends the UML metamodel at the M2[4] level i.e. the metamodel itself. The variant uses the existing architecture of the metamodel for UML and adds concepts (metatypes) to the metamodel. The resulting metamodel is known as a "UML variant".

In a paper of this size it is not possible to give a complete specification of an OML extension/variant to the UML. The focus of this paper is rather to describe the merits of doing so, discuss the relative pros and cons of using the variant or extension mechanism, and to illustrate what form such a variant would take. Following a brief background-setting overview of the OML (Sect. 2), we then give, in Sect. 3, a technical description of some of the metalevel features of UML and OML, and the general nature of the difference between them. We also describe a number of features of the OML which we believe would be particularly beneficial to UML developers. In the following two sections (Sects. 4 and 5), we describe why the UML variant approach seems to make more sense than the UML extension approach for this purpose and also describe, in Sect. 5, how such a variant would be defined. In Sect. 6, we extend the discussion to propose the use of conformant and non-conformant variants.

2   The History of OML and Future Contributions to the UML

OML [2], was published in early 1997 during the standardization process which finally led to the (current) UML Version 1.3 [1]. While contributing to that process, there are still some features of OML which have not yet been adopted into UML. An overview of the comparative features of OML, as compared to UML, was given in [2, 5] and in more detail in [6].

OML has been influenced by pure OO approaches such as RDD [7] and Eiffel [8] but at the same time remains a completely programming language-independent modelling language. OML Version 1.0 (published in 1997) was amended slightly in 1998 [9,10] to bring it into alignment with the, by then OMG-endorsed, UML Version 1.0. Despite the large overlap between the OML and the UML, there are still some useful features of the OML which are not currently adequately supported in the UML and which UML users may find helpful for their modelling work. In this paper, we identify these features and explain how the most important OML-specific features may be reconciled with the OMG standards in the form of a UML variant. By presenting these features as an extension to the UML, we can make these OML features available to developers using UML in their development project. The long-term goal would be to offer these modifications to the OMG for potential inclusion in future versions (e.g. Version 2.0) of the OMG/UML standard.

3 Key OML and UML Metamodel Fragments

3.1 UML

UML is characterized by, and emphasizes, use cases, relationships as reifiable classes, an obvious data modelling heritage, the use of bidirectional associations and rôles on association ends from OMT and an increasing reliance on stereotypes.

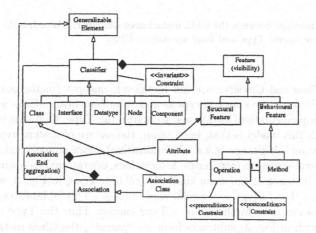

Fig. 2. Main structural elements of the UML metamodel (Version 1.3)

Fig. 2 shows the static architectural model for UML in which there are five major metalevel concepts, collectively called Classifier: Class, Interface,

Datatype, Node and Component. In UML, an *object* is "an instance that originates from a class" [1] (p2-90), but is not defined in the core model package. An
object in UML has only attributes and no operations [1] (p3-53).

Interface is shown as a subtype (using the Generalization relationship) of
Classifier which means, according to the definition of Generalization in [1] (p2-
34) (see also below), that the Interface has all the characteristics of its supertype,
Classifier. Since a Classifier has attributes, methods and operations while an
Interface has only operations, this would appear to be inconsistent with the
axiomatic definitions, meaning that the UML metamodel is not applied correctly
in describing itself.

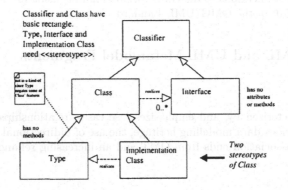

Fig. 3. Relationships between the UML metaclasses Classifier, Class, Interface and the
two Class stereotypes: Type and Implementation Class

While Class and Classifier are full metalevel concepts (metaclasses), there
are other relevant stereotypes: ≪type≫ and ≪implementationClass≫ (Fig. 3)
where the ImplementationClass is said to realize the Type. One area of current
concern with this model is that, once again, the nature of a stereotype is that
of specialization inheritance (a.k.a. Generalization), since stereotyped instances
have "the same structure (attributes, associations, operations) as a similar non-
stereotyped instance of the same kind" [1] (p2-66). Thus, in Fig. 3, we should
be able to say that "a type is a special kind of class". This is, however, not true
since a Class can have Methods but a Type cannot. Thus the Type metaclass
(shown as such in Fig. 3) subtracts from its "parent", the Class metaclass (as
does Interface from Classifier) — this subtraction technique with "inheritance"
was recognized many years ago by Brachman [12] as a bad modelling strategy.

Classifiers are composed of (black diamond notation) features which have
an associated visibility (Fig. 2). Features may be structural or behavioral. Behavioral features are either operations or methods, where methods implement

operations. Structural features are only attributes. Assertions can be included by adding stereotyped Constraints to operations and classifiers as shown.

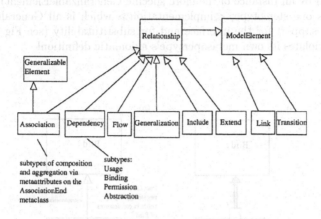

Fig. 4. Metamodel fragment for relationships in UML Version 1.3

Fig. 4 shows the Version 1.3 metamodel for UML relationships. Relationship is an abstract metaclass with no specific semantics which defines "a connection among model elements" [1] (p2-41). Subtypes of Relationship are Association, Dependency, Flow and Generalization. In addition, Include and Extend are metaclasses in the Use Cases package which also inherit from Relationship (in the Core package). Extend and Include are also said to be directed relationships although this does not seem to be enforced anywhere in the metamodel.

Of the metasubtypes, an Association is said to be "a semantic relationship between classifiers" [1] (p2-19) which defines a set of tuples relating the instances of these classifiers. It is bidirectional in nature. A Dependency is a "term of convenience for a relationship other than an Association, Generalization, Flow, or metarelationship" [1] (p2-30). It is a undirectional (or directed) relationship — but the unidirectionality is not enforced. There are four kinds of Dependency: Abstraction, Binding, Permission and Usage (shown with stereotype labels in the notation). Abstraction may be unidirectional or bidirectional and has four predefined stereotypes: Derivation, Realization, Refinement and Trace (also shown with a single stereotype label in the notation). While ≪realize≫ may be unidirectional or bidirectional, ≪derive≫ is unidirectional whereas for ≪trace≫ it is said that the "directionality of the dependency can often by ignored" [1] (p2-19). Binding, Permission and Usage are unidirectional and the last two have several pre-defined stereotypes each. Flow represents a relationship between two versions of an object and is a directed relationship. It has two pre-defined stereotypes.

The Generalization relationship is a "taxonomic relationship between a more general element and a more specific element. The more specific element is fully consistent with the more general element (it has all its properties, members and relationships) and may contain additional information." It is "a subtyping relationship (i.e. an Instance of the more general GeneralizableElement may be substituted by an Instance of the more specific GeneralizableElement)" [1] (p2-34). It has one stereotype, ≪implementation≫ which is all Generalization is *minus* the support for the interface and for substitutability (see Fig. 5(b)). It therefore violates its own metasupertype's axiomatic definition!

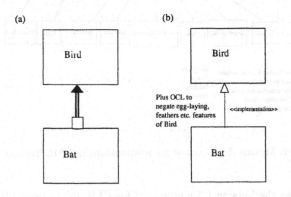

Fig. 5. Modelling Bat as a subclass of Bird because both can fly: (a) using OML, implementation (or white box) inheritance is used directly where this relationship is a subtype of Inheritance and a peer of Generalization; and (b) using UML, implementation inheritance is a stereotype of Generalization which means that, *de facto*, it has the same properties as its superclass, Generalization. Unwanted features of Generalization have then to be negated away by use of appropriate OCL constraints.

Link and Transition are also shown in Fig. 4. While these are not part of the Core package's Relationship hierarchy, they are intimately connected (as are Include and Extend from the Use cases package). A Link is simply an instance of an Association. It is a subtype of ModelElement in UML Version 1.3, not of Relationship. The Transition metatype is also a subtype of ModelElement but in this case it would appear from the UML documents [1] (p2-132) that it should in fact be a subtype of Relationship since it is defined to be "a directed relationship between a source state vertex and a target state vertex" in the state machine metamodel.

Rôles in UML have two meanings: (i) as a label on the AssociationEnd or (ii) as AssociationRole, AssociationEndRole or ClassifierRole in the collaboration diagram. As a label on the AssociationEnd it is "a name string near the end of the path" [1] (p3-61); a concept which does not seem to be supported in the metamodel itself. In the collaboration diagram, the classifier rôle describes how a specific participant (interface) in a collaboration may play a specific rôle. It is effectively a viewpoint on an object in the specific context of the collaboration in question. A ClassifierRole thus defines a set of Features which are themselves a subset of those in the base Classifiers. AssociationRoles and AssociationEndRoles are the corresponding usages of Associations/AssociationEnds in the context of a collaboration. The two different meanings of rôle are described in [13] (p414) as the static and dynamic aspects of rôles.

In summary, the fragments of the UML metamodel[1], described above, to which OML offers extension or modification, relate to the focal points of (a) class/type/interface metamodel structure, (b) responsibilities, (c) relationship hierarchy — especially aggregation relationship, inheritance stereotypes and dependencies and (d) rôle modelling.

3.2 OML

OML is characterized by a balanced use of use cases, responsibilities and rôle modelling. In OML, all relationships are unidirectional to preserve encapsulation/information hiding [15]. Stereotypes are used similarly to UML.

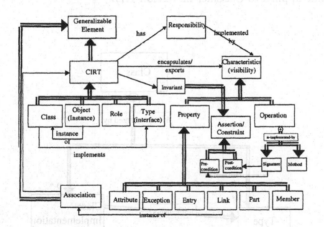

Fig. 6. OML Version 1.1 metamodel — incomplete fragment of the static architecture with the same scope as Fig. 2

Fig. 6 shows the core elements of the static metamodel, laid out in the same way as Fig. 2 for ease of comparison. Ignoring terminological differences, we note strong similarity except that the CIRT supertype is concrete (with its own

---

[1] We restrict our discussion here to aspects of the static, architectural components of both the UML and OML metamodels.

notation) and its subtypes are not identical to those of the UML Classifier. CIRT stands for Class or Instance or Rôle or Type. While Class and Type map roughly to UML's Class and Interface, the three UML subtypes of Datatype, Node and Component are missing (but could easily be added or retained as stereotypes) and OML has two subtypes NOT in the UML metamodel: Instance (Object) and Rôle. In strict metamodelling[2], Instance and Type/Class would not appear in the same Mx layer (Fig. 1). On the other hand, the appearance of Rôle as a subtype of CIRT is purposeful since there is a need to support rôle modelling in the class diagram as well as in the collaboration diagram. Also of note in Fig. 6 is the fact that OML eschews the ideas of AssociationClass and AssociationEnd as distinct metaclasses.

The structure of the Feature/Characteristic hierarchy is also richer in OML. Property has more than the single subtype of Attribute as in UML (Fig. 2). These other subtypes relate to the more extensive use of association and aggregation modelling in OML (see below).

A very important metaclass in OML is that of Responsibility (Fig. 6). Classifiers/CIRTs have high level responsibilities each of which is linked to/realized by one or more features (which may be structural or behavioural). Responsibilities are adopted from the work of [7] and carry significant semantics — unlike the responsibility notion in UML which is a stereotyped comment in Version 1.3 (tagged value applied to Classifier in Version 1.1).

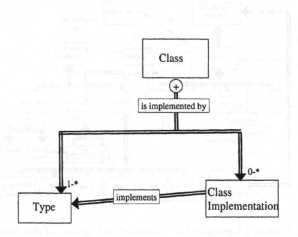

Fig. 7. In OML, the definitional relationship between Class, Type and Class Implementation is that of aggregation not inheritance

In OML, Types and Interfaces are not strongly differentiated in the metamodel. A Type is defined [2] (p17) to be a declaration of visible characteristics

---

[2] Neither UML nor OML employ strict metamodelling.

(= UML Features) that form all or part of the interface. This set of characteristics defining the type is therefore a subset of those defining the interface. In OML, it is not only objects and classes that can have interfaces but also use cases, packages etc. Thus if a class has a single type, type is identical to interface. Since one (type) is a subset of the other (interface), only one concept (metaclass) is retained in the OML metamodel. This is in contrast to UML where a set of operations define a service. This set is called the Interface, which is roughly analogous to Type in OML.

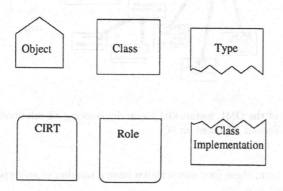

An abstract or deferred class has a dotted outline

Fig. 8. Stereotypes in UML may be given their own graphical icons. Here are some of those suggested in OML

Secondly, since Type in OML is the declaration of the external view or specification, and since the full Interface consists of one or more Types[3], then Interface can be considered as redundant and the totality of the Class is in fact a combination (or aggregation) of this Type and its Class Implementation (Fig. 7). Thus, instead of generalization, OML uses aggregation to link together the concepts of Type (inclusive of Interface), Class and Class Implementation (Fig. 7 compared to Fig. 3). In the OML notation, the various stereotypes of Fig. 3 are all given icons, as permitted within the OMG standard (Fig. 8). These were chosen based on semiotic[4] principles to make learning easier and more intuitive.

The original relationship metamodel for OML made a clear distinction between four sets of relationships: referential and definitional (as used in class diagrams/semantic nets), transitional (state models) and scenario (use cases di-

---

[3] The use of the names Type and Interface in UML and Java is opposite to that in OML

[4] The study of signs and symbols

58

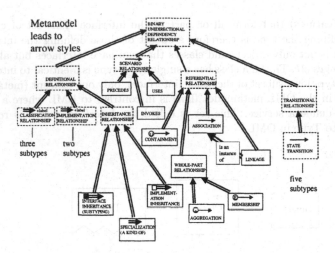

Fig. 9. Fragment of the OML metamodel hierarchy showing all relationships — updated from [2] to Version 1.1 based on [9]

agrams etc.). In turn, these four metaclasses have a number of subtypes (Fig. 9). Of specific interest here are

- all relationships are binary and *unidirectional*
- all relationships are dependency relationships (tying in with their unidirectionality)
- aggregation (Fig. 10), membership and containment are clearly defined subtypes of association
- there are three types of inheritance relationship: generalization/specialization (a-kind-of), interface (blackbox or subtyping) and implementation (whitebox) — all peers.

Since all relationships are unidirectional, they are arrowed to indicate the direction of dependency. Thus associations have a single direction which means that an unarrowed association can be given the meaning of "TBD" (to be decided) — a useful modelling tool when doing rapid analysis and design sketches of the emerging model. Additionally, bidirectional relationships are only a shorthand for a pair of unidirectional relationships that are semi-strong inverses of each other. The iconic representation of the three specific subtypes of Association (Membership, Containment and Aggregation) offers visual differentiation. The black and white box on two of the subtypes of inheritance is another valuable visual reminder (Fig. 5(a)).

OML's specialization inheritance (Fig. 9) is in full agreement with the UML definition of Generalization (see Sect. 3.1). However, generalization is only one kind of "inheritance", the others being interface inheritance and implementation

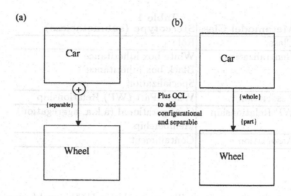

Fig. 10. Modelling the commonest type of whole–part relationship: (a) directly in OML using the configurational symbol (plus in a circle) and the {separable} stereotype; (b) in UML — because neither black nor white diamond can be used to describe a configurational/separable whole–part relationship, it has to be constructed from a regular association to which is added: (i) a {whole} and a {part} constraint, (ii) a navigability arrow and (iii) some OCL constraints to make the association both configurational and (iv) separable.

inheritance (Fig. 5(a)) — although in practice the first two are often purposefully confounded. These three (or pragmatically two) form a partition of an (abstract) superclass in the metamodel (called Inheritance Relationship in Fig.9).

4 The UML Extension Mechanism

The three extension mechanisms available in UML are stereotypes, tagged values and constraints. Some of OML's characteristics could possibly be re-expressed with stereotypes. In particular, Table 1 shows the necessary stereotypes together with the metamodel class which they extend in OML.

Although defined at the model or M1 level, a stereotyped class can thus be thought of as a "virtual" or "pseudo" M2 class which partitions an existing M2 metaclass. In other words, we might describe a stereotype as creating an implicit user-defined metasubtype.

Thus, OML requires a stereotype ≪rôle≫ for Classifiers to permit their use in Class diagrams as well as the existing support in Collaboration diagrams — although a new M2 metasubtype would be much more powerful.

OML has three distinctive kinds of inheritance which could be given stereotypes. The problem here is that specialization, specification and implementation inheritance really create a single partitioning rule i.e. they should be peers, to-

Table 1

Metamodel Class	Stereotype (submetaclass)
Class	Rôle
Generalization	White box inheritance
	Black box inheritance
	Specialization
Association	Whole–Part (WP) Relationship
WP Relationship	Configurational (a.k.a. aggregation)
	Membership
Association	Containment

gether with an abstract supertype. To create this in UML would require a variant not an extension (see Sect. 5).

OML has strong support for "aggregation" (configurational whole–part (WP) association relationship) and "membership" (non-configurational WP relationship). One possible way to represent these using stereotypes would be to add a ≪WPRelationship≫ stereotype to Association (with aggregationKind set equal to none) from which two additional stereotypes of ≪configurational≫ and ≪membership≫ could be created. The third new Association type, Containment, is then a stereotyped Association. On the other hand, a more semantically powerful representation in the metamodel is discussed in Sect. 5 using the idea of a UML variant.

In addition to showing the stereotypes of Table 1 as keywords in guillemets, we recommend the following new icons: rôle: Greek tragedy mask (Fig. 8); kinds of inheritance: white and black box options (Fig. 9); WP relationship: annotations at client end of relationship (Fig. 9); and Containment: annotation at client end of relationship (Fig. 9).

In conclusion, whilst some of OML's constructs can be readily represented as stereotypes, many of these can more cleanly use the variant ideas supported in UML.

5 OML as a UML Variant

In this section, we describe the OML model elements which cannot be simply created at the model level (M1) by judicious use of user-defined stereotypes. Instead, the metamodel (M2) requires modification, thus creating a UML variant.

5.1 Responsibilities

In OML (Fig. 6), a CIRT has Responsibilities which are implemented by Characteristics. In the variant version of UML, we introduce a new metaclass called Responsibility which has a meta-association to Classifier and a meta-association to Feature (Fig. 11 which shows the relevant fragment of Fig. 2, updated in this way.) This replaces the current UML responsibility which is (a) a stereotyped comment with no semantics and (b) confused with the notion of a contract e.g. [11, 16].

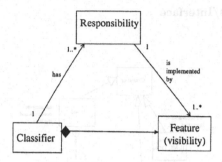

Fig. 11. Addition of Responsibility metatype to UML variant

Notation for responsibilities is already available in UML (adopted from OML: [17]). Responsibilities are documented in a fourth box on the class icon. What is required (see below) are well-formedness rules to ensure that the Class–Responsibility–Operation links are correct.

5.2 Aggregation

Although we have shown in Sect. 4 how whole–part relationships can, to some degree, be represented by the use of stereotypes in a UML extension, a cleaner model can be derived by judicious modifications to the metamodel itself. The following changes would be needed:

- in the AssociationEnd metaclass, the aggregation: aggregationKind meta-attribute should be erased.
- the introduction of a new metaclass called Whole–Part (WP) Relationship. While this can be regarded as a kind of Association, it is important that it inherits from a unidirectional relationship rather than from the bidirectional Association metaclass. This suggests that, despite the clear is-a-kind-of connection between it and the Association metaclass, the new WP Relationship metaclass might best inherit from either Dependency or Relationship.
- the addition of two new subtypes of Whole–Part Relationship metaclass: (a) Configurational and (b) Membership
- the introduction of a new subtype of Association called Containment

Annotation for whole–part and containment relationships are given already in OML. These could be used "as is" (Fig. 9). The Whole–Part Relationship metaclass also needs additional well-formedness rules. These would formally express the mandatory existence of (i) emergent property, (ii) resultant property,

(iii) irreflexivity and (iv) asymmetry ([18]). In addition, careful formal definition of containment, which is *not* a whole–part relationship, is needed — again we encourage the evaluation and derivation of appropriate well-formedness rules.

5.3 Type/Class/Interface

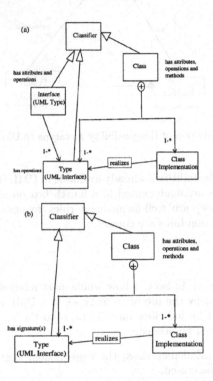

Fig. 12. Aggregation relationships between Class, Type and Class Implementation linked into the UML architecture involving Interface and Classifier metaclasses

The tidiest way to improve the metamodel of Fig. 3 would be a full revision using correct Generalization relationships. In making such a drastic change it might be better to totally revise this fragment of the metamodel. One suggestion is given in Fig. 12(a). It can be seen that a Class is made up from the specification (or Interface) which consists of several Types together with the Class Implementation. If required, an (OML) Interface is then equal to one-to-many Types[5]. However, a better model (Fig. 12(b)) might be one in which Type and

---

[5] This means that an alternative, but equivalent, model could be drawn in which a Class is made up of a single Interface plus one-to-many Class Implementations

Interface are fused together. This metaclass then has one or more signatures (no operations and no attributes).

If an inheritance hierarchy is preferred, then initially it seems possible to make an Interface (UML Type) to inherit from (Generalization relationship) Type (UML Interface) since a UML Interface has operations while a UML Type has not only operations (and their corresponding signatures) but also attributes. However, since Fig. 12 shows that an Interface (UML Type) is equivalent to one-to-many Types (UML Interfaces), a Generalization relationship is clearly inappropriate. Yet the need for a Classifier metaclass remains. However, for it to be a supertype of Interface, Type and Class (as is probably deemed preferable), the definition of Classifier needs to be modified. At present, a Classifier, like a Class, contains attributes, operations and methods. We propose that these elements are deferred to the Class, such that a Classifier is more of a place holder as an abstract class in the hierarchy, representing just that: model elements that represent the abstraction technique known as classification. If the hierarchy is constructed in the way suggested by Fig. 12, then all inheritance relationships are truly generalization and the purity of the metamodel (i.e. being defined using its own rules) is obtained.

It should be noted, however, that in a sophisticated metamodel such as the UML metamodel, there are likely to be other complications resulting from such a change. Nevertheless, so far as the OML variant of UML is concerned, this slight modification to Fig. 3 *does* result in an acceptable and usable definition of Class, Type and Interface — although it should also be noted that the names of Type and Interface in Fig. 12 are UML nomenclature and there is still the terminological argument between "interface" and "type" in the more general OO modelling community.

5.4   All Relationships are Dependencies

While the original (Version 1.0, 1.1) relationship metaclasses were distinct, in UML Version 1.3 a partial unification has taken place as shown in Fig. 4. As well as requiring Transition to inherit from Relationship rather than ModelElement[6] in OML, some discussion of Association and Dependency is needed. In UML, Association is bidirectional and the others (apart from some subtypes) are unidirectional. In OML, all relationships are binary (whereas ternary are permitted in UML's associations), undirectional and dependency relationships. Thus, rather than use Association as the base class, the OML variant will focus on Dependency (which is already unidirectional). From this will be constructed the model elements of Fig. 9. The major elements of OML are shown in Fig. 13. The UML Dependency and Relationship metaclasses are fused together into the root Dependency Relationship metaclass (abstract) and the newly introduced Referential Relationship is an abstract metaclass acting as a place-holder. To avoid

---

[6] We presume this is an error in the Version 1.3 draft documents since the text suggests that in UML V1.3 it was always intended that Transition should inherit from Relationship not from ModelElement as shown in the metamodel.

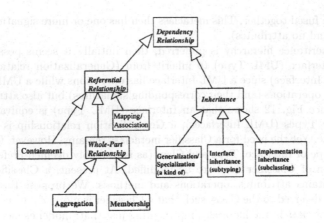

Fig. 13. Suggested OML variant structure for relationships which take the UML Relationship hierarchy of Fig. 4 but refocus on Dependency, ignoring Association and AssociationEnd, and introducing a new Whole-Part Relationship together with its subtypes. Similarly, three subtypes of inheritance are used such that the Generalization metaclass becomes an abstract class, renamed Inheritance for clarity

name clashes in the namespace, OML Association has been renamed Mapping. Association and AssociationEnd are thus not part of the OML variant of UML.

6   Conformant and Non-conformant Variants

The basic premise of a UML variant is that additions should be made at the M2 level. It is implicit that these are *additions* as opposed to changes. Some of the suggestions in Sect. 5 fall into this category. However, other suggestions require changes rather than additions at the M2 level. We may wish to discriminate between these two uses of the word variant by qualifying them. The first we will call a *conformant variant* and the second a *non-conformant variant*. The ideas to improve the relationship and type/class architectures fall into this second class of non-variance. On the other hand, it is feasible to use the existing UML constraint mechanism to eliminate or modify existing metaclasses — in which case the conformant/non-conformant discrimination vanishes. Nevertheless, we will, for the present, label OML as a non-conformant variant of UML — although it contains elements which are extensions and elements which are conformant variants as well.

The possibility of adding the idea of a non-conformant variant to the UML permits evolution in a more flexible fashion than the two current extension mechanisms which both insist that every fragment of the current UML is "correct" and inviolate. The difficulty in adhering to this strict requirement is evidenced by the significant non-conformant changes made to the original UML metamodel by the Revisionary Task Force (RTF) — for instance the welcome addition of the Relationship metaclass in Version 1.3 (not seen in Version 1.1) and the total revision of the use case association types in Version 1.3 (now stereotyped Dependencies rather than the stereotyped Generalizations of Version 1.1).

A major area of concern remains in the use of Generalization relationships in the diagrammatic, and therefore semantic, definition of the UML metamodel. The definition is good but the use of it is clearly seen to be incorrect. It is vital that (non-conformant) checks of *all* uses of Generalization in the UML Version 1.3 metamodel be made (in preference to "patches" involving overwriting constraints) and, once a consistent definition of black and white diamond aggregation has been accepted (see, e.g., suggestions of [20]), the uses of aggregation and composition also need to be carefully examined in the metamodel definition.

Thus the introduction of the idea that suggested changes to the UML metamodel may in fact be non-conformant variants (i.e. improvements to the metamodel which extend and at the same time change/correct it) could be very valuable in both tightening up the UML and also creating a path forward for its future evolution.

7 Summary and Conclusions

A comparison of specific fragments of the published UML and OML metamodels has permitted us to identify the new stereotypes and metamodel changes necessary to permit the OML to be viewed as a UML variant. Finally, we discussed the potential for the introduction of both conformant and non-conformant variants. Non-conformant variants open up the opportunity for true evolution of the UML. OML is one possible step in that direction.

References

1. OMG: OMG Unified Modeling Language Specification (draft), Version 1.3 alphaR2, January 1999 (unpubl.) (1999)
2. Firesmith, D., Henderson-Sellers, B., Graham, I.: *OPEN Modeling Language (OML) Reference Manual*, SIGS Books, New York, 276pp (1997); Cambridge University Press, New York (1998)
3. OMG: UML Extension for Objectory Process for Software Engineering. Version 1.1, 1 September 1997. OMG document ad97-08-06 (1997)
4. Atkinson, C.: Supporting and applying the UML conceptual framework. Procs. ≪UML≫'98 (1998) 1–11
5. Henderson-Sellers, B.: OML: proposals to enhance UML. Procs. ≪UML≫'98 (1998) 319–329
6. Henderson-Sellers, B., Firesmith, D.G.: Comparing OPEN and UML: the two third generation OO development approaches. Inf. Software Technol. **41** (1999) 139–156
7. Wirfs-Brock, R., Wilkerson, B., Wiener, L.: *Designing Object-Oriented Software*, Prentice Hall, Englewood Cliffs, NJ, 368pp (1990)
8. Meyer, B.: *Eiffel: The Language*, Prentice Hall, New York, 594pp (1992)

66

9. Firesmith, D.G., Henderson-Sellers, B.: Upgrading OML to Version 1.1: Part 1. Referential relationships. JOOP/ROAD **11(3)** (1998) 48–57
10. Henderson-Sellers, B., Firesmith, D.G.: Upgrading OML to Version 1.1: Part 2 — Additional concepts and notations. JOOP/ROAD **11(5)** (1998) 61–67
11. Wirfs-Brock, R.J.: Adding to your conceptual toolkit: what's important about responsibility-driven design. Report on Object Analysis and Design **1(2)** (1994) 39–41
12. Brachman, R.J.: "I lied about the trees" or, defaults and definitions in knowledge representation. The AI Magazine **6(3)** (1985) 80–93
13. Rumbaugh, J., Jacobson, I., Booch, G.: *The Unified Modeling Language Reference Manual*, Addison-Wesley, Reading, MA, 550pp (1999)
14. OMG: UML Notation. Version 1.1, 15 September 1997. OMG document ad/97-08-05 (unpubl.) (1997)
15. Graham, I.M., Bischof, J., Henderson-Sellers, B.: Associations considered a bad thing. J. Obj.-Oriented Programming **9(9)** (1997) 41–48
16. Meyer, B.: Applying "design by contract". IEEE Computer **25(10)** (1992) 40–51
17. Booch, E.G.: public communication, Sydney, 19 April 1999
18. Henderson-Sellers, B., Barbier, F.: What is this thing called aggregation?. TOOLS29 (eds. R. Mitchell, A.C. Wills, J. Bosch and B. Meyer), IEEE Computer Society Press (1999) 216–230
19. Booch, G., Rumbaugh, J., Jacobson, I.: *The Unified Modeling Language User Guide*, Addison-Wesley, Reading, MA, USA, 482pp (1999)
20. Henderson-Sellers, B., Barbier, F.: Black and white diamonds. *Procs. ≪UML≫ '99*, Fort Collins, CO, October 1999 (1999), this volume

A Comparison of the Business Object Notation and the Unified Modeling Language

Richard F. Paige and Jonathan S. Ostroff

*Department of Computer Science, York University,
Toronto, Ontario M3J 1P3, Canada.* {paige,jonathan}@cs.yorku.ca

Abstract. Seamlessness, reversibility, and software contracting have been proposed as important techniques to be supported by object-oriented methods. These techniques are used to provide a framework for the comparison of two modeling languages, the Business Object Notation (BON) and the Unified Modeling Language (UML). Elements of the UML and its constraint language that do not support these techniques are discussed. Suggestions for further improvements to both BON and UML are described.

1 Introduction

> ... *There are two ways of constructing a software design: one way is to make it so simple that there are obviously no deficiencies, and the other way is to make it so complicated that there are no obvious deficiencies.*
>
> C.A.R. Hoare, *Turing Award Lecture 1980* [7].

As described by Brooks [1], the key factor in producing quality software is specifying, designing and implementing the conceptual construct that underlies the program. This conceptual construct is usually complex and highly changeable. It is abstract but has many different representations. The complexity of the conceptual construct underlying software is an essential property, not an accidental one. Hence, descriptions of a software entity that abstract away its complexity often abstract away its essence.

A suitable modeling language is needed to describe the conceptual construct, its design and implementation. A satisfactory description of the conceptual construct for an industrial-strength software system prior to its construction is as essential as having a blueprint for a bridge or a large building, before its construction commences.

In 1994, there were between 20 and 50 such languages [14]. Often users had to choose from among many similar modeling languages with minor differences in overall expressive power. But in a landmark meeting in 1994, methodologists and tools producers agreed that users needed a standard. At that moment, the seeds were sown for the UML, and it has since been embraced by leading software developers. From 1997, development of the UML standard – through the Object Management Group – has continued.

This paper conjects that the standardization on UML is premature and perhaps even counter-productive. The reason we use the word 'conject' is that a rational critique of

UML would first require building a theory of software quality and then developing metrics for measuring the quality of software developed via a particular approach. Such a theory and consequent metric is not currently available and we must thus resort to a more qualitative, and hence more subjective, analysis.

In the absence of a theory of quality, our starting point will be the remarks quoted at the beginning of this paper. Our main goal will be simplicity of language. But we will need criteria which will allow us to reject a feature of a language as being excessive.

We define quality software as software that is *reliable* and *maintainable*. Reliable software must be *correct:* it must behave according to specification; and, it must be *robust:* it must respond appropriately to input from outside the domain of the specification. Maintainable software must be *extendible:* that is, easy to change with changing requirements; and *reusable:* that is, it can be re-used in different applications.

Reliability is the key quality requirement. If the software does not work correctly and supply the required functionality it is unusable despite having other qualities. Maintainability is the other key requirement because maintenance often accounts for 70% or even more of the cost of the product.

In order to assess UML's contribution to quality, we will compare it to the BON notation and method. The BON approach to quality is to stress three techniques: *design by contract*, as a contribution towards reliability; and *seamlessness* and *reversibility*, as contributions towards maintainability. These terms are defined as follows.

- **Seamlessness.** Seamlessness allows the mapping of abstractions in the problem space to implementations in the solution space without changing notation. Seamless software development occurs by adding new classes, or by enriching already existing classes with additional features.
- **Reversibility.** Changes made during one stage of development can be automatically reflected back to earlier stages. A modification made to an implementation class can be reflected in changes to a BON design class. CASE tools exist to support such reversibility for BON and Eiffel.
- **Design by contract.** The obligations on and benefits of using features of classes are precisely specified, using assertions, with the class. BON was designed to support design by contract, and this coupled with support in programming languages like Eiffel helps to satisfy the seamlessness and reversibility requirements.

With the above definitions now in place, we will look for the simplest set of features that will allow us to describe the conceptual construct underlying our software. The following will be rejected.

- Any feature that militates against contracting, seamlessness or reversibility.
- Any feature that duplicates one already in the notation. (Note that this does not prevent using different *views* of a model, but, as we discuss in the conclusions, these views should ideally be generated automatically from a single model, to maintain consistency.)
- Any feature that is in the notation merely because a competing notation has it.

The language summary for UML (version 1.3) is 161 pages, whereas the summary for BON is just a few pages [10]. Further, BON has only one classifier (the class), while UML has an additional seven classifiers (e.g., datatype, use case). Among the UML classifiers, a case can be made for redundancy; e.g., datatype and interface can be encompassed by class. We fail to see why all of the UML classifiers are needed. The power of using only the class as a classifier is that it unifies modules (information hiding) with hierarchical subtyping, and thereby abets simplicity and seamlessness.

There are three ways to defeat our arguments.

1. Develop a scientific theory of software quality, and do suitable studies and comparisons to other methods to show the efficacy of UML.
2. Disagree with the notion of software quality, as defined above (although we feel that most developers will want to have reliability and maintainability).
3. Prove UML does at least as good a job as BON at reliability and maintainability.

The rest of the paper will focus on point 3. We hope to show that BON does a significantly better job than UML. We also suggest what changes could be made to UML to better support contracting, seamlessness and reversibility.

Our comparison of BON and UML is founded on the standards for both languages. The BON standard reference is [17]; the UML reference is [16]. We do not consider techniques or extensions beyond the standards. In part, we therefore do not consider UML stereotypes beyond those documented in the standard reference. We refer the reader to [13] for a longer version of this report.

2 Introduction to BON

BON is an object-oriented method possessing a recommended process as well as a graphical and a separate textual notation for modeling object oriented systems. The notation provides mechanisms for modeling inheritance and usage relationships between classes, and has a small collection of techniques for expressing dynamic relationships. The notation also includes an *assertion language*, discussed in more detail in Section 4; the method is predicated on the use of this assertion language. In this sense, BON is based on behavioral modeling. This should be contrasted with UML which is grounded in data modeling. The method is supported by the EiffelCase tool from ISE.

As previously mentioned, BON supports three main techniques: seamlessness, reversibility, and software contracting. As a result, BON provides only a small collection of powerful modeling features that guarantee seamlessness and full reversibility on the static modeling notations.

Early steps of the BON recommended process make use of the informal *chart* (CRC index card) notation for documenting potential classes, clusters of classes, and properties of classes. Intermediate steps rely on the BON static and dynamic modeling notations, which we summarize in following sections. The final step involves mapping a BON model into an OO programming language.

The BON method is not driven by use-cases, unlike UML and its compatible processes. In this sense, we would claim that BON is architecture-centric and contract driven, but not use-case driven. BON does implicitly apply use-cases with its object communication diagrams (they are called 'scenarios' therein), but they are not an emphasized part of the method, and are usually applied after design classes have been discovered.

2.1 What is not in BON?

BON is also distinguished by the so-called 'standard' modeling elements that it omits, in particular *data modeling* (e.g., via some variant of entity relationship modeling) and *state machines*. Using these elements breaks seamlessness and reversibility [17]. The modeling advantages that these elements offer are far outweighed by the advantages of seamlessness and reversibility.

Modeling the behavior of objects using finite state machines introduces an impedence mismatch, which requires translation of finite state machines into code or surrender of the class concept. We also lose seamlessness with data modeling, in part because of its reliance on binary associations, and in part because associations as a modeling concept break encapsulation [3, 4, 17]. It is claimed in [17] that using simple OO primitives, and not binary associations, for class relationships is sufficient for specifying all the interesting relationships between classes, and is guaranteed to maintain seamlessness.

We have previously mentioned that BON does not support use-cases directly, though they are implicitly applied during the later stages of the process where object communication diagrams are developed.

3 Seamlessness and Reversibility

In this section, we outline the basic BON modeling language, concentrating on those aspects of the language that support seamlessness and reversibility. As we shall see, all of the static diagram elements of BON are designed for this purpose. These elements will be compared with equivalents in UML, and we will discuss the support these UML elements provide to the aforementioned techniques.

3.1 Class interfaces

The fundamental construct in BON is the *class*; in UML terminology, the class is the only form of classifier available. A BON class is both a module and a type; it is a possibly partial implementation of an abstract data type. With BON, a class is the only way to introduce new types; this is because of the requirement for seamlessness.

A BON class has a *name*, an optional *class invariant*, and a collection of *features*. A feature may be a *query*—which returns a value and does not change the system state— or a *command*, which does change system state. BON does not include a separate notion of *attribute*. Conceptually, an attribute should be viewed as a query returning the value of some hidden state information.

Figure 1(a) contains a short example of a BON graphical specification of the interface of a citizen class. Class features, with optional behavioral specifications, are in the middle section of the diagram (there may be an arbitrary number of sections, annotated with visibility tags, as discussed later). An optional class invariant is at the bottom of the diagram. The class invariant is a predicate (conjoined terms are separated by semicolons) that must be *true* whenever an instance of the class is used by another object. In the invariant, the symbol @ refers to the current object; it corresponds to this in C++

and Java. Class *CITIZEN* has seven queries and one command. For example, *single* is a *BOOLEAN* query, while *divorce* is a parameterless command. Class *SET* is a generic predefined class with the usual operators (e.g., $\in$, *add*); it is akin to a parameterized class in UML, or a template in C++.

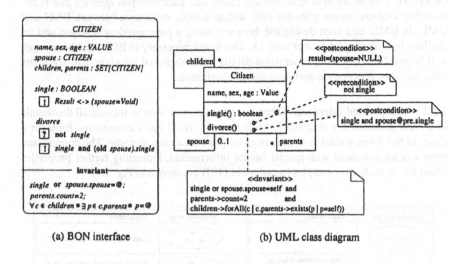

      (a) BON interface   (b) UML class diagram

Fig.1. A citizen class in (a) BON and (b) UML

A textual dialect of BON also exists [17]. It is syntactically similar to the Eiffel programming language.

A UML class diagram for a citizen is shown in Fig. 1(b). It is drawn assuming that we want to represent all details shown in Fig. 1(a) in UML (OCL is used for writing constraints); later, we discuss how the class diagram can be simplified.

Let us discuss the fundamental differences between the diagrams. First, consider the types of the attributes. In UML, attributes are intended to be used to represent data types (i.e., primitive types and enumerations, perhaps with simple multiplicities). A citizen class thus is not used as a type of an attribute in a class. So, *spouse, children*, and *parents* from the BON class interface must be modeled as associations in the UML class diagram, thus making the UML diagram more complicated and making seamlessness difficult to support. In BON, any class may be used in an interface. This leads to simpler models, abets seamlessness, and allows modelers to visually emphasize the most important relationships in their diagrams, thus aiding readability.

A second difference between the UML diagram and the BON diagram is with the behavioral specifications. We shall return to this point in detail in Section 4, but for now we mention that pre- and postconditions and invariants can be modeled using notes, extra boxes, and stereotypes. This clutters the diagram, as Fig. 1(b) shows. For this reason, behavioral details for classes are frequently omitted from diagrams and instead are presented using a textual assertion language, such as the OCL, separate from the diagram. Separation introduces the potential for maintenance and consistency problems.

3.1.1 Features Each class in BON has a collection of features, which may be queries or commands. All features of an object are accessed by standard dot notation. Identical syntax is therefore used to access attributes, and parameterless queries; this is the so-called *uniform access* principle [10], and is a clear difference between BON and UML. In UML, one must distinguish between using a parameterless function and an attribute by suffixing the former with ( ). This is not necessary in BON, and because of it, it is possible to hide implementation details from clients of the class, and allow the redefinition of functions to attributes under inheritance.

3.1.2 Compressed Interfaces Often, specifiers do not want to include all the details of a class interface in a diagram. For this purpose, BON has a *compressed form* for a class. In this form, a class is written as an ellipse containing its name. The compressed form can be annotated with special header information, indicating further properties about the class. Some examples (more are in [13]) are shown in Fig. 2.

Graphical form	Explanation	Graphical form	Explanation
NAME[G,H]	Class is parameterized.	• *NAME*	Class is (potentially) persistent.
* NAME*	Class is deferred. It has no instances, and is used for classification purposes.	▲ *NAME*	Class is interfaced with the outside world; some feature encapsulates external communication.

Fig.2. Compressed views and headers in BON

The ellipse notation in BON is equivalent to the rectangle in UML. The * header in BON, for a *deferred* class, roughly corresponds to an *abstract* class in UML. A deferred class has at least one unimplemented feature. The correspondence between deferred and abstract class is not exact. In UML, classes where all operations are implemented can still be marked as abstract, while this is not possible with BON. Deferred classes are also not the same as UML interfaces, since the former can contain attributes and behavioral specifications, while the latter cannot. Thus, the deferred class notion encompasses the UML notion of interface, and the most common uses of abstract class as well. It is not clear why both a notion of abstract (or deferred) class and interface need to be present in UML.

3.1.3 Visibility Visibility of features in BON is expressed by sectioning the feature part of the class interface, and by use of the feature clause. By default, features are accessible to all client classes that would use them. This is almost the same as public visibility in UML, except that in BON no client class can *change* the value of any query (that is, BON features are externally read-only). This restriction is necessary if we want proper information hiding.

More restrictive visibility of features can be expressed by writing a new section of the class interface and prefixing the section with a *list* of client classes. For example, a section prefix of `feature{A, B}` indicates that only classes A and B may access the features in the section. This form of visibility is directly implementable in Eiffel, and can be mapped directly to C++ via `friend` features and classes.

This should be contrasted with the mechanism supported by UML, which by default permits the C++/Java style of public, private, and protected features, via tagging each feature with a symbol. Tagging can be applied at both the class and the package level. The BON visibility mechanism is more flexible and general. It is also very helpful in the analysis and design phase, when class communication and coupling is being developed [10, 17].

3.2 Static architecture diagrams

BON provides a small, yet powerful selection of *relationships* that can be used to indicate how classes in a design interact. These relationships work seamlessly and reversibly with those that are supported by modern OO programming languages–especially Eiffel. There are only two ways that classes can interact in BON.

- **Inheritance:** one class inherits behavior from one or more parent classes. Inheritance is the subtyping relationship. It corresponds to generalization in UML: everywhere an instance of a parent class is expected, an instance of a child class can appear. There is only one inheritance relationship in BON; however, the effect of the relationship can be varied by changing the form of the parent classes (e.g., making parents deferred), and by using feature modifiers, e.g., *rename* and *redefine*.
  In BON, renaming mechanisms can be used to resolve multiple inheritance problems. By contrast, UML provides no mechanism for resolving such conflicts. According to [14], it is the responsibility of the designer to resolve class conflicts in multiple inheritance, for example, based on some provided programming language mechanism. This approach increases generality, but breaks seamlessness.
- **Client-supplier:** a client class has a feature that is an instance of a supplier class. There are two basic client-supplier relationships, association and aggregation, which are used to specify the *has-a* or *part-of* relationships between classes, respectively (the difference between the relationships is defined in Section 3.2.1). Both relationships are uni-directed; there is no undirected association in BON. These two relationships correspond to *singly navigable* associations and compositions, respectively, in UML, or to usage dependencies. There is no equivalent to UML's aggregation in BON. Client-supplier relationships can also be bidirectional and self-directed; we provide examples later.

Fig. 3 contains a non-trivial architectural diagram using BON, demonstrating examples of both inheritance and association. Thin vertical arrows (e.g., between *EXP* and *SD*) represent inheritance. Double-line arrows with thick heads (e.g., between *FTS* and *TRANSITION*) represent association. On the associations, names and optionally types of client features that use the supplier class can be specified, e.g., feature *events* on the association between *CLOCKCHART* and *EVENT*. The type of *events* is *generic*; *events* is a set of instances of *EVENT*. The BON naming notation for client-supplier relationships roughly corresponds to the UML notation for roles.

74

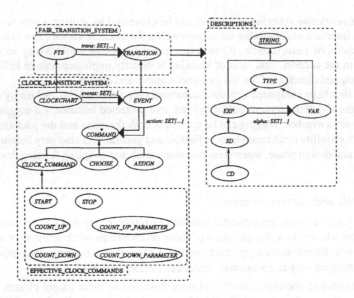

Fig.3. BON architectural diagram for fair transition systems

3.2.1 Client-supplier relationships Client-supplier relationships in BON are between *classes*, and constrain classes. The BON relationships can be mapped directly to attributes or functions in Eiffel and Java, and can be reversibly generated from Eiffel and Java programs.

Associations and aggregations in BON have no object multiplicities; class invariants can be used to express such constraints. In this manner constraint details are kept solely within classes, and thus it is easier to maintain them and to understand their relationships. Multiplicities are just one of many different kinds of constraints that one might want to write on a relationship. Instead of providing a multiplicity notation, BON provides a single, uniform and expressive notation to express all kinds of constraints on relationships.

BON also provides a notion of *aggregation*, which is commonly used to represent the 'part-of' relationship. Aggregation has a precise semantics in BON: it corresponds to the notion of *expanded* type [10]. A variable of expanded type is not a reference. An implication of this is that in BON, aggregates are created and are destroyed with the whole. This most closely corresponds with UML's notion of *composition*.

3.2.2 Clustering In Fig. 3, dashed boxes are *clusters*, which encapsulate subsystems. In BON, clusters are a purely syntactic notion. They can be used to present different views of a system. Clusters roughly correspond to the notion of package in UML, but there are several differences.

The first difference pertains to the extension of BON's relationships to clusters. With BON, inheritance and client-supplier relationships are recursively extended to be applicable to clusters as well as classes, as the figure shows. Precise rules for such extensions can be found in [17]. The extension for client-supplier relationships is similar in meaning to package dependencies (via the <<imports>> stereotype) in UML. A difference arises with inheritance. UML supports generalization between packages, but it differs in meaning from inheritance involving BON clusters. In UML, package generalization defines a substitutability relationship among packages; in BON, it simply means that everything in the child cluster inherits from something in the parent cluster.

The second important difference between clusters and packages is that UML packages introduce import and export facilities. Things inside a UML package cannot see out of the package by default. Further, things outside of a package cannot see inside the package. This can be changed by the specifier by introducing visibility tags on things inside a package. Packages can also explicitly import visible components of other packages, via the <<imports>> stereotype. BON supports none of these features; visibility and accessibility is determined and specified by the modeler at the class level. Clusters provide no namespace control, visibility control, and import/export facilities. All of these features are *only* provided at the class level, because of the requirement for seamlessness.

The limitation with the UML approach is that it makes it difficult to express fine-grained visibility for specific features of classes; this is discussed more in [13].

3.3 Dynamic diagrams

BON provides a simple, uniform expressive notation for specifying message passing and object interactions. This dynamic notation presents a complementary view to that of a static model.

We view the BON dynamic notation as useful for producing *rough sketches* of system behavior [8]. Rough sketches provide informal details of how elements in a system interact. There are two categories of dynamic BON notations: the charts, and the object communication diagram. The charts are an informal card-based notation for describing system events and scenarios. The object communication diagram models objects and the messages that are passed between objects. Objects are represented as rectangles enclosing the name of their class, perhaps with an object name qualifier. Messages are depicted as dashed arrows, optionally annotated with sequence numbers representing order of calls. Sequence numbers can be cross-referenced to entries in a *scenario box*.

The object communication diagram corresponds most closely to UML's collaboration diagram (though it is also semantically equivalent to UML's sequence diagram); both forms of diagram share the ideas of sequence numbers and using two dimensions to express collaborations. The BON and UML syntax for these diagrams is so similar that we omit examples, though some can be found in [13]. We point out that BON provides only one diagram for dynamic modeling, whereas UML provides several.

4 Design by contract and assertion languages

We now turn to the second major technique supported by BON (and supportable by UML), namely design by contract [10]. In doing so, we explain how the technique is used in BON and UML, and discuss the respective constraint and assertion languages.

The notion of design by contract is central to BON. Contracts are used to specify the behavior of features, of classes, and of class interactions. Each feature of a class may be given a contract, and interactions between the class and *client* classes must be via this contract. The contract is part of the official documentation of the class; the class specification and the contract are never separated. This substantially aids readability and specification simplicity.

The contract of a feature places obligations on the clients of the feature (they must establish the precondition) and supplies benefits to the clients of the feature (they can rely that the feature will establish the postcondition). Both BON and UML offer constraint languages than can be used to precisely specify behavioral details about classes, features, and entire systems. BON has a simple assertion language based on first-order predicate logic; the method was designed around the use of the assertion language. By contrast, UML has its Object Constraint Language, which was added to UML in version 1.1, after much work on the modeling language had been completed.

4.1 Assertions in BON

Contracts, and thus class behavior in BON, are written in a dialect of predicate logic. Assertions are statements about object properties. These statements can be expressed directly, using predicate logic, or indirectly by combining boolean queries from individual objects. The basic assertion language contains the usual propositional, predicate, and set theoretic operators and constructors, as well arithmetic operators and constants.

The assertion language can also be used to refer to the prestate in the postcondition of a routine. The **old** keyword, applied to any expression *expr*, refers to the value of *expr* before the routine was called. **old** can be used to specify how values returned by queries may change as a result of executing a command. Most frequently, **old** is used to express changes in abstract attributes. For example, *count* = **old** *count* + 1 specifies that *count* is increased by one.

A formal semantics for contracts in BON, as well as a collection of re-engineered rules for reasoning about BON contracts, can be found in [12].

4.2 The Object Constraint Language

The Object Constraint Language (OCL) is roughly the equivalent of the BON assertion language in UML; a difference is that the OCL is not based on standard predicate logic. The OCL also fixes problems inherent in the UML metamodel. Requirements for the OCL include: precision; a declarative language; strong typing; and, being easy to write and read by non-mathematicians. As a result, OCL syntax is verbose, replacing common mathematical operators and terms with a more programming language-like syntax. To developers experienced with the use of a constraint language, the OCL will appear cumbersome and difficult to use—especially for reasoning.

4.3 Comparison

While the BON assertion language and OCL are similar in terms of how they are intended to be used, there are significant differences between the two languages.

The first difference is in terms of the rôle that the constraint languages play in the modeling language. The assertion language is fully integrated into BON; the graphical (and textual) notation and the process have been designed to use it. With UML, the constraint language is an add-on, and there are syntactic and semantic issues that remain to be considered after OCL's addition [5], such as connecting finite state modeling with constraints.

The BON assertion language provides both a familiar, concise, expressive mathematical notation – in its graphical form – as well as a textual form that may be preferable to inexperienced constraint language users. The graphical BON assertion language is far superior for reasoning, either with a tool or without, than the OCL; even simple proofs, e.g., the kind needed to show totality or satisfiability of a constraint, will be large and difficult to do with OCL's syntax. An example of using the BON assertion language for reasoning can be found in [12].

Another significant difference is that OCL is a three-valued logic; an *expression* may have the value *Undefined*. BON possesses a notion of *Void*, which reference types may take on. However, this is not the same as OCL's *Undefined*, as only a reference variable (and not, e.g., a *BOOLEAN* variable) can take on value *Void*. Three-valued logics need more extensive rules for reasoning than standard predicate calculus. A case for making the OCL a two-valued logic can be found in [5]. Techniques for reasoning about references can be found in [9].

BON also defines the effect of inheritance on constraints: they are all inherited, and may be refined by the child class. With OCL, this approach is suggested, but not required. It is not clear what value there is in not requiring the inheritance of contracts.

4.4 Contextual information

In BON, constraints are written in class interfaces and are never separated from the interface to which they apply, and therefore maintaining constraints and ensuring their consistency with respect to the attributes and queries of a class is straightforward.

With OCL, it is recommended that constraints not be included in the class diagrams [18], in part because doing so clutters the UML class diagram. Constraints are instead written textually, separate from the diagram. For example, to express that an attribute *age* of a class *Customer* is always at least 18, we would write

Customer

$age \geq 18$

Since constraint and diagram are separate, there is increased likelihood of inconsistency, especially without tool support. Even with tool support, separating constraint and class can make it difficult to use existing constraints for further development. Part of the value of using constraints with classes is that we can use existing constraints when writing new ones. This is not easy to do when constraints are not kept in one place.

4.5 Software contracting with OCL

OCL support for software contracting comes in the form of class constraints (which are equivalent to BON's class invariant), and optional pre- and postconditions. These contracts are not, by default, inherited by a child class, though they may be. Here are two example contracts in OCL. They are translated from the BON class *CITIZEN* in Fig. 1(a). On the left is an OCL contract for function *single*, and on the right is the contract for procedure *divorce*.

Citizen :: single()

pre : — none

post : result = (spouse = NULL??)

Citizen :: divorce()

pre : not single()

post : single() and spouse@pre.single()

In the BON contract for *single*, *spouse* is compared with the *Void* reference; a citizen is single if and only if their *spouse* attribute refers to the *Void* object. [18] makes no reference to a *Void* or *NULL* object that can be used with reference (or object) types. We use *NULL??* here for illustration, but a careful consideration of object types, *Void* references, and their effect on the type system of OCL and UML, is necessary.

In the postcondition of *divorce*, the value of attribute *spouse* before *divorce* is called is referred to by using of the @pre notation. @pre can only be applied in postconditions to attributes or associations. This should be contrasted with **old**, which serves a similar purpose in BON. **old** can be applied to any expression. **old** makes specification of certain features very straightforward and convenient. There is no valid technical reason to restrict use of @pre to attributes and associations.

4.6 Using the constraint languages

OCL provides a number of built-in types, including basic types like integers, and collection types like bags, sequences, and sets. Methods of collection types (defined in [18]; examples include *collect, select, forAll* and *exists*) are accessed via the arrow notation $\rightarrow$; methods of basic types are accessed by the standard dot notation. It has been suggested that the arrow notation in OCL is counter-intuitive [2], in part because of its confusion with the pointer dereference syntax of C and implication of logic. A simplifying modification to OCL would be to obey the uniform access principle, and to use dot notation to access both methods and attributes.

The definition of OCL states that collections are flattened [18]; that is, collections cannot contain other collections. Nestings of collections are not permitted because they are considered to be complex to use and explain; however, they are a very useful modeling tool. Further, flattening makes formalization of a theory of collections difficult [5], requires non-standard reasoning about collections, and significantly reduces the modeling power of the notation. We agree with [5] that flattening collections is unnecessary, and it reduces the expressive power of the OCL significantly.

Consider the following illustration of the use of built-in types, taken from [18], that uses the OCL *forAll* operation. Suppose we have a collection (e.g., a set) of customers in a class *LoyaltyProgram* and want to specify that all customers are no more than 70 years old. In OCL, a constraint is

LoyaltyProgram

$$\text{self.customer} \rightarrow \text{forAll}(c : \text{Customer} \mid c.\text{age}() \leq 70) \tag{1}$$

This specification is not very readable. It also contains many unnecessary elements: the →, the empty parentheses, and the type of c. The corresponding BON specification is an invariant of class *LOYALTY_PROGRAM*, which possesses a set attribute *customer*. The constraint is

$$\forall c \in customer \bullet c.age \leq 70$$

It is difficult to argue that the OCL constraint (1) is easier to write and read than this BON constraint.

An alternative OCL specification of (1) is given in [18]. The alternative is, in fact, more concise, and is as follows.

LoyaltyProgram
self.customer → forAll(age() ≤ 70)   (2)

This is easier to read than (1), but it introduces a new problem. age() is an operation of the class Customer. The constraint (2) belongs to LoyaltyProgram. The use of age() in (2) is untargeted; the object to which the call applies is not provided. The OO paradigm clearly states that all operation calls must be targeted, either implicitly to the current object self or to a specified object. Neither case applies to the use of age() in (2), so we must reject use of such constraints for OO modeling.

See [13] for a discussion of OCL's *allInstances* operation, and how the BON assertion language can be used for everything that it can do. A special *allInstances* feature is unnecessary if a single, expressive assertion language based on standard typed set theory and predicate logic is provided.

5 Limitations of BON and UML

In this section, we briefly discuss some limitations that we have identified, with both BON and UML.

5.1 Improvements to BON

Two inadequacies with BON were identified and discussed in detail in [12]: tool support and handling of real-time systems. There does not exist a wealth of tool support for BON; EiffelCase, a CASE tool from ISE, supports the static diagram and interface notation, as well as round-trip engineering and code generation. There is no analytic tool support, e.g., for reasoning about contracts and classes. Work is underway on providing such support, as detailed in [12]. Better tool support is needed for BON in general.

Currently, BON provides no support for real-time specification (concurrency can be expressed using object communication diagrams). UML, by comparison, has a real-time dialect. A long-term direction of research will be to study how to provide real-time features that integrate with BON's behavioral modeling techniques. This could go hand-in-hand with further study and development of dynamic modeling notations in BON. Any extensions to BON will have to maintain seamlessness and reversibility.

5.2 Improvements to UML

The UML has been constructively criticized by others, e.g., [4, 11, 15]. Our comparison of BON with UML has led us to the following suggestions for improvements with the UML.

- **Design by contract.** Design by contract can be supported in UML through the OCL, but it is not a core part of the modeling language. Full support for design by contract in UML would an excellent way to rationalize existing techniques for specifying constraints, and would significantly improve the UML's capabilities for building reliable, robust software. This, however, may be difficult: the visual modeling notation may require changes in order to better integrate design by contract capabilities, and the semantics, particularly with respect to state diagrams, may have to be changed to accommodate contracts.
- **OCL.** As it currently stands, we believe the OCL is too informal and too verbose for behavioral modeling and for reasoning about said models. A formal semantics for the OCL, as well as a less verbose syntax, needs to be developed. Work is underway along these lines [6]. A number of decisions in the design of OCL are also worth revisiting. As discussed earlier, and elsewhere [5], making the OCL a three-valued logic, and requiring the flattening of collections, are questionable decisions and impact on the modeling power of the notation.
  We question whether it is feasible to develop a constraint language that meets all the requirements placed on the OCL. The goals of precision and non-expert understandability seem to be mutually exclusive. A better approach, as is used in the formal methods application area, might be to use a formal contract language for modeling and specification, and to thereafter paraphrase it into natural language.
- **Rationalization.** With the UML, there are typically several ways to write a model. In part, this is an artifact of unification and the desire to make it as easy as possible for users of the unifying methods to move to UML. With the addition of the OCL, a number of modeling concepts, e.g., or-constraints, subset constraints, etc., can be considered redundant. Further rationalization could be done. Alternatively, restrictions of the UML could be examined, e.g., removing those graphical modeling concepts that become redundant upon addition of a precise constraint language. This is discussed more in [13].

6 Conclusions

BON and UML are languages that can be used to model object-oriented systems. BON is founded in behavioral modeling and emphasizes seamlessness, reversibility, and the use of design by contract. It is simple, easy to teach, and scales up to large systems. UML is a data modeling language that emphasizes use-cases, architectural modeling, and expressiveness. It is supported by a constraint language that is optional for developers to use. It is large, general-purpose, and extensible.

One of our goals in writing this paper was to better understand UML and BON, and to potentially identify limitations and aspects for improvement with each notation. With BON, we have identified limitations with respect to real-time specification and tool support. With UML, our main conclusion is that its development is not complete. UML has unified three different approaches to modeling; that is a useful first step. A next step for UML development should be rationalization.

A second goal of this paper was to understand how UML supports, or fails to support seamlessness, reversibility, and software contracting. We believe that these are vital techniques for an OO modeling language to support. BON has been designed to support these techniques, but UML has not. If it is desired to use UML and to support the techniques of seamlessness, reversibility, and contracting, we suggest the following.

- **Seamlessness.** We can treat dynamic diagrams as rough sketches [8], and make contracts the fundamental specification element. State diagrams should be used minimally, and ideally as an automatically generated view for a class (e.g., as is done with SOMA [4]).
- **Reversibility.** Navigable associations should be used. Non-standard stereotypes should not be used. Contracting should be considered for use as a technique that further supports reverse-engineering.
- **Contracting.** The OCL should be carefully formalized, and a precise definition of the effect of contracts on inheritance should be specified. Collapsing of collections should not be carried out.

Suppose that the UML was used in this manner. It is still very questionable whether the UML applied in this way is the best approach for developing software seamlessly and reversibly, using design by contract. The most significant difference between BON and UML is that the former satisfies what we term the *single-model principle*. In BON, there is precisely one model for a class. All information associated with the class, e.g., contracts, invariants, signatures, is always kept in that single model. When we design, we add information to the class model, and as necessary we produce different views of the model. But these views are always based on the single model for the class.

UML does not satisfy the single model principle. Information about a class need not be kept in one place; its contracts and invariants are written in OCL, and are not part of the diagram. Information about attributes that are not 'simple' is kept outside of the class. The semantics of a class may be given using a state machine. There is no single model for a class written in UML, and this may lead to consistency and communication problems as the class is reused or maintained, and as the system evolves.

References

1. F. Brooks. *The Mythical Man Month*, Addison-Wesley, 1995.
2. D. D'Souza and A. Wills. *Objects, Components, and Frameworks with UML: The Catalysis Approach*, Addison-Wesley, 1998.
3. I. Graham, J. Bischof, and B. Henderson-Sellers. Association considered a bad thing. *Journal of Object-oriented Programming* 9(9), February 1997.
4. I. Graham. *Requirements Engineering and Rapid Development*, Addison-Wesley, 1998.
5. A. Hamie, F. Civello, J. Howse, S. Kent, and R. Mitchell. Reflections on the Object Constraint Language. In *Proc. UML'98*, Springer, 1998.
6. A. Hamie, J. Howse, and S. Kent. Interpreting the Object Constraint Language. In *Proc. APSEC'98*, 1998.

7. C.A.R. Hoare. The Emperor's Old Clothes. Turing Award Lecture 1980. *ACM Turing Award Lectures*, ACM Press, 1987.
8. M. Jackson. *Software Requirements and Specifications*, Addison-Wesley, 1995.
9. S. Kent and I. Maung. Quantified Assertions in Eiffel. In *Proc. TOOLS Pacific 1995*, Prentice-Hall, 1995.
10. B. Meyer. *Object-Oriented Software Construction*, Second Edition, Prentice-Hall, 1997.
11. B. Meyer. UML: The Positive Spin. *American Programmer*, March 1997.
12. R.F. Paige and J.S. Ostroff. Developing BON as an Industrial-Strength Formal Method. In *Proc. World Congress on Formal Methods (FM'99)*, Springer-Verlag, September 1999.
13. R.F. Paige and J.S. Ostroff. A Comparison of BON and UML. Technical Report CS-1999-03, York University, www.cs.yorku.ca/techreports/1999/CS-1999-03.html, May 1999.
14. J. Rumbaugh, I. Jacobson, and G. Booch. *The Unified Modeling Language Reference Manual*, Addison-Wesley, 1999.
15. A. Simons and I. Graham. 37 Things that Don't Work in Object-Oriented Modeling with UML. In *Proc. ECOOP'98 Workshops*, TU-Munich Report 19813, 1998.
16. *Unified Modeling Language Specification*. Object Management Group, 1998. www.omg.org.
17. K. Walden and J.-M. Nerson. *Seamless Object-Oriented Software Development*, Prentice-Hall, 1995.
18. J. Warmer and A. Kleppe. *The Object Constraint Language*, Addison-Wesley, 1999.

Formalizing the UML Class Diagram Using Object-Z

Soon-Kyeong Kim and David Carrington

Department of Computer Science and Electrical Engineering
The University of Queensland, Brisbane, Australia
Email: soon@csee.uq.edu.au, davec@csee.uq.edu.au

Abstract. To produce a precise and analyzable software model, it is essential for the modeling technique to have formality in the syntax and the semantics of its notation, and to allow rigorous analysis of its models. In this sense, UML is not yet a truly precise modeling technique. This paper presents a formal basis for the syntactic structures and semantics of core UML class constructs, and also provides a basis for reasoning about UML class diagrams. The syntactic structures of UML class constructs and the rules for developing a well-formed class diagram are precisely described using the Z notation. Based on this formal description, UML class constructs are then translated to Object-Z constructs. Proof techniques provided for Object-Z can be used for reasoning about these class diagrams.

1. Introduction

Informal Object-Oriented (OO) specification techniques based on graphical notations, such as OMT [12] and Booch [2], are often used to analyze and specify user requirements. Their visually appealing and simple notations, such as boxes, circles, and lines, provide specifications that are easy to use and understand. However, without a precise semantic basis for the notations used in most informal OO techniques, the scope for rigorous analysis is limited. The simplicity of the notations also limits the precise expression of user requirements.

Recently, UML [15] was developed as a standard OO modeling notation. Unlike its predecessors, the syntax and semantics of the notations are provided in terms of a meta-model expressed in a combination of natural language descriptions (English), UML notation and Object Constraint Language (OCL) [16]. However, as Evans [6,7] and others [8,9] have pointed out, these natural language descriptions are not sufficient to express the semantics of the UML notation precisely. Moreover, since OCL is used to express complex constraints which the UML notations cannot express alone, semantic analysis and refinement of any UML model should take these constraints into account. Currently, UML does not support rigorous analysis of UML models.

This paper presents a formal basis for the syntactic structures and semantics of UML class constructs, and also provides a basis for reasoning about class diagrams. The syntactic structures of UML class constructs and the rules for developing a well-formed class diagram are precisely described using the Z notation [14] in section 2, extending the approach of Evans [6,7]. Based on this formal description, UML class constructs are translated to Object-Z [4] constructs. Since these Object-Z specifications are developed as formal models of class diagrams, any reasoning about these class diagrams can take place on the corresponding Object-Z specifications using proof techniques provided for Object-Z. Readers may refer to [13] for reasoning techniques with Object-Z specifications.

The advantage of using Object-Z for the translation is that since Object-Z uses the same OO paradigm, its underlying concepts are very similar to those of UML. This reduces a number of complex problems in translating OO models using non-OO formal specification notations such as Z. For example, in the work done by France et al. [9], each class is associated with a given type denoting a set of object identifiers of all instances of that class. Instances of a class can access their attributes and operations via functions. In contrast, a class in Object-Z encapsulates its attributes and operations like UML. Semantically, a class maps to a set of object identifiers. Therefore, each instance of a class is implicitly identified by its object identifier and can access its attributes and operations directly. Moreover, since a class encapsulates its operations as well as its attributes, subclasses inherit not only the static features of their superclasses such as attributes and relationships with other classes but also their dynamic behaviors such as operations. However, these features cannot be expressed concisely using Z. The advantages of using Object-Z for translating informal OO models have been shown in the work done by Dupuy et al. [5] and Araujo [1], who translate informal OO models to Object-Z.

The structure of the rest of this paper is as follows. In section 2, we present formal descriptions for most class constructs and class diagrams. In section 3, we briefly describe Object-Z notation and present translations of UML class constructs to Object-Z constructs. In section 4, we conclude and discuss future work.

2. Formalizing syntactic structures of UML class diagrams

A UML class diagram shows static aspects of a system in terms of the classes of objects in the system, relationships between these classes and constraints on the relationships. Associations represent relationships between classes. Classes can be further classified in terms of generalizations. Syntactically, a UML class diagram is a collection of these class constructs. In this section, we provide a precise description for the syntactic structure of class constructs in UML such as class, association, association class, and generalization and the static semantics of these class constructs. Based on this description, class diagrams are formally described.

2.1. Classes

In UML, a class is defined as a descriptor of a set of objects with common properties in terms of structure, behavior, and relationships [15]. Syntactically, a class is a rectangle in which a name, attributes and operations are stated. Attributes have names and types. Operations have names and parameters. Each parameter of an operation has a name and a given type. Following the approach of Evans [6], our abstract formal description for the syntactic structure of a class defines two given sets, *ClassName* and *Name*, from which the names of all classes and the names of all attributes, operations and operation parameters can be drawn, respectively. We also define *Type*, a meta type that is partitioned into all possible types in UML such as object types, basic types (integer and string) and so on. Each class name is associated with an object type. A schema named *ClassDecl* denotes the components of a class: a finite set of attributes and a finite set of operations. A partial function *attrstate* is declared to map attributes to their types. A partial function *opsigs* is defined to map operations to their parameters and also to map each parameter to its type.

[*ClassName, Name, Type*]

$$
\begin{array}{|l}
\hline
\text{—ClassDecl—} \\
\quad attributes: \mathbb{F}\ Name \\
\quad operations : \mathbb{F}\ Name \\
\quad attrstate : Name \nrightarrow Type \\
\quad opsigs : Name \nrightarrow (Name \nrightarrow Type) \\
\hline
\quad attributes = dom\ attrstate \\
\quad operations = dom\ opsigs \\
\hline
\end{array}
$$

Class names should be unique in the enclosing name space. Thus, the set of classes is defined as a partial function from *ClassName* to *ClassDecl*.

$$
\begin{array}{|l}
\hline
\text{—UMLClass—} \\
\quad classes : ClassName \nrightarrow ClassDecl \\
\hline
\end{array}
$$

2.2. Associations

In UML, relationships between classes are represented as associations. Associations in UML can be classified into three kinds: common association, aggregation, and composition. Associations can be reflexive or transitive. A binary association is a relationship between exactly two classes. In most cases, associations in a class diagram are binary. Moreover, aggregation and composition are always binary relationships. For these reasons, only binary associations are considered in this paper.

Syntactically a binary association is represented as a link between two classes with an association name and two association ends. Each association end has a role name, a multiplicity constraint and a class to which the association end is attached. For aggregation and composition, an aggregation indicator, a diamond, is added to one of the association ends.

We extend the semantics of set *Name* to include all possible association names and role names. A schema named *AssociationEnd* denotes the components of an association end: a role name, a multiplicity constraint and an attached class. A multiplicity constraint in UML denotes the range of allowable cardinalities of instances that may be associated with a single instance of the class attached to its opposite association end. A multiplicity is a sequence of non-negative integer intervals in the format *lower-bound..upper-bound*, where the upper bound can be unlimited (syntactically represented by using the star character *). Based on this information, we define a variable *multiplicity* as a set of non-negative integer values, where each element in the set denotes a valid cardinality constraint for instances of the class to which the multiplicity is attached. When a multiplicity comprises the star symbol *, it maps to the whole infinite non-negative integer set, which is represented as $\mathbb{N}$ in Z. The variable *aggregation* denotes whether or not the attached class is an aggregate. This variable can take the values *none, aggregate,* or *composite.* The constraint in the predicate part states that a multiplicity cannot have the value zero for both its lower and upper bounds.

$$assockind :: = none \mid aggregate \mid composite$$

┌─*AssociationEnd*────────────────────────────
rolename: Name
multiplicity : $\mathbb{P} \, \mathbb{N}$
attachedclass: ClassName
aggregation : *assockind*
├──────────────────────────────────
multiplicity $\neq \{0\}$
└──────────────────────────────────

An association name can appear more than once in a class diagram and a binary association has exactly two association ends. The set of binary associations is defined as a power set of tuples of *Name* and a pair of *AssociationEnd*. We assume that *e1* is the composite or aggregate.

┌─*UMLAssociation*────────────────────────────
associations : $\mathbb{P} \, (Name \times (AssociationEnd \times AssociationEnd))$
├──────────────────────────────────
$\forall$ *n: Name; e1, e2: AssociationEnd* | $(n, (e1, e2)) \in$ *associations* •
    e1.rolename $\neq$ *e2.rolename* $\neq$ *n*
    e1.aggregation $\in \{aggregate, composite\} \Rightarrow$ *e2.aggregation = none*
    e1.aggregation = composite $\Rightarrow$ *e1.multiplicity* $= \{1\}$
$\forall$ *n1, n2: Name; e1, e2, e3, e4: AssociationEnd* | $\{e1, e2\} \neq \{e3, e4\} \land$
    $\{(n1, (e1, e2)), (n2, (e3, e4))\} \subseteq$ *associations* •
    $\{e1.attachedclass, e2.attachedclass\} = \{e3.attachedclass, e4.attachedclass\} \Rightarrow$
        $n1 \neq n2$
└──────────────────────────────────

The constraints in the predicate state the core properties of association:
- An association name must be different from both role names and each role name also must be different.
- For aggregation and composition, there should be an aggregate or a composite end and the other end is therefore a part and should have the aggregation value of *none.*
- For composition, the multiplicity of the composite end must equal one.
- All associations must have a unique combination of name and associated classes [16]. Thus, if attached classes are the same, their association names should be different.

2.3. Association classes

Association classes are similar to associations except that they have class-like properties in terms of attributes and operations. That is, association classes have properties of both classes and associations. The schema, *AssocClassDecl,* inherits *Class-Decl* and includes two association ends. The constraints in the predicate state that the two role names must be different and the *aggregation* value of both association ends is *none.*

```
┌─AssocClassDecl──────────────────────────────────────────
│ ClassDecl
│ e1, e2: AssociationEnd
├──────────────────────────────────────────────────────────
│ e1.rolename ≠ e2.rolename
│ e1.aggregation = none
│ e2.aggregation = none
└──────────────────────────────────────────────────────────
```

The set of association classes is declared as a partial function from *ClassName* to *AssocClassDecl.*

```
┌─UMLAssocClass───────────────────────────────────────────
│ assocClasses : ClassName ⇸ AssocClassDecl
├──────────────────────────────────────────────────────────
│  ∀ c: ClassName; ac: AssocClassDecl | c ↦ ac ∈ assocClasses •
│            c ∉ {ac.e1.attachedclass, ac.e2.attachedclass}
│        ac.attributes ∩ {ac.e1.rolename, ac.e2.rolename} = ∅
└──────────────────────────────────────────────────────────
```

The constraints describe well-formedness rules for association classes:
- an association class cannot be defined between itself and something else and,
- the role names and the attribute names do not overlap [16].

2.4. Generalizations

In UML, generalization is the taxonomic relationship between objects, in which objects of the superclass have general information and objects of the subclasses have more specific information [15]. Two variables, *superclasses* and *subclasses* are declared to express relationships between classes involved in generalizations and constraints on them. The variable *superclasses* is defined as a finite set of *ClassName* denoting all superclasses. The variable *subclasses* is defined as a relation between values of type *ClassName.* Its domain is the set of superclasses and its range is the set of subclasses. A class cannot be a superclass of itself (reflexive inheritance) or any of its ancestors. These constraints are described in the predicate.

```
┌─UMLGeneralization───────────────────────────────────────
│ superclasses : 𝔽 ClassName
│ subclasses : ClassName ↔ ClassName
├──────────────────────────────────────────────────────────
│ dom(subclasses) = superclasses
│ id ClassName ∩ subclasses⁺ = ∅
└──────────────────────────────────────────────────────────
```

2.5. Class diagrams

A UML class diagram is a collection of classes including classes in generalizations and association classes, and associations between these classes. The following Z schema represents UML class diagrams.

```
┌─UMLClassDiagram──────────────────────────────────────────
  UMLClass
  UMLAssociation
  UMLAssocClass
  UMLGeneralization
├──────────────────────────────────────────────────────────
  dom(assocClasses) ∩ dom(classes) = ∅
  ∀ n; Name; e1, e2: AssociationEnd | (n, (e1, e2)) ∈ associations •
     {e1.attachedclass, e2.attachedclass} ⊆ dom(classes) ∪ dom (assocClasses)
     e1.rolename ∉ classes(e2.attachedclass).attributes
     e2.rolename ∉ classes(e1.attachedclass).attributes
  ∀ n: dom(assocClasses); e1, e2: AssociationEnd |
    e1=assocClasses(n).e1 ∧ e2=assocClasses(n).e2 •
     {e1.attachedclass, e2.attachedclass} ⊆ dom(classes) ∪ dom (assocClasses)
     e1.rolename ∉ classes(e2.attachedclass).attributes
     e2.rolename ∉ classes(e1.attachedclass).attributes
  dom(subclasses) ∪ ran(subclasses) ⊆ dom(classes) ∪ dom (assocClasses)
└──────────────────────────────────────────────────────────
```

The constraints describe that:

- Classes and association classes are disjoint.
- Classes that are involved in associations or association classes should be classes in the diagram.
- For an association or an association class, the role name at an association end should be different from the attribute names of the class attached to the other end.
- Classes involved in generalizations should be classes in the diagram.

3. Formalizing the semantics of UML class diagrams

Once a class diagram is developed, the next step is to analyze the semantics of the diagram. For this semantic analysis, it is essential that the individual class constructs represented in the diagram should have a precisely defined semantic basis. Also there should be a sound mechanism to check and prove properties captured in the diagram. In this section, we describe the semantics of UML class constructs precisely using Object-Z. We first give an overview of Object-Z and an example class diagram that is used in this section. We then describe translation of UML class constructs to Object-Z constructs. Transformation rules described in this section are based on the formal description given for class constructs in the previous section.

89

3.1. Object-Z Overview

Object-Z is an object-oriented extension to Z specifically to facilitate specification in an object-oriented style. The detailed notation and the semantics of Object-Z are given by Duke et al. [4]. In this section, we summarize features of Object-Z needed in this paper.

Class

A class in Object-Z is a template for objects that have a common state and operations. The state is composed of attributes. Syntactically, a class in Object-Z is a named box, optionally with generic parameters. A class box is composed of constant definitions, a state schema, an initial schema and operation schemas. The following specification is an Object-Z stack class [4].

[*Item*]

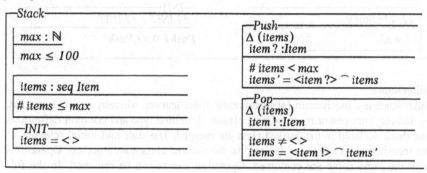

- A **constant definition** is an axiomatic definition for constants used in the class. The example declares *max* to be a constant whose value is less than or equal to 100.
- The **state schema** is a nameless box consisting of state variable declarations and a predicate involving the state variables and constants. The variables and constants comprise the attributes of a class. Instances of a class are in a valid state only when the values of their attributes satisfy the predicate of that class. The state schema of class *Stack* has an attribute, *items,* denoting a sequence of elements of type *Item.*
- The **initial schema**, named *INIT*, consists of a predicate that involves the attributes and other accessible quantities. This predicate is conjoined with the class invariant to define the initial condition. An object can be considered in its initial state whenever the values of its attributes satisfy the initial condition. In the example, an initialized stack has no items.
- Each **operation schema** is a named box specifying transitions that an object of that class can undergo. An operation schema has a Δ-list of state variables whose values may be changed by the operation. The values of state variables not in the Δ-list are unchanged by the operation. The *Stack* class has two operations, *Push* and *Pop*. Operation *Push* appends a given input *item?* to the existing

sequence *items* if the stack has not reached its maximum size. Operation *Pop* outputs a value *item!* which is the first item of the sequence *items* and reduces *items* to the remainder of the sequence. Since the variable *items* is listed in the Δ-list, it is subject to change.

Object instantiation

Objects can be instantiated by declaring them as attributes. The values of such attributes are identities of the objects. The following specification is an example of object instantiation. The class *StackPair* has two individually named references (object identifiers) to stacks. It is intended that the stacks *s1* and *s2* be distinct, hence the references are explicitly distinguished. It is also intended that initially each stack be initialized, hence the explicit initialization. Operation *Push1* promotes *s1's Push* operation to be an operation of *StackPair*.

```
┌─StackPair──────────────────────────────────────────────
│  ┌──────────────────────────┐  ┌─INIT──────────────────
│  │ s1, s2 : Stack           │  │ s1.INIT ∧ s2.INIT
│  ├──────────────────────────┤  │ Push1 ≙ s1.Push
│  │ s1 ≠ s2                  │  │ ...
```

Inheritance

Inheritance is a mechanism for incremental specification, whereby new classes may be derived from one or more existing classes. Inherited type and constant definitions and those declared in the derived class are merged. The state and initial schemas of the inherited class and those declared in the derived class are conjoined. Operations with the same name are conjoined. Operation names can be renamed. In the following example, class *IndexedStack* inherits from class *Stack*. A new operation *IndexedItem* is defined that returns the item *n?* below the top item.

```
┌─IndexedStack─────────────────────────────────────────────
│ Stack
│  ┌─IndexedItem────────────────────────────────────────
│  │ n? : ℕ
│  │ item !:Item
│  ├─────────────────────────────────────────────────────
│  │ n? < # items
│  │ item ! = items(#item − n?)
```

3.2. A library system

The example class diagram used in this paper is shown in Figure 1. The class diagram models part of a library system. The diagram consists of three major entity classes *Reader, Copy* and *Publication*. Class *Publication* is further classified into two subclasses such as *Periodical* and *Book*. An association class *Loan* represents a relationship between class *Reader* and *Copy* and has its own attributes and operations. The multiplicity constraints given for the *Loan* represent that one reader can borrow at most five copies and one copy can be loaned to at most one reader at any point in time. One publication can map to many copies or none. However, one copy maps to exactly one publication. The diagram represents most UML class constructs, namely class, association, association class, and generalization. However, it does not contain more complex relationships such as aggregation and composition. Instead, we give other examples to describe these class constructs in this paper.

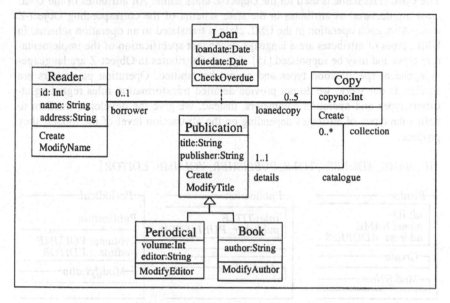

Figure 1. A UML class diagram for the Library system.

3.3. Translating UML class constructs to Object-Z

3.3.1. Semantics of a class

Semantically, a class in UML has two aspects. First, a class defines a set of objects with the same properties, such as attributes, operations, methods and relationships with other objects. This interpretation characterizes the semantics of a class as a type. On the other hand, when a class is interpreted as a component of a UML class diagram, the class represents a set of currently existing instances of that class at some point in time. In our work, these two semantic interpretations of a class are formalized separately:

- Each class (including association classes) in a UML class diagram is translated to an Object-Z class.
- Each class is also instantiated as a set in the system class of the Object-Z specification corresponding to the class diagram. Syntactically, there is no denotation to distinguish the system class from other classes in Object-Z. However, the intention of the system class differs from other classes. The system class captures the behavior of objects as a group, and relationships and interactions between

the objects. In this sense, the Object-Z system class is a formal model of the complete class diagram.

Transformation rules for classes:
The UML class name is used for the Object-Z class name. All attributes of the UML class are declared as attributes in the state schema of the corresponding Object-Z class. Also, each operation in the UML class is translated to an operation schema. In UML, types of attributes are a language-dependent specification of the implementation types and may be suppressed [15]. Types of attributes in Object-Z are language-independent specification types and cannot be omitted. Operation parameters are similar. In our work, we do not provide detailed transformation rules regarding attribute types and operation parameters. Instead, we give the developer freedom to define the types of attributes depending on the abstraction level of the models they produce.

[ID, NAME, ADDRESS, TITLE, PUBLISHER, VOLUME, EDITOR]

```
┌─Reader─────────────┐   ┌─Publication──────────┐   ┌─Periodical────────────┐
│ id: ID             │   │ title: TITLE         │   │ Publication           │
│ name: NAME         │   │ publisher: PUBLISHER │   │                       │
│ address: ADDRESS   │   │                      │   │ ┌───────────────────┐ │
│ ┌─Create─────────┐ │   │ ┌─Create──────────┐  │   │ │ volume: VOLUME    │ │
│ │                │ │   │ │                 │  │   │ │ editor : EDITOR   │ │
│ └────────────────┘ │   │ └─────────────────┘  │   │ └───────────────────┘ │
│ ┌─ModifyName─────┐ │   │ ┌─ModifyTitle─────┐  │   │ ┌─ModifyEditor──────┐  │
│ │                │ │   │ │                 │  │   │ │                   │  │
└─┴────────────────┴─┘   └─┴─────────────────┴──┘   └─┴───────────────────┴──┘
```

The above specification develops Object-Z classes for class *Reader*, *Publication* and *Periodical* from Figure 1. The capitalized attribute names are declared as given types and used for the types of these attributes. Since class *Periodical* is an inherited class from class *Publication*, the Object-Z class *Publication* is declared as its super-class.

The semantics of a class as a set of currently existing instances of that class:
For each class in a class diagram, the corresponding Object-Z class is instantiated as a set in the Object-Z system class to represent existing instances of that class (i.e., in the library example, we assume that class *Library* is the system class).

When a class is defined as a superclass, instances of its subclasses are also instances of that class. The symbol ↓ when prepended to a class name represents the type corresponding to the class and all its subclasses. Constraints should be added to denote that any subclass instances are included in any more general set of instances.

In the *Library*, the Object-Z classes of the five classes such *as Reader, Copy, Publication, Periodical,* and *Book* are instantiated as individual sets (i.e., it is assumed that class *Copy* and *Book* are also translated to Object-Z classes). Since class *Publication* is a superclass of class *Periodical* and *Book*, the symbol ↓ is added to the type declaration for the attribute *publications*. The constraint in the predicate denotes that instances of *periodicals* and *books* are also instances of *publications*. If the *Publication* class is abstract, a constraint, *periodicals* ∪ *books* = *publications*, should be added.

Library

readers: ℙReader
copies : ℙCopy
publications : ℙ↓Publication
periodicals : ℙPeriodical
books : ℙBook

---
periodicals ∪ books ⊆ publications

3.3.2. Semantics of association

Associations in UML can be classified as common associations, aggregations and compositions. An association represents a reference relationship between objects of the classes it associates. However, as Rumbaugh et al. claim [12], at the modeling stage, information represented by an association cannot be buried as attributes of a single class. On the other hand, aggregation and composition represent a part-of relationship between objects. That is, an object is composed of other objects. Therefore, this component information can be part of the composite objects. For this reason, in our work, each type of association is formalized differently.

Common association:
Semantically, an association is a set of pairs of object identifiers of the classes it associates. That is, each instance of an association is a pair of object identifiers. However, it is arguable whether each instance of an association is a separate object, which requires an object identifier to distinguish it from other instances. In addition, associations in UML are bi-directional, unless they are directed to a particular direction with an arrow attached to one of their association ends. For these reasons, we translate each association to a schema rather than a class, in which the association is formalized as a relation between the associated classes. This schema is then instantiated as a single object in the Object-Z system class.

Transformation rules for association:
The association name is used as the schema name. The two role names combined with a hyphen (-) is used for the relation name. The relation is between the corresponding Object-Z classes of the associated classes. The domain and the range of the relation are restricted by the multiplicity constraints given for each association end.

Schema *Catalogue* is declared for the association *catalogue* in Figure 1. The relation *details-collection* represents a relationship between class *Publication* and *Copy*. The multiplicity given for the association end *details* denotes that for all instances of class *Copy*, there should be only one instance of class *Publication*. This is specified in the predicate. However, for all instances of class *Publication*, there can be any number of instances of class *Copy*. Thus, no constraint is given for the publications involved in the set *details-collection*.

94

─Catalogue───────────────────────────────
details–collection : Publication ↔ Copy
───
∀ c: ran(details–collection) • # (details–collection ▷ {c}) = 1
───

A variable of type *Catalogue* is declared in the *Library* class. The association name is used for the variable name. Since an association links existing objects of the classes it associates, objects involved in the set *catalogue* must be existing instances of class *Publication* and *Copy* respectively. This is stated in the predicate.

─Library──────────────────────────────────
│ copies : ℙCopy
│ publications : ℙ↓Publication
│ catalogue : Catalogue
│ ...
│───
│ catalogue.details–collection ⊆ publications × copies
│ ...
──

Aggregation and Composition:
Unlike association, aggregation and composition represent an ownership between objects. That is, component objects are parts of their composite objects (aggregate). Based on this semantics, aggregation and composition are translated to an attribute of the corresponding Object-Z class of the aggregate class.

Transformation rules for aggregation and composition:
The role name attached to the part class is used as the attribute name. The attribute type is a power set of the corresponding Object-Z class of the part class. For composition, no part object can be contained by two composite objects and cyclic relationships are not allowed. Thus, the symbol © which is part of Object-Z, is added to the type of the attribute to represent these properties. For aggregation, cyclic relationships must be avoided. However, a part object can be shared by more than one aggregate. The symbol ⑨ is added to the type of the attribute to represent the shared containment [3]. Depending on the multiplicity constraint given to the part class, the set size can be further restricted.

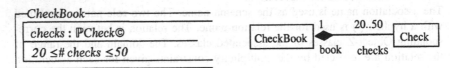

─CheckBook─────────────────
│ checks : ℙCheck©
│──────────────────────────
│ 20 ≤# checks ≤50
───────────────────────────

Figure 2. A composition example

Figure 2 is an example of composition. Class *CheckBook* has a composition relationship with class *Check*. It is assumed that there is an Object-Z class named *Check* for class *Check*. Since class *CheckBook* is an aggregate, class *Check* is translated as an attribute in the state schema of class *CheckBook*. The symbol © is added at the end of the attribute name to represent that instances of class *Check* cannot have any cyclic containment relationship with other objects and also they cannot be shared by more than one instance of *CheckBook*. The multiplicity constraints given describe that each instance of class *CheckBook* should have at least 20 checks and at most 50 checks. This property is described in the predicate part.

Instantiating aggregate classes:
Classes involved in aggregation or composition are translated like other classes. However, additional constraints are given:

- Like association, aggregation and composition link existing instances of the classes they associate. Therefore, objects that are parts of other objects should be existing objects of their classes.
- For composition, the lifetime of component objects is dependent on their composite object. That is, a component object can only exist while its composite object exists. With aggregation, a part can be shared by more that one aggregate. Thus, although one aggregate is destroyed, the part must exist until all its aggregates are destroyed. For this reason, the lifetime constraint is only applied to composition.

The following specification is an Object-Z system class for a bank system. In the state schema, class *CheckBook* and *Check* are instantiated as sets.

$$
\begin{array}{l}
\rule{12cm}{0.4pt} \\
\textit{BankSystem} \\
\rule{12cm}{0.4pt} \\
\quad checkbooks \ : \ \mathbb{P}CheckBook \\
\quad checks : \mathbb{P}Check \\
\rule{12cm}{0.4pt} \\
\quad \forall \ cb{:}checkbooks \bullet cb.checks \subseteq checks \\
\quad \forall \ c \ {:} Check \bullet \\
\qquad \forall \ cb{:} \ CheckBook \bullet c \in cb.checks \implies (cb \in checkbooks \iff c \in checks) \\
\rule{12cm}{0.4pt}
\end{array}
$$

The constraints describe that:

- Instances of class *Check* that are contained by any instance of class *CheckBook* must be existing instances of class *Check* in the system.
- For all instances of class *Check* that are contained by instances of class *CheckBook*, if the composite exists in the system, then the component also must exist in the system.

However, the specification allows that there can be existing instances of class *Check* that are not contained by any instance of class *CheckBook*.

3.3.3. Association class

Since an association class is the case where an association has class-like properties, as well as association-like properties, an association class is formalized in two ways. That is, the class-like properties of an association class are translated to an Object-Z class as described earlier and the association-like properties are translated to a schema as for an association. Since the name of an association class in UML also represents the association name, we add *Cls* to the Object-Z class name and *Assoc* to the schema name to distinguish the two constructs clearly.

The following specification is an Object-Z specification for the association class *Loan* in Figure 1. Its class properties are translated to an Object-Z class *LoanCls* and its association properties are translated to a schema *LoanAssoc*. In the schema *LoanAssoc*, the relationship represented by the class *Loan* is formalized as a partial function from the *LoanCls* to a pair of the corresponding Object-Z classes of class *Reader* and *Copy*. The constraints in the predicate state the multiplicity constraints given for the *Loan*. In the *Library* class, the class *LoanCls* is then instantiated as a set denoting existing instances of that class. The schema *LoanAssoc* is also instantiated as a single object. Since each instance of an association class should map to a pair of instances of the classes it associates, the domain of the variable *loan* is restricted to the set *loans*. An association class links existing instances of the classes it associates. Thus, instances of class *Reader* and *Copy* that are involved in the *loan* should be existing instances of their classes. This is also stated in the predicate of the *Library*.

$[DATE]$

```
┌─LoanCls──────────────
│ ┌──────────────────
│ │ loandate : DATE
│ │ duedate  : DATE
│ ├──────────────────
│ ┌─CheckOverdue─────
│ │
```

```
┌─LoanAssoc──────────────────────────────
│ borrower-collection: LoanCls ⇸ (Reader × Copy)
├────────────────────────────────────────
│ ∀r:dom(ran(borrower-collection) •
│        #({r} ◁ ran(borrower-collection)) ≤ 5
│ ∀c:ran (ran(borrower-collection) •
│        #( ran(borrower-collection) ▷ {c}) ≤ 1
```

```
┌─Library───────────────────────────────────────────────
│ ┌───────────────────────────────────────────────────
│ │ readers : ℙReader
│ │ copies  : ℙCopy
│ │ loans : ℙLoanCls
│ │ loan: LoanAssoc
│ │ ...
│ ├───────────────────────────────────────────────────
│ │ dom(loan.borrower-collection) = loans
│ │ ran (loan.borrower-collection) ⊆ readers × copies
│ │ ...
```

4. Conclusions

In this paper, we present a study that provides a formal basis for the syntax and the semantics of core UML class constructs and class diagrams. The syntactic structures are precisely described using the Z notation. This description is then used as a basis for translation rules from UML class constructs to Object-Z constructs. The Object-Z specifications are formal models of class diagrams. Thus, semantic analysis of these class diagrams can take place on these Object-Z specifications using proof techniques provided for Object-Z. Any inconsistency or error discovered during the analysis provides feedback on the class diagrams. The Object-Z specifications introduced in this paper are type-checked using Object-Z type checker (wizard) and they can be reasoned about by proof techniques introduced in [13].

The approach taken in this paper is an integration of the two common approaches in this area: translating informal OO models to formal specifications [5,9,10] and providing a formal basis for informal modeling constructs at the meta-level [6,7,8,11]. The translation approach provides a mechanism to check and prove the semantics of informal OO models using analysis techniques provided by formal specification techniques. However, it requires the user to have a certain level of fluency with the notations used in combination. The second approach can increase the possibility of producing a precise and understandable OO model. However, it assumes that there are sound techniques that allow developers to reason about and verify the models they produce at the same level of abstraction and representation as their models, but this remains to be demonstrated. These reasons force us to adopt an integrated approach for UML, which provides a formal basis for the syntax of UML class diagrams, and which translates class diagrams to Object-Z specifications to check and verify the semantics of the diagrams. A precise description for the syntax and the rules for constructing well-formed class diagrams should increase the possibility of producing more precise class diagrams. Also translating class diagrams to Object-Z specifications provides a sound mechanism to reason about the semantics of class diagrams.

Dupuy et al. [5] describe a similar translation of OMT object models to Object-Z. In their work, each class is associated with two Object-Z classes that together denote the two aspects of a class: a type and a set of existing instances. In contrast, we translate each class to an Object-Z class and then the semantics of existing instances is formalized by instantiating the Object-Z class as a set in the Object-Z system class. Our approach leads to a more conventional style of Object-Z specifications. There are also differences in the way associations and generalizations are translated. In a similar work, Araujo [1] integrates *Metamorphosis*, a formal object oriented method for analysis and specification, with Object-Z. However, none of this work provides a formal basis for the abstract syntax and the static semantics of the notation used.

Future work
The approach taken in the paper should extend to other diagrams in UML such as the Statechart diagram and the Collaboration diagram. Information provided by these diagrams can then be used to produce a complete Object-Z specification, which is essential for semantic analysis of the Object-Z specification. For example, information provided by Statechart diagrams can be used to complete operation schemas of individual Object-Z classes. Information provided by collaboration diagrams can be used to specify interactions between objects, which are usually specified in the system class of Object-Z specifications. For practical reasons, developing an automatic tool to help generate Object-Z specifications from class diagrams is desirable.

98

References

[1] J. Araujo, *Metamorphosis: An Integrated Object-Oriented Requirements Analysis and Specification Method*. PhD thesis Lancaster University 1996, ftp://ftp.comp.lancs.ac.uk /pub/reports/ThesisJA.ps.Z

[2] G. Booch, *Object-oriented analysis and design with applications*, Benjamin/Cummings, 1994.

[3] J. S. Dong and R. Duke. The Geometry of Object Containment, *Object-Oriented Systems*, vol. 2(1), pp. 41-63, Chapman & Hall, 1995.

[4] R. Duke, G. Rose, and G. Smith. Object-Z: A specification language advocated for the description of standards, *Computer standards & Interfaces*, vol. 17, pp. 511-533, 1995.

[5] S. Dupuy, Y. Ledru, and M Chabre-Peccoud, Integrating OMT and Object-Z, *Proceedings of BCS FACS/EROS ROOM Workshop*, technical report GR/K67311-2, Department of Computing, Imperial College, London, UK, 1997.

[6] A.S. Evans. Reasoning with the UML, *Proc. Workshop on Industrial-Strength Formal Specification Techniques (WIFT'98)*, IEEE Press, 1998.

[7] A. S. Evans and A.N.Clark. Foundations of the unified modeling language. *In 2nd Northern Formal Methods Workshop*, Ilkley, electronic Workshops in Computing,. Springer-Verlag, 1997.

[8] R. B. France, A. Evans, K. Lano, and B. Rumpe, Developing the UML as a Formal Modeling Notation, *Computer Standards and Interfaces*, No 19, pp. 325-334, 1998.

[9] R. B. France, J.-M., Bruel, M. M. Larrondo-Petrie, and M. Shroff. Exploring the Semantics of UML type structures with Z, *Proc. 2nd IFIP conference, Formal Methods for Open Object-Based Distributed Systems(FMOODS'97)*, pp. 247-260, Chapman and Hall, London, 1997.

[10] K. Lano, *Formal Object-Oriented Development*, Springer 1995.

[11] K. Lano and J. Bicarregui. Formalizing the UML in Structured Temporal Theories, *Proc. second ECOOP Workshop on Precise Behavioral Semantics*, pp. 105-121, Springer-Verlag, 1998.

[12] J. Rumbaugh, M. Blaha, W. Premerlani, F. Eddy, and W. Lorensen. *Object-oriented modeling and design*, Prentice-Hall, 1991.

[13] G. Smith, Extending *W* for Object-Z, *ZUM'95: The Z Formal Specification Notation*, pp. 276-295, Springer, 1995.

[14] J. M. Spivey. *The Z Notation: A Reference Manual*, Prentice Hall, 2nd edition, 1992.

[15] The UML group, *UML Notation Guide*, Version 1.1, Rational Software Corporation, Santa Clara, CA-95051, USA, January, 1997, http://www.rational.com.

[16] The UML group, *UML Semantics*, Version 1.1, Rational Software Corporation, Santa Clara, CA-95051, USA, January, 1997, http://www.rational.com.

A Formal Approach to Collaborations in the Unified Modeling Language

Gunnar Övergaard

Royal Institute of Technology, Stockholm, Sweden

Abstract. In this paper we give a formal definition of the collaboration construct in the Unified Modeling Language (UML). We also state what it means that a use case is realized by a collaboration, and what the relationship is between the specification part and the realization part of a subsystem in UML.

1 Introduction

In an object-oriented model a class describes a set of objects that have the same sets of attributes and operations. However, a class does not specify how its objects may be used in different situations, i.e. what different roles the objects may play in the system. This can be defined by *role models*, or so-called *collaborations*. A collaboration defines what properties, such as attributes and operations, each object must have to play a specific role, as well as what communication must take place between the objects for a specific task to be performed. Together, the collaborations express what requirements must be fulfilled by the objects in a system, whereas the classes describe an implementation of these requirements so that their objects will be able to play the different roles. The contribution of this paper is to present a formal definition of the constructs used in collaborations in the Unified Modeling Language.

The *Unified Modeling Language* (UML) [10] is a new, general-purpose language for modelling object-oriented systems which has emerged from the standardization efforts within the *Object Management Group* (OMG). UML originates from the Booch method [1], OOSE [7] and OMT [14], but has also been influenced by other sources, like ROOM [15], OOram [13] and Statecharts [6]. The language consists of a set of constructs common to most object-oriented modelling languages as well as a set of constructs useful for modelling large systems. UML also contains constructs for specification of collaborations.

The specification of UML is "semi-formal", i.e. certain parts of it are specified with well-defined languages while other parts have been described informally in English. The abstract syntax of the different language constructs in UML is specified with the graphical notation of class diagrams in UML itself, while the well-formedness rules of UML are given in an object-oriented constraint language. Ordinary English is chosen for describing the semantics of UML. This makes the structure of the language rigorous whereas the semantics of the language is still quite informal.

However, it is commonly accepted that a language needs a formal specification to be unambiguous. Furthermore, the semantics of the language must be precise if tools are to perform intelligent operations on models expressed in the language, like consistency checks and transformations from one model to another.

In this paper we present some of the results we have achieved when formalizing UML. We focus on the Collaboration construct and some of its related constructs in UML. Using our results it will be possible to state with mathematical precision what it means e.g. that a Use Case is realized by a Collaboration, or that a Subsystem is sound and complete, in the sense that its realization exactly captures the specification of the Subsystem. The ideas are present in the current definition of UML, and our definitions are intended to capture this in a formal way. Without a formal definition, these kinds of statements are ambiguous and are likely to lead to misunderstandings. We also regard our work as a first step towards tool support in checking if models are consistent in this sense. Apart from taking active part in the development of UML, our work aims at giving UML a detailed and precise semantics by providing a formal specification of the language, see e.g. [11, 12]. Other attempts to give UML a formal definition, as in [4, 5, 9, 2, 3, 16], do not give a complete or in-depth definition of the language. Thus, many of the aspects of the different constructs of the language have often been simplified or omitted.

2 Collaborations

It is possible to describe a system in several different ways using different Classifier constructs, e.g. by means of Use Cases, Subsystems or Classes; which kind is chosen depends on the intended focus of the description. In this paper we choose Classes to present the Collaboration construct. This section describes the Collaboration construct, which specifies a structure of roles played by Objects. How the Objects cooperate with each other when playing these roles is specified with Interactions, presented in the following section. Section 4 gives a formal interpretation of these constructs.

A Class describes a set of Objects that have the same sets of features, like Attributes and Operations. Each Object is fully described by its Class. Together, the Classes describe the whole system and all its properties, and they constitute a basis for e.g. the implementation of the system. However, a Class does not specify how its Objects may be used in different situations, i.e. what different roles the Objects may play in the system. In one situation one subset of an Object's Operations will be applied to the Object, and these Operations will depend on some of the Attribute Links (links from an instance to its attribute values) and Links of the Object. In another situation the Object may involve another (not necessarily disjoint) subset of the Operations.

Hence, a Class provides a full description of its Objects; it does not express what roles an Object can play and which of its features are used in each role. However, sometimes it is necessary to describe what features are used in different situations, not only for clarity but also e.g. to facilitate change analysis. One way to describe a particular usage is by defining a Collaboration. A *Collaboration* specifies how the Objects are linked to each other and which of their features are used when they together perform a specific task. (How the Objects actually interact with each other is specified with an Interaction; see Section 3 below.)

Each role in a Collaboration is expressed with a *Classifier Role*, which specifies a projection of a Class. A Classifier Role states which of the features declared in the Class are required in the Collaboration. An Object playing a specific role must conform to the Classifier Role specifying the role, i.e. the Object must offer the Operations stated by the Classifier Role as well as contain Attribute Links corresponding to the Attributes of the Classifier Role. The Object may, of course, contain other features as well, but those will not be used in the specific role.

A Collaboration also specifies which subset of the Associations between the Classes is used in the specific task. Moreover, in a particular Collaboration the Association might be used in a restricted way, e.g. it may state that the multiplicity of the Links is reduced in comparison with the multiplicity of the Association, or that a bi-directional Association is traversed in only one direction. An *Association Role* specifies the role an Association plays in a Collaboration, as well as any restrictions on the Association.

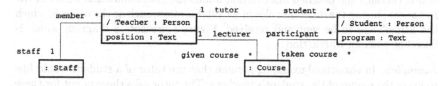

Fig. 1. The diagram presents two named roles, *Teacher* and *Student*, both of them aspects of the Class *Person*, and two unnamed roles, which are aspects of the Classes *Staff* and *Course*, respectively.

Example 1. In a school, teachers and students interact with each other. A teacher gives courses in different subjects, and a student studies different subjects by taking a set of courses. Each course involves one teacher and a set of students. Moreover, each student has a teacher as a tutor for guidance and help, and one teacher may be the tutor of several students. Finally, a teacher belongs to a staff whereas the students do not. In Figure 1, two roles are presented to be played by Instances of *Person*, namely *Teacher* and *Student*, and also two un-named roles, each to be played by Instances of *Staff* and *Course*, respectively.

Figure 2 contains the corresponding Classes and Associations. The diagram shows that a person may belong to a staff: the Association between *Person* and *Staff* has the multiplicity of {0..1} on the *Staff* side. However, the fact that only teachers and not students belong to a staff is not stated in the model presented in this diagram. Moreover, the relationship between a student and the tutor has

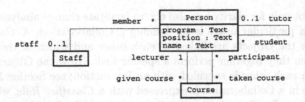

Fig. 2. The underlying set of Classes and Associations of the Collaboration in Figure 1.

been collapsed into an Association between persons; it does not say that only a person who is a teacher may act as a tutor, nor that students should have tutors. It is quite obvious that the school example cannot be expressed with only the Classes and Associations, since they do not specify how they are to be used in this particular example. □

3 Interactions

A Collaboration specifies the structure of the participants when they perform a task, but it does not specify how the participants interact; their communication is defined by an Interaction. An *Interaction* is defined within a Collaboration, and it contains the detailed information about the communication between the Instances playing the different roles in a Collaboration, such as the order in which the Stimuli are to be sent and received. Figure 3 presents a diagram which is based on such an Interaction.

Example 2. In our school example, assume that the tutor of a student would like to know the names of the student's teachers. The tutor asks the student for these names, and the student checks the name of the lecturer of each course taken by the student.

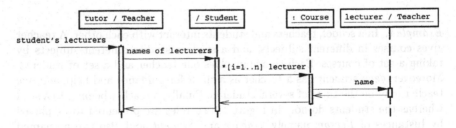

Fig. 3. A sequence diagram presenting an Interaction defined within the Collaboration in Figure 1.

The interaction between the tutor and the student is presented in Figure 3. In this diagram there is one Object called *tutor* who plays the role of *Teacher*, a second (anonymous) Object playing the role of a *Student*, a set of *Course* Objects, and a set of *Teachers* being the lecturers of the student. The tutor sends a Stimulus to the student Object, asking for the names of the lecturers. The student Object iterates over its set of course Objects and sends one Stimulus to each asking for the name of the lecturer. Each course Object retrieves the name of the lecturer and returns it back to the student, who returns the list of the lecturers names to the tutor. □

In this diagram, time proceeds downwards, and Objects playing the different roles are represented as vertical lines. An arrow between two lines denotes a Stimulus sent between those two Objects; the arrow points from the sender to the receiver.

More precisely, an Interaction is defined in the context of a Collaboration. It specifies a collection of Messages between the various Classifier Roles of the Collaboration. Each Message specifies one specific kind of communication, while a communication which actually takes place is expressed with a Stimulus. A set of cooperating Objects playing the roles in the Collaboration interact according to the Messages of the Interaction by sending Stimuli to each other – the Messages define the roles played by the Stimuli that are sent. The Messages of an Interaction are partially ordered, stating in which order the Stimuli are to be sent.

4 Formalizing Collaborations

In this section we present a formal interpretation of Collaborations, i.e. we state what it means that a set of Objects conforms to a Collaboration.

Our formalization of Collaborations is based on sequences. A *sequence* is a totally ordered (finite or infinite) set of elements, such as Actions or Stimuli. We use sequences to represent collections of Actions that are to be executed in specific order or histories of Stimuli that have been sent.

Definition 1. A Σ sequence is a finite or infinite set of elements from a domain Σ together with a total order on the set.

Let $p \leq q$ imply that p is a prefix of q, i.e. p is equal to the sequence obtained from the first n elements in q, where n is the number of elements in p. Furthermore, let $p \subseteq q$ imply that p is a non-contiguous sub-sequence of q, i.e. q contains the same sequence of elements as p, possibly with some extra elements inserted.

In UML a Collaboration consists of a set of (at least one) Classifier Roles, a set of Association Roles and a set of Interactions (see Figure 4). The set of Classifier Roles states what features Objects must have if they are to take part in the cooperation, while the Association Roles specify the subset of connections used in the specific Collaboration. As we are focusing on behaviour, our presentation does not include the Association Roles. The Associations (as well as the Attributes) are implicitly required by the Operations, as the Actions will make

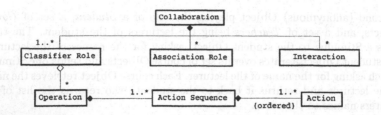

Fig. 4. The diagram uses the UML notation to show that the Collaboration construct consists of three sub-constructs: at least one Classifier Role, a possibly empty set of Association Roles and a possibly empty set of Interactions. A Classifier Role references a possibly empty set of Operations, each with a collection of sequences of Actions describing its behaviour.

use of them. Figure 4 also shows that a Classifier Role consists of a set of Operations, each containing a collection of (non-empty, ordered) Action sequences. (We treat the different paths through the body of an Operation as different sequences.)

In this paper we assume that the descriptions of the Operations are done in terms of Actions. We recognize that there are other description techniques for Operations, like preconditions and postconditions, but we have chosen to use Actions as they give the most detailed description of behaviour in UML. We do not describe the different Action constructs in any detail, as our formalization depends only on their existence and not on their exact semantics.

UML defines that an Object *conforms* to a Classifier Role if the Object has the properties specified by the Classifier Role, i.e. the Attribute Links and the Links of the Object match all the Attributes and Associations specified by the Classifier Role, and all Operations specified by the role may be applied to the Object. The Object may, of course, include more Links etc. than specified by the Classifier Role. This implies that all possible sequences of Actions an Object can perform must include the sequences required by the role [10, Section 2.4.10].

Before we give the formal definition of conformance, we define the semantics of Classifier and Object used in this paper. We assume that each Class contains a set of Operations, denoted *C.operation*, and each Operation contains a name and a set of sequences of Actions, denoted *op.name* and *op.actionSequence*, respectively.

Furthermore, we assume that an Object originates from one or several Classes, denoted *O.classifier*, and the Operations that can be applied to the Object are declared in these Classes. Let *opr (O)* denote the set of Operations that can be applied to the Object O, i.e. $opr(O) = \{op \mid \exists c \in O.classifier \,.\, op \in c.operation\}$.

We also assume an operational interpretation of the execution of an Object, i.e. its semantics is given by the Action sequences it can perform. Let $tr^*(OS)$ be the set of sequences of Actions that the set of Objects OS can jointly perform when executing together. Note that the execution includes synchronization, sequencing etc. of the objects.

We can now formalize the notion of conformance of an Object to a Classifier Role as follows (note, in UML a Classifier Role is a kind of Classifier):

Definition 2. An Object O conforms to a Classifier Role CR, denoted $O \sqsubseteq^c CR$, if

$$\forall co \in CR.operation . \exists oo \in opr(O) .$$
$$co.name = oo.name \wedge co.actionSequence \subseteq oo.actionSequnce$$

The set of Action sequences an Object can perform must include those sequences specified by a Classifier Role, if the Object is to conform to the role.

If the Object conforms exactly to the Classifier Role, i.e. if all behaviour that can be performed by the Object is specified by the Classifier Role, we say that the Object is a *minimal* Object that conforms to the role.

Definition 3. An Object O is a minimal Object that conforms to a Classifier Role CR, denoted $O \sqsubseteq^c_m CR$, if

$$CR.operation = opr(O)$$

Note that Definition 3 implies that for each Operation in one of the sets there must be a corresponding Operation in the other set which has the same name and the same set of Action sequences.

We can now state what it means that a set of Objects conforms to a collection of Classifier Roles.

Definition 4. A set of Objects OS conforms to a set of Classifier Roles CRS, denoted $OS \sqsubseteq^{cs} CRS$, if both of the following hold

- $\forall O \in OS . \exists CR \in CRS . O \sqsubseteq^c CR$
- $\forall CR \in CRS . \exists O \in OS . O \sqsubseteq^c CR$

In other words, each Object in the set must play a role, and each role must be played by an Object. Hence, the relationship $\sqsubseteq^{cs}$ on $OS \times CRS$ is total and surjective. It needs not be injective because several Objects may conform to the same Classifier Role. We use the notation $OS \sqsubseteq^{cs}_m CRS$ to denote that a collection of Objects OS is a set of *minimal* Objects that conforms to a set of Classifier Roles CRS.

The communication between the Objects performing the task of a Collaboration is specified by Interactions. Each Interaction contains a partially ordered set of Messages (see Figure 5).

A *Message* is a specification of a communication between a sender and a receiver. The Message specifies the roles played by the sender Object and the receiver Object, and it states which Operation should be applied to the receiver by the sender. Moreover, the set of Messages in an Interaction is partially ordered. Each Message M specifies which Message activated the Operation from which M is sent, as well as what other Messages are sent from this Operation before M is sent. (In UML, a Message is associated with an Action, which, when executed, dispatches a Stimulus conforming to the Message. The Action defines both the Operation which will be invoked by the Stimulus, and how the arguments are to

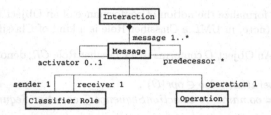

Fig. 5. The diagram shows the Interaction construct, which contains at least one Message. The set of Messages is partially ordered using the *activator* and *predecessor* relationships.

be evaluated. In this paper we have simplified this by associating the Message directly with the Operation, since how the arguments are determined is not of importance in our formalization.)

More formally, an Interaction is defined by $\langle M, \leq \rangle$, where M is a set of Messages and $\leq$ is a partial order on M. A Message is defined by a tuple: $\langle sender, receiver, operation \rangle$, where *sender* and *receiver* are two Classifier Roles to be played by the sending and the receiving Objects, and *operation* is the Operation to be invoked.

Two Objects communicate by sending a Stimulus from one Object to the other. A Stimulus is also a tuple: $\langle sender, receiver, operation, arguments \rangle$, where *sender* and *receiver* are the sending and the receiving Objects, *operation* is the Operation which is invoked on the receiver by the Stimulus, and *arguments* is the sequence of arguments to be used when the Operation is invoked.

If the Objects participate in the performance of the task specified by a Collaboration, part of their communication conforms to an Interaction defined within the Collaboration, i.e. some of the communicated Stimuli must conform to the Messages specified in an Interaction of the Collaboration. Note that the Objects may also communicate due to what other roles they play at the same time, i.e. not all Stimuli sent between the two Objects must conform to the Interaction. We first define what it means that a Stimulus conforms to a Message.

Definition 5. A Stimulus S conforms to a Message M, denoted $S \sqsubseteq^s M$, if all three of the following hold

- $S.sender \sqsubseteq^c M.sender$
- $S.receiver \sqsubseteq^c M.receiver$
- $S.operation = M.operation$

The definition states that the sender and the receiver Objects of a Stimulus must conform to the corresponding roles specified by the Message, and that the Operation invoked by the Stimulus is the same as the one being specified by the Message.

We can now define what it means that a sequence of Stimuli conforms to a sequence of Messages: each Stimulus in the sequence must conform to the corresponding Message. This is formalized as:

Definition 6. A sequence of Stimuli $SS = S_1 \ldots S_n$ conforms to a sequence of Messages $MS = M_1 \ldots M_m$, denoted $SS \sqsubseteq^{ss} MS$, if both of the following hold

- $n = m$
- $\forall i \in \{1..n\} . S_i \sqsubseteq^s M_i$

Before continuing we present two auxiliary functions needed in our definition. Let *send (S)* be the set of sequences of Stimuli being sent by executing the Actions in the sequence S. Moreover, let *lin (MS)* be the set of all possible linearizations of a set of partially ordered Messages, i.e. the set of all sequences of Messages that fulfil the partial ordering.

We are now ready to give a formal specification of what it means that a set of Objects conforms to a Collaboration (what in [10] is called a *Collaboration instance*). First, the Objects and the Classifier Roles in the Collaboration must conform to each other. Secondly, the Objects must be able to perform all the communications specified by the Interactions of the Collaboration, i.e. all possible sequences of Messages defined in the Interactions of the Collaboration must have a corresponding sequence of Stimuli that are communicated between the Objects. Note that this does not preclude the Objects from participating in other communications, as well. This is expressed by the following definition:

Definition 7. A set of Objects OS conforms to a Collaboration C, denoted $OS \sqsubseteq C$, if both of the following hold

- $OS \sqsubseteq^{cs} C.classifierRole$
- $\forall int \in C.interaction . \forall ms \in lin(int.message) .$
  $\exists t \in tr^*(OS) . \exists s \in send(t) . \exists ss \subseteq s . ss \sqsubseteq^{ss} ms$

An Object may communicate with other Objects, even if the communication is not specified in an Interaction of the Collaboration, as the Objects may contain other features that are not involved in the collaboration. For example, if the Collaboration includes an Interaction which specifies the Messages sequence $\langle a, b, c \rangle$, the set of Objects may send the Stimuli sequences $\langle a, b, c \rangle$ and $\langle a, d, e, b, c, f \rangle$, but the sequence $\langle a, e, b \rangle$ does not conform to the Interaction.

We use the notation $OS \sqsubseteq_m C$ to denote that OS is a set of *minimal* Objects conforming to the Collaboration.

5   Generalization between Collaborations

In this section we define the meaning of a Generalization relationship between two Collaborations. Such a relationship is useful e.g. when specifying a task which is a specialization of another task that has already been specified.

Let a Collaboration (the child) have a Generalization relationship to another Collaboration (the parent). The child Collaboration defines an extended and perhaps specialized version of the parent. The set of roles contained in the child includes those of the parent but may also contain some additional roles. However, if the definition of Generalization between Collaborations is to be useful, it must be possible to specialize already existing roles. If not, all new behaviour must be performed by new roles. Hence, the roles defined in the parent Collaboration may be extended to capture additional features required in the child Collaboration, i.e. a role of the parent Collaboration may be replaced in the child Collaboration by a specialization of that role. As the Classifier Role construct is a kind of Classifier, Generalization relationships are used for expressing specialization of Classifier Roles.

We first define what it means that a Classifier Role is a specialization of another Classifier Role:

Definition 8. A Classifier Role S is a specialization of another Classifier Role G, denoted $S \leq G$, if both of the following hold

- $S.name = G.name$
- $\forall go \in G.operation$. $\exists so \in S.operation$.
  $go.name = so.name \land$
  $\forall t \in go.actionSequence$. $\exists s \in so.actionSequence$. $t \subseteq s$

The definition states that all the behaviour sequences of the parent are included in the set of behaviour sequences of the child, although possibly with some additional Actions inserted. With this definition, we can state that a Generalization from a Classifier Role C to another Classifier Role P implies that $C \leq P$.

Furthermore, since Classifier Roles that are specializations of roles of the parent Collaboration, may be extended with additional behaviour, whereas none of the Actions in the original Classifier Roles may be removed, all communication defined in the parent Collaboration must also appear in the child. This implies that the communication specified by the Interactions of the parent Collaboration must also take place according to the Interactions of the child.

This leads to the following definition of specialization of Collaborations:

Definition 9. A Collaboration S is a specialization of another Collaboration G if both of the following hold

- $\forall gr \in G.classifierRole$. $\exists sr \in S.classifierRole$. $sr \leq gr$
- $\forall gi \in G.interaction$. $\forall gis \in lin(gi.message)$.
  $\exists si \in S.interaction$. $\exists sis \in lin(si.message)$.
  $gis \subseteq sis$

Note that this definition does not require that each role in the general Collaboration is extended with additional Actions or Action sequences in the specialized Collaboration.

We can now state that a Generalization between two Collaborations implies that the child Collaboration, including its inherited roles, is a specialization of the parent Collaboration.

6 Use Cases

In the following two sections we give a formal definition of two usages of Collaborations in the definition of UML: the realization of Use Cases, and the correspondence between the specification part and the realization part of a Subsystem. This section first presents the Use Case construct in UML; it is followed by a specification of what it means that a Use Case is realized by a Collaboration. In the next section there is a presentation of the Subsystem construct.

The Use Case construct models the functionality of a system. A *Use Case* specifies a service provided by the system, i.e. one way of using a system, without revealing the system's internal structure. This makes Use Cases suitable for defining functional requirements in the early stages of system development, when the inner structure of the system has not yet been defined.

A Use Case specifies a set of sequences of Actions which the system can perform. Each sequence is initiated by the environment of the system, and it includes the interaction between the system and its environment as well as the system's response to these interactions. The description of a Use Case can be made in many different ways. Formally, a Use Case is specified with a set of Operations or with a State Machine. Both the transitions between the states in a State Machine and the Operations specify the Action sequences which are performed by Instances of the Use Cases. However, in practice ordinary text is often used, although other description techniques are also employed. (In this paper we will not discuss how Use Cases are used in system development, but refer to e.g. [7, 8].)

Instances of Use Cases communicate with their environment by sending and receiving Stimuli, and after receiving a Stimulus a Use-Case Instance performs a sequence of Actions. In view of the scope of this paper no other semantics of Use-Case Instances is required.

So far, we have discussed Use Cases only at the system level, but Use Cases can also be used at subsystem level and at class level. At system level, Use Cases specify how the system may be used by the environment of the system. Similarly, Use Cases at subsystem level and at class level specify how a Subsystem or a Class may be used by other parts of the system. This means that the technique for specifying the behaviour of the entire system can be used also when specifying the behaviour of Subsystems and Classes. This technique is used in the following section when we present the Subsystem construct.

As all behaviour of an implemented object-oriented system is (in principle) performed by Objects, the behaviour specified by a Use Case is performed by Objects in the running system. Therefore, each Use Case is mapped onto a set of Classes in the system. Since a Use Case can be seen as a task performed by the system (it specifies a service offered by the system), Collaborations are well suited for the specification of the mapping of the Use Case onto other Classifiers. We can therefore say that a Use Case is realized by a Collaboration, as Instances of these Classifiers will jointly cooperate in performing all the behaviour specified by the Use Case. However, the Collaboration must not include more behaviour than the Use Case; if so the Use Case will be only a partial specification of the Collaboration's task.

Moreover, as a Use Case is not dependent on the internal structure of the system, it does not specify any Actions that have to do with the organization of the system. Therefore, a Use Case does not specify the communication between Objects inside the system, nor does it include creation and deletion of Objects – a Use Case describes only externally visible Actions. For a Use Case, let $ext\ (t)$ be the sequences obtained by omitting all the Actions that are not externally visible from the sequences in t. The realization relationship from a Collaboration to a Use Case is therefore defined as follows:

Definition 10. A Collaboration C is a realization of a Use Case UC, denoted $C \succeq UC$, if both of the following hold

- $\forall UCI$ instance of UC . $\exists OS$. $OS \sqsubseteq_m C \land tr^*(UCI) = ext(tr^*(OS))$
- $\forall uo \in UC.operation$. $\forall ut \in uo.actionSequence$.
    $\exists CR \in C.classifierRole$. $\exists co \in CR.operation$.
    $\exists ct \in co.actionSequence$.
    $uo.name = co.name \land$
    $\exists w \leq ct$. $ext(w) \leq ut \land (len(ut) > 0 \rightarrow len(w) > 0)$

The sequences of Actions specified by a Use Case are the same as those sequences of externally visible Actions that a collection of Objects can perform. These Objects form a set of minimal Objects conforming to the Collaboration which is a realization of the Use Case. Moreover, each Operation that is defined in the Use Case must have a corresponding Operation in one of the Classifier Roles of the Collaboration, i.e. the Operations must have the same name and the same initial sub-sequences of Actions.

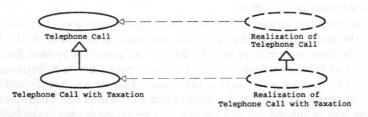

Fig. 6. A Generalization between Use Cases can be realized by a Generalization between Collaborations.

Two Use Cases may have a Generalization relationship between each other. This means that the child Use Case contains all sequences of Actions of the parent, and may also add new Action sequences as well as insert Actions into the inherited sequences [10, Section 2.11.4]. Assume that the parent Use Case is realized by one Collaboration and that the child Use Case is realized by another Collaboration (see Figure 6). Clearly, it must be possible to define the

Collaboration of the child Use Case as a specialization of the parent's Collaboration (although it need not necessarily be a specialization, for the child Use Case might also be realized in a way that is completely different from that of its parent). According to Definition 9, this is also the meaning of a Generalization between the two Collaborations. Hence, a Generalization between two Use Cases can be realized by a Generalization between the Collaborations realizing the Use Cases. This implies that the meaning of the Generalization relationship stated in Definition 9 is both useful and desired.

7 Subsystems

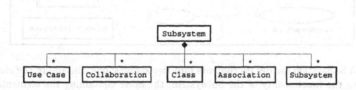

Fig. 7. The diagram shows the Subsystem construct and a subset of the constructs that may be contained in a Subsystem.

In UML a *Subsystem* contains a collection of elements, such as Classes, Associations and other Subsystems, which together constitute a *realization* of the part of the system that the Subsystem models. Apart from these realization elements, a Subsystem contains a set of elements which together form an abstract specification of its behaviour. This *specification* describes the behaviour jointly performed by Instances of the realization elements, e.g. Objects, (recursively) contained in the Subsystem without revealing the structure of the Subsystem [10, Section 2.14.4]. In this section we present the Subsystem construct (see Figure 7), and define the meaning of Subsystems using Collaborations to keep the specification parts and realization parts synchronized.

The specification part of a Subsystem can be expressed in two different ways: by Operations on the Subsystem or by a subset of the model elements contained in the Subsystem. These elements, often Use Cases, are declared as specification elements within the Subsystem. We will here focus on Use Cases; the usage of Collaborations as the realizations of a Subsystem's Operations is performed in a similar way.

It should be noted that a Subsystem does not have any behaviour of its own; all behaviour offered by a Subsystem is in fact performed by Instances of Classifiers within the Subsystem. In other words, a Use Case of a Subsystem specifies a sequence of Actions performed by Instances of the elements realizing the Subsystem. The users interacting with instances of such a Use Case are either users external to the system or elements inside the system but external to the Subsystem. Likewise, the behaviour performed when an Operation declared

112

by the Subsystem is invoked is actually performed by Instances of realization elements contained in the Subsystem.

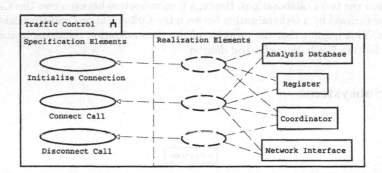

Fig. 8. A Subsystem divides its contents into two subsets: one containing the elements specifying the functionality of the Subsystem, the other containing the elements realizing the functionality. The mapping between the two subsets can be established by a collection of Collaborations.

Example 3. In a telecommunication exchange there is a subsystem, called *Traffic Control*, which is responsible for routing the traffic. The two different views of the Subsystem Traffic Control can be seen in Figure 8: the abstract specification (to the left) and the contained realization (to the right). The former contains three Use Cases:

Initialize Connection	Initialize a session, which includes initializing access to the database containing the analysis information, and reservation of digit storage.
Connect Call	Receive and store a digit and, after receiving the required number of digits, connect the call in the network and send ring signals.
Disconnect Call	Receive and keep track of the hook signals; when both subscribers have hung up, disconnect the call.

The realization of the Subsystem contains four Classes/Subsystems and three anonymous Collaborations:

Analysis Database	Database which e.g. maps a number to an outgoing connection.
Coordinator	Coordinates the Instances which participate in the call.
Network Interface	The interface to the network.
Register	Keeps track of the number being dialled.

Collaboration   Each Collaboration is a realization of a Use Case
                                 in the specification part of the Subsystem and con-
                                 tains a set of roles played by Instances of the at-
                                 tached Classes/Subsystems.

The Use Cases specify what the Subsystem can do, while the Classes and the
Subsystems provide the decided realization of the specification.   □

For a Subsystem to be consistent, all the behaviour offered in its specification part must be realized and implemented by Classifiers (typically Classes or
Subsystems) inside the Subsystem. Otherwise, the Subsystem will not fulfil all
functionality it promises in its specification. Moreover, a Subsystem cannot contain more functionality than what is specified, since this may cause unwanted side
effects on the Subsystem's environment. Therefore, a Subsystem is well-formed
if all its specified functionality, and only that functionality, is also realized in the
Subsystem. We can formalize these requirements in the following way:

Definition 11. A Subsystem SS is *sound* if both of the following hold

$-\ \forall CL \in SS.collaboration\ .\ \exists UC \in SS.useCase\ .\ CL \succeq UC$
$-\ \forall C \in SS.realization\ .\ \forall O$ instance of C .
$\qquad \forall oo \in opr(o)\ .\ \forall t \in oo.actionSequence$.
$\qquad\qquad \exists CL \in SS.collaboration\ .\ \exists CR \in CL.classifierRole$.
$\qquad\qquad\quad \exists co \in CR.operation$.
$\qquad\qquad\qquad oo.name = co.name\ \wedge\ t \in co.actionSequence$

Definition 12. A Subsystem SS is *complete* if both of the following hold

$-\ \forall UC \in SS.useCase\ .\ \exists CL \in SS.collaboration\ .\ CL \succeq UC$
$-\ \forall CL \in SS.collaboration\ .\ \exists OS$.
$\qquad OS \sqsubseteq CL\ \wedge\ \forall O \in OS\ .\ \exists C \in SS.realization\ .\ O$ instance of C

Definition 11 says that all sequences of (externally visible) Actions performed
by Objects originating from Classes realizing a Subsystem, appear in the specification of the Subsystem, while Definition 12 says that all sequences of Actions in
the specification of a Subsystem can be performed by Objects originating from
Classes realizing the Subsystem. A Subsystem which is both complete and sound
is said to be well-formed.

Both Definition 11 and Definition 12 include the notion of *Instance of a Classifier in the realization part of the Subsystem*. If such a Classifier is a Class, the
Instance is an Object. However, if the Classifier is a Subsystem it is not one
Instance but a set of Instances that will act as the Instance of the Classifier, i.e.
the Subsystem. Each Instance in the set originates from a Classifier contained
in that Subsystem. This is due to the fact that a Subsystem does not have any
behaviour of its own; all its behaviour is offered by Instances of Classifiers contained in the Subsystem. This set of Instances will together act as one Instance
and as such it will conform to a Classifier Role and perform sequences of Actions.

8 Concluding Remarks

In this paper we have presented a more formal and more detailed semantics of the Collaboration construct than that given in the definition of UML. We have also shown how Generalization relationships between Collaborations are defined, i.e. how one Collaboration can be a specialization of another. Furthermore, we have given a formal meaning of the realization of Use Cases with Collaborations, as well as stated what it means that a Subsystem has consistent specification and realization parts.

We hope that the present deficiencies in the definition of UML will be removed in future versions of the language. We believe that a good way of finding these deficiencies and explaining new semantics of the language is by providing a formal specification of the language.

The results presented in this paper will facilitate the development of tools that perform intelligent operations on models expressed in UML, like checking if the elements realizing a Subsystem fulfil the specification of the Subsystem, and checking if a Collaboration conforms to the Interactions of its parent Collaborations. Without a formal definition of UML these kinds of operations are not possible.

In this paper we do not include the UML Signal construct, which implies that Operations are the only means for communication between Objects. This simplification has no impact on our results of formalizing Collaborations. The reception of a Signal will cause an Instance to perform a transition and, when doing that, the Instance will perform the Actions attached to the transition, as described by a State Machine of the Instance's Class. Since both the sending of Signals and the invocation of Operations will cause a sequence of Actions to be performed, we have chosen to focus on Operations. An interesting continuation of our work will be to describe the relationship between State Machines used for describing Classifiers and Classifier Roles.

References

1. G. Booch. *Object-Oriented Design with Applications*. Redwood City, 1991.
2. R. Breu, U. Hinkal C. Hofmann, C. Klein, B. Paech, B. Rumpe, and V. Thurner. Towards a Formalization of the Unified Modeling Language. In M. Aksit and S. Matsuoka, editors, *Proceedings of the 11th European Conference on Object-Oriented Programming, ECOOP'97, Lecture Notes in Computer Science 1241*, pages 344–366. Springer-Verlag, June 1997.
3. T. Clark and A. Evans. Foundations of the Unified Modeling Language. In *Proceedings of The Second Northern Formal Methods Workshop*. Springer-Verlag, 1997.
4. A. Evans, R. France, K. Lano, and B. Rumpe. Developing the UML as a Formal Modelling Notation. In P.-A. Muller and J. Bézivin, editors, *Proceedings of the Unified Modeling Language: UML'98: Beyond the Notation, Lecture Notes in Computer Science 1618*. Springer-Verlag, 1999.
5. R. France, A. Evans, K. Lano, and B. Rumpe. The UML as a Formal Modeling Notation. *Computer Standards and Interfaces*, 1998.

6. D. Harel and E. Gery. Executable Object Modeling with Statecharts. *IEEE Computer*, 30(7):31–42, July 1997.
7. I. Jacobson, M. Christerson, P. Jonsson, and G. Övergaard. *Object-Oriented Software Engineering: A Use Case Driven Approach.* Addison-Wesley, 1993.
8. I. Jacobson, J. Rumbaugh, and G. Booch. *The Unified Software Development Process.* Addison-Wesley, 1999.
9. K. Lano and J. Bicarregui. Semantics and Transformations for UML Models. In P.-A. Muller and J. Bézivin, editors, *Proceedings of the Unified Modeling Language: UML'98: Beyond the Notation, Lecture Notes in Computer Science 1618.* Springer-Verlag, 1999.
10. Object Management Group, Framingham Corporate Center, 492 Old Connecticut Path, Framingham MA 01701-4568. *OMG Unified Modeling Language Specification, version 1.3*, June 1999. http://www.omg.org/cgi-bin/doc?ad/99-06-08.
11. G. Övergaard. A Formal Approach to Relationships in the Unified Modeling Language. In M. Broy, D. Coleman, T. S. E. Maibaum, and B. Rumpe, editors, *Proceedings PSMT'98 Workshop on Precise Semantics for Software Modeling Techniques*, pages 91–108. Technische Universitæt, München, Germany, TUM-I9803, April 1998.
12. G. Övergaard and K. Palmkvist. A Formal Approach to Use cases and Their Relationships. In P.-A. Muller and J. Bézivin, editors, *Proceedings of the Unified Modeling Language: UML'98: Beyond the Notation, Lecture Notes in Computer Science 1618.* Springer-Verlag, 1999.
13. T. Reenskaug, P. Wold, and O. A. Lehne. *Working with Objects: The OOram Software Engineering Method.* Manning Publications, 1996.
14. J. Rumbaugh, M. Blaha, W. Premerlani, F. Eddy, and W. Lorensen. *Object-Oriented Modeling and Design.* Prentice-Hall, Englewood Cliffs, 1991.
15. B. Selic, G. Gullekson, and P. Ward. *Real-Time Object-Oriented Modeling.* John Wiley and Sons, 1994.
16. M. Shroff and R. B. France. Towards a Formalization of UML Class Structures in Z. In *Proceedings of Twenty-First Annual International Computer Software and Applications Conference (COMPSAC'97)*, pages 646–651. IEEE Computer Society, 1997.

A Formal Semantics for UML Interactions

Alexander Knapp*

Ludwig–Maximilians–Universität München
knapp@informatik.uni-muenchen.de

Abstract. The UML abstract syntax and semantics specification distinguishes between the statics and the dynamics of collaborations: the rôle context and interactions. We propose a formal semantics of interactions based on the abstract syntax and directly reflecting the specification. The semantics is both parametric in the notion of context and in semantic details that are intentionally left open by the specification, but resolves true inconsistencies. The formalisation uses temporal logic formulae in the style of Manna and Pnueli. We illustrate the flexibility of our semantics by discussing instantiations for a running example; its intuitiveness is substantiated by proving that the temporal formulae give rise to partial orders that also directly can be inferred from interactions.

Introduction

The object-oriented software modelling language UML, the "Unified Modeling Language", supports behavioural modelling, amongst a variety of different techniques, by the stimulus-based notion of interactions in collaborations. In an UML model, collaborations specify how an operation or an use case of the model is realised by a cooperation of several instances of model elements. For the static aspects of such a realisation a collaboration defines a context of class and association rôles describing which features actually participating instances have to show. For the dynamic aspects a collaboration defines an interaction, specifying which actions have to be performed by participating instances, which stimuli to other participating instances these actions have to dispatch, and in which order these stimuli can be sent, sequentially or concurrently.

The UML specification [10] defines the concrete and abstract syntax for collaborations and interactions and gives a description of the intended semantics, but entirely lacks a formal semantics. This omission does not only seriously limit the employment of UML for the construction of analysable and testable software designs in general; for collaborations and interactions the situation is aggravated by a gross vagueness of the specification itself. In particular, the informal semantics falls short of describing how participating instances of an interaction should actually react to an incoming stimulus, when actions are complete, and how local state information is to be treated; for some other, similar, problems see e.g. [9].

---

* This work was carried out during a stay at the Computer Science Laboratory of SRI International as part of the Visitor Exchange Program P-1-3334. It was supported by a DAAD scholarship and partially by the Bayerische Forschungsstiftung.

We therefore propose a formalisation of interactions in UML collaborations that clarifies at least some of the ambiguities of the specification but abstracts from semantic details that are intentionally left open. We claim that our formalisation directly, and intuitively, captures the informal semantics given in the UML specification. In order to reflect the semantic requirements following the specification as closely as possible, the formalisation is based on the UML meta-model. The concurrency constraints of UML interactions are expressed as temporal logic formulas in the style of Manna and Pnueli [8]. These formulae only take into account the sending of stimuli according to actions and the receiving of stimuli by instances. In particular, the formalisation is parametric in the notion of rôle context and in the more detailed execution of actions beyond the sending of stimuli which is under-specified by the UML specification. In order to corroborate our claim that our formalisation captures the UML specification, we provide an alternative approach to the semantics by unfolding interactions into partial orders, using Pratt's pomset framework [11]. We prove that the temporal formulae give rise to these partial orders.

Several formal semantics of interactions have already been investigated, both of the specific notion in UML and of several closely related techniques: Araújo [2] translates a subset of UML sequence diagrams into temporal logic formulas. Gehrke, Goltz, and Wehrheim [6] sketch a translation of UML collaboration diagrams to Petri nets, but do not base their considerations on the meta-model. Wirsing and the author [12] model interaction diagrams of OOSE, one of the main predecessors of UML interactions, by asynchronously communicating finite automata; however, it is unclear whether this approach can be extended to the broader notion of UML interactions. For the equally closely related Message Sequence Charts semantical models based on process algebra, Petri nets, and automata have been investigated, for an overview see e.g. [7]; none of these approaches refers to an object-oriented setting. The semantics of rôle modelling in general is extensively discussed by Andersen [1]; the precise connections to UML collaborations, however, remain to be explored.

The remainder of this paper is structured as follows: In Sect. 1 we summarise UML interactions' abstract syntax and intended semantics. Section 2 presents the generation of temporal formulae from interactions in collaborations that precisely define the semantics of interactions. In Sect. 3 we assign partial orders to interactions and prove that the temporal logic formalisation indeed yields these partial orders. We conclude with an outlook to possible integrations of our semantics with more general approaches.

We assume some familiarity with the UML notation and a superficial knowledge of the UML abstract syntax [10].

1 UML Interactions

We briefly recall the parts of the UML meta-model pertaining to collaborations and interactions and their intended semantics [10] by means of a simple example. Notwithstanding the problems with the mapping from concrete to abstract syntax as described in the UML notation guide we more conveniently present the example in diagram form, but actually discuss the abstract syntax thereby defined. Consider the collaboration diagram in Fig. 1(a); some relevant fragments of its abstract syntax are presented graphically in Fig. 1(b) and Fig. 1(c).

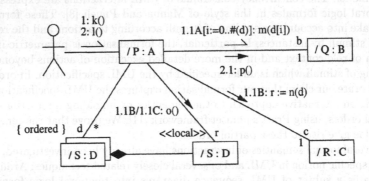

(a) Collaboration Diagram

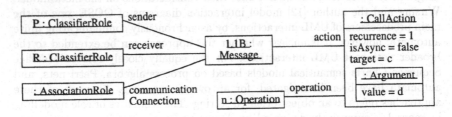

(b) Abstract Syntax of Message 1.1B

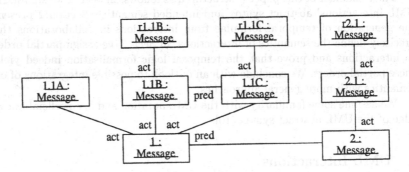

(c) Abstract Syntax of the Interaction

Fig. 1. Example of an UML Collaboration

1.1 Abstract Syntax and Semantics According to the UML Specification

For the static aspects of a collaboration, rôles are specified that have to be filled in order to perform the task of the collaboration, describing the features collaborators have to show. Rôles are based on elements of a surrounding UML model; omitting the details, we require that ClassifierRole[1] P actually has two Operations k and l, that there is an AssociationRole with AssociationEndRoles having type P and Q, and so forth.

The dynamic part of a collaboration is described by Messages gathered in an Interaction. An interaction declares how and in which order stimuli complying to its messages are to be exchanged. Each Message has a reference to a sender ClassifierRole, a receiver ClassifierRole, a communication connection AssociationRole, an Action, a set of predecessor Messages, and, optionally, an activator Message. Semantically, the "activator is the message that invoked the procedure which in turn invokes the current message" [10, p. 2-115]; before a message can be executed all its predecessor messages have to be completed. There are several constraints on these predecessor and activator relations; with M the set of messages of an interaction the following properties have to be satisfied:

i. The graph (M, P) with P the predecessor relation on M is acyclic.
ii. The graph (M, A) with A the activator relation on M is a forest.
iii. For $m', m'' \in M$, if m'' transitively precedes m' then either both have the same activator $m \in M$ or both have no activator.

Executing a message actually means executing its action possibly resulting in the sending of a stimulus that complies to the message and therefore its action. An Action can be a CallAction, a ReturnAction, and of several other kinds, we do not discuss here. Each Action has a target expression that resolves to a set of target instances on evaluation; a recurrence expression that determines how the target set is iterated (sequentially or in parallel); and argument expressions that yield the actual arguments of the Action. Furthermore, an Action can be asynchronous, i.e., on execution no results are awaited, or synchronous, i.e., awaiting results. A CallAction refers additionally to an Operation, a stimulus complying to such an action will call that operation with the evaluated argument expressions as actual parameters. A ReturnAction has no explicit target; it returns the actual arguments to a caller.

According to (our interpretation of) the UML specification the intended semantics of the interaction described in Fig. 1(a) is the following: At start, only stimuli complying to l can occur since it is the only message that has no predecessors and no activators. Such stimuli can only be created by an execution of l's action. Execution of this action means the creation of a single stimulus μ that is sent from the instance playing the actor rôle to the instance playing P and that bears an asynchronous call action of operation k with no actual arguments. Since the action of l is declared as asynchronous, no return stimulus is awaited and

---

[1] As in the UML specification meta-classes are written with initial capital letters.

message 1 is completed; thus, stimuli complying to 2 may now occur. Further-more, on receipt of stimulus μ, 1.1A, 1.1B, and 1.1C are activated; the actions of 1.1A and 1.1B can now be executed concurrently, 1.1C has to wait for the completion of 1.1B. The action of 1.1B calls n with actual argument d; it is synchronous, hence a return stimulus with a value for r is awaited and only on receipt 1.1B is completed. On completion of 1.1B, the action of 1.1C can be executed analogously. However, note that 1.1B and 1.1C share the local link r; this has to be stored such that the return stimulus for 1.1B and the stimulus for 1.1C have access. The action of 1.1A asynchronously calls m as many times as d has elements with actual argument d[i] where i varies with the number of calls; the message is completed when all elements of d are processed. Meanwhile the action of 2 may have been executed in the same vein, activating 2.1. —

Interactions may be composed inside the same collaboration by sharing rôles; or, transgressing the boundaries of one collaboration, by messages sharing actions. In the latter case, execution of an action will generate stimuli complying to all messages sharing this action.

1.2 Problems of the UML Specification

This exposition deviates from or interprets the UML specification in the following points: According to [10, p. 2-115] the activator relationship of messages in an interaction imposes a tree; we relax this condition to forests to allow multiple initial messages. The completion of a message is identified with the completion of its action, which in turn is determined by the termination of its recurrence expression; this remains open in [10]. By [10, pp. 2-99ff.], return actions do not send stimuli; in order to express our semantics more compactly, we require that they do. This also resolves the problem, left open in [10], by which means a synchronous call action learns of the results it is waiting for; we thus require a message with a return action that is activated by the message bearing the call action and that is the final message activated by the call message (see [10, p. 3-128] for a motivation). This leads to the following two additional constraints on the messages M of an interaction:

iv. For $m \in M$, if the action of m is a return then there is no $m' \in M$ such that m activates m'.
v. For $m \in M$, if the action of m is a synchronous call then there is an $r \in M$ such that the action of r is a return, m activates r and m' transitively precedes r for all $m' \in M$ such that m activates m'.

Finally, given our interpretation that shared actions lead to stimuli complying to several messages, the UML specification neglects the consequences of sharing, viz., that shared actions can only be executed if the corresponding messages do not depend on each other. Dependencies in the presence of sharing may be of the form: messages m_0 and m_1 precede messages m'_0 and m'_1, resp., but m_0 and m'_1 share the same action and also m'_0 and m_1 share the same action; none of these action can be executed in this situation. We therefore require that after identifying all those messages of an interaction that share the same action and extending this identification to the predecessor and activator relations, the new predecessor relation has to be acyclic, the activator relation has to form a forest, etc. More precisely, we regard an interaction as a graph (M, R) with the messages M as vertices and the activator and predecessor relations as edges; each vertex m is labelled by $\alpha(m)$, the action of m, and each edge by whether its an activator or a predecessor edge:

vi. The quotient of an interaction (M, R) by α satisfies constraints (i–v).

The UML abstract syntax and its contextual constraints sketched so far serve as a basis for our formalisation of UML interactions. In particular, we will only consider CallActions and ReturnActions, all other kinds may be treated similarly; we do not include the script attribute of Action. Moreover, we do not consider the composition of interactions; we henceforth assume that messages sharing actions are contained in the same interaction.

2 Formal Semantics

We formalise the semantics of UML interactions as formulae in a temporal logic in the style of Manna and Pnueli [8]. We try to argue that this kind of formalisation allows us to capture the informal semantic requirements of the UML specification directly and intuitively. A state variable yields the global system state; functions for local states of actions and stimuli sent and received partially characterise this state. Execution of actions is described by transition systems on the local states of actions. Temporal formulae over the system state constrain the concurrent execution of actions; activations are directly reflected by number sequences. Thus, the global state may change not only subject to the execution of actions in a given interaction, as a collaboration may be embedded in a more comprehensive model. We leave open the representation of instances playing the rôles of an interaction's context and some details of executing actions, like the evaluation of arguments. These are under-specified in the UML specification and we will discuss plausible choices thus demonstrating the flexibility of our approach.

2.1 Semantic Domains

Formally, we require the following semantic domains: A domain Σ of global system states representing the states of instances playing the different rôles in the context of an interaction; for each action a a domain Λ_a of local action states comprising information from the recurrence and the target expression of a; for each action a a domain M_a for stimuli complying to a; and for each message m a domain M_m for stimuli complying to m and the action $\alpha(m)$ attached to it, such that there is a map $\alpha : M_m \to M_{\alpha(m)}$ exhibiting information that is shared by several stimuli. The semantic domain Σ is equipped with maps

$$a_p : \Sigma \to \{\bot, \downarrow\} \uplus (\mathbb{N} \times \Lambda_a) \qquad \text{and}$$

$$m_p : \Sigma \to \{\bot, \downarrow\} \uplus M_m$$

for every action a, every message m, and every $p \in (\mathbb{N} \times \mathbb{N})^+$. The number pair sequences p are used to distinguish different occurrences of actions and stimuli; nesting is reflected by the length of a sequence, the pairs reflect the possibly parallel occurrences on a given nesting level. Intuitively, the map a_p either yields the local action state of the pth occurrence of a together with a "program counter" that is used to create stimuli; or a_p yields that the pth occurrence of a is undefined ($\bot$) in a system state; or has already terminated ($\downarrow$). Analogously, the pth stimulus occurrence for a message m has been sent but not yet received if m_p yields an element of M_m; this stimulus has not been sent if m_p yields $\bot$; and this stimulus has already been received if m_p yields $\downarrow$.

2.2 Transition Semantics for Actions

For each action a we assume that its semantics is given by some initial local action state depending on the system state

$$\lambda_a^0 : \Sigma \to \Lambda_a$$

and transition relations

$$\to_a^\sigma \subseteq (\Lambda_a \times (\{\downarrow\} \uplus \Lambda_a)) \uplus (\Lambda_a \times (\Lambda_a \times \wp M_a))$$

parameterised over the global system state $\sigma \in \Sigma$.

Action occurrences will be created in an initial state; note, however, that such a notion is not required in the UML specification and may therefore be omitted. After creation, an action occurrence may proceed either by terminating and we write $\lambda \downarrow_a^\sigma$ if $(\lambda, \downarrow) \in \to_a^\sigma$; or it may proceed by a silent step changing only its local state and we write $\lambda \to_a^\sigma \lambda'$ if $(\lambda, \lambda') \in \to_a^\sigma$ with $\lambda' \neq \downarrow$; or it may proceed by changing its local state and sending stimuli and we write $\lambda \to_a^\sigma \lambda', M$ if $(\lambda, (\lambda', M)) \in \to_a^\sigma$. An action's recurrence expression may allow for sending several different stimuli in one transition step.

The flexibility offered by such a general semantics for actions does indeed seem to be necessary: Let a be the call action of message 1.1A in our running example (Fig. 1) having: as its recurrence expression i:=0..#(d), as its target expression b, as its operation a reference to m, and as its single argument expression d[i]. Various choices for a formal semantics are possible: For one instance, we could assume that the target expression is only evaluated once, when the action is created, and that the evaluation of the actual argument is atomic. Then we would choose $\Lambda_a = \mathbb{N} \times O$ where O is some semantic domain of object identifiers and $M_a = V \times O$ with an additional semantic domain V for values such that O is a sub-domain of V. The initial state would be

$$\lambda_m^0(\sigma) = (0, [\![b]\!](\sigma))$$

and a transition relation could be defined by

$$(i, o)\downarrow_m^\sigma \quad \text{if } \#[\![d]\!](\sigma) > i,$$

$(i, o) \rightarrow_a^\sigma (i + 1, o), (\llbracket \mathbf{d} \rrbracket(\sigma)(i), b) \quad \text{if } \sharp \llbracket \mathbf{d} \rrbracket(\sigma) \leq i$

where $\llbracket - \rrbracket$ is a function evaluating an expression in a state and $\sharp$ denotes the cardinality function.

However, the evaluation of the actual arguments may require many steps which then has to be reflected in the semantic domain Λ_a. Analogously, the evaluation of the target expression may not be atomic. We simply are not committed to any of these selections.

2.3 Semantics of Interactions

We now turn to the semantics of full interactions and their ordering constraints. We use a linear first-order temporal logic with temporal connectives $\square$ (always), $\lozenge$ (eventually), and W (unless). The underlying state language consists of one flexible state variable σ from the semantic domain Σ, rigid variables, the semantic maps a_p and m_p as function symbols, and $\rightarrow_a^\sigma$ as relation symbols (we also use the various abbreviations introduced above). The semantics of a specification in such a temporal logic is defined to be a transition system whose runs satisfy all formulae of the specification; for more details on this formalism cf. [8].

Let I be an interaction, M its set of messages, A the set of actions that is attached to M, and let $\alpha(m)$ denote the action of $m \in M$. The instantiation of the formula schemes (1–13) defined below according to I define a temporal logic specification of the semantics of I.

First, we embed the transition semantics for actions into temporal logic. Each occurrence of an action $a \in A$ is created in its initial state:

$$\forall p \in (\mathbb{N} \times \mathbb{N})^+ . a_p(\sigma) = \bot \; W \; a_p(\sigma) = (0, \lambda_a^0(\sigma)) . \tag{1}$$

Each occurrence of an action $a \in A$ proceeds as given by its transition semantics:

$$\begin{aligned}
\square(a_p(\sigma) = (i, \lambda) &\Rightarrow (a_p(\sigma) = (i, \lambda) \; W \\
&((a_p(\sigma) = \downarrow \wedge \lambda \downarrow_a^\sigma) \vee \\
&(a_p(\sigma) = (i, \lambda') \wedge (\lambda \rightarrow_a^\sigma \lambda')) \vee \\
&(a_p(\sigma) = (i + 1, \lambda') \wedge (\lambda \rightarrow_a^\sigma \lambda', \{\mu_1, \ldots, \mu_k\}) \wedge \\
&\bigwedge_{\substack{0 \leq j \leq k \\ m \in \alpha^{-1}(a)}} \alpha(m_{p.(i,j)}(\sigma)) = \mu_j \wedge \forall j > k . m_{p.(i,j)}(\sigma) = \bot))) .
\end{aligned} \tag{2}$$

Occurrences of actions that have terminated can not be reactivated:

$$\square(a_p(\sigma) = \downarrow \Rightarrow \square a_p(\sigma) = \downarrow) .$$

Next, stimuli can only be created by appropriate actions:

$$m_{p.(i,j)}(\sigma) = \bot \; W \; (\alpha(m)_p(\sigma) = (i + 1, \lambda) \wedge m_{p.(i,j)}(\sigma) \neq \bot) . \tag{3}$$

Stimuli do not change between sending and receiving:

$$\square((m_p(\sigma) = \mu \wedge \mu \neq \bot) \Rightarrow (m_p(\sigma) = \mu \; W \; m_p(\sigma) = \downarrow)) . \tag{4}$$

Stimuli will be received sometime:

$$\Box(m_p(\sigma) \neq \bot \Rightarrow \Diamond m_p(\sigma) = \downarrow) . \tag{5}$$

Stimuli that have been received can not be resent:

$$\Box(m_p(\sigma) = \downarrow \Rightarrow \Box m_p(\sigma) = \downarrow) . \tag{6}$$

Finally, we treat the order constraints of the interaction I. If a message m is preceded by a message m'

$$\Box(\alpha(m)_p(\sigma) = (0, \lambda) \Rightarrow \alpha(m')_p(\sigma) = \downarrow) , \tag{7}$$

saying that whenever the pth occurrence of message m's action is ready to be executed the preceding message m' must have terminated.

If message m is activated by message m'

$$\Box(\alpha(m)_{p.(i,j)} = (0, \lambda) \Rightarrow m'_{p.(i,j)}(\sigma) = \downarrow) , \tag{8}$$

$$\Box(m'_{p.(i,j)}(\sigma) = \downarrow \Rightarrow \Diamond \alpha(m)_{p.(i,j)} = (0, \lambda)) , \tag{9}$$

saying that when a stimulus with number i adhering to the pth occurrence of the action of message m' has been received, the $p.i$th occurrence of message m will be ready for execution some time later on, but that execution can not start prematurely.

If message r is the return message of message m

$$\Box(r_{p.(i,j).(k,l)}(\sigma) = \downarrow \Rightarrow \alpha(m)_p(\sigma) = (i + 1, \lambda)) , \tag{10}$$

saying that when a stimulus with number (k, l) adhering to the $p.(i, j)$th occurrence of message r's action has been received the pth activation of message m's action is (still) in state $i + 1$.

We additionally designate an initial state. Let N be the messages without an activator in I. Only occurrences of actions of messages in N may exist initially.

$$\forall m \in M . \forall p \in (\mathbb{N} \times \mathbb{N})^+ \setminus \{(0, 0)\} . \alpha(m)_p(\sigma) = \bot \tag{11}$$

$$\forall m \in M \setminus N . \alpha(m)_{(0,0)}(\sigma) = \bot \tag{12}$$

$$\forall m \in N . \Diamond \alpha(m)_{(0,0)}(\sigma) = (0, \lambda_a^0(\sigma)) \tag{13}$$

The *semantics* of I is defined to be all runs (models) of the temporal logic specification yielded by instantiating formula schemes (1–13) according to I.

Example. For our running example as depicted in Fig. 1(c), denoting the action of message m by am,

$$\Box(\text{a2}_p(\sigma) = (0, \lambda) \Rightarrow \text{a1}_p(\sigma) = \downarrow)$$

$$\Box(\text{a1.1A}_{p.(i,j)} = (0, \lambda) \Rightarrow 1_{p.(i,j)}(\sigma) = \downarrow)$$

$$\Box(1_{p.(i,j)}(\sigma) = \downarrow \Rightarrow \Diamond \text{a1.1A}_{p.(i,j)} = (0, \lambda))$$

125

$\Box(\texttt{a1.1B}_{p.(i,j)} = (0,\lambda) \Rightarrow 1_{p.(i,j)}(\sigma) = \downarrow)$

$\Box(1_{p.(i,j)}(\sigma) = \downarrow \Rightarrow \Diamond \texttt{a1.1B}_{p.(i,j)} = (0,\lambda))$

$\Box(\texttt{a1.1C}_{p.(i,j)} = (0,\lambda) \Rightarrow 1_{p.(i,j)}(\sigma) = \downarrow)$

$\Box(1_{p.(i,j)}(\sigma) = \downarrow \Rightarrow \Diamond \texttt{a1.1C}_{p.(i,j)} = (0,\lambda))$

defines the ordering constraints for the messages 1, 1.1A, 1.1B, and 1.1C according to (7–9). A simple model construction will be discussed in the next section.

It may be noted that we assume that for shared actions several different stimuli complying to each of the corresponding messages occur. These may be comprised into only one stimulus complying to several messages and the shared action.

3 Partial Orders from Interactions

In order to provide evidence that our formalisation captures the intuitive semantics of interactions as described in the specification, we investigate the partial orders of stimuli that can be produced by executing an interaction. These partial orders are derived again directly from the abstract syntax. They may appear as an even more obvious approach to the semantics of interactions; however, it seems to be non-trivial to integrate them with a semantics of actions. We thus subsequently prove that our temporal logic semantics yields the same partial orders for terminating interactions.

More precisely, we assign a process, i.e. a set of (labelled) pomsets [11], to an interaction, with the labelling from the messages of the interaction. Such a process represents all possible unrollings or executions of the interaction assuming that each action of a message can be executed an arbitrary, but finite number of times; completion of a message, viz. of its action, will be indicated by special labels. The overall plan of the process construction is to assign to each activation nesting level of messages a set of processes in a bottom-up fashion and to insert the pomsets of these processes whenever the start message of a nesting level occurs in the previous nesting level.

The process is built from simple pomsets containing only a single atom m, denoting the possible occurrence of a stimulus complying to the message m, or $\overline{m}$, denoting the completion of the message's action. From these simple processes we proceed to more complex ones by the well-known (total) process operations of sequential composition (;), parallel composition ($\parallel$), sequential iteration ($*$), parallel repetition ($\dagger$), and homomorphisms [11]; two additional partial operations for the treatment of synchronous and asynchronous actions are introduced. We define these operations for pomsets only; they are lifted to processes in the usual way:

Let p and q be pomsets such that q has a unique minimal element. Let $(X, \leq_X, \kappa)$ and $(Y, \leq_Y, \lambda)$ be representing partial orders for p and q, resp., such that X and Y are disjoint; let $m \in Y$ be the representation of the minimal element of q and let $C = \{x \in X \mid \kappa(x) = \lambda(m)\}$.

The *asynchronous insertion* of q in p, written $p \leftharpoonup q$, is given by the pomset represented by the partial order $(Z, \leq_Z, \mu)$ with: $Z = X \cup (Y \setminus \{m\}) \cdot C$ (where $N \cdot M$ denotes the M-fold disjoint sum of N); for every $z, z' \in Z$ define $z \leq_Z z'$ if either $z, z' \in X$ and $z \leq_X z'$, or $z, z' \in Y \times \{x\}$ for some $x \in C$ and $\pi_1 z \leq_Y \pi_1 z'$, or $z \in C$ and $z' \in Y \times \{z\}$; for every $z \in Z$ define $\mu(z) = \kappa(z)$ if $z \in X$ and $\mu(z) = \lambda(\pi_1 z)$ if $z \in Y \times C$.

The *synchronous insertion* of q in p, written $p \leftarrow q$, is given by the pomset represented by $(Z, \leq'_Z, \mu)$ with Z and μ as for the asynchronous insertion and $z \leq'_Z z'$ for $z, z' \in Z$ if either $z, z' \in X$ and $z \leq_X z'$, or $z, z' \in Y \times \{x\}$ for some $x \in C$ and $\pi_1 z \leq_Y \pi_1 z'$, or $z \in C$ and $z' \in Y \times \{z\}$, or $z \in Y \times \{z''\}$ for some $z'' \in C$ and $z'' \leq_X z'$.

Synchronous insertion is obviously a special case of homomorphism. Both synchronous and asynchronous insertion are associative; additionally

$$p \leftarrow q \leftarrow r = p \leftarrow r \leftarrow q \quad \text{and} \quad p \leftharpoonup q \leftharpoonup r = p \leftharpoonup r \leftharpoonup q$$

hold for pomsets p, q, and r such that all the synchronous and asynchronous insertions are defined, respectively.

3.1 Process Construction

To begin with, we construct processes for interactions without shared actions; the general case will be discussed shortly. Again we perceive such an interaction as a graph I with messages as its vertices and edges labelled either "activator" or "predecessor" which has to satisfy conditions (i–v) of Sect. 1. In order to abbreviate notation, we write $m \dashrightarrow m'$, if message m is the activator of m', and $m \rightarrow m'$, if m is a predecessor of m'. For the set of messages M of I fix a set $\overline{M}$, disjoint from M and a bijective function $\overline{\cdot} : M \to \overline{M}$.

We first describe how a process is assigned to each level of activation, i.e., a full subgraph of the interaction having as one vertex an activating message and as the additional vertices all the messages that it (immediately) activates. Each such level would be an interaction if requirement (v) would be dropped. Let C_N be the set of all graphs with vertices from $N \subseteq M$, edges labelled from $\{\dashrightarrow, \rightarrow\}$, fulfilling requirements (i–iv) above, and having exactly one activator vertex. Let P_N be the class of all processes with labels from $N \cup \overline{N}$. Define $\psi_N : C_N \to P_N$ as follows: Let $C = (N, R) \in C_N$, and let $c \in N$ be its unique activator. Define a pomset p represented by the partial order $(N \setminus \{c\}, \dashrightarrow, \mathrm{id}_{N \setminus \{c\}})$ and a homomorphism $\sigma : p \to P_N$ given by $\sigma(m) = m$ if the action of m is a return; $\sigma(m) = m^* ; \overline{m}$ if the action of m is not a return and always sequential; and $\sigma(m) = m^{\dagger*} ; \overline{m}$ otherwise. Then $\psi_N(C) = c ; p\sigma$.

For the overall construction, let I_M be the set of all interactions with vertices M and define P_M as before. Define $\varphi : \prod_M I_M \to P_M$ mutually recursive as follows: Let $I = (M, R)$ be an interaction.

A. If $I = (M, R)$ such that there is no activator in M (i.e. there are no $c, m \in M$ with $c \dashrightarrow m$), then define a homomorphism $\sigma : M \to P_M$ by $\sigma(m) = m^* ; \overline{m}$

if the action of m is always sequential and $\sigma = m^{\dagger *}$; $\overline{m}$ if the action of m may be parallel; define a pomset p represented by the partial order $(M, \rightarrow, \mathrm{id}_M)$ with $\rightarrow$ the predecessor relation of I. Define $\varphi_M(I) = p\sigma$.

B. If $I = (M, R)$ such that there is at least one activator in M, then let $A = \{a_1, \ldots, a_k\}$ be the penultimate vertices of the activator forest in I (i.e. for every $a \in A$ there is an $m' \in M$ such that $a \dashrightarrow m'$ but there are no two vertices $m', m'' \in M$ such that $a \dashrightarrow m' \dashrightarrow m''$; and A is maximally so); for every $1 \le i \le k$ define $I_i = (M_i, R_i)$ as the full subgraph generated by $\{m \in M \mid a_i \dashrightarrow m\} \cup \{a_i\}$; further, define $I_0 = (M_0, R_0)$ as the full subgraph of I generated by $M \setminus \bigcup_{1 < i \le k}(M_i \setminus \{a_i\})$; finally, define $P_0 = \varphi_{M_0}(I_0)$ and $P_i = P_{i-1} \leftarrow \psi_{M_i}(I_i)$ if the action of a_i is synchronous and $P_i = P_{i-1} \leftharpoonup \psi_{M_i}(I_i)$ if the action of a_i is asynchronous. Then $\varphi_M(I) = P_k$.

Let $I = (M, R)$ be an interaction and $\varphi : \prod_M I_M \to P_M$ defined as above. Then, $[\![I]\!] = \varphi_M(I)$ is the *corresponding process* of I.

Example. We illustrate the construction of an interaction's corresponding process by our running example interaction I as depicted in Fig. 1(c). All actions in I are sequential. The graph I fulfils the conditions of (B) and we get

$$I_0 = \begin{array}{cccc} \text{1.1A} & \text{1.1B} \longrightarrow \text{1.1C} & \text{2.1} \\ & 1 \longrightarrow 2 \end{array}$$

$$I_1 = \begin{array}{c} \text{r1.1B} \\ \uparrow \\ \text{1.1B} \end{array}, \quad I_2 = \begin{array}{c} \text{r1.1C} \\ \uparrow \\ \text{1.1C} \end{array}, \quad I_3 = \begin{array}{c} \text{r2.1} \\ \uparrow \\ \text{2.1} \end{array}$$

Omitting the indices of ψ and φ, we have $\psi(I_1) = \text{1.1B}\,;\,\text{r1.1B}$, $\psi(I_2) = \text{1.1C}\,;$ r1.1C, and $\psi(I_3) = \text{2.1}\,;\,\text{r2.1}$. The graph I_0 again fulfils the conditions of (B):

$$I_0' = 1 \longrightarrow 2$$

$$I_1' = \begin{array}{ccc} \text{1.1A} & \text{1.1B} \longrightarrow \text{1.1C} \\ & 1 \end{array}, \quad I_2' = \begin{array}{c} \text{2.1} \\ \uparrow \\ 2 \end{array}$$

Here, $\psi(I_1') = \text{1}\,;\,(\text{1.1A}^*\,;\,\overline{\text{1.1A}} \,\|\, \text{1.1B}^*\,;\,\overline{\text{1.1B}}\,;\,\text{1.1C}^*\,;\,\overline{\text{1.1C}})$, $\psi(I_2') = \text{2}\,;\,\text{1}^*\,;\,\overline{\text{2.1}}$. Now, by (A), $\varphi(I_0') = \text{1}^*\,;\,\overline{1}\,;\,\text{2}^*\,;\,\overline{2}$, and thus $\varphi(I_0) = \varphi(I_0') \leftharpoonup \psi(I_1') \leftharpoonup \psi(I_2')$; a typical pomset of this process is

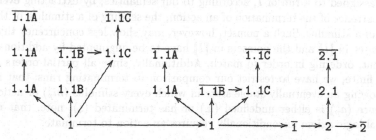

Finally, $\varphi(I) = \varphi(I_0) \leftarrow \psi(I_1) \leftarrow \psi(I_2) \leftarrow \psi(I_3)$; synchronous insertion into the pomset above yields

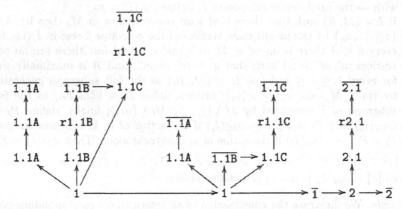

In the general case, actions may be shared by messages. The construction of a process for an interaction without shared actions can be lifted to this situation: Let $I = (M, R)$ be an interaction satisfying conditions (i–vi) of Sect. 1, viewed as an edge-labelled graph as above. Let $\alpha(m)$ denote the action of a message $m \in M$. Define an edge-labelled graph $I/\alpha = (M/\alpha, R/\alpha)$ with M/α the set of equivalence classes $[m]$ of messages quotiented by α, i.e., $[m] = [m']$ if $\alpha(m) = \alpha(m')$; and R/α as $v \dashrightarrow v'$ if $v = [m]$ and $v' = [m']$ and $m \dashrightarrow m'$, and $v \to v'$ if $v = [m]$ and $v' = [m']$ and $m \to m'$. By definition, I/α is the graph of an interaction. Furthermore, for $[m] = \{m_0, \ldots, m_{n-1}\}$, let $\sigma([m]) = m_0 \| \ldots \| m_{n-1}$ and $\sigma(\overline{[m]}) = \overline{m_0} \| \ldots \| \overline{m_{n-1}}$.

Then, $[\![I]\!] = [\![I/\alpha]\!]\sigma$ is the *corresponding process* of I.

3.2 From Runs to Pomsets

There is a precise relation between the formal semantics of an interaction I as given by all runs over a single state variable satisfying the temporal formulae derived from (1–13) in Sect. 2 and the process $[\![I]\!]$ according to the construction above. The pomsets in $[\![I]\!]$ only take into account the sending of messages and the termination of actions; we may additionally interpret an occurrence of a return stimulus in a pomset as the receipt of this stimulus. Conversely, a pomset can be assigned to a run of I, according to our semantics, by extracting every first occurrence of the termination of an action, the sending of a stimulus, or the receipt of a stimulus. Such a pomset, however, may show less concurrency than any pomset in $[\![I]\!]$ and the pomsets in $[\![I]\!]$ have to be augmented by additional, consistent, ordering in order to match. Additionally, since all partial orders in $[\![I]\!]$ are finite, we have to restrict our comparison to terminating runs, that is, runs showing an eventually stable state in which every stimulus (m_p) or action occurrence (a_p) is either undefined ($\bot$) or has terminated ($\downarrow$); note, that no temporal logic axiom unconditionally requires an action to terminate.

More precisely, for a terminating run $(\sigma_i)_{i\in\mathbb{N}}$ of I with messages M define

$$H_i = \{(\overline{m},p) \in \overline{M} \times (\mathbb{N} \times \mathbb{N})^+ \mid \alpha(m)_p(\sigma_i) = \downarrow\} \cup$$
$$\{(m,p) \in M \times (\mathbb{N} \times \mathbb{N})^+ \mid m_p(\sigma_i) \in M_{\alpha(m)}, \; \alpha(m) \text{ no return}\} \cup$$
$$\{(m,p) \in M \times (\mathbb{N} \times \mathbb{N})^+ \mid m_p(\sigma_i) = \downarrow, \; \alpha(m) \text{ return}\}$$

and let $\pi((\sigma_i)_{i\in\mathbb{N}})$ denote the pomset represented by $(H, \leq, \eta)$ with

$$H = \bigcup_{i\in\mathbb{N}} H_i \, ,$$

$$h \leq h' \text{ if } h = h' \text{ or } h \in H_k, \; h' \in H_l \setminus \bigcup_{0 \leq i < k} H_i, \; k < l \, ,$$

$$\eta((\overline{m},p)) = \overline{m}, \; \eta((m,p)) = m \, .$$

It remains to show that these pomsets are augmentations of pomsets in $[\![I]\!]$, i.e., that for every run $(\sigma_i)_{i\in\mathbb{N}}$ there is a $\pi \in [\![I]\!]$ equipped with a bijective homomorphism from π to $\pi((\sigma_i)_{i\in\mathbb{N}})$.

For a proof sketch of this claim, the following observation is decisive: Let $(X, \leq, \lambda)$ be a representative of a pomset in $[\![I]\!]$. Then $x < x'$ only if: $\lambda(x) \notin \overline{M}$ and $\lambda(x) = \lambda(x')$; or $\lambda(x), \lambda(x') \notin \overline{M}$ and $\lambda(x)$ transitively precedes or activates $\lambda(x')$ in I; or $\lambda(x) \in \overline{M}$, and the message of $\lambda(x)$ precedes or activates the message of $\lambda(x')$ in I; or $\lambda(x) \in M$, $\lambda(x') \in \overline{M}$, and the action of $\lambda(x)$ is the return action for the action of the message of $\lambda(x')$. This is easily proven by the construction of $[\![I]\!]$. But these are exactly the orderings that are also at least required by the temporal formulae. Since $[\![I]\!]$ contains all pomsets with an arbitrary number of stimulus occurrences, the claim follows.

The correspondence between the pomset semantics and the temporal semantics may also be extended to non-terminating runs (in the sense explained above), if we drop the requirement that all messages have to be completed in the pomset semantics.

Conclusions and Future Work

We presented a formal semantics for UML interactions. This semantics is given by all runs satisfying certain temporal formulae that can be derived directly from the abstract syntax representation of an interaction; it tries to capture the requirements of the UML semantics specification as intuitively as possible. In particular, the semantics is parametric in both the notion of context of interactions and in a transition semantics for actions, that are under-specified by the UML semantics. Additionally, we investigated the relationship of these models to an event-based construction assigning pomsets over stimuli to interactions; it was shown that temporal runs correspond to augmentations of these pomsets.

It may be interesting to combine the construction of pomset models for interactions with the transition semantics for actions in the style of Cenciarelli et. al. [4]. Such a semantics could form an even more declarative alternative.

Our temporal logic semantics, however, already provides the necessary basis to study the important notion of composition of interactions. We considered sharing of actions as one possibility to combine different interactions in a common contextual collaboration, but much work remains to be done in this respect.

Finally, the contextual parametricity of our semantics allows its smooth integration with existing more general approaches to a comprehensive system model for UML, such as the SysLab [3] or pUML [5].

Acknowledgements. I profitted much from discussions with José Meseguer, Martin Wirsing, and Harald Störrle.

References

1. Egil P. Andersen. *Conceptual Modeling of Objects — A Role Modeling Approach.* PhD thesis, Universitetet i Oslo, 1997.
2. João Araújo. Formalizing Sequence Diagrams. In Luis F. Andrade, Ana Moreira, Akash R. Deshpande, and Stuart Kent, editors, *Proc. Wsh. Formalizing UML. Why? How?*, Vancouver, 1998.
3. Ruth Breu, Radu Grosu, Franz Huber, Bernhard Rumpe, and Wolfgang Schwerin. Systems, Views and Models of UML. In Axel Korthaus and Martin Schader, editors, *The Unified Modeling Language — Technical Aspects and Applications*, pages 93–108. Physica, Heidelberg, 1998.
4. Pietro Cenciarelli, Alexander Knapp, Bernhard Reus, and Martin Wirsing. An Event-Based Structural Operational Semantics of Multi-Threaded Java. In Jim Alves-Foss, editor, *Formal Syntax and Semantics of Java*, volume 1523 of *Lect. Notes Comp. Sci.*, pages 157–200. Springer, Berlin, 1999.
5. Robert France, Andrew Evans, Kevin Lano, and Bernhard Rumpe. The UML as a Formal Modeling Notation. *Comp. Stand. Interf.*, 19(7):325–334, 1998.
6. Thomas Gehrke, Ursula Goltz, and Heike Wehrheim. The Dynamic Models of UML: Towards a Semantics and Its Application in the Development Process. Technical Report 11/98, Universität Hildesheim, 1998.
7. Piotr Kosiuczenko and Martin Wirsing. Towards an Integration of Message Sequence Charts and Timed Maude. In Murat M. Tanik, Jiro Tanaka, Kiyoshi Itoh, Michael Goedicke, Wilhelm Rossak, Hartmut Ehrig, and Franz Kurfueß, editors, *Proc. 3rd Int. Conf. Integrated Design and Process Technology*, Berlin, 1998.
8. Zohar Manna and Amir Pnueli. *The Temporal Logic of Reactive and Concurrent Systems. Vol. 1: Specification.* Springer, New York–etc., 1992.
9. Object Management Group. OMG UML v1.3: Revisions and Recommendations — Appendix A: Issues Database Report. Technical report, Object Management Group, 1999. http://www.omg.org/docs/ad/99-06-11.pdf.
10. Object Management Group. Unified Modeling Language Specification, Version 1.3. Technical report, Object Management Group, 1999. http://www.omg.org/docs/ad/99-06-08.zip.
11. Vaughan Pratt. Modeling Concurrency with Partial Orders. *Int. J. Parallel Program.*, 15(1):33–71, 1986.
12. Martin Wirsing and Alexander Knapp. A Formal Approach to Object-Oriented Software Engineering. In José Meseguer, editor, *Proc. 1st Int. Wsh. Rewriting Logic and Its Applications*, volume 4 of *Electr. Notes Theo. Comp. Sci.*, pages 321–359. Elsevier, 1996.

UML 2.0 Architectural Crossroads:
Sculpting or Mudpacking?
Panel

Moderator:
Cris Kobryn[1]

Panelists:
Michael Jesse Chonoles[2], Steve Cook[3], Desmond D'Souza[4],
Sridhar Iyengar[5] and Guus Ramackers[6]

[1] EDS, [2] Lockheed Martin, [3] IBM,
[4] Computer Associates, [5] Unisys, [6] Oracle

Abstract. As the UML reaches the venerable age of four, both its proponents and critics are scanning the changes in the UML 1.3 revision and the proposed UML 2.0 roadmap. They are asking pointed questions, such as: Has the modeling language matured or bloated during the standardization process? Will UML be able to adapt to the changing requirements of software development? These questions and the strong emphasis of the UML Revision Task Force's final report on architectural issues suggest that UML is approaching an architectural crossroads at the OMG and elsewhere. This panel explores the nature and the extent of the architectural challenges facing UML, and makes constructive recommendations to address them.

1 Discussion Context

As the UML reaches the venerable age of four, both its proponents and critics are scanning the changes in the UML 1.3 revision and the proposed UML 2.0 roadmap. They are asking pointed questions, such as: Has the modeling language matured or bloated during the standardization process? Will the UML be able to adapt to the evolving requirements of software development, such as distributed business components and enterprise architectures for e-business?

These questions and the strong emphasis of the UML Revision Task Force's final report on architectural issues suggest that UML is approaching an architectural crossroads at the OMG and elsewhere [1]. Although the UML appears to have survived, even thrived, under the OMG standardization process, it faces tough new challenges as the UML 2.0 roadmap is planned and driven [2]. These challenges include the following:

- *Learning curve.* UML is a general-purpose modeling language that supports a full range of semantic expression. While UML's basic language constructs can be learned quickly, it requires significant time to master its advanced techniques. It is not surprising then that the learning curve for UML is significantly steeper than it is for CORBA or DCOM IDL.

- *Semantic bloat.* Although it may be argued that UML's size and complexity are inevitable because it is a general-purpose modeling language, it also includes many standard elements that have vague or sparse semantics. UML could be simplified substantially by removing these from the kernel language.
- *Lightweight extensibility.* UML currently provides only lightweight extension mechanisms, such as stereotypes, constraints and tagged values.[1] These mechanisms are challenged by simple extensions (e.g., stereotypes for CORBA IDL interfaces) and are stressed by more demanding extensions (e.g., application frameworks and distributed business components).
- *Metamodeling overemphasis.* Since metamodeling is now recognized as a powerful technique for managing the complexity of distributed architectures, many modelers are eager to apply this new solution sledgehammer to problems where a claw hammer (e.g., a stereotype) or a tack hammer (e.g., a class in a standard model library) would be more appropriate.

It has been suggested that a successful response to these challenges will require that the OMG adopt a sculpting approach (where "less is more") rather than a mudpacking approach (sometimes associated with a "ball-of-mud" pattern) to refine and extend the UML architecture [2]. In particular, modelers need to perform effective triage in order to determine which language extensions should be treated as revisions to the UML kernel language, and which should be handled as separate profiles (e.g., a profile for business process modeling) or standard model libraries (e.g., a model library for healthcare with PatientRecord and Prescription classes).

The issue facing the panel is to assess the nature and extent of UML's architectural challenges, and to make constructive recommendations to address them. Within this general context, the panel asks the following questions:

- What guidelines (if any) do you propose for determining what UML constructs should be defined as metamodel elements (i.e., part of the kernel language), profiles or model libraries?
- What constructs should be added and removed from UML 2.0? (For brevity, select and prioritize the top three of each.)
- Should modelers be able to define their own metaclasses? If so, what will be the ramifications on common semantics and notation?
- To what extent should the UML specification adopt a more formal (or strict metamodeling) approach?

---

[1] A metaclass is considered a heavyweight extension mechanism.

2 Position Statements

Michael Jesse Chonoles

UML has made fantastic penetration in the system development industry but is facing ever-increasing barriers to future adoption. And these barriers are more than just speed bumps. We face barriers of increasing size, complexity, and formality, barriers of untried features, and barriers of mutually unintelligible dialects.

UML overcame some barriers. Before UML was adopted, there were upwards of 40 different OO notations. This was the main barrier to adoption and innovation. Practitioners needed to be trained and re-trained. Expertise and investment in one notation might not be of any value in another. Then UML came along and fixed most of this. Soon tools and training courses were allowing practitioners to have a common intelligible way of expressing their ideas and designs.

UML's success is source of new barriers. The very success of UML attracts interest from many different communities with different needs. While the desire to continuously add special features is understandable, each such addition makes it harder for practitioners to get their arms around UML. Gone are the days of OMT, where one of its features was the ability for non-technical business analysts to embrace it and for developers to employ it almost immediately. Now, despite the fantastic job the Revision Task Force has done clarifying and reconciling inconsistencies, UML is increasingly becoming so complicated and technical that expertise in it requires a dedication that is almost incompatible with having any time to use it. Each proposed addition brings with it specialty notation and interpretations—more barriers.

UML adoption barriers. Imagine a typical 5-day first course in OOAD using UML. Consider that the course must cover some amount of basic OO principles and methodology. How much of UML could it possibly cover? Where should it concentrate? My organization is the leading OO trainer, and our OO course suite is thorough and well designed, but only about one third of UML can be covered and that proportion is decreasing. Not only is UML steadily getting larger and more complicated which each new revision, but also we're way beyond the highly motivated early adopters with advanced backgrounds. To the majority of students, the size, complexity, and formality of UML are true barriers. The true users of UML are not asking for more, they're asking for less!

Recommendations. Modelers should follow good object-oriented practices to break down the barriers:
1. Abstract from the mass of increasing and confusing detail of UML a "core" UML that follows the 20-80-80 rule: The 20% of the notation that 80% of practitioners use 80% of the time. This would be close to the original UML—a merging of OMT and Booch with the addition of use cases. Think of it as the minimum that a

134

practitioner would need to know to claim UML proficiency. It could be the basis for the core UML curriculum and required to be supported by UML-compliant tools. Use the OO principle of *abstraction* to define layers of abstraction that allow for understanding at multiple levels of depth.

2. Encapsulate the UML notation from changes in the metamodel without great provocation. Users already worry that their work will need to be redone as UML evolves. Use the OO principle of *encapsulation* to prevent impacting previously completed work.

3. The notation should hide the formal details of the UML metamodel. For example, most users need not know about the formal differences between types and classes. Use *information hiding* to save us from unwanted theory, especially where it might change.

4. Use *inheritance* properly for extensions by following substitution principles. Limit proposed extensions to not change any existing feature. Some current proposals appear to allow any notation and almost any metamodel to claim to be an extension of UML.

5. Use *configuration management* on proposed extensions. No proposed feature should be officially added until it has been tested. It is much harder to take something out than it is to put it in, and users should not be burdened with untried, attractive-sounding mistakes.

Steve Cook

In my view the most important recent development in UML is the recognition that UML is a family of languages, rather than just one. We distinguish between a core definition, or kernel, and a set of profiles that define specialized extensions of the kernel language. Currently there is a proposal before the OMG for how to define profiles; I believe that this proposal is not yet sufficient in scope, and that the structure of profiles should cover more ground. A key question is what parts of the current UML definition should stay where they are, and which parts should be moved into profiles.

I believe that the current broad scope of UML is about right, but that its definition is not yet sufficiently precise, and that it mixes up levels of abstraction. Perhaps the most significant point of poor definition is the definition of the mapping between the semantics and the notation. The standard document defines, informally, a mapping between notation and metamodel, but for model interchangeability a precisely defined mapping in the other direction is needed. In terms of levels of abstraction, there remain aspects of the current metamodel which are too much influenced by ideas from one specific programming language, such as options for visibility, the fact that structural features cannot have parameters, and the definition of aggregation.

In summary, I think the following changes need to be made in UML 2.0:
• fully recognize the distinction between the kernel definition and the profiles, keeping the structure of the kernel simple, uniform and abstract
• create a composable profile structure which can handle at least the following:

1. extension through new and specialized elements
2. specialized well-formedness rules for diagrams and their relationships for programming language-specific and other bindings
3. dynamic semantics, including models of time, concurrency and communication
4. connecting OCL constraints to diagrams

In answer to the question "Should modelers be able to define their own metaclasses?" I think the question should be re-formulated as whether modelers may define profiles. If we adopt the principle that the interpretation of a design model in a particular case depends on the application architecture, we may envisage a profile which itself represents and encapsulates important aspects of an application architecture. In such a case, the application architect would most definitely want to define his or her own profile.

Desmond D'Souza

UML offers a rich set of standard modeling constructs. Its weakness is that these constructs are not structured on a minimal semantic base, and its extension mechanisms are inadequate to structure the UML itself. Consequently it is semantically bloated, difficult to learn, and often needlessly "extended." Semantically ambiguous extensions spell trouble with widespread re-use of models across teams and tools, and make frameworks like RM-ODP very difficult.

Despite the "Unified" name, we have not reached global agreement on a single modeling language for all purposes. Rule-based systems, compilers, signal-processors, control-systems, etc. all want end-user modeling languages to simplify the expression of their own problems and solutions. But there is even disagreement on the meanings of some UML elements themselves. The "Unified Modeling Language," in fact, must necessarily be a family of modeling languages.

The UML standards groups recently acknowledged this with the concept of "profile" to structure the definition of multiple modeling languages as extensions to the UML. However, the deeper need is for:

- A small and simple core for the UML
- A clean mechanism to package and extend the modeling language(s)
  - Extensions can also specify pre-requisite constraints on usage of their modeling constructs
  - Extensions can specify a translation of convenient syntax into an underlying expanded form
- Using that extension mechanism to re-factor the current UML constructs themselves
- Finally, using that same extension mechanism to structure other UML extensions and architectural styles

UML 2.0 represents an architectural crossroads for the modeling standards effort. Should we proceed with additional widespread and ad-hoc extensions on the current base ("ball-of-mud")? Or should we step back and re-architect the current standard to simplify it so it will scale better to the challenges ahead ("sculpting")?

I believe we should "sculpt," first re-factoring to a minimal base, moving as much as possible into "standard libraries." The minimal base should be layered to minimally impact the semantics of existing modeling elements. Wherever possible, an element should be defined in terms of existing ones (e.g., state as a predicate or boolean attribute). The structuring mechanism must permit different aspects of a modeling element to be defined separately and composed meaningfully. For example, you might impose new constraints on the use of the "inheritance" element, which was itself introduced somewhere else; or, if new elements are defined independently, you might define its translation into existing ones separately from the package which first introduced the element. The resulting structure might resemble this:

- P0: Packages with normal and "instantiation" import mechanism to support all others
- P1: Base 1 package: define object type, and attribute
- P2: Associations - built using P1
- P2.1: Aggregations - built using P2
- P3: Base 2 package: define time, state change, and state-changing actions - built using P1
- P4: State charts: structured states and state transitions - built using P3
- ...
- P10: The refinement relation construct between models
- P11: Particular refinement relations and their rules - built using P10
- P12: Use-cases and their realizations - built using P11
- ...
- P30: CORBA-Component model with its own convenient modeling constructs
- P31: EJB component model with its own convenient modeling constructs
- P32: "Bridge" meta-model with constructs for bridging P30 and P31

Sridhar Iyengar

UML has made a significant impact in the last two years in terms of designers and software vendors using the UML notation. However, the widespread uses of the full modeling language are still limited because it is complex and also because UML is not yet making a fundamental difference to the programmer in his or her daily work. The lack of modeling tool support with good round-trip engineering support has been an inhibiting factor. Another key promise of the OMG standardization of UML, the support of tool interchange, has not yet been realized because the UML specification is not sufficiently precise. The complexity of the modeling language inhibits the use of UML and inhibits the rapid implementation of related technologies. (Compare this to the widespread implementation of XML and Java technologies by almost all

137

vendors—small and large—in the last 3 years). With the UML 1.3 a more practical and concrete representation of UML using OMG XMI (XML Metadata Interchange) is finally available, and this is a step in the right direction.

The generality of UML needs to be tempered by better and more precise guidelines for how to use UML when designing and implementing various aspects of computing—databases, middleware (COM, CORBA, EJB), user interfaces, etc. Work has begun at OMG as well as JavaSoft on standard mappings (or profiles) for EJB, CORBA, etc.

Precision in the definition of both the core UML concepts (the 20% that are most useful) and precise mappings (with flexibility in optimization) to various implementation technologies are essential. The core of UML should be much better integrated with the model management and extensibility constructs (metaclasses as in the MOF, in addition to stereotypes and tagged values). The use of metamodeling to precisely define the structure and semantics of UML as well as the mappings to various implementation technologies is imperative. The UML (and in fact all related metamodels) must adopt a strict metamodeling architecture if there is any hope of true sharing of modeling concepts and more importantly models and frameworks at the semantic level. While stereotypes and tagged values are straightforward mechanisms to use initially for extensibility, they mask the complexity that a developer must face when implementing complex systems that need to be extended. The use of actual metaclasses is recommended over the use of stereotypes—especially in designing and implementing component frameworks. Expert modelers should be allowed to define their own metaclasses, but extensible frameworks that support real metaclassses must support their implementation.

Guus Ramackers

The general themes for UML 2.0 and, I suspect, quite a while beyond will be two-fold. On the one hand, the UML language definition as it stands will be consolidated and integrated, to become more effective in itself, as well as a better basis for future 'families' of extensions for specific domains. On the other hand, UML will need to come down from its 'analysis and design' pedestal and be more closely mapped to, or integrated with, implementation and deployment languages and environments (or 'models' thereof) such as CORBA, Java, SQL and workflow.

The current UML metamodel in principle represents an effective kernel language. Although there are opportunities for pruning and removing detail (e.g., the standard stereotype «friend»), the basic set of diagrams and metamodel elements are appropriately there in a kernel language.

However, in order to map UML to implementation 'models' there are two areas that will need to be revised and extended in the core (i.e., in addition to the many profiles expected), namely in the area of components and architecture ('model management').

The current support in these areas simply does not scale to deal with enterprise-level information systems.

Although the current UML model mostly contains the right set of core concepts, the way in which these constructs are structured and organized needs to be improved. Specifically, this needs to be done in order to be able to define formalizations of the core language. To do this, more integration is needed in the behavioural part of UML (including the anticipated component extensions), i.e., collaborations, state machines, and activity graphs. Model consolidation should then enable (mathematical) formalizations to prove the executability of models (cf. model simulation).

The process of taking the UML standard forward then becomes one of revising the UML metamodel in the areas mentioned (components, model management, behavioural core), and furthermore managing the process of extensions. The basic process for extensions should be a pragmatic, standards based one, driven by RFI and RFP processes. A central consolidation of extensions by an architecture group will be needed to avoid a jungle of overlapping, imprecise extensions, and to occasionally fold common constructs into the core language.

2 Participant Biographies

Michael Jesse Chonoles (HYPERLINKMichael.J.Chonoles@Lmco.com) is the Chief of Methodology for Lockheed Martin Advanced Concepts Center (ACC). An established OO author, columnist, and educator, Michael is responsible for the technological and methodological direction of the ACC's technology transfer and training practice and works with selected clients to surmount barriers to technology adoption and mission success. With over 20 years in the system/software development industry, Michael has been a significant influence on the direction of UML—as the ACC's representative to OMG working with the UML RTF—and as a direct and indirect influence by writing, lobbying, training, and inspiring. Michael has an M.S.E. in Systems Engineering from the University of Pennsylvania and S.B.s in Math and Physics from MIT.

Steve Cook (HYPERLINKsj_cook@uk.ibm.com) is an IBM Distinguished Engineer. He is currently Chief Architect of IBM's Object Technology Practice. He was the lead author of IBM's submission (with ObjecTime Limited) to the OMG's Object-Oriented Analysis and Design Facility standard, which introduced several elements including OCL (Object Constraint Language) into the UML definition. He was the co-author (with John Daniels) of the Syntropy method, a formalized variant of OMT which was the main precursor to OCL. He has been working with object-oriented methods for 20 years.

Desmond D'Souza (HYPERLINKdsouzad@acm.org) is senior vice president of component-based development at Computer Associates, responsible for defining methods, tools, and architectures for effective component-based software engineering.

He is co-author and developer of the CATALYSIS method, published by Addison Wesley, and heads CA's Catalysis Technology Center. Desmond is a frequently invited speaker at companies internationally.

Sridhar Iyengar (HYPERLINKSridhar.Iyengar2@unisys.com) is a Unisys Fellow who leads the technology strategy for object technology products in Unisys. Sridhar's current focus includes the integration of object frameworks, modeling technologies, databases, and metadata repositories with Internet and distributed object technology products. He is the chief architect of the OMG Meta Object Facility (MOF) and the OMG XML Metadata Interchange (XMI), which together with UML forms the core of the OMG Modeling and Metadata architecture. Sridhar represents Unisys at the OMG where he serves on the OMG Architecture Board. He has a master's degree in computer science and is a frequent presenter in industry conferences on topics of repositories, databases, component software and distributed object technology.

Cris Kobryn (ckobryn@acm.orgHYPERLINK) is a Chief Architect in the E.Solutions unit of EDS, where he specializes in software architectures and methods. He has broad international experience leading high-productivity software development teams, and has architected custom systems and commercial packages. Cris has been a major contributor to the OMG UML specification, first as the chair of the UML 1.1 Semantics Task Force and more recently as the chair of the UML 1.3 Revision Task Force. He received a BA degree in geochemistry from Colgate University and a BSCS degree from San Diego State University. His graduate studies at SDSU and UCLA were eclectic, and explored the synergies between linguistics, computer science and artificial intelligence.

Guus Ramackers (gramacke@uk.oracle.comHYPERLINK) works for the CASE development group of Oracle Corporation in Reading, UK and represents Oracle at the OMG OA&D task force. Previously, he developed object reuse processes and tools for Cap Gemini in The Netherlands, after completing a PhD thesis on Petri-net based formalization of object models at Leiden University.

References

[1] UML Revision Task Force. "OMG UML v. 1.3: Revisions and Recommendations," document ad/99-06-10, Object Management Group, June 1999.

[2] Kobryn, C. "UML 2001: A Standardization Odyssey," *Communications of the ACM*, October 1999. To appear.

Core Meta-Modelling Semantics of UML: The pUML Approach

Andy Evans[1] and Stuart Kent[2]

[1] Department of Computer Science,
University of York, York, UK.
andye@cs.york.ac.uk
[2] Computing Laboratory, University of Kent,
Canterbury, UK.
s.j.h.kent@ukc.ac.uk

Abstract. The current UML semantics documentation has made a significant step towards providing a precise description of the UML. However, at present the semantic model it proposes only provides a description of the language's syntax and well-formedness rules. The meaning of the language, which is mainly described in English, is too informal and unstructured to provide a foundation for developing formal analysis and development techniques. Another problem is the scope of the model, which is both complex and large. This paper describes work currently being undertaken by the precise UML group (pUML), an international group of researchers and practitioners, to address these problems. A formalisation strategy is presented which concentrates on giving a precise denotational semantics to core elements of UML. This is illustrated through the development of precise definitions of two important concepts: generalization and packages. Finally, a viewpoint architecture is proposed as a means of providing improved separation of concerns in the semantics definition.

1 Introduction

The Unified Modeling Language (UML) [BRJ98,RJB99] is rapidly becoming a de-facto language for modelling object-oriented systems. An important aspect of the language is the recognition by its authors of the need to provide a precise description of its semantics. Their intention is that this should act as an unambiguous description of the language, while also permitting extensibility so that it may adapt to future changes in object-oriented analysis and design. This has resulted in a Semantics Document [Gro99], which is presently being managed by the Object Management Group, and forms an important part of the language's standard definition. The approach taken in the Semantics Document is to give a meta-model description of the language. This is presented in terms of three views: the abstract syntax, well-formedness rules, and modelling element semantics. The abstract syntax is expressed using a subset of UML static modelling notations. The abstract syntax model is supported by natural language descriptions of the syntactic structure of UML constructs. The well-formedness rules are expressed in the *Object Constraint Language* (OCL). Finally, the semantics of modelling elements are described in natural language.

A potential advantage of providing a semantics for UML is that many of the benefits of using a formal language such as Z [Spi92] might be transferable to UML. Some of the major benefits of having a precise semantics for UML are given below:

Clarity: The formally stated semantics can act as a point of reference to resolve disagreements over intended interpretation and to clear up confusion over the precise meaning of a construct.

Equivalence and Consistency: A precise semantics provides an unambiguous basis from which to compare and contrast the UML with other techniques and notations, and for ensuring consistency between its different components.

Extendibility: The soundness of extensions to the UML can be verified (as encouraged by the UML authors).

Refinement: The correctness of design steps in the UML can be verified and precisely documented. In particular, a properly developed semantics supports the development of design transformations, in which a more abstract model is diagrammatically transformed into an implementation model.

Proof: Justified proofs and rigorous analysis of important properties of a system described in the UML require precise semantics. Proof and rigorous analysis are not currently supported by UML.

Tools: The tools that make use of semantics, for example a code generator or consistency checker, require that semantics to be precise, whether it be expressed as part of the standard or invented in the code by the tool developer.

Unfortunately, the current UML semantics are not sufficiently formal to realise many of these benefits. Although much of the syntax of the language has been defined, and some static semantics given, dynamic semantics are mostly described using lengthy paragraphs of often ambiguous informal English, or are missing entirely. Furthermore, little consideration has been paid to important issues such as proof, compositionality and rigorous tool support. A further problem is the extensive scope of the language, all of which must be dealt with before the language is completely defined.

This paper describes work being carried out by the precise UML (pUML) group and documented in [pG99,FELR98,EFLR98,ARKB99]. pUML is an international group of researchers and practitioners who share the goal of developing UML as a precise (formal) modelling language. The paper reports on work that aims to strengthen the existing meta-model semantics of UML. In section 2 a formalisation strategy is described (developed through the experiences of the group) that is being used to precisely describe the semantics of UML. This aims to give a precise denotational semantics to the core elements of UML. Section 3 identifies a core semantics model for UML, and in sections 4 and 5 an illustration is given of the formalisation of two core concepts: generalization and package.

Finally, section 6 proposes a 'model-instance viewpoint architecture' (MVA), as a route towards integrating the core semantics within the UML semantics.

2   The pUML approach

In this section, we briefly present some of the key objectives of the pUML group's approach to formalising the UML. The formalisation strategy that is currently adopted by the group is also described. A detailed discussion of the approach and formalisation strategy can be found in [ARKB99].

2.1   Working with the standard

An important aim of the pUML approach is to work firmly in the context of the existing UML semantics. The reasons for taking this approach (as opposed to developing our own semantic model) are as follows:

1. We recognise that UML is a standard and that considerable time and effort has been invested in the development of its semantics. It cannot be expected that radically different semantic proposals will be incorporated in new versions.
2. We believe that the existing UML semantics documentation and the metamodelling approach already provide a good foundation for a precise semantics. As described below, the use of denotational semantics is the key to describing the semantics of UML precisely.

Thus, the pUML approach aims to identify and make precise areas of ambiguity and/or missing semantic details within the current UML meta-model.

2.2   Core semantics

To cope with the large scope of the UML it is natural to concentrate on essential concepts of the language to build a clear and precise foundation as a basis for formalisation. Therefore, the approach taken in the group's work is to concentrate on identifying and formalising a core semantic model for UML before tackling other features of the language. This has a number of advantages: firstly, it makes the formalisation task more manageable; secondly, a more precise core will act as a foundation for understanding the semantics of the remainder of the language. This is useful in the case of the many diagrammatical notations supported by UML, as each diagram's semantics can be defined as a particular 'view' of the core model semantics. For example, the meaning of an interaction diagram should be understandable in terms of a subset of the behavioural semantics of the core.

2.3 Adopting a denotational approach

One of the best known (and most popular) approaches to describing the semantics of languages is the denotational approach (for an in-depth discussion see [Sch86]). The denotational approach assigns semantics to a language by giving a mapping from its syntactical elements to a meaningful representation. For example, an association may denoted by a set of links between objects, while a class may be denoted by a set of objects.

UML already partially adopts the denotational approach to describe aspects of the language. The meta-modelling approach used in the UML semantics naturally supports the description of denotational relationships between model elements: model elements and their denotations can both be abstracted as conceptual classes, and the relationships between them can be formalised by associations and OCL constraints. It consequently makes good sense to continue using the denotational approach in the formalisation strategy. The distinguishing feature of the pUML approach is its emphasis on obtaining *precise* denotational descriptions of UML modelling elements.

2.4 Review and feedback

Constructing a semantics for a language as large and complex as UML is clearly not a simple task. Thus, obtaining feedback and reviews of semantic proposals is a key goal of the pUML approach. This is currently being achieved through publications, open collaborations and the group's web-site. Future aims of the group are to develop semantic tests, which can be used to validate the correctness of new semantic proposals. The use of formal notations to gain a alternative view of a semantic proposal is also used.

2.5 Tool support

Tool support is essential if the benefits of a precise semantics are to be realised. Sophisticated analysis and design tools (that support verification and refinement) require a meta-model semantics that can be implemented efficiently, and which supports sophisticated automation by tool vendors.

2.6 Formalisation strategy

In order to implement the pUML approach it is necessary to develop a strategy for formalising the UML. This is intended to act as a step by step guide to the formalisation process, thus permitting a more rigorous and traceable work program. The formalisation strategy consists of the following steps (a more detailed account can be found in [ARKB99]):

1. Identify specific modelling element/s that contribute to a core semantic model.

2. Iteratively examine the element/s, seeking to verify their completeness. Here, completeness is achieved when: (1) the modelling element has a precise syntax, (2) is well-formed, and (3) has a precise denotation in terms of some fundamental aspect of the core semantic model.
3. Use formal techniques to gain better insight into the existing definitions as shown in [FELR98,EFLR98].
4. Where in-completeness is identified, we attempt to address it by strengthening the existing model, or extending it in the most conservative way possible.
5. Feed the results into the UML meta-model, and disseminate to interested parties for feedback.

In the next section, we identify some core concepts for UML before showing how the strategy can be used to formalise their semantics.

3 The UML core

What parts of the UML semantics should be included in the core semantics? This question is already partially answered in the UML semantics document. It identifies a 'Core - Relationships' package and a number of 'Common Behaviour' packages. The Core Relationship package defines a set of model elements that are common to all UML diagrams, such as relationship, classifier, association and generalization. However, it only describes their syntax.

The Common Behavior package gives a partial denotational meaning to the model elements in the core package. For instance, it describes an association between classifiers and instances. This establishes the connection between the representation of a classifier and its meaning, which is a collection of instances. The meaning of an association (a collection of object links) is also given, along with a connection between association roles and attribute values. Finally, the Common Behaviour package also introduces the notion of an action and stimulus. These specifically relate to the modelling of behaviour in UML.

To illustrate the scope, and to show the potential for realising a compact core semantics, the relevant class diagrams of the two models are shown in the Figures 1 and 2. Well-formedness rules and some classes are omitted for brevity.

An appropriate starting point for a formalisation is to consider these two models in isolation, with the aim of improving the rigor with which the syntax of UML core elements are associated with (or mapped to) their denotations (core instances).

4 Generalization/Specialization

This section presents a precise definition of the meaning of generalization and specifically how it relates to instance conformance. The presentation is more structured and detailed than above due to the greater number of omissions in this part of the UML semantics document.

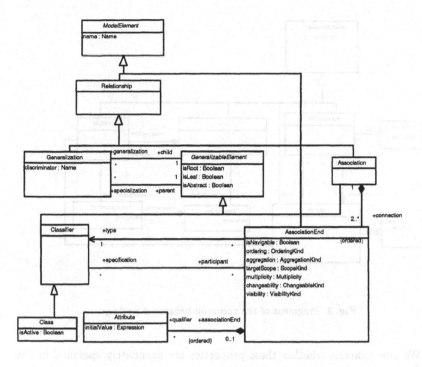

Fig. 1. Fragment of the core relationships package

4.1 Informal description

In UML, a generalization is defined as "a taxonomic relationship between a more general element and a more specific element", where "the more specific element is fully consistent with the more general element (it has all of its properties, members, and relationships) and may contain additional information" [Gro99] (page 2-35) .

Closely related to the UML meaning of generalization is the notion of direct and indirect instances: This is alluded to in the meta-model as the requirement that "no object may be a direct instance of an abstract class, although an object may be an indirect instance of one through a subclass that is non-abstract" [Gro99] (page 2-59).

UML also places standard constraints on subclasses. The default constraint is that a set of generalizations are disjoint, i.e. " (an) instance of the parent (class) may be an instance of no more than one of the given children .." [Gro99] (page 2-36). Abstract classes enforce a further constraint, which implies that no instance can be a direct instance of an abstract class.

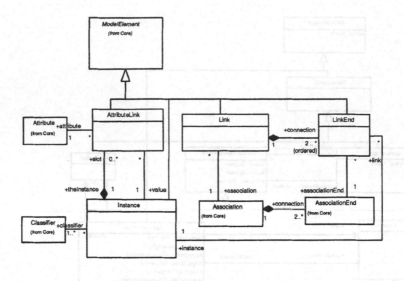

Fig. 2. Fragment of the common behaviour package

We now examine whether these properties are adequately specified in the UML semantics document.

4.2 Existing formal definitions

Bruel and France [BR98] have defined a formal model of generalization. Classes are denoted by a set of object references, where each reference maps to a set of attribute values and operations. Generalization implies inheritance of attributes and operations from parent classes (as expected). In addition, class denotations are used to formalise the meaning of direct and indirect instances, disjoint and abstract classes. This is achieved by constraining the sets of objects assigned to classes in different ways depending on the roles the classes play in a particular generalisation hierarchy. For example, assume that Ai is the set of object references belonging to the class A and that B and C are subclasses of A. Because instances of B and C are also instances of A, it is required that $Bi \subseteq Ai$ and $Ci \subseteq Ai$, where Bi and Ci are the set of object references of B and C.

This model also enables constraints on generalisations to be elegantly formalised in terms of simple constraints on sets of object references. In the case of the standard 'disjoint' constraint on subclasses, the following must hold: $Bi \cap Ci = \emptyset$, i.e. there can be no instances belonging to both subclasses. For an abstract class, this constraint is further strengthened by requiring that Bi and Ci partition Ai. In other words, there can be no instances of A, which are not instances of B or C. This is formally stated as $Ai = Bi \cup Ci$.

We will adopt this model in order to strengthen the existing meta-model definition of generalization as it applied to classifiers and classes.

4.3 Syntax and well-formedness

The abstract syntax of the Generalization model element is described by the meta-model fragment shown in Figure 3. This is taken from the core relationships package of the UML Semantics Document.

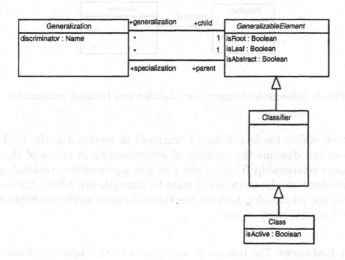

Fig. 3. Syntax of Generalization/Specialization

The most important well-formedness rule which applies to this model element, that is not already ensured by the class diagram, is that circular inheritance is not allowed. This constraint is decribed using the Object Constraint Language (OCL). Assuming 'allParents' returns all the parents of GeneralizableElement, then it must hold that:

```
context GeneralizableElement
invariant
not self.allParents -> includes(self)
```

Here, the *context* of the OCL expression is any instance of a GeneralizableElement.

4.4 Semantics

The completeness of the semantic formalisation versus the desired properties of generalizations is now examined. We restrict ourselves to examining whether the properties of instance conformance, identified in section 4.2, are preserved by the meaning of Generalisation/Specialisation in the UML semantics. Therefore the most important denotational relationship to be examined is that between a classifier and instance. This relationship is formalised in the existing UML semantics by the meta-model fragment shown in Fig 4. This denotes the fact that a Classifier is described by the set of objects that may be instantiated from it. Note that the more generic term for a class in UML is the classifier.

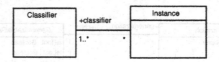

Fig. 4. Meta-model fragment for Classifier and Instance relationship

However, unlike the formal model described in section 4.2, the UML meta-model does not describe the meaning of generalization in terms of the Classifier/Instance relationship. Thus, in order to give a precise denotational meaning to a Generalization, the meta-model must be strengthened with additional constraints on the relationship between the Generalization model elements and the Classifier-Instance relationship.

Indirect instances The first constraint relates to the meaning of indirect instances.

An Instance of a Classifier is also an *indirect* Instance of its parent Classifiers. This is specified as follows:

```
context c : Classifier
invariant
    c.generalization.parent -> forall(s : Classifier |
    s.instance -> includesAll(c.instance))
```

Instance identity Unfortunately, this constraint does not guarantee that every instance is a direct or indirect instance of a related classifier (it only states that generalization implies the existence of indirect instances).

Thus, an additional constraint must be added in order to rule out the possibility of an instance being instantiated from two or more un-related classifiers. This is the *unique identity* constraint:

```
context i : Instance
invariant
    i.classifier -> exists(direct : Classifier |
        direct.allParents -> union(Set{direct}) = i.classifier
```

149

This states that the *only* Classifiers that an object can be instantiated from are either the Classifier that it is directly instantiated from or its parents [1].

Direct instances The meaning of a direct instance can now be precisely defined:

```
context i : Instance
isDirectInstanceOf(c : Classifier) : Boolean
isDirectInstanceOf(c) = c.allParents -> union(Set{c}) = i.classifier
```

A direct Instance directly instantiates a Classifier and indirectly instantiates its parents.

Disjoint constraints Once direct and indirect Instances are formalised, it is possible to give a precise description to the meaning of constraints on generalizations (for example the disjoint constraint).

The disjoint constraint can be formalised as follows:

```
context c : Class
invariant
    c.specialization.child -> forall(i,j : Classifier |
        i <> j implies i.instance ->
            intersection(j.instance) -> isEmpty)
```

For any two children of a Classifier, i and j, the set of instances of i will be disjoint from the set of instances of j. Note that the disjoint constraint is only applied to Classes in UML, not Classifiers.

Abstract classifiers Finally, the following OCL constraint formalises the required property of an abstract classifier that it cannot be directly instantiated:

```
context c : Classifier
invariant
    c.isAbstract implies
        c.specialization.child.instance -> asSet = c.instance
```

Note, the result of the specialization.child path is a bag of instances belonging to each subtype of c. Applying the asSet operation results in a set of instances. Equating this to to the instances of c implies that all the instances of c are covered by the instances of its children. This, in conjunction with the disjoint property above, implies the required partition of instances, and completes the formalisation of this concept.

---

[1] The UML standard does in fact state that static/dynamic multiple classification of instances is permitted without generalization being present. However, the conditions under which this is permitted are not defined, and we therefore defer consideration of this aspect for now.

5 Package Instances

Links, link ends and objects do not generally appear in isolation. A UML object diagram represents a *snapshot* [DA98] of the state of the system. Yet there is no corresponding concept in the meta-model. A reason for this is given in the accompanying English semantics, [Gro99] (page 2-179):

> The purpose of the package construct is to provide a general grouping mechanism. A package cannot be instantiated, thus it has no runtime semantics.

This is a particularly implementation-oriented perspective. Object diagrams can be drawn to show instances of specification models (think of tools which simulate that model) as much as instances of implementation models. As has been identified in the discussion on abstract classes, whether something is instantiable or not at execution is captured not by whether it can or can not have instances, but whether those instances are only instances of that thing.

So, in this section, we develop the idea of instances of packages, corresponding to object diagrams. This also caters for instances of models and subsystems. We believe that this concept is essential for formalising the semantics of behavioural constraints specified in a package. Again, the model-instance view is emphasised in our formalisation.

Although a package is defined to be a collection of model elements (which include instance as well as design elements), it is constrained [Gro99] (page 2-175, [1]) only to include design elements. This suggests a new class is required to capture the concept of a package instance. The relevant class diagram is given in Fig 5.

Classes have also been added to make the distinction between design and instance element clearer, although only some of the different kinds of design and instance elements have been shown on the diagram. The new classes make the OCL constraint in [Gro99] (page 2-175, [1]), redundant.

There is another issue that needs to be considered – but not here as it is beyond the scope of this paper. The allContents associations from Package and PackageInstance, respectively, are examples of the recursive composite pattern. In that pattern there is usually another association to collect together all primitive (as opposed to composite) elements that are contained in the composite either directly, or indirectly through other composites that are elements of the composite. The meta-model has nothing to say about what it means for a package to be contained within another, or the relationship of that concept to package imports. Some hints are provided in the accompanying English semantics, though the description given for packages in general seems at odds with the description given for models. A similar issue arises for package instances.

The association 'accessed' from Package is an attempt at capturing the notion, mentioned in the informal semantics [Gro99] (page 2-173), that elements from other packages may be accessed from a package. The association 'accessible' then represents all those elements which are accessible from a package, i.e. those contained within it and those outside which it is able to access:

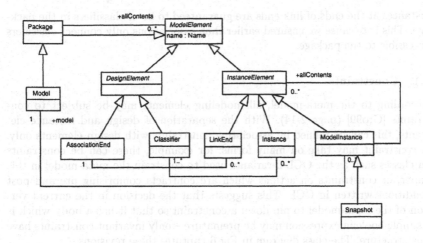

Fig. 5. Meta-model fragment for packages and their instances

```
context p : Package
invariant
    p.accessible = p.accessed->union(p.allContents)
```

At least one further constraint is required, that associations only refer to classifiers accessible to the package.

There can be many kinds of package instance. For example, to give the semantics for dynamic behavioural constraints, it is likely that an idea of a *trace* (in the formal sense of the word) or *filmstrip*, corresponding to an instance of the execution of some sequence of actions or operations will be required. In this paper we restrict ourselves to *snapshots*, instances that correspond to object diagrams. Snapshot is a subclass of PackageInstance. It is only associated with Instance's and Link's.

Of course it is not the case that any snapshot can be an instance of any package. Specifically, the links and instances must be of associations and classifiers, respectively, accessible to that package:

```
context s : Snapshot
invariant
    s.package.accessible->select(oclIsKindOf(Association))
        ->includesAll(s.allContents->select(oclIsKindOf(Link)).association
    and s.package.accessible->select(oclIsKindOf(Classifier))
        ->includesAll(s.allContents->select(oclIsKindOf(Instance)).classifier
```

If a link is of an association accessible to the package then that guarantees that link ends are of association ends accessible to the package. Similarly, instances at the ends of link ends are guaranteed to be of classifiers in the package. This is because we ensured earlier that associations only connect classifiers accessible to the package.

5.1 Constraints

According to the meta-model, all modeling elements may be subject to constraints [Gro99] (page 2-14). With the separation of design and instance elements, this can be refined by associating constraints with design elements only. A constraint may take on many forms: for example, there can be constraints on classes such as the OCL invariants used to constrain the meta-model in this paper, or constraints on actions which are contracts comprising pre and post conditions written in OCL. This suggests that the decision in the current version of the meta-model to pin down a constraint so that it has a body which is a simple boolean expression may be premature – only invariant constraints have this structure. The class diagram in Fig 6 captures these revisions.

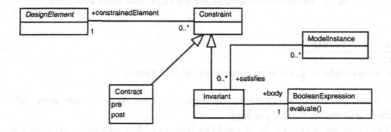

Fig. 6. Meta-model fragment for constraints

The diagram focuses on a single kind of constraint, namely invariants on packages.

6   A Model-Instance Viewpoint Architecture

In terms of future work, it is important to understand how the pUML approach should fit within the overall architecture of the UML semantics. Currently, the UML semantics adopts a rather ad-hoc approach, in which its various elements are distributed throughout a number of different packages, for example, there are a state-machine, model-management and use-case packages. Each of these packages has some overlap with each other, but because this is not made explicit in the architecture, understanding and maintaining the model is difficult.

An alternative architecture is one that makes a clear distinction between the core semantics, which describe the essential concepts and meaning of a language, and viewpoints, which describe the different ways in which the core concepts can be viewed by the modeller. For example a static modelling view encompasses classes and objects. Furthermore, viewpoints can be directly related to the diagrammatical notations that are used to visually represent models, for example class and association icons, etc.

The viewpoint architecture has been successfully applied in the development of the RM-ODP standard [ISO96], where it is used to describe multiple views of open distributed systems. The advantage of adopting a viewpoint-oriented architecture is that it places clear boundaries on the roles that different parts of the semantics plays. While the core semantics makes clear the meaning of the language, the views and diagram elements specify the syntactical features of the language (which is essential to tool designers).

Figure 7 gives an overview of an architecture that both supports a viewpoint-oriented model of the UML semantics, and which also places emphasis on precisely documenting the relationships between model elements and instances (denotations).

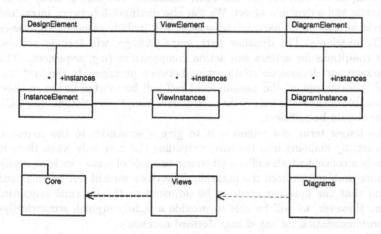

Fig. 7. Model-Instance View Architecture (MVA)

Here, the semantics is to divided into three main packages: the core, views and diagrams-packages. Within these packages, the relationship between model elements and their denotations (instances) are given by the 'instances' associations. This makes explicit the fact that most denotations in UML consist of mappings from generic concepts to the instances which they represent. In the core-package, elements might include modelling concepts such as associations and classes, while their instances are objects and links. In the view-package, views map view elements and instances to appropriate core elements and instances. A behavioural view, for example, would provide a mapping to only behavioural modelling elements in the core, such as operations and actions. Finally, the diagrams-package provides a link between diagram meta-models and their instances (class icons, etc.) to elements in the core model (through possibly many viewpoints).

7   Further Work

This paper has described an approach to the semantics of UML which builds upon the meta-model defined in the UML standard semantics document. Fragments of the semantics have been shown here, specifically the semantics of associations and generalisation, and the introduction of snapshots which will play a pivotal role in the semantics of behavioural constraints. A model-instance viewpoint architecture has been proposed as a way of integrating the core semantics into the complete UML meta-model.

Our immediate goal is to complete this semantics for a core notation set, seeking compliance with the current UML standard. The core is likely to include a static and a dynamic aspect. We can also distinguish between intra- and extra-package. The static part, intra-package, will include associations, classes and OCL invariants. The dynamic part, intra-package, will include actions, pre/post conditions for actions and action compositions (e.g. sequences). The extra-package part focuses on relationships between packages. In the best tradition of "bootstrapping" the meta-model itself will be written in the smallest subset possible of the static part of the core-classes, simple associations and OCL invariants should be sufficient.

In the longer term, our intention is to give a semantics to the complete notation set, by mapping into the core, extending the core only when there is not already a concept which suffices. Of course one role of semantics is to clarify and remove ambiguities from the notation. Therefore we will not be surprised if we find that the notation needs to be adjusted or the informal semantics rewritten. However, we will be able to provide a tightly argued, semantically-based recommendation for any change deemed necessary.

Some consideration also needs to be given to quality assurance. There are at least three approaches we have identified:

1. peer review and inspection
2. acceptance tests
3. tool-based testing environment

So far the only feedback has come from 1. Since a meta-model is itself a model, acceptance tests could be devised as they would be for any model. Perhaps "testing" a model is a novel concept: it at least comprises devising object diagrams, snapshots, that the model must/must-not accept. Better than a list of acceptance tests on paper would be a tool embodying the meta-model, that allowed arbitrary snapshots to be checked against it.

155

References

[ARKB99] A.S.Evans, R.B.France, K.C.Lano, and B.Rumpe, *Towards a core meta modelling semantics of UML*, Behavoral Specifications for Businesses and Systems (Haim Kilov, ed.), Kluwer Press, 1999, To appear.

[BR98] J-M. Bruel and R.B.France, *Transforming UML models to formal specifications*, UML'98 - Beyond the notation, LNCS 1618, Springer-Verlag, 1998.

[BRJ98] G. Booch, J. Rumbaugh, and I. Jacobson, *The Unified Modeling Language user guide*, Addison-Wesley, 1998.

[DA98] D. D'Souza and A.C.Wills, *Objects, components and frameworks with UML*, Object Technology Series, Addison-Wesley, 1998.

[EFLR98] Andy Evans, Robert France, Kevin Lano, and Bernhard Rumpe, *Developing the UML as a formal modelling notation*, UML'98 Proceedings (Jean Bezivin and Pierre-Allain Muller, eds.), Springer-Verlag, LNCS 1618, 1998.

[FELR98] R. France, A. Evans, K. Lano, and B. Rumpe, *The UML as a formal modeling notation*, Computer Standards & Interfaces **19** (1998).

[Gro99] Object Management Group, *OMG Unified Modeling Language Specification, version 1.3beta. found at: http://www.omg.org*, 1999.

[ISO96] ISO/IEC, *Reference model of open distributed processing - part 1-5, ISO/IEC DIS 10746-1*, Tech. report, 1996.

[pG99] The pUML Group, *The precise UML web site: http://www.cs.york.ac.uk/puml*, 1999.

[RJB99] J. Rumbaugh, I. Jacobson, and G. Booch, *The Unified Modeling Language reference manual*, Addison-Wesley, 1999.

[Sch86] D. A. Schmidt, *Denotational semantics: A methodology for language development*, Allyn and Bacon, 1986.

[Spi92] J.M. Spivey, *The Z reference manual, 2nd edition*, Prentice Hall, 1992.

A Metamodel for OCL

Mark Richters and Martin Gogolla

University of Bremen, FB 3, Computer Science Department
Postfach 330440, D-28334 Bremen, Germany
{mr|gogolla}@informatik.uni-bremen.de,
WWW home page: http://www.db.informatik.uni-bremen.de

Abstract. The Object Constraint Language (OCL) allows the extension of UML models with constraints in a formal way. While the UML itself is defined by following a metamodeling approach, there is currently no equivalent definition for the OCL. We propose a metamodel for OCL that fills this gap. The benefit of a metamodel for OCL is that it precisely defines the syntax of all OCL concepts like types, expressions, and values in an abstract way and by means of UML features. Thus, all legal OCL expressions can be systematically derived and instantiated from the metamodel. We also show that our metamodel smoothly integrates with the UML metamodel. The focus of this work lies on the syntax of OCL; the metamodel does not include a definition of the semantics of constraints.

1 Introduction

The UML provides a rich set of language constructs for describing structural, dynamic and functional aspects of a system [16]. For most parts of the language a graphical notation facilitates the construction of UML models. However, the graphical notation is inherently limited when specifying complex constraints. Often a textual constraint language is more appropriate for this task. The Object Constraint Language (OCL) has been designed to extend UML models with constraints in a concise and precise way [15, 21].

Since UML version 1.1, the OCL is an integral part of the UML standard. The application of OCL is not limited to user models, it is also used by the standard itself for specifying well-formedness rules in context of the UML semantics definition [17]. In [20, 10], the authors report on considerable improvements by using OCL instead of English text. Several items of the UML specification which previously allowed ambiguous interpretation could be identified and subsequently clarified or corrected.

The main reference for UML is the semantics document [17]. This document defines the language to be used for specifying well-formed UML models. Language constructs are defined by (1) their abstract syntax, (2) a set of well-formedness rules, and (3) an informal description of their intended semantics. The abstract syntax is defined as a metamodel in form of UML class diagrams. This style of presentation has been chosen for most UML parts but not for the

OCL. In order to achieve a uniform presentation for the complete UML *including* OCL, we therefore propose a definition of the OCL following the style of [17]. Our main contribution to achieve this goal is the definition of an OCL meta-model. The metamodel presented in this paper is not just a different kind of presentation of the OCL. We rather consider it a complementary work to [15]. The OCL document concentrates on the concrete syntax and application aspects of the Object Constraint Language. Our metamodel, on the other hand, tries to define the abstract syntax of OCL. Another benefit resulting from having a metamodel is that it is quite easy to derive an implementation. For example, we currently use the metamodel to build an OCL interpreter.

The UML metamodel is defined as one of the layers of a four-layer meta-modeling architecture [17, p. 2-4]. In this architecture the UML metamodel is an instance of the meta-metamodel of the OMG Meta-Object Facility (MOF) [14]. For an integration of the OCL metamodel with this architecture there are basically two options. We can place the OCL metamodel at the same level (the M2 level) as the UML metamodel and use the MOF as the definition language (provided by the M3 level). The result is a general MOF-compliant metamodel of OCL. The second option is to lift the OCL metamodel to the M3 level. One reason is that OCL may obviously apply to other metamodels than just UML. Furthermore, the MOF also uses OCL for specifying constraints at the M3 level. Consequently it would make sense to define the OCL metamodel as part of the MOF. This way our work can be seen as a contribution to the meta-metamodel level of the standard OMG architecture. Note that there are no differences resulting from these two options with respect to the technical details of the OCL metamodel. In both cases the OCL metamodel is defined in terms of the MOF meta-metamodel. Since OCL cannot be used without a concrete modeling language, we have chosen the first option and placed the OCL metamodel at the same level as the UML metamodel in order to achieve a tight integration.

Related work in context of OCL can be summarized as follows. A general assessment and analysis of the Object Constraint Language has been presented in [9, 7]. In [13], the expressiveness of OCL has been investigated. Extensive applications of OCL can be found in the UML semantics document and, e.g. in [6], where it has been used to define additional constraints for equivalence and transformations rules between UML class diagrams. We consider the development of an OCL metamodel a step towards a formal model of the Object Constraint Language [8, 18]. This is part of a greater effort towards formalizing the Unified Modeling Language (e.g., see [1, 4, 12, 19, 11, 3, 2]).

The presentation of the OCL metamodel is structured in the following way. The metamodel is organized as a set of three UML packages containing a total of 56 classes. Each package describes a different aspect of the language, namely types, expressions, and values. For each package, we first present a class diagram defining the abstract syntax of a concept. Then we give an informal explanation of the diagram contents. Concrete examples are used to illustrate the application of the metamodel. Finally, a set of well-formedness rules specifies additional constraints on the metamodel. These rules are probably not complete, but they should give an idea of how OCL rules given in [15] can be translated and applied to the metamodel. Due to space limitations we do not give OCL expressions for the well-formedness rules but in principle this could be done with reasonable effort.

The rest of the paper is organized as follows. In Sect. 2 we illustrate the package structure of the OCL metamodel. Since OCL is primarily used for specifying constraints on UML models, there exists a strong relationship between both metamodels. A link between OCL and the UML core model will be established in Sect. 3. The *Types* package of our metamodel is presented in Sect. 4. The central part of the metamodel is the definition of *Expressions* which are discussed in Sect. 5. The result of evaluating expressions are *Values* which are not strictly part of the abstract syntax but are included in Sect. 6 for completeness. We close with a summary and draw some conclusions for future work.

2 Structure of the Metamodel

The OCL metamodel will be defined in a package called *OCL*. Fig. 1 shows how this new package is related to the existing UML packages *Core*, *Data Types*, and *Common Behavior* [17, p. 2-7].

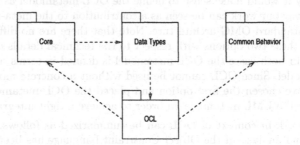

Fig. 1. Dependencies among UML packages and OCL

There are several dependencies among these packages. The *Core* package depends on *Data Types* where a general notion of expressions is defined. These expressions are used for modeling the concept of constraints in the *Core*. The *Data Types* package depends on OCL as soon as we use the Object Constraint Language for defining expressions. On the other hand, OCL depends on model elements (classes, attributes, etc.) of the *Core* package for building expressions. Finally, evaluating OCL expressions results in values or instances. Both concepts are defined in the *Common Behavior* package, hence we have another dependency from *OCL* to *Common Behavior*.

There are more UML packages which use expressions and which are not included in Fig. 1. For example, statecharts may contain boolean expressions for the specification of guard conditions. Of course, these packages will also indirectly (via *Data Types*) depend on (and benefit from) the *OCL* package when they use OCL for expressions.

The *OCL* package is further refined into three separate packages which model different aspects of OCL. Fig. 2 shows the packages *Expressions*, *Types*, and *Values* as part of the OCL metamodel.

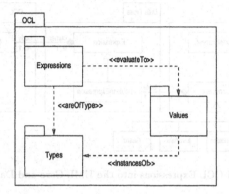

Fig. 2. Package Structure of the OCL Metamodel

We have used special stereotypes for the dependencies to emphasize their different roles. In OCL, every expression has a well-defined type. Instances of types are values which result from the evaluation of expressions. Each of these three packages will be discussed in its own section in the following. But first, we will have a closer look at the relationship between the UML core and OCL.

3 Constraints

Constraints can be used to specify restrictions on UML model elements. The concept of constraints is defined in the *Core* package of the UML metamodel. A constraint may be attached to all kinds of model elements. For example, a class invariant can be defined with a constraint by specifying a boolean expression that has to be true for all instances of the class. The UML does not prescribe the language or formalism for the boolean expression. In general, a constraint can be specified in natural language, OCL, or some other language.

The class diagram in Fig. 3 provides the link between the UML metamodel and the OCL metamodel. It shows partial views of the *Core* package and the *Data Types* package [17, p. 2-13 and p. 2-75]. Only classes and relationships that are important for our purposes are displayed. *Constraints* are *ModelElements* that specify restrictions on other model elements. A *BooleanExpression* forms the body of a *Constraint*. *BooleanExpressions* are just a special kind of *Expression* having a boolean result type. An expression can be defined by using

OCL. In the UML *Data Types* package, **Expression** has attributes *language* and *body* where the body attribute keeps the textual representation of the constraint. We replace the *body* attribute by an optional component relationship to **OclExpression**. It is now possible to use OCL for defining expressions, but we are not forced to do so. Other languages could be integrated similarly.

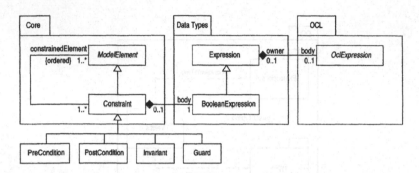

Fig. 3. Integration of OCL Expressions into the UML Core and Data Types Packages

The abstract class **OclExpression** defines the set of all legal expressions in OCL. It is the top-level element of the OCL metamodel and will be further refined in Sect. 5. All kinds of expressions are specializations of **OclExpression**.

Constraints can appear in different contexts. They may be used to specify pre- and postconditions, invariant, and guards. Therefore, in Fig. 3, we have specialized the **Constraints** class into corresponding subclasses **PreCondition**, **PostCondition**, **Invariant**, and **Guard**. The distinction is necessary because some OCL constructs are allowed only in certain contexts, e.g. the @pre modifier makes sense only in postconditions. Alternatively, instead of modeling the different kinds of constraints as subclasses, we could also model this distinction as an enumeration in the **Constraint** class itself. However, our goal was to avoid changes to the original UML metamodel where possible.

The following well-formedness rules refer to some concepts that will be introduced in later sections.

Well-formedness rules:

1. When a **BooleanExpression** is used in a **Constraint** and it is defined by an **OclExpression**, the **Type** of the **OclExpression** must be an instance of **BooleanType**.
2. **PreConditions** and **PostConditions** may be attached only to **Operations** or **Methods** (both are part of the **Core** package).

4 Types

OCL is a typed language. Each expression has a type which is either explicitly declared or can be statically derived. Evaluation of the expression yields a value of this type. Therefore, before we can define expressions, we have to provide a model for the concept of type. A metamodel for OCL types is shown in Fig. 4. Note that instances of the classes in the metamodel are the types themselves (e.g. *Integer*) not instances of the domain they represent.

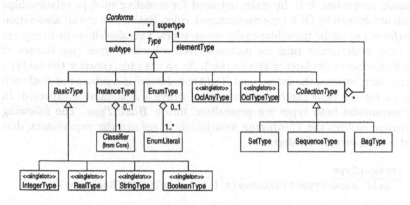

Fig. 4. Metamodel for OCL Types

All available types in OCL are modeled as specializations of the abstract class *Type*. Concrete types are classified in the metamodel as follows.

- basic types: *IntegerType*, *RealType*, *StringType*, and *BooleanType*
- model types: *InstanceType*
- enumeration types: *EnumType*
- special types: *OclAnyType* and *OclTypeType*[1]
- collection types: *SetType*, *SequenceType*, and *BagType*

The first group models the basic types *Integer*, *Real*, *String*, and *Boolean*. We have marked the corresponding elements in the metamodel with a stereotype ≪singleton≫ to indicate that there is exactly one instance for each of these classes, e.g. the one and only instance of *IntegerType* is the type *Integer*.

InstanceTypes are used to refer to *Classifiers* defined by users in a UML class model, e.g. *Person*, *Company*, etc. The meaning of special types is explained in [15]. *EnumTypes* are defined by a list of distinct literals, e.g. *Color:{red, green, blue}*. *CollectionTypes* are parameterized by an element

---

[1] The strange names are a result of our naming scheme: For each OCL type we create a corresponding metamodel element by attaching the word *Type* to its name.

type, e.g. *Set(Integer)*. Those familiar with design patterns [5] will recognize the application of the composite pattern for modeling collection types. In our metamodel, there is no limit on the depth of nesting of collection types. Note that OCL tries to avoid complex collections by a process called "automatic flattening". However, complex values may nevertheless result from the evaluation of some expressions (e.g. by navigating more than one association with multiplicities greater than one). Thus, *before* "flattening" a complex value we need to be able to specify its precise and complete (possibly nested) type.

The specialization of *Type* in the metamodel is used for classifying types by common properties. It is, however, not used for modeling subtype relationships which are defined by OCL type conformance rules. Rather, we use an association *Conforms* to model the subtype relation on types. This also allows us to express the type conformance rules for parameterized collection types (see the set of well-formedness rules later in this section). As an example, consider the subtype relationship between the *Integer* and *Real* type: *Integer* is a subtype of *Real* such that an *Integer* value can be used anywhere where a *Real* value is expected. In the metamodel both types are generalized into a *BasicType*. The following constraint utilizes the *Conforms* association to enforce the requirement that *Real* is a supertype of *Integer*.

```
IntegerType
    self.supertype->includes(t | t.oclIsTypeOf(RealType))
```

Example. We apply the metamodel to the OCL type *Set(Person)*. Figure 5 shows the representation of this type as a UML object diagram. The diagram is an instantiation of the *Types* metamodel. Person is a class defined in a user model. The class induces a corresponding instance type that can be used in OCL. This instance type is used to parameterize the set type.

Fig. 5. Object Diagram for the Collection Type *Set(Person)*

In Fig. 6 we continue the example and add another type *Set(Employee)*. In this example, class Person is a generalization of class Employee. We have also included in this diagram links of the *Conforms* association indicating type conformance. Since Person is a generalization of Employee, the instance types have a similar relationship in OCL. Following from the type conformance rules for collections (see [15, p. 7-20] and items 6 and 7 of the well-formedness rules below), *Set(Person)* is also a supertype of *Set(Employee)*.

Most of the following well-formedness rules can be found as textual descriptions in the OCL documents [15, 21]. With the metamodel, they now can (in principle) be formalized as OCL constraints.

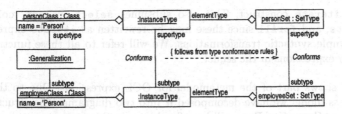

Fig. 6. Type Conformance between Types *Set(Person)* and *Set(Employee)*

Well-formedness rules:

1. *OclAnyType* is the supertype of all *Types*.
2. *IntegerType* conforms to *RealType*.
3. All *EnumerationLiterals* of an *Enumeration* are distinct.
4. *Types* are equal if they have the same textual representation.
5. Type conformance (represented by the association *Conforms*) is transitive, reflexive and anti-symmetric.
6. *CollectionTypes* *C1(T1)* and *C2(T2)* conform to each other when their element types *T1* and *T2* do conform.
7. An *InstanceType* *I1* conforms to a type *I2* when their associated *classifiers* *C1* and *C2* have a generalization relationship, i.e. *C2* is a supertype of *C1*.

5 Expressions

In this section, we define the metamodel for OCL expressions. Perhaps surprisingly, it is quite difficult to say what actually constitutes an expression in OCL. In [15], the term is used informally in a number of places and different contexts. We suggest the following classification based on careful reading and analysis of the description and examples given in [15]. Our definition of expressions has been introduced more formally in our work on developing a formal syntax and semantics for OCL (see [18]). An OCL expression is defined to be one of the following.

1. The `self` keyword.
2. The `result` keyword in a postcondition.
3. A variable.
4. The application of an operation. We distinguish between
   - predefined operations: e.g. +, -, *, <, >, `size`, `max`, ...
   - attribute operations: e.g. `self.age`
   - operations defined by a classifier: e.g. `p.income(...)`
   - navigation by role names: e.g. `self.employees`
   - constants: e.g. 25, `'aString'`, ...

5. The `iterate` construct. This also subsumes `select`, `reject`, `collect`, `exists`, and `forAll` since these may be rewritten as `iterate` expressions by simple syntactic transformations. We will refer to all these functions as *query* expressions in the sequel.

Fig. 7 shows part of the metamodel for OCL expressions. Due to the size of the class model we have decomposed it into two diagrams. The structure of operations (*OperationExp*) will be refined later in this section.

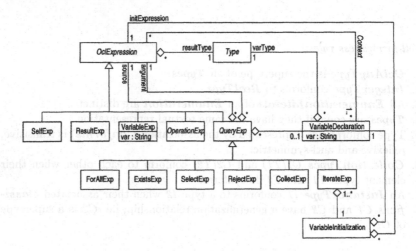

Fig. 7. Metamodel for OCL Expressions

All *OclExpressions* have a well-defined result *Type*. The structure of *Types* has been shown in Fig. 4. *OclExpressions* are specialized into five fundamental expression classes according to the definition given above. A `self` expression is an instance of *SelfExp*, a `result` expression is an instance of *ResultExp*, a variable an instance of *VariableExp*, etc.

Every OCL expression needs a context for evaluation. The context is given by the name of an instance type, e.g. *Person*. All occurrences of `self` expressions are then bound to an instance of this type during evaluation. Furthermore, variables can be explicitly declared in a context, e.g. `p:Person`. A context for a pre- or postcondition includes the signature of the operation to be constrained. Parameters of this operation introduce variable bindings that can be used just as any other variable in an expression. Variable bindings are represented by instances of *VariableDeclaration*. Each declaration associates a variable name with a *Type*.

In the metamodel, the association *Context* connects each expression with a non-empty set of *VariableDeclarations*. There must be exactly one declaration with an empty *var* attribute. This anonymous declaration will be used for binding `self` expressions.

The predefined query expressions `iterate`, `select`, `reject`, `collect`, `exists`, and `forAll` differ from ordinary operations because they may contain language constructs for declaring and – in case of `iterate` – initializing variables that have special meaning in context of their argument expression. We use the simple example model in Fig. 8 to illustrate this.

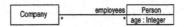

Fig. 8. Example UML Model

The following expression selects from the set of all persons working for a given company those being older than 45.

<u>Company</u>
```
  self.employees->select(p : Person | p.age > 45)
```

The query uses a select expression where a variable p of type *Person* is declared as part of the expression. This variable will be bound implicitly to each element of the source collection (`self.employees`) as the argument expression (`p.age > 45`) is evaluated for each of the collection's elements. In the metamodel, the expression can be represented as a **SelectExp**. We repeat the expression below and mark different components of the expression with braces. Labels refer to the names used in Fig. 7.

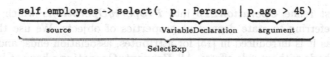

The metamodel class **QueryExp** represents common properties of query expressions. They all have a source expression resulting in a collection which forms the input of the query, and they have an argument expression which is evaluated for each of the source collection's elements. Furthermore, they optionally may have a **VariableDeclaration** introducing an identifier which may be referred to as variable expression (**VariableExp**) in the argument expression.

The more general `iterate` construct is a query expression (an **IterateExp** in the metamodel) which additionally has a mandatory variable initializer. A **VariableInitialization** is split into a declaration part and an initializing expression. The following example illustrates the possible structure of `iterate` expressions.

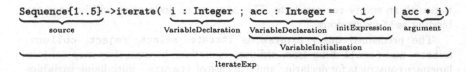

A large group of expressions can be classified as applications of operations. The metamodel element **OperationExp** in Fig. 7 is an abstract class representing all kinds of operations in OCL. A refinement of **OperationExp** is shown in Fig. 9.

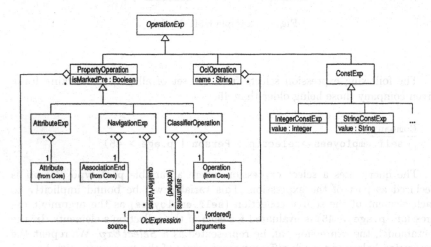

Fig. 9. Metamodel for OCL Operations

We first refine **OperationExp** into **PropertyOperations** which are operations referring to state dependent properties of objects. We use the term property as it is introduced in [15] for attributes, association ends, and operations/methods without side effects. All **PropertyOperations** have at least one argument: an **OclExpression** determining the source object. Furthermore, it is possible in postconditions to refer to a previous value of a property by postfixing it with the @pre keyword. The boolean attribute *isMarkedPre* indicates the reference to a previous value in a **PropertyOperation**.

An **AttributeExp** like self.age is a function with one argument: an **OclExpression** resulting in an instance which owns the attribute. For the expression self.age the owner of the age attribute is given by an instance of **SelfExp**. The name of the operation is the name of the **Attribute** itself.

A **NavigationExp** utilizes associations originating from a classifier. An associated classifier is specified by its role name. The role name is part of an

AssociationEnd in the *Core* package. When using qualified attributes on an *AssociationEnd*, a *NavigationExp* may also specify *qualifierValues*.

ClassifierOperations refer to *Operations* which are defined as classifier features in a user's application model, e.g.:

```
Employee::getSalary(d : Date) : Real
```

They may be used in OCL expressions, if they do not have any side effects, i.e. the *isQuery* attribute of *BehavioralFeature* in the *Core* package is true [17, p. 2-25]. *ClassifierOperations* may require any number of argument expressions.

Predefined operations on values are modeled by the element *OclOperation*. These are characterized by an operation name and an argument list. Examples for predefined operations are: `+`, `-`, `*`, `<`, `>`, `size`, `max`, etc.

Finally, a further group of operations are those that produce constant values. We have modeled these for each of the basic types. For example, an *IntegerConstExp* simply contains an attribute *value* specifying the value of the constant to be created. The ellipsis in the diagram indicates the existence of further classes for the basic types *Real* and *Boolean*. Constant collection values like `Set{1,2}` can be modeled as *OclOperations*, e.g. a function makeSet($e_1 : T, \ldots, e_n : T$) : Set(T) creates a set with n elements of type T. These functions are not visible to the user since they are only used for making the meaning of the literal notation explicit in the metamodel.

Example. We apply the metamodel for OCL expressions to the following query.

<u>Company</u>
```
self.employees->select(p : Person | p.age > 45)
```

Figure 10 shows the abstract syntax of the above expression as an instantiation of the metamodel presented in Fig. 7 and Fig. 9.

The object diagram basically shows an abstract syntax tree. The root of this tree is the `select` expression. It has three child branches. First, the source of the `select` operation is a collection resulting from the navigation expression `self.employees`. The second branch models the variable declaration `p:Person`. Finally, the third branch represents the expression `p.age > 45`. We do not show result types of expressions and the context information (*Company*) in Fig. 10 in order to keep the diagram readable. However, adding this information would be straightforward.

Well-formedness rules:

1. A *ResultExp* may only be used in a *PostCondition*.
2. A modifier `@pre` is only allowed in a *PostCondition*.
3. The variable name used in a *VariableExp* must be part of a *VariableDeclaration* in an enclosing expression (or a parameter if the expression is attached to an operation as a pre- or postcondition). With other words, a variable must be declared before it may be used in an expression.
4. The source expression of a *QueryExp* must have a collection type.

168

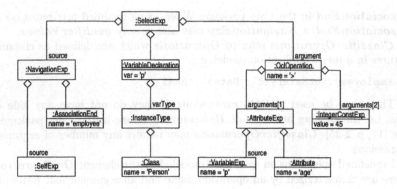

Fig. 10. Example Instantiation of the Expressions Metamodel

6 Values

In the previous sections, we have defined a complete metamodel for the abstract syntax of OCL. Another important aspect is the interpretation of OCL expressions. The result of evaluating an expression is a value. Since our focus is on meta-modeling issues, we do not consider the semantics of expressions here. A complete semantics would have to define a mapping from expressions to values [18]. In the following, we will concentrate on the structure of values. A metamodel for values suitable for representing results from evaluating OCL expressions is shown in Fig. 11.

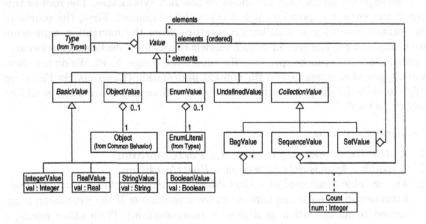

Fig. 11. Metamodel for Values

The general structure of values is very similar to the metamodel for types in Fig. 4 since each kind of value represents the domain of a corresponding type. *Value* is the abstract base class of all kinds of values. Each value has a *Type* defining its properties. Values of basic types are *IntegerValue*, *RealValue*, *StringValue*, and *BooleanValue*. An *ObjectValue* is just a placeholder for an *Object* which is defined as part of the UML *Common Behavior* package. An *EnumValue* is exactly one of the *EnumLiterals* used to define the corresponding type. Since the result of an OCL expression may be undefined (e.g. following from a division by zero), a special *UndefinedValue* is required for these cases.

We also need structures for the collection types *Set*, *Sequence*, and *Bag*. Values of these types may include other values as elements. For *SetValue* we can use an ordinary aggregation, a *SequenceValue* requires the ordered constraint on its elements, and a *BagValue* associates a number (*Count*) with each of its elements.

Example. We apply the metamodel to the value resulting from the OCL expression Set{1,2}. Figure 12 shows the set value containing the two integer values. Each value has a link to its defining type on the left side. In this example, the hierarchical structure of values reflects the hierarchical structure of composite types.

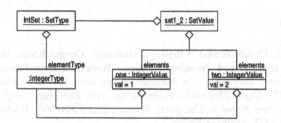

Fig. 12. Object Diagram for the value Set{1,2}

7 Conclusion

In this paper, we have presented a metamodel for the Object Constraint Language OCL. We also have demonstrated the application of the metamodel to concrete OCL expressions. The metamodel has been defined as a set of class diagrams together with well-formedness rules. This style of presentation closely follows the style of the UML semantics document. The metamodel could therefore be easily integrated with the existing UML specification providing a uniform style of presentation for all parts of the Unified Modeling Language.

The metamodel delivers a more precise and detailed view of the OCL. As a consequence from using only well-known modeling concepts of UML which are compliant with the Meta Object Facility (MOF), the metamodel can easily be read by everybody familiar with UML. The decomposition into packages for types, expressions, and values emphasizes the basic structure of OCL. Some issues that remained unclear from the original documentation had to be resolved and solutions for these issues have been integrated into the metamodel. For example, we have presented a concrete approach for connecting the OCL with the UML core concepts. Furthermore, we have given an interpretation for several important OCL concepts, e.g. the context of an expression, undefined values, and result values of expressions.

Finally, we have to note that, although the metamodel provides a precise description of the abstract syntax of OCL expressions, a complete description of OCL still requires a semantics definition. In our framework, a mapping from expressions to values would be necessary. A proposal for this has been made in [18]. It would be interesting to see, whether the UML modeling features are sufficient to express the OCL semantics within the metamodel.

Acknowledgments

We thank the anonymous referees for their helpful comments.

References

[1] Ruth Breu, Ursula Hinkel, Christoph Hofmann, Cornel Klein, Barbara Paech, Bernhard Rumpe, and Veronika Thurner. Towards a formalization of the unified modeling language. In Mehmet Aksit and Satoshi Matsuoka, editors, *ECOOP'97—Object-Oriented Programming, 11th European Conference*, volume 1241 of *Lecture Notes in Computer Science*, pages 344–366, Jyväskylä, Finland, 9–13 June 1997. Springer-Verlag.

[2] Jean-Michel Bruel. Transforming UML models to formal specifications. In Luis Andrade, Ana Moreira, Akash Deshpande, and Stuart Kent, editors, *Proceedings of the OOPSLA'98 Workshop on Formalizing UML. Why? How?*, 1998.

[3] Andy Evans. Making UML precise. In Luis Andrade, Ana Moreira, Akash Deshpande, and Stuart Kent, editors, *Proceedings of the OOPSLA'98 Workshop on Formalizing UML. Why? How?*, 1998.

[4] Robert France, Andy Evans, and Kevin Lano. The UML as a formal modeling notation. In Haim Kilov, Bernhard Rumpe, and Ian Simmonds, editors, *Proceedings OOPSLA'97 Workshop on Object-oriented Behavioral Semantics*, pages 75–81. Technische Universität München, TUM-I9737, 1997.

[5] Erich Gamma, Richard Helm, Ralph Johnson, and John Vlissides. *Design Patterns*. Addison Wesley, Reading, MA, 1995.

[6] Martin Gogolla and Mark Richters. Equivalence rules for UML class diagrams. In Pierre-Alain Muller and Jean Bézivin, editors, *Proceedings of UML'98 International Workshop, Mulhouse, France, June 3 - 4, 1998*, pages 87–96. ESSAIM, Mulhouse, France, 1998.

[7] Martin Gogolla and Mark Richters. On constraints and queries in UML. In Martin Schader and Axel Korthaus, editors, *The Unified Modeling Language – Technical Aspects and Applications*, pages 109–121. Physica-Verlag, Heidelberg, 1998.

[8] Ali Hamie. A formal semantics for checking and analysing UML models. In Luis Andrade, Ana Moreira, Akash Deshpande, and Stuart Kent, editors, *Proceedings of the OOPSLA'98 Workshop on Formalizing UML. Why? How?*, 1998.

[9] Ali Hamie, Franco Civello, John Howse, Stuart Kent, and Richard Mitchell. Reflections on the object constraint language. In Pierre-Alain Muller and Jean Bézivin, editors, *Proceedings of UML'98 International Workshop, Mulhouse, France, June 3 - 4, 1998*, pages 137–145. ESSAIM, Mulhouse, France, 1998.

[10] Anneke Kleppe, Jos Warmer, and Steve Cook. Informal formality? the object constraint language and its application in the UML metamodel. In Pierre-Alain Muller and Jean Bézivin, editors, *Proceedings of UML'98 International Workshop, Mulhouse, France, June 3 - 4, 1998*, pages 127–136. ESSAIM, Mulhouse, France, 1998.

[11] Kevin Lano. Defining semantics for rigorous development in UML. In Luis Andrade, Ana Moreira, Akash Deshpande, and Stuart Kent, editors, *Proceedings of the OOPSLA'98 Workshop on Formalizing UML. Why? How?*, 1998.

[12] Kevin Lano and Juan Bicarregui. Formalising the UML in structured temporal theories. In Haim Kilov and Bernhard Rumpe, editors, *Proceedings Second ECOOP Workshop on Precise Behavioral Semantics (with an Emphasis on OO Business Specifications)*, pages 105–121. Technische Universität München, TUM-I9813, 1998.

[13] Luis Mandel and María Victoria Cengarle. On the expressive power of the object constraint language OCL. Technical report, Forschungsinstitut für angewandte Software-Technologie (FAST e.V.), 1999.

[14] OMG, editor. *Meta Object Facility (MOF) Specification, Version 1.3 RTF, 2 July 1999*. Object Management Group, Inc., Framingham, Mass., Internet: http://www.omg.org, 1999.

[15] OMG. Object Constraint Language Specification. In *OMG Unified Modeling Language Specification, Version 1.3, June 1999* [16], chapter 7.

[16] OMG, editor. *OMG Unified Modeling Language Specification, Version 1.3, June 1999*. Object Management Group, Inc., Framingham, Mass., Internet: http://www.omg.org, 1999.

[17] OMG. UML Semantics. In *OMG Unified Modeling Language Specification, Version 1.3, June 1999* [16], chapter 2.

[18] Mark Richters and Martin Gogolla. On formalizing the UML object constraint language OCL. In Tok Wang Ling, Sudha Ram, and Mong Li Lee, editors, *Proc. 17th Int. Conf. Conceptual Modeling (ER'98)*, pages 449–464. Springer, Berlin, LNCS 1507, 1998.

[19] Bernhard Rumpe. A note on semantics (with an emphasis on UML). In Haim Kilov and Bernhard Rumpe, editors, *Proceedings Second ECOOP Workshop on Precise Behavioral Semantics (with an Emphasis on OO Business Specifications)*, pages 177–197. Technische Universität München, TUM-I9813, 1998.

[20] Jos Warmer, John Hogg, Steve Cook, and Bran Selic. Experience with formal specification of CMM and UML. In Haim Kilov and Bernhard Rumpe, editors, *Proceedings ECOOP'97 Workshop on Precise Semantics for Object-Oriented Modeling Techniques*, pages 167–171. Technische Universität München, TUM-I9725, 1997.

[21] Jos Warmer and Anneke Kleppe. *The Object Constraint Language: Precise Modeling with UML*. Addison-Wesley, 1998.

Tool-Supported Compression of UML Class Diagrams

Ferenc Dósa Rácz, Kai Koskimies

Nokia Research Center
Box 422, FIN-00045 NOKIA GROUP, Finland
Email: ferenc.dosa@research.nokia.com

Software Systems Laboratory
Tampere University of Technology
Box 553, FIN-33101 Tampere, Finland
Email: kk@cs.tut.fi

Abstract

Techniques for tool-supported compression of UML class diagrams are
developed. These techniques allow abstract representations of class diagrams by
effacing (less essential) parts of the diagram. The hidden parts can be made
again visible at selected points. The user can start examining a class diagram
with only few main classes visible and refine the diagram gradually to the
interesting directions, proceeding from abstract view to details. The proposed
techniques help in managing large class diagrams and in extracting high-level
views from object-oriented legacy systems, thus supporting the understanding
of the overall architecture of the system. The construction of the compressed
form of a class diagram can be either automatic or it can be controlled by a
human. An algorithm is given for managing compressed class diagrams, and a
prototype implementation is described.

1 Introduction

Class diagrams are a central diagram for modelling the static structure of software in
UML [RJB99]. However, although a class represents an abstraction, it is usually too
small a unit to provide support for understanding a large system consisting of, say,
hundreds of classes: more coarse-grained abstraction facilities are required for
structuring the model into logical parts.

In UML, packages can be used for grouping several related items. General grouping
concepts like packages are useful for the system designer who can exploit them to
divide the model into logical parts. However, such general structuring concepts are of
little use for understanding a large existing class diagram. Such a diagram can be the
result of automated reverse engineering of legacy software (Fig. 1.), or it may simply
be a part of an object model, which the designer has not been able to divide into
manageable pieces. It should be noted that even a modest class diagram of, say, some

tens of classes may be difficult to grasp for a person who is not previously familiar with the model or the application domain.

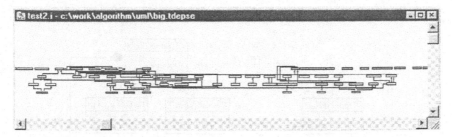

Fig. 1. Part of a complex UML class diagram extracted automatically from a legacy system

Consider for example the class diagram of Fig. 2. For a person familiar with modeling a teller machine this diagram may be immediately understandable, but suppose someone who has never heard of a teller machine sees this diagram. How can she pick up the central classes needed for a more high-level, abstract view which is essential for understanding the model as a whole?

In Fig. 3, another version of the same example is given, but with only the most important classes and associations. This diagram provides a more high-level view of the same system, making the essential concepts of the system clearly visible. Obviously, this version of the model is much more understandable even for an uninitiated reader. Could it be possible to provide tool support for obtaining Fig. 3 on the basis of Fig. 2, so that the user could start with Fig. 3 and refine the diagram as needed? Could the same tool be used to present large automatically generated class diagrams like the one in Fig. 1 in a compact, abstract form?

In this paper we study tool-supported compression methods for UML class diagrams. Ideally, one should be able to start examining a class diagram through a view which shows only a few main classes and their relations, and one could then open the interesting parts gradually, proceeding in a top-down manner from a coarse, abstract view to the details. In this way the amount of visible information is always reasonable, and irrelevant information (with respect to the current needs of the user) does not obscure the diagram. The construction of the compressed form of a class diagram can be either completely automatic or it can be controlled by a human.

A compression technique has been previously proposed for sequence diagrams [KoM95]. In this technique a method call appearing in a diagram is compressed into a single arc from the caller to the the the callee. The call can be opened by mouse-clicking, so that all the internal calls are shown, again in the compressed form. In this way the user can navigate into those parts of the event trace she is really interested in. This compression technique is particularly simple because it naturally complies with the concept of a method call, and because the strict layout conventions of event trace diagrams make it easy to produce automatically different views of the event trace. Although compressing class diagrams is not as straightforward, we argue - and show in this paper - that similar techniques can be applied for them as well. The proposed techniques have been implemented in a class diagram editor prototype.

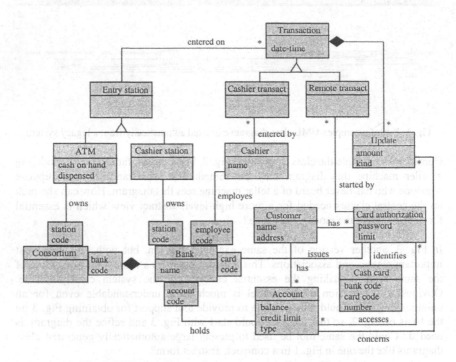

Fig. 2. UML class diagram for an automatic teller machine system

To establish a systematic basis for compression graph-like structures, we first define a general compression technique for directed labeled graphs. In Section 3 we discuss several issues arising when the general compression technique is applied to UML class diagrams. In Section 4 we demonstrate how the compression technique works in the example case, using the prototype implementation. Finally, some concluding remarks concerning remaining problems and future work are presented. An early version of the technique has been presented in [Kos98]. This research is carried out in the context of FAMOOS project [FAM99].

2 Compression and expansion on labeled directed graphs

Here a labeled graph is a representation of multiple relations among a set of nodes: a labeled edge denotes a relation between two nodes. The intuitive idea of graph *compression* is to move certain nodes "inside" another node (*container* node) so that the moved nodes become invisible (*contained* node). All the nodes, which are not

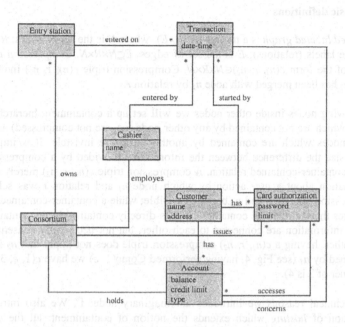

Fig. 3. An abstract version of the class diagram for an ATM system

contained by any other node are visible. To guarantee that the compression somehow reflects the relational structure of the graph, we will compress relations rather than individual nodes and edges. We make the assumption that the relations presented in a graph are transitive, implying that a compression operation must be able to handle the transitive closure of some relation, with respect to a certain node. The compression then effaces this closure from the graph, making the graph in this way (hopefully) more abstract. This can be compared to sequence diagram compression where those arcs belonging to the "call closure" of a particular method call can be collapsed. More precisely, when node n is compressed with respect to relation r, all those nodes that can be accessed from n using r edges are compressed into n.

Expansion reverses the effect of a compression operation. However, since the user may be interested to see in detail only a particular branch in the closure, an incremental expansion operation is provided as well. This operation reveals only those compressed nodes that have direct connection of the relation in question to the container node. In this way the user can expand the graph one level at a time, possibly even changing the followed relation in each step. This corresponds to the expanding of a sequence diagram one nested call at a time. The idea sounds simple, but the technique becomes more complicated when nested compression operations are allowed: containers can be contained by other nodes. We must maintain the containment hierarchy and keep track of the compression and expansion operations. In the following we develop a formal basis for graph compression operations.

2.1 Basic definitions

Directed labeled graph is a triple $G(N, R, E)$, where N is the set of nodes, R is the set of edge labels (relations), E is the set of edges. $E \subseteq N \times R \times N$. *Compression triple* is a triple of the form $c(n_1, r, n_2) \in N \times R \times N$. Compression triple $c(n_1, r, n_2)$ indicates that node n_2 has been merged with node n_1 by relation r.

By moving nodes inside other nodes we will set up a containment hierarchy. Those nodes which are not contained by any other node (i.e. are not compressed) are visible, those nodes which are contained by another node are invisible. It is important to emphasize the difference between the information provided by a compression triple and a container-contained relation. A compression triple $c(n_1, r, n_2)$ merely stores the information about a user action by which node n_1 and relation r was selected for compression operation and n_2 became invisible; while a container-contained nodepair indicates the fact that the contained node is directly contained by its container node. These information are connected to each other, but not necessarily consequences of each other: having a $c(n_1, r, n_2)$ compression triple does not imply that n_2 is directly contained by n_1 (see Fig. 4, having performed <u>Comp</u>(1, e) we have $c(1, e, 3)$, and the container of 3 is 4).

For technical reasons we introduce an imaginary node: Γ. We also introduce the definition of *hosting* which extends the notion of containment: all the container-contained pair will be referred as *host-hosted* nodes. Those nodes, which are not compressed will be hosted by Γ. Therefore, every node will have a host node. Formally, for $\forall n \in N \cup \{\Gamma\}$ we define

$$host(n) = \begin{cases} \Gamma, \text{if } n = \Gamma \\ \text{the container of node } n \end{cases}$$

Compressible graph is a 5-tuple $G(N, R, E, ND, CD)$, where N is the set of nodes, R is the set of relations, E is the set of edges, ND is the *node descriptor table*, $ND \subseteq (N \cup \{\Gamma\}) \times (N \cup \{\Gamma\}), \forall n \in N$ $(n, host(n)) \in ND$, and CD is the *compression descriptor table*, $CD \subseteq N \times R \times N$. The purpose of ND is to represent the host function, and CD will serve as a storage for compression triples.

For $\forall n \in N \cup \{\Gamma\}$ we define the Hosts set, which always contains the symbol Γ and node n itself.

$$Hosts(n) = \begin{cases} \{\Gamma\}, \text{if } n = \Gamma \\ \{n\} \cup Hosts(host(n)), \text{otherwise} \end{cases}$$

For $\forall n \in N$ we define the root of a node as follows: $root(n) = n'$ so that $n' \in Hosts(n) \setminus \{\Gamma\}$ and $host(n') = \Gamma$.

For $\forall(n, r)\in N\times R$ we define Closure$(n, r) = \{n'|$ there is a r labeled edge path from node n to node $n'\}\setminus\{n\}$. Note that n is never included by Closure(n, r) for any r even if it has a reflexive edge.

In addition, we say that two nodes are *compressible*, if their roots are different.

Our aim will be to keep the hosting hierarchy (practically *ND*) and the compression triples (*CD*) always in a consistent state, and define only consistency-preserving operations i.e. the operation must be able to be carried out and it must leave the hierarchy *consistent*. We will omit the exact definition of consistency here, but the informal explanation is as follows: at each compression operation we create a compression triple and set a new containment, and at each expansion operation we destroy one compression triple and reset an appropriate containment relation, i.e. the containment relations and compression triples correspond to each other in a consistent hierarchy.

2.2 Operations

In the sequel, ND(a, b) will denote a pair included in *ND* set.

2.2.1 Constructing an initial compressible graph
Mapping a labeled graph to a compressible graph is quite straightforward: assuming a labeled graph $G(N, R, E)$ we set up the corresponding initial compressible graph as $G'(N, R, E, ND, CD)$, where $ND=\{(n, \Gamma)|n\in N\}$, and $CD=\varnothing$. This initial graph is clearly consistent. All the following operations are defined on compressible graphs. For any of the operations, if the graph is consistent before the execution of the operation, the graph is consistent after the operation as well. We omit the proof here.

2.2.2 Setting a hosting relation

```
SetHosting(n1, r, n2)
do
    assert(compressible(n1, n2))
    CD := CD∪{(n1, r, n2)}
    ND(root(n2), Γ) := ND(root(n2), n1)
od SetHosting.
```

2.2.3 Resetting a hosting relation

```
ResetHosting(n1,r,n2)
do
    assert(compression triple c(n1,r,n2)∈CD)
    CD := CD\{(n1,r,n2)}
    variable actnode
    actnode := n'|host(n')=n1 and n'∈Hosts(n2)
```

```
    ND(actnode, n1) := ND(actnode, Γ)
    od ResetHosting.
```

2.2.4 Compression and expansion
These are the operations called directly by the user, through an appropriate interface.

```
Comp(n, r)
do
        for ∀nodes n'∈Closure(n,r)
        do
            if compressible(n, n')
            then SetHosting(n, r, n')
            od
od Comp.


Exp(n, r)
do
        for ∀ nodes n'∈Closure(n,r)
        do
            if compression triple c(n,r,n')∈CD
            then ResetHosting(n,r,n')
            od
od Exp.
```

A node can be compressed into another node even if it is already hosted by some other node (see the Comp(1,e) transformation in Fig. 4). In that case the uppermost host of the to-be-hosted node becomes hosted by the new host node. For example, if node n_1 is hosted by n_2 and n_2 is the uppermost host (that is, node n_2 is visible), operation Comp(n, r, n_1) moves n_2 and all its hosted nodes into n. Essentially, this corresponds to compression along so called virtual edges (see Section 2.2.6.), since there must have been a virtual edge between n and n_2.

2.2.5 Incremental expansion
If a long chain of classes connected by a certain relation has been compressed, expanding the whole chain is often not sensible for a user: it might be that the user is actually interested only in a particular class directly related to its container, and in the relations this class has with the rest of the system. In this case the possibly large set of uninteresting classes only disturbs the user who very likely compresses them again to be able to focus on the interesting classes. This is analogous to compression sequence diagrams [KoM95]: when expanding a compressed call, it would not be sensible to expand all the nested calls as well, because the user is possibly interested only in particular actions occurring during the compressed call. Instead, the calls are opened incrementally so that the internal calls appear first in the compressed form, allowing the user to open only the interesting nested call(s).

We aim at a similar functionality in class diagrams. This can be achieved using the following revised expansion algorithm. The basic idea is to perform, directly after expansion, a new compression operation for each class directly associated to its container with the relation in question. Hence, after expansion only one level of new classes appear, the rest being compressed to these classes. For example, if the root class of an inheritance hierarchy has been compressed with the inheritance relation, and then expanded, only the immediate subclasses will appear. If the user wants to go deeper into the hierarchy, she must select a particular subclass and expand it.

```
IncExp(n, r)
do
      for ∀e(n,r,n')∈E
      do
        if compression triple c(n,r,n')∈CD
        then
        do
          ResetHosting(n,r,n')
          for ∀n"∈Closure(n',r)
          do
            if compression triple c(n,r,n')∈CD
            then
            do
              ResetHosting(n,r,n")
              SetHosting(n',r,n")
            od
          od
        od
      od
od IncExp.
```

2.2.6 Displaying a compressible graph
Displaying a compressible graph means showing a view on the compression state of the graph. The visible nodes are $\{n \mid host(n)=\Gamma\}\subseteq N$. We have to give a visual representation of the original edges between nodes with different roots (with the exception of reflexive edges). If the representer edge connects different nodes than the orginal one, the edge will be called *virtual*. From the point of view of graphical representation a virtual edge should be distinguished, e.g. using dashed lines.

The code for this operation is straightforward but lengthy, we omit the pseudocode and the details here.

2.2.7 Example
Fig. 4 describes the modifications of a compressible graph during a sequence of compression operations. Virtual edges are presented with dashed lines. Host nodes are shown with thick border lines.

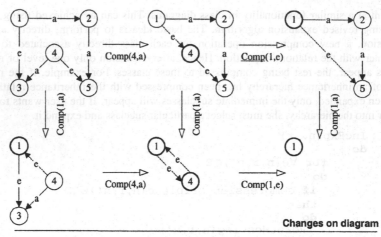

Changes on diagram

Changes on hosting hierarchy and compression triples

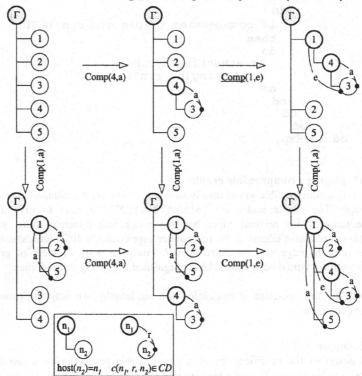

Fig. 4. Compression on a graph

The lower part of Fig. 4. shows the configuration of the hosting hierarchy for each state in the upper part. Each transaction between different states has an opposite transaction having method name "Exp" instead of "Comp", with the same parameters. Our first compression operation Comp(4, a) effaces node 3, creates the virtual edge from 1 to 4, and effaces the edges associated with 3. The second compression operation Comp(1, e) effaces node 4 (because it is the host node for 3 which has e-edge with 1), leaving 1, 2 and 5 visible.

Note that 5 is not effaced because it is not accessible from 1 using normal e-edges. The edge between 4 and 5 becomes invisible, but a corresponding virtual edge from 1 to 5 is created (and the previous virtual edge is deleted). The third compression operation Comp(1, a) effaces nodes 2 and 5, and all the remaining edges. Hence only node 1 is left visible.

From the fully compressed state we have reached, Exp(1, e) restores node 4. Since there is a node compressed in 1 (5) that has an e-edge with 4, there is a virtual edge from 4 to 1. On the other hand, there is also a node compressed in 4 (3) that has an e-edge with 1; hence there is another e-edge from 1 to 4. Performing Exp(4, a) restores 3, and 4 is no more a host. Edges connected with 3 are restored as well. Finally, Exp(1, a) restores the original appearance of the graph.

It should be noted that our definitions of the compression and expansion operations are certainly not the only viable ones, but they lead to a relatively simple model of compression which is logical and simple to implement. In particular, the operations could behave in different ways regarding overlapping or nested groupings.

3 Applying graph compression to class diagrams

3.1 User interface

In their basic form, class diagrams can be viewed as graphs (we ignore for a moment certain special features of class diagrams) in which nodes are classes and edges are relationships (associations or inheritance relations) between classes. Hence we can readily apply the techniques above to compress class diagrams. The user interface for such a facility consists simply of a context-sensitive menu which shows, for each visible class, the relations that can be used to compress or expand the class diagram with respect to the chosen class and relation. A class has a hierarchical context sensitive menu: on the upper level there are menu items "Compress" and "Expand" (the latter only for host nodes), and on the second level there are those relations which can be used for compressing or expanding that particular class. For compression, all the names of the relations attached to that class (and so far not compressed) appear in the menu, while for expanding only those relation names included in the compressed relation set R_n of the node descriptor (n, R_n, h) appear in the menu.

Compressed classes must be distinguishable from ordinary classes, because the user should be able to immediately see which classes can be expanded. In this paper we indicate compressed classes by thick borderlines. Similarly, virtual associations (corresponding to virtual edges) must be distinguishable from "real" associations, because they are only reminders of actual relationships and therefore have different meaning. Here virtual associations will be denoted by dashed lines.

3.2 Compression as abstraction

Above we have considered compression as a general "syntactic" technique for simplifying a graph by effacing all the nodes and edges associated with a particular relation, regardless of the meaning of the relation. If a host node is visually distinguished from normal nodes (which is anyway necessary), this syntactic interpretation of compression may be useful as such. This implies that the user should not understand the resulting diagram as a real UML diagram anymore, but as a diagram that has been compressed using certain conventions. A host node is not a class symbol, but simply a representative of a set of related classes. Similarly, a virtual edge associated with a host node is not an association for the host class, but a mark denoting that there is an association with some of the classes compressed into the host. In this case an appropriate symbol for a host might be a package.

However, our original aim was not to develop an arbitrary compression mechanism but rather to allow the abstraction of a class diagram, either manually or automatically. In this approach the essential question is: when does compression yield in some sense a valid "abstraction" of the original diagram? Clearly, to decide this we must take into account the semantics of relations as well.

Strictly speaking, a compression is semantically correct only if the hosted class can be replaced by the host class without making the diagram semantically invalid. This is essentially a subtype relation between the host class and the hosted class: the host class should be a subtype of the hosted class. Indeed, one of relations in UML fulfils the subtyping requirement: inheritance relation. This means that the superclasses can be effaced by compressing them into a subclass. Clearly the resulting diagram is semantically valid: it simply means that subclass has inherited all the associations of the superclasses, which it does anyway; now the inheritance has only been made explicit. Unfortunately, although this might be a nice feature, too, it is not at all what we are aiming to. In fact, to make the diagram more abstract we should compress along the inheritance relations in the reverse direction: the host (visible) class should be the more abstract class, the superclass.

Hence we must give up strict semantic valentss of compression. Instead, we should look for relations which somehow mean abstraction, going from more general concepts to more specific concepts or vice versa. There are at least three standard relation types of this character in UML: inheritance (but going to the reverse direction with respect to subtyping), aggregation and implementation (of an interface). In all these cases, details can be effaced by compression. We call such relations *abstraction relations.* However, it should be emphasized that although the resulting compressed view is in some sense an abstraction of the original diagram, the virtual edges do not represent actual relations as they would do in the case of subtype compression. Obviously some of the user-defined relations are abstraction relations as well, but there is no way to decide this automatically without additional information. A tool might allow the user to specify this property for a relation. Assuming that all relations are in this way marked either as abstraction relations or non-abstraction relations, the tool could allow compression operations only according to the abstraction relations. A good candidate for a user-defined abstraction relation is one in which the other participant is connected only by this relation to the rest of the diagram; in this case the participant can be effaced without effacing any other relational information.

In the case of reverse engineering, it would be desirable to obtain an abstract view of the class diagram automatically from an arbitrary source, without any additional information concerning the used relationships. In this case we can exploit only the standard abstraction relations. The reverse engineering tool should automatically perform compressions for inheritance, aggregation and/or implementation relations, depending on the user's choice. The user could then see first a view of the class diagram that is as abstract as possible, perceive the system as a whole, and expand the interesting details as necessary.

Another usage scenario is to use predefined, recorded paths of expansion for tutorial purposes. A designer may define a sequence of expansion steps (by recording) and store them as "films" [Mös97] that can be selected and run - step by step - by the user. In this way the designer can offer guided tours through a large class diagram so that the user can look at certain details and understand their relations to the rest of the diagram.

Whether compression should be allowed only for abstraction relations is an open question. The answer depends on the desired character of the tool. If abstraction is not required, the tool should be considered as a "relation navigator" rather than as an abstraction tool. Since the tool should anyway make a clear visual distinction between actual and virtual relations, the view that results from compression never lies. However, if the relation used for compression is not an abstraction relation, the resulting view may be somewhat misleading if the user does not clearly understand the role of virtual edges. In the prototype tool presented in the next section we have taken the syntactic approach which allows the use of compression for various purposes rather than just for abstraction.

3.3 Handling multiplicity and qualification of associations

Assume that a virtual edge originates from an association with multiplicity or qualification specifications. Should these specifications be copied for the virtual edge as well? Generally, no: a virtual edge does not describe a relation with the host but with some class merged with the host. On the other hand, it would useful if as much information concerning the associations was retained as possible, as long as it is not misleading. We can achieve this by applying the following simple principle: if a virtual edge represents an association, and the other end is associated with a non-host class, then the multiplicity and qualification attached to this end in the original association will be copied also for the corresponding end in the virtual edge.

3.4 Alternative presentation options for multiple edges

Since the edges connected to hidden classes are preserved by their container classes, the latter will be often associated with an increased number of edges between the same pair of classes; thus the resulting class diagram may be still messy, even having a reduced number of visible nodes. A possible solution for displaying a group of edges between two classes could be substituting the group of edges with a single edge. The user could then detach chosen associations from the group as separate edges. This could be considered as a kind of shallow compression on edges with straightforward implementation.

4 Prototype

We have implemented a prototype for experimenting with class diagram compression techniques. The prototype is a class diagram editor with very limited editing capabilities, equipped with a compression/expansion facility described in this paper. The system is implemented in Windows/Visual C++, and it is freely availabe for evaluation from the authors.

The view in Fig. 5 could be the one that a person unfamiliar with the application first sees. She could then open the pop-up menu for Consortium and learn that there are invisible classes that are in aggregation relation with Consortium. Having selected relation "aggregation" for expansion, she can obtain a diagram shown in Fig. 6. In this way she can navigate in the diagram as needed, without losing the understanding of the model as a whole.

5 Open problems and conclusions

In the example above, we have more or less retained the original layout when compressing, creating thus empty space within the diagram. This simple technique might be sufficient for class diagrams of modest size, say few tens of classes, but if the diagram is really sizable, or if even the original class diagram is produced automatically (e.g. as an extraction from a legacy system), we need automated layout techniques. An essential requirement for such a technique is that the layout algorithm should retain the general structure of the previous version of the diagram, because otherwise it is very difficult for the user to orient herself according to the new layout. In compressing this means that the relative positions of remaining visible classes should be roughly the same as before. In expansion this means that the new visible classes should roughly take their original relative positions (if they have such).

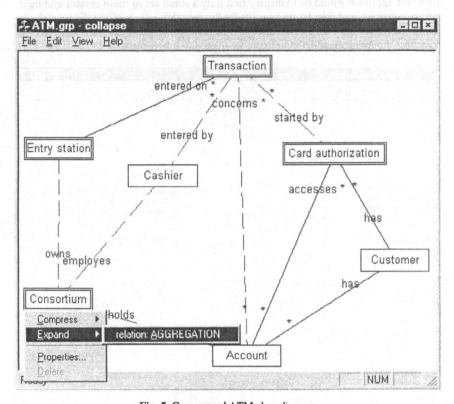

Fig. 5. Compressed ATM class diagram

The prototype is capable of using ISOM (Inverted Self Organizing Map) graph layout algorithm [Meyer98] during expansion. ISOM is practically a reused Kohonen's method. However, our first experiments indicate that it should be possible to explore a compressed graph without actual expansion, but using nested inner windows instead - producing the complete uncompressed diagram is often unnecessarily heavy operation if one only wants to navigate in the structure.

In the context of reverse engineering, an obvious open problem is how to generate a sensible class diagram from source code in the first place: only one type of relations is clearly visible in the code, inheritance. All other relations, including aggregation, should be inferred from the types of reference attributes, parameters etc. This problem is associated with reverse engineering in general.

On the other hand, we believe that class diagram compression is particularly useful in reverse engineering. Various kinds of class diagram extractors are common components of modern OO case tools, but if the results are huge unstructured collections of classes, the benefits of such tools may be questionable. A more attractive approach would be to display first only a small set of main classes and their mutual relations, and then let the user gradually open those parts of the diagram she is really interested in. This can be achieved be letting the class diagram extractor to compress the initial class diagram to the smallest possible form.

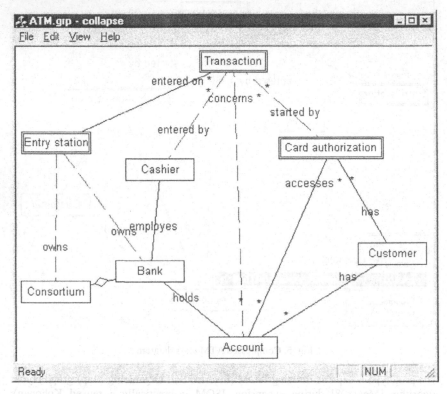

Fig. 6. Partially compressed ATM class diagram

Obviously, similar compression techniques can be used for any graph-like diagrams that tend to become large and hard to understand, assuming that they contain relations that can be regarded as abstraction relations. We anticipate that the same techniques can be used for graph-like visual representations in various tools and systems (say, configuration visualizers, network management systems, etc.).

We are currently implementing the technique for a proprietary UML-based software design environment used within Nokia, called TDE (Telecom Design Environment).

References

[RJB99]   Rumbaugh J., Jacobson I., Booch G.: The Unified Modeling Language Reference Manual. Addison-Wesley 1999.

[FAM99]   FAMOOS home page: http://www.sema.es/projects/FAMOOS.

[Kos98]   Koskimies K.: Extracting high-level views of UML class diagrams. Proc. of NOSA '98 (First Nordic Workshop on Software Architecture), Research Report 14/98, Department of Computer Science, University of Karlskrona/Ronneby, August 1998.

[KoM95]   Koskimies K., Mössenböck H.: Scene - Using Scenario Diagrams and Active Text for Illustrating Object-Oriented Programs. Proc. of the 18th Int. Conf. on Software Engineering (ICSE), Berlin, 1996, 366-375.

[Mös97]   Mössenböck H.: Films as Graphical Comments in the Source Code of Programs. TOOLS USA '97 (Technology of Object-Oriented Languages and Systems), Santa Barbara, July 1997.

[Meyer98] Bernd Meyer: Competitive Learning of Network Diagram Layout. International IEEE Symposium on Visual Languages, Halifax, Canada, September 1998.

A Pragmatic Approach for Building a User-Friendly and Flexible UML Model Repository

Mariano Belaunde

France Telecom CNET,
Technopole ANTICIPA, 2, avenue Pierre-Marzin
22307 Lannion, France

mariano.belaunde@cnet.francetelecom.fr

Abstract. In France Telecom research center in Lannion (France) we have been working for three years on OO modeling as a promising technology for unifying the representation of data. This has led us to develop a Model Repository Tool, which offers, as its default configuration, a full support for the UML 1.3 meta-model. The tool enables the manipulation of models by means of a Java or Python API. It provides a rich and flexible registration capability based on an explicit identification, relying on a two-leveled hierarchical naming space. The paper focuses on the design aspects of the repository tool and highlights its similarities and differences with the design principles of OMG Meta Object Facility specification.

1 Introduction

An Object Oriented Model Repository tool is basically a software that offers facilities to access and store data models that conform with a generic object oriented representation of data (metamodels).

This kind of tools is currently emerging because OO modeling standardization efforts have received a significant acceptance by the IT industry, allowing thus, the possibility for sharing very high-level information between people and between tools themselves.

In France Telecom research center in Lannion (France) we started in 1996 the development of a Model Repository Tool in order to provide support for a home-made metamodel, which we intended to use for our internal projects. As we had to be aware of managing metamodel evolutions, we have built very soon an architecture based on a meta-metamodel level, that allows generating automatically the implementation classes.

In 1998, we decided to implement the UML metamodel [UML]. At the same time, when we compared our work to the OMG MOF specification [MOF], we noticed that most of the principles were identical. Nevertheless, regarding other aspects, like the internal details of the interface definition, or the lack of support for *association classes*, we found that the MOF specification was not entirely satisfactory as to welcome a full alignment with it.

We have achieved two distinct implementations of the repository tool. The first has Python [PYTHON] as the OO target programming language, while the second has Java [JAVA]. Except for some subtle differences, due to the distinct capabilities of the two languages, the design principles are identical for both implementations.

In this paper we will present all along the pragmatic design principles that we have chosen in order to obtain a flexible and user-friendly Model Repository Tool. By the way, we think that this work can be used as a useful material for evaluating and improving the MOF specification.

1.1 UML Graphical Notation Support Versus Metamodel Support

In end users' point of view the Unified Modeling Language is above all a graphical notation that allows people to communicate by means of very intuitive diagrams.

But for the software industry another important point is that the syntax and the semantics of this graphical notation has been, as better as possible, formalized in the form of an object oriented metamodel containing software abstractions such as *metaclasses*, *inheritance relationships*, *association links* and *constraints*.

This leads to the possibility of developing really open model repositories, meaning that they can offer access, storage and navigation capabilities which are strictly based on a non proprietary standard for model representation.

Most current commercial CASE tools are nowadays more concerned with supporting the graphical notation than explicitly supporting the UML metamodel. The situation is likely to evolve as soon as the OMG standard for model interchange [XMI] will be implemented by tool vendors.

2 Meta-metamodel Support

Our first concern was to develop a Model Repository based on a common and unique OO metamodel. As UML arises as a standard, we indeed adopted it. Even if, at first, we were interested in supporting a single metamodel, for practical reasons, we felt the need to have a meta language for describing similar metamodels, so that we could easily manage any metamodel evolution, and handle automatic production of the metamodel's specific parts of the implementation.

2.1 The Four-layer Metamodel Architecture

The four-layer architecture (see Fig.1), is a useful structuring concept which allows the semantics of a modeling process to be clarified [MOF]. The M3 level describes a language for specifying metamodels (the MOF model), the M2 level defines a language for specifying a model (the UML metamodel), the M1 level defines a language to describe an information domain (end-user models), the M0 level describes data attached to a specific information domain (end-user data).

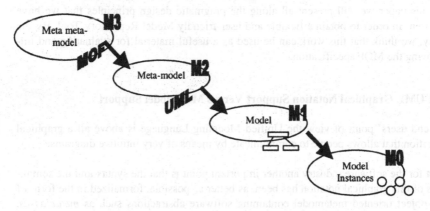

Figure 1 - The four-level metamodel architecture

In a generic model repository tool, however, all these levels can be manipulated in a *uniform way*. The following clarifications are useful.

– We need to keep in mind the distinction between the concrete way to formalize the meta-metamodel abstractions (the *metaentities' abstract syntax*), and the concrete language that is used to enter the definitions of a metamodel (the *metamodel entry notation*).

– In practice, an ordinary *UML class diagram* is often used as the *entry notation*. Therefore, the knowledge of what is the concrete *abstract syntax* being supported is not so significant[1]. For a repository tool, a M3 support is not an end in itself, it is just a way to achieve a generic metamodel support. The UML class diagram graphical notation can hide to end-users the details associated with the M3 level[2].

---

[1] For instance, representing a class inheritance by means of a Generalization entity or by means of a association 'generalizes' is meaningless here.

[2] There are many possibilities for formalizing the graphical representation of a class diagram : use of the MOF, use of Microsoft' RTIM model, direct use of UML, etc.

The theoretical four-level metamodel architecture exhibits only top-down instantiation relationships. However, in practice, a repository tool can get real benefits from exploiting also down-to-top dependencies, such as, model self-definition (reflection) and model to metamodel transformations. The concrete illustration of this, is that, depending on the user's point of view, a model stored in a repository may represent a meta-metamodel (M3), a metamodel (M2), a model (M1) or, even, as we will see later, a model instance (M0).

2.2 The Bootstrap Meta-metamodel

In order to capture any *UML-like* metamodel description, we defined a textual notation reflecting the minimal abstractions that were used to describe *the UML logical metamodel* [UML, Section 2] and which were *significant* for handling automation of the implementation.

The selected abstractions were *packages, classes, attributes, associations, associations ends* and a few basic relationship constructs like *inheritance* and *reification*. The diagram in Fig.2 describes the *abstract syntax* for this textual notation (for clarity *packages* have been omitted). Notice that the underlying meta-metamodel (which will be referred to as the *boot model*) is equivalent to a subset of UML and equivalent to a superset of a subset of the MOF model[34].

We say that this is a *boot model* because its only purpose is to allow repository tool initialization. By the way, after having generated the implementation classes for UML, any user repository may use the *class diagram* UML notation directly[5], as an alternative option for entering new metamodel definitions into the repository, and for generating the corresponding implementation classes [6].

---

[3] The *MOF model* formalizes many modeling aspects not covered by the *boot model*, such as, name scope, object referencing, etc.

[4] The *boot model* can be treated as a simple MOF *projection*, except for the *reification* construct.

[5] As stated before, the use of UML diagrams (M2 level) does not question the need for a M3 level support; it just hides from end-users some of its formalization details.

[6] This relies on an import facility from a UML CASE tool and on a converter program from a *UML class diagram* to the *boot model*.

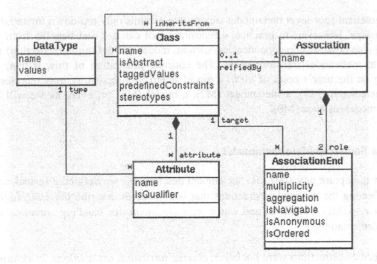

Figure 2 - The *boot model* abstract syntax

2.3 Providing a Partial MOF Support

We have recently entered the OMG MOF model into the repository and generated the implementation classes for it. Therefore, any MOF compliant metamodel can be directly represented and stored in the repository as a MOF *instance*[7]. Notice that this does not mean that we fully support the MOF specification, since this standard describes as well the programmatic interfaces to access the MOF compliant models.

An important problem regarding the MOF, used as a meta-metamodel language, is that it lacks support for the *reification* construct. Metamodels that were initially designed with *association classes*, have to be transformed. As a consequence, one has to maintain two distinct metamodel descriptions (a *logical metamodel* and a MOF compliant *physical metamodel*). This is an important reason for not being aligned with the MOF programmatic interface specification.

2.4 Representing Model Instances in the Repository

After having modeled the static structure of an end-user application (M1), it may be useful to have a way for representing data instances (M0), directly in the terms of the

---

[7] A *UML class diagram* can still be used as the *entry notation*, since it can be interpreted in terms of the MOF abstractions.

model. Such instance-level representation can indeed be used for checking model consistency, prior to software development.

To achieve this, a UML static model stored in the repository tool, may be processed as if it were an ordinary metamodel (M1 to M2 transformation). The generic *API generation program* (the same one that was used for generating the UML implementation) can be applied to produce all the classes described in the end-user model. As a result, the user will be able to create the model instances of his application and store them in the repository. Moreover, these model instances can be externalized and published in the WEB using an XML based format (for instance XMI).

Notice that another alternative to represent instances in the repository is to use the specific UML metaclasses like *Instance, Link, Object, DataValue*, etc. Nevertheless, in this case, instance representation is not as direct as the one that uses the abstractions defined in the user model.

2.5 Towards a Generic Model Repository

To conclude this section, there are many advantages to provide a meta-metamodel level support. Apart from the fact that this helps handling metamodel evolution and enables automatic processing, we have made much more than implementing a UML Model Repository: at the end we have built a generic OO Model Repository.

3 Metamodel Support

3.1 Mapping the Metamodel Abstractions into Language Constructs

An important requirement when building the repository tool was to make it fully accessible from a standard OO language. Most commercial tools offer a specific and proprietary language (like *Rose Basic* from *Rationale*) for accessing the models. Moreover, the *internal metamodel* supported by these tools often deviates from the standard UML.[8]

We were interested in providing a *close* mapping of the metamodel abstractions into the programming constructs.

The natural approach to achieve this was to try the following procedure:

---

[8] Since no standardization exists regarding the way to encode graphical properties, CASE tools provide many metamodel enhancements to manage the graphical rendering of models.

- For each metaclass, create a programming class,
- Supply the metaclass inheritance hierarchy using the inheritance mechanism supported by the target OO language,
- In each metaclass, define a set of operations that allow access to the attributes owned by the class, and a set of operations to allow navigation across *the associations ends* that have the metaclass as a source.

This task is feasible in class based programming languages like Java, C++ and Python. For Java, multiple inheritance at metamodel level can be translated into multiple inheritance of *interfaces*.

3.2 The Tailored Interfaces

The tailored interfaces provide, for each metaclass, a set of methods to access all the attributes and all the roles that are potentially accessible from that class. This interface only depends on the metamodel definition and on the generic rules used to derive it.

We have applied the following rules:
- For each attribute, define a reader access method '*get<Attr>*', a writer access method '*set<Attr>*' and a cleanup operation '*reset<Attr>*'. If the attribute type is of *boolean* type, the reader access method is named '*is<Attr>*' instead of '*get<Attr>*'.
- For each single-valued role, define a reader access method '*get<Role>*', a writer access method '*set<Role>*' and a remove operation '*remove<Role>*'.
- For each multiple-valued role, define a global reader access method '*get<Role>s*', an individual reader access method '*get<Role>*', an add operation 'add<Role>' and an individual remove operation '*remove<Role>*' and a global remove operation '*remove<Role>s*'.
- For each single-valued role that refers to an *association class* in the metamodel, define a reader access method '*getlink<Role>*' to the class instance representing the association. Add an argument to the '*set<Role>*' operation to allow passing the association class instance.
- For each multiple-valued role that refers to an *association class* in the metamodel, define an individual reader access method '*getlink<Role>*' and a global reader access method '*getlink<Role>s*' operation to access the class instances reifying the association. Add an argument to the '*add<Role>*' operation to allow passing the association class instance.
- For each *association class* in the metamodel, define two inverse access methods to get both targets of the reified association: '*getrlink<Role1>*' and '*getrlink<Role2>*'.

Notice that this procedure is quite similar to the one used in the MOF to derive the IDL *tailored* interfaces from a MOF compliant metamodel. One important difference comes from the lack of support for *association classes* in the MOF (no *getlink* or *getrlink* methods). Other noticeable differences will be detailed below.

The MOF compliant CORBA facility specification [UML, section 5] defines an IDL interface for each association in the metamodel. We didn't find it useful to be aligned with this. Navigation across links is performed directly from class instance to class instance. Moreover, access to any link property (like directionality) can be obtained from both targets of the association, by means of convenience methods defined in the *reflective* interface.

Regarding the general policy for avoiding name clashes, some of the IDL interface names, that conflict with the OMG Object Model reserved names, are prefixed, while the majority is not. We think that this is not a very accurate solution since the user needs to be aware of which metaclasses have been renamed. A better approach, according to our proper experience, is either prefix systematically all the metaclasses or not insert prefixes at all. The IDL mapping as proposed in the MOF specification is not as language neutral as expected. For instance, the IDL syntax naming conventions are not well suited to ordinary naming conventions used in Java (*getOwnedElement* instead of *get_owned_element*).

In MOF compliant interfaces, the reader access methods are not prefixed at all (the access method for the 'name' attribute is 'name()'). We think that adding a prefix (such as 'get') is preferable in order to avoid clashes with any other generic operation that could be inherited (such as access to an internal identifier, to status flag, etc).

Another noticeable point concerns the association links defined from a source metaclass to a destination metaclass belonging to another package. In the CORBA facility specification, it is stated that such links are treated as unidirectional. We are not convinced at all with the reasons provided to argue on this. We think that the reasons for people to partition the classes among different packages are, in general, unrelated to the decision to make unidirectional links.

The approach we have, is that navigability is a property that can be specified at meta-metamodel level. In addition, a potentially bi-directional meta-association, still, can be created as unidirectional when it is instantiated. We think that this approach provides the maximum of flexibility required to match any kind of situation.

3.3 The Reflective Interface

The *tailored* interfaces are not well suited for generic applications such as repository browsers, or formatters that may need to manipulate the models independently from the metamodel. For this purpose a generic interface, also known as *reflective* interface can be used. In practice, a *tailored* interface will use the generic interface to implement itself.

The four generic classes that we have defined are *RefPackage*, *RefObject*, *RefLink* and *RefLinkObject*, to represent the basic *object instance* concepts. The *RefLink* and *RefLinkObject* are private classes used for representing internally link instances. *RefObject* is the base class for all the classes generated when implementing a metamodel. It contains general purpose access methods (such as *getRole*, *getAttr*), externalisation methods, and specific methods to access the internal state of the instance.

A *reflective* interface is, of course, also defined in the MOF. The generic classes are *RefObject*, *RefAssociation* and *RefPackage*. Since there is no direct support for association classes, no *RefAssociationObject* is defined. The methods supported by the generic MOF interfaces reflect the specific design choices that were made in the MOF.

We believe that, while agreement on the *tailored* interfaces is feasible (because these interfaces only depend on the metamodel structure), agreement on the *reflective* interfaces will be very hard to obtain, because, it is very difficult to hide at this level the design choices that every tool provider may want to adopt. The *reflective* MOF interface is indeed part of a standard, but this does not ensure that this part will be effectively put to work in the industry.

4 Repository Design Issues

The repository main objective is to achieve naming and storage for each constituent of a registered model. At this level many choices can be made.

There follows a list of typical decisions that have to be taken:

- What kind of naming space will be provided for internal identifiers? Would that be a flat one (single level) or a hierarchical one?
- Will the object identifiers be set explicitly or automatically assigned?
- Are association links always bi-directional? Can we enforce bi-directional meta-associations to be *instanciated*, when needed, as simple unidirectional links?
- Can we enforce ordinary meta-associations (*unspecified aggregation*) to be *instanciated*, when needed, as composition links (*strong aggregation*)?

Since we aim at developing a really *flexible* repository tool, we have chosen, in general the decisions that made the repository the most flexible.

4.1 The Logical Storage Name Space Structure

The main design decision is that the name space for object identifiers be a hierarchical one and that these identifiers can be assigned explicitly, permitting thus, if we want, to have *human readable* identifiers. In other words, the *logical storage name space* in our repository behaves as the name space of an ordinary file system (such as the Unix's file system).

The second important thing is that directory organization in the *logical storage name space* be not enforced to conform to the composition links that could exists inside models. The *storage name space* is orthogonal to the *model name space*[9].

4.2 Repository Encapsulation

All the interactions with the repository (such as loading, registering and storing) are encapsulated in a single interface (the *Repository Session* Interface). This means that the storing and the naming capabilities of the repository can be implemented in many different ways (file system, relational databases, OO databases, etc). The current implementation is based on a file system but this is not compulsory.

4.3 Object Identifiers

As in a file system, an Object Identifier (OID) is made of a *context path* part (a list of successive context identifiers) and a *local object identifier* part (ID). The separator character used between context identifiers and local identifiers is the '/' character.

The local identifier is owned by the metaclass instance. It need not be necessarily provided since it can be automatically computed from the value of an owned attribute that *acts as a qualifier* (for instance the *ModelElement.name* and the *TaggedValue.tag* attribute).

4.4 The Double-layered Naming Hierarchy

A drawback of having explicit naming could be that a user application would have to find a name for each fine grained component of a model, and then, register each item with the chosen name.

To avoid this awkward and not user-friendly constraint, our repository manages a double-layered hierarchy. The first level is provided by the *context* (directory) as we stated before. The second level comes from the distinction between root-objects and

---

[9] This provides flexibility in respect to the different containment policies that may be supported by metamodels. Notice that model containment in UML is not always very well defined.

sub-objects. Root-objects are those that are explicitly registered (with an object identifier containing the full path to reach the object). Sub-objects, on the other hand, are not registered explicitly. They become registered at the moment they are added as a child of a registered parent object.

A sub-object can be attainable as any other object by using its unique OID: this OID is computed automatically on the basis of its top root-object, and on the path required to reach it (a list of 'roleName-localIdentifier' items).

In practice, a file-based implementation will store in a file a root-object and all its sub-objects. This design leads to significant time-consume optimizations for loading and storing objects.

4.5 Stereotype Support

Stereotypes in UML allow extending the metamodel in a controlled fashion. A stereotype that is *based* on a metaclass is an abstraction sharing the structure of the base metaclass but with some added semantics. According to the metamodel a *stereotyped instance* has the type of the base metaclass, and has a link pointing to the Stereotype instance.

To implement this in the repository tool, we have stored the UML metamodel itself as a UML model[10]. UML standard stereotypes are thus represented as registered *Stereotypes* instances with assigned OIDs:

```
TopLevel      /meta/stereotypes/Package_TopLevel

Invariant     /meta/stereotypes/Constraint_Invariant
```

Notice that since each stereotype instance has its own OID, we can distinguish stereotypes that have the same name. User-defined stereotypes are described in the same way, apart from the fact that the OID assignment is the user's application responsibility.

4.6 Default Values for Attributes

One question that a designer of a model repository tool has to solve is the one related to default values of the attributes. Is assignment of all attributes mandatory? At which level default values can be defined: in the metamodel or in the application?

As we want to be as flexible as possible our approach is:

---

[10] This is achieved using a *boot model to UML* converter.

- To make it possible for the user to assign default values at metamodel level,
- To enable the possibility for assigning multiplicity for attributes at metamodel level (a zero, as the lower multiplicity limit, means optional),
- Also, to enable attaching a set of default values to a *repository database* (the default values apply to any model stored in the database) .

Attributes that were not assigned and that did not have a default value are simply *undefined* (in Java the *'get'* method returns the *null* value). A checker utility will, indeed, output errors when detecting unassigned mandatory attributes.

4.7 Proxies Management : Partial Loading of a Model

Model Repositories are expected to maintain a large set of related models. Our repository tool allows incremental loading of models in a transparent way (the parts of a model loaded into the memory are only those which contents are addressed). It allows referring to metaclass instances not already defined (*forward declarations*) so that we could manage models that are still under construction. This is achieved by implementing *proxies*.

A proxy instance, that represents an existing instance in disk space not already loaded into the memory, has no contents at all. A proxy instance, that represents a *forward declaration*, contains all the links being inserted by all the objects that have referenced the *forward declaration*.

Since the responsibility for adding a link always comes from a source object (by applying a method *set<Role>* or *add<Role>*), the implementation can manage a hidden mark to know the source object that creates the link. Therefore, an object that is linked with other objects can distinguish links that are *owned (direct links)* from links that are not owned (*inverse links*). This hidden information is the basis for allowing the definition of forward object declarations, as well as the re-definition of objects (*inverse links* are automatically reinserted into the new definition).

5 Textual Representations of Model Elements

At the beginning of the project, we needed to have a textual representation that could be:
- generic (independent from the contents of the metamodel specification),
- human readable (so that we could accurately inspect the object registered in the repository),
- a formal notation which could be used as an input format.

We have invented, for this purpose, many equivalent textual notations, that only differ on the kind of separators that are used (Python based notation, Java-like notation, XML based notation). Object containment is showed by *inlining* the sub-object definition. As the XMI notation was proposed as a standard for stream based exchange of models, we indeed also provided support to it.

5.1 The *JMI* Textual Notation

The Java-like Metadata Interchange (JMI) is a user-friendly human readable notation, that uses JAVA-like separators.

```
Class /meta/classes/Constraint {
    name = "Constraint";
    feature :
        Attribute body {
            name = "body";
            type :
                Ref DataType /meta/types/BooleanExpression;
        };
    generalization :
        Generalization ModelElement {
            parent :
                Ref Class /meta/classes/ModelElement;
        };
}
```

Figure 3 – A *JMI* textual notation sample

5.2 *XMI* and *XMI+* Textual Notations

The XML Metadata Interchange (XMI) is the stream based interchange format that was standardized by the OMG. This standard makes it possible for models to be exchanged as XML documents. The syntax depends on a DTD file generated from a MOF based metamodel, according to rules defined in the XMI specification.

The metamodel used for generating the standard XMI UML DTD was the *physical UML metamodel* [UML, section 6], which differs on subtle details with the *logical UML metamodel* [UML, section 2]. The XMI rendering program, implemented in the repository, performs all the needed transformations to ensure compliance with the standard DTD.

The XMI+ textual notation is an improved XMI format where links reified by association class instances can be directly expressed (addition of *XMI.reify* and *XMI.reifying* elements). In contrast with the standard XMI for UML, no model transformations are needed. The XMI+ textual notation proposes also some changes to improve readability and ease of use (unqualified names, string representations for data values, *uuid references* that are not attached to a XML document).

```
<Class xmi.uuid="u:/meta/classes/Constraint">
    <name>Constraint</name>
    <feature>
        <Attribute xmi.uuid="u:/meta/classes/Constraint.feature.body">
            <name>body</name>
            <type>
                <DataType xmi.uuidref="u:/meta/types/BooleanExpression"/>
            </type>
        </Attribute>
    </feature>
    <generalization>
        <Generalization xmi.uuid="u:/meta/classes/Constraint.generalization.ModelElement">
            <parent>
                <Class xmi.uuidref="u:/meta/classes/ModelElement"/>
            </parent>
        </Generalization>
    </generalization>
</Class>
```

Figure 4 - A *XMI+* rendering sample (with XML headings skipped)

6 Using the Repository Tool

There are at least two typical main uses for a Model Repository tool. The first concerns its capability to provide a centralized and uniform access to a big amount of data. The other typical usage concerns the capability to perform any kind of *added value* from a model stored in the repository : code generation, automatic test checks, etc.

We will briefly describe some of the applications developed in the France Telecom research center that use the repository tool.

6.1 A TMN Specifications Repository

Our research team was involved in the development of software support for TMN (Transport Management Network) specification technology. We have developed a WEB based repository for TMN specifications compliant to the ITU-T G.851-01 methodology [G.851]. This standard provides a top-down approach based on ODP concepts. It defines a set of templates for entering an *enterprise viewpoint* specification, an *information viewpoint* specification and a *computation viewpoint* specification.

The TMN repository has been implemented on top of the model repository tool. For end users, this is of course not visible. TMN specifications, after being translated into UML, have been registered in the model repository tool. An *enterprise* specification was transformed as a *use-case diagram*, while *class diagrams* and *state diagrams* were used to capture the *information* and *computation* specifications.

The benefits from having built the TMN repository upon a UML repository, were to allow a fine-grained control over the data. All added value computations, such as HTML rendering of the original templates or the search engine for TMN definitions, were achieved by using the navigation capabilities of our repository tool.

An ultimate advantage to use the UML metamodel as the internal data representation, is that the graphical counterpart can be easily derived from it (after importing the models into a graphical CASE tool). In fact, TMN specification teams seem to be more and more interested in using directly the UML notation. Efforts for defining a standardized *UML profile* [Profile] for this is under preparation [Cornily].

6.2 A UML to SDL Translator

The Specification Definition Language (SDL) is a formal language used for defining real-time systems. System behavior is basically described by using flow diagrams. The tool *Object Partner* from *Verilog* contains a powerful environment for simulating the behavior of a SDL system. We intended to derive a complete SDL program from a UML state diagram. To achieve this, it was necessary to constrain the usage of UML in order to avoid any non-formal statement (such as actions or conditions specified in terms of natural language).

We used the repository tool for storing the initial UML *state diagram*. The translator program, written in Java, used the UML tailored interface to traverse the state diagram and generate the SDL files.

7 Conclusions

The increasing acceptance of UML as a standard for OO modeling, reflects the fact that OO technology has reached a good degree of maturity. While agreement on the graphical notation is the basis for helping IT designers to better communicate with each other, agreement on the abstract syntax and the underlying semantics, will significantly help software tools to cooperate better.

As we have seen, the implementation of the model repository tool we developed is based on OO up-to-date modeling techniques. The way it has been designed provides flexibility for creating and storing models. Most of the principles we have used have their counterpart in the Meta Object Facility OMG standard. This includes, a meta-metamodel level support, the definition of a *reflective* and a *tailored* API for manipulating metaclass instances, the automatic production of the *tailored* API. The design differences are mostly due to the fact that we were more interested in being UML compliant rather than being MOF compliant. By the way, we have identified certain deficiencies in the MOF specification, that may be taken in consideration for its improvement.

203

References

[UML] UML Revision Task Force, OMG UML specification v.1.3, Object Management Group, June 1999

[MOF] Meta Object Facility Specification, Object Management Group, January 1997

[XMI] XML Metadata Interchange, Object Management Group, October 1998

[Profile] White Paper on the Profile Mechanism – Analysis and Design Task Force, Object Management Group, April 1999.

[G.851-01] ITU-T Recommendation G.851-01 : Management of Transport Networks – Application of the RM-ODP framework.

[Cornily] *"Specifying distributed object applications using the RM-ODP and the UML language"* – J.M.Cornily, M.Belaunde, EDOC'99, September 1999

[Python] Python language site at http://www.python.org/

[Java] Java language site at http://www.sun.com/java/

A web site presenting the repository tool is accessible at http://universalis.elibel.tm.fr/

Modeling Dynamic Software Components in UML

Axel Wienberg, Florian Matthes[1], and Marko Boger[2]

[1] Software Systems Institute (AB 4.02)
Technical University Hamburg-Harburg, Germany
http://www.sts.tu-harburg.de
[2] Distributed Systems Group, University of Hamburg, Germany
http://vsys-www.informatik.uni-hamburg.de

Abstract. UML provides modeling support for static software components through hierarchical packages. We describe a small extension of UML for modeling *dynamic* software components which can be instantiated at runtime, customized, made persistent, migrated and be aggregated to larger components. For example, this extension can be used to describe systems built with JavaBeans, ActiveX-Controls, Voyager Agents or CORBA Objects by Value. With our extension, the lifecycle of a dynamic software component can be expressed in terms of UML. We can not only describe a system at design time, but also monitor its runtime behaviour. A re-engineering tool is presented that exploits our UML extension for a high-level visualization of the interaction between dynamic components in an object-oriented system.

1 Introduction and Motivation

In academia and industry, the concept of a *software component* [Szy98,Gri98] as a self-contained, persistent, customizable and large-grain building block for (possibly distributed) application systems has attracted a lot of interest.

In our research work on orthogonally persistent and mobile object systems [MS94,MMS96,BWL99], we encountered the need to model and to visualize the state and the behavior of a (possibly distributed) system which consists of a large number of objects, some of which are aggregated to *dynamic software components*. These software components are dynamic, because they can be instantiated at runtime, customized, made persistent and be migrated within a dynamic component hierarchy. We successfully applied UML for this modeling task and we were also able to develop a visualization tool (a debugger extension) to monitor the state and the behavior of such systems by means of UML diagrams.

However, it turned out that the existing notations of UML for static components (e.g., packages, components in deployment diagrams) are not suited for this task and that we had to extend UML slightly by notations for component links, component boundaries (see Sec. 2.1 and Sec. 2.2) and for component aggregation (see Sec. 3.4) to smoothly integrate components into UML object, class, collaboration and sequence diagrams.

Our minimal UML extensions have been chosen deliberately to capture the common semantics of the growing number of industrial component models (e.g., JavaBeans, ActiveX-Controls, Voyager Agents, CORBA with Objects by Value) while leaving room for their differences: Components may or may not consist of further components internally (defining a hierarchic structure as in Java beans or a flat structure as in CORBA); they may be able to migrate as in Voyager or have a fixed location as in CORBA; components may or may not be first class objects, allowing parameter passing of components and substitution of objects for components; and components may be active (running their own thread of control) or passive (waiting for incoming messages), depending on the model used.

The common characteristics of these components are their ability to communicate by sending and receiving messages, the possibility of having multiple components of the same type (class) in a system, and the requirement that a component is the unit of co-location, i.e. that all objects of a component reside completely on one node. Finally, the state of a component consists of the attributes and objects aggregated by the component.

In Section 2, we first introduce the notion of component links and component boundaries to identify components and component hierarchies in object diagrams. The gain in modeling power through this extension in collaboration diagrams, sequence diagrams and class diagrams is discussed in Section 3. As a practical application, in Section 4 we present a re-engineering tool that visualizes component behaviour by monitoring component boundaries at runtime. Section 5 briefly discusses ways to identify component links in existing component systems. After relating our work to that of others, we conclude with a summary.

2 Modeling Component Configurations with UML

In this section, we introduce a notation for dynamic components. A dynamic component is a runtime entity based on objects, and is therefore shown in an object diagram. The configuration of a dynamic component includes its relation to other components, as well as the set and structure of objects internal to the component. Using our notation, these aspects can be expressed in UML.

2.1 Component Links

The fact that a component is made up of certain objects which implement its functionality implies an aggregation of the objects to a component. We shall model this specific form of aggregation using *component links*.

Definition *component link* A form of aggregation link between two objects, a *component* and a *subcomponent*, with the additional semantics that each object is an immediate subcomponent of at most one component. Therefore, the graph of objects and component links forms a forest. A component link may change over time and implies no dependence of lifetimes.

Notation In an object diagram, a component link is represented using the stereotype ≪component≫, displayed graphically as a link with a half-filled diamond on the side of the component. Fig. 1 gives an example.

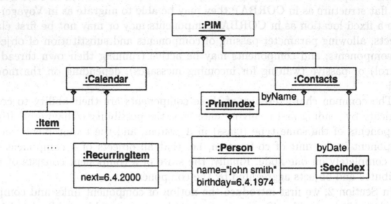

Fig. 1. A UML object diagram with component links

Fig. 1 shows the example of a personal information manager component (PIM) consisting of two independent subcomponents, a contacts database and a calendar, possibly implemented by different vendors. Internally, the contacts database stores its person records in a primary index by name, so that the person object is a subcomponent of the primary index. The secondary index only holds an association link to the person object.

2.2 Component Boundaries

Using component links emphasizes the relation between a component and its immediate subcomponent. However, we also want to show which subcomponents belong together to implement a given component.

This grouping of peer objects cannot be denoted easily through links. Instead, we introduce a slight notational extension, analogous to sytem boundaries in use case diagrams.

Definition *component boundary* A graphical element enclosing exactly the transitive subcomponents of a component.

Notation A dotted boundary with rounded corners is drawn around the set of subcomponents. The object representing the component itself (the *primary object*) is positioned on the boundary, to show its role as the primary interface of the component. Fig. 2 augments the personal information manager example from Fig. 1 with component boundaries.

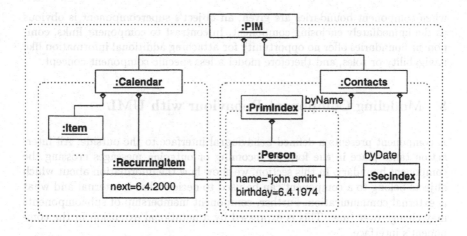

Fig. 2. A UML object diagram with component boundaries

Since the subcomponents of a component may in turn rely on internal subcomponents to implement their functionality, we naturally gain a hierarchical decomposition. This nesting of component boundaries can be used to depict logical as well as physical object spaces. It offers a unified notation for objects within arbitrary levels of components within nodes [Sto97], which in turn might be aggregated to local networks and to administrative domains [CG98].

The hierarchical structure can also be exploited to reduce the level of detail in an object diagram. By collapsing a component bubble and keeping the primary object, the overall structure is preserved, but irrelevant detail (internal to the component) is omitted. Fig. 3 gives a coarser view on our example, which was generated systematically from the detailed view.

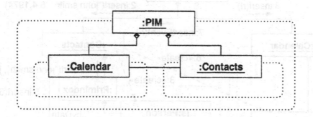

Fig. 3. A coarser view on the PIM.

Component links and component boundaries are essentially two views on the same concept, namely the grouping of objects to components. When a component link is given, the subcomponent will be inside the component's boundaries;

when component boundaries are given, an object's supercomponent is obvious as the immediately enclosing component. In contrast to component links, component boundaries offer no opportunity for attaching additional information like navigability or roles, and therefore model a less specific component concept.

3 Modeling Component Behaviour with UML

A component presents a defined behavioural interface to the outside. All interaction takes place in the form of incoming or outgoing messages crossing the component boundary. In this section, we show how the information about which objects belong to a component can be used to decide what is internal and what is external communication. Further, component membership of subcomponents may change over time. This is expressed by a notational extension in the component's interface.

UML offers two kinds of diagrams for interaction: collaboration and interaction diagrams. We will begin by investigating the influence of components on the former.

3.1 Components in Collaboration Diagrams

So far, we have described component boundaries in object diagrams. Collaboration diagrams are basically object diagrams overlaid with message flow information, so component boundaries can be drawn in a collaboration diagram as well.

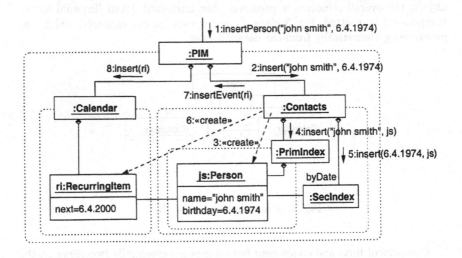

Fig. 4. A collaboration diagram including component boundaries.

When a new person is registered in our personal information manager example, some interactions take place in order to update the indices and to register the person's birthday in the calendar. Fig. 4 gives the whole detailed sequence. When the personal information manager (PIM) receives an insertPerson message (1), it delegates the message to the contacts database (2). There, a new person record is created (3, depicted using the ≪create≫ stereotype [RJB98]) and inserted into the primary index under the person's name (4). A link to the record is also stored in the secondary index (5). The contacts database then creates a calendar item for the person's birthday (6) and informs the personal information manager about it (7). The personal information manager inserts this calendar item into the calendar (8), finally resulting in the state already shown in Fig. 2.

The coarsening that has taken place in the object diagram from Fig. 2 to Fig. 3 can be applied to this collaboration diagram as well, stripping some of the distracting details. When the sender and the receiver of a message are members of the same component, and this component has been collapsed, the message will be abstracted from, and will not be shown. Fig. 5 gives the resulting diagram.

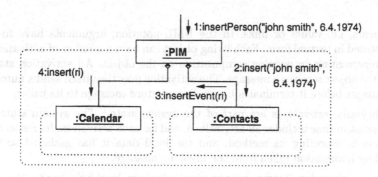

Fig. 5. A high-level collaboration diagram created from Fig 4.

3.2 Components in Sequence Diagrams

In this section, we shall briefly describe sequence diagrams with an emphasis on the concept of activation (which we will need in the next section), and will then consider the influence of components on sequence diagrams.

Sequence diagrams show the interaction of a number of objects, distributed horizontally, over time. Time flows from top to bottom. We will only consider the instance form here, which describes one actual sequence without branches, conditions or repetition. Fig. 6 gives an example.

At each point in time (represented by a horizontal cut), a number of living objects are represented by lifelines. Interactions, such as message send and return message, are depicted as horizontal arrows. Each kind of message can carry

210

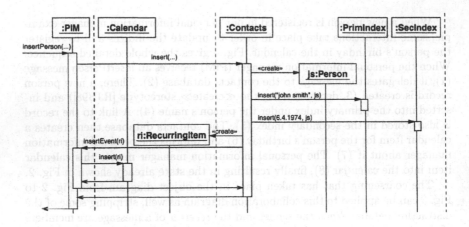

Fig. 6. The sequence diagram corresponding to Fig. 4.

arguments, i.e. values or links. In the UML notation, arguments have to be represented in textual form. Each living object can have a number of activations, each representing an ongoing computation on the object. An activation starts when the object receives a message. The activation may then itself send a number of messages before it terminates by sending a return message to its caller.

Obviously, activations are part of the system state.[1] The system state at some point in time includes all activations, and for each activation, its respective progress in executing its method, and the local data it has gathered so far, including information taken from the invoking message.

In a collaboration diagram or in an object diagram, local links can be depicted using the «local» stereotype [RJB98], but the activation itself cannot be seen: the link originates at the active object to which the activation belongs. In a sequence diagram, activations are visible, but links are not.

We introduce a component boundary notation for sequence diagrams, too: A vertical dotted line, similar to the swimlanes used in UML activity diagrams, separates objects of different components. Messages crossing the boundary are easily recognized. However, since objects only have a one-dimensional (horizontal) position, drawing nested components is awkward.

A high-level sequence diagram is created by subsuming the lifelines of all objects inside a component under that of the primary object, as shown in Fig. 7. Internal messages are abstracted from, as well as internal activations. Note that the resulting interaction is the same as that depicted in Fig. 5.

---

[1] The designers of the programming language BETA even went so far as to unify the concepts of object and activation. We do not follow that trail here.

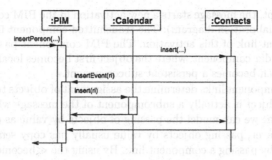

Fig. 7. A high-level sequence diagram created from figure 6.

3.3 The Lifecycle of a Component

In the example we have shown, a subcomponent (the recurring birthday item) is created in one component (the contacts database), and then migrated to another component, where it is made persistent by storing it in a database (the calendar). At a later point in time, e.g. when the person is removed from the contacts database, the birthday item will be deleted from the calendar, and the lifetime of this subcomponent ends. This is what we call the lifecycle of a component.

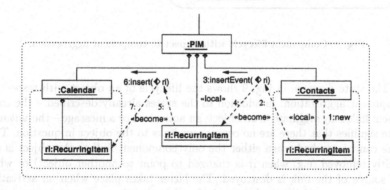

Fig. 8. Migration and persistence of a subcomponent.

Fig. 8 shows this process in greater detail. After creation of the birthday object, the creator (the contacts database) holds a local component link to the created object. The created object is then sent off in a message to the PIM component. As we have mentioned, links in UML messages have to be represented in textual form. In order to indicate that the message carries a component instead of an object reference, we have added a half-filled diamond before the object name. The symbol can be transcribed as the keyword component.

Upon receipt, the message starts a new activation in the PIM component (not visible in a collaboration diagram). The transmitted component link becomes a local component link of this activation. The PIM component passes the object on to the calendar component, where the object first becomes locally bound (not shown) and then becomes a persistent subcomponent.

Because component links determine the assignment of objects to components, the birthday object is actually a subcomponent of the message while in transit. This means that we can model the passing of objects by value as e.g. in RMI or CORBA. However, passing objects by value usually has copy semantics, which is not implied by passing a component link. By using the ≪become≫ stereotype [RJB98], we state that the object's identity is maintained across the migration. The ≪become≫ arrows indicate that the different birthday objects shown in the diagram are actually versions of the same object at different points in time.

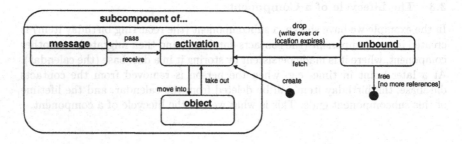

Fig. 9. Component lifecycle with respect to component aggregation.

The state diagram in Fig. 9 shows the lifecycle of an object with respect to component aggregation. In addition to the states already described – the component link may originate in an object, an activation or a message – the *unbound* state signifies that there are no component links to the object in question. This state can be reached when either the only component link to the object is explicitly removed (e.g. when it is changed to point to another object) or when the source of the link is destroyed. The latter takes place when an activation ends or when an object is destroyed. An object's subcomponents are not necessarily destroyed along with the component; the subcomponents may continue an individual existence even after the bubble has burst.

3.4 Components in Class Diagrams

A component as a group of collaborating objects is a concept at the object level, not at the class level. Therefore, dynamic components are not visible directly in a class diagram; especially, it makes no sense to draw component boundaries in a class diagram. For grouping classes, UML provides the concept of packages.

However, there are two impacts of components on class diagrams, as can be seen in Fig. 10. Firstly, an aggregation between two classes can use the ≪component≫ stereotype, turning all instances of this component aggregation into component links. As in the general case of aggregation, the component aggregation in the class diagram may include recursion in order to describe hierarchical structures. An example are the Java AWT user interface beans, where a Container is a kind of Component that may include further instances of the class Component, so there is a cycle between Component and Container. Of course, every concrete user interface hierarchy only has finite depth, so there are no cycles in the object diagram.

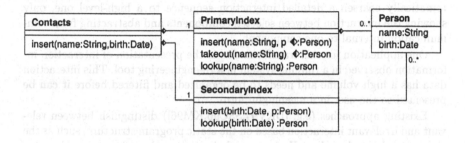

Fig. 10. A class diagram describing components (see text).

Secondly, method parameters and results can be labeled with a half-filled diamond (transcribed as the keyword component) to indicate that component links are expected or returned. When a method takes a component parameter, the object supplied in the actual message is migrated to the message's receiver. Using this notation, the interface of a database can state whether the database will store the object itself or only a link to the object. In Fig. 10, the primary index aggregates the person objects, while the secondary index is only associated with them. So the insert operation of the primary index requires a component link to the person, while the secondary index takes a non-component link.[2]

As a second example, an object factory can state whether it passes the responsibility for its created objects to the client, or whether it manages the set of created objects itself and only passes out association links.

When component boundaries are interpreted as denoting physical location (similar to a deployment diagram), the interface specifies which arguments are to be migrated to the receiver's node, and the component links in those arguments specify which other objects are to be moved along with the primary objects. If a

---

[2] The question of interface compatibility, e.g. whether there may be a common superclass for both primary and secondary index, and the question of parameterization, i.e. whether both classes may be instantiations of the same template, is discussed further in [Wie99] in the context of a strongly typed programming language.

migrated component contains activations, this models a variant of the migrating threads described in [MMS96].

When the component boundary is interpreted as the border between volatile and persistent storage, the annotations state which objects are to be made persistent together. Again, the model covers persistence of active components, corresponding to persistent threads [MS94].

4 Visualizing Component Behaviour at Runtime

As we have shown, the dynamic component structure can be employed to automatically coarsen a detailed interaction sequence to a high-level one, only showing the interaction between selected components and abstracting from communication internal to one component.

One application for this transformation is the presentation of interaction information observed in a running system in a re-engineering tool. This interaction data has a high volume and needs to be organized and filtered before it can be presented to the user in a meaningful form.

Existing approaches (e.g. [DKV94,SSC96,KM96]) distinguish between relevant and irrelevant interaction based on the static program structure, such as the class of sender and receiver. However, we believe that when observing a dynamic phenomenon like interaction, the dynamic system configuration has to be taken into account. For example, using static information only, it becomes impossible to observe the interaction between different complex instances of the same static component, unless the observer reverts to individual objects.

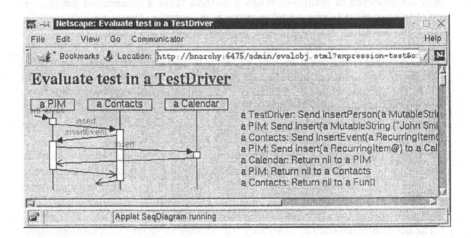

Fig. 11. A high-level sequence diagram corresponding to Fig. 7, automatically created by tracing an annotated program.

By exploiting the component structure, the visualizations produced by the re-engineering tool come closer to the abstract design-phase model, which facilitates understanding as well as validation [LN95]. The semantic gap between design and implementation is narrowed, and architectural properties can be observed in the running system.

For example, due to the hierarchic component structure, it becomes possible to distinguish between up-calls and down-calls. In the personal information manager example in Fig. 5, the PIM performs a down-call towards the contacts databases, which calls back up via an insert event in order to install the birthday item in the calendar. The PIM then does a down-call to the calendar on behalf of the contacts database. Clearly, the PIM functions as a mediator.

In our approach, when examining a running program, the user has to specify the dynamic components whose boundaries [s]he wishes to monitor. This is achieved by *marking* their primary objects. Monitored components need not be disjoint, i.e. a marked object may be a transitive subcomponent of another marked object. The objects to be marked could be specified by an arbitrary predicate, e.g. "all instances of class A", or can be selected individually.

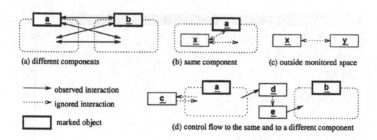

Fig. 12. Types of interaction relative to component boundaries.

Our tracing tool examines all interaction between different objects. For the source as well as the destination of an interaction, the innermost enclosing monitored component is determined. If the objects belong to different monitored components, then we have an interaction crossing both component boundaries (Fig. 12a). This interaction will be presented to the user.

If source and destination belong to the same monitored component (Fig. 12b), or if both are outside of monitored space (Fig. 12c), the interaction is completely ignored.

If only the source object belongs to a monitored component, this means that the action leaves monitored object space. Control may flow through several unmonitored objects before it re-enters a monitored component. The leaving and entering messages are only reported if control flows between different monitored components, not if it re-enters the same component (Fig. 12d).

We have implemented this filtering strategy in a research prototype described in [Wie99]. The sequence diagram in Fig. 11 was created automatically by this tool, from running an annotated program. Individual objects sending and receiving messages are displayed in the textual listing on the right hand side of the window; the sequence diagram itself only includes lifelines for the monitored components.

Besides visualizing the components' behaviour, the research prototype also allows browsing the components' structure, as in Fig. 13. For now, the component hierarchy is displayed as a simple *explorer*-style hierarchy, but we hope to employ the component boundary notation in a future version. Note the difference between reference links (shown as arrows) and component links (shown as expandable folders) in the graphic; for example, the contacts database has a reference link to the enclosing personal information manager (for sending upcalls), but it cannot hold a component link, because this would consitute a cycle. An important consequence of the acyclic component graph is that the object structures displayed by the tool always have finite depth and show each object at most once.

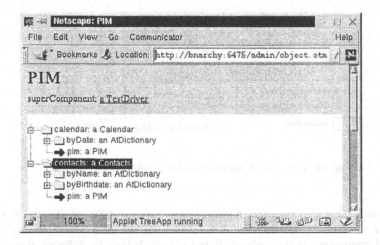

Fig. 13. Exploiting component information for object store browsing.

5 Identifying Component Links in an Implementation

In our research prototype, the observed program is prepared by annotating which variables (instance variables, local variables, parameters and method results) hold component-links and which variables hold non-component links. The component structure is deduced from this information.

If a modification of the examined program is not feasible, an existing distinction between different kinds of links can be mapped onto component and non-component links.

For example, distributed programming languages like Emerald [Jul89] and DOWL [Ach93] offer primitives for specifying a *migration group*, i.e. a set of objects to be migrated together. A variable attribute is used to determine which references are to be followed when computing the migration group. Such an object group can be interpreted as one component; the variable attribute then designates component links. In a similar manner, Java uses the volatile attribute to specify the boundaries of a group of objects to be made persistent together.

In Voyager, remote references can be interpreted as non-component links, and local Java references as component links. The disadvantage of this approach is that local references allow sharing between different components, precluding a clear assignment of objects to components for purposes of visualization, migration, persistence etc. This weakness is fixed in the Dejay system [BWL99] by grouping objects in *virtual processors*.

Many programming languages include the UML concept of composition (e.g. as member objects in C++, expanded objects in Eiffel [Mey97] or static links in BETA [MMPN93]). Due to the inflexibility of composition however, the object groups are usually quite small, and component membership cannot change over time.

If component links are not distinguished in the program source, an object can still be assigned to the component that created it. This creates a relation with the required formal properties. However, migration is not expressible.

As a last possibility, component membership can be assigned manually inside the re-engineering tool. When manipulating a sequence diagram, a command like "merge the lifelines of these objects" could be used. Such a command could easily be integrated in existing visualization tools, and would complement operations like navigation along the call graph [KM96]. The obvious disadvantage is that the grouping has to be specified each time the tool is used, instead of specifying it once and for all during design and implementation.

6 Related Work

Our concept of dynamic components is a generalization of the concepts of component instances and of nodes in UML. We have chosen the term *dynamic* component to stress their twofold dynamic nature: Firstly, they reside on the level of object diagrams as opposed to class diagrams; and secondly, they can migrate, so that the component structure itself is dynamic. Dynamic components also differ from component instances in that their primary object is treated like any other object, i.e. it is treated as first-class. Primary objects could be differentiated from other objects using a stereotype, but only if required. The same is true for nodes, which become another stereotype of a dynamic component, with the additional semantics of a fixed, distinct location.

Civello [Civ93] talks about different kinds of aggregation. He lists several orthogonal properties used to specify an aggregation more exactly. In his terms, our concept of component aggregation is only restricted in that it excludes sharing; apart from that, any aggregation may be labeled as a component aggregation. Especially, Civello says that "it must be possible to model the migration of objects from one composite to another", which is possible in our model.

The importance of object configurations as opposed to static relations is stressed in [GL96], where the concept of environmental acquisition is presented. In the model by Gil and Lorenz, objects acquire properties from their ancestors in a hierarchical aggregation structure similar to the component structure described here. Gil and Lorenz also propose programming language mechanisms for distinguishing between component and non-component links (there called aggregation and nonaggregation links).

The hierarchical grouping of runtime objects to larger units has also been investigated in a number of papers dealing with aliassing in object-oriented programming languages, the latest of which is [NVP98]. The aim of that paper is to enhance encapsulation, which has not been considered in our work. However, the linguistic mechanisms developed in [NVP98] are also applicable for designating component links.

7 Summary

In this paper, we have introduced a small UML extension for denoting nested component boundaries in object and interaction diagrams, and have demonstrated its usefulness for systematically creating high-level diagrams of a component system. The component boundaries are defined based on component links, a form of aggregation between a component and its internal objects.

Secondly, we have shown how migration of components across component boundaries can be specified through an extension of the UML method signature notation.

Thirdly, a re-engineering tool has been presented that automatically creates meaningful sequence diagrams from a program, based on the dynamic component structure of the object graph.

Acknowledgements

This work has been supported in part by the project ISC-CAN-080 CIS of the European Communities.

References

[Ach93]   Bruno Achauer. The DOWL distributed object-oriented language. *Communications of the ACM*, 36(9):48–55, 1993.

[BWL99]   Marko Boger, Frank Wienberg, and W. Lamersdorf. Dejay: Unifying concurrency and distribution to achieve a distributed Java. In *Proceedings of TOOLS Europe '99*, Nancy, France, June 1999. Prentice Hall.

[CG98]   Luca Cardelli and Andrew D. Gordon. Mobile ambients. In *Proceedings of FoSSaCS '98*, volume 1378 of *Lecture Notes in Computer Science*, pages 140–155. Springer-Verlag, March 1998.

[Civ93]   F. Civello. Roles for composite objects in object-oriented analysis and design. In *Proceedings of OOPSLA '93*, pages 376–393, San Jose, California, October 1993.

[DKV94]   Wim DePauw, Doug Kimelman, and John Vlissides. Modeling object-oriented program execution. In *Proceedings of ECOOP '94*, pages 163–182, Bologna, Italy, July 1994. Springer-Verlag.

[GL96]   Joseph Gil and David H. Lorenz. Environmental acquisition – a new inheritance-like abstraction mechanism. In OOPSLA [OOP96], pages 214–231.

[Gri98]   Frank Griffel. *Componentware: Konzepte und Techniken eines Softwareparadigmas*. dpunkt-Verlag, Heidelberg, 1998.

[Jul89]   E. Jul. *Object Mobility in a Distributed Object-Oriented System*. PhD thesis, Department of Computer Science, University of Washington, 1989.

[KM96]   Kai Koskimies and Hanspeter Mössenböck. Scene: Using scenario diagrams and active text for illustrating object-oriented programs. In *International Conference on Software Engineering (ICSE '96)*, Berlin, 1996.

[LN95]   D. B. Lange and Y. Nakamura. Interactive visualization of design patterns can help in framework understanding. In *Proceedings of OOPSLA '95*, pages 342–357, Austin, Texas, USA, October 1995. ACM.

[Mey97]   Bertrand Meyer. *Object-oriented Software Construction, 2nd edition*. Prentice Hall, 1997.

[MMPN93]   Ole Lehrmann Madsen, Birger Møller-Pedersen, and Kristen Nygaard. *Object-Oriented Programming in the BETA Programming Language*. Addison-Wesley, 1993.

[MMS96]   B. Mathiske, F. Matthes, and J.W. Schmidt. On migrating threads. *Journal of Intelligent Information Systems*, 8(2), 1996.

[MS94]   F. Matthes and J.W. Schmidt. Persistent threads. In *Proceedings of the Twentieth International Conference on Very Large Data Bases, VLDB*, pages 403–414, Santiago, Chile, September 1994.

[NVP98]   James Noble, Jan Vitek, and John Potter. Flexible alias protection. In *Proceedings of ECOOP '98*, number 1445 in Lecture Notes in Computer Science, pages 158–185, Brussels, Belgium, July 1998. Springer-Verlag.

[OOP96]   ACM. *Proceedings of OOPSLA '96*, San Jose, California, October 1996.

[RJB98]   James Rumbaugh, Ivar Jacobson, and Grady Booch. *The Unified Modeling Language Reference Manual*. Addison-Wesley object technology series. Addison Wesley Longman, December 1998.

[SSC96]   Mohlalefi Sefika, Aamod Sane, and Roy H. Campbell. Architecture-oriented visualization. In OOPSLA [OOP96], pages 389–405.

[Sto97]   David Petrie Stoutamire. *Portable, Modular Expression of Locality*. PhD thesis, University of California at Berkeley, December 1997.

[Szy98]   Clemens Szyperski. *Component Software: Beyond Object-Oriented Programming*. Addison-Wesley, 1998.

[Wie99]   Axel Wienberg. Dynamic components in an object-oriented programming language - model, language implementation and visualization. Diploma thesis, computer science department, University of Hamburg, Germany, March 1999. *in German*.

Extending UML for
Modeling Reflective Software Components

Junichi Suzuki and Yoshikazu Yamamoto

Department of Computer Science,
Graduate School of Science and Technology,
Keio University
Yokohama City, 223-8522, Japan.
+81-45-563-3925 (Phone and FAX)
{suzuki, yama}@yy.cs.keio.ac.jp
http://www.yy.cs.keio.ac.jp/~suzuki/project/uxf/

Abstract. This paper describes our extension of the UML metamodel for specifying reflective software components. Reflection is a design principle that allows a system to have a representation of itself in the manner that makes it easy to adapt the system to a changing environment. It has matured to the point where it is used to address real-world problems in various areas. We describe how to document reflective components in the framework of UML. Our work allows for recognizing and understanding reflective components in the upper levels of abstraction at an earlier stage of the development process. It leverages the documentation, learning, visual modeling, reuse and roundtrip development of metalevel designs. We also demonstrate the seamless model exchange between different development tools and model continuity across development phases with application-neutral interchange formats.

1 Introduction

The Unified Modeling Language (UML) [1] has been widely accepted as a standard modeling language in the software engineering community. It defines semantics and their notations of model elements required for documenting object oriented software. It is a single and universal language that can be used with any design methodology. UML provides nine diagrams with fine level of abstraction to specify object models for a given problem. Complex systems can be modeled through a small set of nearly independent diagrams. UML has been used for representing a variety of software models such as real-time systems, hypermedia, business processes, engineering design and multiagent models.

This paper describes an extension of the UML metamodel for supporting reflective software components, which allows software to be highly configurable and extensible, and how to specify them with UML. Supporting reflective components in the design level means that we can recognize them in the upper levels of abstraction at an earlier stage of the development process. We also address a description language based on XML (eXtensible Markup Language) for describing reflective components as textual representations. It increases the model continuity across development phases in more precise manner and provides for the interchangeability of reflective model information between different development tools.

The remainder of this paper is organized as follows. Section 2 overviews reflective software and its general constructs; it also describes the benefits of capturing reflective components in the design phase and expressing them with XML. Section 3 describes our extension to the UML metamodel and shows some examples. Section 4 presents some applications using UXF (UML eXchange Format), our XML-based model description language, and XMI (XML Metamodel Interchange) format. We conclude with our current project status and future work in Sections 5 and 6.

2 Reflective Components in the Design Phase

This section overviews Aspect-Oriented Programming. It then considers some aspects of reflective software and describes our motivation to capture its components in UML. We also describe an interchange format for reflective components.

2.1 Aspect-Oriented Programming: Separation of Concerns

Today's software is becoming more and more complex, and has to deal with an greater variety of computing concerns simultaneously, e.g. concurrency, object interaction, persistence, distribution, fault tolerance and realtime constraints. The notion of *separation of concerns* is proposed for managing software complexity well and improving its quality by separating different concerns and introducing clear and minimal dependencies between them at both the conceptual and implementation levels [2].

Aspect-oriented Programming (AOP) is a paradigm for facilitating separation of concerns, which has been proposed by a research group of the Xerox PARC [3]. AOP introduces a new unit of software modularity, called an *aspect*, which represents a computing concern described above. Each aspect provides a better handle for managing *cross-cutting* problems. Cross-cutting is a problem found in many object-oriented software systems; some features of a system, i.e. aspects in the sense of AOP, tend to affect or require the collaboration of groups of objects. They are naturally spread within a whole system and cross-cut the primary decomposition of objects. Typically, non-functional concerns make it difficult to understand and evolve the system.

Aspects are handled to fulfill certain application requirements (e.g. persistence, distribution, real-time and fault tolerance) or manage and optimize underlying computational algorithms (e.g. concurrency and object interaction). Isolating aspects allows them to be:

- abstracted to a higher level,
- easier to understand because an aspect's code is not cluttered with the code of other aspects, and

– coupled loosely with each other; thereby the flexibility and reusability of an aspect is increased.

The process of combining an aspect with other aspects or other portions of a system is called *aspect weaving*. A tool for weaving aspects is called *aspect weaver*. An aspect is expressed by encoding the aspect support as a conventional library, designing a separate language for the aspect, or designing a language extension for the aspect [4]. Aspect weaving is performed by source code transformation, component composition or reflection [4]. For example, AspectJ [5], an aspect weaver extending Java, represents an aspect by designing a language extension and weaves aspects by source code transformation. AOP/ST [6], an aspect weaver extending Smalltalk, represents an aspect by designing a separate language for the aspect and weaves aspects by reflection.

2.2 Reflection

We are using the reflection mechanism to separate aspects and keep them loosely coupled. Alternative approaches are adaptive programming [7] and component composition mechanisms [8,9].

Reflection is a design principle that allows a system to have an explicit representation of itself in a manner that makes it easy to adapt the system to a changing environment [10]. It was originally introduced by 3-Lisp [11]. After that, it has been studied within various programming languages such as CLOS [12], Smalltalk [13], C/C++ [14] and Java [15], in order to extend the language syntax and semantics by providing language constructs as self-representations [12]. Recently, it has been applied to more generic system designs such as databases, concurrent/parallel computing, operating systems, virtual machines, distributed computing, security and agent-based intelligent systems. Reflection has matured to the point where it is used to address real-world problems. In fact, it is identified as a pattern of software architecture (POSA) [16].

In object-oriented systems, the base unit of computation is object (or *baseobject*). Through the interaction among objects, a system computes a certain task. Reflection introduces the notion of object/metaobject separation. A *metaobject* (or *metalevel object*) is an object that contains information about the internal structure and/or behavior of one or more baseobject. In other words, metaobjects can track and control certain aspects (i.e. structure and/or behavior) of baseobjects. A set of metaobjects is called a *metalevel*, and a set of baseobjects is called a *baselevel*.

Reflection is the ability of a program to manipulate as data something that represents the state of the program [17], and to adjust to changing requirements. The goal of reflection is to allow a baseobject to reflect on its own execution state and eventually alter it to change its meaning. In contrast to reflection, reification [18] is the process of making something accessible that is normally unavailable in the baselevel or is hidden from the programmer. For the execution of a baseobject to be supervised, it must first be reified into the corresponding

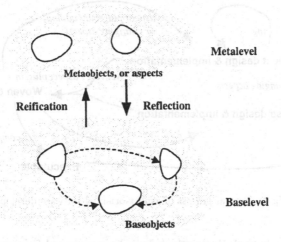

Fig. 1. Typical reflective architecture

metalevel. A set of interfaces through which a baseobject interacts with its metalevel is called *Metaobject Protocols* or *MOPs* [12]. The relationship among the constructs described above is illustrated in Figure 1.

In general, a metaobject protocol establishes the following interactions [19]: (1) attachment of baselevel and metalevel objects, that can be static or dynamic, and in one-to-one or many metaobjects to one baseobject basis; (2) reification of structural and/or behavioral features within a baseobject; (3) execution, which consists of metalevel computation that interferes with the baselevel behavior transparently through the interception and reification mechanisms; (4) modification, which is the capability of the metaobjects of changing behavior and structure of baseobjects.

There are various approaches to achieve reflection, and they can be classified into: compile-time and runtime reflection [14], structural and behavioral reflection [13], and introspection and intercession [20].

In our work, a metaobject is considered as an entity representing an aspect in the sense of AOP. The separation of concerns is performed by separating metaobjects from baseobjects and keeping metaobjects isolated. Aspects and metaobjects are called reflective components.

2.3 Benefits of Capturing Reflective Components in the Design Phase

Reflective components can be identified at the design and implementation phases, though the cross-cutting problem tends to occur at the implementation/coding phase [4]. When a reflective component is identified, or emergent, at the implementation phase, developers often add it to a system and change existing components manually (see Figure 2). Reflective components are maintained only at

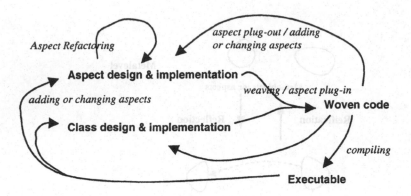

Fig. 2. Typical process of aspect-oriented software development

the source code level. Few methods have been proposed for expressing them at the design level. Supporting them at the design phase streamlines the process of reflective system development (Figure 2) and increases the productivity of metalevel designers by facilitating:

- Documentation and Learning
  Supporting reflective components as design constructs allows developers to recognize them in the upper level of abstraction at an earlier stage of the development process. Metalevel designers can easily document and understand reflective system design.
- Visual modeling
  Metalevel designers can understand reflective components in more intuitive way by visualizing them using CASE tools or model viewers.
- Reuse of metalevel design
  The above characteristics leverage the reuse of the metalevel design. A promising example is pattern catalogs that collect well-known and feasible design of reflective components.
- Roundtrip development
  The incremental and roundtrip development of reflective software is possible with the metalevel source code to design model translation, model to metalevel source code translation and metalevel refactoring (see also Figure 2).

2.4 Benefits of Describing and Interchanging Metalevel Design Models with XML

In the software design phase, model interchange is a very important capability, because there are few application-neutral model exchange formats between development tools. To solve this problem, we developed the UML eXchange Format (UXF) [21], which is similar in some respects to the XML Metamodel

Interchange (XMI) format [1], standardized by the Object Management Group. Both UXF and XMI are based on XML and allow model semantics to be described explicitly and transferred precisely. Such application-neutral interchange format facilitates:

- Interchangeability and reuse of metalevel descriptions:
  Software models change dynamically in the analysis, design, implementation and maintenance phases. Software tools used in each phase usually employ their own proprietary formats to describe model information. For example, current aspect weavers use their own language to describe aspects. An application-neutral format allows aspect information to be interchangeable and reusable between a wide range of different development tools with different strengths, throughout the lifecycle of software development (see also Figure 2). This seamless tool interoperability increases our productivity for designing the metalevel.
- Intercommunication between metalevel designers:
  An application-neutral metalevel description format serves as a communication vehicle for designers. They can communicate their modeling insights, understandings and intentions on a metalevel design with each other. This capability simplifies the circulation of aspect models between aspect designers.

We use UXF basically to describe metalevel design information, because it preceded XMI at the time when our project began. However, our project is slowly migrating to use XMI by providing a UXF-XMI converter and extending XMI. Note that due to space limitations, the basics and benefits of using XML as an interchange format are not covered here. Please see [?] for a more in-depth discussion.

3 Modeling Reflective Components with UML

This section describes our extensions to the UML metamodel for supporting reflective components and how to specify them with UML.

3.1 Aspects

The semantics of an aspect is defined as a UML metamodel element derived from Classifier, which describes behavioral and structural features [1]. As shown in the left of Figure 3, Classifier is a parent element of Aspect, Class, Interface, Node and Component.

An aspect can have a set of attributes, operations and relationships because a classifier has them. An operation of an aspect is considered as the aspect's weave declaration. A weave is a program that centralizes the code affecting (crosscutting) diverse portions in a system. An attribute of an aspect is used by one or more weaves. Relationships of an aspect include generalization, association and

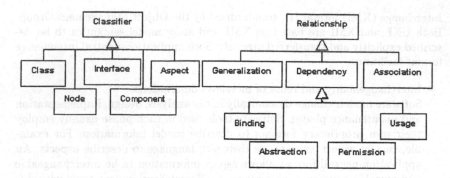

Fig. 3. Aspect as a metamodel element derived from Classifier (left), and some kinds of the Relationship metamodel element(right)

dependency (see Figure 3). If an aspect language and weaver supports multiple kinds of weaves, e.g. introduction and advice weaves in AspectJ [5], they are specified as stereotypes corresponding to their kinds.

The notation of an aspect is a class rectangle with stereotype «aspect» as shown in Figure 4. In this figure, the Singleton aspect represents the Singleton design pattern. Because the aspect encapsulates the instantiation and instance management policies, it is possible for a class to vary the policies. The operation list compartment of the rectangle means the list of weave declarations. Each weave is displayed as an operation with the stereotype «weave». A signature of a weave declaration shows its designator, specifying which model elements (e.g. classes, methods and variables) are affected by the weave.

The Singleton aspect in Figure 4 is defined based on the AspectJ language. It has two introduction weave declarations specified by the stereotype «introduction weave». Singleton introduces a static method GetInstance() and a private constructor in SingletonClass. A static attribute Instance, which is stereotyped with «introduction» is also introduced in SingletonClass.

3.2 Aspect-Class Relationship

UML defines three primary relationships derived from the Relationship metamodel element: Association, Generalization and Dependency (Figure 3). The relationship between an aspect and the classes affected by the aspect is a kind of dependency, because the behavior of a class is constrained by an aspect. The dependency relationship states that the implementation or functioning of one or more elements requires the presence of one or more other elements [1]. The derived metamodel elements of Dependency are Abstraction, Binding, Permission and Usage (Figure 3). The aspect-class relationship is classified as a kind of the abstraction dependency. The abstraction dependency relates two elements that are the same concept at different levels of abstraction or from different viewpoints [1]. UML defines three stereotypes for the abstraction de-

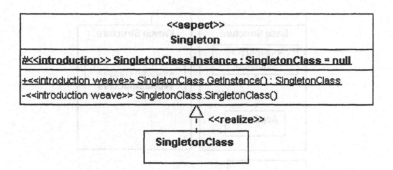

Fig. 4. A simple aspect example. The `Singleton` aspect reifies the implementation policies of the Singleton design pattern.

pendency: *derivation, realization, refinement* and *trace*. The aspect-class relationship is best-suited to the abstraction dependency with the stereotype realization, ≪realize≫. A realization is a relationship between a specification model element and a model element that implements it. The implementation model element is required to support the declaration of an specification model element.

UML defines the notation for an abstraction dependency with the ≪realize≫ stereotype as a dashed generalization arrow. Figure 4 shows a single aspect-class relationship between `Singleton` and `SingletonClass`.

3.3 Woven Class

Aspect and class code are combined using an aspect weaver, and then a woven class is generated (Figure 2). The woven class structure depends on the aspect weaver and programming language used. For example, AspectJ replaces an original class with the generated woven class. AOP/ST generates a woven class derived from the original class [6]. There are other alternative composition strategies, e.g. composition using the Mediator or Decorator design patterns [22].

We introduced the stereotype ≪woven class≫ into the `Class` element in order to represent a woven class. Figure 5 shows two different woven class structures using different aspect weavers, AspectJ and AOP/ST. In general, application developers may not use this stereotype because they do not have to recognize how a woven class is structured. However, since aspect or metalevel designers often need to know the implementation details of weavers, they can use it especially when they remove an aspect and modify the aspect and/or classes to debug them (see also Figure 2).

3.4 Example: Reifying the Observer design pattern in the metalevel

Figure 6 shows an example of an aspect-oriented variant of the Observer design pattern [22] based on the AspectJ language. Reifying design pattern con-

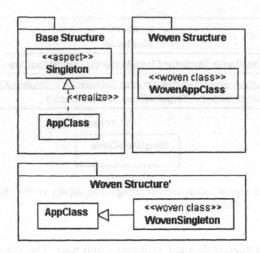

Fig. 5. An aspect-class structure and two different woven class structures using AspectJ (upper) and AOP/ST (bottom).

structs to the metalevel allows for centralizing a variety of implementation policies in a pattern, as proposed in several places [15, 23, 24]. In Figure 6, the aspect SubjectObserverProtocol reifies the behavior of the pattern to define the behavior between the class Subject and Observer. SubjectObserverProtocol can be implemented independently of Subject and Observer because it localizes a policy of the protocol implementation involving several objects in a single aspect rather than spreading multiple code fragments throughout them.

SubjectObserverProtocol defines seven introduction weaves and two attributes. The following definitions:

```
#<<introduction>> Observer.subject: Subject=null
+<<introduction weave>> Subject.notify(arg: Object):void
```

are mapped to the AspectJ aspect definition below:

```
aspect SubjectObserverProtocol{
  introduction Observer{
    protected Subject subject = null;
    ...
  }
  introduction Subject{
    public void notify(Object arg){
      ...
    }
    ...
}
```

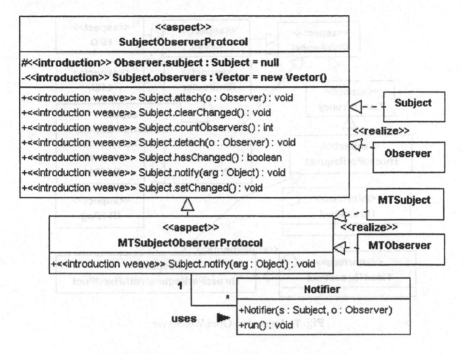

Fig. 6. Aspects reifying the Observer design pattern.

SubjectObserverProtocol has a sub-aspect, MTSubjectObserverProtocol, which is a thread-safe aspect. The implementation policy of a design pattern can be modified by extending an aspect. MTSubjectObserverProtocol is used when a subject and observers are executed on different threads. It is designed to avoid a potential deadlock caused when a subject issues a change notification in a thread while an observer is trying to check the observable's instance variable [25]. MTSubjectObserverProtocol uses an instance of Notifier to issue an event in a new different thread. Notifier is created for a single event notification to an observer. A new thread spawned to execute run() of Notifier does not possess the synchronization lock on a MTSubjectObserverProtocol. Note that MTObserver and MTSubject are not subclasses of Observer and Subject. They can support any other arbitrary implementation and/or interfaces.

3.5 Example: An Aspect-Oriented Web Server

The next example shows an aspect layer (i.e. metalevel) in our adaptive web server named OpenWebServer [26]. OpenWebServer contains a metalevel that supports a wide range of aspects of web servers to allow itself to continuously evolve beyond the static and monolithic servers of today. It has a collection of aspects including Concurrency, Cache, Protocol, RequestHandler,

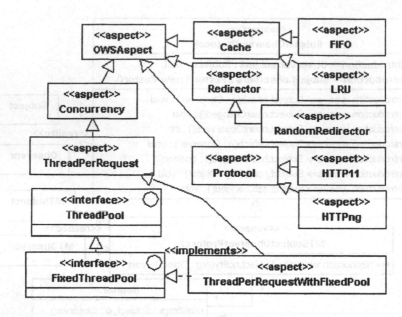

Fig. 7. Aspects in OpenWebServer

`ContentFinder`, `Logger`, `Redirector` (see Figure 7). Each aspect has one or more implementation policies. OpenWebServer can change its behavior at runtime corresponding to a given situation and/or requirement. The system adaptation is achieved with the reflection mechanism. In our project, project members are using the semantics and their notations described above to specify aspects and keep them loosely coupled in both conceptual and implementation levels.

3.6 Metalevel, Baselevel and Metaclasses

All of metalevel, baselevel, metaclasses and baseclasses can be represented using the predefined UML model constructs.

The metalevel and baselevel are expressed with the `Package` element. A metalevel is described with a package with the stereotype ≪metamodel≫. A class stereotyped with ≪metaclass≫ represents a metaclass. The relationship between a class and its metaclass is described with the `InstanceOf` relationship.

Figure 8 shows an example of a reflective implementation of the proxy design pattern, described in [27]. This model defines a *proxy* object [22] that hides the reference to an object and plays the role of its placeholder to control access it. This pattern is generic enough to provide various kinds of proxies including virtual proxy, cache proxy, remote proxy, protection proxy, synchronization proxy, counting proxy and firewall proxy [28]. Figure 8 defines local and remote proxies (`ProxyForX` and `ProxyForY`) for original (or target) objects (`X` and `Y`). Both proxies accept a method invocation and redispatch it to corresponding

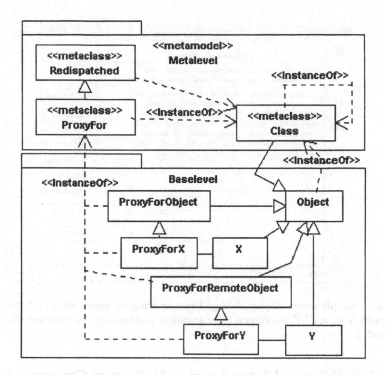

Fig. 8. A reflective implementation of the proxy design pattern

target objects. A remote proxy creates the illusion that the remote object is a local one. Redispatched provides this redispatching capability by placing a redispatch stub in each method table entry with its redispatch(). Therefore, the metaclass for proxies, ProxyFor, is a subclass of Redispatched. The method redispatch() must be overridden in base proxy classes, ProxyForObject and ProxyForRemoteObject, for implementing redispatch stubs according to the kinds of proxies. Object is a base class for all classes and Class is a base class for all metaclasses. Note that, as in the example in Figure 6, X is not a parent of ProxyForX; instead ProxyForX merely has to support the interface of X [27].

The notation of the InstanceOf relationship is a dashed arrow with its tail on a class and its head on a metaclass. The arrow has the keyword ≪instanceOf≫. We introduced alternative notations for the relationship between a class and its metaclass, ≪reflect≫ and ≪reify≫. ≪reflect≫ is a directed association with its tail on a metaclass and its head on a class. ≪reify≫ is a directed association with its tail on a class and its head on a metaclass. These associations are useful when a programming language does not support metaclasses directly as language constructs.

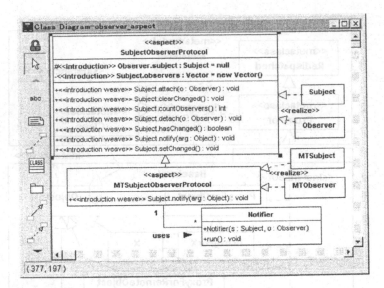

Fig. 9. A sample screen display of MagicDraw showing an aspect model information generated from a UXF description (The graphical positions of the icons are changed manually).

4 Describing and Interchanging Reflective Components with UXF and XMI

As described in Section 2.4, we are using UXF basically to describe metalevel model information. We developed a UXF DTD (Document Type Definition) for metalevel description and then merged it with existing UXF DTDs.

We have developed a translator that converts an aspect source code of AspectJ into a UXF description and vice versa. For translating an aspect code to a UXF description, the translator parses AspectJ code using the Doclet toolkit included in Java Development Kit. For translating a UXF description to an aspect code, the translator parses UXF code with an XML parser and generates AspectJ code. The aspect model interchange can be performed at the aspect design/coding, class design/coding and woven class verification phases (see Figure 2). It helps the forward/reverse engineering in roundtrip aspect development.

We have also developed a tool to convert a UXF description into native formats of Rational Rose [1] and MagicDraw [2], which are popular commercial CASE tools. Figure 9 is a screen display of MagicDraw showing an aspect model generated from the UXF description, which is in turn translated from an AspectJ code (see also Section 3.4 and Figure 6). This example shows the tool interoperability and aspect model interchangeability between an aspect weaver and a CASE

---

[1] http://www.rational.com/rose/

[2] http://www.nomagic.com/

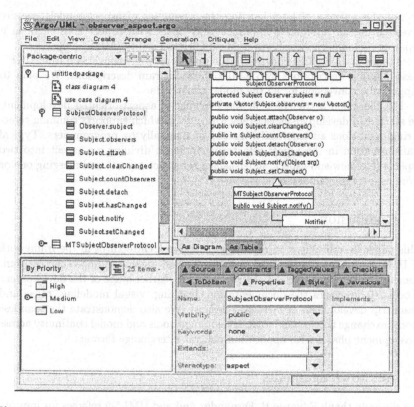

Fig. 10. A sample screen display of Argo/UML showing an aspect model information generated from our UXF-XMI converter (The graphical positions of the icons are changed manually).

tool. In addition, we are developing a UXF-XMI converter for XMI-enabled tools to use UXF descriptions. Figure 10 is a screen display of Argo/UML [9], which is an XMI-enabled CASE tool, showing an XMI formatted model description converted from UXF. This example shows the tool interoperability and model interchangeability between different CASE tools.

5 Current Project Status and Future Work

Our project is using AspectJ as an aspect language and Java as a programming language. We are now evaluating the interoperability of aspect model information between weavers.

---

[3] http://www.ics.uci.edu/pub/arch/uml/

For the purpose of interchanging the semantics of aspect models, we are investigating an alternative approach that deals with aspect constructs in a separate metamodels based on the Meta Object Facility (MOF) [29].

As for our UXF-XMI converter, it supports only the conversion from UXF's class diagram descriptions into XMI's class diagram descriptions. We plan to support the conversion for all the model constructs.

For the smooth transition in the lifecycle of aspect-oriented development, we started to develop an aspect refactoring tool, which supports common refactoring operations automatically, instead of manually by programmers. Typical transformation in the aspect refactoring includes dividing an aspect into two aspects (e.g. new sub-aspect, super-aspect, abstract aspect) and merging one or more aspects into an aspect.

6 Conclusion

This paper describes how to represent reflective architectures in the framework of UML. Our work allows us for recognizing and understanding reflective components in the upper levels of abstraction at an earlier stage of the development process. It leverages the documentation, learning, visual modeling, reuse and roundtrip development of metalevel designs. We also demonstrate the seamless model exchange between different development tools and model continuity across development phases with application-neutral interchange formats.

7 Acknowledgements

We sincerely thank Eduardo B. Fernandez and and UML'99 referees for improving this paper with their careful reading and invaluable comments.

References

1. Object Management Group. *Unified Modeling Language Specification version 1.3.* OMG document number: ad/99-06-08, 1999.
2. W. L. Hursch and C. V. Lopes. Separation of Concerns. Technical report, NU-CCS-95-03, Northeastern University, 1995.
3. G. Kiczales, J. Lamping, A. Mendhekar, C. Maeda C. Lopes, J-M. Loingtier, and J. Irwin. Aspect-Oriented Programming. In *Proceedings of ECOOP'97.* Springer LNCS 1241, 1997.
4. K. Czarnecki. *Generative Programming: Principles and Techniques of Software Engineering based on Automated Configuration and Fragment-based Component Models.* Ph.D. Thesis, Technische Universitat Ilmenau, Germany, 1998.
5. Xerox PARC AOP research group. The AspectJ Primer. available at www.parc.xerox.com/spl/projects/aop/aspectj/primer.
6. K. Bollert. Implementing an Aspect Weaver in Smalltalk. In *Proceedings of STJA'98*, 1998, available at www.germany.net/teilnehmer/101,199268/.
7. K. J. Lieberherr. *The Art of Growing Adaptive Object-Oriented Software.* PWS Publishing Company, 1995.

8. M. Aksit, K. Wakita, J. Bosch, L. Bergmans, and A. Yonezawa. Abstracting object-interactions using composition-filters. In R. Guerraoui O. Nierstrasz and M. Riveill, editors, *Object-based Distributed Processing*. Springer LNCS, 1993.

9. Y. Ichisugi and Y. Roudier. The Extensible Java Preprocessor Kit and a Tiny Data Parallel Java. In *Proceedings of ISCOPE'97*. Springer LNCS 1343, 1997.

10. S. Sonntag, H. Haertig, O. Kowalski, W. Kuehnhauser, and W. Lux. Adaptability using Reflection. In *Proceedings of 27th. Annual Hawaii Int. Conf. on Sys. Sci.*, pages 383–392. Springer LNCS 1616, 1994.

11. B. C. Smith. Reflection and Semantics in Lisp. In *Proceedings of ACM POPL '84*, pages 23–35, 1984.

12. G. Kiczales, J. Rivieres, and D. G. Bobrow. *The Art of the Metaobject Protocol.* MIT Press, Cambridge, MA, 1991.

13. J. McAffer. Engineering the Meta Level. In *Proceedings of Reflection '96*, 1996.

14. S. Chiba. A Metaobject Protocol for C++. In *Proceedings of OOPSLA'95*, 1995.

15. M. Tatsubori and S. Chiba. Programming Support of Design Patterns with Compile-time Reflection. In *Proceedings of Reflective Programming in C++ and Java Workshop at OOPSLA'98*, 1998.

16. F. Buschmann, R. Meunier, H. Rohnert, P. Sommerlad, and M. Stal. *A System of Patterns: Pattern-Oriented Software Architecture*. WILEY, 1996.

17. P. Maes. Concepts and Experiments in Computational Reflection. In *Proceedings of OOPSLA '87*, pages 147–155, 1987.

18. D. P. Friedman and M. Wand. Reification: Reflection without Metaphysics. In *Symposium on LISP and Functional Programming*, 1984.

19. M.L.B Lisboa. A New Trend on the Development of Fault-Tolerant Applications: Software Meta-Level Architectures. In *Proceedings of Internetional Workshop on Dependable Computing and its Applications (IFIP'98)*, 1998.

20. D. Bobrow, R. Gabriel, and J. White. CLOS in Context -The Shape of the Design Space. In A. Paepcke, editor, *Object-Oriented Programming- The CLOS Perspective*, page Chapter 2. MIT Press, 1993.

21. J. Suzuki and Y. Yamamoto. Making UML Models Interoperable with UXF. In J. Bezivin and P-A. Muller, editors, «UML» '98: Beyond the Notation. Springer LNCS 1618, 1999.

22. E. Gamma, R. Helm, R. Johnson, and J. Vlissides. *Design Patterns*. Addison-Wesley, 1995.

23. C. Maros, M. Campo, and A. Pirotte. Reifying Design Patterns as Metalevel Constructs. In *Journal of the Argentine Society for Informatics and Operations Research*. August 1999.

24. L. L. Ferreira and C. M. F. Rubira. The Reflective State Pattern. In *Proceedings of PLoP'98*, 1998.

25. D. Lea. *Concurrent Programming in Java: Design Principle and Patterns*. Addison-Wesley, 1997.

26. J. Suzuki and Y. Yamamoto. OpenWebServer: an Adaptive Web Server Using Software Patterns. In *IEEE Communications Magazine, Vol.37, No.4*, April 1999.

27. I. R. Forman and S. H. Danforth. *Putting Metaclasses to Work*. Addison-Wesley, 1998.

28. H. Rohnert. The Proxy Design Pattern Revisited. In J. Vlissides J. Coplien and N. Kerth, editors, *Pattern Languages of Program Design 2*. Addison-Wesley, 1996.

29. Object Management Group. *Meta Object Facility*. OMG document number: ad/97-08-14, November 1997.

Nine Suggestions for Improving UML Extensibility

Nathan Dykman   Martin Griss   Robert Kessler

University of Utah   HP Laboratories   University of Utah
ndykman@cs.utah.edu   griss@hpl.hp.com   kessler@cs.utah.edu

In this paper we suggest nine improvements that address issues in UML extensibility and UML based tools. These improvements were suggested by work on
SAGE, an extension to Rational Rose that automates the component generation
process in a large scale CORBA financial enterprise framework called Sea-
Bank. SAGE makes extensive use of UML extensibility features and it was
noted that these features were underspecified in the UML standard and under-
supported in current CASE tools. This paper proposes some ideas that would
greatly improve UML extensibility, which we believe is critical for wider
adoption of UML and for next generation domain-specific and UML-based
component development and tools

1   Introduction

SAGE was developed as a UML-based tool to improve development with the Sea-
Bank framework developed at HP Labs[1] and to see how effective UML and CASE
tools would be as a platform for creating domain-specific development tools. SAGE
works by translating annotated UML models into SeaBank component customization
files.

SeaBank is a component framework, built around a set of highly customizable
components and is supported by CORBA-based services for notification, transactions,
persistence and workflow. Early during the design and use of SeaBank, it was envi-
sioned that specialized tools would be needed to assist in the construction, customization and integration of components.

Initially, SeaBank components and their interconnections were sketched by hand
using "box and wire" notation, and then components were built manually using C++
class libraries and C++ code. It was natural to consider a visual tool in which models
of component templates and customization shapes could be used to create components
and wired together into applications.

Several versions of these visual tools were explored using an experimental visual
programming environment, CWave [2] and a commonly available drafting tool, Visio.
[3] SAGE was devised as an improvement over these tools, as it was built as an extension to a powerful CASE tool, Rational Rose [4, 5], which provides support for
modeling the entire application and integration with other tools.

UML was an obvious choice for SAGE, given that models were used in SeaBank
from its inception. However, UML is more than a simple visual notation; it is a complete modeling language with semantics as well as a visual syntax. UML also contains
extensibility features that allow modelers to capture additional semantic information

in a standard manner. In addition, most CASE tools use or will use the UML as a standard modeling notation.

As development of SAGE progressed, it became clear that the semantics of UML extensibility features were incomplete and unclear. Few, if any examples existed in the UML 1.3 (Beta 1) standard on the advanced features of UML extensibility. In addition, Rational Rose and other CASE tools evaluated did not fully support advanced UML extensibility features like stereotype hierarchies and tagged value inheritance and overriding. This made it difficult to use the UML extensibility features consistently and correctly. The lack of examples for these features combined with the lack of tool support required that a number of compromises be made in the development of SAGE. The compromises did impact the usability and effectiveness of SAGE and complicated its design and implementation.

SAGE is not a complex tool. SAGE addresses only one aspect of development within the SeaBank framework. Even though SAGE is a simple tool, the limitations of the UML extensibility semantics and tool support were quickly reached. We believe that for large-scale UML component and framework development tools to be successful, the semantics of UML extensibility must be clarified, extended and supported by all tools.

SAGE was an effective experiment that evaluated the effectiveness of UML [6-11], Rational Rose [4, 5] and SeaBank [1] for automating component development. In this paper, we focus on the problems that arose in using UML and Rational Rose while developing SAGE, and suggest potential improvements to UML and UML tools that arose directly from the work in developing SAGE. SAGE is presented in more detail in [12].

2 Background

The first section presents the SeaBank framework in more detail, giving the necessary background to understand the component generation process and the motivation for developing SAGE. The second section discusses how UML is used to model SeaBank components. The last section discusses how SAGE is implemented.

2.1 The SeaBank Framework

SeaBank, developed at HP Labs, is a banking-domain CORBA-based framework that provides components for task management, reporting, business logic, and database access as they relate to the transfer and management of monetary instruments. SeaBank provides mechanisms for distribution, controlled access and transactional management of these components.

SeaBank applications are built from SeaBank framework components that are customized to match the application requirements. Each component is created from a customized set of component parts. These parts represent a collection of classes that implement various aspects of the component functionality. This customization process is done by creating files that contain the information that specifies the structure of the component and its parts. This file is fed to a frame-based code generator [13] that creates large sections of application-specific code.

SAGE automates the tedious task of creating these customization files by allowing designers to create a UML model that contains all the necessary customization information. For enterprise-scale applications, the customization files quickly become very difficult to manage. In comparison, the UML models scale extremely well.

SAGE concentrates on automating the development of SeaBank Task Model components. Task Model components coordinate various high-level actions in the application to perform a certain business task. The SeaBank Task Model component, like all SeaBank components, consists of one or more parts. These parts represent portions of functionality that are combined together to implement the component functionality.

SeaBank Task Model components have the following parts: Actions (required to complete the task), Viewers (how task progress is presented to users), Business Objects (business rules and actions the task must use), and Task Model (controls the other parts to perform a task).

It was decided that the UML would be a perfect choice for creating models of Task Model components and applications, and that the UML extensibility features would allow these models to be easily annotated with all the information required to generate new Task Model components. Stereotypes were used to denote UML classes and components as Task Model components and parts. Tagged values were used to store additional Task Model component information.

2.2 Modeling SeaBank Components

Task Model components are modeled as UML components with the stereotype «TaskModel». Task Model components contain one tagged value, IssueAnswerDirectory, which contains the directory in which the component customization file is to be placed. Figure 1 shows a Task Model component named TransferFunds.

Task Model component parts are modeled as UML classes. The type of part is denoted by a stereotype that is the name of the SeaBank part. For example, an Action part has «Action» as its stereotype.

Each Task Model component part has a set of items that represent the features or parameters that each part imports or exports. These features are modeled as attributes of the class. The meaning of these attributes depends on the part. For example, for a Viewer part, the attributes represented the data that the viewer presents to users, as shown to Figure 2.

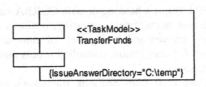

Fig. 1. TransferFunds Task Model Component in UML

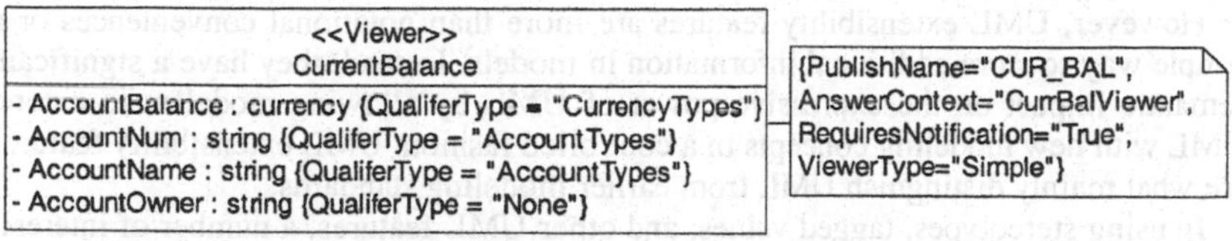

239

```
            <<Viewer>>
          CurrentBalance
- AccountBalance : currency {QualiferType = "CurrencyTypes"}
- AccountNum : string {QualiferType = "AccountTypes"}
- AccountName : string {QualiferType = "AccountTypes"}
- AccountOwner : string {QualiferType = "None"}
```

{PublishName="CUR_BAL",
AnswerContext="CurrBalViewer",
RequiresNotification="True",
ViewerType="Simple"}

Fig. 2. SeaBank Viewer Part in UML

Each component part can be assigned to one or more Task Model components by connecting the classes to components by a UML realization relationship. Each Task Model component can have zero or more Action, Viewer or Business Object parts assigned to it and each component must have one and only one Task Model part to be a well formed component model. These constraints on Task Model components are expressed in UML Object Constraint Language (OCL) [8-10, 14] in Figure 3.

```
self.stereotype = 'Task Model' and
self.oclIsKindOf(Component) implies
        (self.realizes->forAll(r |
                    r.stereotype = 'Action' or
                    r.stereotype = 'Viewer' or
                    r.stereotype = 'Business Object' or
                    r.stereotype = 'Task Model'))
        and
        (self.realizes->select(t |
                t.stereotype = 'Task Model')->size = 1)
```

Fig. 3. OCL Constraints on SeaBank UML Models

2.3 Implementing SAGE in Rational Rose

SAGE was implemented as a small, easily maintained extension to Rational Rose. Rose has a well-defined extensibility interface that allows developers to make third-party extension that integrate with Rose. SAGE is implemented as a Rose addin, which is a COM server that supports interfaces for communicating with Rose.[4, 5]

By developing SAGE as a Rose extension developers can take full advantage of all the features Rose provides for development. Rose was also used to develop SAGE as well. SAGE was modeled in UML, and implemented in Visual Basic from a Rose generated code skeleton. Additionally, a generation metamodel was created in UML and Rose that served as a reference for how SeaBank Task Models were to be structured. [12]

It is clear that UML contains a number of new features that were indispensable in the creation of SAGE; earlier modeling languages and tools would not have been as easy to use. UML stereotypes and tagged values allow models to be annotated with information that is required for SeaBank component generation without compromising the core semantics of UML.

However, UML extensibility features are more than notational conveniences or a simple way to store additional information in models. Instead, they have a significant semantic impact on the expressive power of UML by allowing modelers to extend UML with new modeling concepts in a controlled fashion. UML extensibility features are what mainly distinguish UML from earlier modeling standards.

In using stereotypes, tagged values, and other UML features, a number of interesting problems and shortcomings in the UML standard and tool support became apparent. We look at these problems and suggest potential solutions in the next section.

3 Suggestions for Improving UML Extensibility

The UML metamodel and semantics are considered by some to be a mostly academic concern that has little impact on the average user of UML. [15] Currently, that may be true. However, as UML is used to support advanced component development and re-use of software designs, the metamodel and semantics of UML becomes much more useful and the clarity, conciseness, and preciseness of UML semantics becomes critical.

For an industrial-scale component and framework development and reuse tool to be successful, the issues raised by SAGE and other issues concerning the UML semantics must be resolved. In this case, simplifying and clarifying the UML metamodel and semantics is not an academic exercise, but is an important step in the development of next-generation UML-based CASE tools, components, framework and other model-based reusable software.

Not every developer will need to understand the intricacies of the UML metamodel to develop software. However, developers using the UML to create tools, components and frameworks will need knowledge of the UML semantics to ensure that the domain-specific UML-based tools match the domain components and frameworks, and that these tools support rapid development using the available reusable assets.

The next sections each contain brief proposals on how the UML semantics could be potentially improved to better support advanced UML modeling and component development, and to clarify some existing features in the UML 1.3 standard. [9]

3.1 Distinction Between Keywords and Stereotypes is Unclear

SAGE models Task Model component parts as classes. However, these parts are more properly modeled as interfaces, since parts represent the external interfaces that the component exports in SeaBank. SAGE does not model parts as interfaces because Rational Rose did not allow interfaces to be stereotyped. In Rose, interfaces are modeled as stereotyped classes, and Rose does not support stereotype hierarchies. Visio 5.0 Enterprise Edition supports a limited form of stereotype hierarchy, allows interfaces to be stereotyped, and allows tagged values and constraints to be associated with stereotypes. [3]

In the UML 1.3 standard, interfaces are not stereotyped classes, but are core elements in the UML metamodel. However, in the UML notation specification, interfaces appear exactly as stereotyped classes. They are marked by the keyword «interface» and have an optional appearance as a small circle. However, «interface» is not a

241

stereotype, even though it appears just like a stereotype. The notation is the same for two semantically different aspects of UML; keyword elements are core elements in the metamodel and stereotyped elements are not. This confusion in the notation versus the semantics makes it unclear how interfaces and other keyword elements should be supported by UML tools, and many tools fail to distinguish between keywords and stereotypes, and implements them all as stereotypes.

3.2 Semantics of Stereotype Hierarchies Are Underspecified

If Rational Rose and other tools supported stereotype hierarchies, SAGE could have modeled parts as a stereotype of «interface» by creating a stereotype hierarchy with the «interface» stereotype as the root of the tree. However, Rose does not support stereotype hierarchies, and their semantics are unclear.

Even if the confusion of modeling keyword elements as stereotyped classes was resolved, the UML still contains a number of predefined element stereotypes. For example, UML 1.3 has a predefined component stereotype «file» that notes that the component is a source or other type of file. It is clear that modelers may wish to have stereotypes that specify what type the file is, if it is a C file, Java class file, ActiveX control, and other specific packaging mechanisms. It is more semantically correct to have these stereotypes further stereotype «file», and not be stereotypes of component.

Clearly, stereotype hierarchy is an important aspect of UML, as these hierarchies allow modelers to stereotype any element in the UML, and to model complicated semantic hierarchies of domain concepts and elements that occur in large software systems and domains.

The UML standard states that stereotype hierarchies are supported by creating special UML models. Each stereotype is represented as a class with the stereotype «stereotype», and generalization relationships show the extended stereotype hierarchy. The UML standard states that both the user and the modeling tool must distinguish stereotype-hierarchy models from regular models, but it does not state a standard mechanism for doing so, nor are any mechanisms suggested.

The UML specification and other UML books contain no examples of stereotype hierarchies, and little information is available on how to properly model them. Furthermore, no mention is made on how UML core elements are to be included in a stereotype hierarchy model. This is important, as core elements serve as the root class of every stereotype hierarchy. It is likely the class used to represent the core element in the metamodel is used, but this is not explicitly stated.

Extending the semantic hierarchy of UML via stereotypes is a critical feature for capturing the additional semantics of software domains, frameworks and components. The standard needs to be clarified with respect to forming stereotype hierarchies to enable effective modeling of large-scale software domains. The UML standard should contain a number of examples of how stereotype hierarchies are modeled, how stereotype hierarchies extend the UML metamodel, and the semantics of inheritance as it applies to stereotype hierarchy must be described in more detail as well.

The UML standard also notes that supporting stereotype hierarchies is an optional feature, and tools need not support it. SAGE presents a specific counterexample to this argument. If stereotype hierarchies were supported in Rational Rose, SAGE would have modeled component parts as stereotyped interfaces. This would allow components to be associated with the interfaces in the component model, and the interfaces appear directly in the component models.[1] In addition, SAGE could also model Task Model components as a stereotype of UML «file» component, which more closely matches the semantics of the customization files. We believe these features should be mandatory, not optional.

3.3 Tagged Value Inheritance and Specification are Underspecified

SAGE makes extensive use of tagged values to store information about SeaBank parts and components. However, Rational Rose did not support tagged value inheritance and therefore specific tagged values could not be associated with SeaBank stereotypes. Instead, the Rational Rose extension of allowing multiple tagged value sets for core UML elements had to be used. One set was created for each stereotype for Task Model classes, and the modeler must manually select the proper set, or generation may fail. Visio 5.0 Enterprise does allow tagged values and constraints to be associated with newly defined stereotypes using the Visio UML stereotype editor, though not all constraints and required and default tagged values associated with the stereotype appear to be properly carried over to the stereotyped class.[3]

The UML 1.3 specification states that tagged values are "pseudoattributes" of core or stereotype classes in the metamodel, which we refer to here as metaclasses. Tagged values denote the additional semantic information required by any instance of a metaclass. In addition, it is possible to define a set of required tags for any core or stereotyped element in UML. Finally, stereotyped elements inherit required tags from base classes and can additionally override the default values of tagged values of parent elements

However, like stereotype hierarchies, no examples are given to how these features are modeled and how tagged value specification, inheritance or overriding works. The exact semantics of tagged-values and how they interact with stereotype hierarchy modeling is not specified. It is unclear if the semantics match the semantics of generalization relationships in UML, or if additional semantics must be attached to generalization to match the mechanisms needed to properly model tagged-values semantics in stereotype hierarchies.

Again, most tools do not support these tagged value semantic features. The standard notes that they are to be optionally supported by tools, and again SAGE provides a clear counterexample to this argument. If the proper semantics for stereotyped values were supported, the proper values would always be available, and modelers would not have to manually choose the proper set to insure proper code generation in SAGE. We again believe that the support for these features should be made mandatory.

3.4 UML Should Support Tagged Value Types

Rational Rose provides a useful extension to the semantics of UML tagged values: types. Tagged values in Rose can have the following types: Boolean, integer, real, string, and enumerated. These types restrict the values a tagged value can assume, and is used to insure that tagged values contain proper data. SAGE makes good use of

---

[1] Rational Rose does not show realization relationships in component diagrams. The relationships are implicitly managed by the tool.

243

Rational Rose support for tagged value types to insure users enter the proper data for the values.

It seems that adding a type field to tagged values would be a valuable addition to UML. The UML specification states that the interpretation of tagged values is beyond the scope of the UML. However, having types for UML tagged values would serve as a useful hint as how the values are to be interpreted by tools. In addition, it is possible that the UML specification could provide a predefined set of tagged value types and encodings that all tools would support. This would improve model interoperability.

While UML probably should not provide a complete type system for tagged values, it may make sense to define a small set of basic types and encodings that all tools should understand. A good basic set of types might be the types supported in the OCL specification and the basic types used in the metamodel. .

3.5 Support Tagged Values as Metaattributes in the UML Metamodel

The previous sections noted the problems SAGE had associating tagged-values with stereotypes and how SAGE took advantage of tagged-value types in Rose. It makes sense to model tagged values as attributes in the metamodel, or metaattributes, thereby making it much easier to specify the tagged values that core elements or stereotype elements require and to support tagged value types. To avoid confusion with other attributes in the metamodel, tagged value metaattributes could have the stereotype «metaattribute» . To specify the tagged values in the metamodel or stereotype hierarchy models, all that is required is to add metaattributes to the model classes.

It is important to note how well the semantics of inheritance match the semantics of tagged-values. The only non-trivial case is when a stereotype metaclass wants to override the default value of its parent. This could be simply modeled by having the stereotyped class include an attribute with the same name as the parent tagged-value, with a different default value. The standard rules of inheritance seem to match with the intended semantics, as attribute resolution occurs by traversing up the hierarchy, which means that the overriden value would be found first. Certain attributes could also be marked as final, or not overridable. It makes sense to use the default semantics of inheritance to model tagged-values inheritance, as it is well understood and standardized.

We noted we could distinguish tagged values in the metamodel by using the stereotype «metaattribute». However, it may not be necessary to distinguish tagged values with a stereotype at all. It makes sense to view all attributes in the metamodel as if they were tagged values of elements. Having all attributes in the metamodel be tagged values may improve model exchange with mechanisms like the MOF, XMI or repositories. This would eliminate the need for having two mechanisms: one for storing metamodel attributes, and one for storing tagged values.

3.6 Tools Should Support the OCL

SAGE extracts information from UML models for various pieces of information by using the Rose Extensibility Interface. However UML has a powerful constraint and query language, OCL. OCL allows modelers to query models for information, and to specify additional constraints on model elements. However, many tools do not currently support the OCL, and additional tools are required to explicitly check constraints or query models for information [8, 9, 14]

One key feature of CASE tools is the enforcement of constraints at design time. Modelers are forbidden to draw a diagram that violates the semantics of the modeling language. For example, a UML CASE tool would forbid a modeler to associate a class with a component, but would allow an association with an interface, or a realizes relationship with a class. These constraints are enforced as the design is created.

In the modeling of SeaBank components, SAGE has additional constraints on models that are outside the constraints on regular UML models. These constraints were expressed in OCL in Figure 3. However, SAGE can not enforce these additional constraints at design time. The constraints are only explicitly checked during generation by the SAGE tool. If the constraints are not meet, generation can fail.

Support for the OCL is a useful feature that is missing from CASE tools. Full-featured support of OCL would allow modelers to create and check additional constraints implicitly at design-time or explicitly by running an OCL checker. Also, OCL can be used to create queries against UML models to gather information about a model. Just as SQL queries can be ported from database to database, OCL queries could be ported from CASE tool to CASE tool. Additionally, repositories or XMI or MOF compliant tools could use OCL as the standard query language, allowing software to use the OCL to query for metadata, workflow information or other resources. [16-18]

SAGE must currently allow developers to create improper models of Task Model components, because Rose has no support for adding design-time constraints. Support for implicit constraints via the OCL could prevent modelers from creating improper models, greatly reducing errors. Also, SAGE currently uses the Rose Extensibility Interface to gather information about models. By using OCL queries instead, SAGE could be more easily ported to any tool that supports OCL queries.

It is an open question if OCL can represent all the constraints that modelers may wish to enforce at design time. A number of constraints in the UML metamodel can not be expressed in OCL (e.g. the well-formedness rules on constraints) and so these constraints do exist, and must be hard-wired into tools. Tools could also support constraints by allowing the tool to be extended with arbitrary scripts or code that enforces or checks constraints, thereby circumventing the limitations of OCL.

3.7 Tools Should Comply to the UML Standard

SAGE is a not a portable tool. It relies extensively on the Rose Extensibility Interface to interact with UML models. A better version of SAGE would use standards like OCL, and XMI to be more portable. However, many tools do not support all the UML standards or features, making portability much more difficult.

Compliance to a standard is a critical feature for any tool. Most programming language have a precise standard that give enough detail so tool developers can support every feature of the language. Many languages have a set of standard test-suites that insure compliance to the standard.

UML is a fairly new standard, and has not yet evolved to point of a mature language. If UML is to evolve to the level of maturity that is present in programming languages, it is critical that: the UML standard be precise enough that developers know the semantics of all UML features, UML-based tools support all the standard features in the UML specification, and UML tool compliance be verified from tests. At least, tools should precisely identify what they do and do not support, so one can make informed decisions when considering UML tools. SAGE is a not a portable tool, and is highly tied to a specific CASE tool. Better support for UML in tools would greatly improve the portability of these types of tools

Currently, tools vary in their level of support for UML features, from very basic support, to almost full-featured support. The lack of a standard set of features that all tools provide serves as an impediment to a wider adoption of UML and UML tools. While complete portability between tools may not be possible, the current inability to use any models between tools is a significant problem. Standards like the MOF, XMI, and tools like the Microsoft and Unisys repositories [18, 19] address these issues, but support for these standards or tools also vary from tool to tool. [3-5, 20]

It appears that a set of standard compliance tests would be a critical part of ensuring that tools are compliant with the UML standard. Currently, many tools can claim to support the UML, but their support for UML varies, requiring tools to support all of the standard features of the UML specification before they could claim to be UML compliant would encourage a convergence of tools towards a standard set of features.

SAGE chose Rational Rose as its base UML tool. Porting SAGE to other tools would be a non-trivial task, and SAGE is not a large program. As the size and investment in third-party tools increases, portability between tools and support for multiple tools becomes a critical issue and tool support of UML standards also becomes critical.

3.8 The UML Metamodel Should Be Self-contained

We noted earlier that it makes sense to model tagged values as attributes in the metamodel to support various UML features. If tagged values are to be modeled as meta-attributes, it is natural to ask what metamethods could represent, if anything. It is important to ask if the semantics of UML contain information that could be modeled as methods in the metamodel.

If we view methods of a class as the contract that preserves the integrity of a class, a reasonable use of metamethods suggests itself. Metamethods could be used to represent the well-formedness rules and constraints in the metamodel. In other words, the set of metamethods completely describes how a UML element can be used.

It appears reasonable to have metamethods represent the constraints and well-formedness rules in the metamodel. Each metamethod could contain predefined tagged values that hold the text or OCL for each rule and any additional explanatory documentation. Embedding constraints as metamethods would also allow advanced modelers to model complicated metamodel extensions entirely as UML models.

Any extension to the UML metamodel could contain all the documentation that any core UML element currently has. By embedding both tagged values and constraints directly into the UML metamodel, it is possible that the entire UML standard could be generated directly from the metamodel using documentation automation tools. For example, SAGE could be based entirely on a UML metamodel extension. All the proper tagged values would be specified, and all the additional constraints of the framework would be documented in OCL as well. Enough information would be available in the metamodel extension to implement the modeling of SeaBank components.

Also, having the UML metamodel be self-contained insures that standards like the MOF, XMI, and other model exchange tools would exchange all the information a UML model could contain, including documentation and constraints on stereotyped elements added by modelers. This information is critical to help document and understand domain specific information captured in UML models.

3.9 Tools Should Be Metamodel Based

SAGE could have taken advantage of more powerful extensibility features. Having a self-contained metamodel allows developers to model extensions to UML that are as complete as the UML core metamodel. To implement extensibility, tools should be entirely based on UML-based metamodels. These metamodels would contain all the information needed to describe models and model constraints. As changes occur in the UML metamodel, or as extensions to the metamodel are added, the tool can parse and use the updated metamodel without having to change the tool at all. Platinum Paradigm Plus [20] is believed to be an example of a tool that is currently UML metamodel based.

Metamodel-based tools would allow metamodel extensions to have the same level of support as core UML elements. This would allow advanced modelers and tool developers to extend the semantics of the tool via metamodel extensions. New constraints could be added, new stereotypes, new visual elements, etc. could all be added to the tool without changing or extending the tool itself. This level of flexibility adds a great deal of power and flexibility to CASE tools, and could give CASE tools many, of the features that metaCASE tools support, while retaining all the advantages of using a standard modeling specification and core metamodel.

This level of integration in the UML metamodel of course raises a number of important issues. We believe that an integrated metamodel would allow a number of important advances in UML modeling and UML development tools. Having a self-contained core metamodel and self-contained extensions to that metamodel would allow tools to derive their behavior directly from the metamodel. As the UML standard and metamodel evolved, tools would not longer have to maintained and modified to represent changes in the UML specification, but would merely require changing the metamodel the tool uses to specify new behavior.

Many programming languages are powerful enough to be metalanguages, or powerful enough to easily create tools that execute code in that language. While UML does partially describe itself, it is desirable that the UML metamodel be self-descriptive and self-contained, given proper meta-metamodel or M2 support.

Languages also thrive on precise semantics. This allows different tool vendors to each create tools that provide the same core functionality, and also developers to move code from tool to tool. The same features are important for UML models. The semantics of UML models should be clear, and models should be interchangeable or have the same semantics from tool to tool. Basing all tools on a self-contained UML metamodel would greatly help insure a standard level of compliance and a new level of extensibility.

4 Conclusion

This paper has presented a tool SAGE, that uses the UML and Rational Rose to model components in the SeaBank framework. Not only is SAGE a useful development tool, it served as a case study into the effectiveness of UML and Rational Rose for domain-specific modeling and component generation. While SAGE is a relatively simple tool, it raises a number of important issues in UML.

UML extensibility features are a very important and powerful addition to OO modeling languages. These features allow modelers to extend the UML metamodel in a controlled fashion to capture additional semantics that are not found in the UML core. SAGE uses these features to capture additional information required to create Sea-Bank components in a easy to use manner,

In using the UML extensibility features in SAGE, it became apparent that a number of issues existed in the current UML standard. A number of important features like stereotype hierarchies and tagged value inheritance were not clearly described in the specification. Additionally, tools did not support these features. SAGE would have been simplified in a number of important aspects if these extensibility features were more clearly specified and better supported by tools.

Tools vary on their support for UML standards. Many tools do not support the UML metamodel, OCL or other important UML features. The specification states that in some cases, tools do not have to support some UML features. SAGE serves as an example against this statement. SAGE is a simple tool, and yet it exposed limitations in the UML standard and tool support.

We made nine suggestions as how the UML could be improved or clarified to address these issues. We suggested that examples of how stereotype hierarchy and tagged value inheritance be presented in the standard, and the statement that tools are not required to support these features be removed. We discussed how tagged values could be modeled as attributes in the metamodel classes and how metamodel constraints could be modeled as methods in the metamodel classes. We also noted that making the UML metamodel self-contained would allow tools to support a new level of flexibility by supporting the metamodel. Since the metamodel is self-contained, all the required information on behavior could be extracted for the model. In addition, advanced modelers could create extensions to the metamodel that would add support for domain-specific modeling, without requiring tools to be extensively modified.

Finally, we noted that as the UML standard progresses, it is important that tools that UML all support a standard set of features, to ensure better tool interoperability and portability of models between tools. A number of standards currently exist that address the issues of interoperability, but they are not widely supported. A suite of compliance tests would encourage tools to support a more unified set of UML features, and basing tools on the UML metamodel would encourage compliance as well.

The UML is a very exciting development in the areas of software design, component modeling and software reuse. The UML is not just a new notation, but it is a full modeling language, complete with semantics, a self-describing metamodel, and extensibility features. If the UML is to become more pervasive in all aspects of software development and be used as more than just a notation, it must be based on a solid, consistent and powerful semantic foundation.

References

1. HPLabs: SeaBank Architectural Specification, . 1997, HP Laboratories.
2. Mueller-Planitz, C. and R. Kessler: CWave: A Visual Agent Workbench, . 1999, Dept. of Computer Science, University of Utah.
3. Visio: Visio Enterprise Edition, 1999. http://www.visio.com
4. Rational: Rational Rose 98, . 1998, Rational Software: Menlo Park, CA.
5. Quatrani, T.: Visual Modeling with Rational Rose and UML. The Addison-Wesley Object Technology Series. 1998, Reading, MA: Addison Wesley. xvii, 222.
6. OMG: UML 1.1 Notation Guide, . 1997, Object Management Group.
7. OMG: UML Semantics, . 1997, Object Management Group.
8. OMG: Object Constraint Language Specification, . 1997, Object Management Group.
9. OMG: UML 1.3 Specification, Beta 1, . 1999, Object Management Group.
10. Booch, G., J. Rumbaugh, and I. Jacobson: The Unified Modeling Language User Guide. Addison-Wesley Object Technology Series. 1998, Reading MA: Addison-Wesley. . cm.
11. Rumbaugh, J., I. Jacobson, and G. Booch: The Unified Modeling Language Reference Manual. The Addison-Wesley Object Technology Series. 1999, Reading, MA: Addison-Wesley. xvii, 550.
12. Dykman, N.: SAGE: Generating Applications With UML and Components, in Department of Computer Science. 1999, University of Utah: Salt Lake City.
13. Bassett, P.G.: Framing Software Reuse. Yourdon Press Computing Series. 1997, Upper Saddle River, NJ: Yourdon Press. xvii, 365.
14. Warmer, J.B. and A.G. Kleppe: The Object Constraint Language : Precise Modeling With UML. Addison-Wesley Object Technology series. 1999, Reading, MA: Addison Wesley Longman. xii, 112.
15. Zamir, S.: Taking UML from Innovation to Usage, in Component Strategies. August 1998.
16. OMG: XML Metadata Interchange (XMI), . 1998, Object Management Group.
17. OMG: Meta Object Facility (MOF) Specification, . 1997, Object Management Group.
18. Bernstein, P., et al.: The Microsoft Repository. in VLDB Conference. 1997.
19. Unisys: Universal Repository Technical Overview. 1996: Unisys Corp.
20. Platinum Software: Platinum Paradigm Plus, . 1998. http://www.platinum.com

A Classification of Stereotypes for Object-Oriented Modeling Languages

Stefan Berner[1], Martin Glinz[1] and Stefan Joos[2]

[1]University of Zurich, Winterthurerstr. 190
CH-8057 Zurich, Switzerland
{berner, glinz}@ifi.unizh.ch
[2]Robert Bosch GmbH, Postfach 30 02 20,
D-70469 Stuttgart, Germany
stefan.joos@de.bosch.com

Abstract. The Unified Modeling Language UML and the Open Modeling Language both have introduced stereotypes as a new means for user-defined extensions of a given base language. Stereotypes are a very powerful feature. They allow modifications ranging from slight notational changes up to the redefinition of the base language. However, the power of stereotypes entails risk. Badly designed stereotypes can do harm to a modeling language. In order to exploit the benefits of stereotypes and to avoid their risks, a better understanding of the nature and the properties of stereotypes is necessary.

In this paper, we define a framework that classifies stereotypes according to their expressive power. We identify specific properties and typical applications for stereotypes in each of our four categories and illustrate them with examples. For each category, we discuss strengths and weaknesses of stereotypes and present a preliminary set of stereotype design guidelines.

1 Introduction

Since about 1990 a broad variety of object-oriented modeling languages have been developed [1][2][3][4][9][11][13][14][17]. These languages are used to describe the requirements and the design of a software system. Since 1996, various attempts have been made to unify different methods and languages. As a result of this endeavor, two languages have been developed: the Unified Modeling Language UML [12] and the Open Modeling Language OML [6]. Both UML and OML introduce a distinctive new feature: they allow users to extend or even to modify the base language in order to adapt the language to specific situations or needs. The language construct that is used to implement this feature is called a *stereotype*.

In the context of object-oriented modeling, the notion of stereotypes was introduced before by Rebecca Wirfs-Brock [16]. Her principal idea is to provide a secondary clas-

This work was partially supported by the Swiss National Science Foundation under grant N° 20-47196.96.

sification for objects: stereotypes classify objects according to their use, independently of the primary classification by classes and class inheritance. This classification helps to better organize a model and improves system understanding. Wirfs-Brock uses a fixed number of stereotype classes, namely: domain-, design-, control-, delegation-, structure-, service-, and interface-objects as well as information objects.

UML and OML both generalize Wirfs-Brock's notion of stereotypes from a secondary classification to a concept that allows for general extensions of the base language. A stereotype in UML and OML can add new properties to elements of the underlying language or can modify existing ones. (In the following, we will call the underlying language that is being stereotyped the *base language*.) For example, a stereotype can *extend* UML classes to include a property that designates a class to belong to the model, the view or the controller in a model-view-controller pattern. On the other hand, in older versions of UML an Actor was expressed using a stereotype that redefined the language element Class.

In this paper we discuss the UML/OML kind of stereotypes in a general context of object-oriented modeling languages. However, as the notion of stereotypes is not limited to object-oriented approaches, we define a stereotype independently of object-orientation as follows:

DEFINITION. A *stereotype* in a modeling language is a well-formed mechanism for expressing user-definable extensions, refinements or redefinitions of elements of the language without (directly) modifying the meta-model of the language.

Stereotypes provide language users with limited metamodeling capabilities without giving them (direct) access to the metamodel of the language. This is a very powerful mechanism. However, as is frequently the case with powerful features, stereotypes have both a bright and a dark side. On the bright side, stereotypes can lead to modeling languages which are more flexible and expressive and which are better adaptable to specific problem types and application domains. On the dark side, unsystematic or excessive use of stereotypes can lead to a proliferation of incompatible dialects of a language and can make a language both difficult to handle and to understand. Thus, unconsidered use of stereotypes can do more harm than good.

Like language design in general, designing good stereotypes is a difficult and demanding task. 'Good' in this context means:

- Every stereotype is properly defined. The definition balances formality and understandability in an optimal way. Intuitive semantics and formal definition of semantics (if existing) are congruent. The stereotype does not introduce inconsistencies into the language.

- Every stereotype is useful. That means (1) it provides a concept or feature that the base language does not have and (2) it eases the application of the language in a given context or application domain.

- The set of all stereotypes is consistent. A stereotype must not be inconsistent with other stereotypes, unless a stereotype is explicitly declared to be mutually exclusive with other, contradictory ones.

- The set of stereotypes is orthogonal. When a stereotype introduces a new distinctive feature or concept into a language, then this feature should not be provided by another stereotype, too.

The task of designing good stereotypes would become considerably easier if we had a proper design methodology or at least a set of design guidelines for their creation. However, neither a methodology nor guidelines presently exist. In order to develop guidelines (and finally arrive at a methodology), a deeper understanding of the nature of stereotypes and of the implications of their use is necessary.

In this paper, we contribute to the solution of the stereotype design problem. We introduce a classification of stereotypes according to their expressiveness – that means according to their potential to alter the syntax and semantics of the base language. Every class represents a related set of purposes for using stereotypes and has specific stereotype design requirements associated with it. For each class, we give a first-cut set of guidelines for designing and defining stereotypes of this class.

To our knowledge, there is almost no related work on classification and design of stereotypes. Eriksson and Penker [5] classify the standard (i.e. predefined) stereotypes of UML according to the language concept they are applied to. For user-defined stereotypes, however, there is no classification. In the literature on UML and OML (e.g. [12][5][6]), the design and definition of stereotypes is treated in a very vague and superficial fashion only. In an earlier paper in German [10], we have introduced the idea of our classification framework along with some examples and a preliminary discussion of weaknesses and benefits.

The rest of the paper is organized as follows: Following the introduction we present our classification and illustrate it with examples. In the next section, we identify strengths and weaknesses of stereotypes in each of our four categories. In section 2, we demonstrate how a rigorous definition of a non-trivial stereotype should look like and present a preliminary set of guidelines for the design of stereotypes. The paper concludes with some remarks about our achievements and future research directions.

2 A Classification of Stereotypes

The extent to which stereotypes alter a language ranges from mere notational variations to a complete redefinition of the language.

Simple stereotypes typically change the notation (i.e. the concrete syntax and/or visual representation) of a language element and/or introduce new features that serve as a kind of 'structured comment'. Powerful stereotypes, on the other hand, impose semantic restrictions on the added language elements or even redefine the semantics of language elements. This can go up to a complete syntactic and semantic redefinition of the base language. We classify stereotypes according to their expressiveness into four categories:

- *Decorative stereotypes* vary the concrete syntax of a language.

- *Descriptive stereotypes* extend the syntax of a language such that additional information can be expressed.

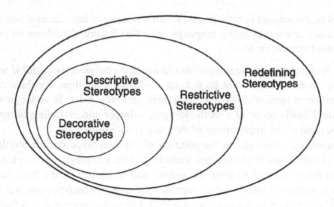

Figure 1. Classification of stereotypes according to their expressive power into four categories and their inclusion hierarchy

- *Restrictive stereotypes* extend the syntax of a language and impose semantic restrictions on these extensions.
- *Redefining stereotypes* modify the (core) semantics of a language element.

Note that our classification forms an inclusion hierarchy, not a partition. The more powerful categories include all the potential of the less powerful ones (see Figure 1). In the following subsections, we describe the four categories more in detail and provide examples.

2.1 Decorative Stereotypes

DEFINITION. A *decorative stereotype* modifies the concrete syntax of a language element and nothing else.

Thus, decorative stereotypes vary the way in which a language element is visually represented. They do not introduce any essential additional information or new concepts into the base language. The represented model and the essence of the language that expresses the model remain unchanged.

A decorative stereotype improves the understandability of a model in the same way that a good illustration improves the understandability of a text. However, it can also worsen the understandability in the same way that an erratic illustration does. As the interpretation of signs depends on the personal and cultural background of the viewer, it can become quite tricky to decide whether a decorative stereotype eases or hampers understandability. Some form of usability testing can act as a decision-maker here. But the potential benefit of the new stereotype – remember it is 'just' a decorative one – must be promising enough to justify the testing effort.

Decorative stereotypes are typically used to adapt the notation of a language or of some of its elements to some given standard or to personal preferences.

Examples

- In textual representations, a decorative stereotype can be used to change key words (like ENTITY_TYPE instead of CLASS, etc.), and if accessible, text/character attributes (font, style, size, color, etc.) or other formats (indents, alignments, line spacing, etc.).

- In graphic representations, a decorative stereotype can change graphic symbols or other attributes like colors, line width, etc. Figure 2 depicts a typical application of a decorative stereotype: In UML the graphic notation of the class symbol is replaced by that of OOAD [1]. Note that this is a pure cosmetic change. Both the abstract syntax and the semantics of UML classes remain untouched.

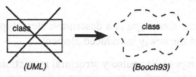

Figure 2. An example for a decorative stereotype

2.2 Descriptive Stereotypes

DEFINITION. A *descriptive stereotype* extends or modifies the abstract syntax of a language element and defines the pragmatics of the newly introduced element. The semantics of the base language remains unchanged. Additionally, a descriptive stereotype may modify the notation (the concrete syntax) of the stereotyped language element.

Thus, descriptive stereotypes are on a pure syntactic level. They do not impose any semantic restrictions on the extended or modified syntax. The persons who use a descriptive stereotype must rely on the description of the stereotype pragmatics in order to use and interpret the stereotype properly.

As a descriptive stereotype can also modify the concrete syntax of the stereotyped language element, it includes the expressiveness of a decorative stereotype. When compared with simple comments, descriptive stereotypes have the advantage of a well-defined syntactic structure, which makes some formal checking and analyses possible.

Secondary classifications (in the sense of Wirfs-Brock's stereotypes [16]) and standardized annotations are typical applications of descriptive stereotypes. Standardized annotations have the form keyword <value> where <value> must conform to a given type.

Examples

- A classification of objects according to Jacobson's OOSE [9] into Entity-, Interface-, and Control classes or distinguishing different kinds of relationships (association, usage, part-of, ...) in a language which provides only 'primitive' relationships are examples of secondary classifications.

- An extension of the definition of a UML class with a stereotype ConfigInfo that consists of Author : string, Created : date, LastModified : date is a standardized annotation which allows to specify the author and the creation/modification dates of a class in a controlled way. Note that string and date are static data types that can be checked on a purely syntactic level. Semantic restrictions would lead to restrictive stereotypes (see section 2.3).

  It should also be noted that a *Tagged Value* in UML is equivalent to a simple descriptive stereotype. So, in UML an alternative to this stereotype ConfigInfo would be the usage of three user-defined tags: author, created and lastModified.

2.3 Restrictive Stereotype

DEFINITION. A *restrictive stereotype* is a descriptive stereotype that additionally defines the semantics of the newly introduced element.

Typically, the semantics impose compulsory structural restrictions on the newly introduced language element – hence the name restrictive stereotype. A restrictive stereotype does not change the semantics of the base language – it only extends it by the semantics of the stereotype.

The concept of restrictive stereotypes allows for a fully formal definition of the stereotype. However, in practice the definition will frequently be semi-formal only. This is no surprise when considering that most contemporary specification languages (including UML) do not have a completely formal definition of their semantics.

Restrictive stereotypes are first-class members in the language they are added to. They have the same expressive power and can be defined with the same degree of rigor as the elements of the base language themselves. Restrictive stereotypes are typically used to add missing features to some elements of a language, to strengthen weak features or to introduce a metalanguage on top of a given language.

Examples

- A class category construct (borrowed from OOAD [1]) can be introduced into UML in order to strengthen UML's rather weak package construct. In order to behave like the original class categories of OOAD, the stereotype must fulfill the following restrictions:

  - A class category must only contain classes or other class categories.

  - Any relationship between two class categories must be a uses-relationship.

  - Inside a class category a class may be tagged for being exported using instances of another stereotype, namely a descriptive stereotype export on classes. Classes outside a class category may have relations only with those components of a class category that are tagged for being exported.

- The support for requirements tracing is very weak in UML. The existing Trace dependency is quite unspecific. Its use is optional and lacks detailed semantics

([12] p. 56). Thus, mandatory requirements tracing that is enforced by a tool is almost impossible using Trace. A significantly improved requirements tracing feature can be introduced into UML by first introducing a restrictive stereotype requirementId : int with a restriction unique. Using this stereotype, every element of an UML requirements model can be tagged with a unique number identifying it. Now we introduce another restrictive stereotype traceBackToRequirements : array of int with the restriction that these integers must be existing requirement identifiers. Using this stereotype, we can trace design elements back to their requirements by attaching a set of requirement identifiers to every element of an UML design model. Note that neither a descriptive stereotype nor the UML Trace would suffice to accomplish this task, because they would also accept values other than valid requirement identifiers.

- A meta-language for the identification of design patterns [7] in a model can be introduced on top of a modeling language with restrictive stereotypes: the elements of a design model can be tagged with the pattern name and the role they play in this pattern (see Figure 4). Structural restrictions that the pattern imposes on the elements that instantiate the pattern can be enforced by defining restrictions for the pattern language stereotypes accordingly (see Figure 3).

The last example from the above list also demonstrates a situation where it is clearly preferable to define a feature with stereotypes instead of including it in the base language: patterns evolve and are frequently application-dependent. Thus, the set of those patterns that can be used and have to be documented in a model should be definable on the level of projects or organizations, and, hence, not be a part of the base language. Restrictive stereotypes are the appropriate means for doing so.

2.4 Redefining Stereotype

DEFINITION. A *redefining stereotype* redefines a language element, changing its original semantics. Concerning syntax, a redefining stereotype behaves in the same way as a restrictive one.

With decorative, descriptive and restrictive stereotypes, instances of the stereotype remain valid instances of the stereotyped language element. For redefining stereotypes, this is no longer true. A redefining stereotype can introduce a new language element that is no longer related to the element of the base language that it stereotypes.

Using redefining stereotypes, deep and radical changes can be imposed to a language. New language concepts can be introduced. In its extreme, redefining stereotypes can embed another language in a given base language.

Examples

- In earlier version of UML, some model elements that do not belong to the core of the language were defined as redefining stereotypes, e.g. Use Case and Actor were stereotypes of Class.

- A subset of the specification language Z [18] can be embedded in UML using a redefining stereotype Scheme for UML classes. Instances of Scheme are no longer classes, but valid Z schemes.

3 Strengths and Weaknesses of Stereotypes

In this section, we discuss the pros and cons of stereotypes both in general and for our four categories in particular.

The main general advantage of stereotypes is that they make a language *flexible* and *adaptable*. When used properly, they improve a modeling language, making models easier to express and to understand.

On the other hand, there are two general drawbacks and risks.

- Working with stereotypes requires effort for designing and maintaining them, and for training all the users and readers of a stereotyped language how to use and interpret the stereotypes.

- Badly designed stereotypes and the use of an excessive number of stereotypes both turn the potential benefit of stereotypes into its contrary: they harm a language, making it more difficult to use and to understand.

The potential benefit of stereotypes as well as the drawbacks and risks grow with increasing power of the stereotypes. In the sequel, we identify and discuss specific strengths and weaknesses of the stereotypes in the four categories of our classification.

Decorative Stereotypes. The definition of decorative stereotypes is easy and requires little effort. Furthermore, it is easy to incorporate the features required for the definition and use of decorative stereotypes into a tool. When used to adapt the language syntax to a given standard, decorative stereotypes improve the understandability of the language for all persons working with this standard.

On the other hand, decorative stereotypes threaten the very purpose of any language – that is, communication among people. When everybody uses her or his own decorative stereotypes, we quickly end up in the same situation as the ancient people of Babel when attempting to build their tower... Furthermore, a decorative stereotype does not add any real power to a language, it is mere 'syntactic sugar'.

Descriptive Stereotypes. Descriptive stereotypes enrich a language in a controlled way by introducing structured annotations (comparable to tagged values in UML) and secondary classifications. This is clearly better than providing the same information with comments only, because descriptive stereotypes can be made mandatory and are amenable to syntactic checking and analyses. The definition of descriptive stereotypes is easy and requires little effort: only the syntax and a short natural language statement on the pragmatics of the stereotype are required. These properties make descriptive stereotypes a favorite candidate for adapting a language to company-wide or project-specific documentation standards.

On the negative side, the power of descriptive stereotypes is very limited, because they are purely syntactical. This is frequently insufficient, in particular for structured annotations. As soon as we want to impose semantic restrictions on the application of a stereotype and/or on the values of its variables, a descriptive stereotype no longer works and a restrictive one is required instead. Thus, a descriptive stereotype cannot improve bad or weak features of the base language. Neither can it supply a real feature (that means one with significant semantics) that is missing in the base language. Finally, in the same way that too many comments can 'drown' the code of a program, using too many descriptive stereotypes can yield clumsy and hardly readable models, where the essence of the model disappears in a flood of annotations.

Restrictive Stereotypes. Restrictive Stereotypes add real power to a language. Their capabilities go far beyond those of descriptive or decorative stereotypes. Not only can they add structured annotations with semantic restrictions, but also essential features that are missing in the base language. Restrictive stereotypes also can improve bad or weak features of the base language. Another powerful feature of restrictive stereotypes is their capability for defining a metalanguage on top of the base language (see the pattern examples in sections 2.3 and 4). In contrast to descriptive stereotypes, it is possible to check models for their compliance not only with syntax rules, but also with given semantic restrictions. When these restrictions are formally defined, checking can be automated and built into a tool.

Finally, a well-defined language forces its users to adhere to a basic set of modeling and design principles that the language is founded upon. Restrictive stereotypes can do the same for the features they introduce into the base language.

On the negative side, restrictive stereotypes have the following disadvantages and limitations.

- Designing restrictive stereotypes is expensive. The designers require a profound knowledge of: the desired properties of the stereotype to be designed, the base language, the general principles of good language design, and the metalanguage that is used to specify the semantics of the stereotype. Ignoring these requirements leads to bad (i.e. incomprehensible, contradictory or simply wrong) stereotypes. Badly designed restrictive stereotypes damage the base language instead of improving it.

- In order to define restrictive stereotypes properly, the metamodel of the base language should provide a mechanism for a formal specification of the semantics of the stereotype. This is a demanding requirement, which, for example, is only partially met by UML. Only a formal specification makes it possible to automatically check restrictions defined by a stereotype (of course, only those that can be automatically checked at all). Without this capability, restrictive stereotypes lose much of their power and finally become similar to descriptive ones.

- Defining too many restrictive stereotypes has the same disadvantages as described for descriptive stereotypes.

Redefining Stereotypes. Redefining stereotypes give the users full control over the base language. In contrast to restrictive stereotypes, a redefining stereotype does not merely extend the semantics of the base language, it modifies it. Instances of a redefining stereotype do no longer need to be instances of the stereotyped base language element. This capability has two major advantages:

- The base language can be extended by new features that have nothing to do with the element of the base language that is being stereotyped. It is even possible to (virtually) delete the stereotyped element from the base language by restricting the allowed number of non-stereotyped instances to zero. Of course it is impossible to really delete an element from the base language, because a stereotype cannot delete elements from the metamodel. However, by restricting the number of allowed instances of a language element to zero accomplishes the same effect as a real deletion.

- Starting from a common general-purpose language, highly specialized languages can be derived for every specific problem or application domain. This can help to keep the base language simple an clean.

On the negative side, all the disadvantages and limitations listed above for restrictive stereotypes also apply to redefining stereotypes, but to an even stronger degree. Three problems deserve special attention.

- The introduction of every redefining stereotype creates a new dialect of the base language that is semantically different. If everybody creates her or his own redefining stereotypes, this is a deadly threat to the very essence of any language: enabling communication between people that have to share information.

- Among all kinds of stereotypes, redefining stereotypes require the highest effort for creation, training and maintenance.

- Making redefining stereotypes is language design, a task requiring special knowledge and experience that typical users of a language do not have. Letting these people nevertheless do language design bears a high risk of failure. And, by the way, modelers are usually employed for creating models, not modeling languages.

4 Defining Stereotypes

If we want to use stereotypes properly, we must have them properly defined first. However, this is not quite easy. The UML and OML reference manuals [12][6] are both considerably vague about the definition of stereotypes. The UML metamodel [15] is somewhat more precise. According to this model, the definition of a stereotype consists of a name, the host (language element(s) being stereotyped), an (optional) new graphic symbol and optional sets of UML tagged values and constraints. However, the UML metamodel still gives the stereotype designer (too) many degrees of freedom: Arbitrary notations may be used for the definition of tag data and constraints. Thus, anything goes, ranging from a primitive (name, host) declaration up to a highly sophisticated definition of the syntax and semantics of a stereotype.

The support for the definition of stereotypes in current tools tends towards the minimum: it is mostly restricted to the modification of graphical symbols – that is to say, to the definition of decorative stereotypes.

In order to define stereotypes of all categories properly, the base language should provide a framework for the definition of stereotypes that is well adapted to these categories. The required elements are shown in Table 1.

Table 1. Elements of a proper stereotype definition

	Decorative Stereotypes	Descriptive Stereotypes	Restrictive Stereotypes	Redefining Stereotypes
SYNTAX:				
Unique name for the stereotype	yes	yes	yes	yes
Host (the language element(s) that can be stereotyped)	yes	yes	yes	yes
Concrete syntax: graphic symbol, etc.	yes	yes	yes	yes
Pragmatics (concept behind the stereotype, for what it shall be used), given with text	(yes)	yes	yes	yes
Syntactic properties of the stereotype (a set of: keyword type [initial value])	no	yes	yes	yes
SEMANTICS:				
Restrictions that have to be fulfilled by all model elements that are instances of the stereotype*	no	no	yes	yes
Formally defined semantics, including restrictions for model elements that are not instances of the stereotype	no	no	no	yes

* Restrictions that shall be formally analyzable and checkable must be formally specified, for example with a constraint language like OCL [8]. Any other restrictions may be stated semi-formally or informally.

In Figure 3, we give a short example how the definition of a restrictive stereotype should look like. We define a stereotype Observer which supports the application of the observer pattern [7].

```
stereotype Observer {                              /* the name of the stereotype */
    host Class;                                    /* Classes can be stereotyped by this stereotype */
    properties {                                   /* declaration of the syntactic properties
        String id;                                      the stereotype introduces */
        ['Observable', 'Observer'] role;
    }
    restrictions {                                 /* restrictions for a restrictive stereotype */
        ( role = 'Observable' ) implies            /* Observable/Subject must have specific methods */
```

```
( ( exists a, d, n in self.operations |
        (a = "attach") and (d = "detach") and (n = "notify") )
    and ...    )
and
( role = 'Observer' ) implies              /* Observer must have update method and ... */
    ( ( exists u in self.operations |
            u = "update")
        and ( self.subject -> notEmpty )   /* there must be an observable for this observer ... */
        and ( exists c in self.subject |       /* ... this means, an associated class ... */
            exists s in c.stereotypes |         /* ... with a matching stereotype ... */
                (s = Observer) and (s.role = 'Observable') and (s.id = id) )
        and ...    )
    }
}
```

Figure 3. Example for the definition of a restrictive stereotype Observer. When this stereotype is assigned to a class, an id and a role must be specified. The restrictions prescribe that depending on the role the class must have some specific methods and that an observer must have a corresponding observable/subject.

Figure 4 gives an example where the stereotype Observer is applied to three classes. It should be noted that the latest released version of UML Semantics ([15] p. 53) does allow a model element to be assigned more than one stereotype, whereas the Language Reference Manual [12] forbids this, so that the definition of class B in Figure 4 would be *illegal* according to [12] but *legal* according to [15]. The authors of [12] motivate their decision with the argument of simplicity ([12] p. 450). In our view, this is a bad decision that seriously restricts the applicability of stereotypes.

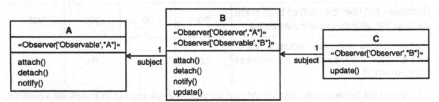

Figure 4. Example for the application of the stereotype Observer. In this example, Class B observes Class A and Class C observes Class B. Thus, Class B is acting both as an observable/subject and as an observer.

5   Guidelines for Stereotype Design

As mentioned earlier, designing stereotypes is a demanding task and the potential benefit of stereotypes heavily depends on taking the right design decisions.

From our experience with stereotypes we have assembled a preliminary set of guidelines for stereotype design.

General Advice

- Define a stereotype policy and enforce it: who (identify roles) has the right to define stereotypes of which category (e.g. according to our classification) for which purpose and with which scope (e.g. individual, project, department-wide, and company-wide).

- Make sure that every stereotype is properly defined and documented.

- Have every stereotype definition reviewed prior to using it.

- Avoid the creation of stereotypes when its scope is below the level of a project.

- Whenever you define a new stereotype, make sure that you will be able to maintain it in the scope and for the duration of its use.

- Make the stereotype definitions available to all people who need to know them and train these people how to apply and how to interpret them, respectively.

- Define less stereotypes and apply the existing ones more uniformly and with a wider scope.

Guidelines for Decorative Stereotypes

- Thoroughly examine the need for a decorative stereotype. Does the new symbol increase the semiotic value of a language element significantly? If you are not sure that there is a benefit in terms of increased understandability, then do not waste time with decorative stereotypes.

- Do not try to be an artist. Do not ever create a decorative stereotype because you have the feeling that a model looks better then.

- Use decorative stereotypes only to adapt the concrete syntax of a notation to a compulsory company or customer standard.

Guidelines for Descriptive Stereotypes

- Before defining a stereotype for an annotation or secondary classification, thoroughly determine that this is a required standard information with some concept behind it. A model style guide helps to decide.

- Decide whether it is better to include an annotation in the model or to document it separately. If the latter is true, do not define a stereotype.

- When annotations or classifications neither are a required standard nor have a clear concept behind them, use *notes* or other commenting constructs instead. *Notes* indicate that the interpretation of the given information is completely up to reader whereas a descriptive stereotype points out that there is a specific syntax and pragmatics in place.

- Do not try to cure symptoms. Descriptive stereotypes will not cure an unstructured model or the wrong choice of language element, diagram type, etc.

Guidelines for Restrictive Stereotypes

- Thoroughly investigate and discuss the need for the language extensions or modifications that shall be introduced with a restrictive stereotype.

- Never define restrictive stereotypes on the fly.

- Leave the definition of restrictive stereotypes to language and method specialists. For example, state in your stereotype policy (see above) that restrictive stereotypes may only be defined by your software methods group and that a formal validation and approval process has to be employed.

- Take care that a restrictive stereotype is really a restrictive stereotype and not a redefining one. State the semantics of the stereotype as precisely and formally as possible.

Guidelines for Redefining Stereotypes

- Do not do it. If you must do it, do it with extreme care and let only experienced language and method specialists do it.

- Explicitly forbid the definition of redefining stereotypes by individual engineers or within projects.

- Organize the definition of a redefining stereotype as a project of its own. Impose a rigorous design, validation and approval process.

6 Summary and Conclusions

Stereotypes are powerful, but care and experience is required to harness this power. Our classification helps to better understand the nature of stereotypes and to control their application. Every category in this classification represents a typical kind of applications for stereotypes. Hints for proper definition of stereotypes and preliminary design guidelines have been presented.

The discussion in section 3 and the guidelines in section 5 make clear that using decorative and redefining stereotypes both is highly problematic. Variations of the concrete syntax or the style of representation as well as a fundamental redefinition of the semantics of the base language should be done very restrictively only. Hence, descriptive and restrictive stereotypes are the most important ones in practice. Stereotypes from these two categories are especially useful to:

- make models more expressive by augmenting them with additional information in a standardized way

263

• compensate for deficits and weaknesses in a given modeling language in order to make it better adapted to some classes of problems or to given domains.

Stereotypes are no silver bullet. Their application does not automatically result in 'better' models. They always increase the complexity of the base language and introduce overhead for definition, training and maintenance. So, before introducing language extensions or modifications on the basis of stereotypes, it should be made sure that these are clearly beneficial.

Our further research on stereotypes will be directed towards empirical work on the usefulness of our classification, on better capabilities for the definition of stereotypes and on improving and validating the design guidelines. Finally, this endeavor should end up in a sound methodology for stereotypes.

References

[1]   Booch, G. (1994): *Object-Oriented Analysis and Design with Applications*, 2nd ed. Redwood City, Ca.: Benjamin/Cummings.

[2]   Champeaux de, D., Lea, D., Faure, P. (1993): *Object-Oriented System Development*. Reading, Mass., etc.: Addison-Wesley.

[3]   Coad, P., Yourdon E. (1991): *Object-Oriented Analysis*. Englewood Cliffs, N. J.: Prentice Hall.

[4]   Embley, D.W., Kurtz, B.D., Woodfield, S.N. (1992): *Object-Oriented Systems Analysis*. Englewood Cliffs, N. J.: Prentice Hall.

[5]   Eriksson, H.-E., Penker, M. (1998): *UML Toolkit*. New York: John Wiley & Sons.

[6]   Firesmith, D., Henderson-Sellers, B. H., Graham, I., Page-Jones, M. (1998): *Open Modeling Language (OML) – Reference Manual*. SIGS reference library series. Cambridge, etc.: Cambridge University Press.

[7]   Gamma, E. (1995): *Design Patterns: Elements Of Reusable Object-Oriented Software*. Reading, Mass., etc.: Addison-Wesley.

[8]   IBM et al. (1997): *Object Constraint Language Specification*; Version 1.1. [http://www.software.ibm.com/ad/oc]

[9]   Jacobson, I., Christerson, M., Jonsson, P., Övergaard, G. (1992): *Object-Oriented Software Engineering – A Use Case Driven Approach*. Reading, Mass., etc.: Addison-Wesley.

[10]   Joos, S., Berner, S. Glinz, M. (1998): Stereotypen und ihre Verwendung in objektorientierten Modellen – eine Klassifikation [Stereotypes and Their Usage in Object-Oriented Models – A Classification (in German)]. In: K.Pohl, A. Schürr, G. Vossen (eds.): *Proceedings of the GI workshop Modellierung'98*. TR 6/98-I, University of Münster, Germany. 111-115.

[11]   Rumbaugh, J., Blaha, M., Premerlani, W., Eddy, F., Lorensen, W. (1991): *Object-Oriented Modeling and Design*. Englewood Cliffs, N. J.: Prentice Hall.

[12]   Rumbaugh, J., Jacobson, I., Booch, G. (1998): *The Unified Modeling Language Reference Manual*. Reading, Mass., etc.: Addison-Wesley.

[13]   Selic, B., G. Gullekson, P. T. Ward (1994): *Real-Time Object-Oriented Modelling*. New York: John Wiley & Sons.

264

[14]   Shlaer, S., Mellor, S.J. (1988): *Object-Oriented Systems Analysis: Modelling the World in Data*. Englewood Cliffs, N. J.: Prentice Hall.

[15]   Unified Modeling Language Specification (1998): *UML Semantics v1.1*. Framingham, Mass: Object Management Group. [http://www.omg.org/techprocess/meetings/schedule/Technology_Adoptions.html#tbl_UML_Specification]

[16]   Wirfs-Brock, R., Wilkerson, B., Wiener, L. (1994): Responsibility-Driven Design: Adding To Your Conceptual Toolkit. *ROAD* 1, No. 2; (July-August 1994), 27-34.

[17]   Wirfs-Brock, R., Wilkerson, B., Wiener, L. (1993): *Designing Object-Oriented Software*. Englewood Cliffs, N. J.: Prentice Hall.

[18]   Wordsworth, J.B. (1992): *Software Development with Z*. Reading, Mass., etc.: Addison-Wesley.

First Class Extensibility for UML –
Packaging of Profiles, Stereotypes, Patterns

Desmond D'Souza, Aamod Sane, Alan Birchenough
{ desmond.dsouza | aamod.sane | alan.birchenough } @platinum.com

Abstract. We discuss a first-class extensibility mechanism for the UML based on Catalysis packages and frameworks [3]. Packages define and structure meta-model extensions for different modeling language "profiles". Package frameworks support lightweight extensions like stereotypes as well as heavyweight extensions. OCL can be used to define constraints and rules for profiles and frameworks. Our approach rationalizes and consolidates some core concepts within the UML standard, uses a simple general mechanism for layering facilities onto that core in a precise and well-defined way, and offers a way to simplify and re-factor the UML specification.

1   Introduction

The UML has become a family of modeling languages, each with its own specialized stereotypes and other extensions. This is understandable, since rule-based systems, compilers, signal-processors, control-systems, etc. all want the end-user modeling language to simplify the expression of their own problems and solutions. The OMG "green" paper document ad/99-04-07 titled "White Paper on the Profile Mechanism" [1], proposes a mechanism named "Profile" to add to UML 1.3 to structure UML extensions and provide:

- Specialization, visibility, and selection of meta-models
- Context for stereotypes, tagged values, and constraints
- Heavy-weight extensions to the meta-model
- Miscellaneous extensions, such as rules and notations.

In this paper we show how to consolidate current UML facilities to improve model structuring and management, satisfying the requirements for meta-model "profiles" while providing first-class extensibility, with a simple application of:

- the existing UML package and package-import/generalization mechanism
- frameworks - to describe recurring patterns in models and meta-models
- OCL - to express constraints on the meta-model

The argument is quite simple:

- Use packages to define and structure both the meta-model and its extensions: this is how the meta-model is already structured. Packages and import (or its variant, package *generalization*) provide all that is required to control granularity of element grouping and visibility; profiles are just packages.

- Utilize package imports (or package *generalization*) so an importing package can extend the definition of an imported element: there are good reasons why different aspects of a meta-element should be defined in different packages e.g. Profiles. UML 1.3 ostensibly supports this for package generalization.

- Define the "join" rules by which such extended definitions are combined: if a given meta-element is extended in two different profile packages, then we need clear rules for combining these two extensions.

- Use meta-model level frameworks to express common structural patterns in models: For example, a framework can express structural patterns like JavaBeans, lightweight extensions like stereotypes, relationships such as "instantiation" and its implied constraints on the graphs of model elements, as well as heavyweight meta-model extensions.

- Use OCL at the meta-model level to describe constraints: For instance, to describe a constraint that the "Java" profile does not allow multiple inheritance, add OCL expressions to the Generalization meta-model element to forbid multiple inheritance.

2   Why use UML Packages as the Base?

All UML modeling is done in some package. Package import (via generalization) define visibility of elements, so elements from another package are only visible if that package is transitively imported (or at least transitively used via package generalization relationship). The granularity of a package determines the granularity of visibility control. The UML meta-model is already defined in packages, so we already have a mechanism for controlled granularity and visibility of meta-models. If the granularity is not adequate it simply means that the UML meta-model should be re-factored into smaller packages.

3   Packages for Structured Meta-Model Extension

A profile allows an element of the UML meta-model to be selected and specialized. Figure 1 shows how a *Real Time* profile may add new properties to the model element named *Activation*, such as an attribute *duration*; and a *Load Balancing* profile may associate a *Node* with an activation. These profiles define extensions to the same meta-element, *Activation*, rather than defining sub-classes. In particular, if you work with a *Load Balancing + Real-Time* profile, every Activation would have both a *duration* as well as a *Node*.

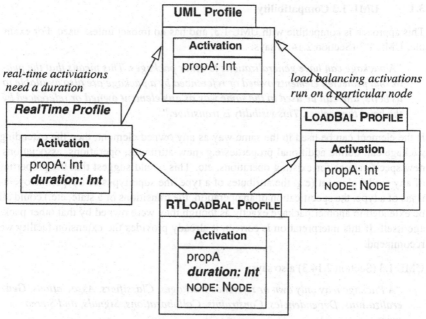

Figure 1: Package extension to define multiple "profiles"

A given (meta) model element — *Activation* in this case — may be introduced in one package, and then extended in other packages that have a generalization relation to it. As an analogy, note that *class* extension lets you *extend* a method (the mechanics might require "calling" the superclass method, or may offer an automatic mechanism such as before:after: declarations); common sub-typing specification practices let you *extend* an operation spec by adding more pre/post conditions. We want, at the package level, to *extend* any model element: same class with new attributes, operations, or invariants; same operation with new pre/post specs; etc.

When this feature is used, there is a difference between these two queries:

(a) *"What are the attributes of type Activation?"*

and:

(b) *"What are the attributes of Activation in package RealTime (or LoadBal)?".*

Naturally, (b) would return { *propA, duration* } for RealTime; and { *propA, NODE* } for LoadBal. We might choose for (a) to return { *propA* }; or it could be an undefined query. Thus, for a user, the result would depend upon which profile package was visible to him at that time.

3.1 UML 1.3 Compatibility

This approach is compatible with UML 1.3, and has no impact unless used. For example, UML 1.3 (Section 2.14.4) says:

> *"A package can have generalizations to other packages. This means that the public and protected elements owned or referenced by a package are also available to its heirs, **and can be used in the same way as any element owned** or referenced by the heirs themselves. This visibility is transitive."*

If the element can be used in the same way as any owned element, then the extending package can define additional properties e.g. new attributes, operations, associations, new specifications for existing operations, etc. This would suggest that the properties of any model element (e.g. the attributes of a type, the supertypes of a type, the operations of a type, the specification of an operation, the transitions of a state, etc.) could all be extended in another package exactly as though they were owned by that other package itself. If this interpretation is correct, it already provides the extension facility we recommend.

UML 1.3 (Section 2.14.3) also says:

> *"A Package may only own or reference Packages, Classifiers, Associations, **Generalizations**, Dependencies, Constraints, Collaborations, Signals, and Stereotypes."*

If a package owns generalizations, then package P3 can import package P2 and own a generalization between classes P2::A and P2::B; hence some features become associated with class P2::A only in package P3, by inheritance from P2::B via this generalization. This implies the UML must already deal with package-scoped properties of a class. UML 1.3 also already permits the same model element to be defined incrementally (with overlaps) across multiple diagrams in a given package, and so already has to compose incremental definitions from multiple places.

3.2 Contrast with Subclassing Approach

UML recommends subclassing as the default mechanism for extensibility. Contrast these two ways of accomplishing this in Figure 2: the one on the left is based on our approach, where each package describes a separate perspective on the (meta) classes; the one on the right uses subclasses with explicit multiple inheritance, and becomes unmaintainable very quickly. In fact, the subclassing approach becomes far worse if you consider extensions not just of (meta) classes, but of attributes, associations, operations, etc. due to the combinatorial explosion of artificial sub-classes and explicit multiple-inheritance. The example in Figure 1 would not work correctly if two separate subclasses were created for the RealTime and Load Balancing properties, unless you used explicit multiple-inheritance. Moreover, our approach supports an explicit subclass if needed.

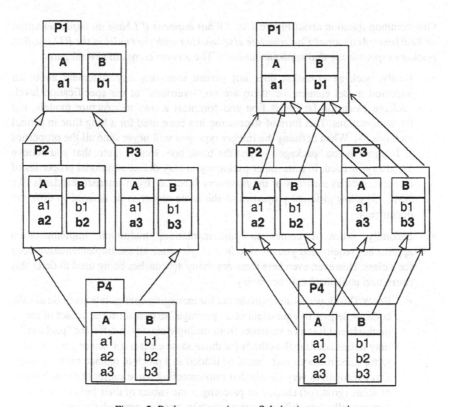

Figure 2: Package extension vs. Subclassing approaches

3.3 But <my favorite language> Works Differently

Let's start off by addressing a typical question:

> *"C++, Java, IDL don't let you define a class/interface in more than 1 place; why do this with UML?".*

Implementation languages impose different limitations on physical structure of code. Source-file based systems might require a complete class implementation to be in a contiguous section of a single file; modern-day IDEs offer direct manipulation of distinct program entities such as data members, methods signatures, and method bodies, with no underlying flat-file structure. Models and specifications, by their very nature, separate out different requirements or aspects of the same implementation unit. e.g. *"all remote calls must be synchronous"* and *"database calls must be remote"* and *"the place_order method must update the database"* are three distinct specification fragments that all influence the *place_order* method body. While certain UML packages might correspond to implementation units like files, namespaces, or modules, it is too restrictive to equate every model package with a particular language unit since you then lose much of the flexibility of factoring and re-using models and specifications.

One common question about Figure 1 is: *"What happens if I have an implementation of RealTime::Activation? Can someone else working with the model in the RTLOADBAL package expect to use that implementation?"* The answer comes in several parts:

- Firstly, package extension does not permit removing any statement about an imported model element i.e. there are no "overrides" at the specification level. Package extension facility is first and foremost a way to structure models, not implementations. This form of structuring has been used for a long time in formal methods [4]. When defining the Integer type you will never state all the properties of Integers in one "package" — just the basic ones like +, -, etc. that you believe are always needed. In some other package you may define additional properties of integers, such as *statistics* or the *fibonacci* function. Even Internet standards like XML [5] allow different properties of the same object to be defined in different documents.

- Secondly, when it comes to implementation, traditional implementation approaches require that you provide *in a single place* all the implementation code for a class. However, even here there are many approaches being used to defer this restriction until absolutely necessary.

  - ENVY/Developer — an environment for managing configurations of Smalltalk code — defines its equivalent of a "package" to contain some subset of the methods and instance variables from multiple classes; a different "package" may define additional methods for those same classes; or newer versions of those same methods that should be loaded as a patch to replace existing ones. Envy recognizes very clearly that implementing some subset of the behavior of some (group of) classes, or patching some subset of their behaviors, is a semantically very different operation than creating new subclasses.

  - *Aspect-oriented programming* is based on the assumption that you want to implement different *aspects* of a class (or a method, a collection of classes, an object, etc.) in different places to improve separate evolution and maintainability of those aspects. There is a stage where these different aspects are "woven" together into the combined implementation — in our terms, a package where the different implementations aspects are "joined".

  - The usual (IDL and C/C++) facility of *#include* is one implementation counterpart of package structure and imports. It is actually a very simple *"join"* based on textual concatenation. There is no reason to restrict ourselves to this, except at those selected package boundaries which must necessarily correspond to language boundaries (e.g. IDL modules).

- Thirdly, if you do have an implementation of *RealTime::Activation* and a client wanted to use *RTLoadBal::Activation* with that existing implementation, then you might use a delegation / wrapper around that implementation. Remember that the full name for any type is *PackageName::TypeName*.

3.4   Define "Join" Rules to Combine Element Definitions

Since 2 profiles may each "say more" about the same meta-model element, X, and we need to be able to combine profiles, when combined, you want meta-model element X to be automatically "joined". These join rules must be defined carefully. For example, when we are joining the two definitions of *Activation* into the RTLOADBAL package, the rule may include:

> *Join two type specs, the resulting attribute set is the <u>union</u> of the two attribute sets.*

> *Join two operation specs, the resulting spec is <u>pre1 & pre2</u> with <u>post1 & post2</u>.*

Note the following points regarding "join":

- The stereotype / sub-classing approach will not support this automatic "join" without combinatorial multiple inheritance; our "extension" approach will automatically perform the "join" (see Figure 2).

- The UML approach to multiple diagrams already requires "join". UML 1.3 permits a given model element to be defined across multiple appearances on multiple diagrams; hence the UML 1.3 already needs to combine these appearances (the 1.3 spec unfortunately omits these rules). We would utilize this behavior for definitions that were introduced across more than 1 package.

- A "join" facility is already needed for UML patterns, supported notationally in UML 1.3. Since several patterns can be applied to a given model element, substituting for types, attributes, etc, you end up with parts of multiple patterns defining a type, attribute, operation etc. These must be "joined" in the resulting model element.

- It is not possible to insist on global consistency across all profiles ever defined, or to require that every pair of extensions be disjoint. Hence it is possible that the "joined" versions across two profiles will result in an inconsistent meta-model. There are many different strategies to dealing with this situations, and we will not explore them here.

3.5   What can this be used for?

This facility can be used in many ways (subject to "join" rules, in Section 3.4).

- Tags: Separate tags into an importing package so basic model elements remain independent of those tags.

- Notations: Separate the notational aspects (e.g. stereotypes, specialized graphical or textual syntax) from the core model constructs they represent.

- Views: Support end-user structuring of models into separate views or perspectives, without having to create combinatorial (and often semantically confusing) subclasses and multiple-inheritance.

- OCL extension: those using OCL today quickly reach its limits in readability. For example, if you only have a fixed set of operations on *Sequence* it becomes very

272

awkward to define something like *shortestInitialSubsequence* repeatedly in terms of the predefined set. However, in our approach all of the OCL is defined in a package (using packages as frameworks we routinely define generic types such as *Seq(X)* and instantiate them with framework application — Section 4). You simply define your own *OCL Extension* package and extend the definition of *Sequence*; and you can continue to re-use packages that were developed in terms of the original OCL package. Subclassing would not work correctly here either: since there are existing packages which may use the original *Sequence* type, creating a separate subclass *MySequence* will not help with using those existing packages.

4   Frameworks — Light and Heavyweight Extensions

Frameworks are packages that capture recurring patterns. We suggest that they should be used at modeling as well as meta-modeling levels. The following sections show how frameworks help realize JavaBeans in a Java profile, stereotypes, and meta-level patterns like instantiation.

4.1   JavaBeans for a Java Profile

JavaBean models are based on properties, methods, and event. For example, a Stock object might have a *price* property; this means that the Stock class should have two methods, *getPrice()* and *setPrice()*, and an instance variable, *price*. In the JavaBean profile, you don't want to work at the level of individual getters and setters, but directly use the concept of property. This can be captured as the *Property* framework in Figure 3, in which we have used the UML pattern notation to show framework application. You import the *Property* subpackage of the *JavaBean* package (visible because you have imported *JavaBean*), and substitute Stock for Bean, Dollar for T, and Stock::price for Bean::property. The methods *get<price>* and *set<price>*, with their corresponding OCL specifications, are generated as a result of framework application. Note that the JavaBean Profile package provides the context for the Property pattern; the same approach can be used to define a profile for the Corba Component model, containing patterns such as *Ports, Facets, Receptacles,* and *Events*.

4.2   Stereotypes as Framework Application

A simple, novel, and semantically-rich way to define stereotypes is as a direct shorthand for framework application. Frameworks can simulate the default UML mechanism of implicit meta-model subclasses, or patterns on existing meta-model classes.

In UML, a stereotype is a meta-model element that is a subclass of a standard meta-model element. When you apply a stereotype <<property>> to an Attribute, that Attribute Instance is an instance of the *Property* subclass of Attribute. Frameworks already provide the capability to define a stereotype, and framework application lets us use a stereotype. We can make the *Property* stereotype definition in the traditional

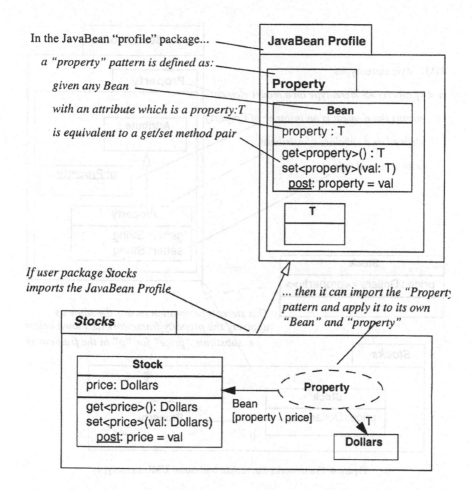

In the JavaBean "profile" package...

a "property" pattern is defined as:

given any Bean

with an attribute which is a property:T

is equivalent to a get/set method pair

JavaBean Profile

Property

Bean
property : T
get<property>() : T set<property>(val: T) post: property = val

T

If user package Stocks
imports the JavaBean Profile

... then it can import the "Property
pattern and apply it to its own
"Bean" and "property"

Stocks

Stock
price: Dollars
get<price>(): Dollars set<price>(val: Dollars) post: price = val

Property

Bean
[property \ price]

T

Dollars

Figure 3: JavaBean profile can be defined with packages and frameworks

UML form explicit in a framework, as in Figure 4. Note that the JavaBean profile, imported by *Stocks*, provides the context for the definition of the Property stereotype.

Rather than implicit sub-classes and tagged-values, we can apply stereotypes as we would apply any modeling pattern: the stereotype serves as a short syntax for an expanded form of that pattern, and additionally defines its semantics as a translation. For example, suppose we define *Stock* with the stereotype *property* as in Figure 5. We can regard this as syntactic sugar for the framework application in Figure 3. This is our preferred means to give semantics to stereotype definition and stereotype application, as framework application is grounded in very clear semantics. Still, frameworks can also support the subclassing style.

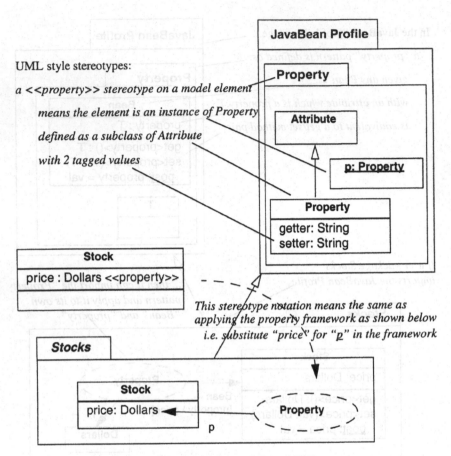

UML style stereotypes:

a <<property>> stereotype on a model element

means the element is an instance of Property

defined as a subclass of Attribute

with 2 tagged values

This stereotype notation means the same as
applying the property framework as shown below
i.e. substitute "price" for "p" in the framework

Figure 4: Frameworks can realize traditional UML stereotypes

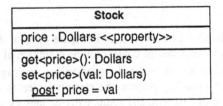

Stock
price : Dollars <<property>>
get<price>(): Dollars set<price>(val: Dollars)   post: price = val

Figure 5: Stereotype as syntax for pattern application

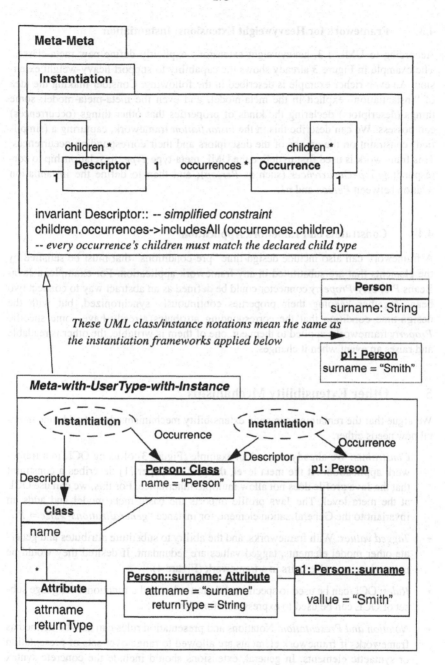

The following text appears within the figure:

Meta-Meta

Instantiation

children *
Descriptor 1   occurrences *   **Occurrence** children *
1   1   1

invariant Descriptor:: -- *simplified constraint*
children.occurrences->includesAll (occurrences.children)
-- *every occurrence's children must match the declared child type*

Person
surname: String

These UML class/instance notations mean the same as
the instantiation frameworks applied below

p1: Person
surname = "Smith"

Meta-with-UserType-with-Instance

Instantiation   Instantiation

Occurrence   Occurrence

Descriptor   Descriptor

Person: Class   **p1: Person**
name = "Person"

Descriptor

Class
name

*

Attribute
attrname
returnType

Person::surname: Attribute   **a1: Person::surname**
attrname = "surname"
returnType = String   value = "Smith"

Figure 6: Some frameworks apply across user, meta, and meta-meta levels

4.3 Framework for Heavyweight Extensions: Instantiation

According to UML 1.3, heavyweight extensions explicitly define new meta-classes. The example in Figure 3 already shows the capability to support heavy-weight extension. An even richer example is described in the following. Consider making the idea of "instantiation" explicit in the meta-model, and even the meta-meta-model: some things (descriptors) declaring the kinds of properties that other things (occurrences) can possess. We can describe this in the *Instantiation* framework, capturing a (simplified) constraint on the graphs of the descriptors and their corresponding occurrences. This framework is used first to define the UML meta-type *Type*, its relationship to corresponding *TypeOccurrences* (such as *Person*); and then to define the instantiation relation between *Person* and *p1*.

4.4 Constraints on Framework Application

A framework can also include design-time "pre-conditions" that must be satisfied by the elements that are substituted in any framework application. For example, a Java-Beans *Property-Property* connector could be defined as an abstract way to connect two beans together keeping their properties continuously synchronized, but with the design-time condition that the corresponding attributes already have some specific *Property* frameworks applied to them e.g. one of them is writable, the other is readable and raises an event when it changes.

5 Other Extensibility Mechanisms

We argue that the remaining semantic extensibility mechanisms do not add fundamental new needs either.

- *Constraints*: We already showed an example (Figure 3) of using OCL in a framework application. At the meta-level, the profiles paper [1] describes a constraint that the Java profile does not allow multiple inheritance. For this, we can use OCL at the meta level. The Java profile imports the basic meta model and adds an invariant to the Generalization element, for instance "*generalizations->size = 1*".

- *Tagged values*: With frameworks, and the ability to substitute attributes and generate other model elements, tagged values are redundant. If desired they would be untyped string-value pairs in a framework (Figure 4).

- *Rules*: OCL can be used to specify validation rules, and a functional-language subset of OCL can be used to express transformation rules.

- *Notation and Presentation*: Notations and presentation rules can also be defined as frameworks if framework elements are allowed to range over sets of presentation or syntactic elements. In general, extensions should include the concrete syntax that defines each newly introduced modeling construct. However, some new machinery may still be required to deal with this style of incremental syntax definition and syntactic ambiguities.

277

6 Proof of Concept

The mechanisms described in this paper are well understood and used in the Catalysis approach to using the UML [3] and implemented successfully in modeling tool prototypes. In Catalysis, all modeling and meta-modeling notions, extensions, and specializations, are defined in a structure of packages and frameworks. For example, see Section 9.8.2, Section 9.9.3, or Section 9.9.4 of the Catalysis book.

7 Conclusions

We have shown how packages and frameworks can provide a first-class extensibility mechanism for UML, and to structure UML extensions into "profiles". The UML 1.3 metamodel is already quite large and is known to contain some inconsistencies. Rather than add yet more facilities to it for extensibility or profiles, our approach is to rationalize and consolidate some core concepts within the standard, and then use a simple general mechanism for layering facilities onto that core in a precise and well-defined way. In fact, the UML itself could be simplified considerably if re-factored, and our approach makes it possible to re-factor the UML to a cleaner specification, while retaining the original semantics of the UML constructs where needed.

References

[1] "White Paper on the Profile mechanism", Version 1.0, OMG Document ad/99-04-07. Also available at http://uml.shl.com/u2wg/default.htm

[2] "OOAD and Corba/IDL - A Common Base", D. D'Souza and A. Wills, http://www.iconcomp.com/papers/omg-ooa-d/OMG-OOA-D-rfi.frm.html. This paper outlined a semantic scheme for extensibility and a layered meta-model to accomplish the goals that the Profiles paper raises.

[3] "Objects, Components, and Frameworks with UML - the Catalysis Approach", D. D'Souza and A. Wills, Addison-Wesley, 1998.

[4] "Larch — Languages and Tools for Formal Specification", http://www-the-ory.dcs.st-and.ac.uk/~mnd/larch.html

UML-Based Fusion Analysis

Shane Sendall and Alfred Strohmeier

Swiss Federal Institute of Technology
Department of Computer Science
Software Engineering Laboratory
1015 Lausanne-EPFL
Switzerland

email: *{Shane.Sendall, Alfred.Strohmeier}@epfl.ch*

ABSTRACT In recent times, there has been an increased requirement for software to be distributed. The well-known Fusion development method, however, can only be used to develop sequential reactive systems, and certain restricted kinds of concurrent systems. In contrast, the Unified Modeling Language (UML) provides a rich set of notations that can be used to model systems that are distributed. In addition, UML provides the ability to introduce rigor into diagrams through its constraint language OCL. In this paper, we present a UML-based Fusion analysis phase by way of a simple bank case study, and we discuss some enhancements that were made in addition to a mapping of notations; our proposal is the first step towards providing a Fusion-based analysis phase which supports high-level modeling of distributed systems.

KEYWORDS Fusion, UML, Analysis, Object-Oriented Software Development.

1 Introduction

The Fusion object-oriented software development method [4] is an integration of a variety of OO methods, including OMT [13], Booch [2] and CRC [18], in a single method framework. It provides a systematic process that takes the developer from requirements analysis to implementation. A downside of Fusion is that it can only be used to develop sequential reactive systems, and certain restricted kinds of concurrent systems.

The Unified Modeling Language (UML) [3] is a rich set of notations, brought about to unify the large number of notations that exist in the object-oriented development method community. It does not prescribe how its notations should be used, and thus methods are free to use it in any way that they deem fit; nevertheless, the target method must adhere to the well-defined semantics of UML.

The UML semantics is defined by a meta-model, which relies heavily on natural language and the Object Constraint Language OCL [17] to define well-formedness. UML's adoption by OMG in November 1997 has seen it since become the defacto standard for industry and academia alike.

The primary purpose for introducing UML into Fusion is to support the modeling of distributed systems, while retaining the systematic process, and separation of concerns of Fusion. Also, UML provides a vehicle to remove some of the known problems from Fusion models, such as unresolved interpretations of the object model [10], the inability of the life cycle model to express certain operation sequences [8], and a general difficulty in using regular expressions for modeling the system life cycle [5].

In this paper, we present a UML-based Fusion analysis phase by way of a simple banking case study and discuss some enhancements that were made in addition to the translation. In essence, we make four contributions to Fusion analysis above and beyond the result of a translation from Fusion to UML notations:

- establishment of precise definitions for actors and system entities and the communication between the two.
- formalization of the style and language for writing operation schemas via the introduction of OCL and usage rules.
- the ability to model concurrent interactions with multiple users.
- the ability to model multiple users as instances of actor classes and to identify such instances in order to interact with them (a known problem of Fusion in describing multi-user systems [15]).

This paper is organized in the following way: section 2 presents an overview of the Fusion analysis phase and the proposed mapping between diagrams; section 3 presents and discusses the UML enhanced analysis models for the Bank case study; section 4 is a discussion of some of the issues that arise from this work and compares it with work by other authors; finally section 5 sums up the main results presented in the paper.

2 Fusion analysis and UML

2.1 Fusion analysis

The analysis phase of Fusion abstracts from the internal working of the system, placing a particular emphasis on defining the system interface and the concepts related to the system domain. The dependencies between the models of the Fusion analysis phase (from the static point of view) is shown in figure 1. It is noteworthy that the flow of the elaboration process goes generally in the opposite direction of the dependencies. The concepts of the problem domain are described by the object model. The object model later evolves into the system object model. Neither the classes inside the object model nor inside the system object model have methods assigned to them. The procedure of attaching system operations to classes is deferred until the design phase. Thus, during analysis, the system can be viewed as a functional black-box.

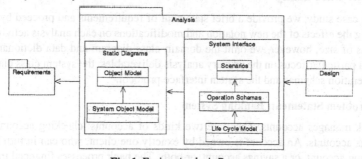

Fig. 1. Fusion Analysis Process

During analysis, Fusion elaborates a comprehensive set of specifications for the system level operations, described by operation schemas. An operation schema specifies the changes in system state and the production of output events by pre- and postconditions. Fusion also specifies temporal sequencing between system operations by regular expressions in the life cycle model. The operation schemas and the life cycle model complement each other well. Scenarios of system usage are used to find the right granularity for system operations and to aid the discovery of the life cycle model. Also, the use of a data dictionary is part of Fusion and it is used throughout the development life cycle.

2.2 Mapping to UML

UML provides a rich set of graphical notations which reflects the wide background and motivations of the UML Partners Consortium. Hewlett Packard's inclusion in the consortium and Fusion's roots in well-known methods has seen the incorporation of most of Fusion's models in UML. Thus, the high-level mapping for the diagrams in common is straightforward. Figure 2 displays the mapping, where the 'Model Name' column indicates the name that we have given to the resulting model.

Operation schemas do not have an equivalent UML diagram because all diagrams in UML are graphical. The developers of UML explicitly omitted textual models from the standard because they were considered the concern of CASE tools. Operation schemas are integrated into UML by attaching them to the corresponding operation declarations.

Classic Fusion Model	UML Diagram	Model Name
Object model	Class diagram	Domain class diagram
System object model	Class diagram	System class diagram (SCD)
Operation schema	— (use of OCL)	Operation schema
Scenario	Sequence diagram	Scenario
Life cycle model (reg. expressions)	Statechart diagram	System interface protocol (SIP)
System context diagram	Collaboration diagram	System context diagram

Fig. 2. Fusion to UML notation mapping

3 Bank Case Study

In this case study, we provide a brief statement of requirements and proceed by highlighting the effects of the new notation and modifications on each analysis activity. For reasons of size, however, we omit the domain class diagram and data dictionary. We place a particular focus on the primary analysis deliverables: the system class diagram, the operation schemas and the system interface protocol.

3.1 Problem Statement: Banking System

A bank manages accounts. There are two kinds of accounts: checking accounts and savings accounts. An account is owned by exactly one client, who can in turn own a checking account, or a savings account, or both. The bank processes financial transactions and enquiries: get the balance of an account, withdraw cash from an account, deposit cash into an account, and transfer money from one account to another. A transaction can, therefore, involve one or two accounts. The transactions are recorded, because at the end of each month statements are sent to all clients showing all transactions performed on their accounts during the last period.

We will suppose that all transactions and enquiries are performed by the clerks of the bank on behalf of the clients. The clients are, however, informed directly about the outcome of these transactions and enquiries, for instance by a receipt.

3.2 Class Diagrams

As one would imagine, the UML notation for class diagrams is very similar to the one used by Fusion for its object models. Nevertheless, there are slight differences in the definitions, such as associations and aggregations, but their explanation is outside the scope of this paper.

The domain class diagram, replacing the object model, defines the concepts and their associations related to the application and its domain by defining the classes, attributes, and the relationships between the classes. The Bank domain class diagram is not included here, due to its similarity to the system class diagram.

The system class diagram (SCD), replacing the system object model, may filter or extend the domain class diagram to represent the state space of the system. In the SCD we will use stereotypes to indicate specific semantics for some elements. These new stereotypes are system, actor, and id, as described in figure 3. Note that the term actor corresponds roughly to an agent in Fusion. For instance, a human agent could act as a customer or car driver.

Keyword	Applies to element	Well-formedness rules & rules for general use	Meaning
system	classifier	1. There is exactly one element that is stereotyped 'system' for each SCD. 2. All associations with the system (and those that originate from within the system itself) that go outside the system entity are connected to actors. 3. These associations (defined by 2) are not navigable unless stereotyped 'id' (described below). 4. The system element has a single instance.	It defines an element that is equivalent to the system boundary of Fusion's system object model, except that it can have associations of its own, i.e. the association is attached to the system entity itself. From a behavior point of view the system is reactive.
actor	classifier (stick-figure icon)	1. An actor element can only communicate with the system if an association exists between itself and the system, the system entity itself or an entity within the system. 2. If such an association exists (defined by 1), it implies that the actor can only communicate with the system through the system interface, and thus cannot directly access the internals of the system. 3. An actor element has implicitly zero or many instances.	It defines an entity that provides stimuli to the system. It exhibits active behavior. It does not have visibility into the internal structure of the system.
id	association	1. An 'id' association must be between an actor and the system (either the system itself or an entity within the system). 2. This stereotype places an obligation on the implementation to realize a mechanism which determines a unique actor instance given an object which is part of the system and participates in the association.	An 'id' association serves the purpose of allowing the system to identify the actor on the other end of the association.

Fig. 3. Stereotypes on UML Elements of the System Class Diagram

3.2.1 Bank System Class Diagram

The Bank SCD, shown in figure 4, defines accounts, customers, transactions and a calendar as entities of the system, and clerks, clients and a clock as actors. The calendar and clock are present, as they represent time in the system and its continual update. Transaction is a generalization of deposit, withdraw and transfer transactions. Account is a generalization of the Checking and Savings accounts. Constraints and invariants can be placed in UML notes and attached to elements of the SCD, including the system.

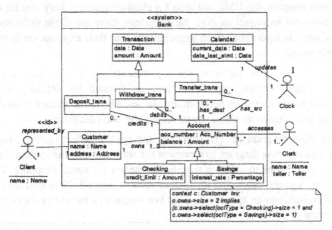

Fig. 4. Bank System Class Diagram

The Bank SCD displays a system invariant, written in OCL in a UML note; it states that any customer who owns two accounts must have one checking account and one savings account. In the OCL expression, the dot operator stands for navigation, for instance from an individual object to a set, e.g., if c is a customer then c.owns is the set of accounts owned by customer c. The select operator is a predefined collection operator of OCL that designates all the elements satisfying a given expression, e.g. select(oclType = Checking). Postfix notation is used for denoting operations on collections, e.g., ->size is predefined and yields the size of the collection which precedes the arrow.

3.3 System Interface

The system interface is defined by the system context diagram, the collection of operation schemas and the life cycle model. Scenarios of system use are constructed in all but the most simple applications; they aid the discovery of the principal models.

3.3.1 Bank System Context Diagram

The system context diagram is a model that has found its way into the method via the recommendation of several authors [16, 12]. It provides a clear view of the system interface, by showing all events exchanged at the system interface, input and output. The Bank system context diagram is shown in figure 5, showing the input and output events at the Bank system interface. We model it in UML using a descriptor-level collaboration diagram, where the half-arrow lines indicate asynchronous events exchanged between the system entity and actor entities. In this diagram, actors represent classes rather than individual instances. For the purpose of conciseness, event parameters are omitted from this diagram.

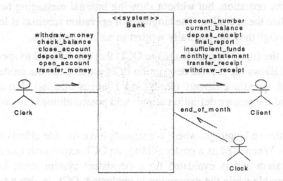

Fig. 5. Bank System Context Diagram

3.3.2 Bank Scenario

Scenarios are used to discover and check the system interface. They are indispensable when the sequencing of the system operations is not intuitive or the system interface contains complex interactions. Also, they provide a tremendous aid in finding the right granularity for the system operations. The scenario in figure 6 illustrates a possible series of interactions between the system, SwissBank, and two clients, Josh and Sophia, speaking to their respective clerks. The clients ask to open accounts and perform transactions on their respective accounts, and the system responds to these requests. At the end of the month, the system sends a statement to the two clients. Normally, many scenarios will be produced, but we show only one here.

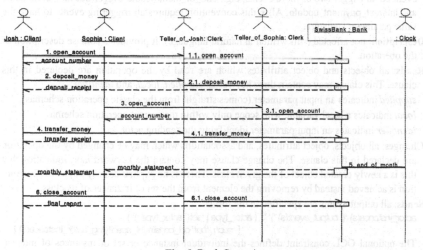

Fig. 6. Bank Scenario

3.3.3 Bank Operation Schemas

An operation schema defines the functionality of a system operation. A schema is written in a declarative style, which details all objects and associations that are read or updated by the operation, events generated by the operation, and the pre- and postconditions ruling the operation, but without showing internal messaging between objects required to realize the operation. The language of operation schemas is left unspecified by Fusion, although they are generally written in natural language.

UML provides the (semi-)formal language OCL that can be used to specify operation schemas. OCL allows the precise description of constraints on the model (invariants), the mention of specific elements (reads and changes clauses) through navigation expressions, and the precise definition of pre- and postconditions (assumes and results clauses).

OCL is a declarative language, where constraints have no side effects and are evaluated atomically. When used as a postcondition, an OCL expression can not change the state of the system and is evaluated on a consistent system state, i.e., no system changes are possible while the expression is evaluated. OCL is also a typed language where types are any kind of Classifier, the basic building block of UML. OCL also provides the benefits of precision and unambiguity associated with a formal language.

For the format of operation schemas we will keep the one used for Fusion. For their contents, however, we will use OCL together with some additional keywords and operators. The clauses of an operation schema have the following meanings and contents:

Operation: displays the operation name. The *triggered_by* keyword is used when the input event that triggers the operation is of a different name—not possible in classic Fusion. If this keyword is not supplied, the triggering event is assumed to have the same name as the operation. The *triggered_by* keyword is followed by a simple or-expression indicating that the operation can be triggered by different events on different executions. Note that any execution of the operation is only triggered by a single event. Also, note that it is perceivable that a single event could trigger two or more events, e.g., end_of_month could trigger account_statements and interest_payment_update. Also, this convention requires all triggering events to have the same parameter signature.

Description: not affected (still written in natural language). It provides a concise description of the operation.

Reads: all objects and object attributes which are read by the operation are declared in this clause. This clause may contain three keywords: *supplied, local,* and *identifier*:
supplied indicates an input parameter (comes straight from the classic operation schemas).
local indicates a local variable with scope only within the current operation schema.
identifier indicates an input parameter identifying the sending actor.

Changes: all objects, object attributes, and associations which may be changed by the operation are declared in this clause. The change clause may contain the keyword *new*, indicating that this is a newly created object of the given class. There is no explicit destroy keyword. Destruction is achieved instead by removing the element from the set of instances of its class.

Sends: all output events are specified in the form:
actor_instance(s) '{' output_event(s) '}' ':' (actor_type | 'set(' actor_type ')')
['*such_that*' ocl_constraint_specifying_actor_instance(s)]
The optional OCL constraint defines the individual instance or set of instances of interest. Note that in classic Fusion it is not possible to send an event to a set of actors.

285

Assumes: OCL precondition, a boolean expression.

Results: OCL postcondition, a boolean expression including the if-then-else[1] conditional construct (highlighted with bold font) and the keyword *sent*, which indicates that an event has been sent to an actor.

Here are some additional comments applying to operation schemas:

- The additional keyword *such_that* can be used in the Reads, Changes, and Sends clauses; it is used to select the exact instance(s) of interest described by an OCL expression.
- There is an implicit logical *and* between blocks within assumes and results clauses.
- bank.customer designates all instances of the class Customer, which can also be written bank.customer.allInstances. Because all expressions are evaluated in the context of the system, the prefix 'bank' can be omitted, i.e. customer is equivalent to bank.customer. As a consequence, class names should not be used as instance names, even with an initial lowercase letter.
- We allow the use of side-effect free functions defined in the data dictionary. A function description consists of a function signature separated by the '==' symbol, followed by the body, which is an OCL expression returning a value compliant with the signature.

We now present two operation schemas of the Bank system: transfer and account_statements.

Operation: transfer_money
Description: Transfer an amount of money from a given account to another, only if the amount of money does not exceed the source account's limit.

Reads: *supplied* src_a: Account
supplied dest_a: Account
supplied amount: Amount
calendar.date : Date
transaction -- all transactions, equivalent to Transaction.allInstances
local src_has_suff_funds: Boolean =
(src_a.oclType = Savings *and* src_a.balance >= amount) *or*
(src_a.oclType = Checking *and* src_a.balance + src_a.credit_limit >= amount)

Changes: *new* t : Transfer_trans
src_a.balance, dest_a.balance: Amount
has_src, has_dest: (Transfer_trans, Account)

Sends: client {transfer_receipt, insufficient_funds} : Client *such_that*
src_a.owns.represented_by->includes(client) -- client owner of src account is sent the event

Assumes:

Results: **If** (src_has_suff_funds) **then**
src_a.balance = src_a.balance@pre - amount
dest_a.balance = dest_a.balance@pre + amount
t.oclIsNew -- OCL property which evaluates to true if the object is created during the operation
t.date = calendar.date
t.amount = amount
src_a.has_src = src_a.has_src@pre->including(t) -- the new transaction is linked to its source account
dest_a.has_dest = dest_a.has_dest@pre->including(t) -- the new transaction is linked to its dest account
sent client transfer_receipt(src_a, dest_a, amount)
else -- not src_has_suff_funds
sent client insufficient_funds(src_a, amount)
endif

Fig. 7. Transfer_money operation schema

---

1. This is in addition to the OCL *if-then-else* which returns a value when evaluated.

The transfer_money operation schema is shown in figure 7. It describes the action taken to transfer money from one account to another provided that the source account has sufficient funds. The reads clause contains a local boolean declaration, which is used to improve the readability of the schema. In the results clause, if the source account has sufficient funds then the money is transferred from the source to the destination account, a new transaction is created and linked to the involved accounts, and a receipt is sent to the source account owner; otherwise the source account owner is sent an insufficient funds event.

Operation: account_statements
     triggered_by end_of_month
Description: Statements are sent to all account owners, detailing their respective transactions over the last period (i.e., since last statement).
Reads: calendar.current_date: Date
    account – all accounts, equivalent to Account.allinstances
    transaction – all transactions, equivalent to Transaction.allinstances
Changes: calendar.date_last_stmt : Date
Sends: clients {monthly_statement} : set(Client) = customer.represented_by – sent to all clients
Assumes:
Results: Account->forall(a | *sent* a.owns.represented_by monthly_statement(a,
last_period_transactions(a, transaction, calendar.date_last_stmt@pre)))
        – a statement (details all transactions of an account since the last period) is sent to the owner of each account
    calendar.date_last_stmt = calendar.current_date

-- defined in Data dictionary
function
last_period_transactions(a:Account, trans: Set(Transaction), earliest: Date): set(Transaction) ==
    trans->select(t | t.date >= earliest *and* (t.credits = a *or* t.debits = a *or* t.has_src = a *or* t.has_dest = a))

Fig. 8. Account_statement operation schema

The account_statements operation schema is shown in figure 8. It describes the action of sending out account statements at the end of the month. The *triggered_by* keyword in the operation clause indicates that the name of the triggering event is end_of_month whereas the name of the operation is account_statements. The first line in the results clause states that a statement is sent out for each account to its owner; the statement details all the transactions since the last statement.

3.3.4 Bank Life Cycle Model

The Fusion life cycle model defines the behavior at the system interface. It abstracts all possible scenarios by describing all possible sequences of operations at the system interface. The usefulness of the life cycle model should not be underestimated. The life cycle model becomes very important in systems that exhibit complex interactions, especially when there are too many possible scenarios.

The regular expression format of life cycles is not found in UML. The closest UML equivalent is the statechart diagram used as a protocol specification. We prefer statechart diagrams over activity diagrams, the other possible choice, because the former are more concise when modeling event-level interactions. A protocol specification captures all the external interactions of the system, leaving all internal transitions invisible.

The typical use of state diagrams as low-level behavior descriptions of objects is, therefore, not the intended use here. In short, we will describe the system interface protocol (SIP) with UML statechart diagrams used as protocol specifications. Note that the SIP offers additional expressiveness over regular expressions, which we will demonstrate in what is to follow.

```
Bank: ((Initialization | Transaction | Enquiry | Finalization)* || Statements)
Initialization = open_account . #account_number
Transaction = withdraw_money . (#withdraw_receipt | #insufficient_funds)
              | deposit_money . #deposit_receipt
              | transfer_money . #transfer_receipt
Enquiry = check_balance . #current_balance
Statements = account_statements . #monthly_statement*
Finalization = close_account . #final_report
```

Fig. 9. Classic Life Cycle for Bank

Notation	System interface protocol terms and constraints
event-name '(' parameters ')' '/' action(s) '^' send-clause A B when '(' bool-expr ')'	• Input events are mapped to transition triggering events. Parameters are optional, and should generally be omitted for the sake of readability, since they are found in the operation schemas anyway. • System operations are mapped to actions on transitions. *Convention*: no action clause implies that the operation name is the same as the input event name. • Output events are placed in the send-clause. A send-clause is a special action on transitions, indicating the destination of the event. • A simple state (e.g. A) is interpreted as a wait state for the next event. • Initial and Final states are shown with filled circles and bulls-eyes respectively. • A change events (**when**) is triggered when its boolean condition becomes true.
B C A	A diamond symbol represents a branching transition. Branches are strictly used to differentiate possible output events of a system operation. The conditions for the branches are not shown in SIPs to avoid redundancy with the operation schemas.
A B	Concurrent state machines (so called AND-states) are denoted by a thick dotted line dividing the parent state machine, which means the parent is divided into two or more concurrent submachines whose initial transitions are triggered at the same time as the parent state machine is entered.
A * A A	A star, or more generally a multiplicity string, in the upper right hand corner denotes a dynamic state machine. It specifies the number of concurrently executing identical state machines. These concurrent state machines can start execution at different times, however for completion they have to wait for each other. The multiplicity of the dynamic state machine is resolved when entering the state and is equal to the number of argument lists supplied to it. Dynamic state machines are used in UML activity diagrams and are not strictly speaking part of UML statechart diagrams.

Fig. 10. System interface protocol mapping

Lets now contrast the old model and notations against the new ones. We illustrate this by the classic Fusion Bank life cycle model (figure 9) and a more expressive SIP that is able to express concurrent account activities (figure 11). The life cycle model (figure 9) displays the sequencing of the bank operations and output events by way of four operators: sequence '.', alternative '|', optional repetition '*', and interleaving '||'. The prefixed '#' symbol indicates an output event. Subexpressions, e.g. Enquiry, are used as a structuring mechanism, so that the life cycle model is more readable and manageable.

Before we display and discuss the SIP model, we show in figure 10 the notations used in UML statechart diagrams, the restrictions we apply on them, and their relations to Fusion terms.

Figure 11 shows the SIP for the bank system. The system comprises two concurrent state machines, Account_activity and Statement_activity. Account_activity in turn is a dynamic state machine. Thus, many account activities can operate concurrently to each other.

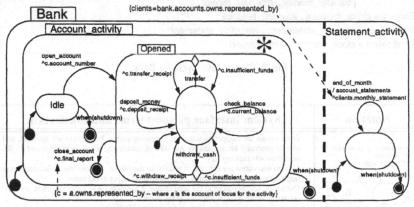

Fig. 11. Bank system interface protocol

According to the definition of a dynamic state machine, the multiplicity of its concurrent state machines must be resolved upon entering it, i.e., the number of state machines that are spawned, when the dynamic state machine is entered, must be known. It is, however, not possible to predict how many accounts will be opened during the lifetime of the bank. A possible, not ideal, solution is to work with a fixed upper limit for the multiplicity and therefore the maximum number of accounts.

As explained in figure 10 we can exit the dynamic state machine Account_activity only when all its submachines have completed. This means that we have to take care of the cases where the account is never opened, which are handled by the transition when(finished) exiting from the Idle state.

In Statement_activity, we can see that it is possible to differentiate the triggering event name from the executing operation name; end_of_month/account_statements is such an input event/operation pair. It is also important to note that the send-clause shows that an event of type monthly_statement is sent to each client of the bank, the set of these clients being defined by an OCL expression as the account owners.

4 Related Work and Discussion

4.1 Related Work

Atkinson [1] gives a broad overview of a Fusion-based method that uses UML notations, called FUML. He proposes a split in the analysis phase: conceptualization and detailed analysis. In our approach, we do not propose an explicit split but, as we will discuss below in section 4.3, we make a difference in the level of detail between models within analysis. Also, FUML proposes the use of a design class diagram in the design phase models, a proposal we are in agreement with. A design class diagram allows one to make a clean separation between analysis and design classes. There is no explicit design class diagram in classic Fusion. The method uses the analysis class diagram as the basis for design, allowing only limited additions to the diagram, an approach which does not really scale well to large complex systems.

Lano et al. [12] make an interesting comparison between Fusion and Syntropy and roughly outline a method out of the two, which includes a few supplementary models. Although, they do not go into the details of their method, the goal of their proposal relates closely to ours: retaining the systematic approach of Fusion while injecting rigor; they achieve this rigor via Syntropy, which uses an early form of OCL.

Catalysis [9] is a method that has roots in Fusion [8], and advocates rigor through constraints and refinement, and reuse through frameworks.

We claim that the proposals by Atkinson and Lano et al. do not provide the same level of detail as we do in this paper, whereas the Catalysis approach indeed offers a direct alternative to our approach.

4.2 System Interface Protocol

As we have shown, the addition of dynamic state machines is not well suited to model the behavior of multiple concurrent users, such as clients opening, managing and closing their accounts independently of one another. Because of the importance of dynamic actors in distributed systems, this problem is worth to be solved.

As stated we use UML statecharts with limited capabilities because we want only to model the interface protocol of the system. The problems related to the dynamic creation of activities are due to the constraint not to expose the internal working of activities, since without that constraint it would be possible to dynamically create active objects.

In fact, this is the approach taken by Catalysis. Catalysis goes even further by attaching statecharts to types instead of classes[1]. The claimed advantage of attaching statecharts to types is that as design evolves and classes are added and changed the analysis types containing the statecharts are unaffected. However, this approach requires one to assign behavior to types, which in many cases may be premature. Thus, one loses the flexibility that one would have when making the SIP independent of any type or class.

---

1. In short, a type is an analysis level class that will typically be refined into a design class at design time, but is not obligated to. Catalysis uses type models rather than class models in analysis.

Together with Fusion we believe that a clear separation between analysis and design activities promotes cleaner and more effective models. We agree with Fusion which claims that responsibilities should be allocated during design rather than analysis.

We, therefore, believe the most effective approach is to model the SIP independent of analysis classes/types. Thus, a proposal for enhancing UML statecharts with dynamic creation of activities is required. Such a proposal is the subject of our current work.

4.3 Operation Schemas

The approach we propose for describing the contents of operations schemas provides a precise and consistent notation that will allow them to play a greater role in verification and testing. The replacement of natural language with OCL does, however, make operation schemas harder to produce. This can be remedied by the intermediate development of use cases between functional requirements capture and work on operation schemas. This is consistent with the findings of Coleman [5]. An expansion on this idea is presented in [11], and is an area of our current research.

Catalysis promotes a generalized notion of operation schema, named action: a localized action is equivalent to an operation schema and a joint action to a use case. However, we are not sure that these additional concepts make it easier to specify system operations. We claim that the language and guidelines that we use for specifying operations are more precise than the ones used by Catalysis, almost only implicitly defined by examples. Finally, it might be important to know that Catalysis uses some terms and definitions that are not compatible with the UML/OCL standard.

4.4 In Perspective

Since our work is focused on modeling distributed systems during analysis, this paper defines the initial ground work that we will build upon to fulfil our goal of producing a comprehensive set of guidelines and techniques for modeling distributed systems, while endeavoring to provide a solid foundation for supporting the explicit description of software architectures.

5 Conclusion

A UML-based and enhanced Fusion analysis phase has been demonstrated on a simple bank case study. This new approach resolves many of the problems of the original Fusion analysis phase thanks to the introduction of UML notations and changes to the method itself, and it allows one to take advantage of the rich tool support available for UML.

In essence, we have made four contributions to the Fusion analysis phase above and beyond the result of a translation from Fusion to UML notations:

* establishment of precise definitions for actors and system entities and the communication between the two.
* formalization of the style and language for writing operation schemas via the introduction of OCL and usage rules.
* the ability to model concurrent interactions with multiple users.

- the ability to model multiple users as instances of actor classes and to identify such instances in order to interact with them.

Acknowledgments

The authors would like to thank Mohamed Kandé, Didier Buchs, Jörg Kienzle and the anonymous referees for their helpful comments. This work has been supported partially by the Swiss National Science Foundation, grant no. 20-50635.97.

References

[1]   C. Atkinson. *Adapting the Fusion Process to Support the UML*. Object Magazine, Sigs Publications, Nov 1997.

[2]   G. Booch, *Object-Oriented Analysis and Design with Applications*, second edition, Addison-Wesley, 1994.

[3]   G. Booch, J. Rumbaugh and I. Jacobson. *The Unified Modeling Language User Guide*. Addison-Wesley 1998.

[4]   D. Coleman, P. Arnold, S. Bodoff, C. Dollin, H. Gilchrist, F. Hayes and P. Jeremaes. *Object-Oriented Development: The Fusion Method*. Prentice-Hall 1994.

[5]   D. Coleman. *Fusion with Use Cases - Extending Fusion for Requirements Modelling*. OOPSLA Conference Tutorial Slides 1995 (available at http://www.hpl.hp.com/Fusion/me_books.html).

[6]   S. Cook and J. Daniels, *Designing Object Systems: Object-Oriented Modelling with Syntropy*. Prentice Hall 1994

[7]   T. Cotton. *Evolutionary Fusion: A Customer-oriented Incremental Life Cycle for Fusion* in Object-Oriented Development at Work: Fusion in the Real World. pp. 179-202, (eds. R. Malan, R. Letsinger and D. Coleman) Prentice-Hall 1996.

[8]   D. D'Souza and A. Wills. *Extending Fusion - Practical Rigor and Refinement* in Object-Oriented Development at Work: Fusion in the Real World. pp. 314-359, (eds. R. Malan, R. Letsinger and D. Coleman) Prentice-Hall 1996.

[9]   D. D'Souza and A. Wills. *Objects, Components and Frameworks With UML: The Catalysis Approach*. Addison-Wesley 1998.

[10]   G. Eckert. *Improving the Analysis Stage of the Fusion Method* in Object-Oriented Development at Work: Fusion in the Real World, pp. 276-313, (eds. R. Malan, R. Letsinger and D. Coleman) Prentice-Hall 1996.

[11]   Hewlett Packard. *Engineering Process Summary (Fusion 2.0)*, http://www.hpl.hp.com/Fusion/, 1998.

[12]   K. Lano, R. France and J-M. Bruel. *A Semantic Comparison of Fusion and Syntropy*. Object-Oriented Systems, 1998; To be published, http://www.univ-pau.fr/~bruel/publications/oos98.ps .

[13]   J. Rumbaugh, M. Blaha, W. Prembalini, F. Eddy and W. Lorensen. *Object-Oriented Modeling and Design*. Prentice Hall, 1995.

[14]   J. Rumbaugh, I. Jacobson and G. Booch. *The Unified Modeling Language Reference Manual*. Addison-Wesley 1999.

[15]   S. Sendall, N. Guelfi and A. Strohmeier. *Fusion Applied to Distributed Multimedia System Development: the Easy Meeting Case Study*. Technical Report EPFL-DI No 99/304, Swiss Federal Institute of Technology in Lausanne.

[16]   A. Strohmeier. *The Fusion Method, with Implementation in Ada 95*; SIGAda'98 Tutorial Notes, Washington, DC, USA, 1998, ACM Press, 1998.

[17]   J. Warmer and A. Kleppe. *The Object Constraint Language: Precise Modeling With UML*. Addison-Wesley 1998.

[18]   R. Wirfs-Brock, B. Wilkerson and L. Wiener. *Designing Object-Oriented Software*. Prentice-Hall, 1990.

Using UML for Modelling the Static Part
of a Software Process

Xavier Franch[1], Josep M. Ribó[2]

[1] Universitat Politècnica de Catalunya (UPC),
c/ Jordi Girona 1-3 (Campus Nord, C6) E-08034 Barcelona (Catalunya, Spain)
franch@lsi.upc.es
[2] Universitat de Lleida
P. Víctor Siurana 1, 25003 Lleida (Catalunya, Spain)
josepma@eup.udl.es

Abstract. We study in this paper the use of UML as a tool for modelling the process of software construction. As a case study, we deal with the process of building a library of software components. UML is used in order to define the static part of the process, i.e., the elements that take part on it and their structural relationships. We think that our approach supports some interesting properties in the field of software process modelling (e.g.: modularity; expressivity in model construction; sound formal basis; and flexibility in model enactment). Besides showing the adequacy of UML for modelling the static part, the paper outlines also some drawbacks concerning the description of the dynamic behaviour of the process using only UML, and some possible solutions to them.

1 Introduction

A model for a software development process (i.e., a *software process model* [5], SPM for short) is a description of this process expressed in some *process modelling language* (PML). The process can be viewed as the cooperation of many *tasks* (e.g.: requirements elicitation, component testing) that use and develop some *documents* (e.g.: specification, test plan) with the help of some *tools* (e.g.: CASE-tools, debuggers) and using some *resources* (e.g.: data bases, computer networks). Tasks involve many *agents* (e.g.: people, hardware media) which play specific *roles* (e.g.: programmer, manager) and which coordinate through some *communication* media (e.g.: e-mail, fax).

Hence, the definition of a SPM must state all the elements just mentioned, and also the way in which this model must be executed (*enacted*). This idea leads to the notion of *static* and *dynamic* parts of a model. The static part is given by means of a conceptual model that defines the elements that take part in the SPM. On the other hand, the dynamic part consists of a description of the way in which the model is enacted (e.g.: ordering of tasks). The systematic description of both parts not only helps in understanding software development, but also allows the construction of systems for supporting automation of the process up to an acceptable level.

This topic has drawn a special attention within the scientific community and, as a result, several PMLs have been developed (see [6] and [5] for a survey). Currently a second generation of PMLs is coming into existence ([22], [23]) trying to fix some common drawbacks of most of former approaches: difficulty to express complex processes and to understand the resulting model; non-visual models or too naïve visual ones; difficulty to formalize the many facets of the process; use of one single language paradigm; etc. However, even taking this last generation into account, it is the case that most of them fail in a central issue, the use of standard, widespread notations and tools. It is a fact that this is one of the main reasons for which none of existing approaches has been widely adopted by the software engineering community.

We are developing a PML called PROMENADE (**PRO**cess-oriented Modelling and **ENA**ctment of software **DE**velopments) intended to be a part of this second generation of languages. PROMENADE, among other features (e.g. modularity, expressivity, formality and flexibility), aims at taking advantage of standard languages and tools in software engineering. One of such emerging languages is UML [20], which has acquired a great deal of interest in the last few years and which is becoming a standard *de facto* in software engineering (both in industry and in academia). This fact, together with its adequacy for modelling the structural elements of a software process, makes UML playing an important part in PROMENADE to describe the static part of a SPM.

In this article we propose, as a case study, the modelling of the construction process of a library of components in PROMENADE. We have chosen this particular process because it plays a central role in our research project, *ComProLab* (a Component Programming Laboratory, see [8]). We will focus on the description of the static part of such a process, made with UML.

In ComProLab, a *component* is defined by means of a *specification*, which includes two parts: a *functional specification*, stating how does the component behave, and a *non-functional specification*, that declares additional requirements referred to some operational *non-functional attributes* (as efficiency and reliability); these attributes are defined in *property modules*, which are imported in non-functional specifications. Once the specification is complete, an *implementation* may be built for this component, which is required to fulfill the properties stated in both parts of the specification. Implementations include a description of their *non-functional behaviour*, which determines the values that the operational attributes take in this implementation, possibly stating some additional constrains over implementations of the imported components. Fig. 1 shows the whole picture.

The rest of the paper is organised as follows. Section 2 presents both the PROMENADE metamodel and reference model, which acts as the basis on which any other PROMENADE model will be expressed. Sections 3 and 4 show the most relevant aspects of the static part of the modelled process (documents and tasks). Section 5 sketches some limitations of UML in order to model the dynamic part of a software process and outlines some possible solutions. Finally, section 6 provides the conclusions and some related work.

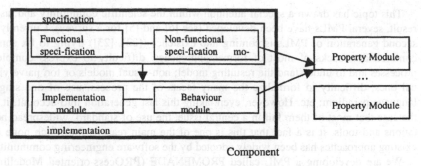

Component

Fig. 1. Organization of a component in the ComProLab approach

2 The Metamodel and the Reference Model

The PROMENADE metamodel is built as an extension of the UML metamodel [18], adding a metaelement for each one of the core elements that take part in any PROMENADE model (e.g., it adds a *MetaTask* metaelement for tasks, a *MetaDocument* metaelement for documents, a *SPMetamod* metaelement for the model itself, etc.). The model elements (e.g., the class *SpecifyComponent* for the task of specifying a component) are seen as instances of the metaelements (e.g., *MetaTask*) whose (meta)features (metaattributes, metaoperations, etc.) have been given a value. Therefore the process of building a model in PROMENADE will consist mainly in creating instances for the metamodel classes and give values to its metafeatures.

The PROMENADE approach to process modelling defines a universal or *reference model* (an instance of the PROMENADE metamodel) that constitutes the basis on which any other process model will be built (as done for instance in [3]). Hence, this *reference model* is the common and extensible kernel shared by all processes modelled in PROMENADE. It is responsible for defining the core elements that will be a part of any model described in PROMENADE (e.g., the classes *Task, Document* and *Model* are defined by the reference model). The features associated to these elements are the ones needed to characterize any instance of them. Notice the difference between the metaelement features (which are used to characterize model classes) and the element features (which are used to characterize model class instances). Sometimes we may refer to the former group as *class features* and to the latter one as *class instance features*.

2.1 The Metamodel

The PROMENADE metamodel extends the UML one with the following elements (see figures 2 and 3):

- The classes *MetaDocument, MetaTask, MetaRole,* as *Class* subclasses (being *Class* defined in the UML metamodel). These three classes are the ones whose instances really characterize particular SPM.

- The class *SPMetamod* (which stands for *Software Process metaModel*), as *Model* subclass (again *Model* refers to the UML metamodel element).
- The classes *Precedence* and *Trigger*, that are used in the specification of the dynamic behaviour of process models.

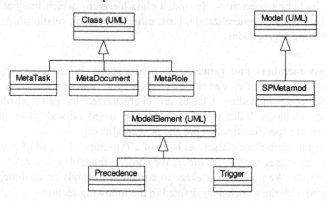

Fig 2. Elements and generalizations of the PROMENADE metamodel

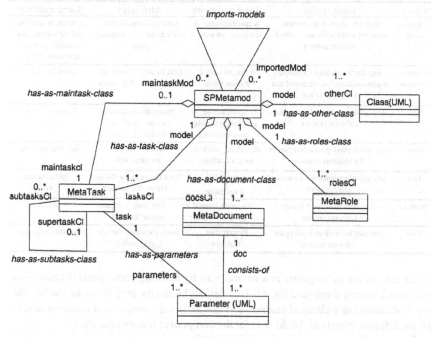

Fig 3. Associations of the PROMENADE metamodel

2.2 The Reference Model

The reference model is built upon three kinds of information, which yield to several complementary UML class diagrams. First, the individual information of the classes themselves, including constraints. Second, a class hierarchy which integrates all the documents by means of generalization. Last, other association relationships between classes (including aggregations).

Classes, class members and generalization hierarchy. In our O.O. approach, a generalization hierarchy of classes is the natural way to represent the many concepts involved in process modelling. Classes are characterized by many attributes and support many methods. Valid value attributes are stated through class invariants, while methods are specified through pre and post conditions.

Table 1 summarizes these classes. As heirs of a *Type* superclass, all of them share a few common attributes, such as *identifier*. We show in the table a few relevant attributes and methods. We do not refer either to structural methods or attributes that are directly related with the associations defined in the following section.

Table 1. Predefined classes

Class	Description	Some instances	Attributes	Some methods
Docu-ment	Any container of information involved in the software development process	A specification; a test plan; an e-mail	Link to contents; relevant dates; version; status	Document updating with or without new version creation; document edition
Commu-nication	Any document used for people communication; can be stored in a computer or not	A fax; an e-mail; human voice	Link (if any) to contents; transmission date; status	*Send* and *Read*
Task	Any action performed during the software process	Specification; component testing; error-reporting	Precondition; status; success condition; deadline (if any)	Changes of task status
Agent	Any entity playing an active part in the software process	Myself; my workstation; a compiler	Profile; location; humans skills	Just structural ones
Tool	Any agent implemented through a software tool	A compiler; a navigator	Root directory for the tool; binary file location	Just structural ones
Resource	Any help to be used during software process	An online tutorial on Java; a Web site	Platform requirements; location	*Access*
Role	Any part to be played during the software process	Programmer; manager	Tasks for which the role is responsible	Just structural ones

These classes are put together in a natural way by defining some generalization relationships between them (see fig. 4). This default hierarchy may be extended by adding new classes in a classical manner, and this allows the creation of concrete models given different criteria as we do next for the component library case study.

Some predicates appear in the static part. One the one hand, *invariants*; that is, consistency predicates bound to classes that express constraints concerning attributes and relationships between them that should be kept at any time during the enaction of the model. On the other hand, *pre* and *post conditions* associated to class methods

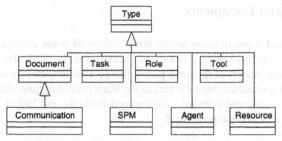

Fig. 4. Default generalization hierarchy in the PROMENADE reference model

which will be enforced respectively at the starting and ending point of such methods. We express these elements in UML as stereotyped constraints (with the predefined stereotypes *invariant, precondition* and postcondition, respectively) put in UML notes. We express the constraints in OCL as we consider it to be a natural, convenient and close to standard language to express constraints.

Association relationships. It is clear that the classes presented above must be related beyond the generalization relationship, and this is done by means of the UML association relationships (*associations* for short). Using associations, we can state which documents are manipulated by which tools, which tasks use which resources, etc. One significant kind of such associations are the *aggregation relationships* which relate a class with its components (whole-part relationship), as usual.

Fig. 5 shows the associations bound to the default classes. We highlight the existence of an association (in fact, a UML *aggregation*) from tasks to tasks to catch the concept of task decomposition.

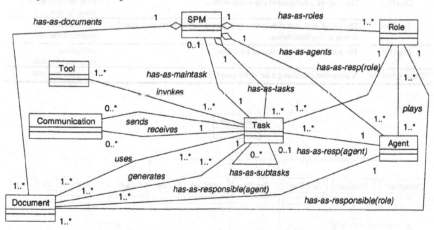

Fig. 5. Association relationships between classes

3 Static Part: Documents

Sections 3 and 4 are devoted to the presentations of some excerpts of a particular process, the construction of a library of components. In particular they present the most important aspects of the static part of this process (documents and tasks).

The main documents that take part in the library building process are shown in table 2, along with some relevant attributes and methods, and linked together through generalization relationships in fig. 6.

Table 2. Document classes

Class	Description	Attributes, methods
Component	A package of software provided with functional and non-functional specifications, code and non-functional behaviour. It may be reused and customized. It must be stored in a library.	stored flag; tested flag ; structural methods
Library	A collection of elements that may be reused when building software. Two kinds of libraries are considered: component libraries and property module libraries.	directory of contents; statistics; store, retrieve
SpecDoc	The specification part of a component. It is subdivided in functional and non-functional specification documents.	structural methods
ImplDoc	The implementation part of a component. It is compounded of a code document and a behaviour one.	structural methods
FSpec	The component functional specification document. It may be of several kinds: informal, formal, etc.	component signature; specification itself; tested flag
NFSpec	The component non.-functional specification document. It defines a list of non-functional (NF) requirements.	NF-requirements list; structural methods
Behaviour	The behaviour of the implementation with respect to the component NF-requirements.	behaviour description; structural methods
CodeDoc	The code of a component implementation.	code file; tested flag; compile, link
TestPlan	The list of tests that should be applied on a formal functional specification or an implementation.	list of tests; structural methods
EvalDoc	The result of the application of a test plan.	evaluation; structural methods
PropertyModule	A document containing a list of NF properties.	structural methods

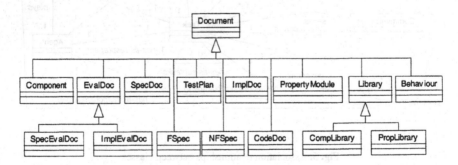

Fig 6. Generalization relationships for the component library model

The whole-part relationships between documents are shown by means of the aggregation relationships presented in figure 7. As an example, we depict in figure 8 the UML definition of one of these *Document* classes: *FSpec*. Notice the inclusion of the document invariants expressed in OCL.

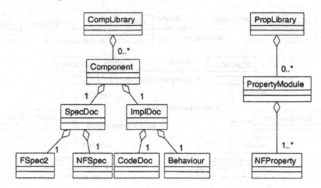

Fig 7. Aggregation relationships concerning documents

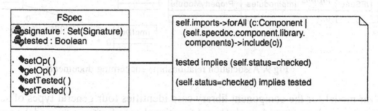

Fig 8. The UML description of *Fspec* document

Some important association relationships between documents are shown in figure 9. Among others, note that *SpecEvalDoc* and *ImplEvalDoc* are defined as association classes within the association relationships between *SpecDoc* and *SpecTestPlan*, on the one hand; and between *ImplDoc* and *ImplTestPlan*, on the other hand. Therefore, a document of the class *ImplEvalDoc* will contain the results of testing a specific component implementation by means of a given implementation test plan document. Notice also that the class documents *NFSpec*, *Fspec* and *ImplDoc* need to import components and that a specification document may be implemented by means of different implementation documents belonging to different components. Finally, a *SpecDoc* may be implemented by means of several *ImplDoc* documents belonging to different components.

Clearly, one crucial aspect concerning documents is the thorough modelling of the specific document classes that appear in table 2 which will lead to the extension of the hierarchy by defining new classes aiming at modelling (describing) finer aspects of the process. Let's focus, for instance, in the specification side of components.

There are several ways in which a component may be functionally specified. Each one of them generates a functional specification document (*FSpec*) with some specific features (i.e., attributes). Since we want to allow the coexistence of different specification techniques for different components, it becomes necessary to define a subhierarchy of functional specification documents in order to state precisely which kind of document will be generated by each functional specification choice.

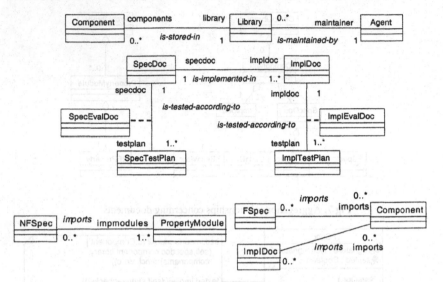

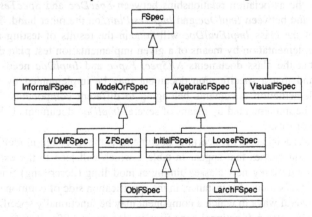

Fig 9. Association relationships concerning documents

Our model for the component library case identifies four general types of such specifications, which range from informal functional specifications to model-oriented ones (see fig. 10). But this is just a starting point; these classes should be further decomposed to introduce more concrete specification methods (e.g., initial or loose algebraic specifications) down to classes bound to concrete specification languages (e.g., *Z* or *VDM* model-oriented specifications; Larch and OBJ algebraic ones).

Fig 10. Functional specification document subhierarchy: an excerpt

Two aspects are intimately related with the coexistence of several functional specifications in the library of components: the definition of a different task refinement to specify functionally a component for each specification approach (see section 5), and the existence of a different *test plan* document class for each functional specification approach.

Non-functional specification documents are in charge of defining the non-functional requirements that should be fulfilled by any implementation of that component. As in the functional case, many notations can be used here, and the resulting hierarchy is similar to the one of fig. 10. Our usual choice is the NoFun language [7] defined as part of the ComProLab project, although other approaches can be followed [14,16].

4 Static Part: Tasks

From the PROMENADE point of view, two key concepts are relevant in order to describe tasks involved in process modelling: *task description* and *task decomposition*. The first one focuses on the statement of the dynamic behaviour of the task (mainly stating different kinds of precedence relationships between subtasks), and it is outlined in section 5. Concerning the second one, tasks may be decomposed along two dimensions that turn out to be orthogonal: task decomposition *by refinement* and *by aggregation*. This section focuses in these two aspects, which are part of the static model.

4.1 Task decomposition by aggregation

Tasks in PROMENADE may be of two kinds according to the complexity of their behaviour: *atomic* tasks and *composite* ones. Atomic tasks cannot be decomposed in other tasks. Their enactment is performed by means of a call to an external tool or in a manual way. On the other hand, composite tasks may be further decomposed into *subtask classes*, which may be in turn atomic or composite. Therefore, a composite task enactment may involve the enactment of several *subtasks* in some suitable order (determined in the dynamic part of the model).

Some of these aggregation relationships concerning task classes are shown in figure 11. This figure shows the task classes that are involved in the upper levels of the construction of the component library, which should be completed by tasks concerning lower levels. Specifically, some subtasks of *BuildComponent* and *SpecifyComponent* are shown. On the other hand, table 3 shows the parameters (documents) used and generated by some of these tasks along with a brief description. Notice in this table that in addition to the usual parameter modes (input, output and input/output), a feedback mode is introduced to represent the flow of information when a task ends in failure.

Table 3. Task classes

Task	Parameters	Description
BuildComponent	b: **i/o** Library, c: **out** Component	Construction of a component
SpecifyComponent	c: **out** Component, sd: **in** SpecDoc	Specification of a component (both functional and non-functional)
ImplementComponent	c: **i/o** Component, id: **out** ImplDoc, evd: **in fbk** ImplEvalDoc, b: **in** Library	Implementation of a component.
GenerateTestPlans	sd: **in** SpecDoc, stp: **out** seq(ImplTestPlan)	Generation of plans for testing a component.
TestComponent	id: **in** ImplDoc, evd: **out fbk** ImplEvalDoc, stp: **in** seq(ImplTestPlan)	Test of a component according to its test plans.
Store	c: **in** Component, b: **i/o** Library	Storing a developed component into a library.
FSpecifyComp	c: **i/o** Component fsd: **out** FSpec evd: **in fbk** SpecEvalDoc b: **in** Library	Functional specification of a component (different task refinements may be considered.
ModelOrFSpecify	c: **i/o** Component fsd: **out** ModelOrFSpec evd: **in fbk** SpecEvalDoc b: **in** Library	One of the task refinements of the functional specification of a component.

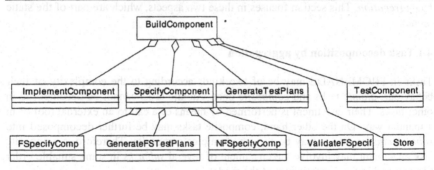

Fig 11. Aggregation relationships between tasks

4.2 Task decomposition by refinement

The behaviour of a composite task is encapsulated in PROMENADE by means of *task refinements*. Intuitively, a task refinement is a concrete way to perform a task. A bit more formally, a task refinement of a composite task class *T* is a task class that expresses one specific way in which *T* may be decomposed into subtasks and the precedence relationships that should be kept among them at enactment time. Since, in general, it is possible to think of several ways to perform a task, it makes sense to define several task refinements for a specific composite task.

Task refinements are modelled in PROMENADE by means of generalization rela-tionships between a task class and the set of task classes that refine it. In this way, the subclasses of a task class T represent its possible refinements. Notice that task re-finements and aggregations are two complementary mechanisms which will appear tightly intertwined in class diagrams: a task can be refined in many ways, and for each task refinement a particular decomposition into subtasks may exist, and those subtasks can be in turn refined in different ways, and so on.

Some examples of how a task class may be refined are given in figure 12. There are two ways to carry out the functional specification of a component (*FSpecify-Comp*): *model-oriented functional specification* (by which a formal model is associ-ated to a component and its operations are specified according to that model) and *informal functional specification* (which describes in natural language the component and its operations). Each one of these functional specification methods is considered to be a new task refinement in PROMENADE and it is represented by means of a generalization relationship.

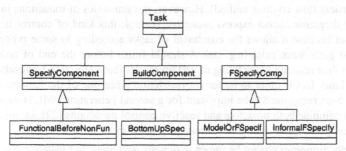

Fig 12. Generalization relationships between tasks (task refinements)

Usually, the selection of a particular task refinement enforces the use of some specific documents. This enforcement is established in the parameter definition of the refine-ment. For example, the selection of the *ModelOrFSpecify* task refinement (for the task *FSpecifyComp*) must result in the automatic selection of a *ModelOrFSpec* as functional specification document and a *ModelOrSpecTestPlan* as specification test plan and this is reflected in the *ModelOrFSpecify* parameters (see table 3). In general, a task refinement for task A may use as parameters some subclasses of the parameter classes of A.

The behaviour of a task refinement is described by means of a specific control flow which is depicted in a PROMENADE activity diagram (see section 5). Other task refinements may come up in the activity diagram that describes a specific task refinement for a given task class T (including other T's refinements).

Task refinements help in achieving *expressivity* (since various different behaviours may be given to a single activity; hierarchies of task refinements are also allowed), *modularity* (task refinements are encapsulated into tasks; therefore it is easy to add /remove task refinements to/from a model), and *model flexibility* (since we can decide at enactment time, instead of modelling time, which specific task refinement is to be enacted; this allows the enactment of incomplete models).

5 Drawbacks Concerning the Dynamic Part Description with UML

While the static part of a SPM (its structural view) may be well described using class and object UML diagrams, this does not seem to be the case of its dynamic part. The dynamic part of a SPM deals with the description of the control flow of the modelled process. This control flow is usually split into composite and atomic activities (tasks) whose behaviour should be described somehow. Among the diagrams provided by UML to cope with the behavioural view of the system (sequential diagrams, collaboration diagrams, statechart diagrams and activity diagrams), activity diagrams have been reported to be the most suitable ones to describe SPM [1, 15], but it has also been pointed out that they suffer from several limitations, remarkably:

- *Lack of expressivity*. UML activity diagrams are a sort of event diagrams which are useful to express event-driven (*reactive*) control allowing *swimlanes* (partition of activities according to their responsible), *branching* (conditional flows), *forking* (control split into several independent flows) and *joining* (independent flow controls unified). However, the semantics of transitions in activity diagrams cannot express *proactive* control; this kind of control is important because it allows the enactment of tasks according to some predetermined precedence rules (e.g. task *A* should finish before the end of task *B*) rather than reacting to the rising of certain events (basically, task finalization). This latter fact is quite an important restriction since one of the features that have been recognised to be important for a second generation PML is its ability to combine both proactive and reactive control paradigms [22], in order to get less prescriptive and more expressive process models.

- *Activity properties cannot be shown in activity diagrams*. For instance, activity deadlines and duration, roles related to an activity (who should be informed about it...), used resources and tools... Needless to say that expressing all these aspects is crucial in both software process modelling and workflow management.

PROMENADE extends the expressivity of UML activity diagrams by providing (a) proactive control (provided by means of several kinds of precedence relationships) which may be combined in the same diagram with the usual event-driven control, and (b) visual access to activity properties (which may be depicted in the activity diagram). The description of the PROMENADE dynamic model is beyond the scope of this article. We refer the interested readers to [9] and [10].

6 Conclusions and Related Work

This paper has presented a case study in the field of software process modelling using UML for modelling the static part this process. Specifically, we have presented a model for building a library of software components involving the specification of non-functional requirements and the statement of the behaviour of a particular implementation with respect to those non-functional requirements.

The way to construct a software process model in our approach is based on the extension of a reference model which describes the hierarchy with the most important concepts related to process modelling, their structure, behaviour, constraints and the association relationships between them (task responsible, task attributes and resulting documents...). This reference model is described using several class and object UML diagrams. Any other specific model will be considered as an extension of the reference one (extending the generalization default hierarchy with new classes and new associations between them). Thus, we have defined the model for constructing a library of components by defining new documents (component, library, specification document, functional specification document...), new tasks (specify functionally a component, implement it...) and new associations. All these elements concern the static part of the model and have been described in UML (and the constraints in OCL) which has proved to be a suitable and powerful notation to deal with such structural part, and a very important improvement to our former approach using OOZE [9] (an object-oriented dialect of the specification language Z).

This has not been the case with the dynamic (behavioural) part. Activity diagrams seem to be the UML diagram that fit better into the control flow description that is needed in software process and workflow modelling. Unfortunately they seem not to be expressive enough to deal with all the elements required in such environments. We have suggested the highlights of some extensions for UML activity diagrams which have been included in PROMENADE, our process modelling language: the depiction of the required task properties (responsible, generated documents...) in the diagrams and the addition of proactive declarative control (by the definition of several kinds of precedence relationships) which do not enforce a strict execution order triggered by some events (like the ending of a task) but just the declarative enumeration of the essential precedence requirements to be kept during enactment. Our experiences seem to confirm that the event-driven transitions combined with proactive control (implemented by means of several types of precedence relationships with a declarative policy) provide a much more flexible, realistic and expressive process modelling. As we have pointed out, the combination of both controls is encouraged for second generation PMLs [22].

We are not aware of almost any other PML using UML to describe the static part of the SPM. For the sake of giving some examples of well-known PMLs: APEL uses OMT-like diagrams [4]; E3 defines its own notation [11]; MERLIN describes the structural aspects with extended entity-relationships and statecharts [19]; APPL/A and JIL use textual constructs strongly based on Ada (which has been extended with some additional features like relations between objects) [21, 22]. The Rational Software Corporation *et al.* have developed a UML extension for objectory process for software engineering [17]. Essentially, it extends some metamodel classes by means of stereotypes. Neither structure nor behaviour are given to those *stereotyped* classes; no integrity constraints are defined; and no means to improve the UML features in order to deal with the dynamic process are provided. Therefore, this proposal seems not to be adequate to meet the requirements of SPM.

[12] presents an approach which describes with UML the dynamic part of the model using class diagrams with stereotyped associations for showing the control and data flow. The metamodel is defined by attaching stereotypes to model elements. In our opinion, other UML diagrams (for instance, activity diagrams) are better suited for the description of the dynamic part of a model. On the other hand, stereotypes and the other UML extension mechanisms suffer from several limitations in order to define a metamodel (being expressivity and comprehensibility two of them).

Task refinements, defined as different implementations of a task that may coexist in a single model, also appear in [13]. In this approach, a task definition consists of a task interface and, potentially, various task bodies (in the form of workflow definitions) attached to it. At enactment time a decision is taken for each task about which task body is to be enacted. PROMENADE broadens this feature by allowing the definition of task refinement hierarchies along with the substitution of a task by any of its offsprings, either at modelling time or at enactment time and by providing a powerful way to describe a task behaviour as have been outlined in section 4.

Last, we are not aware of any approach intended to improve the capacity of UML in order to model the behaviour of a software process. But there are several proposals in the related field of business processes. [1], for example, also states that UML diagrams are not sufficient for business process modelling. The authors propose to integrate a well-known process modelling formalism (EPC, Event-driven process chains) with UML diagrams. [15] proposes the use of the *stereotype* mechanism of UML to extend activity diagrams in the context of business process modelling. The new diagrams can express the required activity properties (computer support to the activity, duration...). In both cases, no new control paradigm is provided.

References

1. Allweyer, T; Loos, P: Process Orientation in UML through Integration of Event-Driven Process Chains. Proceedings of UML 98' Workshop, Ecole Superioeure des Sciences Appliquées pour l'Ingénieur-Mulhouse Université de Haute-Alsace (1998), 183-193

2. Bandinelli, S.; Fuggeta, A.; Ghezzi, C.; Lavazza, L.: SPADE: An Environemnt for Software Process Analysis, Design and Enactment. In [6] (1994), 223-247

3. Conradi, R.; Larsen, J.; Minh, N.N.; Munch, B.P.; Westby, P.H.: Integrated Product and Process Management in EPOS. Journal of Integrated CAE, special issue on Integrated Product and Process Modeling (1995)

4. Dami, S.; Estublier, J.; Amiour, M.: APEL: a Graphical Yet Executable Formalism for Process Modeling. E. di Nitto and A. Fuggetta (eds.), Kluwer Academic Publishers (1998)

5. Derniame, J.-C.; Kaba, B.A.; Wastell, D. (eds.): Software Process: Principles, Methodology and Technology. Lecture Notes in Computer Science (LNCS), Vol. 1500. Springer-Verlag, Berlin Heidelberg New York (1999)

6. Finkelstein, A.; Kramer, J.; Nuseibeh, B. (eds.): Software Process Modelling and Technology. Advanced Software Development Series, Vol. 3. John Wiley & Sons Inc., New York Chichester Toronto Brisbane Singapore (1994)

7. Franch, X.: Systematic Formulation of Non-Funcional Requirements of Software. Proceedings 3rd International Conference on Requirements Engineering (ICRE), Colorado Springs (USA), IEEE Computer Society Press, Los Alamitos (1998), 174-181.

8. Franch, X.; Botella, P.; Burgués, X.; Ribó, J.M.: ComProLab: A Component Programming Laboratory. Proceedings 9th Software Engineering and Knowledge Engineering Conference (SEKE), Knowledge Systems Institute, Skokie (1997), 397-406

9. Franch, X.; Ribó, J.M.: A Structured Approach to Software Process Modelling. Proceedings 24th EUROMICRO Conference, IEEE Computer Society Press, Los Alamitos Washington Brussels Tokyo (1998), 753-762

10. Franch, X.; Ribó, J.M.: PROMENADE: A Modular Approach to Software Process Modelling and Enaction. Research Report LSI-99-13-R, Dept. LSI, UPC (1999)

11. Jaccheri, M.L.; Picco, G.P.; Lago, P.: Eliciting Software Process Models with the E3 Language. ACM Transactions on Software Engineering and Methodology (1999)

12. Jäger D., Schleicher A., Westfechtel B.: Object-Oriented Software Process Modeling. To appear in the proceedings of the 7th European Software Engineering Conference (ESEC), Toulouse, September 1999.

13. Joeris G., Herzog O.: Towards a Flexible and High-Level Modeling and Enacting of Processes. Proceedings of the 11th. Conference on Advanced Information System Engineering (CAISE), LNCS 1626, pp. 88-102, 1999.

14. Landes, D.; Studer, R.: The Treatment of Non-Funcional Requirements in MIKE. Proceedings 5th European Software Engineering Conference (ESEC), Barcelona (Catalunya, Spain). Lecture Notes in Computer Science, Vol. 989. Springer-Verlag (1995)

15. McLeod, G: Extending UML for Entreprise and Business Process Modeling. Proceedings UML 98' Workshop, Ecole Superioeure des Sciences Appliquées pour l'Ingénieur-Mulhouse Université de Haute-Alsace (1998), 195-204

16. Mylopoulos, J.; Chung, L.; Nixon, B.A.: Representing and Using Nonfunctional Requirements: A Process-Oriented Approach. IEEE Transactions on Software Engineering, Vol. 18, N. 6 (1992), 483-497

17. Rational Software Corporation: UML extension for Objectory Process for Software Engineering. http://www.rational.com/uml

18. Rational Software Corporation et al.: UML Semantics. http://www.rational.com/uml

19. Reimar, W.; Schaefer, W.: Towards a Dedicated Object-Oriented Software Process Modelling Language. Workshop on Modeling Software Process and Artifacts, held at 11th ECOOP, Jyvaskyta (Finland) (1997).

20. Rumbaugh, J.; Jacobson, I.; Booch, G.: The UML Reference Manual. Addison Wesley (1999)

21. Sutton S.M.; Heimbigner D.; Osterweil L.J.: APPL/A: A Language for Software Process Programming. ACM Transactions on Software Engineering and Methodology. Vol 4. N. 3, July 1995. 221-286.

22. Sutton, S.M.; Osterweil, L.J.: The Design of a Next-Generation Process Language. Proceedings of ESEC/FSE '97, Lecture Notes in Computer Science, Vol. 1301, M. Jazayeri and H. Schaure (eds.). Springer-Verlag, Berlin Heidelberg New York (1997), 142-158

23. Warboys, B.C.; Balasubramaniam, D. et al: Instances and Connectors: Issues for a Second Generation Process Language. Proceedings of the 6th European Workshop in Software Process Technology, LNCS 1487, V. Gruhn (ed.). Springer-Verlag (1998)

Framework for Describing UML Compatible Development Processes

Pavel Hruby

Navision Software
Frydenlunds Allé 6
DK-2950 Vedbæk, Denmark
ph@navision.com

Abstract. Have you ever tried to specify an accurate development process for your organization and later faced difficulties with the complexity of the description? Instead of describing a specific process, it might help to describe a process framework and reuse it by creating specific processes for specific needs. This paper describes the object-oriented framework of a development process, which considers software development artifacts as objects and evolution as collaborations between the objects. Such an object-oriented process definition can deal with the complexity of a development process in a better way than a traditional description based on workflow. This paper discusses features of such a process framework with an eye towards approaches such as Fusion, OPEN and the Rational Unified Process.

Acknowledgement. I would like to thank to Prof. B. Henderson-Sellers of the School of Computing Sciences, University of Technology, Sydney, Australia, for "shepherding" the article, and for his useful suggestions and comments. I do, of course, take full responsibility for any omission or errors.

1 Introduction

"The software development process, as actually performed, is so complex that we cannot write it down accurately, and if we could, no one could read that description and learn to perform it," (Alistair Cockburn, panel discussion, ECOOP'98 [2]). In this paper, this problem is solved in the following way. Instead of writing down a concrete process scenario, we specify a process framework that describes all allowable processes. Such a framework is abstract yet precise. To meet the demands of specific development problems, the framework is reused by creating specific development processes. The concrete development processes can be represented at the necessary level of accuracy. This solution significantly simplifies the description of concrete development processes, because the complexity is localized in the abstract form in the process framework.

This paper is structured in the following way. The second section explains the traditional specification of software development processes and its drawbacks. The next four sections explain the main ideas of the object-oriented process specification, and the structure of the object-oriented framework is outlined in the section 6: The Object-Oriented Specification of Development Processes. The next section compares the object-oriented specification with the original specification of three contemporary design methods and methodological frameworks.

2 Traditional Specification of Software Development Processes

The purpose of this section is to clarify the terminology used throughout the paper because different authors define these terms differently in different contexts.

The traditional specification of a development process is typically illustrated with a graph of tasks, techniques, software development artifacts and activities. *Tasks* are small behavioral units that usually result in a software development artifact. Examples of tasks are construction of a use case model, construction of a class model and writing of code. *Techniques* are formulas for performing tasks, for example, object design using CRC cards, functional decomposition and programming in Visual Basic. *Software development artifacts* are final or intermediate products resulting from software development, for example, a use case model, a class model, or source code. *Activities* (in this paper) are units that are larger than task units. Activities typically include several tasks and software development artifacts. Examples of activities are requirement analysis, logical design and implementation.

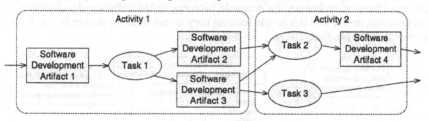

Fig. 1. Traditional specification of a development process. It can be compared with the object-oriented specification illustrated in Fig. 10

In general, the traditional specification of a development process cannot cover all the possible combinations of activities and software development artifacts without becoming overly complex. This paper will show that in the specification of development processes, the object-oriented approach deals with complexity of the development process better than the traditional approach illustrated in Fig. 1. This is similar to experience from within software development, where the object-oriented approach can deal with the complexity of large software solutions better than the traditional structured approach.

3 Basic Features of the Product-Focused Object-Oriented Process Specification

Software development and management artifacts produced during a software development process are considered objects with various methods and attributes. Evolution during software development is represented as collaborations between software development artifacts, management artifacts and users of the method.

The object-oriented specification of the software development process distinguishes between artifacts and their representations. A *software development artifact* determines information about a *software product*. Examples of software development artifacts are use cases, software architecture, object collaborations and class descriptions. A *management artifact* determines information about a *management product*, such as a project and a team. Software development artifacts can be very abstract, such as the vision for the software system, or very concrete, such as the source code. The *representation* determines how the artifact is presented. For example, a use case model is represented by a use case diagram; a state model can be represented by a statechart diagram, an activity diagram or a state transition table. The object interaction model can be represented by a set of sequence diagrams or a set of collaboration diagrams. Various design methods typically recommend a suitable representation of each software development artifact. However, the choice often depends on the concrete situation, and it is sometimes advisable to leave the final decision about the representation to a user of the method.

An artifact has a representation, properties, responsibilities, methods, relationships to other artifacts and attributes, all of which are discussed later. Consistency check, process phase or technique, for example, are not called software development or management artifacts in this article because they do not describe design of a software or management *product*.

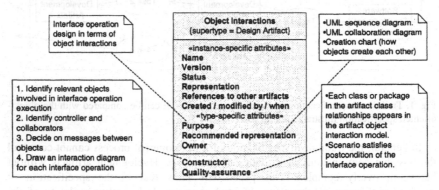

Fig. 2. Specification of the artifact type object interaction model

The object-oriented specification of the development process distinguishes between the artifact *type* and the artifact *instance*. *Types of artifacts* specify properties, attributes and methods of various kinds of software development and management artifacts. *Instances of artifacts* are concrete software development and management products produced during software development. An example of a software development artifact type is a use case model. An example of a software development artifact instance is a concrete set of use cases, actors and their relationships, represented by a use case diagram.

Artifact types have two kinds of methods:

- *Constructors*, which are methods describing how to create an artifact.
- *Quality-assurance methods*, such as completeness and consistency checks.

Artifacts have instance-specific *attributes*: name; version; representation, which typically contains a diagram, a table or a text; status, such as draft, completed, tested; references to other software development and management artifacts; and attributes such as who created and modified the artifact and when. In addition, artifact types have type-specific *attributes*: the purpose, the recommended representation and the owner of the artifact type. Artifacts may have other additional attributes and methods than those mentioned above. Fig 2. illustrates the object-oriented specification of a software development artifact type with attributes and methods. The inheritance diagram of the artifact types is in Fig. 3.

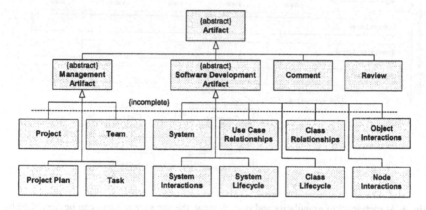

Fig. 3. The inheritance diagram illustrating types of artifacts used by the development process. The fact that the inheritance tree is incomplete allows for flexibility throughout the process and for creating new artifacts to match different kinds of development processes

4 Static Structure of Software Development and Management Artifacts

This section specifies the static relationships between software development and management artifacts. The static structure is based on the pattern for structuring project repositories with UML design artifacts [11]. This section outlines how the pattern is applied to create a static structure of the framework for describing UML compatible development processes.

A software system can be represented from various viewpoints and at various levels of granularity, see Fig 4. Examples of views[1] and levels of granularity are discussed later. In each view and at each level of granularity, a UML compatible system can be described by four artifacts: static relationships between classifiers[2], dynamic interactions between classifiers, classifier responsibilities and classifier lifecycles.

The artifact called *classifier relationships* specifies static relationships between classifiers. The artifact called *classifier interactions* specifies interactions (an instance of a scenario) between classifiers. The artifact *classifier* identifies the classifier and specifies classifier responsibilities and other static classifier properties, for example, a list of classifier operations with preconditions and postconditions, and a list of classifier attributes that can be read and set. The *classifier lifecycle* specifies classifier state machine and dynamic properties of classifier interfaces, for example, the allowable order operations and events.

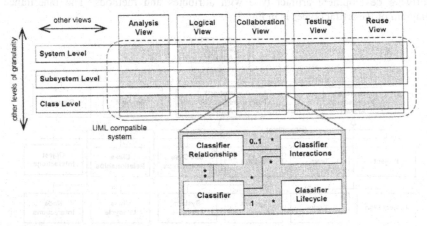

Fig. 4. At each level of granularity and in each view, the software product can be described by four types of software development artifacts. Each software development artifact identifies specific information about the software product (After ref. [10].)

Examples of views are the logical view, the collaboration view, the deployment view and the analysis view. The *logical view* describes the logical structure of the product in terms of subsystems and classes and their responsibilities, relationships and interactions. The *collaboration view* identifies types of collaborations with actors of the system, subsystems, classes, components and nodes. The *deployment view*

---

[1] In this article, I use the term "views" to mean complete "slices" through a model of a software system across different levels of granularity from various viewpoints. This meaning is different from the term "view", as used in reference [12], where it means a non-complete set of significant elements.

[2] Classifiers represent static entities in a system model. In UML, classifiers are class, object, interface, datatype, use case, subsystem, component and node. Management artifacts, such as team and project, are mapped to UML as stereotyped classes.

describes the physical structure of the system in terms of hardware devices and their responsibilities, relationships and interactions. The *analysis view* describes the logical structure of the product in terms of analysis subsystems, objects and their responsibilities, relationships and interactions. The analysis view differs from other views in the way that the software entities in the analysis view do not specify the software system precisely. The purpose of the analysis view is to record preliminary or alternative solutions to design problems, and to record requirements or user's view of the system. Analysis objects may – but do not always – correspond to logical or physical software entities existing in the product.

Examples of levels of granularity are the system level, the subsystem level and the class level (see Fig. 4). The *system level* of granularity describes the context of the software system. The system level specifies the responsibilities of the system being designed and the responsibilities of the other systems that collaborate with it, responsibilities of physical devices and software modules outside the system, and static relationships, along with the dynamic interactions between them and the system being designed. The *subsystem level* of granularity describes subsystems, software modules and physical devices inside the system, along with their static relationships and dynamic interactions. The *class level* of granularity describes the detailed design of the subsystems in terms of classes and objects, and their relationships and interactions.

The software product can be represented by additional views, such as the *testing view* and the *view of reusable elements*. The software product can be specified at additional levels of granularity, such as the *tier level* for systems with layered architecture and the *organizational level* for business systems. At each additional level of granularity and in each additional view, the software product is specified by static relationships between classifiers, dynamic interactions between classifiers, classifier responsibilities and classifier lifecycles. The semantics of these additional software development artifacts are out of the scope of this paper. See paper [10], *Structuring Design Artifacts with UML,* for details about these artifacts.

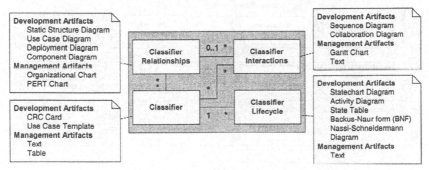

Fig. 5. Typical representation of software development and management artifacts

As mentioned in section 3, the object-oriented model distinguishes between the information itself (called software development or management artifact in this article) and its representation. Software development artifacts can be represented by UML diagrams, tables or text, see Fig. 5. The artifact *classifier relationships* is represented by a UML static structure diagram, a use case diagram, a deployment diagram and a component diagram, if classifiers are classes, use cases, nodes and components, respectively. The artifact *classifier interactions* is represented by a UML sequence diagram and a collaboration diagram. The UML Notation Guide describes only interaction diagrams in which classifiers are objects; it does not describe interaction diagrams in which classifiers are use cases, subsystems, nodes or components. These diagrams are discussed in [10]. The artifact *classifier* is represented by a CRC card, use case template, structured text or table. The artifact *classifier lifecycle* is represented by a UML statechart diagram, activity diagram, a state table, a Backus-Naur form and a Nassi-Schneidermann diagram.

Management artifacts can be represented by project diagrams, tables or text (see Fig. 5). The artifact *classifier relationships* is represented by an organizational chart, if the classifiers are roles or teams; or by a PERT chart, if the classifiers are tasks and projects. Overeager UML users can use class diagrams, in which the classes have a stereotype «task» and «project». The artifact *classifier interactions* is represented by a Gantt chart if the classifiers are projects or by text if the classifiers are roles and teams. The artifacts *classifier* and *classifier lifecycle* are represented by table or text.

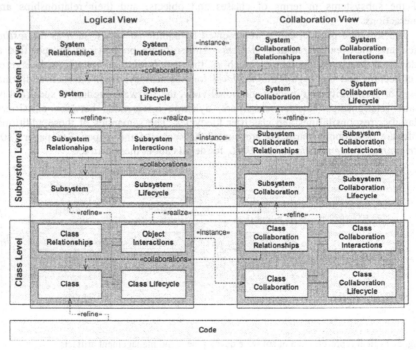

Fig. 6. Static structure of software development artifact types at the system, subsystem and class levels of granularity and in the logical and collaboration views (After ref. [10].)

Fig. 6 specifies the static structure of software development artifact types at three levels of granularity and in the logical and collaboration views. Each software development artifact type specifies certain information (discussed in paper [10]) about the software system. The structure uses the pattern described above, and the result is a regular structure, which allows consistent customization of the design specification. In simple cases, the specification consists of only a small subset of the software development artifacts identified in Figs. 5 and 6. Conversely, if software designers have to specify something unusual or unexpected, such as the things not covered by the method, the specification is extended by adding additional views and additional levels of granularity.

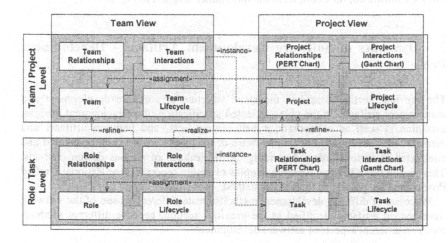

Fig. 7. Static structure of management artifact types in the team and project views

Fig. 7 illustrates the application of the pattern for structuring management artifacts. The pattern can be used to structure two kinds of management product: teams and projects. Management artifacts can be shown as stereotyped classes in UML, such as «team», «role», «project» and «task». The artifacts *team relationships* and *role relationships* specify the organizational structure at two levels of granularity. The artifact *team* specifies the responsibility of the team and the artifact *role* specifies the role of the team member. Examples of roles are developer, program manager, product manager, user education and logistics. The artifacts *team interactions* and *role interactions* specify scenarios - interactions between teams and team members, which are responses to various events. The artifacts *project* and *task* specify static properties of projects and tasks. The artifact *project relationships* and *task relationships* specify static relationships between projects and tasks. These artifacts can be represented by a PERT chart. The PERT chart shows the task dependencies, which are the most important static relationships between tasks. The artifact *task interactions* specifies a project scenario in terms of starting and finishing tasks. Accordingly, the artifact *project interactions* specifies the project scenario in terms of starting and finishing projects. These artifacts are typically represented by Gantt charts, but they might be represented by UML sequence diagrams as well. Gantt charts show the task constructors, which are the most important messages between tasks.

It can be noted that every project generates a number of artifacts not captured by the pattern. Examples of such artifacts are a glossary, minutes of meetings, reviews, comments and notes. These artifacts do not describe a *product*, and therefore they cannot be structured using the pattern. These artifacts can be related to any other software development or management artifact. For example, the glossary is related to the artifact *system*, the minutes of meetings are related to the artifact *project*. In order to reuse them in a consistent way, they have specified types in the process repository. They are illustrated, for example, in the inheritance diagram in Fig. 3.

5 Dynamics of Software Development Artifacts - Development Processes

The previous section described the static structure of software development and management artifacts. In the object-oriented specification of a development process, evolution is seen as collaborations between artifacts, and between artifacts and members of a development team. Software development artifacts can be created and completed in various orders depending on the features of various design methods. This section describes two typical examples of design processes: the Rational Unified Process and the process of the Fusion method.

Processes of different development methods create different subsets of the software development artifacts identified in the previous section, because different methods focus on different aspects of software development.

5.1 Rational Unified Process

The object-oriented specification of the Rational Unified Process [12], [15] is illustrated in Fig. 8. The figure illustrates the scenario of the requirements, analysis and design workflows of the Rational Unified Process. The figure shows the evolution of the software as a number of interactions between the worker (such as the system analyst, the use case specifier, the architect and the designer) and the software development artifacts of the Rational Unified Process.

A worker who uses the Rational Unified Process is responsible for calling the constructors of the software development artifacts in the order illustrated in the Fig. 8. Constructors and quality assurance methods generate various messages between the objects.

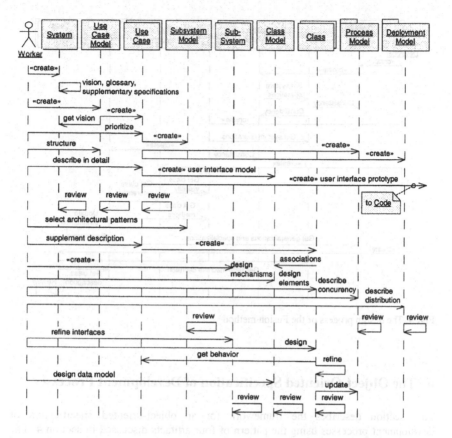

Fig. 8. The process of the requirements, analysis and design workflows of the Rational Unified Process

5.2 The Process of the Fusion Method

The object-oriented specification of the development process of the Fusion method [3] is illustrated in Fig. 9. The figure illustrates the Fusion development scenario in terms of interactions between software development artifacts, and between software development artifacts and the developer.

The Fusion method, as described in reference [3], contains more details than the Rational Unified Process 5.0, as described in reference [12]. For the sake of comparison with the Rational Unified Process (to keep the Figs. 8 and 9 at the same level of detail), Fig. 9 shows only high level interactions of the Fusion method. Note that this figure is meant as an illustration of the idea, not as a detailed description of the entire method.

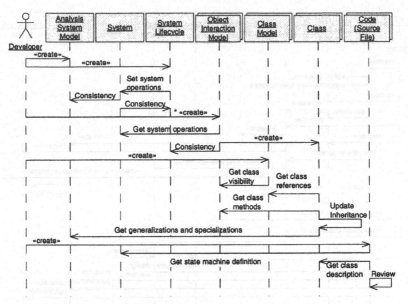

Fig. 9. The design process of the Fusion method

6 The Object-Oriented Specification of Development Processes

This section describes the framework for an object-oriented specification of development processes using the pattern of four artifacts discussed in section 4. The structure of the process specification framework is illustrated in Fig. 10.

The artifact called *artifact relationships* specifies the static relationships between software development and management artifacts. It can be represented by a UML static structure diagram, where classes and objects have stereotypes, such as «class relationships», «class lifecycle», «project» and «team». The artifact called *artifact interactions* specifies the development scenario. This artifact can be represented by a UML sequence diagram or a UML collaboration diagram. In the Rational Unified Process [12] these scenarios are called *workflows*. The artifact called *artifact type* specifies the purpose, constructor, quality assurance, and specific attributes of software development and management artifacts. The artifact called *artifact lifecycle* specifies the artifact states and the events that change the artifact state, such as creation, completion and approval. The lifecycle shown in Fig. 10 is an illustrative example; different software development and management artifacts have different and often more complex lifecycles.

The artifact relationships are discussed in section 4 ("Static Structure of Software Development and Management Artifacts"). The artifact interactions are discussed in section 5 ("Dynamics of Software Development Artifacts"). The artifact types are discussed in section 3 ("Basic Features of the Object-Oriented Process Specification").

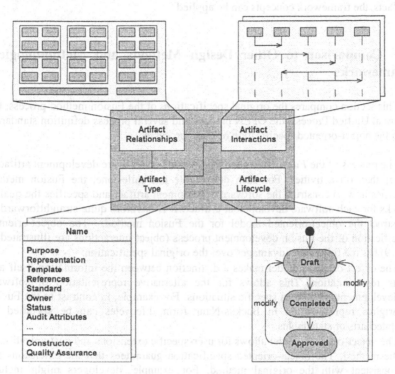

Fig. 10. Structuring the object-oriented specification of a development process

It can be noted, however, that the framework described in this article is not a so-called document-based process. Document-based processes focus on the documents delivered. The documents often have a form, which is well-defined by templates and checklists. Document-based processes have a bad reputation, because the success of a project has often been measured in terms of the documents being delivered on time. For software projects, this encourages problems, such as bureaucracy, slow delivery and requirements drift.

Many software development and management artifacts in the framework, such as class lifecycle, are too small to be useful documents delivered separately, for example, at a project milestone. For pragmatic reasons, software development and management artifacts can be combined together to make useful documents. Examples of such aggregated software development artifacts are in ref. [10].

The framework specified in this article can be characterized as information-focused, rather than document-focused. Software development and management artifacts are essentially pieces of information (see the artifact definition in section 3). The framework strictly distinguishes between the information itself (called software development or management artifact in this article) and its representation (called deliverable or document in the traditional approach). Sometimes the information is not represented physically and might exist only as a mental model. Even for such artifacts, the framework concepts can be applied.

7   Comparison to Other Design Methods and Methodological Frameworks

This section compares the original specifications of the Fusion method process, the Rational Unified Process, the OPEN process and several process definition standards, with the object-oriented specification of these processes.

The process of the *Fusion method* [3] is focused on software development artifacts, rather than on activities. For each deliverable and milestone, the Fusion method specifies how to construct the software development artifact and specifies the quality checks for each software development artifact. It is therefore quite straightforward to construct the object-oriented model for the Fusion method. The object-oriented specification of the Fusion development process (object interactions are illustrated in Fig. 9) has the following advantages over the original specification:

- The object-oriented model makes a distinction between the information itself and its representation. This allows for the alternative representation of software development artifacts in specific situations. For example, in contrast to the Fusion original representation in Backus-Naur form, lifecycles can be specified by statecharts or state tables.
- The object-oriented model allows for user-specific extensions and customization of the method. The object-oriented specification guarantees that the extensions are consistent with the original method. For example, developers might include specifications of subsystem lifecycles, class lifecycles and use cases in the original process. The object-oriented specification makes it easy to define relationships between these new artifacts and the original Fusion deliverables.
- The object-oriented model allows for consistent mapping between management artifacts and software development artifacts. In particular, the object-oriented model defines relationships between software development artifacts and the project and task, along with relationships between software development artifacts and the team model, which specifies which team roles are responsible for which deliverables.

The Rational Unified Process [12] is specified by a workflow model focused on activities. The original Rational Unified Process 5.0 [15] is complex, but it is made more manageable by viewing the process in different ways, such as by notation, workflow, documentation, artifacts and workers. It is also made more manageable by necessary configuration (adaptation) to meet specific needs. The on-line representation of the process helps significantly in dealing with the complexity of the process. The object-oriented specification of the Rational Unified Process (object interactions are illustrated in Fig. 8) has the following advantages over the original specification:

- The object-oriented model is simpler than the original one. If the set of artifacts included in the demo version of the process is used, the same information is described in 26 different documents in the demo version and in 11 documents in the object-oriented specification. The demo version was available at Rational's Web site during 1997-1998.
- The Rational Unified Process, rewritten in an object-oriented manner, provides a consistent framework: it makes inconsistencies in the original process noticeable and draws attention to missing information.
- Each artifact of the object-oriented specification has quality defined by quality-assurance methods. For comparison: the Rational Unified Process 5.0 [15] defines the quality of only about 30% of artifacts.

The Fusion process, the Rational Unified Process and their customized versions, are instances of the framework discussed in the previous section. Using the framework, these processes can be compared against each other by identifying their software development and management artifacts and specifying the order in which the artifacts are created, updated and completed, as it is shown in Figs. 8 and 9. Furthermore, the constructors, quality-assurance methods and artifact lifecycles of various processes can be compared against each other using the framework in this article.

Methodological frameworks such as OPEN and various process definition standards can be compared against the framework in this article by finding mappings between their elements. Such mappings are not analyzed in detail in this paper.

OPEN [8] is a process-focused object-oriented framework for software development methods. OPEN provides a range of activities, tasks and techniques, which can be tailored specifically to each individual organization or individual project. The OPEN process is an object-oriented framework that regards activities as objects and tasks as their methods. The execution of the objects (activities) is guarded by pre- and postconditions on their methods (tasks). The postconditions include testing requirements and deliverables. OPEN has a well-defined metamodel [4], [7] consisting of projects, activities, tasks, deliverables, techniques and sequencing rules. OPEN addresses the same problem as the framework in this article – instead of specifying a single process, it specifies the process framework and instantiates it (that is, it creates a concrete method) for specific situations. Although the aim is similar, both approaches differ in the following details.

- OPEN is a process-focused framework; this article describes a product-focused framework. OPEN objects (the first class citizens in the object-oriented process model) are *activities*. In contrast to this, objects of the framework discussed in this paper are *products*.
- Modularity and encapsulation of OPEN are at the level of *activities*. For example, OPEN allows an *activity* to be outsourced to an external organization; the contract between activities becomes a business contract between the two organizations (ref. [8], page 45). The modularity of the framework in this article is at the level of *products* (software development and management artifacts). The *product* can be obtained from an external organization, and the constructor and quality-assurance methods become a business contract between the two organizations.

- Both OPEN and the framework in this article can create specific methods by choosing the method components that meet specific demands in specific situations. A concrete method based on OPEN is created by selecting the *activities* (with tasks) necessary to perform the project; the framework in this article is instantiated by selecting the *software development and management artifacts* (pieces of information) necessary to make a final product. If the situation requires artifacts not specified in the framework, the pattern discussed in section 3 guarantees that new artifacts are added in a consistent manner.
- OPEN can be transformed into the framework in this article and the other way around. Although the artifacts in this article are smaller than OPEN deliverables, each of the OPEN deliverables can be mapped to one or more software development or management artifacts in this article, and all OPEN tasks and techniques can be expressed as the constructors of artifacts and quality-assurance methods. The OPEN activities can be mapped to collaborations between software development and management artifacts, but this mapping is only possible in certain cases. This is because the links between OPEN objects and their methods are specified in terms of probabilities (deontic certainty factors) that allocate which tasks are needed for which activities [9]. The framework in this article does not have such a probabilistic mechanism. A classifier (such as collaboration) either does or does not have its elements. Therefore, the mapping between the collaborations in this framework and OPEN activities is only possible in cases in which the values of the deontic factors change to a bimodal distribution (0 or 1).
- The mapping between OPEN and the framework in this article might be useful for the further development of OPEN. The framework in this article has a succinct structure of software development and management artifacts, and has "placeholders" for additional deliverables and tasks not included in the OPEN process specification [8]. Moreover, the framework in this article provides a pattern for extensions of the process in a consistent manner. The new improved version of the OPEN metamodel [7] indicates that OPEN process specification can easily be extended to cover software development and management artifacts from the framework in this article.

Several *standards related to process definition*, such as the Capability Maturity Model [13], ISO 9000 and the Alistair's Cockburn's VW-Staging [1] define quality criteria and the key practices that concrete development processes should meet. The object-oriented specification of a development process can be directly evaluated against the standards simply by comparing the quality-assurance methods of the software development and management artifacts with the key practices and quality criteria defined by the process definition standard. However, the process definition standards are not supported by any well-defined metamodel. The OMG Process Working Group White Paper [14] suggests a metamodel using OMG Meta Object Facility, but without regular structures and guidelines for specifications of the artifacts.

Unlike the object-oriented framework presented in this article, none of the standards mentioned describe patterns for structuring the software development and management artifacts. A development process is therefore more difficult to reuse and extend in a consistent manner. The process instantiated from the metamodel in the

OMG Process Working Group White Paper is customized off-line by means of project profiles that reflect, for example, the organizational cultures, industry domains and technology types. The idea of the profile is certainly useful. However, the object-oriented framework discussed in this paper allows for process customization at a much lower level of granularity; the processes can be customized on-the-fly to fit various *problems* being solved.

8 Summary

This paper discussed the product-focused object-oriented framework for the specification of software development processes. The software development and management artifacts are modeled as objects with constructors and quality-assurance methods, along with a number of specific attributes. The object-oriented specification of a development process is simpler and more consistent than traditional specification based on tasks and deliverables.

References

1. Cockburn, A.: Using "V-W" Staging to Clarify Spiral Development, available at: http://members.aol.com/acockburn/papers/vwstage.htm
2. Cockburn, A.: ECOOP 98 panel discussion on Software Development and Process, Brussels, Belgium, July 1998.
3. Coleman, D. Arnold, P. Bodoff, S. Dollin, C. Gilchrist, H. Hayes, F., Jeremaes ,P.: Object-Oriented Development: the Fusion method, Prentice Hall, 1994
4. Henderson-Sellers B.: A Methodological Metamodel of Process, JOOP, 11(9): 45-55, February 1999.
5. Henderson-Sellers B.: Instantiating a Process Metamodel, JOOP, 12(3): 51-57, June 1999.
6. Henderson-Sellers B.: Mellor S. J.: Tailoring Process-Focused OO Methods, JOOP, 12(4): 40-44, July/August 1999
7. Firesmith D., Henderson-Sellers B.: Improvements to the OPEN Process Metamodel, JOOP, 12(6), October 1999
8. Graham I., Henderson-Sellers B., Younessi H.: The OPEN Process Specification, Addison-Wesley, Harlow, 1997
9. Graham I.: Message in the mailing list OTUG, Subject: RE: (OTUG) Unified Process ????, 17 September 1998, 01:09 GMT.
10. Hruby, P.: Structuring Design Artifacts with UML, in: <<UML>>'98: Beyond the Notation, Bezivin J., Muller P.A. (editors), Springer Verlag LNCS 1618, 1999.
11. Hruby, P.: The Pattern for Structuring UML Based Repositories, OOPSLA'98, Vancouver, Canada, 1998.
12. Kruchten, P.: The Rational Unified Process, Addison-Wesley, 1998.
13. Paulk M. C. et al.: Capability Maturity Model for Software version 1.1, CMU/SEI-93-TR-024
14. OMG White Paper on Analysis & Design Process Engineering, Process Working Group, Analysis and Design Platform Task Force, OMG document ad/98-07-12, July 1998.
15. The Rational Unified Process 5.0, Rational Corporation, 1998.

On the Behavior of Complex Object-Oriented Systems

David Harel
The Weizmann Institute of Science, Rehovot, Israel
harel@wisdom.weizmann.ac.il
and
I-Logix, Inc., Andover, MA 01810

Over the years, the main approaches to high-level system modeling have been *structured-analysis* (SA), and *object-orientation* (OO). The two are about a decade apart in initial conception and evolution. SA started out in the late 1970's by De Marco, Yourdon and others, and is based on 'lifting' classical procedural programming concepts up to the modeling level and using diagrams [CY, D]. The result calls for modeling system structure by functional decomposition and the flow of information, depicted by hierarchical data-flow diagrams. As to system behavior, the mid 1980's saw several methodology teams (such as Ward/Mellor [WM], Hatley/Pirbhai [HP] and our own Statemate team [H+]) enriching the basic SA model with means for modeling behavior: state diagrams, or the richer language of statecharts [H]. A state diagram or statechart is associated with each function to describe its behavior. Obviously, careful behavioral modeling is especially crucial for embedded, reactive, and real-time systems. A detailed description of the way this is done in the SA framework appears in [HP]. The first commercial tool to enable model executability and code synthesis of high-level models was Statemate from I-Logix in 1987 (see [H+, IL]).

OO modeling started in the late 1980's. Here too, the basic idea for system structure was to 'lift' concepts from object-oriented programming up to the modeling level, and to do things with diagrams. Thus, the basic structural model for objects in Booch's method [B], in OMT [R+], in the ROOM method [SGW], and in many others (e.g., [CD]), has notation for classes and instances, relationships and roles, and aggregation and inheritance. Visuality is achieved by basing this model on an enriched form of entity-relationship diagrams. As to system behavior, most OO modeling approaches, including those just listed, adopted the statecharts language for this. A statechart is associated with each class, and its role is to describe the behavior of the instance objects.

However, here there are subtle and complicated connections between structure and behavior, that do not show up in the simpler SA paradigm. Classes represent dynamically changing collections of concrete objects, and behavioral modeling must address issues related to their creation and destruction, the delegation of messages, the modification and maintenance of relationships, aggregation, true inheritance, etc. These issues were treated by OO methodologists in a broad spectrum of degrees of detail — from vastly insufficient to adequate. The test, of course, is whether the languages for structure and behavior and their inter-links are defined sufficiently well to allow full model execution and code synthesis. This has been achieved only in a couple of cases. One is the Objec-Time tool, which is based on the ROOM method of [SGW]. The other is the Rhapsody tool (see [IL]), which is based on the executable object modeling work of [HG]. The latter was originally intended as a carefully worked out language set based on Booch and OMT object model diagrams, driven by statecharts, and addressing the issues above in a way sufficient to lead to executability and full code synthesis.

In a remarkable departure from the similarity in evolution between the SA and OO paradigms for system modeling, the last three years have seen OO methodologists working together. They have compared notes, have debated the issues, and have finally cooperated in formulating the UML, which was adopted in 1997 as a standard by the OMG (see [UML]). This sweeping effort, which in its teamwork is reminiscent of the Algol'60 and Ada efforts, has taken place under the auspices of Rational Corp., spearheaded by Booch, Rumbaugh and Jacobson. Version 0.8 of the UML was released in 1996 and was rather open-ended and vague, lacking in detail and well thought-out semantics. For about a year, the UML team went into overdrive, with a lot of help from methodologists and language designers from outside Rational Corp. Our team contributed quite a bit too, and the languages underlying Rhapsody [HG, IL] are indeed the executable kernel of the UML. The version of the UML adopted by the OMG is thus much tighter and more solid than version 0.8. With some more work there is a good chance that the UML will become not just an officially approved standard, but the main modeling mechanism for the software that is constructed according to the object-oriented doctrine. And this is no small matter, as more and more software engineers are now claiming that more and more kinds of software are best developed in an OO fashion.

The recent wave of popularity that the UML is enjoying will bring with it not only the UML books written by Rational Corp. authors (see, e.g., [RJB]), but a true flood of books, papers, reports, seminars, and tools, describing, utilizing, and elaborating upon the UML, or purporting to do so. Readers will have to be extra-careful in finding the really worthy trees in this forest. Despite this, one must remember that right now UML is a little *too* massive. We understand well only parts of it; the definition of other parts has yet to be carried out in sufficient depth as to make clear their relationships with the constructive core of UML (the class diagrams and the statecharts). Moreover, there are still major problems in the general area of behavioral specification and design of complex object-oriented systems that await treatment. These still require extensive research.

Here are brief discussions of two examples of research directions that seem to me to be extremely important. One has to do with message sequence charts (MSCs) and their relationship with state-based specification, and the other has to do with inheriting behavior.

As to the first one, there is a dire need for a highly expressive MSC language, with a clearly defined graphical syntax and a fully worked out formal semantics. Such a language is needed in order to construct semantically meaningful computerized tools for describing and analyzing use-cases and scenarios. It is also a prerequisite to a thorough investigation of what might be *the* problem in object-oriented specification: relating *inter*-object specification to *intra*-object specification. The former is what engineers will typically do in the early stages of behavioral modeling; namely, they come up with use-cases and the scenarios that capture them, specifying the inter-relationships between the processes and object instances in a linear or quasi-linear fashion in terms of temporal progress. That is, they come up with the description of the scenarios, or 'stories' that the system will support, each one involving all the relevant instances. A language for scenarios is best used for this. The latter, on the other hand, is what we would like the final stages of behavioral modeling to end up with; namely, a full behavioral specification of each of the processes or object instances. That is, we want a complete description of the behavior of each of the instances under all possible conditions and in all possible 'stories'. For this, a state-machine language such as statecharts [H] appears to be most useful. The reason the state-machine intra-object model is what we want as an output from the design stage is implementation: ultimately, the final software will consist of code for each process or object. These pieces of code, one for each process or object instance, must — together — support the scenarios as specified in the MSCs. Thus the "all relevant parts of stories for one object" descriptions must implement the "one story for all relevant objects" descriptions.

Now, there are several versions of MSC's, including the ITU standard (see [ITU]), and the sequence diagrams adopted in the UML. However, both versions are extremely weak in expressive power, being based essentially on a simple way of constraining the partial order of events. Nothing much can be specified about what the system will actually do when run. A particular troublesome issue is the need to be able to specify 'no-go' scenarios, or what I like to call *anti-scenarios*, ones that are not allowed to occur. In short, there was a serious need for a more powerful language for sequences. In a recent paper [DH], we have addressed this need, proposing an extension of MSCs, which we call *live sequence charts* (or *LSCs*). One of the main extensions deals with specifying *"liveness"*, i.e., things that must occur. LSCs allow the distinction between possible and necessary behavior both globally, on the level of an entire chart, and locally, when specifying events, conditions and progress over time within a chart. (In so doing it makes possible the natural specification of forbidden behavior.) LSCs also support subcharts, synchronization, branching and iteration. It is not clear yet whether this language is exactly what is needed: more work on it is required, experience must be gained using it, and of course an implementation is needed. Nevertheless, it does make it possible to start looking seriously at the two-way relationship between the aforementioned dual views of behavioral description. How to address this *grand dichotomy of reactive behavior* is a major problem. For example, how can we synthesize a good first approximation of the statecharts from the LSCs? Finding efficient ways to do this would constitute a significant advance in the automation and reliability of system development. One must realize that this is a far more doficult problem than synthesizing from conventional MSCs, which are simple partial-order based, and are existential in nature. There have number of attempts to address that problem. LSCs are much richer, and it is their universality, for example their ability to specify anti-scenarios, that makes synthesis a lot more problematic. In recent work, we propose a first-cut at devising synthesis algorithms from LSCs, and at analyzing their complexity, albeit in a slightly restricted object model [HKg].

The second direction of research involves inheriting behavior. Inheritance is one of the key topics in the object-oriented paradigm, but when working on the analysis and design levels (rather than in the programming stage) it is not at all clear what it exactly means for an object of type B to be also an object of the more general type A. In virtually all approaches to inheritance in the literature, the is-a relationship between classes A and B entails a basic minimal requirement of *protocol conformity*, or subtyping, which roughly means that it should be possible to 'plug in' a B wherever an A could have been used, by requiring that what can be requested of B is consistent with what can be requested of A. In addition, *structural conformity*, or subclassing, is often requested, to the effect that B's internal structure, such as its set of composites and aggregates, is consistent with that of A.

Nevertheless, these form only weak kinds of subtyping, and they say little about the *behavioral conformity* of A and B. They require only that the plugging in be possible without causing incompatibility, but nothing is guaranteed about the way B will actually operate when it replaces A. Thus we don't have full behavioral substitutability, but merely a form of consistency. In fact, B's response to an event or an operation invocation might be totally different from A's. Here we are concerned with investigating the plausibility (and indeed also the very wisdom) of guaranteeing full behavioral conformity. In practice. behavioral conformity is often too stringent; many times one does not expect the inheritance relationship between A and B to mean that anything A can do B can do too and in the very same way. They are often satisfied with guaranteeing that anything A can do, B can be *asked* to do, and will look like it is doing, but it might do so differently and produce different results.

In recent work [HKp], we have obtained preliminary results that show that on a suitable schematic, propositional-like level of discourse there are strong connections between questions of inheritance and well-known semantic notions of refinement between specifications (such as trace containment and bisimulation). The rather high-level computational complexity of detecting behavioral conformity is also established in [HKp]. However, here too there is still much research to be done, including the discovery of restrictions on behavioral specification that would guarantee behavioral conformity, and algorithms for finding out if given models satisfy such restrictions.

Many other significant challenges remain, for which only the surface has been scratched. Examples include the true formal verification of software modeled using high-level visual formalisms, automatic eye-pleasing and structure-enhancing layout of the diagrams in such formalisms, satisfactory ways of dealing with hybrid object-oriented systems that involve discrete as well as continuous parts, and much more.

It is probably no great exaggeration to say that there is a lot more that we *don't* know and *can't* achieve yet in this business than what we do know and can achieve. Still, the efforts of scores of researchers, methodologists and language designers have resulted in a lot more than we could have hoped for ten years ago, and for this we should be thankful and humble.

References

[B]   Booch, G., *Object-Oriented Analysis and Design, with Applications* (2nd edn.), Benjamin/Cummings, 1994.

[CY]   Constantine, L. L., and E. Yourdon, *Structured Design*, Prentice-Hall, Englewood Cliffs, 1979.

[CD]   Cook, S. and J. Daniels, *Designing Object Systems: Object-Oriented Modelling with Syntropy*, Prentice Hall, New York, 1994.

[DH]   Damm, W., and D. Harel, "LSCs: Breathing Life into Message Sequence Charts", *Proc. 3rd IFIP Int. Conf. on Formal Methods for Open Object-based Distributed Systems (FMOODS'99)*, (P. Ciancarini, A. Fantechi and R. Gorrieri, eds.), Kluwer Academic Publishers, 1999, pp. 293–312.

[D]   DeMarco, T., *Structured Analysis and System Specification*, Yourdon Press, New York, 1978.

[H]   Harel, D., "Statecharts: A Visual Formalism for Complex Systems", *Sci. Comput. Prog.* **8** (1987), 231–274. (Preliminary version appeared as Tech. Report CS84-05, The Weizmann Institute of Science, Rehovot, Israel, Feb. 1984.)

[HG]   Harel, D., and E. Gery, "Executable Object Modeling with Statecharts", *Computer* (July 1997), 31–42. (Also, *Proc. 18th Int. Conf. Soft. Eng.*, Berlin, IEEE Press, March, 1996, pp. 246–257.)

[HKg]   D. Harel and H. Kugler, "Synthesizing Object Systems from Live Sequence Charts", in preparation.

[HKp]   D. Harel and O. Kupferman, "On the Inheritance of State-Based Object Behavior", submitted 1999.

[H$^+$] Harel, D., H. Lachover, A. Naamad, A. Pnueli, M. Politi, R. Sherman, A. Shtull-Trauring, and M. Trakhtenbrot, "STATEMATE: A Working Environment for the Development of Complex Reactive Systems", *IEEE Trans. Soft. Eng.* **16** (1990), 403–414. (Preliminary version appeared in *Proc. 10th Int. Conf. Soft. Eng.*, IEEE Press, New York, 1988, pp. 396–406.)

[HP]   Harel, D., and M. Politi, *Modeling Reactive Systems with Statecharts: The STATEMATE Approach*, McGraw-Hill (258 pp.), 1998. (Early version titled *The Languages of STATEMATE*, Technical Report, I-Logix, Inc., Andover, MA, 1991.)

[HP]   Hatley, D., and I. Pirbhai, *Strategies for Real-Time System Specification*, Dorset House, New York, 1987.

[IL]   I-Logix, Inc., products web page, http://www.ilogix.com/fs_prod.htm.

[ITU]   *ITU-TS Recommendation Z.120: Message Sequence Chart (MSC)*. ITU-TS, Geneva, 1996.

[RJB]   Rumbaugh, J., I. Jacobson and G. Booch, *The Unified Modeling Language Reference Manual*, Addison-Wesley, 1999.

[R$^+$] Rumbaugh, J., M. Blaha, W. Premerlani, F. Eddy and W. Lorensen, *Object-Oriented Modeling and Design*, Prentice Hall, 1991.

[SGW] Selic, B., G. Gullekson and P. T. Ward, *Real-Time Object-Oriented Modeling*, John Wiley & Sons, New York, 1994.

[UML] Rational Corp., documents on the Unified Modeling Language, http://www.rational.com/uml/index.jtmpl.

[WM] P. Ward and S. Mellor, *Structured Development for Real-Time Systems* (Vols. 1, 2, 3), Yourdon Press, New York, 1985.

UML-RT as a Candidate for Modeling Embedded Real-Time Systems in the Telecommunication Domain

Dominikus Herzberg*

Ericsson Eurolab Deutschland GmbH
Ericsson Allee 1
52134 Herzogenrath
Germany
Dominikus.Herzberg@eed.ericsson.se

Abstract. UML-RT offers a set of extensions to the UML, which are a basis for modeling real-time systems. Some investigations have shown that UML-RT provides concepts, which are close to the constructs used at Ericsson for designing mobile switching systems in the GSM (Global System for Mobile Communication) telecommunication domain. Telecommunication systems are some of the most challenging systems regarding size, complexity, and real-time constraints. This article describes the main constructs used at Ericsson, which display a robust framework for building real-time systems; and shows the mapping to UML-RT. An addition to UML-RT is presented; it allows UML-RT to be a candidate for modeling embedded real-time systems in the telecommunication domain.

1 Introduction

Telecommunication systems, especially switching systems, are among the most complex embedded real-time software systems and pose many challenging software engineering problems. To get an idea of the real-time requirements which must be fulfilled consider that up to several hundred-thousand lines and some ten-thousand calls are processed in parallel under restrictive time constraints. Additionally, a switching system has to be fault-tolerant and highly reliable; customers will only tolerate a downtime of a few minutes per year.

So far, the Unified Modeling Language (UML) [8,9] supports the object-oriented paradigm, but it is too general to support real-time modeling aspects. In its present state the UML extension for real-time (UML-RT) [7] just adds three new concepts to the UML (capsules, ports, connectors) and claims to satisfy real-time needs. Although these three concepts do not cover any timing aspects, the claim has to be interpreted in such a way that capsules, ports, and connectors are a result of best-practice concepts in the real-time domain. While hard to prove in general, this article shows that UML-RT in fact offers concepts,

---

* Special thanks to Stefan Sandh from dpart.com for all the inspiring discussions about architectural modeling.

which are similar to the ones identified by Ericsson for real-time design. However, note that UML-RT aims for architectural modeling [7]. According to one of its inventors [5], architecture is "the organization of significant software components interacting through interfaces, those components being composed of successively smaller components and interfaces". This definition will not be questioned for the purpose of this paper, but it is agreed to be necessary to distinguish between architecture and design.

Section 2 introduces the main concepts used at Ericsson for real-time design. In section 3, these concepts are compared with the concepts offered by UML-RT. The comparison also results in a proposal on how to extend the UML-RT for modeling of priorities on an architectural level. Some words about the Specification and Description Language (SDL) [3] and its relation to UML/UML-RT in section 4 round off the article. Finally, section 5 summarizes the results.

2 Three Basic Concepts for Real-Time Design

The experience and expertise in real-time design for switching systems[1] is reflected in Ericsson's in-house programming language called PLEX (Programming Language for EXchanges) and the underlying hardware. Both complement each other – and when building software the hardware is taken into consideration. PLEX mirrors a lot of concepts found in the machine language of the processor but on a more abstract, user-friendly level. The benefit is that a programmer can easily estimate how much time is consumed by a specific part of code, which is essential for real-time programming; the drawback is that some constructs common to high-level languages are missing.

Three key concepts can be extracted out of this language/hardware synergy: blocks, signals, and job buffers. Beside this, a set of well-defined design rules and guidelines build the foundation of Ericsson's success in real-time design for over 20 years now. The design guidelines include patterns and techniques for real-time design. But one factor should not be underestimated: it is the human intellect. Design rules, guidelines, and methods support the designer, but they do not and are not intended to replace the creative individual. Within the software construction process the designer is responsible to ensure that the real-time constraints are met. The knowledge, the skills, the expertise, and the genius of a human being cannot be substituted by any methodology or modeling language.

2.1 Blocks

A *block* is a self-contained unit, which encapsulates data and code. All variables are local to a block and hidden to others. Blocks are in general comparable to the concept of objects, but they lack object-oriented features like inheritance due to a conscious decision made by Ericsson. Physically, a block is restricted

---

[1] The article refers to the software development of Ericsson's Mobile Switching Center (MSC) in the GSM System.

in size: a specific amount of code and data cannot be exceeded. This restriction forces a designer not to overload a block in functionality and to be careful in design decisions. This is a very simple but effective method in controlling the complexity of a block and the amount of time granted to design a block.

Internally, blocks are mostly implemented by state machines. Although not explicitly offered as a construct in PLEX, design guidelines include patterns, which instruct the designer on how to implement state machines. Once familiar with the design rules and guidelines, it is easy to read the code and understand the underlying concepts used.

2.2 Signals

The only way to exchange data between blocks is via *signals*; it is the sole communication mechanism between blocks. A signal consists of a signal name, a sender and a receiver, and optionally of a set of data attached to the signal. If several blocks are allowed to receive a specific signal, the signal sending statement includes a reference to the receiving block to avoid ambiguity. Whenever a signal is received by a block, this signal is the entry point to code execution in the block.

Different types of signals exist. Buffered signals go to a signal stack. The signals are forwarded to the receiving block on a first-in first-out basis. Buffered signals are used for implementing soft real-time requirements. Hard real-time requirements like time critical operations can be handled by direct signals; they by-pass the signal buffer and go immediately to the receiving block.

It is allowed to change the signal type by attaching an attribute to the signal sending statement. Thereby, a signal pre-defined as buffered can be sent directly to another block.

Furthermore, a signal can be either a single or a combined signal. Combined signals are pairs of signals, that means when block A sends a signal to block B, block A expects a responding signal back from block B. In contrast to that, single signals do not depend on any other signal. The crucial point is that the time consumed between the forward and backward signal of a combined signal pair must not exceed a critical limit. Therefore, a designer of a block, who uses combined signals, can rely on that the environment of the block to which it communicates will fulfill the timing constraints; otherwise the control system restarts the block during execution.

Figure 1 summarizes all the signal types discussed.

2.3 Job Buffers

The processor used as a platform for PLEX code execution works with so-called *job buffers*. A job buffer is a set of registers plus a signal buffer; this hardware's specific characteristics are used as a concept when designing software. The fact that makes the Ericsson platform a real-time system is the existence of several job buffers; and each of the job buffers is associated with a different priority in execution. The underlying control system specifies the job buffer handling and triggers code execution in a deterministic way.

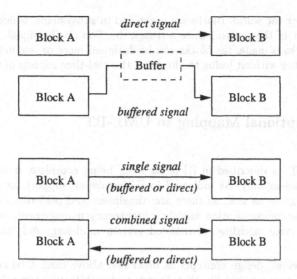

Fig. 1. The different types of signals

In principle, the code in one and the same block can be executed with different priorities. In practice, the design rules advise the designer to be careful on this point. Hence, a block is typically associated with a prefered job buffer level.

A change of priority is initiated only by signals. That means a signal addresses a specific job buffer level. The job buffer puts the signal either on the signal stack as buffered, or executes it directly. Depending on the priority of the job buffer, a direct signal interrupts execution on lower levels.

To ease the software design of a real-time system, certain activities are assigned to specific job buffer levels. For example, all operation and maintenance activities run on their own job buffer level, traffic-related activities run on another job buffer level with a higher priority. Priority levels allow telephone calls to have top priority in a switching system. In the background system tracings or interactions with the operator may be handled without negatively impacting the real-time constraints on the traffic.

2.4 The Implications of the Concepts Introduced

The block/signal view of a system is quite natural for telecommunication systems, because the communication between nodes in a telecommunication network is very much protocol driven. But this is only one aspect.

Blocks and signals are more than domain-specific design elements. They are also organizational prerequisites for a multi-site and distributed development of a software system. Ericsson makes heavy use of this "effect". It enables Ericsson to develop software with hundreds of designers located in different subsidiaries spread all over the world. Blocks are organized in subsystems, which constitute a logical part of the system. Once a design decision on the signals to be used between blocks is made, the blocks can be designed more or less independently from each other without losing the focus on the real-time aspects of the system.

3 Conceptional Mapping to UML-RT

The UML-RT as described in [7] is far from being complete, it covers architectural modeling only. As indicated in [5], modeling support for other areas needs to be given as well, as there are: timeliness and performance modeling, time-sensitive communication models, concurrency management, resource and quality of service modeling, distributed system modeling, and fault-tolerance techniques.

To be precise, design concepts as described above cannot be compared to architectural concepts just like that. Design and architecture serve different purposes and are therefore different domains. For example, we at Ericsson make a clear distinction between architecture and design since we cannot afford breaking the architecture down to design in a straight forward manner (e.g. by structural decomposition). The architecture, which naturally does not consider all aspects – like real-time constraints – sufficiently, would thereby predetermine design decisions as well as the structural design of our system. This would result in an unacceptable loss in characteristics. Putting it to the extreme, such a predetermined design might conflict with real-time requirements. The point is that the design has to fulfill the requirements and the specifications; only in the ideal case can the design also fulfill the architecture. That is simply an industrial insight.

For some reason it is stated that capsules and ports have to be physical [7]; furthermore UML-RT is seen primarily as an instrument for modeling the software architecture as mentioned above. However, in practice UML-RT is used and promoted in both domains; architecture and design. This leads to a loss of formal rigor for the discussion in this paper and is a symptom of the problem to capture the characteristics of a domain formally.

Nonetheless, it is beneficial to map Ericsson's design concepts to the concepts offered by UML-RT as a language. In the next step one may ask if a design concept is applicable on an architectural level and vice versa. For instance, the requirement that capsules have to be physical is less critical on a design level but questionable on an architectural level as explained previously. One has to be aware of the domain one is in and how the domains relate to each other.

Therefore, the task needs to be slightly re-formulated: Do the basic design concepts introduced (blocks, signals, job buffers) have a counterpart in UML-RT? If yes, the UML-RT is probably a good candidate for modeling embedded real-time systems in the telecommunication domain; if not, the question is, if there is an aspect missing in UML-RT or if this aspect is only specific to Ericsson.

3.1 Blocks, Capsules and Ports

The concept that corresponds to blocks is capsules in UML-RT. Capsules are complex and (on a design level) physical like blocks are. In addition to blocks, capsules can be nested and inherited.

Moreover, ports are the part of a capsule that mediate the interaction of the capsule to the outside world. This is a remarkable extension compared to the concept of a block. Conventionally, blocks (like objects in object-orientation) only specify an outside-in context by the signals they can receive and not an inside-out context. This is done by ports, which specify the capsule's environment and are the only communication interface for the capsule. This makes capsules highly reusable; this is especially important for architectural modeling. Consequently, ports can be interpreted as specifications for an architectural component, which can then be put into effect by one or more blocks.

Similar to blocks, state machines can be used in the context of capsules. As long as capsules are used either on an architectural or on a design level the use of them is not problematic. But if UML-RT is used in both domains for the same system the risk is quite high of mixing purposes. Does the state machine specify the architectural component on an abstract level or is it meant to be a first description of the behavior of a design component, which just needs to be further refined and eventually implemented? The same argumentation is valid for ports.

3.2 Signals and Connectors

Connectors in UML-RT realize protocols, which define a flow of information between different ports; connectors capture the key communication relationships between capsules. Therefore, a set of signals that represents the information exchange between two blocks can be regarded as an instance of a connector.

When talking about signals it has to be noted that the UML does not know signals in the sense as they were introduced in the context of blocks [1, 8, 9]. For architectural purposes connectors are a suitable abstraction. But if the designers of the UML-RT intend to give modeling support for purposes other than architectural modeling, it is recommended to take into account the signal concept as described above. Signals are a key concept for successful real-time design.

3.3 Job Buffers and Priority Layers

To reduce the job buffer concept to the essence, job buffers stand for different priority layers. In ROOM (Real-time Object-Oriented Modeling), from which UML-RT is derived, events and event priorities are known, but a corresponding architectural concept is missing [6].

As explained above, job buffers or rather priority layers are essential in grouping activities according to their importance in execution and should be regarded as a design pattern. It is proposed to import this pattern as a language construct in UML-RT. This adds another but important dimension to the model of a real-time system and makes the priority associated with specific activities easier to understand.

A graphical presentation of priority layers for capsules is suggested in figure 2. It shows different capsule layers related to different priority layers. Additionally, a small symbol in the corner may specify the priority layer. This makes it easy to assess rapidly if a capsule works on different priority layers, and if so, on how many and on which. It is then up to a modeling tool to present, for example, a view for one specific priority layer. For design or specification purposes it makes sense to associate a separate state machine for each priority layer of a capsule.

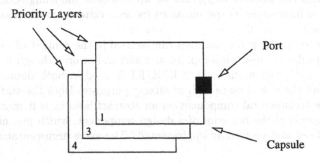

Fig. 2. An extension to capsules for priority layers

Designing with priority layers in mind has been good practice for over two decades in Ericsson now. All the details of priority handling, signal queuing, and scheduling can be defined by and hidden in an execution model. Since priority layers as such are stable and rarely change over time, this seems to be an argument to use this extension of the UML-RT also for architectural modeling. However, the question is whether a design concept is also valid as an architectural concept. From a pragmatic viewpoint we think that priority layers are of use in the architectural domain and add a quite interesting aspect to the architectural model, independent of the discussion of whether architectural components need to be physical or not. The lack of a formal approach for describing domain-specific characteristics makes it quite hard to formally prove the consistency of a concept domain relation.

4 Design and Modeling in the Telecommunication Domain

There is a clear preference in the telecommunication domain to use SDL (Specification and Description Language) as a specification and design language [2, 3].

For instance, the standards for the 3rd generation systems in mobile telecommunication are proposed to be specified in SDL. In fact, SDL offers a lot of concepts that match the block/signal view.

To be precise and efficient in development, one has to distinguish carefully between design and architecture as well as recognizing the relationship between the two. Different domains may require different approaches and concepts, because they serve different purposes.

SDL addresses design purposes. It is a mature and stable language standard and is unlikely to disappear from the telecommunication world over the next 10 years. UML-RT, on the other hand, originally addresses architecture purposes, an area, where standards lack a solid proof of industrial use. UML-RT has the potential to fill this gap for real-time modeling. The discussion demonstrated that it offers suitable concepts for that purpose.

All this requires that UML/UML-RT and SDL can co-exist and form a powerful pair for modeling and designing real-time systems. If there is a clear separation between the architecture and the design domain the combined use of SDL and UML-RT is less critical; coupling the domains requires an adjustment of the concepts and interfaces of the two languages. Only a combined use within the same domain causes severe problems: UML-RT and SDL are based on different paradigms, which makes adaptation difficult. This requires further studies and research. Currently, activities are ongoing to specify an ITU standard that brings UML and SDL closer together [4]. The success and acceptance of these efforts assumed, we can predict a future in which modeling real-time systems in the telecommunication domain is more a reality than a theory.

5   Summary and Conclusion

For real-time modeling, UML-RT extends the UML by three concepts, namely capsules, ports, and connectors. According to Ericsson's experience in designing switching systems, UML-RT lacks a facility: it does not offer support for modeling priority layers. Therefore, an extension to UML-RT is proposed that introduces one or more layers of a capsule, which are associated with a specific priority in execution. With this extension UML-RT is a candidate for modeling real-time systems in the telecommunication domain.

However, we do not agree that a capsule in UML-RT has to be physical; we cannot afford allowing a modeling language to put requirements on what has to be implemented in the system – at least not on an architectural level, which has an impact on the real-time characteristics. We speculate whether this statement is generally applicable for real-time systems. Furthermore, we believe that architecture and design are different domains, which are not necessarily strictly separated and that some kind of overlap is to be expected. If a modeling language is supposed to be used in both domains with the very same language concepts, we reflect on what kind of conclusions we have to draw. Does it blur the distinction in domains? Or does the language mediate optimally between the two domains? Further investigations in that area seem to be needed.

338

References

1. Booch, G., Rumbaugh, J., Jacobson, I.: The Unified Modeling Language User Guide. Addison Wesley, 1999
2. Ellsberger, J., Hogrefe, D., Sarma, A.: SDL: Formal Object-oriented Language for Communicating Systems. Prentice Hall, 1997
3. ITU Recommendation Z.100: Specification and Description Language (SDL), 1996
4. ITU Recommendation Z.109: SDL Combined with UML, 1999
5. Selic, B.: UML-RT: Using UML for Modeling Complex Real Time System Architectures. Presentation given at Rational in Kista (Sweden), 8 Oct 1998; see also http://www.objectime.com/otl/technical/umlrt.html
6. Selic, B., Gullekson, G., Ward, P. T.: Real-Time Object-Oriented Modeling. John Wiley and Sons, 1994
7. Selic, B., Rumbaugh, J.: Using UML for Modeling Complex Real-Time Systems. White paper available at http://www.objectime.com/otl/technical/umlrt.html
8. UML Notation Guide: Version 1.1 (Sep 1997). Object Management Group, http://www.omg.org
9. UML Semantics: Version 1.1 (Sep 1997). Object Management Group, http://www.omg.org

Modeling Hard Real Time Systems with UML
The *OOHARTS* Approach

Laila Kabous and Wolfgang Nebel

University of Oldenburg
Department of Computer Science
Postfach 2503
26111 Oldenburg
Germany
{Laila.Kabous|Wolfgang.Nebel}@informatik.uni-oldenburg.de
http://eis.informatik.uni-oldenburg.de

Abstract The use of object oriented techniques and methodologies for the design of hard real time systems appears to be necessary in order to deal with the increasing complexity of such systems. Recently many object oriented methods have been used for the modeling and the design of real time systems. Due to their complex nature, such systems are extremely difficult to specify and to implement. The use of an appropriate design method that supports both the modeling of the functional and the non-functional requirements of such systems is a crucial prerequisite for the successful modeling of such systems. In this paper we propose our approach, which enables the development of hard real time systems using UML and other notations enabling both the analysis of the functional requirements as well as the analysis of the non-functional requirements. **Keywords** : Hard real time systems, Object Orientation, UML, schedulability analysis.

1 Introduction

Real time applications vary in size and scope from microwave ovens to factory automation, nuclear power plant control, telecommunication and avionics systems. Such systems are in continual interaction with their execution environment. They react to external stimuli coming from their environment in a specified amount of time. The correctness of such systems depends on the time within they respond to external events.

The one common feature of all real time systems is timeliness. However, this ostensibly simple property characterizes a vast spectrum of very different types of systems. There are two main classes of such systems: hard real time systems, which we discuss in this paper, are incorrect if their responses do not meet the specified deadline and soft real time system. The response times of soft real time systems are also important, but the system will still function correctly if deadlines are occasionally missed. The quality of soft real time systems is however regarded worse with increasing violations of the soft timing constraints.

The most important stage in the development of any hard real time system is the generation of a consistent design that satisfies the specification of the requirements. Real time systems in general are no different from other computer applications. However, hard real time systems do differ from traditional data processing systems in that they are constrained additionally by certain non-functional requirements (reliability, dependability, timing, etc.). These requirements and the constraints imposed by the execution environment, must be considered throughout the system development cycle. Hard real time systems and real time systems in general include specific constraints which make them harder to develop and test than other systems. A design methodology for hard real time system must handle some specific requirements:

- − Specification of timing constraints,
- − Characterization of the execution environment,
- − Support of concurrent activities,
- − Priority assignment,
- − Validation of the models designed in early phases,
- − Simulation of the system models,
- − Schedulability analysis,
- − Reuse,
- − Implementation independence.

During the last ten years over 40 object oriented design methods have been developed. In recent years object oriented methodologies have been also employed in the construction of real time systems. The use of object oriented technologies is of great interest. Some examples of them which are used in the development of real time systems, are given below. For a comparison of some design methodologies for real time systems see [5], [6].

ROOM [17] Real Time Object Oriented Method support the object oriented concepts (e.g. abstraction, inheritance, polymorphism) and appropriates for the validation of the designed models. Its behavior is based on RoomCharts [9]

Real Time UML [7]: The method is the application of the Unified Modeling Language UML [16] to the design of real time systems. It represents the system with structural and behavioral models. These models are class diagrams and statecharts. The design modeling includes the architectural, mechanistic and the detailed design.

HRT-HOOD [1]: Hard Real Time Hierarchical Object Oriented Design method HRT-HOOD is an extension of the HOOD [10] method for the development of hard real time systems. The method allows the designer to describe the system architecture. Schedulability analysis can be used to validate the hard real time requirements of the system. HRT-HOOD is an object based method since the method does not support some object oriented ideas like polymorphism and inheritance.

OCTOPUS [3]: The method can be used for the development of soft real time systems. OCTOPUS is based on a combination of the object oriented Method

OMT [12] and Fusion methodologies. It provides a guide to integrate object oriented concurrency in real time software. Hard real time requirements are not approached.

UML-RT [15]: UML-RT is co-developed by ObjecTime and Rational Corporation. The method combines the modeling concepts of the UML1.1 standard and special modeling constructs and formalisms originally implemented in ObjecTime developer and defined in the ROOM language [17]. The UML collaboration diagrams are used to explicitly represent the structural design patterns. In order to support the modeling of real time systems some stereotypes (capsules [1], port [2], connector [3]) have been introduced in to UML.

The methodologies listed above are based on object oriented concepts [4] but for the design of hard real time system they do not satisfy all the requirements mentioned above for a successful modeling of these specialized class of real time systems. The use of object oriented techniques is of great interest, but some properties of real time systems are difficult to specify using these techniques. Many methodologies have been proposed for the development of real time systems, but often the explicit specification of timing properties, that such systems underly, and the validation of these timing requirements are not approached.

For the design of hard real time systems we need a design methodology that provides a framework for the development of such systems with strict timing constraints and which fulfill the topics mentioned above in order to produce reliable and safe systems.

This paper presents a development paradigm for hard real time systems which we call Object Oriented Hard Real Time System *OOHARTS*, and the notations on which it is based. An example of a simple control system is presented in section 2. In Section 3.1 we give an overview of our approach and describe briefly each main step of the *OOHARTS* development process. In the following sections we describe our approach *OOHARTS* which extends the applicability of object oriented design UML to hard real time systems with strict timing constraints. Other notations are also added for describing the behavioral aspects of such systems.

2 An Example Model

To demonstrate the principle of our modeling approach and illustrate the notation, the well known *Mine Pump Control System* has been chosen. It possesses many of the characteristics of real time systems. The system is used to pump mine water, which collects in a sump at the bottom of shaft to the surface. The main safety requirement is that the pump should not be operated when the level of methane gas in the mine reaches a high values, due to the risk of explosion. A

---

[1] correspond to the ROOM concepts of actors
[2] physical part of the implementation of a capsule
[3] capture the communication relationships between capsules
[4] except HRT-HOOD which is an object based method

simple schematic is given in Figure 1, which indicates that if the water collecting at the bottom of a mine shaft rises above a certain limit, the pump should be switched on. The pump should be switched off when the water has been sufficiently reduced. The pump can also be turned on and off by a human operator. Any operator can control the pump when the water level is between high and low sensors, and a specific operator designated the "supervisor" has the authority to control the pump whatever the water level is.

For safety reasons there are sensors monitoring methane (CH4) and carbon monoxide (CO) concentrations, and airflow, and an indication must be given of any critical values. Due to the risk of fire, the pump must not be operated when the atmosphere contains too much methane. All three sensors' values along with the pump status should be periodically (60.0 s) logged. Critically readings from any of the sensors cause alarm messages to be sent to the operator's station. The current status of the system is presented to the operator (within 1.0 s if the methane or carbon level is critically) and the system also records periodically the readings from the sensors and the pump activities on the log.

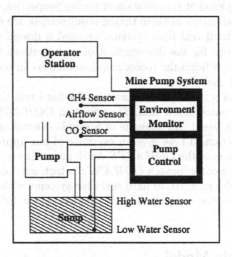

Figure1. The Mine Pump System

3 Overview

OOHARTS is developed for the design of object oriented hard real time systems. It is based on the unified modeling language UML and the hard real time concepts of HRT-HOOD. The development process is based on the essential phases of "water fall" model and comprises the classical main phases of development:

1. Requirements definition
2. Analysis
3. Design
4. implementation

We use in our development process UML notations and diagrams. For the design of hard real time systems we add some specific notations and rules to UML [5]. In the next section we give a short presentation of our approach.

3.1 *OOHARTS* Development Process

The Object Oriented Hard Real Time System *OOHARTS* development process is based on an iterative and incremental development process (Figure 2).

Requirements Definition: The first development phase consists of the definition of the system requirements. During the requirements phase the main question that needs to be addressed is "What is desired?". The user and the analyst must discover which features are required from the system. During this design phase the functional (object identification, etc.) and non-functional requirements of the system must be addressed.

Hard Real Time Analysis: The main aim of the analysis is to identify the problem in a clear manner. The defined requirements during the last development process step will be analyzed in order to understand the requirements fully and describe them in sufficient detail to model the problem. The analysis phase is decomposed into three phases.

The first phase consists of building a "System Environment Interaction SEI" model in order to capture the external events and messages flowing between the execution environment and the system. It shows the interactions among the system and its environment [6].

The next design step during the analysis is to define the structure of the system. In this sub-step of the analysis we use the UML basic class diagram, which we call "Object Class" (**OC**) model, to identify and define the objects, classes and their relationships.

The third analysis phase is described with a specialized form of statecharts in order to define the behavior of each object and class. We call it "Object Behavior Chart" (**OBC**).

Hard Real Time Design: This phase is the process of specifying an implementation that is consistent with the results of the analysis phase. The objects defined during the analysis will be refined during this design step and the design models will be simulated in this step in order to detect errors before we begin with the implementation.

The design phase comprises also the following activities: Assignment of objects to tasks using task diagrams [7], Process or task allocation to processors and process scheduling to ensure that all processes residing on the processor(s) will meet their deadlines.

---

[5] Without modification of the UML-Standard
[6] UML use case diagrams can be also used at this design step

Implementation: During this design step the system will be implemented. The software part can be implemented e.g. in the object oriented programming language C++ or Ada95. The implementation of the hardware can be realized in Objective VHDL [14], which is the object oriented extension to the hardware description language VHDL.

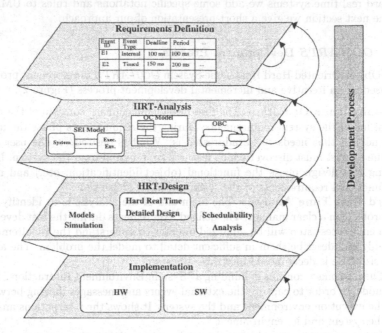

Figure2. The Development process of Hard Real Time Systems

4 Requirements Definition

Once the problem has been clearly stated, it is important to understand the parameters, requirements and the constraints that influence the problem. We do not consider in our approach *OOHARTS* at this point any implementation constraints, only those that belong to the problem area. We can extract from the problem description potential parameters, events and timing constraints. This informations given in the problem statement can be represented in a requirements table as given in Table 1.

5 Hard Real Time Analysis

In this section we present a framework for describing both the structure and the behavior of hard real time systems based on UML diagrams and the new intro-

	Event ID	Event Type	Periodic/aperiodic	Deadline	...
CH4-Sensor	CH4-level	Environmental	Periodic	1.0s	...
CO-Sensor	CO-level	Environmental	Periodic	1.0s	...
...	...	...	...	...	...

Table1. Defining the requirements of the system

duced notations. The system is designed as a collection of interacting objects. The structure of the system is described as a set of object oriented models, which include the description of identified classes, objects and their relationships. In this section we illustrate our approach and its applicability using the mine pump system case study (see Sect. 2).

5.1 System Environment Interaction Model

Since the responses of hard real time systems depend heavily on its environments, we use in the *OOHARTS* method a new model called System Environment Interaction model SEI-Model in order to be able to represent the events and messages flow between the system and its execution environment. To illustrate this we give in Figure 3 the mine pump system SEI-model. The lines are annotated with arrows identifying the events flow. The event flow is marked with a special mark in the case of periodic events (see Figure 3).

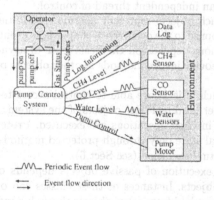

Figure3. The System Environment Interaction Model

5.2 Defining the Object Class Model

Objects in hard real time systems differ in their nature. We consider in *OOHARTS* various object types that can be found in hard real time systems. These objects can specify cyclic, aperiodic, passive, protected or environment objects.

Based on the objects defined in HRT-HOOD, which are common in hard real time systems, and the UML stereotypes, we describe in this section a notation that is suitable for hard real time systems with strict timing constraints.

The UML stereotypes allow the description of classes and their relationships. A stereotype establishes a specific semantic of a class including the kind of relations to other objects and the method calls. We have identified a set of stereotypes with a precise defined semantic to represent these hard real time objects and the relationships that can be specified between them. The stereotypes allow the description of the different hard real time objects. The structure of the system can be specified using these objects and classes.

HRT-HOOD extend the types defined in the software design method HOOD [10] and defined several types of objects [7] e.g. passive, active, cyclic, sporadic, protected. The development process does not comprises an analysis phase. It consists of a design phase (logical and physical), an implementation [8] and a test phase. The objects are described using the object description skeleton ODS and the object control structure OBCS which have a BNF structure.

Our approach *OOHARTS* differs from the object based HRT-HOOD method, since we use another development process, consisting of an HRT-analysis (see Sect. 5), HRT-Design (see Sect. 6) and implementation (see Sect. 7) phase. We have added a new object type to those defined in HRT-HOOD. Following classes are defined in *OOHARTS*:

Cyclic class: These art of real time objects represent cyclic behaviors. The activation of the methods of a cyclic class depends on the passage of a given period. They have an independent thread of control.

Aperiodic Class: Such objects represent aperiodic activities. They execute some actions in response to the occurrence of aperiodic events. The execution of aperiodic operations depends on the occurrence of asynchronous events. They may spontaneously invoke methods in other objects by the occurrence of an event.

Protected Class: Such objects are able to control access to resources which are used by the other hard real time objects of the system. They are able to control when their invoked operations are executed. Protected objects can change their internal state only through protected methods. Their methods are executed in mutual exclusion (see Sect 6).

Passive Class: The execution of passive objects depends only on the calls made from other objects. Instances of these classes can only change their state when that is required from other objects through a method invocation. Passive objects do not spontaneously invoke methods in other objects and have no control over when invocations of their methods occur.

Environment Class: Since a hard real time system is in continual interaction with its environment, this class type is needed in order to model sources and destinations of external events.

---

[7] HRT-HOOD does not distinguish between classes and objects
[8] for software systems in ADA

The stereotypes of the Unified Modeling Language UML permit us to describe the classes and objects that are specific for hard real time systems in *OOHARTS*. Following stereotypes «cyclic », «passive », «protected », «aperiodic », and «environment » are added to UML. The structure of real time objects can be specified and defined using these classes given above in the object class model based on the UML class diagram. To give an example of the use of these defined hard real time objects in *OOHARTS* the system structure of the mine pump system given in Sect. 2 can be defined as follows (See Figure 4).

Many strategies can be used to identify objects and classes in the class di-

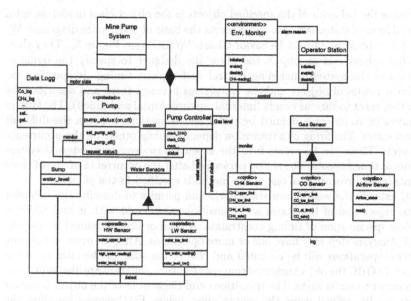

Figure4. The Object Class Model of the Mine Pump Control System

agrams e.g. underlying the nouns, identifying the physical devices and so on. The set of real time class types that are used, are based on the new defined UML stereotypes. The identified objects and classes have attributes (*real world attributes*[9] and *real time attributes*[10]) and methods (services) that allow them to fulfill their responsibilities. Classes have relationships to each other. These may be associations among class instances, like aggregation or they may be relationships between classes, like generalization. Objects use these associations to communicate by sending messages to each other. The identified classes with their attributes and methods are represented in the "Object Class" model (OC). After specifying the system structure which contains different objects and classes

---

[9] They represent the states of objects
[10] specified for real time objects

we assign for the cyclic and aperiodic objects additionally to *real-world* attributes a number of *real time attributes* like:

- the period for each cyclic object,
- minimum arrival time and the maximum arrival time[11],
- deadlines for all aperiodic and cyclic objects.

5.3 Describing the Behavior using the Object Behavior Chart in *OOHARTS*

To define the behavior of the specified objects in the object class model we use a special form of statecharts [9], which forms the basis of UML state diagrams. We call this extension "Object Behavior Charts" (OBC) (see Figure 5). They show the full behavior of an object and allow the designer to specify the dynamic aspects of the system, including control and explicit timing. It specifies the different modes of objects and the transitions between them and describes the way they react to various events (internal, environmental and timed). The system behavior of an object is defined by states and transitions between the different object states. The firing of a transition depends among others on the occurrence of events. Those can be events from the execution environment, internal system events or timed events. The action performed after the occurrence of one of these events is the invocation of one of the methods specified in the object.

OBC extends the well-known statecharts and permits to describe the behavior of an object and of the states and transitions associated with it and allows a explicit specification of timing constraints. A class or object defined during the HRT-Analysis step may have one or more operations. After an event occurs one of these operations will be executed and change the state of the class or of the object. In OBC the object oriented concepts e.g. encapsulation are also supported in contrast to statecharts. The transitions and the actions of the object behavior chart can be defined using the syntax given below. Furthermore we allow the specification of timing constraints. The syntax of transitions is w.r.t. the object oriented paradigms and shows how those can be specified.

```
Transition_Trigger := Event/Method_Invoke
Event := Internal | Environmental | Timed
Method_Invoke := Object_Name.Operation_Name(Arguments)[RT_Constraint]
RT_Constraint := [Temp_Req, Priority, Importance]
Temporal_Req:= [Deadline, Period, Min_Arrival_Time, Max_Arrival_Time]
Importance := [Hard, Soft]
```

We explain this using the simplified example given below. A complete specification of the mine pump control system can be found under [11]. This example shows the object behavior chart of the pump consisting of two different states on and off of the pump. The pump is at first in the default state off. If the methane level and the carbon monoxide are safe, the pump change to the state on. If the gas level are not safe, the pump must change to the state off after 1.0 s. Since

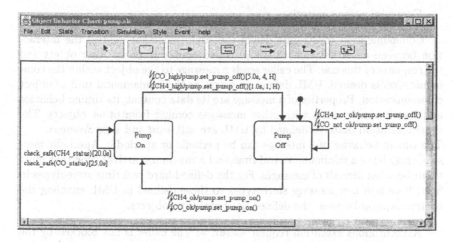

Figure5. Object Behavior Chart to describe the System Behavior

they are safe (unsafe), the pump is still on (off). We allow in *OOHARTS*, the expression of timing constraints (see sect. 5) in order to engineer hard real time systems. This real time constraints can be added to transitions to define object behavior. Such constraints are for example deadlines, to specify the amount of time, after which the object must change its state. This is the case for example if the pump is on and the methane is to high, so the pump must change to the state off in 1.0 s in order to avoid explosions. Other kinds of real time constraints which we allow to specify are periods of execution, minimum- and maximum-arrival time for aperiodic objects. Additinionally we allow the designer to set a priority to transitions, so that the transition with the higher priority will be at first executed. The designer can also define an importance constraints to differentiate between hard and soft real time objects. The priority and the importance constraints are very helpful, also later during the detailed design.

6 Hard Real Time Design

6.1 Refinement of Behavioral Aspects

The behavior of the specified objects describes the various modes, operations and conditions of the objects. We distinguish between constrained operations and unconstrained operations. Unconstrained operations can be called without special care. They are non-sensitive to any external events. On the other hand constrained operations can provide their services if certain conditions are met. These constraints imposed on operations are limited to:

– Concurrency constraints,

---

[11] only for aperiodic objects

– Communication constraints.

Communication constraints define the conditions that control the interaction between an object that calls a provided operation of an other object and the response of this one. The caller sends a message to the object within the communication is desired. UML defines messages as the fundamental unit of object communication. Properties of a message are its data content, its timing behavior and how it synchronizes with other messages coming from other objects. The synchronization patterns defined by UML are *call*, *wait* and *asynchronous*.
The timing behavior of a message can be periodic or aperiodic. Aperiodic messages may have a minimum arrival time and a maximum arrival time that must occur between arrivals of messages. For the defined hard real time stereotypes in Sect. 5 we add new message stereotypes to those defined in UML enabling the communication between the defined hard real time objects.

- Asynchronous execution request «**aser** »: The caller is not blocked by the request[12]
- Loosely synchronous execution request «**lser** »: The caller is blocked by the request until the called object is ready to provide the service.
- Highly synchronous execution request «**hser** »: The caller object is blocked by the request until the called object has serviced the request.
- Time-Out-lser «**tm-lser** » and Time-Out-hser «**tm-hser** »: A timeout constraint added either to "lser" or "hser" allows a caller to request and check that an operation is executed within a given period of time. The timeout is the maximum elapsed time before the request is taken into account or executed by the server.

We have defined different types of objects. The objects send messages to each other for data exchange. One of the major problems using for example synchronous communication is that objects may be blocked *deadlock*. It can happen when two objects simultaneously invoke each other synchronously. In that case, since both objects are blocked, neither is capable of replying to the other. Another form of deadlock can happen, when there is a circular chain of invokes (e.g. the pump controller invoke the operation CH4-reading of the environment monitor, which in turn invoke the operation alarm of the operator, which finally invoke the operation set-pump of the pump object). Such circularities can be difficult to detect and must be avoided. One common way to get around these problems is to provide the designer with synchronization patterns, which can be used to specify and constrain the communication and the synchronization depending on the object type. The timeout synchronization mechanisms specifies a maximum time interval during which the reply must be received. If the reply is not received within this interval, the synchronous communication is aborted and an exception is released or a change of mode is needed. This can be the case in the communication between the controller and the CH4-Sensor given in the example below. The Sensor must service the request from the controller to

---

[12] similar to the asynchronous modus in UML

check the methane level in the specified time of 1.0 s otherwise the pump should be set off and an exception is raised. Furthermore we allow to set priorities to messages. This is very useful if two objects send asynchronously a message to a third object, so the message with the highest priority, will be serviced.

We add these annotations to UML sequence diagrams which show the sequence of messages between objects. The graphical syntax for the specialized sequence diagrams is shown in Figure 6. This figure shows a simplified sequence diagram for the mine pump system. The horizontal line shows messages. We add to each message additional synchronization features mentioned above.

The other kind of constraints are *Concurrency Constraints*. They define the

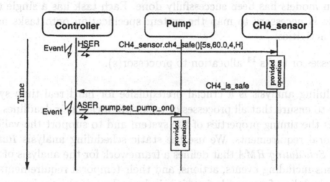

Figure6. Sequence Diagram

conditions that allow the access to a method of an object by several objects at the same time. Following stereotypes are needed to allow a successful operation access and the execution of concurrent operations in *OOHARTS*:

- Mutual exclusion execution «**mutex**»: The operation is executed in mutual exclusion. If a object calls an "mutex" operation while some other objects are currently executing it, it is held until the operation are completely executed. The operation is protected. This concurrency constraint is particularly needed for protected real time objects.
- Write execution request «**wer**»: The operation is defined as possibly modifying the object state to which it belongs. If an object thread is executing a "wer" constrained operation, no other object thread can execute a constrained operation from the same object.
- Read Execution Request «**rer**»: Means that the operation is allowed to access the state of the object, but performs no changes of global variables that would require mutual exclusion. A "rer" operation can be simultaneously executed by several client threads.

Before we continue with the detailed hard real time design the design models must be evaluated at this stage before coding. One of our research aims is the development of a tool (simulator) that allows us an early feedback of functional and non-functional behavior of the system and that reduces the iteration loop between the design and the implementation of hard real time systems. Simulation allows us to check the temporal behavior of design models and allows us the detection of the non fulfillment of temporal requirements.

The aim of hard real time design is to specify an implementation that is consistent with the hard real time analysis results.

Hard real time systems often execute several threads[13] of control simultaneously. The entirety of these threads can be defined as a task. The system tasks are identified and represented in a system task diagram [7] under consideration of the constraints defined during the analysis phase and after the simulation of the design models has been successfully done. Each task has a single thread of control. It is necessary to map the system specification onto tasks and to do following activities:

- Processes or tasks [14] allocation to processor(s).

- Scheduling analysis is a crucial prerequisite for hard real time systems in order to ensure that all processes or tasks will meet their deadlines, to guarantees the timing properties of the system and to support the validation of temporal requirements. We use the static scheduling analysis *Rate Monotonic Scheduling RMA* that defines a framework for the analysis of real time systems including events, actions, and their temporal requirements. We use this scheduling framework since it is independent of the application and the development domain.

By defining task diagrams to perform an *RMA* analysis, we consider the messages, the timing requirements, and the constrained communication between the objects defined in the sequence diagrams and how each object reacts internally, (specified in the object behavior chart), on the occurrence of events. All these informations must be taken into account to undertake schedulability analysis. We do not detail this step, since it goes beyond the scope of the conference.

7 Implementation

OOHARTS uses at this stage the UML deployment diagrams which allows us to specify the physical relationship among hardware and software components in the hard real time system. The system consists of a hardware part and of a software part. Both system parts will be implemented during this design step. The software part can be implemented e.g. in the object oriented programming language C++ or Ada95. The hardware part can be translated in to Objective VHDL [13] which is an object-oriented extension to the hardware description language VHDL. Automatic code generation is scheduled on future work.

---

[13] is a set of sequentially executed actions
[14] the terms are used interchangebly in this paper

For a successful design of a hard real time systems a tool is needed to support the *OOHARTS* methodology. The tool set is now under development in the programming language Java. It comprises the design of the object structure, the design of the object behavior using the new defined object behavior chart OBC, a simulator to execute the designed models and a scheduling analyzer based on the rate monotonic scheduling analysis.

8 UML Extensions Summary

The *OOHARTS* methods use an extension of the unified modeling language UML notations enabling the development of hard real time systems as introduced above. We give in this section a summary of these extensions made to UML. These extensions have been introduced in order to address the needs for the development of hard real time systems. We do not detail the entire process, but emphasize the most significant extensions defined.

Hard real time class stereotype: The new class stereotypes are applied to define and identify the different common objects types found in hard real time systems. Several real time stereotypes have been introduced into UML «cyclic», «aperiodic», «protected», «passive», «environment» (see Sect 5).

Modeling the object behavior: For definition of the object behavior we have introduced a specialized form of UML statecharts, which we call object behavior chart OBC. The OBC shows the full behavior of an object, containing states and transitions between states. The transition trigger consists of events and actions to perform after the occurrence of this events and can be defined using the syntax given in section 5 w.r.t. object oriented concepts e.g. encapsulation. This specialized form of UML statecharts are used to describe the full behavior of the specified hard real time objects with explicit timing constraints e.g. deadline, period.

Constraining the object synchronization: In UML Messages are the fundamental unit for object communication. The synchronization patterns supported by UML are call, wait and asynchronous. We have introduced in the *OOHARTS* approach other communication stereotypes «aser», «lser», «hser», «tm-hser», «tm-lser» to constrain the communication and synchronization explicitly in dependence on the type of hard real time objects. We specify the interactions between objects using sequence diagrams, in which the kind of synchronization is identified for each object. Some objects e.g. protected objects are not allowed to call *blocking* operations in other objects. The communication and synchronization between the system objects must be explicitly described in order to avoid that objects may block each other "*deadlock*".

Constraining the objects concurrency: In UML the concurrency problem is taken into account through an attribute on operations named concurrency. This

354

attribute can take following activities:

sequential: Callers must coordinate and synchronize with each other, so that nothing is guaranteed by the system if concurrent calls occurs to a same object.

guarded: This means that if multiple calls from concurrent threads occurs simultaneously to one instance, only one is allowed to commence. The others are blocked until the performance of the first is completed. The designer must ensure that deadlocks do not occur due to the simultaneous blocks.

concurrent: Concurrent calls are allowed to be executed concurrently with any other method of the object, that is, with concurrent, guarded or sequential operation.

In UML the concurrency attribute on operations can take one of the above mentioned values. This attribute is set from the callers side. This implies that concurrency control is transfered outside of the called object. Since each hard real time object in *OOHARTS* is responsible for its encapsulated data, we have introduced new concurrency stereotypes «mutex», «wer», and «rer» to UML and maintain through them that concurrency control is inside the object that might be target of concurrent calls.

«mutex»: Stereotype implies that the operation is executed in mutual exclusion. Through this constraint the operation is protected and facilitate particularly the access to operations of protected objects.

«wer»: A «wer» declaration implies that the operation is defined as possibly modifying the object state. No other object thread can execute a «wer» constrained operation, if another object thread is executing it.

«rer»: This declaration allows that several calls from concurrent threads can occur simultaneously and will be executed concurrently.

9 Conclusion

The design of hard real time systems in *OOHARTS* includes different types of objects. These objects are used to specify the most common hard real time structure including cyclic , aperiodic, passive and environment objects. The stereotypes of the Unified Modeling Language UML allowed us to extend the language, without making changes to the UML standard, for the design of hard real time systems. The structure of hard real time systems in *OOHARTS* can be defined using these new defined stereotypes in the UML class diagram, which we call the object class diagram. The behavior of the system can be specified in the object behavior chart. We have chosen the mine pump control case study to show the applicability of our method for the design of hard real time systems. Other more complex case studies, e.g. the autonomous intelligent cruise control, have been also designed using the *OOHARTS* method.

The design models must be evaluated before their implementation. When design errors are not discovered in early design phases they can affect important problems during later design phases. Different evaluation methods are possible to detect errors. Model simulation and scheduling analysis can be used for this purpose. Model simulation allows the evaluation of the system models and shows their timing behavior. Schedulability analysis provides informations about possible timing errors. A tool set that supports these evaluation methods is under development in the Java programming language.

References

1. Burns, A., Wellings, A.: HRT-HOOD A Structured Design Method for Hard Real-Time Systems. Elsevier (1995)
2. Burns, A., Wellings, A.: Real-time systems and Programming Languages. Addision Wesley (1997)
3. Awad, M., Kuusela, J., Ziegler, J.: Object-Oriented Technology for Real-Time Systems: A Practical Approach using OMT and Fusion. Prentice Hall (1996)
4. Booch, G.: Object Oriented Design with Applications. Benjamin/Cummings Publishing Company (1991)
5. Calvez, J.P.: Embedded Real-Time Systems: A Specification and Design Methodology. Wiley (1993)
6. Cooling, J.E: Software Design for Real-Time Systems. Chapmann and Hall (1991)
7. Douglass, B.P.: Real-Time UML: Developing Efficient Objects for Embedded Systems. Addision Wesley (1997)
8. Fowler, M.: UML Distilled-Applying the Standard Object Modeling Language. Addision Wesley (1997)
9. Harel, D., Politi, M.: Modeling Reactive Systems with Statecharts, The STATE-MATE Approach. McGraw-Hill (1998)
10. HOOD Technical Group: HOOD Reference Manual Release 4.0. (1999)
11. Kabous,l: The Mine Pump Control System: A Case Study in OOHARTS Available on the WWW from URL http://eis.informatik.uni-oldenburg.de/~laila/mpcs.html
12. Rumbaugh, J., Blaha, M., Premerlani, W., Eddy, F., Lorenson, W.: Object-Oriented Modeling and Design. Prentice Hall (1991)
13. Nebel, W., Putzke Roemig, W., Radetzki, M.: A Unified Approach to Object Oriented VHDL. Journal of information Science and Engineering 14, pp. 523-545. (1998)
14. Nebel, W., Putzke Roemig, W., Radetzki, M.: OOVHDL: What It is? and why do We Need It?. Asia Pacific Conference on Hardware Description Languages. Hsin-Chu, Taiwan. (1997)
15. Rumbaugh, J., Selic, B.,: Using UML for Modeling Complex Real Time Systems. Available on the WWW from URL http://www.objectime.com
16. Rational Software: Unified Modeling Language UML. Available on the WWW from URL http://www.rational.com/uml/index.shtml
17. Selic, B., Gullekson, G., Ward, P.: Real-Time Object Oriented Modeling. Wiley (1994)

UML Based Performance Modeling Framework for Object-Oriented Distributed Systems[1]

Pekka Kähkipuro

Department of Computer Science, University of Helsinki
P.O. Box 26 (Teollisuuskatu 23), FIN-00014 University of Helsinki, FINLAND
Pekka.Kahkipuro@cs.Helsinki.FI

Abstract. As object-oriented distributed systems, e.g. those based on CORBA, Java, and DCOM, are entering the mainstream of information technology, it is increasingly important to predict and understand their performance characteristics. To support this need, we describe a framework that allows UML diagrams to be used for building performance models for such systems. A mapping is proposed from the high-level UML notation to queuing networks with simultaneous resource possessions, so that the models can be solved for the relevant performance metrics. The main goal of the framework is to support performance engineering and, thus, flexibility and ease of use have been emphasized.

1 Introduction

Object-oriented distributed systems, such as those based on CORBA, Java, and DCOM, are gaining popularity in information technology. As a result, it is important to understand their performance characteristics, e.g. to meet performance related system requirements. While benchmarking [1] and performance heuristics [2] offer short-term solutions, their help is limited to minor improvements. A more comprehensive approach, *performance engineering*, provides a long-term solution. It proposes special development techniques and tools for software engineers so that performance aspects can be considered during all phases of software development. Several reasons are in favor of using performance engineering in most software development efforts. Firstly, end users have always performance expectations. Ignoring them during early stages of development does not make the expectations go away, but it may increase the effort of meeting these expectations. Secondly, detecting potential performance problems early in the development cycle may significantly reduce the implementation cost – it is much cheaper to modify the design compared to modifying the code. Thirdly, the "fix-it-later" approach may sometimes fail altogether, as some architectural mistakes cannot be fixed without a complete redesign. See [3, 4] for more motivations for using performance engineering. The role of performance modeling is essential in per-

[1] This work has been carried out in the CORBA-FORTE project funded by Sonera Oyj, Tellabs Oy, Tietotekniikkayhtiö Tieturi Oy, Tieto Corporation Oyj, and the National Technology Agency under contracts 40032/98 and 40910/98.

formance engineering, as it allows designs to be validated against performance requirements at various stages of the development cycle at a moderate cost.

Traditional performance modeling has concentrated on centralized systems and communication issues in client/server systems [3, 4, 5]. Less attention has been paid to issues like software complexity or server contention that are essential in object-oriented distributed systems. Some attempts have been made [6, 7, 8], but they have not gained wide support. Also, most traditional performance modeling tools use specialized notations and assume a thorough understanding of the underlying modeling techniques. This makes them less attractive in general-purpose software engineering.

In this paper, we present a performance modeling framework that is specifically designed for object-oriented distributed systems. The framework is based on object-oriented concepts and uses the standard Unified Modeling Language (UML) [9] with minor extensions for representing the performance models. As a result, functional modeling and performance modeling can be done in parallel with the same tools and diagrams. We also present a precise textual notation, the Performance Modeling Language (PML), for representing those UML features that are relevant for performance models. The framework provides an algorithm for solving the performance models for the relevant performance metrics. The idea is to transform the UML based models into stochastic queuing networks with simultaneous resource possessions, and to use the method of decomposition (MOD) [10] for solving them. Although we mainly use CORBA related examples, the approach is equally well suited for other object-oriented distributed systems, such as those based on Java or DCOM.

Related work includes a number of graphical approaches for presenting performance models. For example, Smith has developed a special *execution graph* notation for visualizing performance models [3]. She also proposes to use message sequence charts (similar to UML sequence diagrams) as an intermediary notation for transforming object-oriented designs into execution graphs [7]. Waters *et al.* have adopted an object-oriented notation (Booch) for modeling the application structure, but they use a proprietary notation for modeling the workloads and the execution environment [15]. The framework in this paper is the first known approach that is completely based on object-oriented modeling – most likely due to the limitations of pre-UML notations.

The rest of this paper is structured as follows. Section 2 discusses the detailed requirements for the framework, and section 3 proposes an overall architecture. Section 4 briefly presents the foundations: queuing networks and the method of decomposition. Section 5 presents the UML modeling techniques, and section 6 proposes ways to use them in actual modeling work. Section 7 gives an example, and section 8 discusses limitations of the approach and proposes topics for further study. Finally, section 9 concludes the paper.

2   Framework Requirements

The primary goal for our framework is to reflect the needs of performance engineering in the context of object-oriented distributed systems. To support this goal, at least four basic requirements must be met.

First, the framework should produce performance models that can be solved automatically in a reasonable amount of time for the relevant performance metrics, such as throughputs, response times, and queue lengths. We can slightly relax this requirement by allowing the use of approximate solutions. No user intervention should be required for solving the models, and no structural limitations should be imposed for the application design. This way, the framework can be easily integrated into various software development tools and methodologies.

Second, we require the framework to support the usual style of UML modeling as proposed in the UML standard [9] and in the literature, e.g. [11]. This way, it is possible to extend existing functional models into performance models without having to rewrite them. Moreover, it should be possible to develop both models in parallel with as many common diagrams as possible.

Third, we require the framework to clearly distinguish between different architectural aspects of object-oriented distributed systems. The framework should keep application objects, system infrastructure, hardware resources, and network topology in separate UML diagrams. This way, it is possible to experiment with different design alternatives in some parts of the system without having to modify other parts.

Fourth, the framework should support incremental development style, as this is commonly used with object-oriented analysis and design [11]. In particular, it should be possible to build solvable (but inaccurate) performance models already at the analysis phase when the infrastructure and hardware issues are still unknown. Also, it should be possible to upgrade the first tentative performance models into more accurate ones without the need to update the complete set of UML diagrams.

3   Framework Architecture

The architecture of the framework consists of four *representations* that correspond to different ways of expressing performance models. Each representation uses its own notation, and the architecture defines mappings between them, as shown in Fig. 1. The idea is to start from the UML representation and proceed downwards using the mappings. Once the bottom has been reached, the model can be solved for the relevant performance metrics. The mappings also indicate how the metrics can be propagated upwards.

The *UML representation* describes the system with UML diagrams. This representation may contain purely functional elements that are not needed for performance modeling. To reduce the complexity of the diagrams, we assume that the UML representation is divided into separate layers describing, for example, the application, the infrastructure, and the network.

The *PML representation* provides an accurate textual notation for representing performance related items in the UML diagrams. The PML representation has the same layered structure as the UML representation, and the mapping from UML to PML is straightforward. The purpose of this representation is to filter out those features that have no significance for performance modeling, such as graphical UML variations and purely functional parts of the UML model. The PML representation has an important role in our prototype implementation, as it is the input format for the experi-

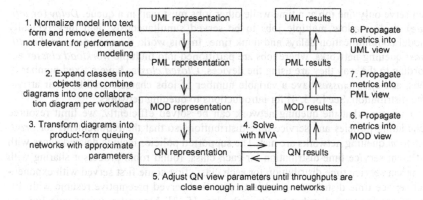

1. Normalize model into text form and remove elements not relevant for performance modeling	UML representation	UML results	8. Propagate metrics into UML view
2. Expand classes into objects and combine diagrams into one collaboration diagram per workload	PML representation	PML results	7. Propagate metrics into PML view
3. Transform diagrams into product-form queuing networks with approximate parameters	MOD representation	MOD results	6. Propagate metrics into MOD view
	QN representation	4. Solve with MVA	QN results

5. Adjust QN view parameters until throughputs are close enough in all queuing networks

Fig. 1. Architecture of the performance modeling framework

mental tool for solving performance models. However, the use of PML is not mandatory, and any other normalized UML representation could be used as a bridge from UML diagrams into solvable performance models.

The *MOD representation* describes the system as a queuing network with simultaneous resource possessions. This format allows us to apply the MOD algorithm for solving the model. The MOD representation is obtained from the PML representation by expanding object classes into separate object instances. The result of this transformation can be visualized with a set of large collaboration diagrams. Each diagram corresponds to a workload with a given number of jobs and a particular way of using the resources.

The *QN representation* consists of product-form queuing networks that are obtained from the MOD representation during the execution of the MOD algorithm. The MOD algorithm generates initially queuing networks with approximate parameters, and iteration is needed for finding a solution with the required level of accuracy [10].

Similar architectures are common in performance modeling [5], but the top representation is often a sophisticated performance modeling notation, such as a variant of Petri nets, to support advanced techniques. In our case, a non-technical top representation is used for hiding most of the underlying performance modeling issues.

4 Technical Foundations

4.1 Queuing Networks

In queuing network based performance models, systems are represented in terms of *devices* that correspond to hardware or software resources, and *jobs* that correspond to software processes requiring access to the resources. A device is characterized by its scheduling policy and service time distribution. The model indicates the *service demands* for the jobs, i.e., the average times the jobs visit each device. A *queuing device* can serve only one job at a time while other jobs must wait in a queue. *Delay* (or *infinite*) *devices* allow multiple jobs to be served simultaneously, and they typically model communication delays and think time. In this work, we consider *mixed multiclass* queuing networks, where jobs are partitioned into multiple *workload classes* according to the way they are using the devices. *Closed classes* have a fixed number of jobs, and *open classes* have a variable number of jobs characterized by their arrival rate distribution. See [5, 3, 4] for introductions to queuing networks.

To ensure that the queuing network can be solved efficiently, we limit resource scheduling policies and service time distributions so that the requirements for *product-form* queuing networks are met. The supported policies include delay devices with different service time distribution for each class, round robin processor sharing with different service time distribution for each class, first come first served with exponential service time distribution, and last come first served preemptive resume with different service time distribution for each class [5, 3]. Moreover, we assume Poisson distribution for arrivals in open classes. Efficient techniques, such as mean value analysis (MVA) approximations, exist for solving product-form queuing networks [12, 13], and different tradeoffs between accuracy and speed can be chosen. The available metrics include throughputs for job classes, average queue lengths for devices, and residence times for each class in a particular device. The main limitation of MVA is the unavailability of distributions for the metrics.

There are similarities between resources in a queuing network and objects in a distributed system. Both are abstractions that hide complex internal implementations and both have a scheduling policy that indicates the way clients are served. In addition, queuing networks are intuitively simple and easy to use. Therefore, system engineers that are not particularly familiar with performance modeling can use them.

However, the fact that a job cannot possess more than one resource at a time in a queuing network prevents their straightforward use in our framework. In object-oriented distributed systems, simultaneous resource possession occurs frequently as the basic communication primitive is a synchronous call that blocks the caller until it gets the result (e.g. a remote procedure call, a Java RMI, a CORBA invocation). Also, it is commonplace to use nested invocations, so that several objects are blocked unless multi-threading is used in the object implementations. Furthermore, most software and hardware resources in object-oriented systems are represented in terms of objects. If a client wishes to access such a resource, it must first acquire the possession of the corresponding object, and simultaneous resource possession may result.

4.2 The Method of Decomposition

The method of decomposition (MOD) improves queuing networks by adding support for simultaneous resource possessions. UML collaboration and sequence diagrams are used for visualizing such networks. Devices are shown as UML objects, and each class of jobs is represented as a separate diagram where properties in operation invocations indicate the service demands for the devices. A distinction is made between *direct use* (as in traditional queuing networks) and a *synchronous call* (an invocation that blocks the calling device until the reply is obtained). The former is modeled as a simple message with an open arrowhead, and the latter is modeled as a synchronous

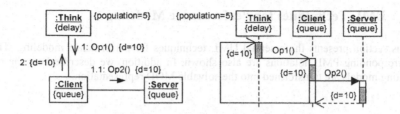

Fig. 2. Collaboration diagram (*left*) and sequence diagram (*right*) illustrating the direct use of a resource (*Op1*) and a synchronous call (*Op2*)

message with a filled arrowhead. Fig. 2 illustrates both types of invocation in sequence and collaboration diagrams. Both diagram types can be used interchangeably.

The property d indicates the service demand of each invocation. In sequence diagrams, gray indicates the operation execution and white indicates queuing or waiting. In both diagrams, we use a feedback arrow to indicate a closed workload that does not allow jobs to leave the system. Open workloads are labeled with the *arrivalrate* property giving the average number of arriving jobs in a unit of time. The *population* property indicates the size of a closed workload population, and the *queue* and *delay* properties indicate scheduling policies.

The MOD algorithm is used for solving queuing networks with simultaneous resource possessions. First, the network is decomposed into *primary* and *secondary* networks. The primary network is obtained by removing synchronous calls and their target devices from the original network. A *surrogate delay device* is added for representing the time spent in the synchronous calls. Secondary networks are created to represent devices that were removed from the primary network. Each secondary network has an *auxiliary delay device* for representing the time spent in the rest of the system. Primary and secondary networks no longer have simultaneous resource possessions and, hence, can be solved efficiently with MVA approximations.

At first, the service demands of the surrogate and auxiliary devices in the primary and secondary networks are not known. Iteration is used to adjust them until the throughputs of all networks are equal. During each round, approximate MVA is applied to the networks. Once the throughputs are close enough, they can be considered to represent the same overall system in a stable state. The performance metrics of the primary and secondary networks directly show the performance metrics of the complete system. Full details of this algorithm are presented in [10].

In principle, the MOD representation could be used for modeling object-oriented distributed systems. In a reasonably accurate model, devices would exist for representing the application, the infrastructure, the hardware, and the network. Even a simple invocation across the network would use most of these resources and, hence, the workload specifications would be complex diagrams mixing application logic with technical details. From the four requirements, this approach satisfies only the first one, solvability.

5   UML Techniques for Performance Modeling

This section presents the essential UML techniques for performance modeling. The corresponding PML notations are also shown. In addition, we describe how the resulting models are transformed into the solvable MOD representation.

5.1   Resource Representation

Software and hardware resources are represented as classes in UML diagrams. We indicate such *resource classes* with the *queue* or *delay* properties. The *queue* property represents queuing resources, such as single-threaded software servers or physical devices. The *delay* property represents delay resources, such as think time, network delays, or multi-threaded software servers that do not impose queuing for their clients.

A resource class or any other class that is contained in a resource class may define service demands for its operations. Service demands are indicated with user named properties. An operation may have a single property for the total service demand, or it may have multiple properties to indicate components of the service demand (e.g. CPU and disk usage). Accurate service demands can only be obtained by measuring real systems, but estimates can also be used to produce approximate performance models. Service demands are not mandatory in class diagrams as they can be also defined in workloads. Fig. 3 illustrates a queuing resource class *CDatabase* with two operations.

Service demands can be specified also in separate interface definitions. Fig. 5 shows an example where the *CHandler* class realizes the separately specified *IHandler* interface. Resource classes can be used like any other UML classes and mixed freely with other classes. For example resource classes may be contained in non-resource classes or *vice versa*. To simplify resource instantiation, we assume the existence of two predefined resource classes, *Queue* and *Delay*.

The mapping of resource classes into the MOD representation is straightforward. Each instance of a resource class is mapped to a queuing network device. The number of instances for a class is specified with UML object and deployment diagrams. Fig. 8 provides an example of such diagrams.

5.2   Workload Representation

Performance workloads are modeled with special collaboration diagrams (*workload diagrams*) that indicate service demands for invocations and show job populations. Workload diagrams resemble those used for MOD representation, but there are two

```
class CDatabase {
  property queue;
  Read() {cpu=100,disk=500};
  Write() {cpu=100,disk=800};
};
```

CDatabase {queue}	
Read()	{cpu=100,disk=500}
Write()	{cpu=100,disk=800}

Fig. 3. An example queuing resource with two operations

363

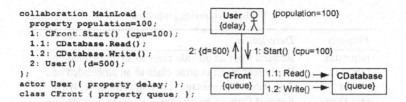

```
collaboration MainLoad {
    property population=100;
    1: CFront.Start() {cpu=100};
    1.1: CDatabase.Read();
    1.2: CDatabase.Write();
    2: User() {d=500};
};
actor User { property delay; };
class CFront { property queue; };
```

Fig. 4. An example workload diagram

differences. First, workload diagrams usually contain classes instead of objects. This way, the defined workloads are generic, and the indicated service demands are divided between the instances of the target class. Second, only application level resources need to be shown since infrastructure and network resources are activated with triggering properties (see section 5.3). Fig. 4 illustrates a simple workload diagram where 100 users are accessing a database through a front-end application.

Service demands need not be specified explicitly in workload diagrams if they are specified elsewhere in class diagrams. However, service demands in workload diagrams override those given elsewhere. Anonymous invocations (e.g. think time) can be used in workload diagrams but they require explicit service demands.

The *thinktime* property can be used for representing think time or any similar delay when there is no need to explicitly present the responsible device (e.g. the end user). For example, we could add the property *thinktime* = 500 to the workload diagram in Fig. 4 and remove the *User* class and the anonymous message 2.

The mapping from workload diagrams into the MOD representation involves several tasks. First, the service demand of each invocation is concluded from workload and class diagrams. If the target is a non-resource class, the service demand is attributed to the closest surrounding resource class. Second, each invocation to a class is expanded into invocations to one or more instances of that class. The details are discussed in section 5.6. Finally, the *thinktime* property is mapped to a delay device.

5.3 Triggering Properties

Triggering properties model application's side effects in the infrastructure and network. These effects are normally excluded from the application's workload diagrams. Examples include network delays and process context switching overhead. Any UML class may specify up to nine triggering properties, as listed in Table 1.

The value of a triggering property must be a reference to a collaboration diagram (*triggering diagram*) describing the actions that take place when the triggering condition is satisfied. We use UML package names to indicate these references, but other referencing mechanisms can also be used. For example, the lower part in Fig. 7 defines a constant context switching delay $d = 2$ for messages passing between entities (typically processes) within instances of *CNode* (representing network nodes).

Triggering diagrams can use the service demands of the operations that satisfy the triggering conditions. Suppose, e.g., that the service demand of each operation is split

Table 1. Triggering properties

Property	Description
requestin	Request from an outside class to an embedded one
replyin	Reply sent by an outside class to an embedded one
msgin	Any incoming message (i.e. requestin or replyin)
requestout	Request from an embedded class to an outside one
replyout	Reply sent by an embedded class to an outside one
msgout	Any outgoing message (i.e. requestout or replyout)
requestpeer	Request sent between first-level contained classes
replypeer	Reply sent between first-level contained classes
msgpeer	Any internal message (i.e. requestpeer or replypeer)

into *cpu* and *adapter* properties. We might then map the *adapter* property to an *Adapter* device with an appropriate triggering diagram, where the service demand for the *Adapter* would be defined as $d = adapter$. Service demands can be adjusted with arithmetic expressions: a fast adapter might be modeled with $d = adapter*0.9$.

Triggering properties help to keep application, infrastructure, and network diagrams separated. This separation disappears when they are transformed into the MOD representation. All invocations that satisfy a triggering condition are expanded into a series of invocations as indicated in the corresponding triggering diagrams. If the expanded invocation is a synchronous call, the calling object remains blocked during the additional calls. Otherwise, direct uses are generated.

5.4 Service Demand Binding

If a service demand property is not used in any triggering condition, the service demand is attributed directly to the object that receives the invocation. In some cases, however, service demands should be bound to a specific resource, such as CPU or disk. This is carried out in two steps. First, the target resource needs to be declared. It is often placed inside a container class (e.g. a node) for indicating the binding scope. Second, a binding property is attached to the resource for indicating which service demands it binds. The lower part of Fig. 7 shows how the *cpu* and *disk* properties can be bound for all requests inside a node to model contention for CPU and disk. Arithmetic expressions can be used in bindings. For example, a slow CPU might be modeled with $d = cpu*1.5$.

Nested binding is allowed. For example, a special resource could bind all application level service demands for file access. The service demand of this resource might then be bound the actual disk. Another useful technique is to bind the same demand more than once. For example, if a node has two disks, half of the service demand for each disk access could be bound to one disk and the other half to the other disk.

Service demand binding generates additional invocations to the bound resources when workload diagrams are converted into the MOD representation. If the bound resource is a queuing device, the binding is transformed into a synchronous call. If the bound resource is a delay device, a direct use is generated.

5.5 Network Connections

In UML deployment diagrams, a connection between nodes is an association that may be refined by a stereotype for identifying the protocol or medium (e.g. «*TCP/IP*» or «*Ethernet*»). An extension to UML is proposed for specifying the details of such association refinements. Essentially, we need the same triggering properties that are available for UML classes. Hence, we define a class stereotype «*connection*» for specifying association stereotypes for network connections. The upper part of Fig. 7 shows the definition for a connection with a delay $d = 5$ for each message that passes between two nodes in a «*LAN*». Fig. 8 shows how this connection can be used in a deployment diagram.

In performance models, association stereotypes representing node-to-node connections are treated as if they were container classes for these nodes. Accordingly, the mapping of network connections to the MOD representation is the same as the mapping for other classes with triggering properties or service demand bindings.

5.6 Class Name Resolution

When transforming workload diagrams into the MOD representation, a special name resolution scheme is used for finding the objects that execute the service demands of a particular class. The idea is to access the closest instances of a service by default.

Class names are resolved in a *resolution context*, such as a composite object or a package. For the first invocation in a collaboration diagram, the resolution context is the current package. For all subsequent invocations, the initial resolution context is the smallest context for the object executing the previous invocation. If there is no match in the initial context, resolution is attempted in the next surrounding context, and so forth until the resolution context is the whole system.

When one or more instances of the target class are found in a resolution context, the service demand is distributed evenly among them, and the subsequent resolution contexts are determined by these instances. Hence, a single invocation in a workload diagram may spawn any number of invocations in the MOD representation. To bypass the name resolution scheme, scoped names can be used for explicitly indicating the intended resolution context.

6 How to Use the Framework

This section discusses how the framework can be used in actual modeling work. To cope with the complexity of distributed systems, we propose to split the models into a number of separate layers. The following partition has been found useful:

- Application layer,
- Interface layer,
- Infrastructure layer,
- Network layer, and
- Instantiation layer.

The *application layer* describes the structure and behavior of applications. The structure is given with normal UML class diagrams. Resources that are relevant for performance modeling are indicated with the *delay* or *queue* properties. Application behavior is modeled with collaboration diagrams in the usual UML style. Diagrams that represent essential workloads must be extended with the framework features, as shown in Fig. 4. Some middleware services, e.g. the CORBA Naming service, can also be modeled at the application layer because they are visible in the application logic. As this layer only requires a few performance related properties, it is possible to maintain a single set of UML diagrams for both functional and performance models.

The *interface layer* describes operations implemented by application layer objects. This reflects the principle of separating interfaces from object implementations [14]. Interfaces can be represented with UML diagrams or with an interface definition language (e.g. OMG IDL). The framework requires interfaces to be extended with service demands for each operation. Certain parts of the service demand (e.g. marshaling time) could be inferred automatically but, as service demand normally depends on the operation's semantics, we assume that it is explicitly given in the model.

The *infrastructure layer* describes the infrastructure support for the applications (e.g. middleware, operating system, hardware). It is represented with class diagrams equipped with triggering properties and collaboration diagrams. Often, only implicit links exist between the application and infrastructure layers (i.e. triggering properties and service demand binding). Explicit links can be created with service demand binding. Elements at this layer are organized into nested classes according to their physical structure. In complex models, this layer can be further divided into sub-layers for threads, processes, nodes, etc.

The *network layer* describes the network topology. The UML notation for network connections has been extended to provide the same features that are available at the infrastructure layer (i.e. triggering conditions and service demand binding). The network layer can also be divided into sub-layers for modeling the network details. For example, LAN and WAN connections might be in separate layers.

The *instantiation layer* specifies the actual configuration in terms of objects, processes, nodes, etc. with deployment diagrams. This layer can be described at different levels of detail. For example, at an early stage of development, it may be useful to instantiate only the application objects. Later, when the effect of network traffic is investigated, network layer objects might be added. Finally, a full model may be specified when the final configuration has to be validated.

7 Prototype Implementation and a Modeling Example

This section briefly discusses our prototype implementation of the framework and illustrates its use with an example. Currently, the prototype has the following parts:

- A *parser* that accepts PLM input files,
- An *expander* that expands the PML model into the MOD representation,
- A *solver* that solves the model with the MOD algorithm, and
- An *output module* for presenting the results in textual form.

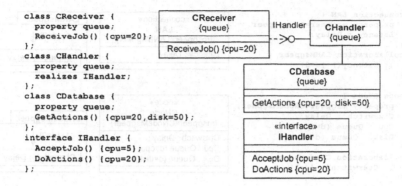

```
class CReceiver {
    property queue;
    ReceiveJob() {cpu=20};
};
class CHandler {
    property queue;
    realizes IHandler;
};
class CDatabase {
    property queue;
    GetActions() {cpu=20,disk=50};
};
interface IHandler {
    AcceptJob() {cpu=5};
    DoActions() {cpu=20};
};
```

Fig. 5. Application and interface layers for the network monitoring system

More work is needed to make the prototype more convenient for software engineering. In particular, we are investigating the possibility to integrate it with an existing graphical modeling tool. A suitable performance engineering methodology is also needed. One possibility is to add object-oriented features to an existing performance engineering methodology (e.g. [4]). Alternatively, we might extend a UML based methodology (e.g. [11]) with performance engineering features.

We now illustrate the framework with an example describing a simplified network monitoring system. In this system, a *receiver object* accepts messages from network elements and forwards them to *handler objects* that are responsible for executing appropriate actions. A *database* maintains descriptions for the actions. Fig. 5 gives class diagrams for the application and interface layers. Service demand estimates are in milliseconds. The model has two workloads (Fig. 6). The background load has an estimated rate of 1 database query per second. The main load models the treatment of network element messages. We initially estimate a rate of 7 messages per second.

The system uses a CORBA infrastructure, for which we give a cut down model. We simply assume an average 3 ms inter-node communication latency and a 2 ms intra-node context switch. A node has two hardware resources: a CPU and a disk. The

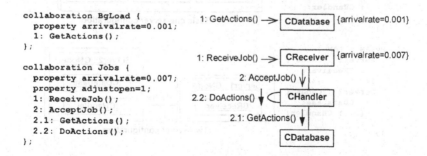

```
collaboration BgLoad {
    property arrivalrate=0.001;
    1: GetActions();
};

collaboration Jobs {
    property arrivalrate=0.007;
    property adjustopen=1;
    1: ReceiveJob();
    2: AcceptJob();
    2.1: GetActions();
    2.2: DoActions();
};
```

Fig. 6. Workloads for the network monitoring system

368

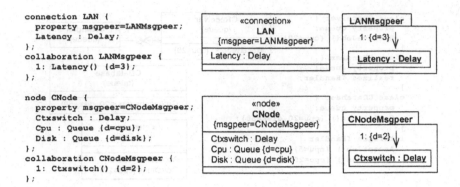

```
connection LAN {
  property msgpeer=LANMsgpeer;
  Latency : Delay;
};
collaboration LANMsgpeer {
  1: Latency() {d=3};
};

node CNode {
  property msgpeer=CNodeMsgpeer;
  Ctxswitch : Delay;
  Cpu : Queue {d=cpu};
  Disk : Queue {d=disk};
};
collaboration CNodeMsgpeer {
  1: Ctxswitch() {d=2};
};
```

Fig. 7. Infrastructure and network layers for the network monitoring system

network and infrastructure layers are illustrated in Fig. 7.

The basic one-server configuration has only one receiver, handler, and database (Fig. 8, top). The tool's report for this configuration is shown in Fig. 9. It summarizes device utilizations and also shows response times, throughputs and the number of jobs in the system for each workload. A more detailed report is available for viewing the expanded MOD representation and the average times spent in each step.

The handler's high utilization (67 %) suggests that it might be a software bottleneck. Additional experiments with two and three handlers in the same node confirm this assumption. Fig. 10 shows average response times of the main load for different configurations. Hence, the program code for the system should be written to support the replication of handlers. Additional improvements can be achieved with two server nodes. The two-server configuration (Fig. 8, bottom) does not significantly improve the throughput but the maximum sustainable load is clearly higher (Fig. 10).

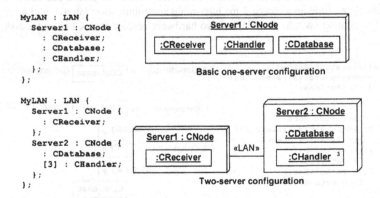

```
MyLAN : LAN {
  Server1 : CNode {
    : CReceiver;
    : CDatabase;
    : CHandler;
  };
};

MyLAN : LAN {
  Server1 : CNode {
    : CReceiver;
  };
  Server2 : CNode {
    : CDatabase;
    [3] : CHandler;
  };
};
```

Fig. 8. Two configurations for the network monitoring system

```
Utilization       Type      Device
-----------       ----      ------
0 %               Delay     MyLAN.Latency
53.3145 %         Queue     MyLAN.Server1.$CDatabase
67.5453 %         Queue     MyLAN.Server1.$CHandler
8.37829 %         Queue     MyLAN.Server1.$CReceiver
32.5008 %         Queue     MyLAN.Server1.Cpu
4.20009 %         Delay     MyLAN.Server1.Ctxswitch
40.0008 %         Queue     MyLAN.Server1.Disk

BgLoad
------
Resp.time: 119.759    Throughput: 0.001    Nbr.in system: 0.119759
Time share: 31.9589 %     MyLAN.Server1.$CDatabase
            11.4127 %     MyLAN.Server1.Cpu
            56.6284 %     MyLAN.Server1.Disk

Jobs
----
Resp.time: 290.412    Throughput: 0.007    Nbr.in system: 2.03288
Time share: 1.88293 %     MyLAN.Server1.$CDatabase
            66.9387 %     MyLAN.Server1.$CHandler
            0.37707 %     MyLAN.Server1.$CReceiver
            18.2065 %     MyLAN.Server1.Cpu
            2.06603 %     MyLAN.Server1.Ctxswitch
            18.0934 %     MyLAN.Server1.Disk
```

Fig. 9. Example report from the framework prototype tool

8   Discussion

In this section, we discuss some limitations of the framework, and consider possibilities for overcoming them. The first limitation is the lack of support for some UML features in collaboration diagrams. In particular, it is not possible to spawn, kill, or synchronize threads (although multi-threading is supported through infinite devices). To have full thread support, the MOD could be extended so that the explicitly created

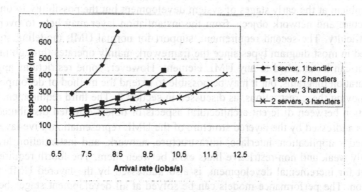

Fig. 10. Response time estimates for different configurations at different arrival rates

threads are transformed into additional classes of jobs in the primary and secondary networks. Synchronization between threads can be achieved by adjusting the propagation of parameters during iteration [10].

The second limitation is related to UML diagram types that are not supported by the framework. We intentionally used a small set of UML elements in order to keep the framework easy to use. However, UML state and activity diagrams can express performance related information in a convenient manner, and future extensions of the framework might use them. For example, state diagrams could be used to represent detailed information on the service demands of operation invocations, and activity diagrams could be used to represent interactions (e.g. synchronization) between workloads.

The third limitation is inherited from the MVA. The available set of workload distributions and scheduling policies is fairly restricted. Also, distributions cannot be obtained for the solved metrics. This limitation can be avoided by solving the model through simulation. This approach would give much more flexibility, but the drawback is the added time it takes the run the simulations. A modeling tool might support both approaches, so that performance modeling during iterative work could be done with the MOD, and final performance estimates could be obtained from simulations.

9   Conclusions

We have described a UML based performance modeling framework that allows standard UML diagrams to be used for specifying performance models of distributed object-oriented systems. Also, a transformation from the UML notation to queuing networks with simultaneous resource possessions is proposed. This way, the method of decomposition can be used for solving the performance models for relevant performance metrics. An experimental prototype has been developed for investigating the usability of the framework for performance engineering.

We identified four requirements for the framework. The first requirement, solvability, is satisfied due to the mapping from the UML diagrams to queuing networks. However, the models can only be solved if the object instances are known. This may be a problem at the early stages of system development but the possibility to omit infrastructure and network objects from the instantiation layer may help to overcome this difficulty. The second requirement, support for normal UML modeling style, is satisfied in most diagram types since the framework mainly operates through properties that can be attached to any UML element. However, some restrictions apply to collaboration diagrams, but it may be possible to extend the method of decomposition for removing these restrictions, as discussed in section 8. The third requirement, clear distinction between different architectural aspects of object-oriented distributed systems, is addressed by the layered structure of the UML representation. Five layers are proposed – application, interface, infrastructure, network, and instantiation layers – and only weak and non-restrictive links exist between them. The fourth requirement, support for incremental development, is also addressed by the layered UML representation. The performance models can be solved at all development stages because some of the layers, e.g. network layer, can be completely or partially omitted.

While the framework meets the requirements, more work is needed to use it up to its full potential. In particular, support for a graphical modeling tool and suitable performance engineering methodologies are needed.

References

1. Object Management Group: Draft Benchmark White Paper. OMG Document bench/99-02-01, Framingham, Massachusetts, USA (1999)
2. Kähkipuro, P.: A survey of techniques and guidelines for improving the performance of CORBA-based distributed systems. Proceedings of the HeCSE Workshop on Emerging Technologies in Distributed Systems, Research Report A 50, Digital Systems Laboratory, Helsinki University of Technology (1998)
3. Menascé, D.A., Almeida, V.A.F., Dowdy, L.W.: Capacity Planning and Performance Modeling. Prentice Hall, New Jersey, USA (1994)
4. Smith, C.U.: Performance Engineering of Software Systems. Addison-Wesley, Reading, Massachusetts (1990)
5. Haverkort, B.R.: Performance of computer communication systems: a model-based approach. John Wiley & Sons, New York, New York, USA (1998)
6. Rolia, J.A., Sevcik, K.C.: The Method of Layers. IEEE Transactions on Software Engineering 21 (8), August 1995, 689-699
7. Smith, C.U., Williams, L.G.: Performance engineering evaluation of object-oriented systems with SPE.ED. In: Marie, R., Plateau, B., Calzarossa, M., Rubino, G. (eds.): Computer Performance Evaluation – Modeling Techniques and Tools. Lecture Notes in Computer Science, Vol. 1245. Springer-Verlag, Berlin Heidelberg New York (1997)
8. Woodside, C.M., Neilson, J.E., Petriu, D.C., Majumdar, S.: The Stochastic Rendezvous Network Model for Performance of Synchronous Client-Server-like Distributed Software. IEEE Transactions of Computers 44 (1), January 1995, 20-34
9. Rational Software: Unified Modeling Language Documentation, version 1.1. California, USA (1997). Available from "http://www.rational.com/"
10. Kähkipuro, P.: The Method of Decomposition for Analyzing Queuing Networks with Simultaneous Resource Possessions. Proceedings of the Communications Networks and Distributed Systems Modeling and Simulation Conference (CNDS'99), The Society for Computer Simulations International, San Diego, California, USA (1999)
11. Eriksson, H-E., Penker, M.: UML Toolkit, John Wiley & Sons, New York, NY (1998)
12. Bard, Y.: Some extensions to multiclass queuing network analysis. In: Performance of Computer Systems, North Holland, Amsterdam (1979)
13. Chandy, K.M., Neuse, D.N.: Linearizer: A Heuristic Algorithm for Queuing Network Models of Computing Systems. Journal of the ACM 25 (2), February 1982, 126-134
14. Kähkipuro, P.: Object-Oriented Middleware for Distributed Systems. Report C-1998-43, Department of Computer Science, University of Helsinki, Finland (1998)
15. Waters, G., Linington, P., Akehurst, D., Symes, A.: Communications software performance prediction. In: Kouvatsos, D. (ed.): 13th UK Workshop on Performance Engineering of Computer and Telecommunications Systems, July 1997. BCS Performance Engineering Specialists Group, UK (1997) 38/1-38/9

Defining the Context of OCL Expressions

Steve Cook [1], Anneke Kleppe[2], Richard Mitchell[3], Jos Warmer[4], Alan Wills[5]

[1]IBM UK Ltd, 79 Staines Road West, Sunbury-On-Thames, Middlesex TW16 7AH, UK
sj_cook@uk.ibm.com
[2] Klasse Objecten, Postbus 3082, 3760 DB Soest, The Netherlands
a.kleppe@klasse.nl
[3]University of Brighton, Brighton BN2 4GJ, UK
Richard.Mitchell@brighton.ac.uk
[4] Klasse Objecten, Postbus 3082, 3760 DB Soest, The Netherlands
j.warmer@klasse.nl
[5]TriReme International Ltd, UK
alan@trireme.com

Abstract. Expressions written in Object Constraint Language (OCL) within a UML model assume a context, depending upon where they are written. Currently the exact nature of this context is not fully defined. Furthermore there is no mechanism for defining the context for OCL expressions in extensions to UML. This paper defines the context of OCL expressions, and proposes precise and flexible mechanisms for how to specify this context.

1. Introduction

Recently more and more software professionals have become acquainted with the object constraint language (OCL), which is an integral part of the OMG standard that defines the Unified Modeling Language (UML). OCL is a simple language designed to express logical constraints on a UML model, especially those constraints which cannot be expressed using UML's diagrammatic notation, such as invariant relationships between different attributes of one or more classes, and pre- and post-conditions on operations. OCL can be used to construct logical expressions that access attributes, invoke operations, navigate along associations, and manipulate collections. Its syntax is rather like Smalltalk, and does not require the use of any special characters outside the normal alphanumeric set. OCL is documented in the UML standard [1], and explained fully in the book by Warmer and Kleppe [2].

OCL has been used in a wide variety of domains, and this has led to the identification of some under-specified areas in the relationship between OCL and UML. In this paper we investigate a specific aspect of this relationship, namely the question of how to define the context of an OCL expression within a UML model in a more precise and more generic way.

This subject was brought up and discussed during a workshop on OCL, which was held in Amsterdam in November 1998. For the complete results of the Amsterdam workshop we refer to 'The Amsterdam Manifesto on OCL' published by the Technical University of Munich [3].

2. OCL expressions and their context

A recent definition of UML (version 1.3 beta R1, April 1999, section 6.3) [1] makes the following statements about the context of OCL expressions:

"OCL can be used for a number of different purposes:
- *to specify invariants on classes and types in the class model*
- *to specify type invariants for Stereotypes*
- *to describe pre- and post-conditions on Operations and Methods*
- *to describe Guards*
- *as a navigation language*
- *to specify constraints on operations"*

Note that this list appears insufficient, as it does not refer (for example) to pre- and post-conditions of transitions, or constraints in sequence diagrams.

"Each OCL expression is written in the context of an instance of a specific type"

"The reserved word self is used to refer to the contextual instance".

The specification goes on to describe how the context of OCL expressions can be stated for just two of the cases where it may be used, namely invariants and pre- and post-conditions:

Invariants. The OCL expression is part of an invariant, which is a UML Constraint stereotyped with «invariant». In this case, *self* by default, or another specified name, denotes the instance of the UML Classifier to which the invariant applies. The context is specified by a declaration of one of the following forms:

```
context Company inv:
self.numberOfEmployees > 50
-- here self is the name of the contextual instance
```

```
context c : Company inv:
c.numberOfEmployees > 50
-- here c is the name of the contextual instance
```

```
context c : Company inv enoughEmployees:
c.numberOfEmployees > 50
-- here enoughEmployees is the name of the constraint
```

Pre- and Post-conditions. The OCL expression is a Constraint stereotyped «precondition» or «postcondition». In this case the Constraint will be associated with an Operation or Method (according to the OCL specification, whereas in the UML specification it says they are only associated with Operations), and the contextual instance *self* is of the type which owns the operation as a feature. In this case there is no way to specify an alternative name for self. The following form of declaration is used for pre- and post-conditions:

```
context Person::income(d : Date) : Integer
```

```
pre:  not self.isUnemployed
post: result = 5000
```

Other uses of OCL. The only guidance given for using OCL for the other purposes suggested in the list above is for General Expressions, where the OCL is "the value of an attribute of the UML class Expression or one of its subtypes". In that case, the specification states that "the semantics section describes the meaning of the expression". It is unclear whether there are any cases of this.

According to the current UML meta-model, valid places for an OCL expression are (1) an OCL specification as body of the metaclass Constraint, in other words, a constraint which may be coupled to any other model element, and (2) an OCL specification as body of an Expression attribute in a metaclass, in other words, any expression.

Summarising, the currently-specified position is that OCL expressions are encouraged for use within UML models, but the places where these expressions can be used, and the valid ways to use OCL in each of these places, remains rather ill-defined and incomplete. What is required is a mechanism for specifying precisely the places where OCL expressions can be used in a UML specification, and the full context in which an expression can be written in each of these places.

The first consequence of such a mechanism would be that the current state of definition as summarised above could be made considerably more rigorous. A second consequence would be to provide a mechanism for defining valid contexts for OCL expressions in extensions to UML. By design, UML is an extensible language, and can be regarded as a family of languages, embracing a variety of syntactical specialisations and meanings suited for different application areas and domains. The mechanisms proposed in this paper give a way to define, for each such family member, exactly where and how OCL expressions may be used.

3. A more general approach

We believe that specifying the connection between OCL specifications and UML models simply by enumerating the cases, as in the current definition, is insufficient. We need a more general way to specify the connection. This has to define:

- The places in a UML model where an OCL specification can validly be attached.
- Given such a place (and maybe some other ingredients such as might be defined in a preface, see below) the set of valid OCL expressions which can be written in that place.
- The UML elements that can be used in these OCL expressions, and the ways they can be used.

Additionally it may be desirable to define some things using OCL which are globally accessible throughout a model. We believe that the way to handle such global issues is through the mechanism of a *preface*, which can be imported into UML documents as part of the process of specifying a UML extension. Detailed proposals for defining such prefaces appear in [4].

For the purposes of this paper we are only concerned with localised OCL expressions and how they are connected to their associated UML model.

We envision an OCL expression in a box, like this:

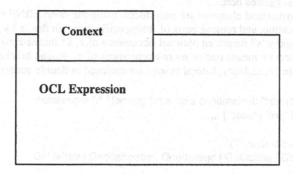

Figure 1. Context for OCL expressions

The large box contains an OCL expression. The smaller box, marked Context, contains implicitly a set of declarations, which establish the names that can be used in the OCL specification, and their types. Note that there is no *export* of newly defined elements from the OCL specification. What exactly appears in the Context depends upon the overall UML model in which the OCL expression is embedded, and whereabouts it is embedded in the model. We may think of the small box as the "hole" through which the UML model appears as the environment for the OCL expression.

This concept is analogous to constructs that appear in many programming languages to control namespaces. In the Java language, for example, the meaning of an identifier in an expression is found by searching, in order, through local variables, method parameters, the enclosing class or interface type, explicitly named imported types, other types declared in the same package, implicitly named imported types, and other visible packages [5]. Similar concepts are found in Pascal, Modula-2, Eiffel etc.

From now on we use the term Context to denote the complete contents of this box. Note that this is a somewhat different meaning of the word *context* from that implied by the **context** declarations in OCL specifications, which may be thought to imply that context denotes solely the contextual instance (*self*).

4. Declaration syntax

The previous section proposed that a Context contains declarations; the remainder of this paper discusses these in detail. In this section we introduce a concrete syntax for these declarations. By introducing this syntax, we extend the language OCL with new elements that go beyond the current heart of OCL: the OCL expressions. The syntax of OCL expressions themselves remains untouched.

In this paper declarations are used solely as concepts for discussion. Nevertheless, declarations could be a useful extension to OCL, particularly for when OCL expressions are used standalone, where no explicit context can be inferred from a diagram. By introducing declarations in this paper we do not intend to imply that such declarations will become an official part of OCL; if they were to do so, mechanisms would be needed to ensure that declarations were more user-friendly than the basic concepts suggested here.

The following syntactical elements are introduced, using the same EBNF syntax as in the OCL specification, and reusing parts of that specification. In this, x | y means a choice between x and y, x? means an optional occurrence of x, x* means zero or more occurrences of x, and x+ means one or more occurrences of x. Round brackets () are used for grouping and readability. Literal strings are enclosed in double quotes "".

```
context-expr ::= "context" declarations expr-kind (name)? ":" expression
expr-kind ::= "inv" | "pre" | "post" | ...


declarations ::= (declaration ";")*
declaration ::= typeD | attributeD | operationD | associationD | variableD


typeD := "OCLType" pathTypeName ( ":" supertypes )?
supertypes := pathTypeName ( "," pathTypeName )*

attributeD := pathTypeName "::" name ":" pathTypeName

operationD := pathTypeName "::" name "(" formalParameterList? ")" ":" pathTypeName
formalParameterList := name ":" pathTypeName ( "," name ":" pathTypeName )*

associationD := singleAssociationD | collectionAssociationD
singleAssociationD := attributeD
collectionAssociationD := pathTypeName "::" name ":"
                         collectionKind "(" pathTypeName ")" "[" min ".." max "]"
min := number
max := number | "*"

variableD := name ":" pathTypeName
```

An example:

```
context
        OCLType Person;                              -- a typeD
        OCLType Company;                             -- a typeD
        OCLType Integer;                             -- a typeD
        self : Person;                              -- a variableD
        Person::age : Integer;                       -- an attributeD
        Person::income(d:Date) : Integer;           -- an operationD
        Person::employer : Set(Company)[ 0..*];     -- an associationD
inv :
        self.age >= 0
```

For the purposes of this paper we have omitted to define syntax for, or otherwise consider, declarations of association types and qualified associations. These could readily be incorporated in a similar manner.

5. Global and local declarations

There will be two kinds of declaration within a given Context:

- Those *global declarations* that are there simply because of the structure of the overall UML model
- Those *local declarations* that are there specifically because of the placing of the OCL expression within the model, attached to a specific model element.

With respect to the first kind, we need to say what is meant by the "overall UML model". According to the UML Reference Manual [6] the "complete model" can be represented by a package with the stereotype «system». But there is no reason why a given assembly of UML needs to have such a package. Furthermore, a model will often be developed "bottom-up", by defining separate packages in different places and subsequently assembling them into an overall system.

On the other hand it does appear that the use of packages within UML models is universal. We propose, therefore, to derive "global" declarations at the package level. This means that within a package a set of declarations are derived from the contents of the package itself and appear in the Context of all OCL expressions within that package. Furthermore, additional declarations will appear as a consequence of the package of interest being enclosed in, importing, accessing, or inheriting from other packages, subject to UML visibility rules.

According to our interpretation of the UML visibility rules, the set of global declarations in a Package P should be the following:

- A type declaration for every Classifier included in P, i.e. every Classifier which is an ownedElement in the namespace of the Package.
- A type declaration for every Classifier not defined in P but visible in P. These include Classifiers explicitly imported into P, Classifiers within an enclosing Package, and those in Packages available via an «access» or «friend» dependency.
- An attribute declaration for every public or protected attribute defined for the Classifiers within P
- An attribute declaration for every public or protected attribute defined for the Classifiers within enclosing Packages
- An attribute declaration for every public attribute defined for the Classifiers within imported or accessed Packages
- Declarations for operations and association ends, similarly to those for attributes

Note that a declaration of the variable "self" does not make sense in a global declaration.

From the perspective of the UML meta-model, if we are willing to allow that an OCL declaration is a valid "BooleanExpression", then OCL declarations can be incorporated in a model as the body attributes of instances of Constraint with stereotypes «OCLTypeDeclaration», «OCLAttributeDeclaration», «OCLOperationDeclaration», and «OCLAssociationDeclaration». For global declarations such Constraints will be associated with (i.e. constrain) a Package.

We move on to local declarations, which are there because of the specific placement of the OCL constraint within the model. In UML, a Constraint can be associated with any set of one or more ModelElements. Within the current UML specification, though, we are interested in just a few of the possible placements of OCL expressions:

- a Constraint associated with exactly one Classifier T, in which case the following local declaration is present within the Context:

```
self : T
-- optionally another name can be used instead of self
```

- A Constraint, stereotyped as «precondition» or «postcondition», associated with exactly one Operation, declared in the global declarations as:

```
T :: op ( param1 : Type1, param2 : Type 2, …) : ResultType
```

where the following local declaration is present within the Context:

```
self : T
```

- Similar arrangements for other valid cases within the current UML definition.

Thus, the earlier example can be broken down into global and local declarations as follows:

```
context
-- global declarations, attached to the enclosing Package
      OCLType Person;                              -- a typeD
      OCLType Company;                             -- a typeD
      OCLType Integer;                             -- a typeD
      Person::age : Integer;                       -- an attributeD
      Person::income(d:Date) : Integer;            -- an operationD
      Person::employer : Set(Company)[ 0..*]; -- an associationD
-- local declarations, attached to the Person Classifier
      self : Person;                               -- a variableD
inv :
      self.age >= 0
```

For cases where UML is extended, new possibilities may arise. For example in Catalysis [7], which may be considered as an extension of UML, constructs called *actions* and *effects* may be defined. A Catalysis action is an abstraction of a model change, which may abstract an entire series of interactions and smaller changes, and a Catalysis effect is a specification of a state change which may be reused within the specifications of several actions. OCL expressions could be used to specify these elements, and the mechanisms in this paper used to specify the contexts for these expressions.

6. Derivation of global declarations

As indicated above, for every Package, global OCL declarations are required for visible Classifiers, Attributes, Operations and Associations. Specifically, for every visible Classifier, we must declare a Type, and similarly for every visible Operation,

Attribute and Association. To do this we need a little bit of special apparatus. To understand this it is essential to note that this apparatus needs to be able to shift meta-levels, i.e. to declare things at the model level using expressions at the meta-model level. Referring back to the syntax developed above, we propose some auxiliary operations on the metaclass Package for creating the BooleanExpressions corresponding to declarations. The specification of these operations is as follows:

```
declareType(typename : String¹, supertypes : Set(String))
                                  : BooleanExpression

-- declares an OclType named typename
-- with supertypes as specified,
-- note that the supertypes need to be declared elsewhere
-- note that the names can be pathnames


declareAttribute(    attrname: String,
                     owningType: String,
                     attrtype: String) : BooleanExpression

-- declares that owningType has an attribute
-- called attrname with type attrtype
-- all types must be declared
-- type names can be pathnames


declareOperation(    opname: String,
                     owningType: String,
                     resultType: String,
                     params: Sequence(Sequence(String)))
                                  : BooleanExpression

-- delares that owningType has an operation called opname
-- with result type resultType
-- and parameter list params, a Sequence of any length of
-- ordered pairs of Strings.
-- (OCL doesn't allow us to define pairs explicitly)
-- again, all type names can be pathnames


declareAssociationEnd(roleName : String,
                     owningType : String,
                     collectionKind : enum{ Set, Bag, Sequence },
                     targetType: String,
                     min : Integer, max: Integer,
                     infinite : Boolean) : BooleanExpression

-- declates that owningType has an association called roleName
-- referring to the type targetType
-- which gives a Set, Bag or Sequence as specified
-- minimum cardinality is min
-- if infinite then maximum cardinality is infinite
-- otherwise it is max
-- type names can be pathnames and all types must be declared
-- note that unary associations are declared as attributes
```

---

[1] We've used Strings instead of Names throughout this treatment to keep things simple, although this is strictly incorrect

Armed with these operations we can specify the global declarations in a Package. In the spirit of the UML specification we assume the existence of an auxiliary Package operation – remembering we are operating at the meta-level - called `allVisibleElements()`, which delivers, for any Package, a Set containing every model element visible within it. Included in this Set are all of the visible Classifiers, Operations, Attributes and AssociationEnds within this and related Packages. Using this, the following OCL assertion – again, remembering we are operating at the meta-level - states that there is a suitable type declaration for every visible Classifier in the package:

```
(self.allVisibleElements()->select
      (el | el.oclIsKindOf(Classifier)))
            ->forall (cl | self.constraint->exists
                 (decl|
                  decl.body = self.declareType(
                          cl.name,cl.parent.name )))²
```

Similar assertions can be developed for declarations of Operations, Attributes and AssociationEnds. These are omitted for readability, as is the definition of `allVisibleElements()`.

It remains to define the auxiliary declaration operations themselves. For this purpose we assume that BooleanExpression is type-equivalent to String. Then:

```
self.declareType(typename, supertypes) =
      "OCLType ".concat(typename). concList(supertypes)
```

using the auxiliary function `concList` to produce a colon appended to a list of comma-separated names for the supertypes (in arbitrary order):

```
concList(names: Set(String)) : String
concList(names, separator) =
      if names->size = 0 then ""
      else
      let nameSeq = names->asSequence in
      let firstName = nameSeq->first in
      let otherNames = nameSeq->subSequence(2,nameSeq->size) in
            concat(" : ").otherNames->iterate
                 (name: String ;
                  res : String = firstName |
                  res.concat(", ").concat(name))
      endif
```

7. Derivation of local declarations

The only local declarations are variable declarations, in the case of the current definition of UML. In this treatment we restrict ourselves to the declaration of local variables called self; the mechanisms can readily be extended to allow other names. Once again, remember when reading this section that we are using OCL at the meta-level. We introduce a variable declaration operation on the metaclass Classifier:

```
declareVariable(varname: String, typename: String) :
```

---

² This is somewhat simplified because it ignores the possibility of aliased names.

```
                                         BooleanExpression
-- declares a variable of name varname and OclType typename,
-- which can be a pathname
```

From the perspective of the UML meta-model, OCL variable declarations can be incorporated in a model as the body attributes of instances of Constraint with stereotype «OCLVariableDeclaration» associated with (i.e. constraining) a Classifier.

Then the following constraint on Classifier asserts the need for the variable declarations:

```
self.constraint->exists
( decl | decl.body = self.declareVariable( "self", self.name ))
```

The definition of the declareVariable operation is simple:

```
self.declareVariable(varname, typename) =
varname.concat(": ").concat(typename)
```

8. Example

As an example we investigate the use of an OCL expression as a condition to a message in a sequence diagram. The example UML class model in figure 2 defines a package P, represented by the outer rectangle. P contains the classes Company, Person and SalaryAccount. The (partial) sequence diagram in figure 3 shows object *comp* of class Company sending a message *deposit(amount)* to object *acc* of class SalaryAccount. This message is conditional. The obvious meaning of this condition is that the company will only transfer salaries to an account owned by one of its employees. This is expressed using the OCL:

```
    comp.employees->includes(acc.holder)
```

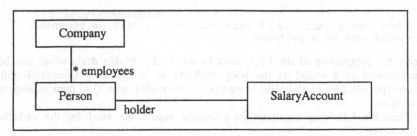

Figure 2: Example class diagram

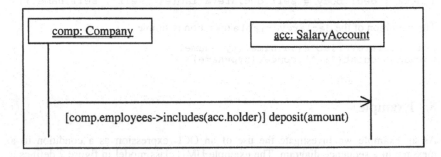

Figure 3: Example sequence diagram

We can use the apparatus developed in this paper to define which UML elements may be used in a condition to a message in a sequence diagram. The condition is specified by an OCL expression associated with an instance of the UML metaclass Message, stereotyped «OCLMessageCondition». In the UML meta-model a Message is associated with two ClassifierRoles representing the source (in this case comp) and target (in this case acc) of the message, each of which is associated with a Classifier. We get the global declarations from the enclosing package for the source Classifier. In this case the source is the object comp, the class of comp is Company, and the enclosing package is P. Therefore we include the global declarations of P as part of the Context for the message condition.

The local declarations that form the rest of the Context declare the names of the object instances used in the sequence diagram, in this case *comp* and *acc*. (If the names were omitted we might introduce default variables called *source* and *target* instead.)

Using the OCL to express the condition in the example, the Context is (incompletely) defined as:

```
context
        -- the global declarations for P
        OCLType Integer                         -- a typeD
        OCLType Person                          -- a typeD
        OCLType Company                         -- a typeD
        OCLType SalaryAccount                   -- a typeD
        SalaryAccount::deposit(amount:Integer)  : Integer
                                                -- an operationD
```

```
Company::employees : Set(Person) [ 0..*]
                                        -- an associationD
SalaryAccount::holder : Person          -- an associationD

-- the local declarations for the message
comp : Company                          -- a variableD
acc : SalaryAccount                     -- a variableD
```

cond :

```
    comp.employees->includes(acc.holder)
```

9. Summary

In this paper we have observed that the current specification of the context of OCL expressions within UML models is both incomplete and inflexible. To remedy this we have proposed mechanisms based on the introduction of explicit declarations embedded within the UML model. There are two types of declarations: global declarations, associated with UML packages, which declare the types, attributes, associations and operations that are visible within the package; and local declarations, associated with Classifiers or elsewhere, which declare the variables that can be accessed from OCL expressions. The overall context of an OCL expression is then understood as the combination of the global declarations from its enclosing package and the local declarations from its associated metaclass instance. We have given formal definitions for the majority of this apparatus.

References

1. UML 1.3 beta R1 draft specification [at uml.shl.com in April 1999]
2. Warmer J and Kleppe A, *The Object Constraint Language: Precise Modeling with UML*, Addison Wesley Longman, Reading, Massachusetts (1999).
3. Cook S, Kleppe A, Mitchell R, Rumpe B, Warmer J and Wills A. *The Amsterdam Manifesto on OCL*. Technical University Munich technical report (1999)
4. Cook S, Kleppe A, Mitchell R, Rumpe B, Warmer J and Wills A *Prefaces: Defining UML Family Members* (in preparation)
5. Arnold K and Gosling J, The Java Programming Language, Addison Wesley Longman, Reading, Massachusetts (1996)
6. Rumbaugh J, Jacobson I and Booch G *The Unified Modelling Language Reference Manual*. Addison Wesley Longman, Reading, Massachusetts (1999).
7. D'Souza D and Wills A *Objects, Components, and Frameworks with UML*. Addison Wesley Longman, Reading, Massachusetts (1999)

Mixing Visual and Textual Constraint Languages

Stuart Kent

Computing Laboratory, University of Kent, Canterbury, UK
sjhk@ukc.ac.uk

John Howse

School of Maths & Computing, University of Brighton, Lewes Road, Brighton, UK
John.Howse@brighton.ac.uk

Abstract. The Object Constraint Language (OCL) is a precise language for notating behavioural constraints on UML models. Constraint diagrams have been proposed as a means of notating similar constraints, but in a visual form. This paper explores the utility of these two notations for depicting constraints, and shows how they can be used effectively together. The goal of this work is to provide more intuitive and expressive languages to support the construction and presentation of rich and precise models.

1. Introduction

The object constraint language (OCL, [10]), which is part of the UML standard [7], has been put forward as a "formal language, that remains easy to read and write" for writing unambiguous constraints that can not be captured completely in UML diagrams. Constraint diagrams (CDs, [4], [3], [6]) is a notation put forward for visualising constraints over UML models.

No doubt there is a discussion to be had on the relative intuitiveness of CDs and OCL. The increasing popularity of diagrams, such as those found in UML, suggests that for some people CDs may be more intuitive; and for others they are probably not.

Certainly, CDs are less expressive than OCL, but it is a moot point whether they could or should be extended to solve this. If they are not extended, then the intuition gained by some people using a visual notation could be countered by a lack of expressiveness. If they are extended, then the simplicity and natural intuitiveness of the diagrams might be lost.

The purpose of this paper is to contribute to the development of more intuitive and expressive languages to support the construction and presentation of rich and precise models, by showing how textual and visual constraint languages, represented by OCL and CDs, respectively, could be used together for notating constraints. The intention is not to favour one above the other, but to explore the strengths and weaknesses of each and indicate how they could be harmonised to play to their strengths.

Sections 2 and 3 overview the OCL and CDs, respectively. Strengths and weaknesses of the two notations are identified, and the relative expressivity of CDs against OCL discussed. Section 4 is the substantive part of the paper. It provides a recipe for mixing OCL and CDs so that one can play to their individual strengths. Specifically, it considers (a) using OCL to annotate CDs, (b) embedding CDs in OCL, and (c) using a combination of (a) and (b) to go back and forth between OCL and CDs.

The fact that these notations can be intertwined, suggests that they are both concrete syntaxes for a uniform set of underlying concepts. The next stage in this work will be to formally define those concepts and mappings from OCL and CDs, with the aim of supporting the development of tools to translate between the different notations, check the well-formedness and consistency of constraints, etc. Section 5, *Further Work,* outlines a meta-modelling approach to this.

2. OCL

OCL is a form of first order predicate logic adapted for use within an object-oriented setting. Expressions in OCL always appear in the context of a class, either as invariants on that class, or as pre or post conditions of operations on that class. This paper restricts itself to invariants, but the ideas are just as applicable to pre and post conditions. This paper uses the newest version of OCL [8].

To illustrate OCL we have selected some invariants which appear in [5], which develops a meta-model for a fragment of UML that encodes semantics. The part we will focus on is the relationship between models and model instances. Specifically, a model is a package which contains just *classes* and *roles* (*association ends* in the latest version of UML). An instance of a model is a *snapshot,* which in UML is depicted by an *object diagram.* A snapshot is made up of *objects* and *links* between objects, which are themselves instances of classes and roles respectively. The class diagram for this meta-model fragment is given in Fig. 1.

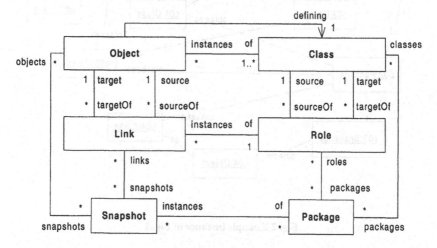

Fig. 1 Fragment of UML meta-model

Additonal constraints are required to ensure that only appropriate snapshots are instances of packages. All the required constraints may be expressed as invariants on the class Snapshot. The first of these is given below.

```
context s:Snapshot inv:                                    (1)
every package that s is a snapshot of, includes all the classes that s.objects
are instances of
s.of->forAll(p |
       p.classes->includesAll(s.objects.of))
```

The context part of an OCL expression indicates the context in which the expression should be interpreted. In this case, the OCL expression should be read as an invariant on the class Snapshot. The variable s is used to denote an arbitrary object of that class.

s.of is a navigation expression returning the set of all objects (in this case of class Package) connected to s by links of the role (or in new UML parlance *association end*) labelled of on the class diagram. For the object diagram in Fig. 2, representing an instance of Fig. 1, if s was bound to s01 then s.of would return the set of objects {p01,p02}.

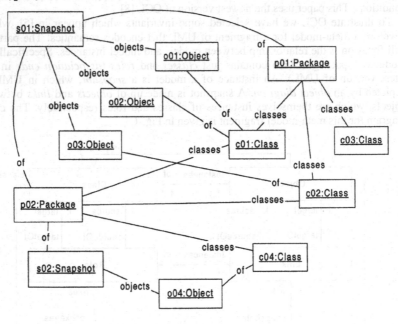

Fig. 2 Example instance of Fig. 1

X->forAll(p | f(p)) is a boolean expression which evaluates to true if f(p) (which must also be a boolean expression) evaluates to true for every subsitution of p for an element in the set X. In this case X is s.of, and f(p) is p.classes->includesAll(s.objects.of).

So evaluating the whole expression in the object diagram Fig. 2 for the same binding of s as above, requires a check that f(p) holds for p bound to p01, and also for p bound to p02, where these are the elements of s.of. The evaluation of f(p) for p substituted with p01 proceeds as follows:

1. p.classes evaluates to {c01, c02, c03}
2. s.objects evaluates to {o01, o02, o03}
3. s.objects.classes evaluates to {o01, o02, o03}.classes which evaluates to {c01}->union{c02}->union{c02} which evaluates to {c01,c02}
4. p.classes->includesAll(s.objects.of) evaluates to {c01, c02, c03}->includesAll{c01, c02} which is true.

The evaluation of f(p) for p substituted for p02 follows similarly.

The second constraint is very similar to 1:

context s:Snapshot inv:
every package that s is a snapshot of, includes all the roles that s.links are instances of   (2)
s.of->forAll(p | p.roles->includesAll(s.roles.of))

The third constraint illustrates a number of OCL constructs:[1]

context s:Snapshot inv: (3)
the links of s respect cardinality constraints for their corresponding role
s.links.of->forAll(r | r.source.instances->
    intersection(s.objects)->forAll(o | let
        linksFromOForR=o.sourceOf->intersection(s.links)
            ->intersection(r.instances)
        ((linksFromOForR->size <= r.upperBound or
            r.upperbound->isEmpty)
        and linksFromOForR->size >= r.lowerBound))
)
)

- let x = exp in P(x) is true if P(x) is true substituting x for the expression exp. In the constraint, the let expression allows linksFromOForR to be used instead of the much longer expression
  o.sourceOf->intersection(s.links)
      ->intersection(r.instances)
  with the obvious advantages.
- intersection is the complement of union.
- A->size returns the number of elements in A

[1] The corresponding constraint given in [4] is wrong - it forgets to consider that the links must be counted for each object in the snapshot of the source class of the role. This has been fixed here.

388

- *Numbers*. OCL includes the basic (value) types, Integer, Real, Boolean and String, in addition to collections such as sets. These are different to object types, which are UML classes.

```
context s:Snapshot inv:                                          (4)
if a link is of a role with an inverse, then there is a corresponding reverse link
s.links->select(l1 | l1.of.inverse->notEmpty)->
    forAll(l | s.links->select(l2 | l2.source=l.target
    & l2.target=l.source & l2.of=l.of.inverse)->size=1
```

This invariant introduces a further construct:

- A->select(a | P(a)) results in a set which is all the elements in A for which P(a) is true. In the above constraint, select is used to select all the links in the snapshot which are instances of roles that have an inverse.

Sets are just one kind of collection. OCL can also deal with *bags* and *sequences*. Sequences are well known, bags perhaps less so. The utility of bags often arises when summing over the result of a navigation expression that returns a collection of numbers. For example, the income of a consultant over a year is the sum of the income of each contract completed by that consultant during the year:

```
context c:Consultant,y:Year inv:                                 (5)
    c.income(y)=c.contracts->select(year=y).earned->sum²
```

Now, suppose in the year there were only two contracts which happened to earn the same, say $2500 apiece. If the navigation expression returned a set of numbers, then the result would be just {2500} which, when summed, gives 2500. Clearly this is the wrong result. However, if the navigation expression returned a bag of numbers, then repeats would not be removed and the result of navigation would be {2500,2500} which, when summed, returns 5000, the correct answer. OCL has operators to coerce one collection into another. For example, co->asSet coerces the collection co into a set. If co was a bag, then the effect of this operation would be to remove repeated elements.

3. Constraint diagrams

Constraint diagrams (CDs) were proposed in [4] as a notation for visualising invariants, and in [6] for visualising pre and post conditions, in an object-oriented modelling context. The CD corresponding to constraint 1 in Section 2 is given in Fig. 3.

---

[2] This is not true OCL syntax, which does not allow more than one class to appear in the context. The true syntax is context c:Consultant inv: Year.allInstances->forAll(y | ...) We feel this is unncessarily messy.

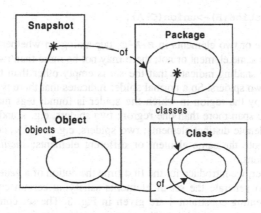

Fig. 3 CD for constraint 1

The main features of CDs are:

- *Venn diagrams/Euler circles.* CDs combine (and significantly extend) the best of Venn diagrams [9] and Euler circles [2] to show relationships between sets. In Fig. 3, the classes (represented by rectangle contours) are disjoint sets, and these each contain other sets obtained by navigating arrows (see below). In the bottom right hand corner, the set at the target of the arrow labelled of is contained in the set at the target of the arrow labelled classes. In the OCL version of constraint 1, the latter corresponds to the subexpression:

  ```
  p.classes->includesAll(s.objects.of)
  ```

- *Spiders.* Spiders are used to depict elements of sets. A spider may be a *wildcard*, representing a universally quantified variable which ranges over the region in which it is located, or *normal*, in which case it represents a specific element, which may be derived, or introduced via an existential quantifier. There are two spiders in Fig. 3, both wildcards. The first spider is contained in the class Snapshot and can be read as "for all elements s in Snapshot, ...". In the OCL constraint, this corresponds to: context s:Snapshot inv: ... ; alternatively, Snapshot.allInstances->forAll(s |...

  Fig. 4 shows spiders which are not wildcards. One of these spiders, z, has *legs*. Two others, u and v, are connected by a *strand*.

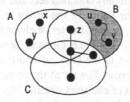

Fig. 4 Various spiders

The meaning of the diagram is as follows: there are two distinct elements, x and y, in the set A-B-C; there is an element, z, in the set

```
A->intersection(B)->union(C-A);
```

there may be one or two elements in B-A-C, depending on whether the spiders u and v denote the same element or not; there may not be more than two elements in this set, because shading indicates that the set is empty other than the element(s) denoted by the two spiders. So a normal spider indicates that there is an element in the set denoted by the region in which the spider is found; legs may be used to allow a spider to span more than one region; two spiders, e.g. x and y, which are not connected, denote distinct elements; two spiders, e.g. u and v, connected by a *strand*, may denote the same element or different elements; *shading* is used to show empty regions.

In this paper, we introduce for the first time the notion of a *generating spider*, which is used to generate the set of elements satisfying certain conditions. The diagrams representing constraint 4 are given in Fig. 5. The set contained in A is generated by the spider appearing on the contour of that set: the set denotes all the elements of A (which are links of some arbitrary snapshot) which are of a role that has an inverse – see below for the interpretation of arrows.

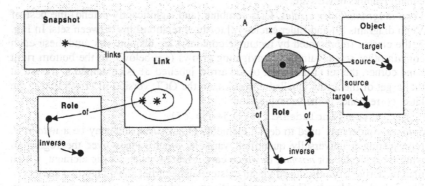

Fig. 5 CDs for constraint 4

- *Arrows*. Arrows are used to constrain relationships between sets/elements. The set or element at the target of an arrow is the result of navigating the named relationship from the element(s in the set) at the source. A CD without arrows is known as a *spider diagram* (it only contains sets and spiders).

In Fig. 3, the set at the target of the arrow labelled of from the wildcard spider in the Snapshot box, denotes the set of elements obtained by navigating the of relationship (association) from the element denoted by that wildcard. This corresponds to the term s.of in the OCL of constraint 1. Similarly for the set at the target of the objects arrow. The set at the target of the of arrow, sourced on this latter set, denotes the set of elements obtained by taking the union of the result of navigating the of relationship from each element in the set at the source of the arrow (i.e. from each object in s). The corresponding OCL expression is
s.objects.of

391

The notation is currently going through a process of improvement and refinement. The semantics of spider diagrams is in place [3], and the semantics of constraint diagrams is being written up. As explained in [3], the semantics is non-trivial for two reasons: every symbol on a CD interacts with every other symbol on the diagram; the order in which symbols are read makes a difference. The challenge has been to develop a semantics that retains one's intuitive interpretation of the diagrams.

Constraint diagrams, at least in their current form, are not as expressive as OCL. This is evident from attempts to construct the CDs for the OCL constraints introduced in Section 2. Constraint 1 was depicted in Fig. 3, and constraint 2 is very similar – no problems so far. Fig. 6 is a partial representation of constraint 3.

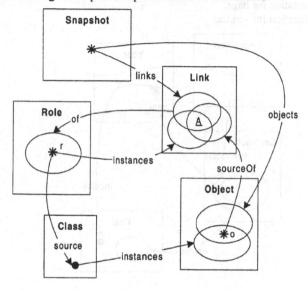

Fig. 6 CD partially representing constraint 3

The constraint diagram is very good at identifying the set of links that need to be counted: the set denoted by the region A. The problem is that there is no way of depicting the size of that set. The constraint is actually that the size of the set denoted by the region A is between r.lowerBound and r.upperBound, or not upper-bounded if the latter is empty. This could be resolved, for example by choosing a different kind of arrow to represent the OCL ->, as opposed to the OCL . operator which arrows on CDs currently represent. Another approach, used in [6], is to annotate contours with numerical expressions, including variables, which are assumed to be equal to the size of the set denoted by the contour. Both these approaches add further complexity to the notation, which might make it more difficult to define and use.

The diagrams representing constraint 4 were given in Fig. 5. This illustrates the use of multiple diagrams to break down the constraint into more comprehensible chunks. The use of multiple diagrams also avoids the problem of interaction between symbols in a complex diagram. Specifically, the labels A and x are defined in the diagram on the left; the diagram on the right then places further conditions on this set and this element. Separating the constraint into two diagrams means that we do not have to worry about the relationship between the subset of A introduced on the left and the subset of A introduced on the right[3]; nor do we need to worry about the relationships between the roles introduced on each diagram.

Fig. 7 is an attempt at depicting constraint 5. This nearly depicts the navigation routes involved in the constraint, which in the OCL version of the constraint are `c.income(y)` and `c.contracts->select(year=y).earned`, respectively. Specifically, the diagram is deficient in two respects:

1. There is no notation for bags.
2. There is no notation for `->sum`.

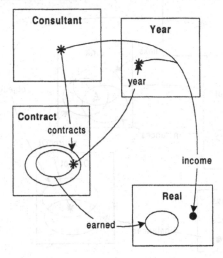

Fig. 7 CD partially representing constraint 5

A solution to (2) has already been suggested: have a special arrow for the OCL `->` operator on sets. A solution to (1) could follow along similar lines, and, therefore, suffer from the same drawbacks: introduce a different style of navigation arrow for generating bags. It is even worse in this case, because the result of this arrow will no longer be a set but a bag; but contours are interpreted as sets.

In summary, constraint diagrams seem to have much potential, and for some constraints they are ideal. However, there are other constraints, where, at some point, they are seemingly let down by a lack of expressive power. It may be possible to fix this, but at the danger of making the notation too complicated, thereby losing its simplicity and intuitiveness.

---

[3] Although, after some reflection, the reader may realise that the subset on the right will always be contained in the subset on the left if it is always the case that the inverse of the inverse of a role is the role itself.

An alternative is to harmonise OCL with constraint diagrams so that they may be freely mixed in the expression of constraints. This would allow modellers the option of using constraint diagrams where they are helpful and can be used, reverting to a textual language when this is not the case.

4. Mixing OCL with constraint diagrams

4.1. Annotating CDs with OCL

Fig. 6 visualises most of the structure of constraint 3, but not the constraint itself. It fails to work because there is no notation for depicting the size of the region A. Similarly, Fig. 7 works until the point at which the navigation expression returns a bag. The solution in both cases is simple: use labels on the diagram to provide links between the diagram and accompanying OCL expressions.

A comprehensive labelling scheme for sets and elements has already been devised for spider and constraint diagrams [3]. It works as follows:
- Labels for contours are plain, with an initial uppercase letter, and are co-located with the contour.
- Labels for regions appear in the region concerned, and are underlind to distinguish them from contour labels. They also begin with an uppercase letter.
- Labels for spiders begin with a lowercase letter and are co-located with the spider.

Thus a diagram may be accompanied or *annotated* by additonal textual constraints that refer to sets and elements in the diagram. For Fig. 6, the constraint is completed with the OCL text:

```
context "Fig. 6":
    (A->size <= r.upperBound or r.upperBound->isEmpty)
        and A->size >= r.lowerBound
```

This text refers to the labels A and r in the diagram. A reference to the diagram has been used in the context section to indicate that the OCL should be read in the context of this constraint diagram.

Similarly Fig. 7 may now be replaced by a correct, annotated CD, which is given in Fig. 8.

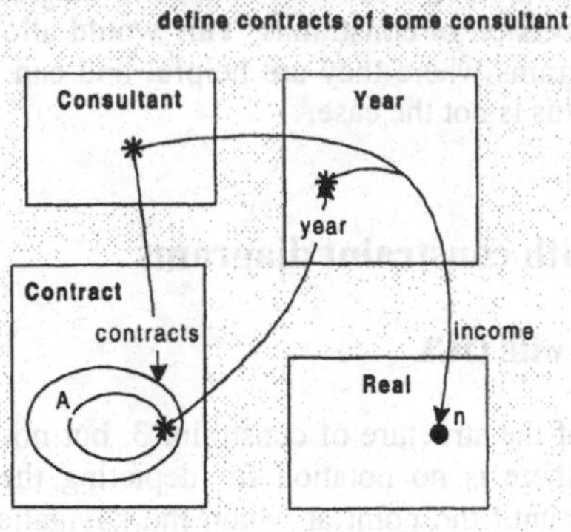

394

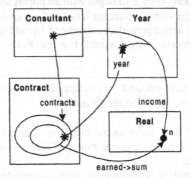

define contracts of some consultant

```
context "define contracts of some consultant":
        A.earned->sum=n
```

Fig. 8 Annotated CD for constraint 5

In this case the annotation can be integrated with the diagram, by labelling an arrow with the navigation expression `earned->sum`, as in Fig. 9.

Fig. 9 CD for constraint 5 with integrated annotation

4.2. Embedding CDs in OCL

So it is possible to annotate CDs with additional OCL constraints. It is also possible to embed constraint diagrams within OCL expressions: labels are again used to connect elements and sets in the diagram to expressions in the OCL (`let` expressions in the latest version of OCL are essential here); diagrams are embedded directly in the OCL expression or referenced by e.g. a diagram name.

For example, consider constraint 6, which is a special case of constraint 3.

```
context s:Snapshot inv:                                        (6)
```
if a role with cardinality 1, then there is one link of that role per object of the
source class of that role
```
s.links.of->select(r | r.upperBound=1 & r.lowerBound=1)->
    forAll(r | r.source.instances
        ->intersection(s.objects)->forAll(o |
            (o.sourceOf->intersection(s.links)
                ->intersection(r.instances))->size=1))
```

Suppose one did not wish to use generating spiders in constraint diagrams: playing
devil's advocate, one could argue that they add an extra layer of complexity to the
notation, which is undesirable. Then, adopting the suggested technique, we could still
use a constraint diagram to represent the main forAll clause of constraint 6, by
embedding it in an OCL expression which provides the select clause, as in Fig. 10.

```
context s:Snapshot inv:
    let RolesOfCardinality1 =
        s.links.of->select(r | r.upperBound=1 & r.lowerBound=1)
    in
```

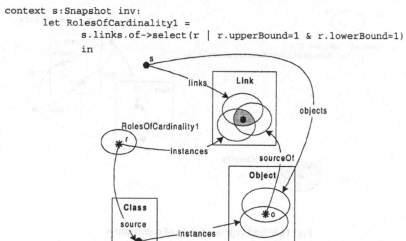

Fig. 10 Constraint 6 using an embedded CD

4.3. Mixing OCL and CDs

OCL annotations and CD embedding can be used in combination to allow arbitrary
mixing of the two notations. One may begin with an OCL expression that embeds a
constraint diagram which itself is annotated by OCL; or one may begin with a CD that
has an OCL annotation, which embeds another CD; and so on.

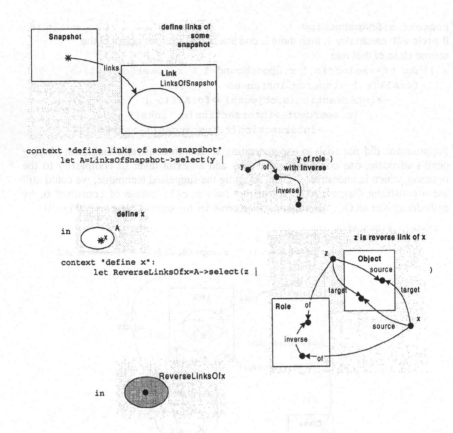

Fig. 11 Mixed representation of constraint 4

This technique is illustrated by constraint 4. Suppose, again, one did not wish to use generating spiders in constraint diagrams. Then, as illustrated by Fig. 11, constraint 4 can be represented by mixture of OCL annotations to CDs and CD embeddings in OCL.

5. Further Work

The paper has presented ways in which OCL and constraint diagrams can be used in combination to notate constraints. This suggests a promising way forward in the development of more intuitive and expressive languages to support the construction and presentation of rich and precise models. To realise this potential, a number of tasks need to be undertaken:

1. Give precise definitions of the concrete syntaxes for each notation.

2. Define a common abstract syntax that supports both notations, and define mappings from concrete to abstract, and vice-versa where possible.
3. Build tools based on 1 & 2, in particular tools that allow constraints to be entered and presented using the mixed approach.
4. Evaluate the mixed approach, versus, say, the purely textual or purely diagrammatic approach, with and without tools and to different target audiences. This does not preclude evaluation by testing the market, i.e. are the techniques widely adopted, does the tool sell?

With regard to 1 & 2 we are following the meta-modelling approach, advocated in the UML standard. Under this approach, the abstract syntax is described using a minimal core of the modelling notation itself. A fragment of the UML class diagram notation plus a constraint language, which may be any combination of (subsets of) the languages presented here, seems to be sufficient. An important difference in our approach is, perhaps, that we have managed to incorporate true semantics into the meta-model [1][5], giving rise to the possibility of tools that can check consistency of models, validate instances of models, generate instances of models (a form of partial simulation), etc. Resources are currently being harnessed to build proof of concept tools.

An enhancement of the ideas presented here, is to introduce a third notation to be mixed with the other two. Specifically, we envisage using object diagrams to show prototypical behaviour of elements. The prototypes can be embedded in the OCL or constraint diagrams: indeed there is a fragment of the constraint diagram notation which is equivalent to object diagrams (e.g. some of the embedded constraint diagrams in Fig. 11 could be replaced with object diagrams). Object diagrams could be further rendered in a domain specific visualisation: for example, an object diagram representing a communications network could be rendered in the notation for a network familiar to communications engineers.

Acknowledgements

The authors acknowledge support from the UK EPSRC under grant number GR/M02606. Thanks are also due to Steve Gaito and Niall Ross of Nortel Networks UK, and Yossi Gil of the Technion, Israel, for many valued discussions and feedback.

References

[1] Evans A. and Kent S. Core Meta-Modelling Semantics of UML: The pUML approach, in Procs. UML'99 (this volume), Springer Verlag, 1999.
[2] Euler L. Lettres a Une Princesse d'Allemagne, Vol 2, Letters No. 102-108, 1761.
[3] Gil, J., Howse J. and Kent S. Formalizing Spider Diagrams. To appear in Procs. VL'99, IEEE, 1999.
[4] Kent S. Constraint Diagrams: Visualising Invariants in Object Oriented Models. In: Proceedings of OOPSLA97, ACM Press, 1997.

[5] Kent, S., Gaito, S. and Ross, N. Putting Semantics into the UML Meta-Model. In: Kilov, H., Simmonds, I. and Rumpe, B., (Eds.) Behavioral Specifications of Business and Systems, Kluwer Academic, 1999.

[6] Kent, S. and Gil, J. Visualising Action Contracts in OO Modelling. IEE Proceedings: Software **145**, 1998.

[7] OMG UML 1.1. Specification. OMG Documents ad970802-ad970809, 1997.

[8] OMG Object Constraint Language. In: *UML Version 1.3 beta R7*, Object Management Group, 1999.

[9] Venn, J. On the Diagrammatic and Mechanical Representation of Propositions and Reasonings. *Phil.Mag.* **123**, 1880.

[10] Warmer, J. and Kleppe, A. The Object Constraint Language: Precise Modeling with UML, Addison-Wesley, 1998.

Correct Realizations of Interface Constraints with OCL [*]

Michel Bidoit [1]   Rolf Hennicker [2]   Françoise Tort [1]   Martin Wirsing [2]

[1] Laboratoire Spécification et Vérification, CNRS & ENS de Cachan, France
[2] Institut für Informatik, Ludwig-Maximilians-Universität München, Germany

Abstract. We present an OCL-like formal notation for interface constraints, called ICL, suited to describe the required observable behavior of any correct interface implementation (provided by some class). The semantics of the ICL notation is defined by a translation to the observational logic institution. For specifying constraints on classes we use a subset of OCL to express invariants and pre- and post-conditions on operations. The semantics of the OCL expressions is defined by a translation into an algebraic specification. Using these semantic foundations we introduce a formal correctness notion for implementation relations between interfaces and classes and we show how to prove implementation correctness by using observational proof techniques.

1 Introduction

1.1 Motivations

Pragmatic software engineering methods and graphical notations as provided by UML are an important means to enhance software quality. The correctness of a program w.r.t. a given specification, however, can only be ensured with the help of formal methods. Therefore it is an important aim to investigate possibilities for combining pragmatic and formal approaches to software development.

A seminal step in this direction was provided by the object-oriented language Eiffel where assertions are an integral (and executable) part of class declarations [M88]. Recently, the Object Constraint Language OCL (see [Rat99], [WK99], [KWC99]) offers a formal notation to constrain the interpretation of modeling elements occurring in UML diagrams. OCL is systematically used for rigorous software development in the Catalysis Approach [SW98]. The OCL notation is particularly suited to constrain classes and associations since OCL expressions allow one to navigate along associations and to describe conditions for object attributes in invariants and pre- and post-conditions of the operations. However, OCL can hardly be used to constrain interfaces since at the interface level no attribute and no association either is available. Moreover, OCL is not designed to express implementation relations and, to our knowledge, there is no

[*] This work is partially supported by the ESPRIT Working Group 29432 CoFI [CoFI] and by the Bayer. Forschungsstiftung.

proof calculi for verifying semantical consequences of constraints and for proving implementations. The aim of this paper is therefore to contribute to a solution to these issues by providing a general method, formal notations and proof techniques which are compatible with OCL and support correct realizations of interfaces with constraints.

1.2 The General Method: an Overview

We focus on UML class diagrams which contain interfaces and corresponding implementations provided by some classes. In order to ensure the correctness of an implementation (also called "realization"), we propose three major steps briefly summarized in the following.

The Interface Level (cf. Section 3). First, the declaration of the operations provided by an interface must be supplemented by a formal description of their semantic properties. For this purpose we adopt a behavioral approach which allows us to specify the observable properties of the operations in an abstract and intuitive way. Technically this is achieved by splitting the operations of an interface into so-called "observer operations" (to be used for observing abstract, non-visible states) and the "other" operations whose behavior can then be specified by describing their effects w.r.t. the given observers. Thus attributes and associations (which are already implementation oriented) are indeed not necessary for constraining the operations. Therefore we propose an interface constraint language, called ICL, with the same syntax as OCL except that the dot-notation of OCL for attributes (and role of association ends) is replaced by a dot-notation for observers. The other changes are that the @pre-notation for attributes is replaced by an @pre-notation for self and that a symbol for expressing "observational equality" of the encapsulated states is added. Semantically, these ideas are realized by translating the invariants and pre- and post-conditions of interface constraints to the observational logic institution [HB99]. The resulting specification is presented in an extension of the algebraic specification language CASL [CoFI98].

The Class Level (cf. Section 4). At the class level we can use attributes and navigations on associations to describe invariants and pre- and post-conditions on the operations. For this purpose we use a subset of OCL which will be supplied with an algebraic semantics by translating OCL class constraints into a corresponding algebraic specification. In particular, this specification provides a precise description of the notion of an object by relating object identifiers and environments. It will also be presented using CASL.

The Implementation Relation (cf. Section 5). Since interface and class constraints are equipped with a well-defined semantics which in each case is given by a class of algebras, it is now straightforward to define formally the correctness of an implementation (realization): we say that a constrained interface I is correctly implemented by a set of classes SC if all algebras belonging to the semantics of (the translation of) SC belong also to the semantics of I. According to the observational semantics of interfaces, this expresses that the implementation has the required observable behavior. For proving the correctness of an implementation we can apply observational proof techniques on the basis of [BH98].[1]

Throughout the presentation we will illustrate our approach with the help of a simple but representative example described in the next section.

2 The Account Example

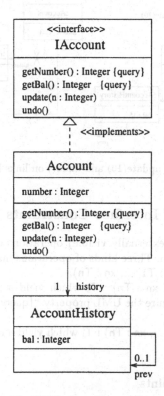

Fig. 1. The UML class diagram for accounts

Figure 1 presents an UML diagram for modeling bank accounts. The IAccount interface shows services that a (simplified) bank account may offer: get the number of the account, get its balance, update the account (with a new

---

[1] These proof techniques are effective in the sense that they lead to proof obligations that can be discharged using existing first-order theorem provers such as, e.g., the Larch Prover.

positive or negative amount, hence update subsumes the usual credit and debit operations), and undo an update.

A class named Account implements the IAccount interface. Its objects are characterized by a number and a history. The history of an account provides its current balance (the attribute bal of the AccountHistory class) and points to its previous balance (via the association prev).

The expected implementation of the update and undo operations is depicted in Fig. 2, where crosses over links mean that these links are deleted after execution of the operations.

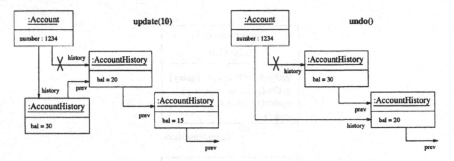

Fig. 2. Impact of update(10) and undo() on links between objects

3 Specification of Interface Constraints

An interface shows the (externally visible) operations of a class. It has no attribute and no association. Three kinds of operations may occur in an interface:
- pure procedures op(x1 : T1, ..., xn : Tn),
- queries op(x1 : T1, ..., xn : Tn) : T, which yield a result and have no side effect (in this case we require the UML property "{query}" to be attached to the operation),
- procedures op(x1 : T1, ..., xn : Tn) : T, which yield a result and may have side effects.

3.1 Interface constraints

In this section, we provide a specification method and a formal notation for interface constraints which allow us to restrict the admissible implementations of an interface to those classes which respect the given interface constraints. Since interfaces do not contain any implementation details, the interface constraints will focus on the "observable behavior" of the operations, i.e., on those properties which are visible "from the outside".

Observer operations. To formally describe observable properties we fix a subset of the interface operations as so-called "observer operations" (or "observers"

for short). The behavior of the other operations can then be constrained by defining their effects w.r.t. the observable experiments determined by the given observers. The observers have to be chosen by the programmer in such a way that they adequately express an indistinguishability relation for objects. (This is quite analogous to the functional programming paradigm where the programmer has to choose the adequate data type constructors.) For instance, for the IAccount interface we choose getNumber(), getBal() and undo() as observers.[2] We have chosen these observers because from the user's point of view two accounts a and b are "observationally equal" if they have the same number and the same balance and if this also holds after performing the same number of undos both for a and b (i.e., if a and b have the same history).[3]

Observational specification method. The declaration of a distinguished set of observers leads to a specification method which shows how to constrain the non-observer operations in a systematic way. Thereby the idea is to specify the observable effects of a non-observer operation by a (complete) case distinction w.r.t. the given observers (see [BH99] for a formal treatment of observer complete definitions which indeed are quite analogous to constructor complete definitions in functional programming). For instance, the observable effect of updating an amount for an account is specified w.r.t. the three observers get-Number(), getBal() and undo() (see the IAccount interface example below). Note that in contrast to OCL, we admit the call of observer operations with side effects such as undo(). This is possible since we translate operations into functions and constraints into declarative axioms, hence the constraints are not executed but interpreted in a declarative logical formalism which provides all possibilities for reasoning about state changes without actually performing them (see below the section on the semantics of interface constraints).

Syntax of interface constraints. To express the interface constraints, we use an OCL-like syntax, called ICL, where attributes and roles are replaced by the following constructs:
- Calls of the form o.op(p1,...,pn), where o is an object, op is an observer, possibly with side effects, and p1,...,pn the actual parameters, are allowed. We deviate on purpose in this point from the recommendations in [WK99] for OCL extensions in the sense that observers can change the state of objects. We assume that the interface can address only objects in the arity of the operation op, i.e., either the object o or one of the parameters xi.
- o@pre where o stands for an object. o@pre can be used in a post-condition of an operation to denote the state of o before the execution of the operation. For example, self@pre denotes the (state of the) object self before the call of the

---

[2] As a graphical notation, the "{observer}" property can be attached to each of these operations in Fig. 1.

[3] undo() is also called an "indirect observer" because it does not directly produce a visible output but nevertheless contributes to distinguish accounts (since after applying an undo(), a "direct observer" getNumber() or getBal() can always be applied).

operation and similarly xi@pre denotes (the state of) an in-out parameter before the call of the operation.

- o ~ ol denotes the "observational equality" of the objects o and ol, i.e., the indistinguishability of their respective states.

The IAccount interface is restricted by the following ICL constraints.

context IAccount
    inv: self.getBal() >= 0

context IAccount::update(n : Integer)
    pre: self.getBal()+n >= 0
    post: self.getNumber() = self@pre.getNumber() and
          self.getBal() = self@pre.getBal()+n and
          self.undo() ~ self@pre

As explained above, in the constraint for update(n), self denotes the current (state of the) account object after applying the update(n) operation and self@pre denotes (the state of the) self before the call of update(n). Moreover, if an update(n) operation followed by an undo() is performed for an account a then the result is not a visible value but rather a new state of the object. Hence the constraint self.undo() ~ self@pre requires that this state is observationally equal to the original state.

3.2 Semantics of Interface Constraints

In the previous section we have intuitively introduced the syntax of interface constraints and their intended meaning. Formally, the semantics of our interface constraint language ICL is defined by associating an observational specification to a given interface. We do this in two steps: first, we construct an algebraic signature (see [W90]) of a given interface. Then, we translate the invariants and the pre- and post-conditions of the operations to axioms written in many-sorted first-order logic over this signature. For defining the signature of an interface I:

- any pure procedure op(x1 : T1, ..., xn : Tn) is translated into an operation symbol $\_op(\_,\ldots,\_) : I \times T1 \times \ldots \times Tn \to I$,
- any query op(x1 : T1, ..., xn : Tn) : T is translated into an operation symbol $\_op(\_,\ldots,\_) : I \times T1 \times \ldots \times Tn \to T$ and
- any procedure op(x1 : T1, ..., xn : Tn) : T with result is translated into an operation symbol $\_op(\_,\ldots,\_) : I \times T1 \times \ldots \times Tn \to Tuple[I, T]$, whereby $Tuple[I, T]$ is a sort which denotes the cartesian product of I and T.[4]

By exception, if the result type of an operation is Boolean, then the operation is translated into a corresponding predicate symbol.

The atomic formulas built over this signature are either predicates (as usual) or, for terms t and r of the same sort s, equations $t = r$ (if s is the name of a

---

[4] Here we assume that the operations have only input-parameters. In the case of in-out parameters additional result sorts would be needed.

basic data type like Integer) or observational equations $t \sim r$ (if s is the name of an interface).

For defining the semantics of a constrained interface, we add to the signature the translation of the ICL constraints into observational logic axioms. This translation is done according to the following scheme. Let us consider an interface constraint of the form:

context I::op(x1 : T1, ..., xn : Tn)
  pre: ψ
  post: ϕ

where the context indicates that op is an operation (in this case a pure procedure) of an interface I, x1, ..., xn are input parameters, and ψ and ϕ are logical expressions defining the pre- and post-conditions on op. Let $Trans_{pre}$ and $Trans_{post}$ be functions translating logical expressions of ICL into first-order formulas, then the above constraint is translated into $Trans_{pre}(\psi) \Rightarrow Trans_{post}(\phi)$.

An invariant constraint of the form:

context I
  inv: ϕ

is translated into the axiom $Trans_{pre}(\phi)$.

The "$Trans_{...}$"functions are defined as follows. The ICL boolean operations not, and, or, etc., are translated into the corresponding logical connectives $\neg, \wedge, \vee$, etc. The ICL equality test "$=$" is kept and considered as an equality symbol to form equations between terms, except for boolean terms where it is translated to logical equivalence $\Leftrightarrow$. The ICL observational equality test "$\sim$" is kept. The ICL operators for collection types are also kept and considered as predefined operation or predicate symbols.

The translation of ICL expressions of the form t.op, where t is a complex expression of the form t1.op1, is recursively defined by $Trans_{...}(\text{t.op}) = Trans_{...}(\text{t}).op$.

The recursion terminates if the translation function is finally applied to a variable which remains unchanged, or if it is applied to self or self@pre. In that cases the translation is more subtle:

• self in an invariant or in a pre-condition denotes the state of the object before the call of the procedure op, it is kept by the translation: $Trans_{pre}(\text{self}) = self$,
• self in a post-condition denotes the state of the object after the call of op, it is translated into $self.op(x1, \ldots, xn)$. Whereas self@pre in a post-condition denotes the state of the object before the call of op, it is translated into $self$: $Trans_{post}(\text{self}) = self.op(x1, \ldots, xn)$ and $Trans_{post}(\text{self@pre}) = self$.

We obtain an observational specification in the framework of the observational logic institution [HB99], whereby the interface constraints are now axioms of the specification. The semantics of such a specification is given by the class of all (observational) algebras which observationally satisfy the axioms. Thereby the observational satisfaction relation is defined in the same way as the usual satisfaction relation of the first order predicate calculus but the symbol "~" is interpreted by a semantically defined observational equality relation between the elements of an algebra (for details see [HB99]). Based on these semantical notions, one can reuse the sound and complete proof system of [HB99] for proving observational theorems of interfaces with constraints.

The observational specification associated to the IAccount interface is shown below. It is written in an extension of the CASL algebraic specification language [CoFI98] to take into account the distinction between observer and non-observer operations.[5] The specification PREDEFINEDOCLTYPES is an algebraic specification of the basic OCL Types, which is not shown here for lack of space.[6]

obs spec IACCOUNT =
PREDEFINEDOCLTYPES **then**
sort *IAccount*;
obs ops
    _.getNumber() : IAccount → Integer;
    _.getBal() : IAccount → Integer;
    _.undo() : IAccount → IAccount;
op   _.update(_) : IAccount × Integer → IAccount;
vars self:IAccount; n:Integer;
axioms
  %% Invariant:
  self.getBal() ≥ 0;
  %% update:
  self.getBal() + n ≥ 0 ⇒
    self.update(n).getNumber() = self.getNumber() ∧
    self.update(n).getBal() = self.getBal() + n ∧
    self.update(n).undo() ~ self;
end

4 Specification of Class Constraints

4.1 Using OCL to specify Class Constraints

As shown in Fig.1, interfaces are implemented by classes which contain attributes and which may be connected via associations. For describing such implementations we use (a subset) of OCL 1.3 [Rat99] to state invariants and pre- and post-conditions on the operations. Thereby, similarly to the observational specification method, operations can now be constrained by describing their effects

[5] We also make an intensive use of the "display annotation" facility of CASL. This facility allows us to use the ICL syntax for operation names, terms and formulas.
[6] All the sorts introduced by PREDEFINEDOCLTYPES are of course observable ones.

w.r.t. the given attributes and associations (in fact, we will use the role names at the appropriate association ends). For instance, the classes Account and AccountHistory are restricted by the following OCL constraints (which, for update and undo, are conform with Fig. 2).

context AccountHistory
  inv: bal >= 0

context Account::getNumber() : Integer
  post: result = number

context Account::getBal() : Integer
  post: result = history.bal

context Account::update(n : Integer)
  pre: history.bal + n >= 0
  post: number = number@pre and
      history.ocllsNew and
      history.bal = history@pre.bal@pre + n and
      history.prev = history@pre

context Account::undo()
  pre: history.prev→notEmpty
  post: number = number@pre and
      history = history@pre.prev@pre

In the above expressions, we have used certain assumptions and predefined constructs of OCL. First, it is implicitly assumed that attribute and navigation expressions like number, history.bal, history.prev and accounts are shorthand notations for self.number, self.history.bal, self.history.prev and self.accounts where self is a special variable for objects whose type can be derived from the context. Moreover, an expression in a pre-condition refers always to a value before execution of the operation while the same expression in a post-condition refers to the value after execution of the operation. However, one can also refer in a post-condition to a previous value by using the symbol "@pre" at the appropriate positions. For instance, history@pre.bal@pre refers to the previous balance of the previous history object of an account. In a post-condition one can also apply the "ocllsNew" construct to an object o to express that after execution of the operation, o is a new object. For instance, after performing update, a new account history object is created.

4.2 Semantics of Class Constraints

The semantics of the class constraints of a class C is defined by a translation into an algebraic specification C_OO written in the algebraic specification language CASL [CoFI98]. For providing the semantics one has to take into account environments, for which we introduce a sort *Env*, and object identifiers, for which

408

we introduce a sort *CId*. Then any object can be represented by a pair $<ev, id>$ where *ev* is an environment and *id* is an object identifier. However, note that conversely not any pair $<ev, id>$ denotes an object. The reason is that we must always be able to create new objects, i.e., to find an identifier which did not represent an object in an "old" environment *ev* but does represent an object in a "new" environment ev'. Therefore the resulting specification C_OO introduces, besides the sorts *Env* and *CId*, an explicit sort *C* and a sort *Tuple[Env, CId]* such that $C < Tuple[Env, CId]$ (i.e., *C* is a subsort of *Tuple[Env, CId]*). The sort *Tuple[Env, CId]* (and its corresponding pairing function $<\_,\_>$ and projections $\_.env$ and $\_.cId$) is defined by means of the "free-type" construct of CASL to denote the cartesian product of *Env* and *CId*; the subsort *C* is used to denote the subset of all pairs $<ev, id>$ which indeed are objects.

In the next step we introduce appropriate operation symbols which still have to be added to obtain the full signature of the specification C_OO. For this purpose we first associate to any attribute a : T of the class C an operation symbol $\_.a : C \to T$ and for any role r at the association end to a class D an operation symbol $\_.r : C \to DId$ (if the role has at most multiplicity 1) and $\_.r : C \to Set[DId]$ (if the multiplicity of r is greater than 1). Note that we have used the sort *DId* (*Set[DId]* resp.) as the result sort of $\_.r$ because values of a role are indeed object identifiers. Finally, we introduce for each operation of the class a corresponding operation symbol in the same way as for the operations of interfaces.[7] From a technical point of view it is also convenient to introduce for the attributes, roles and operations, the overloaded operation symbols working on tuple sorts.

The most important step of the translation concerns the translation of OCL constraints into corresponding axioms. There exists a general translation algorithm which (for lack of space) cannot be detailed here. Nevertheless, let us sketch the basic ideas of the translation scheme by considering a constraint of the form:

context C::op(x1 : T1, ..., xn : Tn)
  pre: ψ
  post: ϕ

where op is a pure procedure (the other cases are quite similar). Then ψ refers to an environment *ev* before execution of op and ϕ refers to an environment ev' after execution of op. Since the constraint must be satisfied whenever op is performed for an object $<ev, self>$ of the class C, we translate the constraint into the following axiom:

$$\forall ev, ev' : Env; self : CId; x1 : T1; \ldots; xn : Tn \bullet$$
$$<ev, self> \in C \wedge ev' = <ev, self>.op(x1,\ldots,xn).env \wedge Trans_{pre}(\psi)$$
$$\Rightarrow <ev, self>.op(x1,\ldots,xn).cId = self \wedge Trans_{post}(\phi)$$

The "$\in$" symbol is predefined in CASL to test membership in a subsort and $Trans_{pre}$ and $Trans_{post}$ are functions which translate logical expressions of OCL

---

[7] Apart from the case where an operation has an object valued parameter. Then the translation is slightly different, but is not explained here for lack of space.

into first-order formulas. The translation of logical expressions and OCL operators is as in ICL (except that now a class C is translated to identifiers of type CId).

Note that a special role is played by the operator $\rightarrow$isEmpty (and similarly by its negation $\rightarrow$notEmpty) which on the one hand can be used to test whether a set is empty (in this case it is translated to the predicate symbol $\_\rightarrow isEmpty$: $Set[CId]$) and, on the other hand, can be used to test whether an object identifier is "nil" (for this test we use the overloaded predicate symbol $\_\rightarrow isEmpty$: CId ; cf. the pre-condition of undo where the negation $\_\rightarrow notEmpty$: CId is used).[8]

The most interesting case is the translation of OCL "properties" whereby we consider expressions of the form t.a where t is a navigation path (in the simplest case just the variable self) and a is an attribute or a role (at an association end). Then $Trans_{pre}$ and $Trans_{post}$ are recursively defined by:

$$Trans_{pre}(\text{t.a}) = <ev, Trans_{pre}(\text{t})>.a\,,$$
$$Trans_{post}(\text{t.a}) = <ev', Trans_{post}(\text{t})>.a\,.$$

The recursion terminates if the translation function is finally applied to a variable which remains unchanged. As an example, we compute:[9]

$$Trans_{post}(\text{history.bal}) =$$
$$Trans_{post}(\text{self.history.bal}) =$$
$$<ev', Trans_{post}(\text{self.history})>.bal =$$
$$<ev', <ev', Trans_{post}(\text{self})>.history>.bal =$$
$$<ev', <ev', self>.history>.bal$$

The above definition of $Trans_{post}$ is not yet complete because in a post-condition one can also refer to previous values by the symbol "@pre" and one can use the "ocllsNew"construct. In these cases, the translation is defined by:

$$Trans_{post}(\text{t.a@pre}) = <ev, Trans_{post}(\text{t})>.a\,,$$
$$Trans_{post}(\text{t.ocllsNew}) =$$
$$\neg(<ev, Trans_{post}(\text{t})> \in C)$$
$$\wedge <ev', Trans_{post}(\text{t})> \in C$$

Finally let us point out that besides the translated class constraints any specification C_OO still contains two additional axioms, one which requires that "nil is not an object"(expressed by $\rightarrow notEmpty$) and one which requires "consistency of environments". Intuitively, this means that in any fixed environment each navigation path that starts from an object (and does not yield "nil") provides an identifier or a set of identifiers which denote indeed objects in this environment. The following specifications show the result of the translation of the class constraints in our example.

spec AccountHistory_OO =
PredefinedOCLTypes **then**

---

[8] Hence, as in OCL, we do not introduce an explicit denotation for "nil".

[9] This computation is used in the translation of the post-condition of the update operation.

sorts *Env, AccountHistoryId*;
free type *Tuple[Env, AccountHistoryId]* ::=
                 <__, __>(__.env : Env; __.accountHistoryId : AccountHistoryId);
%% This CASL construct abbreviates the declarations of:
%% the sort *Tuple[Env, AccountHistoryId]*, a pairing function <__, __>,
%% two projections .env and .accountHistoryId,
%% and the corresponding axioms.
sort *AccountHistory* < *Tuple[Env, AccountHistoryId]*;
ops __.bal : AccountHistory → Integer;
       __.prev : AccountHistory → AccountHistoryId;
%% We need also the overloaded versions on the tuple supersort:
       __.bal : Tuple[Env, AccountHistoryId] → Integer;
       __.prev : Tuple[Env, AccountHistoryId] → AccountHistoryId;
vars *ev:Env; self:AccountHistoryId*;
axioms
         %% "nil" is not an object:
         $<ev, self> \in AccountHistory \Rightarrow self \to notEmpty$;
         %% Consistency of environments:
         $<ev, self> \in AccountHistory \wedge <ev, self>.prev \to notEmpty \Rightarrow$
           $<ev, <ev, self>.prev> \in AccountHistory$;
         %% Invariant:
         $<ev, self> \in AccountHistory \Rightarrow <ev, self>.bal \geq 0$;
end

spec ACCOUNT_OO =
   ACCOUNTHISTORY_OO **then**
   sort *AccountId*;
   free type *Tuple[Env, AccountId]* ::=
                 <__, __>(__.env : Env; __.accountId : AccountId);
   sort *Account* < *Tuple[Env, AccountId]*;
   ops __.number : Account → Integer;
         __.history : Account → AccountHistoryId;
         __.getNumber() : Account → Integer;
         __.getBal() : Account → Integer;
         __.update(__) : Account × Integer → Account;
         __.undo() : Account → Account;
       %% Overloaded versions on the tuple supersort
       %% omitted for lack of space.
   vars *ev, ev':Env; self:AccountId; n:Integer*;
   axioms
         %% "nil" is not an object:
         $<ev, self> \in Account \Rightarrow self \to notEmpty$;
         %% Consistency of environments:
         $<ev, self> \in Account \Rightarrow$
           $<ev, <ev, self>.history> \in AccountHistory$;
         %% history has multiplicity 1, not 0..1:

$$<ev, self> \in Account \Rightarrow <ev, self>.history \rightarrow notEmpty;$$
%% getNumber
$$<ev, self> \in Account \Rightarrow$$
$$<ev, self>.getNumber() = <ev, self>.number;$$
%% getBal
$$<ev, self> \in Account \Rightarrow$$
$$<ev, self>.getBal() = <ev, <ev, self>.history>.bal;$$
%% update
$$<ev, self> \in Account \wedge ev' = <ev, self>.update(n).env$$
$$\wedge <ev, <ev, self>.history>.bal + n \geq 0 \Rightarrow$$
$$<ev, self>.update(n).accountId = self$$
$$\wedge <ev', self>.number = <ev, self>.number$$
$$\wedge \neg(<ev, <ev', self>.history> \in AccountHistory)$$
$$\wedge <ev', <ev', self>.history> \in AccountHistory$$
$$\wedge <ev', <ev', self>.history>.bal = <ev, <ev, self>.history>.bal + n$$
$$\wedge <ev', <ev', self>.history>.prev = <ev, self>.history ;$$
%% undo
$$<ev, self> \in Account \wedge ev' = <ev, self>.undo().env$$
$$\wedge <ev, <ev, self>.history>.prev \rightarrow notEmpty \Rightarrow$$
$$<ev, self>.undo().accountId = self$$
$$\wedge <ev', self>.number = <ev, self>.number$$
$$\wedge <ev', self>.history = <ev, <ev, self>.history>.prev ;$$
end

In addition to the above algebraic specifications, which result from our systematic translation of OCL expressions, it is necessary to take into account the so-called "Global Frame Assumption (GFA)" [L95]. For each non-query operation, it is always assumed that the state of any object not explicitly mentioned in the post-condition is left unchanged. This leads to a further algebraic specification SYSTEM_OO, which is just an enrichment of ACCOUNT_OO by the axioms expressing the global frame assumption for *update* and *undo*.

spec SYSTEM_OO =
  ACCOUNT_OO **then**
  axioms
    %% GFA for update:
    $$\forall self: AccountId; \; ev, ev': Env \bullet$$
    $$<ev, self> \in Account \wedge ev' = <ev, self>.update(n).env \Rightarrow$$
    $$\forall id: AccountId \bullet \neg(id = self) \wedge <ev, id> \in Account \Rightarrow$$
    $$<ev', id>.number = <ev, id>.number$$
    $$\wedge <ev', id>.history = <ev, id>.history$$
    $$\wedge \forall id: AccountHistoryId \bullet <ev, id> \in AccountHistory \Rightarrow$$
    $$<ev', id>.bal = <ev, id>.bal$$
    $$\wedge <ev', id>.prev = <ev, id>.prev ;$$
    %% Similarly for undo ...
end

412

5 Justification of the Implementation Relation

We now have all the necessary ingredients to make precise what it means for the classes Account and AccountHistory to be a correct realization of the IAccount interface.

We will say that a set of classes SC is a correct implementation (realization) of an interface I if, by definition, the algebraic specification SYSTEM_OO(SC) which expresses the semantics of the set of classes SC (together with their OCL constraints) is a correct implementation (in the algebraic sense, see, e.g., [ST97] and [BH98]) of the observational specification OBS_SP(I) which expresses the semantics of the interface I (together with its ICL behavioral constraints).

In our example, the classes Account and AccountHistory are a correct implementation of the IAccount interface if the specification SYSTEM_OO is a correct implementation of the observational specification IACCOUNT.

Once we have a formal correctness notion for implementation relations between interfaces and classes, the next step is to provide concrete means for proving the correctness of a given implementation. To do this, we can simply reuse the existing results from the literature (see, e.g., [BH98] and [HB99]). For instance, to prove the correctness of the implementation in our example, it is enough to show that the axioms of IACCOUNT (after renaming *IAccount* into *Account*) are logical consequences of the specification SYSTEM_OO, enriched by a suitable axiomatization of $\sim$. Here, this axiomatization of $\sim$ is as follows:[10]

$$\forall a, b : Account \bullet$$
$$a \sim b \Leftrightarrow (\bigwedge_{i \in \mathbb{N}} a.undo^i.getBal() = b.undo^i.getBal()$$
$$\land \ a.getNumber() = b.getNumber())$$

The proof of the correctness of the implementation relation can then be performed using standard proof techniques. Thereby it is indeed essential that we have used an observational approach for interface specifications since the implementation satisfies the required observational equation $a.update(n).undo() \sim a$ but does not satisfy this equation if "$\sim$" is replaced by "$=$" (since update creates a new object).

6 Related Work

There is a large amount of work formalizing class diagrams (see, e.g., [SF97], [EFLR99] using Z, [L95] using Z++ and RAL, [BGHRS97] using stream-oriented algebraic specifications, [WK96] using algebraic specification and rewriting logic, [WB98] using algebraic specification and dynamic logic, [GR98] translating TROLL to UML, and [O98] using operational semantics). Most of these approaches propose also formal notions of implementation (see, e.g., [EFLR99],

---

[10] After some obvious simplifications. Infinitary conjunctions are not directly expressible in CASL, and a further step, not explained here for lack of space (see [HB99]), is used to get rid of these.

[L95], [WK96]). However, none of these papers treats interfaces directly (although interface constraints can be expressed using dynamic logic or predicates on streams) and none of the papers uses OCL. To our knowledge, also the use of observational logic for expressing the behavior of interfaces is new, although similar semantic approaches such as coalgebra (see, e.g., [JR97]) and hidden algebra (see, e.g., [GM97]) have been used for modeling objects.

There are two other formal semantics of OCL: [RG98] gives a set-theoretical model of OCL with focus on associations and the iterate construct; the semantics of [BG98] is an algebraic semantics using Larch. In particular, the latter approach is similar to our semantics (although both semantics have been developed independently); the main difference is that we give an explicit representation of the set of existing instances as a subsort. In this way we can use the well-known semantic notions and proof methods of algebraic specifications for modeling the implementation of interfaces by classes. In contrast, [BG98] model the set of existing instances implicitly by a boolean function but they are not concerned with (such) implementations. Other differences are the choice of the specification language (Larch instead of CASL) and w.r.t. [RG98] the modeling of undefinedness: [RG98] introduces a bottom element; in [BG98] and also in our approach undefined elements are "unspecified".

7   Concluding Remarks

In this paper we have presented an approach for specifying UML interface constraints and for proving the correctness of implementation relations between interfaces and classes. Our approach includes a proposal for an interface constraint language OCL-like, a method for developing interface constraints and modular proof methods directly derived from the theory of behavioral algebraic specifications. Inheritance has not been considered due to space limitations. It can easily be included e.g. by using the subsorting approach described in [ACZ99]. Moreover, dynamic behavior described by interaction and state diagrams in UML has not been considered in this paper. More generally, an interface is realized by a subsystem consisting of one or several classes whose dynamic behavior is described by sequence, collaboration and state diagrams [JBR99]. It should be straightforward to integrate our class specifications and the OCL semantics into the stream-based system model of [KRB96]. Extending our behavioral interface specifications by sequence diagrams seems to be more difficult since it requires integrating the observational logic of [BH99] with nondeterministic rewriting (for a possible approach see [D96]) or once more with the system model of [KRB96].

Acknowledgment: Thanks go to the anonymous referees for many constructive remarks which helped us to improve the paper.

References

[ACZ99]   D. Ancona, M. Cerioli and E. Zucca. *A formal framework with late binding.* Proc. ETAPS/FASE'99, Springer LNCS 1577, pp. 30-44, 1999.

[BG98]   M. Brickford and D. Guaspari. *Lightweight analysis of UML*. Draft Technical Report, Odyssey Research Associates, July 1998.

[BH98]   M. Bidoit and R. Hennicker. *Modular correctness proofs of behavioural implementations*. Acta Informatica 35:951–1005, 1998.

[BH99]   M. Bidoit and R. Hennicker. *Observer complete definitions are behaviorally coherent*. Proc. FM'99 (UGM OBJ/CafeOBJ/MAUDE), to appear, 1999.

[BGHRS97] R. Breu, R. Grosu, F. Huber, B. Rumpe and W. Schwerin. *Towards a precise semantics for object-oriented modeling techniques*. Proc. ECOOP'97 Workshop Reader, Springer LNCS 1357, 1997.

[CoFI]   CoFI: *The Common Framework Initiative for algebraic specification and development (WWW pages)*. http://www.brics.dk/Projects/CoFI/.

[CoFI98]   CoFI Task Group on Language Design. CASL – *The CoFI algebraic specification language – Summary (version 1.0)*. http://www.brics.dk/Projects/CoFI/Documents/CASL/Summary/, 1998.

[D96]   R. Diaconescu. *Foundations of behavioural specifications in rewriting logic*. Proc. RWLW96, Electronic Notes in Theoretical Computer Science, Vol. 4, 1996.

[EFLR99]   A. Evans, R. France, K. Lano and B. Rumpe. *Developing UML as a formal modeling notation*. Proc. The Unified Modeling Language. <<UML>>'98: Beyond the Notation, Springer LNCS 1618, 1999.

[GR98]   M. Gogolla and M. Richters. *On combining semi-formal and formal object specification techniques*. Proc. WADT'97, Springer LNCS 1376, pp. 238–252, 1998.

[GM97]   J.A. Goguen and G. Malcolm. *A hidden agenda*. Report CS97-538, Univ. of Calif. at San Diego, 1997.

[HB99]   R. Hennicker and M. Bidoit. *Observational logic*. Proc. AMAST'98, Springer LNCS 1548, pp. 263–277, 1999.

[JBR99]   I. Jacobson, G. Booch and J. Rumbaugh. *The unified software development process*. Reading, Mass.: Addison-Wesley Longman, 1999.

[JR97]   B. Jacobs and J. Rutten. *A tutorial on (co)algebras and (co)induction*. EATCS Bulletin 62, pp. 222-259, 1997.

[KWC99]   A. Kleppe, J. Warmer and S. Cook. *Informal informality? The Object Constraint Language and its application in the UML metamodel*. Proc. The Unified Modeling Language. <<UML>>'98: Beyond the Notation, Springer LNCS 1618, 1999.

[KRB96]   C. Klein, B. Rumpe and M. Broy. *A stream-based mathematical model for distributed information processing systems - Syslab system model*. Proc. FMOODS '96, Chapmann & Hall, 1996.

[L95]   K. Lano. *Formal object-oriented development*. Springer, 1995.

[M88]   B. Meyer. *Object-oriented software construction*. Prentice Hall International, 1988.

[O98]   G. Overgaard. *A formal approach to relationships in the Unified Modeling Language*. Proc. of ICSE'98 (Workshop on Precise Semantics for Software Modeling Techniques). IEEE Computer Society, 1998.

[Rat97]   Rational. *Unified Modeling Language: Semantics, Version 1.1*. Rational Software Corporation. http://www.rational.com/uml/, 1997.

[Rat99]   Rational. *Object Constraint Language, Version 1.3*. Rational Software Corporation. http://www.rational.com/, 1999.

[RG98]   M. Richters and M. Gogolla. *On formalizing the UML Object Constraint Language OCL*. Proc. 17th Conceptual Modeling – ER'98, Springer LNCS 1507, 1998.

[ST97]   D. Sannella and A. Tarlecki. *Essential concepts of algebraic specification and program development.* Formal Aspects of Computing 9:229–269, 1997.

[SF97]   M. Shroff and R.B. France. *Towards a formalization of UML class structures in Z.* Proc. COMPSAC'97, IEEE, pp. 646-651, 1997.

[SW98]   D. D'Souza and A.C. Wills. *Objects, components and frameworks with UML: the Catalysis approach.* Addison-Wesley, 1998.

[WK99]   J. Warmer and A. Kleppe. *The Object Constraint Language: precise modeling with UML.* Reading, Mass.: Addison-Wesley Longman, 1999.

[WB98]   R. Wieringa and J. Broersen. *A minimal transition system semantics for lightweight class- and behavior diagrams.* Proc. of ICSE'98 (Workshop on Precise Semantics for Software Modeling Techniques). IEEE Computer Society, 1998.

[W90]   M. Wirsing. *Algebraic specification.* Handbook of Theoretical Computer Science, North-Holland, pp. 675-788, 1990.

[WK96]   M. Wirsing and A. Knapp. *A formal approach to object-oriented software engineering.* Proc. RWLW96, Electronic Notes in Theoretical Computer Science, Vol. 4, 1996.

Generating Tests from UML Specifications

Jeff Offutt and Aynur Abdurazik *

George Mason University, Fairfax VA 22030, USA

Abstract. Although most industry testing of complex software is conducted at the system level, most formal research has focused on the unit level. As a result, most system level testing techniques are only described informally. This paper presents a novel technique that adapts pre-defined state-based specification test data generation criteria to generate test cases from UML statecharts. UML statecharts provide a solid basis for test generation in a form that can be easily manipulated. This technique includes coverage criteria that enable highly effective tests to be developed. To demonstrate this technique, a tool has been developed that uses UML statecharts produced by Rational Software Corporation's Rational Rose tool to generate test data. Experimental results from using this tool are presented.

1 Introduction

There is an increasing need for effective testing of software for complex safety-critical applications, such as avionics, medical, and other control systems. These software systems usually have clear high level descriptions, sometimes in formal representations. Unfortunately, most system level testing techniques are only described informally. This paper is part of a project that is attempting to provide a solid foundation for generating tests from system level software specifications via new coverage criteria. Formal coverage criteria offer testers ways to decide what test inputs to use during testing, making it more likely that the testers will find any faults in the software and providing greater assurance that the software is of high quality and reliability. Such criteria also provide stopping rules and repeatability.

Although UML provides a powerful mechanism for describing software that is safety-critical or that must be highly reliable, it still is not in widespread use. One advantage of using software description languages such as UML is that they provide a convenient basis for selecting tests. The purpose of this research project is to take advantage of UML to produce highly effective software system-level tests. Although UML is not completely formalized, certain aspects of the language are precise enough to be utilized for test generation, in particular the UML Statecharts.

The research presented in this paper is part of a long term project that is looking at ways to generate tests from specifications. This paper presents formal

---
* This work is supported in part by the U.S. National Science Foundation under grant CCR-98-04111 and in part by Rockwell Collins, Inc.

criteria for developing test inputs from UML statecharts. An overview of software testing is given in the next section, then the criteria are described, algorithms and a proof-of-concept tool are presented, and finally results from an empirical evaluation are given.

2 Software Testing

Software testing includes executing a program on a set of test cases and comparing the actual results with the expected results. Testing and test design, as parts of quality assurance, should also focus on fault prevention. To the extent that testing and test design do not prevent faults, they should be able to discover symptoms caused by faults. Finally, tests should provide clear diagnoses so that faults can be easily corrected [3].

This paper uses the following definitions. A *test* or *test case* is a general software artifact that includes test case input values, expected outputs for the test case, and any inputs that are necessary to put the software system into the state that is appropriate for the test input values. A test specification language (TSL) is a language that can be used to describe all components of a test case. The components considered here are *test case values*, *prefix values*, *verify values*, *exit commands*, and *expected outputs*. Test case values directly satisfy the test requirements, and the other components supply supporting values. A *test case value* is the essential part of a test case, which comes from the test requirements. It may be a command, user inputs, or software function and values for its parameters. In state-based software, test case values are usually derived directly from triggering events and preconditions for transitions. A test case *prefix value* includes all inputs necessary to place the software system into the appropriate state for running the test case values. Any inputs that are necessary to show the results are *verify values*, and *exit commands* terminate the execution of the software. *Expected outputs* are the outputs of the test case on a correct version of the software.

Test requirements are specific things that must be satisfied or covered during testing, for example, reaching statements are the requirements for statement coverage. *Test specifications* are specific descriptions of test cases, often associated with test requirements or criteria. For statement coverage, test specifications are the conditions necessary to reach a statement. A *testing criterion* is a rule or collection of rules that impose test requirements on a set of test cases. A *testing technique* guides the tester through the testing process by including a testing criterion and a process for creating test case values.

Test engineers measure the extent to which a criterion is satisfied in terms of *coverage*, which is the percent of requirements that are satisfied. There are various ways to classify adequacy criteria. One of the most common is by the source of information used to specify testing requirements and in the measurement of test adequacy. Hence, an adequacy criterion can be specification-based or program-based.

A *specification-based* criterion specifies the required testing in terms of identified features of the specifications of the software, so that a test set is adequate if all the identified features have been fully exercised. Here the specifications are used to produce test cases, as well as to produce the program. A *program-based* criterion specifies testing requirements in terms of the program under test and decides if a test set is adequate according to whether the program has been thoroughly exercised. For example, if the criterion of branch testing is used, the tests are required to cover each branch in the program.

There are two main roles a specification can play in software testing [9]. The first is to provide the necessary information to check whether the output of the program is correct [8]. Checking the correctness of program outputs is known as the *oracle problem*. The second is to provide information to select test cases and to measure test adequacy [12].

Specification-based testing (SBT) offers many advantages in software testing. The (formal) specification of a software product can be used as a guide for designing functional tests for the product. The specification precisely defines fundamental aspects of the software, while more detailed and structural information is omitted. Thus, the tester has the essential information about the product's functionality without having to extract it from inessential details.

Formal specifications provide a simpler, structured, and more formal approach to the development of functional tests than using non-formal specifications. One significance of producing tests from specifications is that the tests can be created earlier in the development process, and be ready for execution **before** the program is finished. Additionally, when the tests are generated, the test engineer will often find inconsistencies and ambiguities in the specifications, allowing the specifications to be improved before the program is written.

3   Test Data Generation Techniques Based on UML

Offutt has previously developed criteria for generating test data from SCR specifications [7] and from SOFL specifications [6]. These techniques have been adapted to UML Statecharts, and a tool has been built that automatically generates tests from UML statecharts created using Rational Software Corporation's Rational Rose tool [4].

UML can be used to specify a wide range of aspects of a system. UML statecharts are based on finite state machines using an extended Harel state chart notation, and are used to represent the behavior of an object. As they are the most formalizable aspects of UML, statecharts provide a natural basis for test data generation.

The *state* of an object is the combination of all attribute values and objects the object contains. The dynamics of objects are modeled through transitions among states. The UML syntax for transitions is:

```
event-name (parameters) [guard] / action list ^ event list
```

Event-name is a label for the transition, and parameters are the variables associated with the event that must be triggered for the transition to be taken place. Guard is a precondition that controls whether the transition is taken or not, and action list and event list define changes in the software that occur as a result of the transition.

UML categorizes transitions into five types: *high-level transitions*, *compound transitions*, *internal transitions*, *completion transitions*, and *enabled transitions*. This research is only interested in enabled transitions. The previous testing model was based primarily on predicate satisfaction. In UML, the enabled transitions are similar to transitions that are based on the notion of *predicate satisfaction*. An *enabled transition* is enabled by an event, and it originates from an active state. An enabled transition is triggered when there exists at least one full path from the source state to the target state.

Four kinds of events can be specified in UML: call events, signal events, time events, and change events. A *call event* represents the reception of a request to synchronously invoke a specific operation. A *signal event* represents the reception of a particular (synchronous) signal. A *time event* represents the passage of a designated period of time after a designated event (often the entry of the current state) or the occurrence of a given date and time. A *change event* models an event that occurs when an explicit boolean expression becomes true as a result of a change in value of one or more attributes or associations.

Since change events can be expressed as predicates, these are used as the basis for generating tests. A change event is raised implicitly when the associated predicate becomes true. Change events are different from guards. A guard is only evaluated when an event is dispatched whereas, conceptually, the boolean expression associated with a change event is evaluated continuously until it becomes true. The event that is generated remains until it is consumed even if the boolean expression changes to false. In UML, a change event is modeled by using the keyword *when* followed by a Boolean expression. Figure 1 illustrates a change event. This change event indicates that the function selfTest() is called at 11:49 PM.

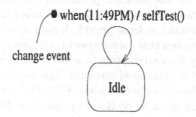

when(11:49PM) / selfTest()

change event

Idle

Fig. 1. Change Events

Change event enabled transitions are used to define four levels of testing: (1) the transition coverage level, (2) the full predicate coverage level, (3) the transition-pair coverage level, and (4) the complete sequence level.

It is possible to apply all levels, or to choose a level based on a cost/benefit tradeoff. The first two are related; the transition coverage level requires many fewer test cases than the full predicate coverage level, but if the full predicate coverage level is used, the tests will also satisfy the transition coverage level (full predicate coverage *subsumes* transition coverage). Thus only one of these two should be used. The latter two levels are meant to be independent; transition-pair coverage is intended to check the interfaces among states, and complete sequence testing is intended to check the software by executing the software through complete execution paths. As it happens, transition-pair coverage subsumes transition coverage, but they are designed to test the software in very different ways.

3.1 Transition Coverage Level

It seems reasonable to expect that to test the software adequately, the tester should at minimum use tests that cause every transition in every statechart to be taken. This level requires just that, by requiring test cases that satisfy each precondition in the specification at least once. In the criteria definitions, T is a set of test cases, and SG is a *specification graph*, a graph that represents the transitions in a statechart. Although the tests are intended to be executed on an implementation of the specification, we say that a test *traverses* a transition to indicate that, from a modeling perspective, the test causes the transition's predicate to be true, and the implementation will change from the transition's pre-state to its post-state.

Coverage Level 1, Transition Coverage: *the test set T must satisfy every transition in the SG.*

3.2 Full Predicate Coverage Level

Small inaccuracies in the specification predicates can lead to major problems in the software. The full predicate coverage level takes the philosophy that to test the software, we should at least provide inputs to test each clause in each predicate. This level requires that each clause in each predicate on each transition is tested independently, thus attempting to address the question of whether each clause is necessary and is formulated correctly. This paper follows the definitions in DO178B [10]. The Boolean operators are AND, OR, and NOT. A *clause* is a Boolean expression that contains no Boolean operators. For example, relational expressions and Boolean variables are clauses. A *predicate* is a Boolean expression that is composed of clauses and zero or more Boolean operators. A predicate without a Boolean operator is also a clause. If a clause appears more than once in a predicate, each occurrence is a distinct clause. (DO178B uses the terms "condition" and "decision", but the more common "clause" and "predicate" are used here.)

Full predicate coverage is based on the philosophy that each clause should be tested independently, that is, while not being influenced by the other clauses. In other words, each clause in each predicate on every transition must independently affect the value of the predicate.

Coverage Level 2, Full Predicate Coverage: *for each predicate P on each transition, T must include tests that cause each clause c in P to result in a pair of outcomes where the value of P is directly correlated with the value of c.*

In this definition, "directly correlated" means that c controls the value of P, that is, one of two situations occurs. Either c and P have the same value (c is true implies P is true and c is false implies P is false), or c and P have opposite values (c is true implies P is false and c is false implies P is true). This explicitly disallows cases such as c is true implies P is true and c is false implies P is true.

Note that if full predicate coverage is achieved, transition coverage will also be achieved. To satisfy the requirement that the *test clause* controls the value of the predicate, other clauses in the predicate must be either **True** or **False**. For example, if the predicate is $(X \wedge Y)$, and the test clause is X, then Y must be **True**. Likewise, if the predicate is $(X \vee Y)$, Y must be **False**. More details, including how this can be accomplished with general predicates, are given elsewhere [6, 7].

3.3 Transition-Pair Coverage Level

The previous testing levels test transitions independently, but do not test sequences of state transitions. This level requires that pairs of transitions be taken.

Coverage Level 3, Transition-pair Coverage: *For each pair of adjacent transitions $S_i : S_j$ and $S_j : S_k$ in SG, T contains a test that traverses the pair of transitions in sequence.*

3.4 Complete Sequence Level

It seems very unlikely that any successful test method could be based on purely mechanical methods; at some point the experience and knowledge of the test engineer must be used. Particularly at the system level, effective testing probably requires detailed domain knowledge. A *complete sequence* is a sequence of state transitions that form a complete practical use of the system. In most realistic applications, the number of possible sequences is too large to choose all complete sequences. In many cases, the number of complete sequences is infinite.

Coverage Level 4, Complete Sequence: *The test engineer must define meaningful sequences of transitions on the statechart diagram by choosing sequences of states that should be entered.*

4 A Rose-based Test Data Generation Tool

It is possible to automate almost all of the steps of generating test data for these criteria. If a machine-readable form of the specifications is available, the transition conditions can be read directly. Test requirements take the form of partial truth tables defined on state transition predicates and pairs of state transition predicates. Given a formal functional specification, most if not all of these test requirements can be generated automatically. The most complicated part of the test case is the test prefix, which includes inputs necessary to put the system into a particular pre-state. Test prefixes are handled by creating a specification graph, which contains all the states and their transition relationships. Prefixes can then be generated automatically by walking through the graph.

To evaluate these criteria, we have developed a proof-of-concept test data generation tool. This tool, UMLTEST, is integrated with the Rational Rose case tool [4]. Rose saves UML specifications in a plain text format, which can be easily read. UMLTEST generates test cases at full predicate and transition-pair levels. Transition coverage is subsumed by full predicate coverage, and complete sequence coverage is not fully automatable.

UMLTEST parses a Rose specification file (called an MDL file) to get the semantic meanings of the specifications. MDL files store specification information from different perspectives. There are two main categories of information, logical and physical. The specification itself is grouped into two packages: *use cases* and *object collaboration diagrams* are packaged into *Use Case Packages*, and *class diagrams* and *state transition diagrams* are packaged into *Logical Views*. Figure 2 shows the internal structure of the class diagram and state transition diagram in an MDL file.

UMLTEST makes several assumptions about the UML specification file input. It relies on the Object Constraint Language (OCL), which is an expression language that enables constraints to be described on object-oriented models and other modeling artifacts [11]. A *constraint* is a restriction on one or more values within an object-oriented model or system. OCL is part of UML and is the standard for specifying invariants, preconditions, postconditions, and other kinds of constraints. UMLTEST makes several assumptions:

- All transitions are triggered by change events.
- Events and conditions are expressed through boolean type class attributes.
- The specification is written strictly following the UML notation. For example, *when* denotes a change event and conditions are in solid brackets ([]). Because there is no way to check whether a specification is well-formed or consistent, this assumption cannot be checked. The OCL does not have a mechanism to enforce its syntactic rules on all parts of the UML specification. Also, Rose does not have a function to write the specification in OCL.
- State transitions are deterministic.

Figure 3 is a UML class diagram describing UMLTEST. Classes are represented as boxes that have three parts, the class name, data members that are declared in the class, and methods of the class. The main entry point (UMLTest)

```
Logical Models
    object Class
        classAttributes

        -------------- State Transition: Logical -------------

        State Machine
            object State /* StartState, Normal, EndState */
                State Transition
                State Machine
                    Object State /* Normal */

        -------------- State Transition: Physical ------------

        State Diagram
            State View /* StartState, Normal, EndState */
            Transition View

    object Association
        object Role
Logical Presentations
    object ClassDiagram /* with grouping and name */
        object ClassView
            Association View
            Role View
            Inheritance View
```

Fig. 2. Structure of MDL File for Class Diagrams and State Transition Diagrams

has three objects, (1) a UML specification parser, (2) a full predicate test case generator, and (3) a transition-pair test case generator.

UMLSpecParser reads a UML specification text file, parses it, and generates state transition table(s) for classes that have state machines. FullPredicate takes a state transition table as an input, generates test cases for the full predicate coverage criterion, and saves the test cases into an ASCII text file. TransitionPair takes a state transition table as input, generates test cases for the transition-pair coverage criterion, and saves the test cases in an ASCII text file.

4.1 Algorithms

This section presents two algorithms used in UMLTEST. Algorithms were developed to generate test cases for the *full predicate coverage, transition coverage,*

424

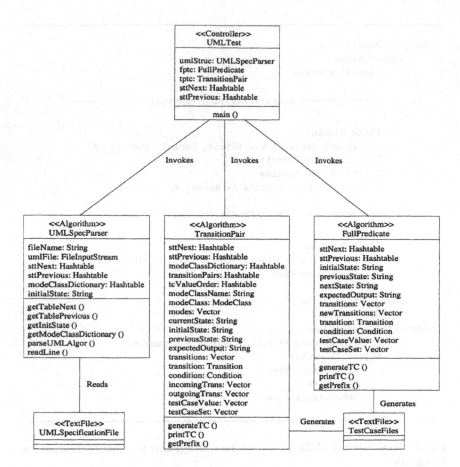

Fig. 3. Class Diagram for UMLTEST **Tool**

and *transition-pair coverage* criteria. A *prefix generation* algorithm was used in test data generation algorithms to create the values necessary to reach a particular state. Because of space limitations, only the full predicate coverage and prefix generation algorithms are presented. The additional algorithms are available in a technical report [1].

Get Prefix Algorithm. Figure 4 gives an algorithm for generating test prefix values from a specification graph. The input is a state (the *test state*) in the graph, and it finds a path from an initial state in the graph to the test state.

Generate Full-Predicate Coverage Test Cases Algorithm. Figure 5 gives an algorithm for generating test cases for the *full predicate coverage* criterion. Algorithm *GenerateFullPredicateCoverageTCs* takes a *state transition table* as

```
algorithm:          GetPrefix (State)
input:              Test state of a transition.
output:             Inputs to get to the given state.
output criteria:    No redundant inputs.
declare:            prefix (s) -- Inputs to reach state s.
                    incomingTrans (s) -- The set of incoming transitions.
                    event (otr) -- Trigger event for transition otr.
                    whenCondition (otr) -- Precondition for otr.
                    nextState (otr) -- Next state for transition otr.
                    expectedOutput -- Post-state after transition.
                    TCValue (otr) -- Value assignments for the trigger
                    event and when condition variables for otr.

GetPrefix (State)
BEGIN -- Algorithm GetPrefix
    s = State
    prefixStates = prefixStates ∪ s
    WHILE (s IS NOT initial state) LOOP
        get incomingTrans (s)
        prefix (s) = EMPTY
        IF (∃ transition itr ∈ incomingTrans (s) such that
        prevState (itr) = initialState) THEN
            s = prevState (itr)
            prefixStates = prefixStates ∪ s
            EXIT
        ELSE
            s = prevState (itr) such that itr ∈ incomingTrans (s) ∧
                prevState (itr) ∉ prefixStates
            prefixStates = prefixStates ∪ s
        END IF
    END LOOP
END Algorithm GetPrefix
```

Fig. 4. The GetPrefix Algorithm

input, and generates test cases for the full predicate coverage criterion. It processes each outgoing transition of each source state, generates a test case that
makes the transition valid, and then generates test cases that make the transition
invalid. When generating a test case, GetPrefix () is used to obtain prefixes
to reach the source state of a transition. Then each variable in the transition
predicate is assigned a test case value. To avoid redundant test case value assignments, those variables that already have assigned values in the prefixes are not
considered in the test case value assignment process. After all test case values
are generated, an additional algorithm is run on the test cases to identify and
remove redundant test cases.

```
algorithm:        GenerateFullPredicateCoverageTCs (STTable)
input:            State transition table.
output:           Test cases for full predicate coverage.
output criteria:  Test cases contain prefix, test case values,
                  and expected output.
assumption:       Clauses are disjunctive.
                  No redundant assignments in prefix and test cases.
declare:          prefix (s) -- Inputs to get to the state s.
                  outgoingTrans (s) -- Set of outgoing transitions.
                  event (otr) -- Trigger event for transition otr.
                  whenCondition (otr) -- Precondition for otr.
                  nextState (otr) -- Next state for transition otr.
                  expectedOutput -- Post-state after transition.
                  TCValue (otr) -- Value assignments for the trigger
                  event and when condition variables for otr.
```

```
GenerateFullPredicateCoverageTCs (STTable)
BEGIN -- Algorithm GenerateFullPredicateCoverageTCs
   TestCaseSet = EMPTY
   FOR EACH source state s in STTable
      prefix (s) = GetPrefix (s)
      get outgoingTrans (s)
      -- Generate one test case for each transition
      FOR EACH outgoing transition otr ∈ outgoingTrans (s)
         expectedOutput = nextState (otr)
         TCValue (otr) = EMPTY
         get event (otr) and whenConditions (otr)
         -- Check for redundancy
         IF (¬ ∃ a condition variable var ∈ prefix (s) s.t.
         var.name = event (otr).name ∧
         var.value = event (otr).value)
            TCValue (otr) = TCValue (otr) ∪
            {(event (otr).name, event (otr).beforeValue)}
         END IF
         -- Assign value for clauses in when condition
         FOR EACH clauseᵢ in whenConditions (otr)
            IF (¬ ∃ a condition variable var ∈ prefix (s) s.t.
            var.name = clauseᵢ.name ∧ var.value = clauseᵢ.value)
               TCValue (otr) = TCValue (otr) ∪
                  {(clauseᵢ.name, clauseᵢ.value)}
            END IF
         END FOR
         TCValue (otr) = TCValue (otr) ∪ {(event (otr).name,
            event (otr).afterValue)}
         TestCaseSet = TestCaseSet ∪ {(prefix (s), TCValue (otr),
            expectedOutput}
```

Fig. 5. The GenerateFullPredicateCoverageTCs Algorithm

```
        -- get test cases for invalid transitions
        expectedOutput = current state s
        FOR EACH variable var in TCValue (otr)
            TCValue (otr) = TCValue (otr) - {(var.name, var.value)}
            var.value = ¬var.value
            TCValue (otr) = TCValue (otr) ∪ {(var.name, var.value)}
            TestCaseSet = TestCaseSet ∪ {(prefix (s),
                TCValue (otr), expectedOutput)}
        END FOR
    END FOR
  END FOR
END Algorithm GenerateFullPredicateCoverageTCs
```

Fig. 5. The GenerateFullPredicateCoverageTCs Algorithm - continued

5 Empirical Evaluation

An empirical study has been undertaken to demonstrate the feasibility of these criteria. The goal was to demonstrate that the specification-based criteria can be effectively used. Tests were created and then measured on the basis of their fault-detection abilities. One moderate size program was used (cruise control [2, 5]), representative faults were seeded, and test cases were generated by UMLTEST.

UMLTEST generated 54 full predicate test cases; after redundant test cases were eliminated, 34 remained to be used during testing. UMLTEST generated 34 test cases for transition-pair coverage.

A model of the cruise control problem was implemented in about 400 lines of C. Cruise has seven functions, 184 blocks, and 174 decisions. Twenty-five faults were created by hand and were inserted into the program by creating separate versions. Most of these faults are small modifications (such as changing a variable name or arithmetic operator), and most were in the logic that implemented the state machine. Four were naturally occurring faults, made during initial implementation.

For comparison, tests were created to satisfy statement coverage. These tests were generated by hand to execute every statement in the (original) implementation at least once, and 27 tests were needed. To eliminate bias, these tests were not generated by either author.

Figure 6 shows the fault coverage percentage for full predicate, transition-pair, and statement coverage criteria test cases. The full predicate tests were able to find all the faults, and the transition-pair tests did significantly better than the statement coverage tests.

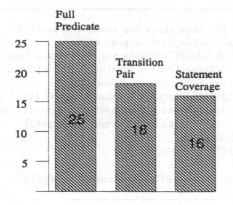

Fault Coverage Results

Fig. 6. Test Data Results

6   Conclusions

This paper presented three results. First, a novel collection of testing criteria for UML was described. These criteria allow software system-level tests to be derived from UML statechart diagrams. As far as we know, this is the first formal testing technique that is based on UML. Second, a proof-of-concept automatic test data generation tool has been developed. This solves one of the most important problems in software testing – developing the actual tests. This tool, UMLTEST, is integrated with the Rational Rose tool and is completely automated. This is the first tool that we know of that can automatically generate tests from UML specifications. Third, empirical results from using UMLTEST to evaluate the testing criteria were given. The results indicate that highly effective tests can be automatically generated for system level testing. This provides strong motivation for using UML, and raises the possibility of developing software that is more reliable than what is being developed by current means.

There are a number of limitations of this current work. The most important is that tests were derived from statecharts and enabled transitions; other aspects of UML have not yet been addressed. By not using other types of transitions, certain states may never be entered. We are currently extending our test criteria to incorporate other types of transitions. We also hope to incorporate other parts of UML if the semantics can be sufficiently formalized, including UML class models. Although the test model is general, the current tool requires all variables to be boolean. Extending the tool to generate values for variables of other types is straightforward, but tedious, so has not been done for this research tool. Finally, the empirical study was limited to one program and one set of tests for each testing technique. We are currently working on extending this study to compare with multiple programs.

429

References

1. Aynur Abdurazik and Jeff Offutt. Generating test cases from UML specifications. Technical report ISE-TR-99-105, Department of Information and Software Engineering, George Mason University, Fairfax VA, 1999. http://www.ise.gmu.edu/techrep.

2. J. M. Atlee. Native model-checking of SCR requirements. In *Fourth International SCR Workshop*, November 1994.

3. B. Beizer. *Software Testing Techniques*. Van Nostrand Reinhold, Inc, New York NY, 2nd edition, 1990. ISBN 0-442-20672-0.

4. Rational Software Corporation. *Rational Rose 98: Using Rational Rose*. Rational Rose Corporation, Cupertina CA, 1998.

5. Zhenyi Jin. Deriving mode invariants from SCR specifications. In *Proceedings of Second IEEE International Conference on Engineering of Complex Computer Systems*, pages 514–521, Montreal, Canada, October 1996. IEEE Computer Society.

6. Jeff Offutt and Shaoying Liu. Generating test data from SOFL specifications. *The Journal of Systems and Software*, 1999. To appear.

7. Jeff Offutt, Yiwei Xiong, and Shaoying Liu. Criteria for generating specification-based tests. In *Proceedings of the Fifth IEEE International Conference on Engineering of Complex Computer Systems (ICECCS '99)*, Las Vegas, NV, October 1999. IEEE Computer Society Press.

8. Andy Podgurski and Lori A. Clarke. A formal model of program dependences and its implications for software testing, debugging, and maintenance. *IEEE Transactions on Software Engineering*, 16(9):965–979, September 1990.

9. D. J. Richardson, O. O'Malley, and C. Tittle. Approaches to specification-based testing. In *Proceedings of the Third Symposium on Software Testing, Analysis, and Verification*, pages 86–96, Key West, Florida, December 1989. ACM SIGSOFT'89.

10. RTCA Committee SC-167. Software considerations in airborne systems and equipment certification, Seventh draft to Do-178A/ED-12A, July 1992.

11. Jos Warmer and Anneke Kleppe. *The Object Constraint Language*. Addison-Wesley, 1999. ISBN 0-201-37940-6.

12. Hong Zhu, Patrick A. V. Hall, and John H. R. May. Software unit test coverage and adequacy. *ACM Computing Surveys*, 29(4):366–427, December 1997.

Formalising UML State Machines for Model Checking

Johan Lilius and Iván Porres Paltor

Turku Centre for Computer Science (TUCS),
Lemminkäisenkatu 14 A, FIN-20520 Turku, Finland,
{Johan.Lilius, Ivan.Porres}@abo.fi

Abstract. The paper discusses a complete formalisation of UML state machine semantics. This formalisation is given in terms of an operational semantics and it can be used as the basis for code-generation, simulation and verification tools for UML Statecharts diagrams. The formalisation is done in two steps. First, the structure of a UML state machine is translated into a term rewriting system. In the second step, the operational semantics of state machines is defined. In addition, some problematic situations that may arise are discussed. Our formalisation is able to deal with all the features of UML state machines and it has been implemented in the vUML tool, a tool for model-checking UML models.

1 Introduction

The Unified Modelling Language (UML) is a standardised notation for specifying object-oriented software systems [1, 2, 15]. A UML model is a set of diagrams describing and documenting the structure, behaviour and usage of a software system. UML is used to model all kinds of software systems, including concurrent and embedded systems. There are commercial modelling tools available on the market to help the designer in creating the UML models and these tools can also generate program code from some diagrams of the model. UML is an expressive and rich language, but UML models must still be verified, since a model may contain behaviours not expected by the designer.

France et al. [4] already discuss the need of formal semantics for UML. In this paper, we discuss the formalisation of UML state machines. The main motivation of this work is to give a clear and unequivocal description of the UML state machines. Such a description is needed as a reference model for implementing tools for code generation, simulation and verification of UML Statecharts.

Our formalisation consists of two parts. The reason for this is that the operational semantics of UML state machines is non-trivial (as can be seen in Figure 7), with lots of features that break the thread-of-control. Thus it makes sense to study the formalisation of the structure of UML state machines and the operational semantics separately. The main contributions of this paper are:

1. a formalisation of the structure of UML state machines that is both simple and declarative, and allows us to easily formulate the transition selection algorithm, and

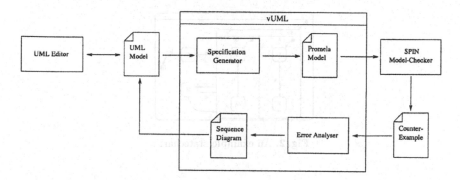

Fig. 1. Using vUML

2. a formalisation of the operational semantics of UML state machines, that to our knowledge, it is the first formalisation that covers all the features of UML state machines.

The semantics presented in this paper has been used to develop the vUML tool [11]. vUML is a tool for model checking[1] UML models that has been designed to be as automatic and transparent to the designer as possible. We wanted a tool that verifies the same UML model that is used to analyse, design, document and implement the software. Figure 1 shows how vUML works. The UML model is first translated into PROMELA [6], the input language of SPIN [7]. SPIN is a widely used model checker that is freely available over the Internet. If SPIN finds an error in the system, it produces a counter-example (an error trace). This counter-example is translated into a UML sequence diagram and then displayed to the user. So, although the tool internally uses the SPIN model checker to perform the verification, the designer does not have to know how to use SPIN or the PROMELA language in order to use vUML. The only requirement is that he must understand the errors that vUML is able to find. Currently these are: deadlocks, reaching an invalid state, violating a constraint on an object, sending an event to a terminated object, overrunning the input queue of an object, overrunning the deferred event queue and livelocks.

The main contents of the paper is in section 2. There we first describe the formalisation of state machines, and then sketch how collaboration diagrams are formalised (a more complete description of this can be found in the full paper [10]). We also want to take the opportunity to discuss some of the unexpected results that may arise in the semantics. We finish the paper with some conclusions and comparison with related work.

---

[1] See [3] for an overview of model checking.

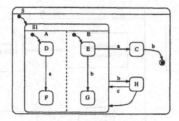

Fig. 2. An example statechart

2 Formal Semantics

In this section, we propose a formalisation of the state machines as defined in the UML standard semantics [1]. In UML, each class has an optional state machine that describes the behaviour of its instances (the objects). This state machine receives events from the environment and reacts to them. The reactions include sending new events to other objects and executing internal methods on the object. The vUML tool analyses the behaviour of a set of interacting objects. The interactions are described with a collaboration diagram that instantiates associations between the objects into links, where all events will be sent along these links.

Our formalisation consists of two parts. We first explain our formalisation of the structure of a UML state machine. This formalisation is needed for the formulation of the transition selection algorithm in the operational semantics of the state machine (c.f. Def. 8). Then we discuss the operational semantics (c.f. Fig. 7), which also defines the execution algorithm that has been implemented in vUML. We continue by describing how a collaboration diagram is interpreted as a network of communicating objects, and close this section with a few observations on the behavior of UML state machines that follows from the semantics. We shall assume that the reader is familiar with UML state machines as described in e.g. [2].

2.1 Formal description of state machines

We assume that we are given a state machine SM. The states Σ of the state machine can be partitioned into three disjoint sets: *compound states* Σ_{cs}, *simple states* Σ_{ss} and *final states* Σ_{fs}, e.g. $\Sigma = \Sigma_{cs} \cup \Sigma_{ss} \cup \Sigma_{fs}$. Final states do not usually have names in the graphical representation, but we shall use a default naming scheme $fs_1, \dots, fs_n$ where $n = |\Sigma_{fs}|$. An *initial state* is graphically depicted as a state with an incoming arrow from a black dot. The set of initial states Σ_{is} is a subset of $\Sigma_{cs} \cup \Sigma_{ss}$. Note that the single top state is always an initial state. The state machine in figure 2 will be our running example. The sets Σ_{cs}, Σ_{ss} and Σ_{fs} are: $\Sigma_{cs} = \{S, S_1, A, B\}$, $\Sigma_{ss} = \{C, D, E, F, G, H\}$, and $\Sigma_{fs} = \{fs_1\}$, while $\Sigma_{is} = \{S, S_1, D, E\}$.

The current state of a UML state machine is defined as a tree of states starting with the single top state at the root down to individual simple states at the leaves. Such as state is called a *state configuration*. We will encode state configurations as terms over a signature, where we will interpret the sets Σ_{cs} and Σ_{ss} as operator symbols. The *arity* of a state is defined as the number of orthogonal regions contained in it: $arity : \Sigma_{cs} \cup \Sigma_{ss} \to \mathbb{N}$. Thus in our example in figure 2 we have:

s	S	S_1	A	B	C	D	E	F	G	H
$arity(s)$	1	2	1	1	0	0	0	0	0	0

Let X be a countable set of variables. Then given the set of symbols Σ, we can form the set of *linear terms* $T_\Sigma(X)$ as the smallest set satisfying:

1. $\Sigma_{ss} \cup X \subseteq T_\Sigma(X)$
2. $op(t_1, \ldots, t_n) \in T_\Sigma(X)$, where $op \in \Sigma_{cs}$, $n = arity(op)$ and $t_1, \ldots, t_n \in T_\Sigma(X)$
3. $\forall t \in T_\Sigma(X), \forall x \in Var(t), \#(x, t) = 1$

Definition 1. *A state configuration of a state machine* SM *is a term in* T_Σ.

The last condition restricts the set of terms to those where each variable occurs only one time. We call such a term *linear*. $T_\Sigma(\emptyset)$ is called the set of *linear ground terms* and simply denoted T_Σ. From now on, we shall use term as a shorthand for linear term.

We will need to reference the set of symbols $symb(s)$ of a term $s \in T_\Sigma$.

$$symb(s) = \{s\} \text{ iff } s \in T_{\Sigma_{ss}} \vee s \in T_{\Sigma_{fs}}$$
$$symb(s(s_1, \ldots, s_n)) = \{s\} \cup_{i=1..n} symb(s_i)$$

Each state in a state machine may have an *entry*, *exit*, and *activity* action associated with it. We shall assume that these are defined in some language $\mathcal{L}_a$. In the case of vUML, this language is PROMELA. Moreover, a state may have a number of *internal transitions*. These concepts will be discussed more in detail in the context of the operational semantics below.

A transition in a UML state machine consists of the following elements: *source:* the originating state vertex, *target:* the target state vertex, *trigger:* The event that fires the transition, *guard:* a boolean predicate that must be true for the transition to be fired (there can be at most one guard), and *effect:* an optional action to be performed when the transition fires. In a UML Statecharts diagram, transitions can be distinguished graphically. In the formalisation we will introduce a set of transition names Tn to be able to make this distinction. State machines communicate by sending events to each other. We shall assume a global set of events E. The set of trigger events $E_t \subseteq E$ of the state machine SM is the union of all trigger events appearing on the transitions of the state machine. The UML standard also allows triggerless transitions in the statechart. Such transitions, called *completion transitions* have an implicit trigger, the *completion event*, which is generated when all transition and entry actions

$$t_1 : S(S_1(A(D), B(X))) \xrightarrow{a} S(S_1(A(F), B(X)))$$

$$t_2 : S(S_1(A(X), B(E))) \xrightarrow{b} S(S_1(A(X), B(G)))$$

$$t_3 : S(S_1(A(X), B(E))) \xrightarrow{a} S(C)$$

$$t_4 : S(S_1(X, Y)) \xrightarrow{b} S(H)$$

$$t_5 : S(H) \xrightarrow{c} S(S_1(A(D), B(E)))$$

$$t_6 : S(H) \xrightarrow{H_c} S(S_1(A(D), B(E)))$$

$$t_7 : S(C) \xrightarrow{b} S(f s_1)$$

Fig. 3. The translation of the state machine in figure 2.

and activities in the currently active state are completed. Completion events are special because they have priority over all other events. Completion transitions will be implemented in our semantics by assigning a special completion trigger to such transitions. The set of completion trigger events is E_c, and the set of events of a state machine $E_{SM} = E_t \bigcup E_c$. In figure 2 the transition without trigger emanating from state H will be labelled H_c in our semantics. We will discuss the generation of completion events in more detail below. From now on, we shall assume that all completion transitions have been labelled with the corresponding completion event.

Definition 2. *A transition is a triple* (s, t, s'), *where* $s, s' \in T_\Sigma(X)$ *and* $t \in Tn$. *The set of transitions of a state machine is denoted by* Φ. *We shall consistently confuse a transition with its name. The projections are given by* $source(s, t, s') = s$, $target(s, t, s') = s'$. *Each transition has a number of attributes: a function* $trigger : \Phi \to E$ *that assigns a triggering event to a transition, a function* $guard : \Phi \to P$ *(the default guard is the predicate* **true***), and a partial function* $effect : \Phi \hookrightarrow E$. *We will use the following notation for transitions:*

$$t : s \xrightarrow{e/g/\hat{}\,a} s',$$

where $e = trigger(t)$, $g = guard(t)$, *and* $a = effect(t)$. *We assume that the predicates are elements of a language* $\mathcal{L}_p$. *An effect is of the form* $a \mid a() \mid link.a \mid link.a()$, *where* a *is an event name,* $a()$ *is a call event, and* $link$ *is the name of the receiving object (if not specified, then the object itself).*

In the case of vUML the language $\mathcal{L}_p$ is the subset of boolean valued expressions of PROMELA. The transitions in figure 2 do not have any guards or effects. The encoding of the transitions are given in figure 3. The completion transition t_6 has the completion event H_c as its trigger, to mark the fact that it will be fired when the activities in state H have been finished. Our notation for states makes the hierarchical relationship of states explicit. The fact that t_1 lives in an orthogonal component is made explicit through the fact that the variable X

appears on both the left-hand-side and the right-hand-side in the same position $B(X)$. This means that firing t_1 does not change the state of the B component in the orthogonal region S_1.

Let us at this point summarise the different elements that a state machine consists of:

Definition 3. *Assume that we are given a languages for describing actions* $\mathcal{L}_a$ *and predicates* $\mathcal{L}_p$ *and a set of events* E. *A UML state machine is a tuple*

$$< \Sigma, entryAction, exitAction, activity,$$
$$\Phi, source, target, trigger, guard, effect >,$$

where

Σ	*is the set of states of the state machine,*
entryAction	*the entry action, is a partial function* $\Sigma \hookrightarrow \mathcal{L}_a$,
exitAction	*the exit action, is a partial function* $\Sigma \hookrightarrow \mathcal{L}_a$,
activity	*the activity, is a partial function* $\Sigma \hookrightarrow \mathcal{L}_a$,
deferred	*the set of deferred events* $\Sigma \to \mathcal{P}(E)$,
Φ	*is the set of transitions* $\Phi \subseteq T_\Sigma \times Tn \times T_\Sigma$,
source	*is the left projection* $\Phi \to \Sigma$,
target	*is the right projection* $\Phi \to \Sigma$,
trigger	*is the trigger event* $\Phi \to E$,
guard	*is an guard of the transition* $\Phi \to \mathcal{L}_p$
	(the default guard is **true**), *and*
effect	*is the side-effect of the transition, a partial function* $\Phi \hookrightarrow \mathcal{L}_a$.

The UML standard defines a number of *pseudostates*, e.g. initial, join vertices, fork vertices, and junctions. These are vertices in the state machine that typically are used to connect multiple transitions into more complex state transition paths. These are extensions the UML state machine syntax, because the primitive constructs are not rich enough to express the wanted behaviour. However, it turns out, that our formalisation allows us to directly express most of these constructs without any extensions. So we will not give explicit translation of these features but instead just indicate how they can be directly expressed in our formalisation.

Initial states: We can treat initial states by requiring that transitions that go to the boundary of a compound states are translated so, that they explicitly reference the initial state (cf. transitions t_5 and t_6 in figure 3.).

Join and fork vertices: A join vertex is used to merge several transitions emanating from the source vertices in different orthogonal regions causing the execution to synchronise, while fork vertices serve to split an incoming transition into two or more transitions terminating in orthogonal target vertices. Join and fork vertices can be translated by making the source and target vertices explicit in the translation. An example is given in figure 4.

Junction vertices: These are a generalisation of join and fork vertices, where any number of transitions may enter or exit the vertex. The transitions may also contain guards. We will not deal with junction vertices explicitly, since

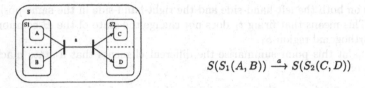

$$S(S_1(A, B)) \overset{a}{\longrightarrow} S(S_2(C, D))$$

Fig. 4. The translation of join and fork vertices

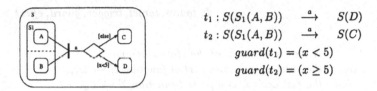

$$t_1 : S(S_1(A, B)) \quad \overset{a}{\longrightarrow} \quad S(D)$$
$$t_2 : S(S_1(A, B)) \quad \overset{a}{\longrightarrow} \quad S(C)$$
$$guard(t_1) = (x < 5)$$
$$guard(t_2) = (x \geq 5)$$

Fig. 5. The translation of junction vertices

they can always be translated into simpler constructs by logically combining the optional guards on the transitions (cf. figure 5).

The standard defines two other pseudostates: *history states* and *synch states*. These cannot be considered as simple syntactic extensions but require a little more work. They have been omitted in the present paper but their formalization can be found in the full paper [10].

2.2 The Run-To-Completion step

In the following discussion, we will try to formalise the informal description of the semantics given in section 2.1.4 of [1]. Recall that to each object we associate only one UML state machine that, according to the standard, "completely defines the behavior of a model element". The execution semantics of a UML state machine is defined in terms of a hypothetical machine whose key components are:

- an *event queue* that holds incoming event instances until they are dispatched,
- an *event dispatch mechanism* for selecting and de-queueing event instances from the event queue, and
- an *event processor* that processes dispatched events.

Each object will execute an algorithm, the *run-to-completion* (rtc) step, that dispatches and executes events on its event queue until the top state generates a completion event and the state machine exits. The standard does not define the semantics of the input queue. We shall implement the event queue as an ADT with operations *enqueue* and *dequeue* for adding events to the end and deleting events from the front respectively. The main task of the event processor is to find out which transitions are *enabled*, e.g. can be fired based on the first event in the queue. The set of transitions is then grouped into a step. The transitions in the step are executed in sequence, but the event processor will only consider the next event in the queue after all transitions in the step have been fired.

First, we need to determine the transitions whose source term matches the current state configuration. We call these transitions *active*:

Definition 4.

$$active(s) = \{t \in T : \exists \rho : X \to T_\Sigma : \rho(source(t)) = s_1\}$$

The substitution ρ is called the mode *of the transition t. We use an auxiliary function $mode(t, s)$ to denote it. A pair (t, ρ) is called a* transition instance *of t.*

The last event that is dequeued from the queue is called the *current event*. Given a current event e we need to find those transitions t in $active(s)$ whose trigger matches t and whose guard evaluates to true.

Definition 5. *An active transition $t \in active(s)$ is enabled iff the current event $e = trigger(t)$, and $guard(t)$ is true.*

The set of transitions enabled by the current event e is denoted by $enabled(s, e)$. If the current event is clear from the context we write $enabled(s)$.

Two transitions in a UML state machine that are enabled at a state s are in *conflict* iff the intersection of the set of states their corresponding transition instances exit is non-empty. E.g. in figure 2 transitions t_1 and t_3 are in conflict at state $S(S_1(A(D), B(E)))$, because both transition instance $(t_3, \{X \mapsto D, Y \mapsto E\})$ and transition instance $(t_1, \{X \mapsto E\})$ in state $S(S_1(A(D), B(E)))$ exit the simple state D. To define the conflict relation we need to calculate the set of states that is exited by firing a transition instance. Assume that transition $t = (s_1, t, s_2)$ is enabled at state s with mode σ. Then the set of states that are exited (and also the set of states that are entered) in this firing mode is given by:

$$exit(t, \sigma) = symb(\sigma(s_1)) - symb(\sigma(s_2))$$
$$entry(t, \sigma) = symb(\sigma(s_2)) - symb(\sigma(s_1))$$

Definition 6. *The conflict operation between enabled transitions is defined as:*

$$conflict(t, t', s) \stackrel{def}{=} exit(t, mode(t, s)) \cap exit(t', mode(t', s))$$

Two transitions are in conflict iff $conflict(t, t', s) \neq \emptyset$.

In figure 2, at $S(S_1(A(D), B(E)))$, transitions t_1 and t_3 are in conflict, because $conflict(t_1, t_3, S(S_1(A(D), B(E)))) = \{D\}$, as are transitions t_2 and t_3, because $conflict(t_2, t_3, S(S_1(A(D), B(E)))) = \{E\}$.

Given two transitions that are in conflict at a state, we can sometimes resolve the conflict with the help of a priority relation. The UML standard [1] defines the priority of a transition based on its source state. If t_1 is a transition with $source(t_1) = s_1$ and t_2 a transition with $source(t_2) = s_2$, then:

- if s_1 is a direct or transitively nested substate of s_2, then t_1 has higher priority than t_2,
- if s_1 and s_2 are not in the same state configuration, then there is no priority difference between them.

The intuition is to give priority to transitions that start "lower" in the hierarchy. E.g. in figure 2 the conflict between t_2 and t_4 can be resolved in favour of t_2 because t_4 starts at the border of S_1 which is clearly higher up in the hierarchy than E. On the other hand, the conflict between t_1 and t_3 cannot be resolved, because they start in states that are not substates of each other.

The relation "direct or transitively nested substate" can be translated into a notion of covering of terms:

$$cover(a,b) \overset{def}{=} \exists \rho : X \mapsto T_\Sigma(X) : \rho(a) = b,$$

e.g. a covers b iff we can find a substitution ρ that allows us to transform a into b. As an example $S(S_1(X,Y))$, the source state of transition t_4, covers state $S(S_1(A(X),B(E)))$, the source of t_2, because by substituting $\{X \mapsto A(X), Y \mapsto B(E)\}$ we can transform $S(S_1(X,Y))$ into $S(S_1(A(X),B(E)))$. On the other hand if we look at the conflict between t_1 and t_3 we see that both $S(S_1(A(D),B(X)))$ covers $S(S_1(A(X),B(E)))$ and $S(S_1(A(X),B(E)))$ covers $S(S_1(A(D),B(X)))$, e.g. the conflict cannot be resolved. Thus the relation "has-priority-over", denoted by $\succ$ is given by:

$$a \succ b \overset{def}{=} cover(b,a) \wedge \neg cover(a,b).$$

So we have:

Definition 7. *The static priority relation on transitions* $t = (s_1, t, s_2)$, *and* $t' = (s_1', t', s_2')$ *is given by:*

$$t \succ t' \overset{def}{=} s_1 \succ s_1'$$

In our running example, if at state $S(S_1(A(D),B(E)))$ the current event is b, then transitions t_2 and t_3 are enabled and in conflict. But since t_3 covers t_2, we will only fire t_2. On the other hand since neither $t_1 \succ t_3$ nor $t_3 \succ t_1$, we cannot resolve the conflict between t_1 and t_3.

We can now define the rule for calculating a step Γ of transitions to fire from a state s assuming a given current event e. Given a state s the set $conflict(s) \subseteq enabled(s)$ is the set of transition instances that are in conflict at s. The set of transitions can be ordered by the relation $\succ$ to resolve some of the conflicts by priority. We denote the remaining set of conflicting transition instances by $conflict_\succ(s)$. The set of fireable transition instances is obtained by deleting all those transitions that cannot fire because of the priority relation:

$$fireable(s) = (enabled(s) - conflict(s)) \cup conflict_\succ(s)$$

```
RTC()
 1   while true
 2   do
 3       if queue ≠ ∅
 4           then currentevent ← queue.dequeue()
 5                enabled ← enabled(currentstate, currentevent)
 6                Select a step Γ based on Def. 8
 7                newstate ← currentstate
 8                for  each γ ∈ Γ
 9                do
10                    newstate = nextstate(γ, newstate)
11                    execute effect(γ)
12                currentstate ← newstate
```

Fig. 6. The first version of the run-to-completion step

Definition 8. *A step Γ is a maximal conflict free set of transitions, Γ ⊆ fireable(s), s.th.:*

$$\forall t, t' \in \Gamma : conflict(t, t', s) = \emptyset$$
$$\wedge \, (t \in fireable(s) \wedge (\exists t' \in \Gamma : conflict(t, t', s) = \emptyset) \implies t \in \Gamma)$$

The next configuration of a state machine given a step is computed as the sequential composition of the *nextstate* function, which is defined as

$$nextstate(s, (s_1, t, s_2)) = (mode(t, s))(s_2).$$

The idea of the run-to-completion step is to ensure that an event can only be dequeued and dispatched after the previous current event has been fully completed. Given the above mathematical tools we can now proceed to give a first version of the algorithm that implements the run-to-completion step as defined in the UML semantics. This is shown in figure 6. In this version we wait for an event to arrive on the queue, calculate the enabled transitions, and execute a step. If no transitions are enabled then the step will be empty, which implements the "implicit consumption" semantics as mandated by the UML semantics.

However, the UML semantics defines a number of notions that we have overlooked in the previous discussion. These are the different *actions* associated to a state; *entryAction, exitAction, activity* ; *completion events, deferred events, call events, internal transitions* and *time events*. The full rtc algorithm is given in figure 7. The rtc algorithm is a while loop that is executed until the top state of the state machine is completed. The rtc step proper is given in lines 4-35. We shall discuss the extensions individually.

State actions and activities: The UML standard attaches to each state a number of activities. Whenever a state is entered, the *entry action* is executed

```
RTC()
  1   finished ← false
  2   while ¬finished
  3   do
  4       if queue ≠ ∅
  5       then currentevent ← queue.dequeue()
  6            enabled ← enabled(currentstate, currentevent)
  7            if enabled = ∅ and currentevent ∈ deferred(currentstate)
  8               then deferredevents.push(currentevent)
  9            Select a step Γ based on Def. 8
 10            newstate ← currentstate
 11            for each γ ∈ Γ
 12            do
 13                    for each s ∈ exit(γ)in "inside-out" order
 14                    do
 15                        abort activity(s)
 16                        execute exitAction(s)
 17                    newstate = nextstate(γ, newstate)
 18                    execute effect(γ)
 19                    if effect(γ) is a method call
 20                       then wait for return of corresponding call event
 21                    for each s ∈ entry(γ)in "outside-in order
 22                    do
 23                        execute entryAction(s)
 24                        start activity(s)
 25            if currentevent is a call event
 26               then send return
 27            currentstate ← newstate
 28       if any activities been completed
 29       then push the corresponding completion event in front of the queue
 30       if all subregions have reached a terminal state
 31       then push the corresponding completion event in front of the queue
 32       if all activity in top state has been completed
 33       then finished ← true
 34       if enabled ≠ ∅
 35       then queue.insert(deferredevents)
```

Fig. 7. The final version of the run-to-completion step

prior to any other action (l. 26). Analogously, when the state is exited, the *exit action* is executed as the final step before leaving the state (l. 17). In the case of nested states the entry actions are executed outside-in, e.g. the entry action of the outermost state is executed first. The exit actions are executed in the opposite order, e.g. inside-out.

A state may also include an *activity*. This is an action that is forked as a concurrent activity directly after the entry action has been completed (l. 28). The activity continues executing while the object remains in the same state. When the state is exited,the activity is aborted before the exit action is executed (l. 15). If the activity ends before, a completion event is generated (l. 28).

Completion events: Completion events are generated when all of the following activities have ended in the currently active state: all subregions have reached a terminal state, all entry actions have been executed, and all activities have been finished.

After each execution of the while loop (note that this does not require the execution of a step!) we check if any activities have been completed and push the corresponding completion events in front of the queue (l. 28-29). Then, if any subregions have reached a terminal state we push the corresponding completion event in front of the queue. Finally, if all activity in the top state has finished we need to stop the rtc loop (l. 32-33).

Deferred events: Usually an event that does not enable any transitions is discarded. Sometimes it is however useful to keep this event waiting until the next state. Such an event is called a *deferred event*. The UML standard allows one to attach a set of deferrable events to a state[2]. We modify the transition selection mechanism so it does not necessarily discard the current event. If the current event does not enable any transitions, but belongs to the deferrable events of the current state configuration (l. 7), it is pushed onto a special list (l. 8), and the main loop is run through once. At the end of the loop (l. 34-35), we insert the deferred event back into the queue. But we can only do this in the case that we have actually fired an event, e.g. changed the object state. Otherwise we would go into an infinite loop, because the deferred event which does not enable any events will be put in the input queue and dispatch it again and again. The change of state can be checked by testing if the set *enabled* is nonempty (l. 34).

Internal transitions: Each state may also have a number of internal transitions. These are transitions whose source and target is the same state. Firing an internal transition does not cause exit and entry actions to be executed, nor does it cause the state activity to be restarted.

Call events: The final touch to the rtc step is the treatment of call events (lines 21-23 and lines 30-32). Call events are special events that represent synchronous invocation of a specific operation. A call event is generated by a method call action. Although the standard does not make this explicit, we have assumed that each call event has a corresponding return event. On the level of a single state machine, when a method call is executed as an action, the whole transition

---

[2] The set of deferrable events of a state configuration is the union of the defer set of the member states.

442

step is only completed when the invoked objects complete their own run-to-completion steps. Thus at line 22 the caller waits for each of method call to end. In line 31 the callee has finished its own rtc step, and must now send the corresponding return events.

Time events: Time events are generated by timers when they expire. The idea behind time events is that they make it possible to model deadlines and thus to express constraints on the *implementation* of the state machine. Thus we can model time events by assuming that each timer is actually an object that runs in an infinite loop sending out time events. Unfortunately, this leads to the so-called state-space explosion problem [17], because the time events will be interleaved with all other events. This is not the case in the implementation, where typically the time events will only occur at certain times in the execution. To the best of our knowledge, nobody has yet proposed a workable solution to the problem of modelling timeouts in an untimed formalism like PROMELA.

2.3 Collaboration of objects

The vUML tool is not used to model check a single state machine, but instead we are interested in the interactions between objects. This means that the above semantics needs to be extended. In practice, this turns out to be relatively easy to do. Recall that a collaboration diagram consists of objects and associations that describe how the objects communicate. In our semantics, each object in the collaboration diagram runs in a different process. Each process is an instance of the run-to-completion step, and it is scheduled asynchronously. Processes communicate by sending events trough links, which are instances of associations of a class. This means that if link l goes from object A to object B in the collaboration diagram, then any message sent along l from A will automatically be routed to B. Note that a link can introduce arbitrary delays when transmitting an event, but it can not lose events. A more detailed description can be found in [10].

2.4 Particular comments on the semantics

At the end of this section, we would like to point out a few problematic situations that may arise due to the way the semantics has been defined in the standard.

Our first example concerns the situation in figure 8. We have two objects A and B that are communicating. A can send events to object B through a link, for simplicity called B. Assume that A somehow receives an event a. This enables both transitions in A and the corresponding effects are executed. Because the order of execution in the rtc step is not defined but is required to be sequential, the object B is able to distinguish between different implementations of A. In the example, if A elects to fire the left transition, machine B will end up in the upper state, while in the other case it will end up in the lower state. This is not necessarily a problem in a single implementation, but if we want to use objects as components then this clearly is a problem. Fortunately, it is possible to detect this situation during the verification.

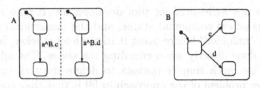

Fig. 8. Machine B is able to observe the execution order of A.

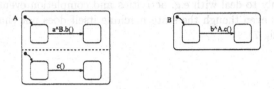

Fig. 9. Machine A will deadlock

Our second example concerns call events. If there is a circular chain of call events from an object back to itself, the object may block. The trivial case of this situation is when an object wants to execute a call action on itself. To illustrate the more complex situation consider the example in figure 9. Assume that object A receives an event a. This trigger the sending of a call event b to machine B, which sends the call event c back to A. Now, even though the event c could be acted upon in the second orthogonal region of A, the semantics of UML forbids this, and thus we have a *deadlock*. A solution to this problem would be to interpret the rtc step recursively and instantiate concurrent instances of the rtc loop for each orthogonal region. Then the call to $c()$ in figure 9 could finish and the system would not block.

Finally, we would like to point out that the fact that since side-effects are allowed to change the attributes of an object, orthogonal regions may not necessarily be independent.

3 Conclusions

We have presented a formalisation of the UML state machines as defined in the UML standard [1]. To our knowledge, it is the first formalisation that covers all the features of UML state machines. The only omission is the creation and destruction of objects. This could lead to unbounded behaviours, which are impossible to verify automatically with a model checker.

Several semantics for Statecharts have been proposed in the literature, e.g. [5, 16, 14, 9, 13, 8]. Most of these are concerned with defining the semantics of Harel's Statecharts. From our point of view [8] is interesting and worth a comparison. There the authors discuss a formalisation of UML state machines that takes hierarchical automata [12] as basis. In our opinion the hierarchical automata representation is not well-suited for tool development. It does not support directly transitions across compound states, and the hierarchical structure must be "flattened" or "unfolded" before using it in a model checker. In our approach the unfolding is done directly when encoding the states. The advantage of this is that we can use much simpler notions to define the structure of the state machine. Another problem of the approach in [8] is that they completely ignore the features of UML state machines that break the flow-of-control in the simple rtc-step of figure 6. Indeed it is difficult to see how their approach could be extended cleanly to deal with e.g. activities and completion events, since these require actions even though the state machine itself does not change state (cf. lines 28-29 of fig 7).

References

1. OMG Unified Language Specification (draft). Version 1.3 alpha R5, March 1999, available at http://www.rational.com/uml/resources/documentation/media/OMG-UML-1_3-A%1pha5-PDF.zip.
2. G. Booch, J. Rumbaugh, and I. Jacobson. *The Unified Modelling Language User Guide*. Addison-Wesley, 1999.
3. E. M. Clarke and R. P. Kurshan. Computer-aided verification. *IEEE Spectrum*, 33(6):61–67, 1996.
4. R. France, A. Evans, K. Lano, and B. Rumpe. The UML as a formal modeling notation. *Computer Standards & Interfaces.*, 19:325–334, 1998.
5. D. Harel and A. Naamad. The STATEMATE semantics of Statecharts. *ACM Transactions of Software Engineering and Methodology*, 5(4) Oct 1996.
6. G.J. Holzmann. *Design and Validation of Computer Protocols*. Prentice Hall, 1991.
7. G.J. Holzmann. The model checker Spin. *IEEE Transactions on Software Engineering*, Vol. 23, No. 5, May 1997.
8. Diego Latella, Istvan Majzik, and Mieke Massink. Towards a formal operational semantics of UML statechart diagrams. In *3rd International Conference on Formal Methods for Open Object-Oriented Distributed Systems*, Boston, 1999. Kluwer Academic Publishers.
9. F. Levi. *Verification of Temporal and Real-Time Properties of Statecharts*. PhD thesis, University of Pisa-Genoa-Udine, Pisa, Italy, 1997.
10. J. Lilius and I. Porres Paltor. The semantics of UML state machines. Technical Report 273, Turku Centre for Computer Science, 1999.
11. J. Lilius and I. Porres Paltor. vUML: a tool for verifying UML models. Technical Report 272, Turku Centre for Computer Science, 1999.
12. E. Mikk, Y. Lakhnech, and M. Siegel. Hierarchical automata as model for Statecharts. In *3rd Asian Computing Science Conference (ASIAN'97)*, volume 1345 of *Lecture Notes in Computer Science*, pages 181–196, Berlin, 1997. Springer.
13. E. Mikk, Y. Lakhnech, M. Siegel, and G. Holzmann. Implementing Statecharts in Promela/SPIN. In *Workshop on Industrial-Strength Formal Specifications Techniques (WIFT'98)*, Boca Raton, FL, USA, 1998. IEEE Computer Society Press.
14. A. Pnueli and M. Shalev. What is in a step: On the semantics of Statecharts. In *Theoretical Aspects of Computer Science*, volume 526 of *Lecture Notes in Computer Science*, Berlin, 1991. Springer-Verlag.
15. J. Rumbaugh, I. Jacobson, and G. Booch. *The Unified Modelling Language Reference Manual*. Addison-Wesley, 1999.

445

16. A. Uselton and S. Smolka. A compositional semantics for Statecharts using labeled transition systems. In *CONCUR'94*, volume 836 of *Lecture Notes in Computer Science*, Berlin, 1994. Springer-Verlag.
17. A. Valmari. The state space explosion problem. In *Advances in Petri Nets*. Springer-Verlag, 1999.

SDL as UML: Why and What
Panel

Moderator:
Bran Selic, ObjecTime Limited

Panelists:
Philippe Dhaussy, ENSIETA, France
Anders Ek, Chief Methodologist, Telelogic
Øystein Haugen, Ericsson AS, Billingstad, Norway
Philippe Leblanc, Verilog, Toulouse, France
Birger Møller-Pedersen, Ericsson AS, Billingstad, Norway

There is a very popular series of movies in which one of the main characters is a blundering French detective who entertains the audience to great comic effect as he speaks English contorted to a French grammatical form. While this may not necessarily be an appropriate analogy for the case of SDL and UML convergence, it highlights one of the principal hazards when merging languages of different lineage.

SDL originated in the telecom industry as a specification language for describing standard telecom protocols in a clear and precise manner. Over time, it expanded its expressive power taking on an object-based and, eventually, an object-oriented form. The computational model of SDL is based on a network of co-operating state machines.

UML, with its support for state machines (a more sophisticated hierarchical form than used in SDL) and object collaborations, certainly has the foundations required for expressing all the essential ideas of SDL and much more. Thus, from a technical perspective, the convergence of the two languages seems feasible.

This leaves two potential sets of issues: motivation and form. Is the convergence something that is beneficial to the user community? If so, what should the result look like?

Clearly, current SDL users would benefit if it meant preservation of their legacy software while allowing them to gain a foothold in the broader UML world. For this group, the primary requirement is one of backward compatibility. To date, SDL has been used mostly in the telecom domain, for protocol specification and realization, despite its potential for broader applicability. (As experience with programming languages teaches us, this is not necessarily due to any technical deficiencies of SDL.)

Given that, perhaps a more important issue is whether the much broader UML community would benefit from using the basic ideas of SDL? The answer to this depends very much on the final form of the merged result. For example, will its users be able to take advantage of the richer and more expressive set of base concepts in UML (relative to SDL)? If not, what is the cost (or benefit) of the lost modeling power? Will users have access to all available UML tools, or will they be confined to those supporting only the merged language? Furthermore, is the integrated form seamless, both notationally and conceptually, so that true iterative and incremental development can take place?

The objective of the panel on SDL and UML convergence is to explore and possibly answer these and related issues. The panel includes distinguished representatives from the two standards bodies, vendors of both UML and SDL tool vendors, as well as users from industry.

Philippe Dhaussy, Joel Champeau, Michel Boulestin:
Using UML and SDL for embedded systems

Context

The research topics of the Control and Real Time Systems group at ENSIETA are oriented towards the design, implementation and validation of advanced robotics control systems. Our group is involved in the development of software for autonomous marine vehicles such as Autonomous Surface Vehicle (ASV) and Autonomous Underwater Vehicle (AUV). The information system of AUV is based on a distributed architecture which comprises : an ethernet network which connects several VME-sites (Power-PC) and another one which uses the CAN technology for controlling several communicated intelligent devices (TMS340 DSP) coupled to actuators and sensors. The hardware supports cooperative subsystems: actuator control, integrated sonar, obstacle avoidance, mission control, navigation and motion control systems.

The development of a real time embedded system such as an AUV is a highly complex process. These critical systems require a high level of maintainability and reliability. In this scope, the advantage of object-oriented analysis and design methods is that the object-based concepts is consistently applied during the whole software life cycle. This paradigm is well suited to implement subsystem frameworks (application frameworks) which we would like to specify and to reuse them for the development of similar subsystems.

Many works currently focus on object oriented methodologies for real time systems [6,2,1]. UML has got much attention when it comes to modeling of requirements. It provides an excellent framework for requirement capture and analysis [5]. Unfortunately, the current UML semantics [8] doesn't give enough support for the development of distributed real-time systems which require the possibility to express unambiguous interfaces and to simulate designs before implementation. And, the validation tools that have to be used continuously during the object-oriented life cycle of the distributed software are not still available [11].

So like some software research groups, we investigate the UML/SDL combination approach to take advantage of the best features of object modeling techniques and the formal language of SDL for a more rigorous description of the dynamic behavior of subsystems. SDL have proved useful for the formal specification of protocols and distributed systems. It is based on communicating, extended finite state machines. It allows the behavior of active objects to be specified formally and to be verified and validated.

In this approach, our motivations are to experiment the actual limits and drawbacks of a object-oriented / formal language co-development. We are examining the conditions to use specific design patterns (communication, state, observer patterns)

[7] and their SDL specifications [9]. We turn our attention to research contributions exploring the iterative design process and continuously validated development process for object-based distributed systems.

Actual software methodology

In our developments, the analysis of a subsystem and early design phases is done with UML. The static structure is described by class diagrams with association, inheritance and dependencies between the classes. The cooperation between the different objects in the system is represented by sequence and collaboration diagrams which are instances of use-case that expressed the requirement analysis. The next step is to identify the active objects. This identification is done from the analysis of the previous diagrams. The collaboration diagrams can combine active and passive objects. One of these diagrams is used to synthesize the relational structure of the active objects. To combine UML and SDL, we use Argo software environment (http://www.ics.uci.edu/pub/c2/). We are implementing translation rules to map the collaboration diagram to an SDL system with some hypotheses on the associations between the different active objects which are interpreted like signal route between SDL processes. These signal routes are the support of asynchronous communications.

Then, we describe the behavior of the active objects with SDL process. We choose a set of essential use cases and define the behavior of the SDL processes and data types that implement these uses cases. SDL process graphs provide a graphical notation for state machines. This activity is done iteratively by starting with an initial set of essential use cases, making that part of the design complete and then testing the design.

With Telelogic-TAU environment [13], we perform the testing with two strategies: simulation using the SDL simulator and MSC verification and automatic testing using the SDL validator. Simulator is very useful to adjust the process specifications. Validation allows an automatic exhaustive verification of SDL specifications by testing all possible runs of an SDL system. During the exploration, a number of rules are checked and violations reported, deadlocks, loops and exceeded queue length.

The drawback of this technique is the well-known state explosion problem. The limit due to the number of states in the model is very quickly reached on calculators with only a reasonable amount of memory even if we focus in testing the subsystem control part. However, the need to specify more and more complex systems implies mastering the specifications with a composing approach [3]. An another problem is the time encoding one [4,12] and the proper capture of real-time constraints in a specification. SDL and MSC specifications do not allow to express real-time bounded response properties [10].

Anders Ek: UML /SDL Convergence

UML and SDL Today

SDL is today a mature but still evolving language, which has been used for application development in numerous telecom and real time projects for many of years. It is a standardized graphical language that has moved from being used mainly for the specification of distributed telecom protocols to a language used for design and implementation of applications characterized by a high degree of concurrency, real time requirements and often a distributed architecture.

When comparing UML and SDL it is useful to keep in mind the differences in scope of the languages. UML covers many relevant aspects of an application including the logical architecture, using e.g. class diagrams and state charts, the requirements structure, using use cases diagrams, and the physical architecture using component and deployment diagrams.

SDL (including MSC) is only intended to cover the logical architecture of the application, i.e. how the application is decomposed into a hierarchy of logical components and a precise definition of the behavior of each component.

Another difference is in the application area coverage. UML is a general-purpose language that can be used in for all kinds of object-oriented applications. SDL is a special-purpose language intended mainly for concurrent, distributed applications with real time requirements.

It is also worth remembering that object-oriented analysis and SDL have been used together for a long time. UML (mainly class diagrams and use cases/scenarios) are used analyze requirements and identify objects. SDL/MSC is then used for the precise definition of the architecture, requirements and behavior of the system.

SDL Characteristics

SDL is characterized by three major features that give focus to the language and that provide substantial benefits to SDL users:

- It is based on concurrent state machines and asynchronous communication.
- It has a precisely defined semantics and a complete definition of behavior.
- It facilitates a precise definition of hierarchical structure, encapsulation and interfaces.

Special concern is thus taken to applications that to a large extent can be described by concurrent state machines that communicate mainly using an asynchronous communication mechanism. This makes SDL a special-purpose language, very suitable for concurrent, distributed real time applications that typically are very efficiently described using the concurrent state machine metaphor.

The precisely defined semantics and the fact that an SDL system is a complete description of an application have some very useful side effects. The tool support for analysis and verification of SDL systems can be very efficient since SDL formally defines the complete dynamic semantics of an SDL description. Consequently, most SDL tools provides extensive support for syntactic and semantic analysis, simulation, formal verification and testing of SDL models.

Another benefit given by the precisely defined and complete SDL definitions is that it is possible to generate applications directly from SDL. Most SDL tools provide code generators that will generate complete application code adapted to various target environments, such as a naked machine with no OS support, or integrated as tasks in a real-time operating system. Code generation together with the high-level nature of SDL provides a very cost efficient way of developing applications based on the communicating, extended finite state machine paradigm.

The focus on architecture and interfaces is in SDL accomplished by special structure diagrams that define the hierarchy, communication paths and interfaces in an SDL application. They make SDL very suitable for large-scale development where a complex application is decomposed into manageable components to be developed by different teams.

Current Trends

Today there are trends both within OMG and ITU that from a UML/SDL perspective are very positive. Within OMG there is work within the Analysis & Design Task Force to define an SDL style action semantics. There is also work within the Real Time Analysis & Design Working Group to extend UML with concepts to handle hierarchical systems that seems to match the SDL concepts very well. Within ITU the definition of SDL-2000 (the next version of SDL) is done with UML harmonization as one of the primary concerns. The result is that there now is a fairly large common subset between UML and SDL. In addition ITU is also in the process of accepting a new standard, Z.109, that defines a mapping between the UML metamodel and SDL.

From a UML & SDL tool vendors point of view this is very good news. It will now be possible to merge the UML & SDL tools and make the combined use of UML and SDL even smoother. From a UML developers point of view this will give him access to SDL benefits like simulation, formal verification, complete application generation and large scale hierarchical system definition directly in his UML tool. From an SDL developer's point of view he can use the strength of object oriented analysis directly in his SDL tool. He can also benefit from the other UML models that are available.

Øystein Haugen: Converging MSC and UML Sequence Diagrams

MSC is a language standardized by ITU since 1992. The current standard supported by tools is from 1996 and a revision is due in November 1999. In summary MSC-2000 adds powerful structuring mechanisms such as roadmaps, decomposition, object orientation with inheritance and redefinition, diagram references, and inline expressions; improve the precision and understanding of time and control flow; and devise a flexible, but formal way of handling alternative data languages.

Purpose of MSC

The purpose of recommending MSC (Message Sequence Charts) is to provide a trace language for the specification and description of the communication behavior of system components and their environment by means of message interchange. Since in

MSCs the communication behavior is presented in a very intuitive and transparent manner, particularly in the graphical representation, the MSC-language is easy to learn, use and interpret. In connection with other languages it can be used to support methodologies for system specification, design, simulation, testing, and documentation.

MSC and similar notations have been widely used in numerous industrial projects. Simple charts are intuitively understood and this boosts the use of the notation such that the ambitions of what to express increase. When the ambitions reach describing major functionality of large products, the need for structuring mechanisms arise. The increased reliance on the sequence diagrams to show the functionality makes it necessary to improve the precision of the notation into a more formal language. MSC can be used as a property language to formulate behavioral properties to be checked against a corresponding executable model.

Position Statement

UML Sequence Diagrams are comparable with basic MSC as standardized in 1992. The feedback from users of MSC-92 resulted in new features in MSC-96. Roadmaps (High-level MSC) were added for improved structure; inline expressions for compactness of description and MSC references for better reuse of charts.

Notations for time and control flow are present in UML Sequence Diagrams, but no precise definition is given. MSC-2000 has incorporated time and control flow based on the notational constructs found in UML. While the underlying execution model for UML Sequence Diagrams seems to be that of alternating (but not concurrent) processes, MSC has as its underlying execution model a set of interacting, concurrent processes. The concepts of time, control flow and data are all strongly affected by this difference in perceived foundation.

MSC-2000 incorporates data in a flexible, but powerful way. Data language is parameter to the MSC document such that the user can apply the data language of his/her liking without losing the powers of formal reasoning.

We believe that the attractive and popular UML Sequence Diagrams will be even more appreciated if they converge with the advances of MSC-2000.

In summary this would add powerful structuring mechanisms such as roadmaps, decomposition, object orientation with inheritance and redefinition, diagram references, and inline expressions; improve the precision and understanding of time and control flow; devise a flexible, but formal way of handling alternative data languages.

One should bear in mind that the MSC community over the last 8 years have designed large industrial systems with reasonable success and MSC-2000 is the most updated answer to their experiences with earlier versions.

Philippe Leblanc: SDL-UML Convergence

Presentation of the Company
CS VERILOG is a CASE tool vendor founded in 1984. We distribute for about 10 years an SDL-MSC toolset called ObjectGEODE (formerly GEODE). More than 3000 licenses are installed all around the world.

The toolset currently supports — totally or partly — the 1996 versions of SDL, MSC, ASN.1, TTCN, and the UML Class diagram and State diagram. It includes graphical editors, SDL-MSC static checker, SDL-MSC debugger, SDL-MSC model checker, TTCN test suites generator, C and C++ code generators for various commercial RTOS (pSOS+, VRTX, VxWorks, Chorus, UNIX, POSIX, Win32...).

Involvement in Standardization Works
The company is involved in the ITU-T Study Group 10, and is actively collaborating to Q6/10 for the definition of SDL-2000. Personally, I am the Q7/10 Rapporteur on Methodology. The company is also ETSI member.

We are a full OMG member, and we participate to the RTAD and RTF works. In particular, we are working on the Action semantics RFP (within the Action Semantics Working Group).

Market Analysis
Our customers come mainly from the Telecom sector (from large public switches to mobile phones), Aerospace (embedded software for aircraft and satellite) and Automotive (electronic equipment). They use SDL as a design language, automating the design, implementation and test phases.

Today there is a significant mind share in favor of UML, as there are a lot of expertise, literature, training courses, tools, etc., publicly available, even if actual benefits have not yet been proven on complex real-time system development. As a consequence, there is a strong demand from the market to give to UML the same support in terms of software productivity gains than those already provided by SDL-based technologies like ObjectGEODE.

To meet these needs, we are working on the convergence of SDL and UML, in order to have UML tools powered by the SDL advanced technology benefiting from features such as: model debugging, model checking, test generation, generation of executable code deployable on RT platforms.

This means 1) aligning the SDL concepts on UML in case of SDL limitations; 2) giving the UML visual representation to the SDL concepts (when relevant); 3) integrating the SDL advantages to UML in a coherent way, mainly regarding formal semantics and visual action language.

Which Benefits Are We Expecting from SDL-2000 and MSC-2000?
Compared to UML 1.x, SDL-96 is a formal specialization of the UML for real-time and distributed systems with a different visual representation and some limitations whose major ones are: for active classes — no dynamic instantiation of container; for passive classes — no redefinition through inheritance; for associations — only

association instances can be specified; for state machines — no composite states, no entry/exit actions.

From our point of view, SDL-2000 is the right answer, SDL deficiencies are removed while its advantages are kept:

- Visually: The UML graphical representation is included in SDL-2000 each time it is relevant. In particular, SDL agents (active classes) and data types can be drawn with the UML Class iconic representation. Associations can also be drawn between agents.
- Conceptually: Container agents can be dynamically instantiated. The concept of interface is completely supported. State machines contain composite states with entry and exit actions. Data modeling supports inheritance, redefinition and polymorphism.

Compared to UML 1.x Use cases and Sequence diagrams, MSC-96 already provides nearly the same visual representation and only the specification of focus of control region is not yet supported. On the other hand, MSC-96 contains much more constructs to specify real-time behavior such as composition operators (to run several MSCs together), timer management, creation/deletion of objects... MSC-2000 will include the concept of control region, thus MSC will be a formal superset of the Use case and Sequence diagram notations (the traditional Use case representation with ellipses and actors is out of the scope of MSC, thus a drawing tool should still be used instead).

Company's Current and Next Activities

CS VERILOG has set up an ambitious development program including standardization activities and tool development:

- We are working in the ITU-T (SG 10) and OMG (RTAD WG, RTF) to collaborate on proposals for a concrete and efficient convergence of SDL and UML.

    At the ITU-T, we have participated to the definition of SDL-2000. Next years, we will participate to the annual revisions of SDL, focused on its convergence with UML. We will lead the methodological work (Q7/10), and an engineering process will be defined allowing a consistent use of SDL, MSC and UML. At the OMG, we are participating to a proposal for the "Action Semantics" RFP. As a result the visual action language defined in SDL would be compatible with the proposed UML action semantics. We will work on the "Complex System Modeling" RFP as soon as it is open.

- With respect to tool development, a very large amount of effort is already scheduled to upgrade our tools according to the SDL-UML convergence. The development plan also includes other advanced topics related to object-orientation and UML, such as CORBA, Java, XML and QoS analysis.

Birger Møller-Pedersen, Thomas Weigert[2]:
Towards a Convergence of SDL and UML

(² Motorola, Schaumberg, Illinois)

The SDL UML profile defined by ITU Z.109 is the first contribution to a convergence between the two dominant modeling languages in the telecom and computing domains, SDL and UML. This note suggests a number of other features of SDL that would benefit a modeling language that leverages the best of both worlds.

Convergence of the telecom and computing worlds has been on the agenda for a while. Being the dominant modeling languages in each of their areas, SDL for real-time telecom applications and UML for general-purpose computing applications, a convergence of these two languages will result in a language that includes the best of both.

While UML aims to be applicable to a wide range of application domains, SDL has focused on the modeling of reactive, state/event driven systems typically found in telecom applications. In order to subsume the possible variances of application domains, UML does not define all language concepts (such as its concurrency semantics) to the level of detail necessary to allow unambiguous interpretation. SDL, on the other hand, gives precise, formal semantics for all its concepts. In addition, UML relies on implementation languages for executable specifications; SDL is a language for specifying executable models independently of an implementation language.

After the emergence of UML, a typical SDL usage scenario is that UML is used for describing the entities of a system and the relationships these entities bear to each other during analysis modeling, while SDL is used for detailed design. User requests for better support of this usage scenario have formed the basis for a first contribution to a convergence: The forthcoming ITU Recommendation Z.109 defines the *SDL UML Profile*. Thanks to this profile, users can smoothly transition from the more abstract UML analysis models to the unambiguous and executable SDL design models. The SDL UML profile allows users to treat an SDL model as a specialization of the generic UML model thus giving more specific meaning to entities in the application domain (blocks, processes, services, gates, channels, etc.).

A number of features have been introduced in SDL which directly support SDL and UML convergence:

- UML-style class symbols provide both partial type specifications and references to type diagrams containing the definition of that type;
- UML-style graphics for SDL concepts such as types, packages, inheritance, and dependencies;
- Composite states that combine the hierarchical organization of Statecharts with the transition-oriented view of SDL finite state machines;
- Interfaces that define the encapsulation boundary of active objects; and
- Associations between class symbols.

While UML has its focus and strength on object oriented data modeling, SDL has its strength in the modeling of concurrent active objects, of the hierarchical structure of active objects, and of their connection by means of well-defined interfaces. As a response to user requirements for more design level concepts as well as for better support of object-oriented modeling of active objects, SDL provides the following concepts, which are also applicable to the convergence of SDL and UML:

- A complete *action language* that makes SDL independent of implementation languages. In line with the rest of SDL, behavior is specified in an imperative style and may be mixed with graphical SDL.
- *Object oriented data* based on single inheritance and with both polymorphic references (objects) and values, even in the same inheritance hierarchy. Type safety is preserved in the presence of covariance through multiple dispatch.
- An *exception handling* mechanism for behavior specified either through state machines or constructs of the action language that makes SDL suitable as a design/high-level implementation language.
- Composite states that are defined by *separate state diagrams* (for scalability); *entry/exit points* are used instead of state boundary crossing (for encapsulation), any composite state can be of a *state type* (for reuse), and state types can be *parameterized* (for even more reuse).
- Object-orientation applied to active objects including inheritance of behavior specified through state machines and inheritance of the (hierarchical) structure and connection of active objects.
- *Virtual types* that allow the redefinition of inherited types.
- *Constraints* on redefinitions in subclasses and on actual parameters in parameterization that afford strong error checking at modeling time.

We maintain that modeling of systems in UML would greatly benefit if these concepts, the need for which has clearly been demonstrated by large user communities, were to be provided also by the UML. These improvements would make SDL a proper subset of the UML. As a benevolent side effect of convergence, tool vendors from the different areas would focus on a common language and thereby compete in the same market space.

References

1. Burns A. and Wellings A.: Hrt_hood: a structured design method for hard real time systems. In Real-time Systems. 1994.
2. Lanusse A., Gerard S., and Terrier F.: Real-time modeling with uml: the accord approach. In UML'98 International Workshop, Mulhouse, France, June 1998.
3. M. Abadi and L. Lamport.: Composing specifications. ACM Transaction on Programming Languages and Systems, 15(1):73--132, 1993.
4. E. Asarin, M. Bozga, A. Kerbrat, O. Maler, A. Pnueli, and A. Rasse.: Data-structures for the verification of timed automata. In HART'97. Springer-Verlag LNCS 1201, 1997.
5. Douglass B.: Real-Time UML : Developing Efficient Objects For Embedded Systems. Adison Wesley, March 1998.
6. B.Selic, G. Gullekson, and P.T Ward.: Real-Time Object Oriented Modelling. John Wiley and Sons, 1994.
7. Gamma E., Helm R., Johnson R., and Vlissides J.: Design Patterns : Elements of Reusable Object-oriented Software. Adison Wesley, 1998.
8. A. S. Evans and A.N.Clark.: Foundations of the unified modeling language. In 2nd Northern Formal Methods Workshop, Ilkley, electronic Workshops in Computing. Springer-Verlag, 1998.
9. B. Geppert, F. Rossler, R.L. Feldmann, and S. Vorwieger.: Combining sdl patterns with continuous quality improvement: An experience base tailored to sdl patterns. In First Workshop of the SDL Forum Society on SDL and MSC, SAM'98, Berlin, Germany, 1998.
10. D. Hogrefe and S. Leue.: Specifying real-time requirements for communication protocols. Technical report, 1992.
11. C. Jard, JM. Jezequel, and L. Nedelka.: An approach to integrate formal validation in an oo life-cycle ofprotocols. In FMOODS'96, Paris, France, March 1996.
12. K.G. Larsen, F. Larsson P. Pettersson, and W. Yi.: Efficient verification of real-time systems : Compact data structure and state-space reduction. In 18th IEEE real-Time Systems Symposium, RTSS'97. IEEE Computer Society press, December 1997.
13. TELELOGIC: TAU/SDT 3.3. TELELOGIC, June 1998.

UML Behavior: Inheritance and Implementation in Current Object-Oriented Languages

Jean Louis Sourrouille

L3I, INSA, Bat. 502,
F69621 Villeurbanne Cedex
FRANCE
E_mail : sou@if.insa-lyon.fr
Phone: (+33) 4 72 43 87 43 Fax: (+33) 4 72 43 85 18

Abstract. The UML dynamic model is described using notions like state, event or active object that current object-oriented languages don't support. When the implementation is not done using a state machine interpreter, these notions had to be translated into the target language. This work aims to study how to translate as automatically as possible UML state diagrams into current object-oriented languages (OOLs), distinguishing sequential and concurrent execution. This translation requires to map UML notions onto OOLs ones, to adapt the abstract state machine, and to add information to state diagrams. Behavior inheritance is a key problem, and both theoretical and practical solutions are examined to ensure behavior substitutability. Then, two main ways for state representation are compared from the inheritance point of view, and automatic code generation is discussed.

1. Introduction

To specify how objects behave during their lifetime, UML ([UML99]) provides state diagrams, sequence diagrams, collaboration diagrams and activity diagrams (dynamic model). These formalisms use notions like active object, event or state machine, most of them being unknown to current typed Object-Oriented Languages like Eiffel [Mey88], C++ [Str92] and Java [Java]. At implementation stage, the developer had to translate the behavior into the target OOL. Implementation using a state machine interpreter is immediate. This article deals with implementation without interpreter and gives insights into the key points of the translation of UML state diagrams into current OOLs.

The principle is to choose a subset of the UML notions, and if needed to limit their semantics to ease translation. The study mainly deals with translation of UML notions into usual OOL notions (section 3), abstract state machine adaptation (section 4), behavior inheritance (section 5), and behavior implementation (section 6).

The proposal follows the spirit of the object-oriented approach, reducing the UML adaptations to the very minimum, rather to clarify and to simplify than to extend. The main contribution of this article is to discuss the whole problems related to UML behavior in the context of current OOLs, and to propose practical and consistent (not always novel) solutions. Original insights concern points like state handling (implicit vs explicit), state inheritance, state dependency, sequential and concurrent abstract state machine definition, safety of the preorder relationship and link with pre/postconditions, behavior inheritance textual rules, or inheritance anomaly-free implementation.

2. Aims

As current OOLs provide very similar notions, it is enough to define a common mapping between UML and OOLs notions, and generic policies to transform notions. The underlying layers (called OS for Operating System in the sequel) will supply notions not provided by the target language (e.g., thread).

The main formalism for behavior description is the state diagram (similar to *Harel statecharts* [Har87]). Its semantics is given in an operational way via an abstract state machine, for instance "the transitions fire one at a time". Under concurrent execution, this definition goes against the decentralized nature of object-orientation and can be enhanced to increase object independence.

Regarding behavior inheritance, the advice given in *UML Semantics* are heuristic rules, UML remaining open to any policy for "state machine refinement". In contrast with heuristic rules that don't ensure any clear property, it would be useful to define formal rules providing well-defined guarantees.

To implement their behavior, objects hold a description either explicit or implicit of their state diagram. Obviously, inheritance anomalies should be avoided ([Mat93] etc.). State representations requiring a class for each state (e.g., [Rum93], *design patterns* [Gam94]) seems to be intricate, so are there other practicable solutions?

To sum up, the aim is to reduce the state diagram notions to a subset easy to translate into current OOLs, to adapt the state machine to concurrency, to define inheritance rules based on formal properties, and to study ways to implement the behavior. State diagrams can be used to describe several UML model elements, but the only concern is class state diagrams. The target language is assumed to provide a minimum "object model". Concurrency is only considered from a behavior point of view.

3. Notions

3.1. Transition

The general form of a transition is $s \xrightarrow{\textit{event (parameters) [guard] / actions}} s'$. In the state s, the arrival of an enabled *event* whose *guard* is true triggers a transition to the state s' when *actions* execution is completed. An object state has duration during witch it is observable, while a transition is instantaneous. This means that a state is not observable during a transition (e.g., temporary forbidden situation). Obviously, a transition has duration but the state can not be computed while a transition is firing.

Event. UML events are the signal receipt (*SignalEvent*), the receipt of a request for a method execution (*CallEvent*), the temporal event (*TimeEvent*), or the event occurring when an expression becomes true (*ChangeEvent*). In current OOLs, the notions closer to event are the message receipt (*CallEvent*) and the raising of an exception (specialization of *SignalEvent*).

In a program, "real" events both come from the outside (user interface, interrupt, etc.) or the inside (e.g., logical event), but they are always conveyed by the OS. These events can be translated into messages (like "*On* event$_i$ *do* object.message$_j$") sent to one or more objects. Temporal event (*after 3ms*) and signal event are included in the events conveyed by the OS, thus only events coming from objects remain to express:

- At the receipt of an exception, the currently running method is interrupted, the code associated to the exception is executed (without return as a signal is always sent asynchronously in UML), then the program resumes at some place in the call stack. Although exceptions are processed differently according to languages, this execution mode is generic enough.

- The expression-change event (*when expression*) seems to require to continually look after the *expression* value, that would be difficult to implement. In fact, any change results from the execution of an action. But any action is atomic and runs to completion before a new transition can fire. A solution is to test the *expression* at the end of any transition capable of modifying it, and when it becomes true to send a message to the concerned object (if *expression* depends on other objects, required links have to be added).

Transitions simplified form. Finally, events amount to request or exception receipt (similar to *protocol state machines* with exceptions in UML 1.3). In the sequel, all events will be considered as requests, exceptions being a particular case. Requests trigger the execution of methods that modify the object attributes and this way change its state. The new transition general form becomes $s \xrightarrow{\textit{request}(\textit{parameters})} s'$. The Fig. 1 gives a sample of transition transformation into requests for the execution of a method m.

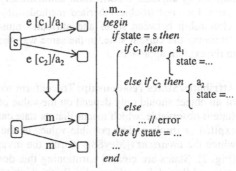

Fig. 1. Transition transformation

Discarded events. In UML, events with no enabled transition (or whose all guards are false) are discarded. As message passing usually does not allow discarding messages, all the received requests should be accepted whatever the current state. To ignore a request it is enough to add a transition that does nothing, so this constraint does not reduce the expressive power. On the other hand, UML allows unlabelled transitions to fire as soon as the state activity is completed ("transition completion"). With the above constraint, unlabelled transitions are forbidden.

3.2. State

Explicit vs implicit handling. To change states, the *Actor* language ([Agh86]) uses the instruction *"become newState"*. Numerous languages have taken up the same mechanism of explicit transition. Unfortunately, explicit state handling introduces inheritance anomalies when states are added in descendants. For instance, the state *Forward* of the class *Vehicle* (Fig. 2) is refined by the substates *Road* and *Town* in the subclass *Car*. The instruction *"become Forward"* written in any method of *Vehicle* is wrong when executed on instances of

Vehicle

invariant	$d'_0 = -20 < speed < 130$	
Off	On	
speed=0	speed ≠ 0	
	Forward	Reverse
	speed>0	speed<0

Car

invariant	$d_0 = d'_0 \wedge power \geq 0$	
Off	Forward	Reverse
	Road	Town
	power≥60	power<60

Fig. 2. State definition domain partitions

Car since only *"become Road"* or *"become Town"* are licit. To cope with this well-known problem, *"become newState"* is replaced by *"become NextState()"*, *NextState()* being a function returning a value characterizing the new object state ([Tom89][Kaf95], etc.)

This approach is close to the implicit one in which the state is not directly handled but deduced from the value of object attributes. For instance to know if a *Car* is in the state *Road*, the attributes *speed* and *power* will be tested (Fig. 2). Transitions are implicit as attribute modification automatically induces state changes. Adding new states does not trouble inherited methods since states don't appear explicitly. The relationship between states and attributes must be declared to make it possible to determine the current state, in the same way the method *NextState()* must be defined in the explicit approach.

Attributes/States relationship. To conform to the encapsulation principle, the state of an object should only depend on the value of its attributes. An object in a (stable) state is observable, which means that its state can be assessed, whether the handling is explicit or implicit. The possible values of the object attributes define a domain in which the *invariant* ([Mey88]) is true: the invariant of *Vehicle* is $-20 < speed < 130$ (Fig. 2). States are created partitioning this domain or the domain of their ancestor state and thus define a hierarchy. Partitioning ensures that any object is always in one and only one leaf of the hierarchy. Each state introduces restrictions to the inherited definition domain. Thus the domain of the state *Forward* is (Fig. 2):

(speed > 0) *and* (speed ≠ 0) *and* (-20 < speed < 130).

The domain of the state *Road* takes into account the invariant restriction of *Car*:

(power ≥60) *and* (speed >0 *and* speed ≠0 *and* (-20< speed <130)) *and* (power ≥0)

The mapping between states and attributes depends on the implementation and is hidden. The state name, by nature abstract, is public and exported. For instance, enabled operations in the state *Empty* of a collection are known.

Domain description with UML. UML provides a way to add properties to model elements (*taggedValue*). To complete the description, a predicate added to each state specifies the domain restriction. The effective definition domain of a state is the conjunction of these restrictions along all its ancestors (like *Forward* above). In the other hand, each class may define an invariant (stereotype «invariant» belonging to constraints). From this information, the function returning the object state can be automatically generated.

Stable state. A state is *stable* when it can not change without the receipt of an event (no hidden transition). Attribute encapsulation is not enough to ensure stability as modifying a shared attribute (in an other object) may trigger a transition without notification. To avoid this problem, a strict rule is applied for the translation of OMT into LOTOS ([Wan97]): any object interaction should go through the provided communication mechanisms (*gates* in LOTOS). The proposal is to allow only *state dependency* between objects, i.e., only object states can appear in the definition domain predicate of other objects. This way, only state changes can trigger hidden transition, and dependencies remain abstract (additional comments below).

3.3. Active objects

An active object is the association of an object with a process or a thread (stereotypes «thread» and «process»). Current OOLs do not provide active object notion, however the only aim is to define a minimum active object model. To allow concurrency and asynchronous messages between active objects, each object holds a request queue. To conform to the encapsulation principle, each object holds its own thread in which it executes its methods. Otherwise, a synchronization mechanism should be added to avoid undesirable simultaneous execution of methods in the same object.

Thus, a simple and basic active object model includes a thread, a request queue and the "sequential" object. Whether the thread belongs to the object or not, only one method can run at a time to avoid attribute sharing conflicts. Numerous concurrent OOLs offer this kind of model ([Pap95]). The ideal would be that an active object component be available in each OOL. It is quite easy to build basic active objects as long as they remain in the same address space.

Message processing. Messages between active objects are sent (maybe asynchronously) via a special method of the active object. In sequential execution, the arrival of a message not enabled in the current state (for instance *Get* to an *Empty* collection) is an error. In concurrent execution, the same policy can not be applied since message order does not always make sense. The advocate solution is to only process messages enabled in the current state, while the other ones remain in the queue. To achieve this, the active object should know its currents state and the messages enabled in each state. This policy comes to consider all the events that are not enabled in a state as *deferred* in UML.

462

4. Abstract state machine

The semantics of the state diagrams is indirectly described by the operations an abstract state machine interpreter would do. This abstract machine includes an event queue common to all the objects. Events are processed one at a time in undefined order (UML 1.3). Each event triggers one or more transitions that should be fully completed before the next event is processed (*run-to-completion* principle). Event serialization ensures that to some extent (depending on the actions on the transition) there are no simultaneous transitions in the system, with the advantage of safety but the drawback of low concurrency. Sequential and concurrent execution are now distinguished.

4.1. Sequential execution

Messages are all assumed to be synchronous. As transitions are transformed into requests for method execution, neither an event queue nor any other mechanism is needed. As a result, the request arrival order for an experiment is always the same. Static type checking ensures objects accept the sent messages. Each object should verify that the invoked method is enabled in the current state and set its new state after method execution (cf. implementation). Requests being processed at once, *deferred* events are forbidden.

Simultaneous transitions and cycles. A transition should be completed before next event processing. But a transition may execute an (atomic) action including synchronous *CallEvents* themselves triggering transitions. To avoid simultaneous transitions, actions would trigger asynchronous events only (as in [Har97]). As requests all are synchronous, several transitions may be in progress simultaneously. This is of no consequence except in case of cycle since an object is then observed during a transition. Fig. 3, the transition *b* sends the message *c* to *o'* that in turn sends *x* to *o*. When $x \equiv d$, the transition is acceptable, but when $x \equiv e$ the transition is forbidden since simultaneous transitions in the same concurrent state subset are not possible (the subset is "frozen"). First, cycles must be avoided if possible, otherwise the developer had to check they don't lead to forbidden situations.

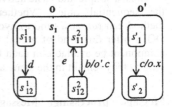

Fig. 3. Requests cycles

Usual execution environments. In present graphical environments, each window is associated with an event queue, a controller, and a thread or process. The controller processes the events and executes the relevant code. This architecture precludes any user wait in the window thread, else actions on the window as moving would be queued. Any concurrent event processing would lead to problems of execution control (priority?). So, the proposal is well adapted to sequential programs running in a window. Additional events are simply inserted into the window event queue.

4.2. Concurrent execution

Decentralized execution better conforms to an object-oriented approach. The abstract state machine is composed of independent abstract state machines associated each one with an object owning its event queue. Events are processed simultaneously in all the objects, but one at a time in each object (intra-object concurrency is out of the scope of this article). Objects choose the processing order, provided that events not enabled in the current state remain in the queue.

This definition implies that several transitions may fire in parallel in different objects, with consequences in case of state dependency (for instance the state of the *Elevator* changes when the state of its component *Door* changes). The state is then not stable and this problem should be carefully managed. Any cycle with synchronous messages is impossible, as it would lead to a deadlock.

5. Behavior inheritance

Behavior inheritance constraints aim to ensure the substitution rule: "An instance of a subtype can always be used in any context in witch an instance of a supertype was expected" ([Weg88]). In practice, this property can not be enforced due to method redefinition and attribute modification in descendants. A weaker relationship is defined, obviously a preorder, and related constraints are detailed. State inheritance is examined apart from transition inheritance, more problematic and for which only a summary of the main points is supplied (in-depth study in [Sou97]).

5.1. State inheritance

The Fig. 2 shows the principle: all the states are inherited, and new states are created partitioning ancestor states. Only invariant strengthening (usual rule [Mey88]) can remove a state making it unreachable. This organization conforms to the substitution rule: an object in a state (*Road*) is also in its ancestor states (*Forward, On*). This property is required since an instance of *Car* in the state *Road* should be substitutable for an instance of *Vehicle* in the state *Forward*.

5.2. Transition inheritance

5.2.1. Definition of a preorder relationship

Traces. It is neither necessary nor sufficient to inherit all transitions for behaviors to be substitutable ([Sou96a]). For an object to be substitutable for an other object, it should first accept the same sequences of requests (*traces* [Bro84]). Let S a set of states, R a set of requests, $\rightarrow \subseteq S \times R^* \times S$, $U \subseteq S$:

$$traces\,(U) = \{\ seq \in R^* \mid u \in U.\ \exists\ v \in S.\ u \xrightarrow{\ seq\ } v\ \}$$

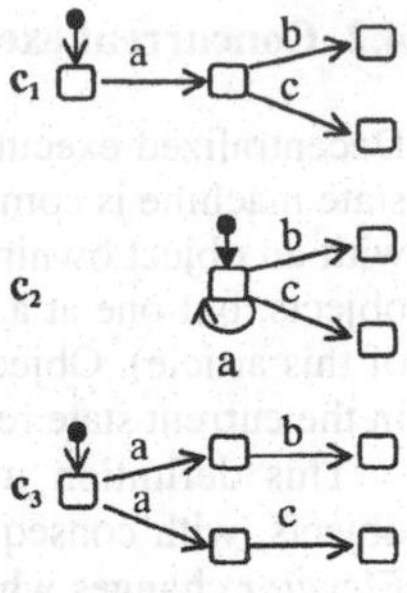

For a behavior c' (initial state I') to be substitutable for a behavior c (I) requires $traces(I) \subseteq traces(I')$. Fig. 4, traces are for c_1 $\{a,ab,ac\}$, for c_2 $\{a,ab,ac,aa,aab,aac,...\}$ and for c_3 $\{a,ab,ac\}$. The traces of c_1 are included into the traces of c_2, and c_2 is actually substitutable for c_1. The traces of c_1 are also included into the c_3 ones, but c_3 is not substitutable for c_1 since after the receipt of the request a, the request b may be refused in c_3 while it was always accepted in c_1. Trace inclusion being not sufficient to deal with non-determinist, *failures* have to be considered.

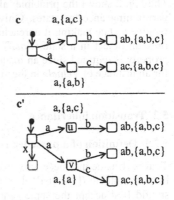

Fig. 4. Behavior samples

Preorder relationship based on failures. With the same definitions as above and $\rightarrow \subseteq S \times R \times S$, the enabled set of requests from a state s is:

$$initial(s) = \{ r \in R \mid s \in S, \exists x \in S. \ s \xrightarrow{\ r\ } x \ \}.$$

The failures are the set of pairs (seq, X) specifying the set of requests *refusals* X after the sequence seq of requests ([Bro84][Hoa85]):

$$failures_R(I) = \{ \ (seq, X) \mid seq \in R^*, s_{init} \in I, \exists s \in S. \ s_{init} \xrightarrow{\ seq\ } s \ , X = R - initial(s) \ \}$$

The substitutability context assumes that in a valid program, an object with behavior c enabling the request sequences $traces(I)$ is replaced by an object with behavior c'. Requests added in c' and more generally any request sequence not belonging to $traces(I)$ don't play any role (else the program would not be valid). To be substitutable for c, c' should have *at the most* the same failures than c (and at least the same traces):

$$traces(I) \subseteq traces(I') \wedge$$

$$\forall \ (seq,X') \in failures_R(I'), seq \in traces(I). \ \exists \ (seq,X) \in failures_R(I). \ X' \subseteq X$$

Fig. 5, the failures comply with this constraint: c' is substitutable for c in the sense of the failure relationship. Fig. 4 for $seq = a$, the refusal of c_1 is $\{a\}$ while the refusals of c_3 are $\{ac\}$ and $\{ab\}$. The refusals of c_3 being not included into the refusals of c_1, c_3 is not substitutable for c_1 (but c_1 is substitutable for c_3).

Limits of the relationship. When the behavior c is non-determinist and despite the failure constraint, a sequence accepted by c still may be refused by c'. Fig. 5 in c' for $seq = a$, when a leads to the state u, the sequence ac is refused while when a leads to v the sequence ac is accepted. This "potential" error results from the fact that c behave the same, i.e., that c can refuse or accept ac after a:

Fig. 5. Failures

This preorder only ensures that behaviors have the same properties.

This failure preorder has already been proposed, particularly in [Nie93] that introduces the notion of "request substitutability" to define a subtyping relationship between objects viewed as CCS processes. Obviously, a stronger relationship would be desirable to avoid errors, but attempts seem to lead to the major drawback that an object is not substitutable for itself when its behavior is non-determinist, so reflexivity is lost. The proposal is pragmatic: when non-determinism is avoidable, substitutability is error free, else potential errors must be accepted and the code had to be written accordingly. Another way is to reduce apparent non-determinism using transition guards ([Sou97]). The failure preorder being independent of any implementation, it can be applied to types like *interfaces* in *Java*.

Checks. UML tools ideally would check constraints so a prototype has been implemented. State diagrams are translated into usual "flat" state machines whose expressive power is the same in this context, but with the advantage that numerous results may be applied to. Inheritance and concurrency are removed making the cartesian product of the concurrent state subsets, and keeping only the leaf states of the hierarchy. This transformation also removes transition synchronization and *history* pseudostates. The prototype verification is based on the algorithm of [Nie93]. In practice, for an average number of n states, the time complexity of is very lower than the maximum $O(2^n)$ ($O(n^2)$ for a deterministic behavior). Efficiency was not the concern, but seemingly faster algorithms could be implemented ([Fer91][Cle93], m being the number of transitions, $O(n^4 \times m)$ and even $O(m^2)$). Until automatic checks are available, rules can be applied manually for small applications.

5.2.2. Sequential execution

UML (Notes in *Semantics*) suggests several "refinement" policies with decreasing constraints: subtyping aiming to guarantee substitutability, strict inheritance to encourage reuse, and simple refinement. Only the first policy falls within the scope of behavior inheritance.

UML subtyping. UML list of advice looks like "Transitions are only added, not deleted" etc. This subtyping is based on the preservation of the pre/postconditions that would guarantee substitutability: when a method is redefined in a descendant, preconditions should be weakened and postconditions strengthened ([Mey88]). In fact, these rules are neither necessary nor sufficient to guarantee substitutability ([Sou96b]). Fig. 6, in (1) the precondition of a is strengthened (definition domain of s_1 included into the s one) and yet c' is substitutable for c. In (2) the postcondition of b is strengthened but the behavior of c' is no longer substitutable.

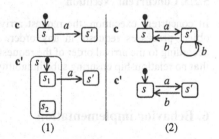

Fig.6. Strengthening precondition (1) and Postcondition (2)

New formulation. Behavior substitutability is separated from pre/postcondition rules. Modifying the ancestor behavior enforcing the rules below guarantees the substitutability in the sense of the failure preorder. The constraint can be formulated: for each sequence *seq* accepted in c_i, c_j is substitutable for c_i if in each state (at least one) reachable following *seq*, c_j accepts at least the requests accepted in any state of c_i reachable following *seq*. Practically, the following rules apply:

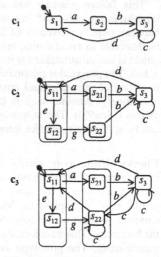

Fig.7. Substitutability

- Adding transitions creating new request sequences (paths in the graph) not accepted in the ancestor is always allowed.

- Removing a transition without removing a path is allowed (decreases non-determinism).
  Fig. 7, c_2 is substitutable for c_1 and c_3 for c_2. In c_2, the two descendants of s_2 inherit the transition $s_2 \xrightarrow{b} s_3$. On the other hand, s_1 is partitioned into two states but only $s_{11} \xrightarrow{a} s_{21}$ is preserved from $s_1 \xrightarrow{a} s_2$, the three other transitions being removed. This removing has no effect on the paths that remain of the form $ab[c]*d$ (*e* plays no role as it is unknown in c_1). In c_3, the transition $s_3 \xrightarrow{c} s_3$ could be removed.

- Adding transitions creating new paths for an existing request sequence *seq* of the ancestor (thus increasing non-determinism) is allowed providing each new state reachable by *seq* accepts at least the paths accepted by any state of the ancestor also reachable by *seq*.
  Fig. 7 in c_3, the transition $s_3 \xrightarrow{c} s_{22}$ introduces non-determinism but the paths $ab[c]*d$ and $egb[c]*d$ remain always allowed. Adding $s_{22} \xrightarrow{d} s_{11}$ does not affect substitutability as the path *egd* was not possible in c_2.

5.2.3. Concurrent execution

In concurrent execution, the request arrival order is not meaningful and the active object processes requests in any order. As an object does not behave the same according to the arrival order of the requests, the context is not favorable, and it seems that no relationship ensuring substitutability has been established.

6. Behavior implementation

The main problems to solve are the choice of a state representation, the control of the transitions and the comparison of states (guard [*in* state] on a transition). First the sample Fig. 2 is completely specified to illustrate the solutions. The Fig. 8 shows the

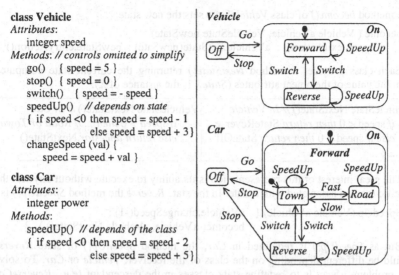

```
class Vehicle
Attributes:
    integer speed
Methods: // controls omitted to simplify
    go()    { speed = 5 }
    stop()  { speed = 0 }
    switch()   { speed = - speed }
    speedUp()   // depends on state
    { if speed <0 then speed = speed - 1
                else speed = speed + 3}
    changeSpeed (val) {
        speed = speed + val }

class Car
Attributes:
    integer power
Methods:
    speedUp()  // depends of the class
    { if speed <0 then speed = speed - 2
                else speed = speed + 5}
```

Fig. 8. Code sample and state diagrams for two classes

state diagrams (fictitious and incomplete!) of the classes *Vehicle* and *Car*. An instance of *Car* is substitutable for an instance of *Vehicle*. The code is just given as a sample, without checks or state tests (algorithmic language with blocs between brackets).

6.1. State representation

Hierarchy of state classes. A first solution is to represent states using a class hierarchy derived either from the actual object ([Rum93]) or from a special state class (*design patterns* [Gam94]) as in Figure 9. Each state class *X* has only one instance (classes without attribute) named *StateX* to make simpler. The attribute *state* in *Vehicle* expresses the association with *VehicleState*, and the method *SetState()* changes this attribute. Objects forward the received messages to the current state, for instance in *Vehicle*:

    Go() { state.Go(self) }

In the state *Off* the method *Go()* calls the suitable object method (lower-case first letter) to change attributes while conforming to encapsulation:

    Go (Vehicle aVehicle)
    { aVehicle.go();
        become(aVehicle, aVehicle.NextState()) }

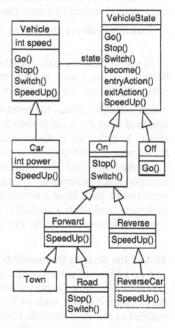

Fig. 9. State representation based on classes

The method *become()* of class *VehicleState* sets the new state:

become (Vehicle aVehicle, VehicleState newState)
      { exitAction() ; aVehicle.SetState(newSate) ; newState.entryAction() }

Each class provides a method *NextState()* returning the current state computed from the value of the object attributes (*StateX* is the instance of *X*):

VehicleState NextState() // In Vehicle	VehicleState NextState() // In Car
{ if speed < 0 then return StateReverse	{ if *speed* > 0 then ... // Road or Town
else if speed = 0 then return StateOff	else return ancestor.NextState()
... }	}

The great interest of this representation is its ability to execute without any test the specific method associated with a state. In the state *Reverse* the method *SpeedUp()* is:

SpeedUp (Vehicle aVehicle) { aVehicle.changeSpeed(-1) ;
                  become(aVehicle, aVehicle.NextState()) }

But as *SpeedUp()* is redefined in *Car*, the method *SpeedUp()* of state *Reverse* should be different depending on the class of the receiver, *Vehicle* or *Car*. To solve this problem, a way is to redefine state classes in the descendant (e.g., *ReverseCar* Fig. 9), but this is an inheritance anomaly. Moreover, the method *NextState()* should be redefined in each class (state *Reverse* replace with *ReverseCar*). Another way, also leading to inheritance anomaly (when *SpeedUp()* is not modified in all the states) but more convenient, is to add a method *SpeedUpCar()* for *Car* in every state it is enabled. Finally, to avoid inheritance anomaly, an obvious way is to execute an object class method instead of a state class method (with the drawback of state test):

SpeedUp (Vehicle aVehicle) { aVehicle.speedUp () ;
                  become (aVehicle, aVehicle.NextState()) }

In such a case, the state class hierarchy is not at all interesting, its major drawback being to increase very quickly the number of classes and methods. To deal with concurrent states, the easier way is to build the corresponding flat state machine, which will still increase the number of classes. The transition *Road* $\xrightarrow{Switch}$ *Reverse* inherited from *Forward* is not allowed in the state *Road*. Hence the method *Switch()* should be redefined in *Road* to catch forbidden use (idem for *Stop*), thus increasing the number of methods.

On the other hand, this solution guarantees that an object in a state is always in its ancestor states (inheritance). Thus it is easy to implement a method finding that an object in the state *Road* is also in the state *On*. Processing of the *entry* and *exit* actions is obvious, and when a method is not allowed, the default error method defined in the root *VehicleState* is executed but it may be redefined in the descendants.

Test of the state in the methods. In the chosen solution the transitions are implicit, and no additional class is required. Each object class has a method *CurrentState()* (similar to *NextState()* above) that returns the current state ID. A table *T* is built from the flat state diagram, such as *T[methodID,stateID]* is true if the method *methodID* is enabled in the state *stateID*. Methods keep always the same ID, while state IDs are modified to remain sequential (*Forward* is no longer a state in *Car*).

As this coding does not preserve the state inheritance relationship, an additional hierarchical code is associated with the states (Table 1). This code is built prefixing the ancestor code (010 for *Forward*) with the code of the state in the class (01 for *Road*). Each class holds a mask (111 for *Vehicle*) that defines the used bits (multiple inheritance requires disjoint masks). The question "Is the state *s* of class *X* an ancestor of *u*" becomes: "u bitAnd maskOfX = s". The comparison of an object current state with the state *aState* of a class (ancestor) *X* is written:

currentState() = aState bitAnd X.Mask() // *Mask is a class method*

Each class holds a mapping table giving the hierarchical code from the sequential code returned by *CurrentState*. Obviously, this implementation requires a tool, and only small tables can be built manually. Unlike the previous one, this solution requires data (tables) but their size is small (method number x state number bits, and state number x 32 or 64 bits for hierarchical state codes).

Sequential execution. Each method is modified according to the template:

```
... M( ... )
{    if EnabledTransition ( M_ID ) then  ... // Normal execution
     else ... // Specific processing of error, optional
}
```

The test at the beginning of the method detects errors (this solution is not affected by hidden transitions). The methods *CurrentState()* and *EnabledTransition()* should be written:

```
boolean EnabledTransition(methodID m_ID)

{    if T[m_ID, CurrentState() ] then return true // T[i,j] true if i allowed in state j
     else ... // Common error processing
          return false
}
```

Transitions being not necessarily inherited, the table *T* had to be redefined in each class (class variable). As in the previous solution, the state is computed once at method entry instead of exit. It can be computed only for calls coming from the outside since intra-object calls are in principle error-free after debugging. An additional table giving the possible states at the end of the methods allows controlling the exit state ([Lec93]). It is not always convenient to add *entry* and *exit* actions at the end of methods, so another way is to gather all actions in one method.

	Mask	States (stateID)			
Vehicle		Off (0)	Forward (1)		Reverse (2)
Hierarchical code	111	001	010		100
Car		Off (0)	Road (1)	Town (3)	Reverse (2)
Hierarchical code	11111	00 **001**	01 **010**	10 **010**	00 **100**

Table 1. Hierarchical coding of states (bits)

Concurrent execution. The tables are here processed differently as requests forbidden in the current state remain in the request queue of the active object. Object methods are not modified. The current state is computed at the end of methods and kept, the *entry* and *exit* actions being executed at the same time. To manage dependencies between object states, a way is to notify all the concerned objects of state changes. The main loop of the active object "controller" looks like (affected by hidden transitions):

```
... EventsLoop( ... )
{
      ...
      message = queue.NextMessage()
      if EnabledTransition ( message.Method_ID() ) then
      {         ... // Normal execution of the method
                UpdateCurrentState()      // exit action, new state, entry action
      }
}
```

7. Conclusion and related works

The mapping between transition and method execution is widely used. The partition of state definition domains in the descendants is rational ([Lec93] [Pae94]). The relationship between state and attributes stay hidden so as not to export class implementation details, while public state IDs allow useful tests (*if state ≠ Empty then...*). The state machine adaptation to increase concurrency fits well with objects running independently. As a result, in case of shared attribute, a state may change without request receipt (non-stable), but this problem can be avoided processing state dependencies (not detailed). Keeping requests into the active object queue until they are enabled solve the problem of the arrival order that generally is not meaningful in concurrent execution.

Behavior inheritance constraints are mostly restricted to pre/postcondition rules, both in formal ([Lis94][Dha96]) and informal approaches ([Coo94][Pae94][UML99]), thus failing to bring the expected guarantees. In [Dha96], pre/postcondition rules are weaker than in [Lis94] (and in [Mey88]), but none of them provides a practical way to express constraints on traces. In [Coo94], state domains (invariant) are defined, but inheritance does not enforce their inclusion (hence a state is not always in its ancestor state), and inheritance rules aim at preserving the inherited behavior and enforcing the pre/postcondition rules (hence actual guarantees are unknown).

Not surprisingly, subtyping and inheritance behavioral constraints are close, since in current OOLs inheritance should ideally imply subtyping. The failure preorder, also used in [Nie93] to define substitutability but in a subtyping context, brings noticeable guarantees, although it is not always safe as errors may remain when the behavior is non-deterministic. One can assume that no simulation relationship is stronger than the failure preorder (in the article context), but this still has to be proved. In the meantime, this compromise preserves the greater expressive power compatible with safety. A tool checks the constraints and guarantees are precisely known.

Generally, proposals to implement states using classes don't take into account inheritance (e.g., [Rum93][Gam94]). Extending these solutions to inheritance leads to serious problems and except cases with many state-specific methods and few method redefinition, it will be easier to test the state in each method. The proposal is particularly well adapted to concurrent execution since processing of queued messages is centralized and methods are not modified. Another essential feature is the ability to compare object states whatever the object class.

A tool is useful to implement this solution for all the transformations to be made automatically. The method that computes the current state is built from the state diagram improved with the state definition domains (*taggedValue*). All the needed data as the table of the enabled methods in each state are also built automatically. The method that checks enabled transitions is always the same, and object methods all are modified according to the same template. Finally, substitutability checking requires a tool as soon as the number of states and paths become important.

A prototype linked to a market UML CASE tool implements the proposal in C++. The present prototype takes care on substitutability checking and a great part of additional code generation, while in the CASE tool code generation had to be modified. The basic classes of the active object also have been implemented to achieve the initial goal. An industrial solution would only require a tool linked with a CASE tool generating the code for different target languages and operating systems, and frameworks of basic active object classes to complete according to requirements.

References

[Agh86] Agha G., *Actors: a Model of Concurrent Computation in Distributed Systems*, MIT Press, Cambridge, 1986
[Bro84] S.D. Brookes and C.A.R Hoare and A.W. Roscoe, "A Theory of Communicating Sequential Processes", *Journal of the ACM*, Vol 31(3), 1984, p.560-599
[Cle93] R. Cleaveland, J. Parrow, B. Steffen, "The Concurrency Workbench: A Semantics Based Tool for the Verification of Concurrent Systems", *ACM Trans. on Programming Languages and Systems*, Vol.15(1), 1993, p.36-72
[Coo94] Cook S., Daniels J., *Designing Object Systems*, Prentice Hall, 1994
[Dha96] K.K. Dhara, G.T. Leavens, "Forcing Behavioral Subtyping Through Specification Inheritance", *Proc. Int. conf. on Software Engineering*, 1996, pp258-267
[Fer91] J.C.Fernandez, L. Mounier, "On the Fly Verification of behavioural Equivalences and Preorders". *Workshop on Computer-Aided Verification*, Springer Verlag, LNCS 575, 1991
[Gam94] Gamma E., Helm R., Johnson R., Vlissides J., "*Design Patterns : Elements of Reusable Object-Oriented Software.*", Addison Wesley, 1994.
[Har87] Harel D., "Statecharts: A Visual Formalism for Complex Systems", *Science of Computer Programming*, 1987, p.231-274
[Har97] Harel D., Gery E., "Executable Object Modeling with Statecharts", *IEEE Computers, July 1997*, Vol 30(7), pp31-42
[Hoa85] C.A.R. Hoare, *Communicating Sequential Processes*, Prentice-Hall, 1985
[Java] Java Documentation, Sun Microsystems
[Kaf95] D. G. Kafura and R. G. Lavender. "Concurrent Object-Oriented Languages and the Inheritance Anomaly", in *Parallel Computers: Theory and Practice*, IEEE Press, 1995, pp.165-198
[Lec93] H. Lecoeuche, J.L. Sourrouille, "Introducing states in the object model.", *Proc. TOOLS'93 USA*, p.69-81

[Lis94]   B.H. Liskov, J.M. Wing, "A Behavioural Notion of Subtyping", *ACM Trans. on Programming Languages and Systems*, Vol 16(6), 1994, p.1811-1841

[Mat93]   Matsuoka S., Taura K., Yonezawa A., "Highly Efficient and Encapsulated Re-use of Synchronization Code in Concurrent O-O Languages", *OOPSLA'93, ACM Sigplan Notices*, 1993 pp. 109-126.

[Mey88]   B. Meyer, *Object Oriented Software Construction*, Prentice Hall, 1988

[Nie93]   O. Nierstrasz. "Regular types for active objects", *OOPSLA'93, ACM Sigplan notices* 1993, Vol. 28(10), p.1-15

[Pae94]   B. Paech, B. Rumpe, "A new Concept of Refinement used for Behaviour Modelling with Automata", *Formal Methods Europe*, Springer Verlag, LNCS 873, p.154-175

[Pap95]   Papathomas M., "Concurrency in Object-Oriented Programming Languages", *In. Object-Oriented Software Composition*, Prentice Hall, 1995, pp.31-68

[Rum93]   J. Rumbaugh, "Controlling code. How to implement dynamic models", JOOP, Vol.6(2), 1993, pp.25-30

[Sou96a] J.L. Sourrouille, "Should subclasses inherit of all states and transitions?", *Report on Object Analysis and Design*, Vol 2(6), 1996, p.19-21

[Sou96b] J.L. Sourrouille, "A framework for the definition of behaviour inheritance", *JOOP*, Vol 9(1), 1996, p.17-21

[Sou97]   Sourrouille J.L., "Une sémantique pour l'héritage de comportement", LMO'97, 1997, pp.131-146

[Str92]   B. Stroustrup, *The C++ Programming Language*, Addison-Wesley, 2nd edition, 1992

[Tom89]   C. Tomlinson and V. Singh. "Inheritance and synchronization with enabled-sets", *OOPSLA'89, ACM Sigplan Notices*, 1989, pp. 103-112

[UML99]   "OMG Unified Modeling Language Specification (draft)", Version 1.3, March 1999.

[Wan97]   E.Y. Wang, H.A. Richter, B.H.C. Cheng, "Formalizing and Integrating the Dynamic Model within OMT", Proc. *ACM ICSE* 1997, p.45-55

[Weg88]   P. Wegner and S. Zdonik, "Inheritance as an Incremental Modification Mechanism or What Like Is and Isn't Like", Proc. *ECOOP* 1988, Springer Verlag, p.55-77

UML Collaboration Diagrams and Their Transformation to Java

Gregor Engels[1], Roland Hücking[2], Stefan Sauer[1], and Annika Wagner[1]

[1] University of Paderborn, Dept. of Computer Science, D 33095 Paderborn, Germany
engels|sauer|awa@uni-paderborn.de
[2] SAP AG, Lo. Dev. PP-PI, Neurottstr. 16, D 69190 Walldorf, Germany
roland.huecking@sap-ag.de

Abstract. UML provides a variety of diagram types for specifying both the structure and the behavior of a system. During the development process, models specified by use of these diagram types have to be transformed into corresponding code. In the past, mainly class diagrams and state diagrams have been considered for an automatic code generation. In this paper, we focus on collaboration diagrams. As an important prerequisite for a consistent transformation into Java code, we first provide methodical guidelines on how to deploy collaboration diagrams to model functional behavior. This understanding yields a refined meta model and forms the base for the definition of a transformation algorithm. The automatically generated Java code fragments build a substantial part of the functionality and prevent the loss of important information during the transition from a model to its implementation.

Keywords: Collaboration diagram, methodical guidelines, code generation, Java, pattern-based transformation algorithm

1 Introduction

The Unified Modeling Language (UML, [8, 9, 13]) provides a variety of diagram types for an integrated specification of both the structure and the behavior of a system. Collaboration diagrams belong to the behavioral diagrams like sequence diagrams, statecharts and activity diagrams.

Tools to support the development of software, so-called CASE tools, often do not only support the analysis and design of systems, but also contain code generators to automatically create code fragments of the specified system in a target programming language. Unfortunately, the capabilities of code generators to transform the design to an implementation are often restricted to produce class definitions consisting of attributes and operation signatures captured in class diagrams, but not the methods to implement the procedural flow within the operations.

Using also behavioral information for code generation prevents the loss of substantial information during the transition of a model to its implementation. Existing approaches in this direction transform statecharts into executable code

[5, 1, 2]. Statecharts are used as object controllers for specifying when an object is willing to accept requests. CASE tools supporting code generation from statecharts are e.g. Statemate [15], Omate [5], and Rhapsody [11].

In contrast, it is our aim to transform the specification of the *functional behavior* of objects into code fragments. The functional model can be described in terms of interactions between objects in an abstract way by UML interaction diagrams.

The only tool known to us that is capable of generating code from interaction diagrams is Structure Builder [16]. Sequence diagrams are used there, but code is not directly generated from them, but from an intermediate representation called Sequence Methods. Sequence Methods are based on the concept of Interaction Graphs [14], resulting from the Demeter project [7], which are directed labeled trees with nodes representing object variables and edges representing actions. They basically resemble a representation of additional information that, in agreement with our approach, needs to be interactively entered by a developer to extend the interaction modeled in UML diagrams. Such information being necessary for the generation of working Java code is e.g. how objects can be accessed, how they are transported between methods, and instantiation of links etc. These details can not be specified in sequence diagrams, but most of them are already captured in collaboration diagrams.

Thus, we selected collaboration diagrams from UML interaction diagrams as the source for the transformation process since, in contrast to sequence diagrams, they do not only supply the message flow information of an interaction, but also the underlying structural information building the context of the interaction, i.e. the links via which messages are sent. Additionally, we stay within the diagram types of UML whereas Sequence Methods are outside the UML.

Java was selected as the target language because it is a purely object-oriented programming language of growing importance and it offers concepts for concurrent programming to extend the transformation mechanisms to parallel flow of control.

The paper is organized as follows: In Sect.2, we introduce the main features of collaboration diagrams and state methodical guidelines for their deployment. A general overview of the transformation approach for collaboration diagrams based on the transformation of class diagrams is given in Sect.3. The next section introduces a refined meta model which forms the basis for a detailed description of the transformation algorithm for collaboration diagrams in Sect.5. The paper ends with some concluding remarks and perspectives.

Further details can be found in an extended version of this paper that is available as a technical report (see [4]).

2 Deploying UML Collaboration Diagrams

In this section, we outline a methodical approach on how to deploy UML collaboration diagrams to model functional behavior. This approach is based on the general UML specification [8, 9], but it extends it by additional pragmatic guidelines and constraints. A systematic usage of this approach will ensure that collaboration diagrams describing the functionality of methods can automatically be translated into corresponding Java code. In the following, we assume that the reader is familiar with the standard UML notations (see [8, 13]).

In general, collaboration diagrams can be used to model system functionality or more precisely the control flow within a system. This is described by sending messages between instances of classes. Collaboration diagrams are feasible to model not only the behavioral, but also the structural context of such an interaction, called a *collaboration*.

In [8], the following two possibilities, among others, of deploying collaboration diagrams in the above sense are introduced:

- Method: specify the implementation of an operation as an interaction,
- Use case: describe the functionality of a main operation of a system on an abstract level.

Both kinds of usage differ not only on their level of abstraction, but also in their main intention. Whereas use cases are deployed in earlier phases of modeling, the method-oriented usage is already close to implementation. Use cases describe scenarios. They are intended to examplify a certain situation, i.e., very often they describe only one possible control flow path. In contrast, within a method specification, the general situation with all posssible control flow paths has to be modeled. As a consequence, collaboration diagrams are used on the instance level in the first case, describing the interaction of different objects with each other. In the second case, they are used on the type level possibly containing iterations or conditional flows [9]. Type level modeling is in accordance with the specification of methods within classes of object-oriented programming languages.

With this background, two main steps can be identified within the development process producing the systems functionality. The first task is stepping from different scenarios to the general situation. And the second task is stepping from a model of the general situation to its implementation. In this paper, we concentrate on the second task, where one type level model for each method, i.e., exactly one collaboration diagram per method, serves as the basis for automatic code generation.

The first task of combining different instance level collaboration diagrams specifying the same operation can not be done automatically in the general case. Collaborations define views on the classes specified in the class diagram. Therefore, problems in combining several collaboration diagrams resemble typical problems of view integration [3]. Input by a developer is necessary to handle conflicts or to specify details of combination like contextual constraints or conditions. This interactive intervention should receive support by code generation tools. Situations where an automatic combination is possible are, e.g., mutual exclusive execution conditions for different occurences of the same operation for branching as well as iteration.

On the other hand, collaboration diagrams are not able to fully model the functionality of an operation. One restriction is their inability to model operations on data types, i.e., primitive base types like Integer, Real or predefined enumeration types like Boolean, whose values do not posses an identity. Thus, collaboration diagrams can not serve as a fully-fledged visual programming language. Moreover, usually not all aspects of a system are completely modeled. Exception handling, for example, will usually be separately specified and added later in the implementation. For these reasons, code generation from collaboration diagrams is by their definition restricted to object interactions. Generating this kind of working code, prevents the loss of information during the step from modeling to implementation and simplifies the task of transition what states our objectives.

Before we start explaining our approach in more detail, we introduce a running example for a system to be modeled. Figure 1 shows the class diagram of an example application where a Company object is related to zero or more Store, Order, and Delivery objects. A Store is related to multiple Delivery objects, which in turn are related to one Customer and one Order. A Customer can place several instances of Order, and one or none Delivery objects belongs to an Order.

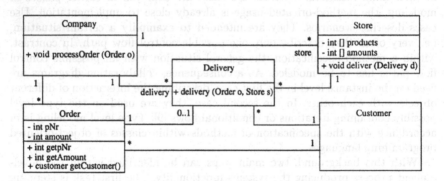

Figure 1. Class diagram of a modeling example

A typical scenario within that setting is the situation where a customer orders a product from the company. On the use case level one would model that scenario by sending an order from a customer to the company, followed by forwarding that order from the company to one of its stores, followed by delivering the ordered products from the store to the customer.

After the step of refining and combining different use cases into a method-oriented specification one might end up with a collaboration diagram for a method processOrder as depicted in Fig.2. Here, the company first obtains the product number pNr and the ordered amount a of that order using defined access functions. It then checks all stores to find one that can supply the requested amount of the demanded product. A delivery is created, and the selected store is called to send it out. Finally, the delivery is added to a container holding all deliveries of the company.

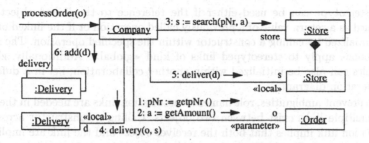

Figure 2. Collaboration diagram for the modeled example

We will now introduce methodical guidelines as a foundation of the later on presented transformation approach. As a consequence of deploying a unique collaboration diagram for specifying the implementation of one operation, two basic model entities build the basis for the forthcoming concepts:

- The *specified operation* is the operation whose implementation is modeled by the collaboration diagram (processOrder in Fig.2).
- The *target object* is that object on which the specified operation is called. The specified operation belongs to the class of the target object, its signature must be declared in the operation compartment of the corresponding class in the class diagram (:Company in Fig.2).

As a result of the refinement and combination of different scenario-oriented collaboration diagrams we obtain a collaboration diagram with a single level of nesting. Thus, we specify which operations are called in the specified operation directly, but we do not consider those that are subordinately called within these nested operations. We consider this to be meaningful when we specify the implementation of an operation, since the subordinately called operations belong to collaboration diagrams for the nested operations. This is alike the definition of procedures and procedural calls in programming languages. As an implication, the target object is the sender of the call message for all operations in a collaboration diagram except for the specified operation.

One end of a stereotyped link must be directly connected to the target object (see Fig.2). Conventional links based on associations can also be indirectly connected to the target object. They can be accessed by traversing along a path of links of which only the first may be a stereotyped link. If a link with the stereotype «parameter» (e.g. to :Order in Fig.2) is used, then a reference to the object on the other end of the link must be transported to the target object as a parameter of the specified operation. The names of objects that are connected to the target object by a «parameter» link must be identical to the parameter names of the corresponding operation in the collaboration diagram.

Stereotyped links of kind «local» (e.g. between :Company and s:Store in Fig.2) depict that the linked objects are locally accessible within the specified operation.

This stereotype can be used either if the reference to the linked object was obtained as a return value of a previously called operation or if the linked object was initialized by calling a constructor within the specified operation. The same restrictions apply to stereotyped links of kind «global». Additionally, global variables can also be initialized within another collaboration, i.e. in a different collaboration diagram.

To prevent ambiguities, role names on association links are needed in the case that multiple links exist between two objects. Calling a constructor across an association link implies that both the receiver object and the link are implicitly {new} (see 4: in Fig.2). Thus, the constraint is optional. In contrast, adding to and deleting from multiobjects (notion for container in UML collaboration diagrams) can be explicitly defined by the modeler in order to specify the exact sequence of messages (see 6: in Fig.2).

Objects may not be marked with the constraint {destroyed} because Java does not contain a predefined destructor. Otherwise, one would have to solve the problem that all references to that object must be deleted to make the garbage collector delete the object, even those references specified in other collaborations.

Further details of the implications of our approach will be shown in Sect.4 where the refined meta model for collaborations is presented.

3 Transformation Approach

In Sect.2, methodical guidelines on how to deploy collaboration diagrams have been explained. Following these, all collaboration diagrams to be translated have a well-formed structure. This is an important prerequisite and enables a systematic translation of collaboration diagrams into corresponding Java code.

The translation algorithm for collaboration diagrams is based on a standard algorithm for translating class diagrams. The underlying idea is to translate class definitions into corresponding Java class definitions and to translate associations into bi-directional references between the two participating classes. This standard algorithm has been refined, e.g., with respect to automatically generated "get" and "put" access operations for attributes or a generic search operation to select certain objects from a set of existing objects. Further details on the refined class translation algorithm can be found in [4].

The basic idea of the overall transformation algorithm from a class diagram and associated collaboration diagrams into corresponding Java code is to identify standard patterns in a given diagram and to translate those patterns into corresponding Java code. This *pattern-based transformation algorithm* will be presented in a technical, formal way in Sect.5. Here, we give two simple examples to sketch informally how this pattern-based translation works.

First, Fig.3(a) shows a part of the collaboration diagram given in Fig.2 where operation getpNr() is sent via a parameter link with role o to an object of class Order. This collaboration diagram is depicted in the lower left part of Fig.3(a), while the corresponding class diagram can be found in the upper left part. The right hand side shows the generated Java code for such a parameter link pattern within a collaboration diagram.

Second, Fig.3(b) shows another pattern taken from Fig.2. Here, the collaboration diagram comprises a pattern consisting of a local link combined with a newly created object of class Delivery. The resulting Java code comprises a definition of a local variable d of type Delivery, as well as the invocation of the constructor of class Delivery in order to create a new instance.

The complete structured and pattern-based transformation algorithm will be explained in Sect.5. In order to be able to describe certain patterns within a class or collaboration diagram, a uniform internal representation of diagrams is an important prerequisite. As known from the UML language definition, such an internal representation can best be defined by a meta model. Therefore, the next section will present an adapted UML meta model, which incorporates the restrictions introduced in Sect.2.

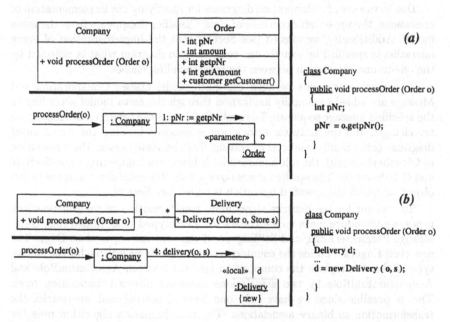

Figure 3. Transformation of (a) parameter and (b) local links into Java code

4 Refined Meta Model

Based on the UML meta model, we present a refined meta model for collaborations, that has been adapted according to the assumptions and restrictions described in Sect.2. The methodical guidelines for deploying collaboration diagrams to model method implementations have been integrated and are thus reflected on the meta model level now. Since the transformation algorithm presented in the next section is based on this meta model representation, the methodical guidelines also affect the code generation. The benefits of this adapted meta model are two-fold. First, the methodical guidelines have become part of the modeling language. Thus, only well-structured collaboration diagrams can be instantiated from this meta model. Second, the adapted meta model shows a granularity which is very well suited as basis for the pattern-based transformation algorithm.

Figure 4 depicts the changes to the original UML meta model. Elements that are replaced or deleted are crossed out, while new or changed meta classes and associations are shaded. Note that associations connected to new classes are also new even if they are not explicitly marked for simplicity. Some classes from other meta model packages of UML have been included, but all changes to existing associations have been marked.

Due to the use of collaboration diagrams for specifying the implementation of operations, the upper left occurence of class Classifier disappears from the meta model. Additionally, we argued (see Sect.2) that the implementation of every operation is specified by exactly one collaboration diagram what is reflected by the one-to-one association between the corresponding classes.

Two new associations between the meta model classes Collaboration and Message are added to simplify navigation through the meta model according to the specified message sequence. The multiplicities on the predecessor association are changed and the activator association is removed because the transformed diagrams contain only one level of nesting. For the same reason, the association to ClassifierRole with the role name sender is bent, now connecting ClassifierRole and Collaboration: The sender for messages within this collaboration is the target object on which the specified operation is called (see Sect.2).

To account for the distinct algorithmic transformation of the different link types, we introduce meta model classes for stereotyped links LocalEdge, GlobalEdge, ParameterEdge, and SelfEdge, and the abstract super class Edge. The new class EdgeEnd builds the counterpart to AssociationEndRole for the stereotyped links. We replace the composition relation between AssociationRole and AssociationEndRole by two associations modeling directed association roles. This is possible since we have only one level of nesting and we restrict the transformation to binary associations. The transformation algorithm uses the roles to and from to traverse association links in the direction of message flow.

New is also the abstract meta class Node as a super class of ClassifierRole. Its purpose is to hold an attribute of type N_T_Kind representing the default constraints {new} and {transient} that can be attached to an object (ClassifierRole) in a collaboration diagram. An equivalent attribute of the super type N_T_D_Kind has been added to the class AssociationEndRole. Due to this extension, the mapping of constraints on the appropriate subclasses of the meta model class Action [8] is no longer needed.

We further introduce a meta model class Expression and subclasses (not shown on the diagram) for data values, operators and their operands, etc. to

481

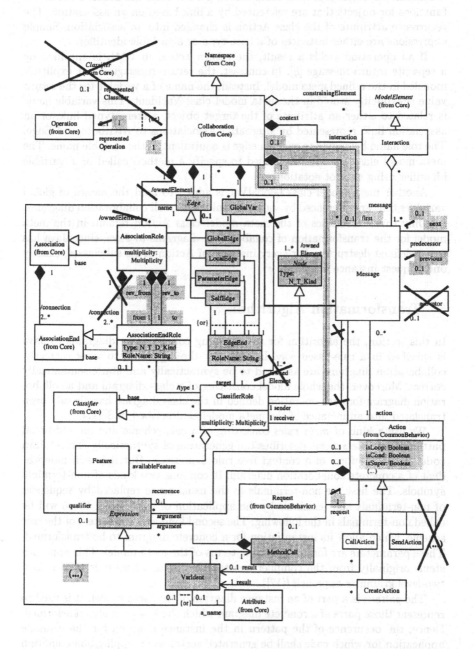

Figure 4. Extended meta model for transforming collaboration diagrams

decompose expressions in their components. This enables the definition of access functions for objects that are referenced by a link based on an association. The recurrence attribute of the class Action is changed into an association. Simple expressions are either instances of a base type or a variable identifier.

If an operation yields a result, the return action in UML is specified by a separate return message [8]. In contrast, the return message is not explicitly modeled in the refined meta model. Instead, the name of a variable for the return value is explicitly stored in the meta model class VarIdent. This variable name is related to either an attribute of the target object, a stereotyped link, or an association link, represented by alternative associations to Attribute and Edge. The role name belonging to such an edge is equivalent to the variable name. The meta model class MethodCall is used to specify a method called on a variable identifier using the dot notation.

Another meta model class GlobalVar is added to hold the names of global variables that are referenced by «global» links within all collaboration diagrams.

Only three subclasses of the meta model class Action remain in the meta model for the transformation of collaboration diagrams to Java, since Java has no predefined destructor. For every instance of Action or its subclasses, exactly one Request instance is linked.

5 Transformation Algorithm

In this section, the algorithm for transforming collaboration diagrams to Java is specified in a rule-based way. In order for the algorithm to work correctly, collaboration diagrams are assumed to be syntactically and static-semantically correct. Moreover, the whole model consisting of a class diagram and a collaboration diagram for each operation defined in the class diagram has already been translated into an instance of the meta model as described in Sect.4.

We use a kind of *meta rules* consisting of a rule scheme and an additional pattern. The *rule scheme* describes the generation of syntactically correct Java code. It has the form of a context free rule expression. But it is still independent of a concrete collaboration diagram. It contains two kinds of non-terminal symbols. The first are non-terminals in the usual sense replaced by sequences of non-terminals and terminals by the application of rules. Only those will be called non-terminals in the following. The second kind are parameters of the rule scheme, which allow its instantiation for a concrete diagram to be transformed. These parameters are formulated using terms of the meta model. This approach stems originally from the compiler construction area, where it is known as a two-level grammar approach ([17]).

The *pattern* is a part of an instance diagram of the meta model. It is used to represent those parts of a concrete diagram which shall be actually transformed. Hence, the occurence of the pattern in the instance diagram for the example application for which code shall be generated serves as an application condition for the whole meta rule to be applied. Moreover, the concrete occurence links together the general code generation possibilities, described by the rule scheme,

483

and the actual elements of the concrete collaboration diagram that has to be transformed. The parameters of the rule scheme occur in the pattern and can hence be replaced by actual values in order to instantiate the rule scheme.

Figure 5 shows two meta rules for the transformation of class diagrams. These meta rules will be used in the following to illustrate how the algorithm is specified in principle. On the left hand side, the part of the class diagram actually translated by the meta rule is shown. In the middle, we give its translation to part of an instance of the meta model, which forms the pattern. On the right hand side, the rule scheme for generating Java code is shown. Words in capital letters denote non-terminal symbols, whereas words in small letters denote terminal symbols if they are underlined, or they denote parameter expressions over the pattern if not. These parameters will be evaluated to terminal symbols as soon as a concrete occurence of the pattern is chosen, leading to an instantiation of the rule scheme for the concrete diagram.

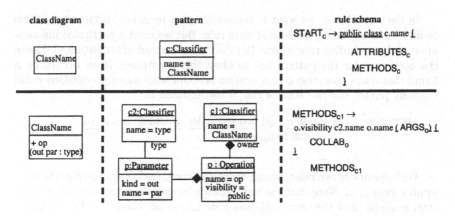

Figure 5. Meta rules for class diagram

The first meta rule shown in Fig.5 allows to transform a single class into the frame for a class declaration in Java. Here, c refers to the instance of classifier which represents the class in the instance of the meta model. Hence c.name is a parameter which will be replaced by the name of the class, i.e., the concrete value of this attribute in an occurence of the pattern. The non-terminal symbols $START_c$, $ATTRIBUTES_c$ and $METHODS_c$ will also be instantiated with more concrete non-terminal symbols. The name c of the classifier object is used to keep track of the concrete classifier object currently dealt with during the next steps of code generation. It already determines partly the occurence of the pattern belonging to the meta rule for replacing this non-terminal. The second meta rule shown in Fig.5 serves for the generation of the method frames for each operation defined in the class diagram in an analogous way. The instantiation of the non-terminal symbol $COLLAB_o$ will be replaced by the code generated for the collaboration diagram of this operation.

484

Note that the meta rules are only applied once for each occurence of the according pattern. Different occurences may overlap. For example, in case of the second rule the same classifier object may occur as owner of an operation and as parameter of another or even of the same operation.

Consider again our example application introduced in Sect.2. The class diagram shown in Fig.1 can be transformed into Java code using the above meta rules in the following way: We search for an occurence of the pattern of the first meta rule in the instance diagram of the meta model. Classifier c is mapped to classifier com, whose name attribute has the value "Company". For this occurence of the pattern we instantiate the rule scheme leading to

$$\text{START}_{com} \longrightarrow \text{public \underline{class} Company \{}$$
$$\text{ATTRIBUTES}_{com}$$
$$\text{METHODS}_{com}$$
$$\underline{\}}$$

In the second step, we want to replace the non-terminal METHODS_{com}. This could be done by using the second meta rule. But we need a particular instantiation of the according rule scheme (METHODS_{com} instead of METHODS_{c1}). Hence the occurence for the pattern has to obey this constraint. If an occurence is found that maps operation o to operation procOrd with name processOrder and visibility public, the rule scheme can be instantiated to

$$\text{METHODS}_{com} \longrightarrow \text{public \underline{void} processOrder (ARGS}_{procOrd}) \{$$
$$\text{COLLAB}_{procOrd}$$
$$\underline{\}}$$

With the above two rules, we can deduce a primitive class frame from the start symbol START_{com}. Note that the instantiation process leads to a set of different start symbols since the generated Java code has to be stored in different files.

We now advance to the transformation of collaboration diagrams. We start with meta rules for replacing the non-terminal COLLAB_o by a sequence of other non-terminals in order to determine the structure of the generated code of the body of a method. First, the local variables have to be declared. Then, we invoke the methods in the order which is indicated in the collaboration diagram by the sequence numbers. Finally, we have to add newly inserted links, which are not used to invoke a method, and to remove links, which are indicated as destroyed. The according meta rule is depicted in Fig.6.

In the sequel, the first two meta rules generated by this substitution are explained in detail. Figure 7 shows the meta rule for declaring local variables. Remember that we also assume that indirectly declared local variables (return values of method invocations referencing objects) are to be represented as local edges in the instance of the meta model. Hence each LocalEdge uniquely represents a local variable, the name of which is stored as the RoleName of its EdgeEnd. The LocalEdge belongs to the collaboration of operation o. The type of a local variable is given by the name of the base (classifier) of the target of the EdgeEnd. This information is represented in the pattern. Moreover, it is used

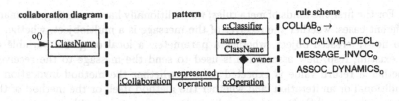

Figure 6. Meta-Rule for splitting of COLLAB

in the rule scheme by the parameters c.name for the type and e.RoleName for the name of the local variable. We add the possibility of declaring more than one local variable within the same operation o by repeating the non-terminal $LOCALVAR\_DECL_o$. Again, different applications of the meta rule imply different occurences of the pattern ensuring that each local variable is declared only once. The meta rule in the lower part of Fig.7 serves for the end of the declaration process. The rule scheme replaces the non-terminal $LOCALVAR\_DECL_o$ by the empty string. It may only be applied, if the upper meta rule is not applicable any more.

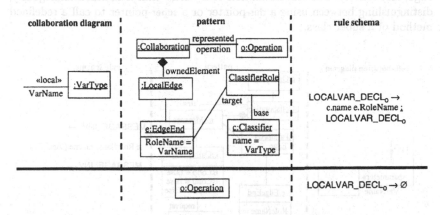

Figure 7. Meta rule for local variable declaration

Now we come to the generation of the real body of an operation, namely the invocation of methods. Generally, we have to generate the method invocation code in the order indicated by the sequence numbers in the collaboration diagram. This order is represented in the meta model by the predecessor edge between messages and by the edge assigning the first message to a collaboration. Hence, we have three kinds of meta rules: The first kind serves for invoking the first method. The second kind traverses the predecessor edge from the previous to the next message. The third kind ends the process. Meta rules of the last kind look like the last one discussed for the local variable declaration above. They are neglected in the following.

For the first two kinds of meta rules, we additionally have to distinguish many different cases: whether the receiver of the message is a multiobject, whether it is a newly created object, whether a parameter, a local or global variable or an existing resp. new association is used to send the message to the receiver, whether a return value is expected or not, whether the method invocation is conditional or an iteration, and whether the method itself or the method of the super class is called. Due to space limitations, we are not able to present all according meta rules in this paper (see [4] for an exhaustive presentation).

Instead, Fig.8 shows as an example a meta rule for invoking a method on a parameter object. It is a rule of the first kind, meaning that the method invocation is the first one in the actually transformed collaboration. A method for operation o is invoked. The kernel of this method invocation is that an operation r.name is called on the parameter object referred to by e.RoleName. The arguments for this call are generated from the non-terminal symbol $ARGS_r$. We omit a more detailed view on that, since we left out the specialization of class Expression in the meta model in Sect.4 that is necessary for this purpose. The meta rules for invoking an operation on a local or global variable look quite similar. Only the parameter edge in the pattern is replaced by a local or global edge, respectively. The transformation of a «self» link is handled analogously, distinguishing between using a this-pointer or a super-pointer to call a redefined method of a super class.

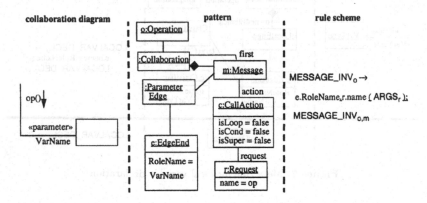

Figure 8. Meta rule for method invocation on parameter object

The pattern of the meta rule for method invocation via an association link differs in that ParameterEdge and EdgeEnd are replaced by AssociationRole and AssociationEndRole, respectively. Other additional requirements on attributes of the CallAction and the receiving ClassifierRole ensure that one deals with the simplest case and not with multiobjects, for instance. Another difference is that the method may not directly be called using the RoleName stored in the AssociationEndRole if sender and receiver are only indirectly linked. Hence we include a non-terminal symbol PATH$_{s,r}$ which has to be replaced by an expression determining the shortest existing link path from the sender s to the receiver r.

In order to allow more than one method invocation in the body of an operation, the rule scheme in the above meta rule generates a new non-terminal symbol MESSAGE_INV$_{o,m}$, distinguished by the differing parameter expression. This second kind of non-terminals for method invocations can be replaced by the second kind of meta rules.

Using the complete set of meta rules as shown for the generation of the code for the class diagram by instantiating the meta rules and reducing the non-terminals to terminal symbols, the following Java code is generated from the collaboration diagram depicted in Fig.2.

```
public class Company {
      ...
      public void processOrder (Order o) {
         Delivery d;
         Store s;
         int pNr;
         int a;
         pNr = o.getpNr();
         a = o.getAmount();
         s = search_stores(pNr, a);
         d = new Delivery(o,s);
         s.deliver(d);
         add_deliveries(d);
      }...
}
```

6 Conclusion and Perspectives

In this paper, we have investigated the modeling of behavior by UML collaboration diagrams and their automatic transformation into Java code. We have introduced methodical guidelines how to deploy collaboration diagrams in a structured way. This formed the basis for the formulation of a transformation algorithm.

The objective of this automatic transformation is to prevent a loss of substantial information during the transition from a model to its implementation. But, this does not imply that UML collaboration diagrams offer a means to specify the behavior of a system completely and that UML can be used as a visual programming language. UML collaboration diagrams focus on the modeling of object interactions, while computations on data values are neglected, and thus have to be added to the generated Java code by hand.

This paper focussed on the transformation of sequential behavior descriptions. The next steps will be to implement the transformation algorithm by extending the often used, commercial tool Rational Rose [10] and to extend the transformation algorithm to the transformation of concurrent behavior as well as of asynchronous and synchronous communication descriptions. The already chosen target language Java will facilitate this development. First results of that extension can be found in [6].

488

Finally, it is intended to investigate whether and how the in this paper reused approach of two-level grammars (cf. [17]) is an appropriate means to specify
and realize easily adaptable code generators for forthcoming versions of UML
and for a visual modelling language in general.

References

1. Ali, J., Tanaka, J.: Generating executable code from the dynamic model of OMT
   with concurrency. In: *Proc. IASTED International Conference on Software Engineering (SE'97)*, San Francisco, 1997, pp. 291–297
2. Ali, J., Tanaka, J.: Implementation of the dynamic behavior of object oriented
   systems. In: *Integrated Design and Process Technology (IDPT)*, Vol. 4, Society for
   Design and Process Science, 1998, pp. 281–288
3. Engels, G., Heckel, R., Taentzer, G., Ehrig, H.: A view-oriented approach to system modelling using graph transformations. In Jazayeri, M., Schauer, H. (eds.):
   Proceedings European Software Engineering Conference (ESEC'97), Zürich, LNCS
   1301, Springer, 1997, pp. 327–343
4. Engels, G., Hücking, G., Sauer, S., Wagner, A.: UML Collaboration Diagrams
   and Their Transformation to Java. Technical Report TR-RI-99-208, University of
   Paderborn, 1999
5. Harel, D., Gery, E.: Executable Object Modeling with Statecharts. *IEEE Computer*, **30** (July 1997) 31–42
6. Hücking, R.: UML Collaboration Diagrams and Their Transformation to Java (in
   German). Master's Thesis, University of Paderborn, September 1998
7. Lieberherr, K.: *Adaptive Object-Oriented Software: The Demeter Method with
   Propagation Patterns*. PWS Publishing Company, Boston MA, 1996
8. OMG: UML Notation Guide, Version 1.1. The Object Management Group, Document ad/97-08-05, Framingham MA, 1997
9. OMG: UML Semantics. Version 1.1. The Object Management Group, Document
   ad/97-08-04, Framingham MA, 1997
10. *Rational Rose 98.* Rational Software Corporation, Cupertino CA, 1998
11. *Rhapsody.* Version 2.1. I-Logix, Andover MA, 1998
12. Rumbaugh, J., Blaha, M., Premerlani, W., Eddy, F., Lorensen, W.: *Object-Oriented
    Modelling and Design*. Prentice-Hall, 1991
13. Rumbaugh, J., Jacobson, I., Booch, G.: *The Unified Modeling Language Reference
    Manual*. Addison-Wesley, Reading MA, 1999
14. Sangal, N., Farrel, E., Lieberherr, K.: Interaction Graphs: Object interaction spec-
    ifications and their compilation to Java. Technical Report NU-CCS-98-11, North-
    eastern University, Oct. 1998
15. *Statemate MAGNUM.* Release 1.2. I-Logix, Andover MA, 1999
16. *Structure Builder.* Version 3.1.5. Tendril Software Inc., Westford MA, 1999
17. A. van Wijngaarden: The Generative Power of Two-Level Grammars. In J. Loeckx
    (ed.): *Automata, Languages and Programming*, 2nd Colloquium, University of
    Saarbrücken, 1974. LNCS 14, Springer, 1974, pp. 9 –16

Towards Three-Dimensional Representation and Animation of UML Diagrams

Martin Gogolla, Oliver Radfelder, Mark Richters

University of Bremen, FB3, Computer Science Department

Abstract. The UML notation is intended to be drawn on two-dimensional surfaces. However, three-dimensional diagram layout and animation may improve comprehension of complex diagrams significantly. The paper concentrates on special UML diagram forms well-suited for advanced visualization. It makes a proposal for representing and animating such UML diagrams in a three-dimensional style.

Keywords: UML Diagram, Visualization, Animation, Graph, Planar Representation, Three-Dimensional Representation.

A journey of a thousand miles must begin with a single step.
Lao-Tzu, The Way of Lao-Tzu.

1 Introduction

The Unified Modeling Language UML [BJR99] aims to become the industrial standard for software development and documentation. But also research projects consider the UML as a serious effort. Work is done in a large number of aspects ranging from questions of semantics to implementation issues.

Describing UML in an over-simplified manner, one could say that the language provides a number of different diagram forms for emphasizing different aspects of software. The diverse diagrams will be typically used differently in the various phases of software development. However, the documents [BJR99, p. 3-7] introducing the UML clearly point out the two-dimensional nature of UML diagrams:

> UML notation is intended to be drawn on two-dimensional surfaces. Some shapes are two-dimensional projections of three-dimensional shapes (such as cubes), but they are still rendered as icons on a two-dimensional surface. In the near future, true three-dimensional layout and navigation may be possible on desktop machines; however, it is not currently practical.

Thus, all diagrams in UML and all examples given in the defining documents possess a two-dimensional nature. This paper supports the vision indicated above stating that three-dimensional layout should be studied. We see two main reasons for this: (1) three-dimensional layout may *enhance* the comprehension of complex UML diagrams, and (2) three-dimensional layout may be *necessary* in situations where the two-dimensional representation is unable to express the intent of the UML diagram. For example, complex diagrams may be structured according to appropriate criteria by pushing uninteresting parts into the background, or graphs, which can only be displayed in the plane with intersecting edges, may be shown in three dimensions without intersection.

There is a number related works with a strong connection to the ideas presented here. One of our starting points was our experience with UML, its formal foundation, and some smaller example diagrams [GR98b,GR98a,RG98]. Animation of OMT scenarios has been studied in [Sys97]. The approach has similarities with the animation of interaction diagrams we propose. In [GK98,KG98] three-dimensional representation of diagrams is studied with respect to software engineering aspects. Information systems seem to be the main influence for the approach in [Koi92,Koi93]. Three-dimensional representation of diagrams and graphs is also well-known in the visual language and programming community [Shu88]. A basic vocabulary of three-dimensional representation elements was studied in [FGJ95], and the general benefits of lay-outing graphs in three dimensions was put forward in [WF94]. Three-dimensional diagrams were employed for distributed objects [VDS93,ZZ95], and algorithm animation [BN93],

The rest of the paper is structured as follows. Sections 2 to 5 study some particular UML diagram examples which benefit from a three-dimensional representation. The sections are devoted to class diagrams, object diagrams, animation of sequence diagrams, and three-dimensional sequence diagrams, respectively. The paper ends in Sect. 6 with concluding remarks and future work to be done.

2 Class Diagrams

We start with a very simple approach: We take a two-dimensional diagram, put it into three-dimensional space as a sheet (i.e., first we take only one layer), and then we push the uninteresting items into the background. Thereby, emphasis is laid on the important things in the foreground. The view may be changed by turning uninteresting things into interesting things and vice versa. The advantage of this approach is, that in three dimensions

- there is a natural notion of making things important, namely by drawing them closer to the viewer, and
- there is a natural notion of taking different perspectives, namely by moving around the scene in question and thereby watching it from different views.

491

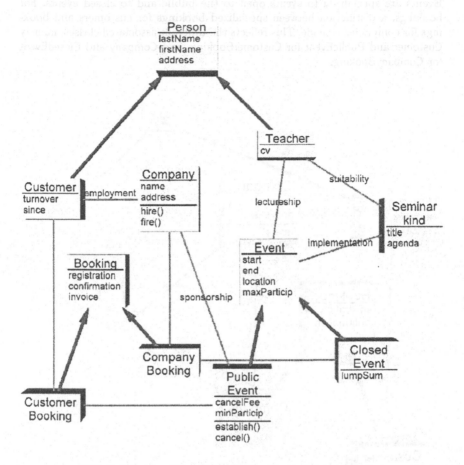

Fig. 1. Complete Class Diagram for Seminar Organization

Figure 1 gives an example of a non-trivial class diagram. Classes are shown as boxes, associations as thin connections, and generalization relationships as directed thick connections with the parent class positioned above the child class. Only a selection of the attributes and operations is shown. The class diagram is still moderate in size and describes aspects of a system for organizing and managing seminars. The original design can be found in [Bal96]. The classes on the right mainly deal with the seminars, the ones on the left with bookings for seminars. For seminars, (1) the class SeminarKind describes a general template, for example, a SeminarKind object may have the title Object-oriented Programming, and (2) the class Event characterizes concrete implementations, for example, the Object-oriented Programming events starting on 02-02-1999 and on 03-03-1999.

Events are specialized to events open to the public and to closed events. For bookings, a distinction between specialized bookings for customers and bookings for companies is made. This reflects the different associated classes, namely Customer and PublicEvent for CustomerBooking, and Company and ClosedEvent for CompanyBooking.

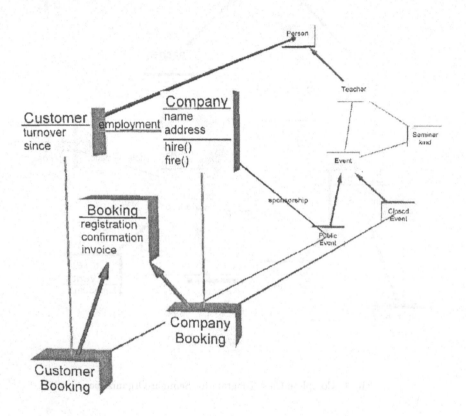

Fig. 2. Class Diagram for Seminar Organization with Emphasis on Bookings

The classes in Fig. 1 are placed as boxes into space in a single plane, but they are viewed from a simple front perspective. In contrast, perspective and emphasis in Fig. 2 is different. Here, one concentrates on the booking aspect of the system by pushing the classes which are *not* important for that aspect into the background. Pushing classes into the background goes hand in hand with hiding the classes' details. Therefore the attributes and operations of unimportant classes are left out. In addition, one can now interactively take other perspectives than a front perspective or interactively rearrange the selected classes. In comparison to Fig. 1, the viewer in Fig. 2 has moved to the right. Changing views is also supported by smooth transitions between the different views, i.e., there is an animation from the first to the second view in which the respective classes are constantly moved until their desired position is reached. Thus the viewer of the animation does not perceive both views as different snapshots (as here in the static two-dimensional presentation of this paper), but by watching the animation, the interesting and important classes like the Booking class can be fixed and its position development can be traced. Analogously to highlighting the booking aspect, another view could emphasize the seminar aspect by focusing on and enlarging the right-side classes Teacher, SeminarKind, Event and its subclasses. Of course in general, any set of diagram elements could be selected to be focussed on. The expected benefit is increased clarity by embedding a small section of a class diagram into the entire system. This small example only demonstrates the principle idea of representing class diagrams in a three-dimensional way. The benefit will become more evident on much larger diagrams.

3   Object Diagrams

The next approach we want show is to take a two dimensional diagram and put the items of the diagram into three-dimensional space by some uniform criterion, for example by putting aggregates above the components and by choosing different planes for different kinds of objects.

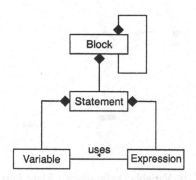

Fig. 3. Class Diagram for Block-Statement-Expression

Figure 3 shows part of a class diagram of a development environment for a programming language. The diagram mainly expresses that blocks may include statements and other blocks as components and that statements may involve variables and expressions as components. The association uses expresses that an expression may use a variable, e.g., by referring to its value. In Fig. 4, an unfinished program with various partly nested blocks, statements, variables, and

494

```
begin               // Block B1
integer X, V1, V2; ...  // Statement S1, Variable V1, Variable V2
for ...             //
  begin             // Block B2
  integer V2, V3; ...   // Statement S2, Variable V2, Variable V3
  X:=V1*V2; ...     // Statement S3, Expression E3
  end; ...          //
X:=V1*V2; ...       // Statement S4, Expression E4
if ...              //
begin               // Block B3
integer V3, V4; ... // Statement S5, Variable V3, Variable V4
X:=V2*V3; ...       // Statement S6, Expression E6
end; ...            //
end                 //
```

Fig. 4. Sample Program to be Represented by an Object Diagram

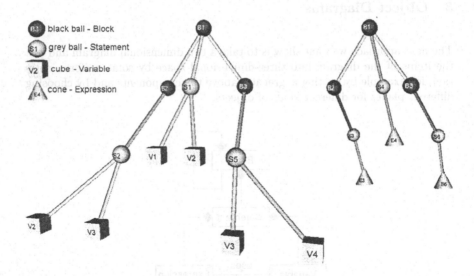

Fig. 5. Left Part: Variable Declarations - Right Part: Expressions

expressions is given. Statements are variable declaration statements or assignment statements referring to previously defined variables.

We now want to show in Figs. 5 and 6 that the example program of Fig. 4 may be represented as a three-dimensional object diagram fitting to the class diagram in Fig. 3. The object diagram will become a kind of pyramid. The pyramid was chosen because in this example we have two tree structures: one for the variables declarations and one for the expressions; one tree structure will become the

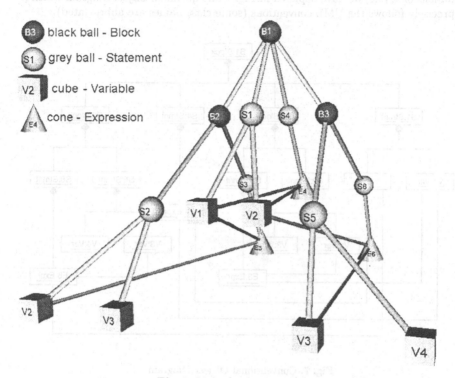

Fig. 6. Links for Association uses

front side of the pyramid, the other the back side. In Fig. 5 we have represented two initial parts of that object diagram: The left part displays the variable declaration statements whereas the right part covers the expression statements. The variable declaration statements lie in the front of the pyramid whereas the expression statements are positioned in the back of the pyramid. The uniform rule for displaying objects is that blocks are shown as black balls, statements as grey balls, variables as cubes, and expressions as cones. Blocks B1, B2, and B3 constitute the intersection of both parts. In both diagrams the aggregates lie above the components, and the aggregates are linked to their components by thick lines. Figure 6 finally puts the two parts mentioned before together and in addition shows with thin lines the uses links from the expressions in the back to the variables in front. For example, expression E6 is linked to (1) variable V2 declared in statement S1 and (2) variable V3 declared in statement S5. The edges representing the association uses are directed from the back to the front and are either parallel to the ground (variable declared in the same block) or climb upwards (variable declared in an enclosing block). In order to encourage a comparison between the advanced object diagram in Fig. 6 and a standard two-dimensional one, we have depicted in Fig. 7 an equivalent object diagram which precisely follows the UML conventions (some class names are abbreviated).

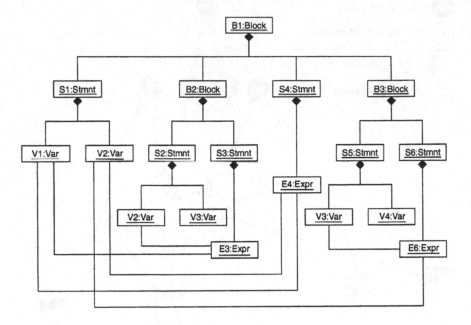

Fig. 7. Conventional Object Diagram

The three-dimensional object diagram strictly obeys the following rules:

- Display objects of different classes in different shapes or different colors (for example, variables as cubes and expressions as cones).
- Position aggregates above components (for example, blocks above statements).
- Arrange different kinds of objects in different planes (for example, variable declarations on the pyramid's front plane, expressions on the pyramid's back plane). Planes must not necessarily be disjoint.
- Show links between objects of different classes as edges connecting one plane to the other plane (for example, links for the association uses between the pyramid's back and front plane).

In this paper, we have to give an impression of the three-dimensional diagram by two-dimensional means. The three-dimensional nature is captured much better if one can analyze the diagram on a screen interactively by mouse movements, turn around the diagram and look at it from different perspectives. Thus, a user can generate her or his own animation as needed. This is possible, for example,

497

with a simple prototype we have built which uses the Virtual Reality Modeling Language (VRML) [CG97,MC97] for visualization and animation.

4 Animating Sequence Diagrams

Up to now we have mainly taken one static UML diagram and have viewed it from different perspectives. Now our intention is different, because we want to point out dynamic aspects of one diagram by animating it. At the same time we would like to be able to move around the diagram. This section also explains the two orthogonal aspects of the title of the paper, namely the aspect concerning three dimensions and the animation aspect.

Figure 8 is taken from the UML notation guide [BJR99]. It shows the establishment of a telephone connection by means of a sequence diagram. Central to the diagram is the communication aspect between the three participating objects caller, exchange, and receiver. These three objects communicate by sending messages indicated as arrows directed from the sender to the receiver of the message.

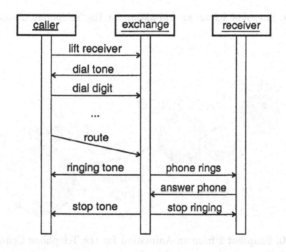

Fig. 8. Sequence Diagram for Establishing a Telephone Connection

In Figs. 9 and 10, two snapshots from a three-dimensional animation of the diagram are captured. The three objects are represented as balls and are put into space on a single line. The connections between the balls show the communication channels. Messages sent from one object to another are displayed in the animation by a ball labeled with the message name. The ball moves from the sender object to the receiver object. The direction of the moving message ball is indicated in Figs. 9 and 10 by a grey arrow (the arrow does not appear in the animation and is used here only to indicate the move direction). The two snapshots are taken at different moments in the sequence, namely the lift-receiver and the ringing-tone/phone-rings moments. Also, two different perspectives are taken: The first perspective emphasizes the caller and the second one the receiver. Emphasizing means that the respective objects are put into the front of the scene. Thereby one process can be perceived from quite different viewpoints.

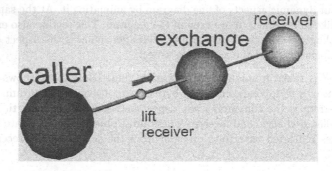

Fig. 9. Snapshot 1 from an Animation for the Telephone Connection

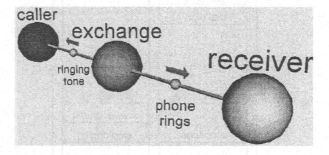

Fig. 10. Snapshot 2 from an Animation for the Telephone Connection

5   Three-Dimensional Sequence Diagrams

As a last example we want to explain how to employ three-dimensional diagrams for phenomena which are not expressible with two dimensions. It is quite well-known, for example, that there are graphs which in the plane do not possess a representation without intersecting edges.

Sequence diagrams in UML are structured with respect to the participating objects. Each object in the sequence diagram has a lifeline, whereas communication between objects is presented by arrows from one lifeline to another lifeline. This means that it is easy and convenient to show communication between lifeline neighbors, but the representation of communication between non-neighbors becomes more involved. Figure 11 shows a sequence diagram with four simultaneous messages sent from one object to four other objects. In such a case, there is no easy to understand two-dimensional representation. Of course, this problem already occurs when one has to show three simultaneous messages. The work-around introduced in the diagram to express the simultaneous sending is a small curve. For example, the curve directly underneath the North object (together with the rest of the arrows) indicates that there are two messages, which go to the left and start at the same moment in the Control object: One message for the West object, the other for the North object. The curve is a feature not mentioned in the UML notation guide.

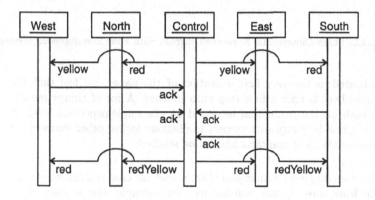

Fig. 11. Sequence Diagram with four Simultaneous Messages

A way out of the dilemma is pictured in Fig. 12 where a three-dimensional representation of the diagram is shown. The lifelines are placed into space using different x- and y-coordinates. Simultaneous occurrence of communication can now easily be expressed, because in three-dimensional space it is possible that one lifeline has more than two neighbors.

6 Conclusion and Future Work

We have given some examples for UML diagrams which were designed to demonstrate that a three-dimensional diagram layout can enhance the comprehension of complex scenes. Our examples used class, object, and sequence diagrams.

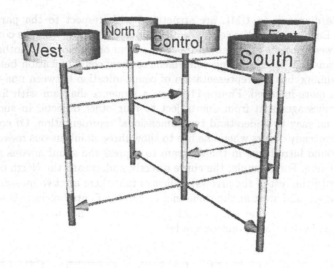

Fig. 12. Three-Dimensional Sequence Diagram with Four Simultaneous Messages

As indicated in the very first quotation of the paper, we feel that the work presented here is only a first step on a journey. A lot of things remain to be done, both on the conceptual level and on the implementation level. Because UML is a rich language with respect to diagram forms, other kinds of diagrams and combinations of diagrams have to be studied:

- Use case diagrams with complex actor and use case relationships might benefit from three-dimensional features, for example, the position of actors in the diagram could reflect their position in the real-world system.
- Object diagrams could be looked at from different views, not only from the aggregation view we mainly used here, but also from a generalization or general association view.
- Activity diagrams with swimlanes can be very similar in their spatial requirements to the ones we mentioned for sequence diagrams.
- Three dimensions can also point out the relationships between different diagrams nicely. This is possibly because the different diagrams could take their own place in three-dimensional space, and relationships between diagrams can be indicated as connections in between like this has been done to some extent in [GK98,KG98].
- Object diagrams and their corresponding class diagrams, i.e., type and instance aspects, can be displayed by providing space for the classes and objects, and by indicating the type-instance relationship with connections. Layout of object diagrams in three dimensional space may follow various paradigms, for example, a real-life-space or real-life-time paradigm. Instead of zooming classes into the foreground (as described in Sect. 2), instances of

classes could grow out of classes and come to the foreground where they form an object diagram which can be animated by a sequence of interactions.

- Sequence and statechart diagrams can be animated together in such a way, that the concrete message sequence turns out to be a run through the general and abstract scheme of the statechart.
- Our animation of sequence diagrams can also be understood as an animation of an equivalent collaboration diagram.
- Two complex sequence diagrams, which are closely connected, for example, through common objects, but which are too complicated to be shown in a single diagram, can be animated together. The simultaneous animation has the advantage, that only the active messages are shown. Thereby, the complexity can be reduced.

References

[Bal96] H. Balzert. *Lehrbuch der Software-Technik: Software-Entwicklung.* Spektrum Akademischer Verlag, Heidelberg, 1996.

[BJR99] G. Booch, I. Jacobson, and J. Rumbaugh, editors. *UML Summary (Version 1.3).* Rational Corporation, Santa Clara, 1999. http://www.rational.com.

[BN93] M.H. Brown and M.A. Najork. Algorithm Animation using 3D Interactive Graphics. In *Proc. 6th ACM Symposium User Interface Software and Technology (UIST'93)*, pages 93–100, 1993.

[CG97] R. Carey and B. Gavin. *The Annotated VRML 2.0 Reference Manual.* Addison Wesley, 1997.

[FGJ95] E. Freeman, D. Gelernter, and S. Jagannathan. In Search of a Simple Visual Vocabulary. In *Proc. 11th IEEE Symposium Visual Languages*, 1995.

[GK98] J. Gil and S. Kent. Three Dimensional Software Modeling. In *Proc. 20th Int. Conf. on Software Engineering (ICSE'98)*, pages 105–114. IEEE Computer Society Press, 1998.

[GR98a] M. Gogolla and M. Richters. Equivalence Rules for UML Class Diagrams. In Pierre-Alain Mullor and Jean Bézivin, editors, *Proc. UML'98 Int. Workshop (UML'98)*, pages 87–96. ESSAIM, Mulhouse, France, 1998.

[GR98b] M. Gogolla and M. Richters. On Constraints and Queries in UML. In Martin Schader and Axel Korthaus, editors, *The Unified Modeling Language – Technical Aspects and Applications*, pages 109–121. Physica-Verlag, Heidelberg, 1998.

[KG98] S. Kent and Y. Gil. Visualising Action Contracts in OO Modelling. In *IEE Proceedings Software, 145*, pages 1–18, 1998.

[Koi92] H. Koike. Three-Dimensional Software Visualization: A Framework and its Applications. In *Visual Computing: Proc. Computer Graphics Int. Conf. (CG'92)*. Springer, Berlin, 1992.

[Koi93] H. Koike. The Role of another Spatial Dimension in Software Visualization. *ACM Transactions Information Systems*, 11(3):266–286, 1993.

[MC97] C. Marrin and B. Campbell. *VRML 2.* Sams Net, Indiana, 1997.

[RG98]   M. Richters and M. Gogolla. On formalizing the UML object constraint language OCL. In Tok Wang Ling, Sudha Ram, and Mong Li Lee, editors, *Proc. 17th Int. Conf. Conceptual Modeling (ER'98)*. Springer, Berlin, LNCS 1507, 1998.

[Shu88]   N.C. Shu. *Visual Programming*. Van Nostrand Reinhold Company, 1988.

[Sys97]   T. Systä. Automated Support for Constructing OMT Scenarios and State Diagrams in SCED. Technical Report A-1997-8, Department of Computer Science, University of Tampere, 1997.

[VDS93]   J.-V. Vion-Dury and M. Santana. Virtual Images: Interactive Visualization of Distributed Object Systems. In *Proc. 8th OOPSLA*, 1993.

[WF94]   C. Ware and G. Franck. Viewing a Graph in a Virtual Reality Display is Three Times as Good as a Two-Dimensional Diagram. In *Proc. 10th IEEE Symposium Visual Languages*, 1994.

[ZZ95]   D.-Q. Zhang and K. Zhang. A Visual Programming Environment for Distributed Systems. In *Proc. 11th IEEE Symposium Visual Languages*, 1995.

Typechecking UML Static Models

Tony Clark

Department of Computing, University of Bradford, West Yorkshire, UK, BD7 1DP
a.n.clark@scm.brad.ac.uk

Abstract. UML static models are expressed using a mixture of class diagrams and OCL expressions. In a well formed static model, the OCL expressions and class diagrams are type consistent. Checking for type consistency of static models involves both inclusion and parametric polymorphism. This paper defines a semantics of type consistency in terms of a type theory for UML static models. The type theory is shown to be correct with respect to a value semantics for OCL. The existence of a consistency checking algorithm for UML static models is established.

1 Introduction

The Unified Modeling Language (UML) is a general-purpose visual modelling language that is designed to specify, visualise and construct the artifacts of an object-oriented system [14]. UML supports the OO development process by providing constructs for describing a proposed software system as a collection of data and behavioural models.

UML defines system states using *static models* that freely describe configurations of objects in terms of classes and associations. Each class declares state variables and methods. Associations between classes declare the potential for message communication between instances.

Expressions in an *object constraint language* (OCL) [15] restrict the freely constructed system states. OCL is used to express class invariants and method specifications (see [8] and [9] for more information on the use of conditions in OO designs). A class invariant is a condition that must hold for any instance of the class. Methods are specified using OCL pre- and post-conditions. In all cases OCL expressions are conditions on the values of object state variables, method calls and associations.

It is possible for the static models and the OCL expressions to be type inconsistent. Type inconsistencies result in references to a non-existent object fields and to the application of operators such as + to operands of incorrect types.

The aim of this paper is to define a set of rules for checking the type consistency of UML static models. To achieve this we define a value semantics for OCL expressions with respect to a UML class diagram. A type theory for OCL expressions is defined and shown to be sound and complete with respect to the value semantics. The type theory is used to define type consistency for UML static models.

2 UML Static Models

A UML static model denotes a collection of legal system states and method calls. Each method call is a pair of system states (a pre-state and a post-state), a method, a list of arguments and a return value. The model is expressed using UML class diagrams and OCL expressions. Both class diagrams and OCL expressions denote collections of system states and method calls; the legal system states are the intersection of the two collections. This section defines the components of a static model.

2.1 Requirements and Class Diagram

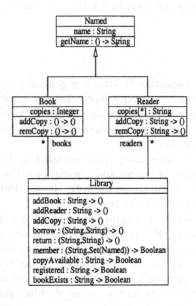

Fig. 1. A Static Model for a Library System

Software to control a library is required. The library contains a number of copies of books and has a number of registered readers. Two different books cannot have the same name. Two different readers cannot have the same name. Readers borrow and return copies from the library. The system must record the number of copies currently shelved and on loan. The user may query whether a copy is available and whether a reader is registered.

Figure 1 shows a UML class diagram for this system. Since both readers and books have names, the class **Named** is defined as a shared super-class of **Reader** and Book. An instance of Book contains an integer copies being the number of copies currently shelved. An instance of Reader contains a set of strings copies being the names of books currently borrowed (assuming a limit of one copy per book).

The declaration of copies[*]:String in Reader corresponds to the OCL type Set(String). The operator Set is a *type constructor* whose argument defines the type of set elements. Library is the source of two associations, readers and books; the cardinality constraint * indicates that a single library is associated with many instances of Reader and Book respectively.

A meaning for a class diagram such as that shown in figure 1 is a freely constructed collection of object states. Class diagrams alone are not sufficiently expressive; they cannot express exactly the required set of object states. For example, the class diagram shown in figure 1 permits more than one reader with the same name.

OCL is used to restrict the possible states in terms of *invariants* and *method specifications*. An invariant is a condition that must hold at all times for each instance of a class, referred to as self. For example:

```
Library:
    self.readers->forAll(r1 |
        self.readers->forAll(r2 |
            not(r1 = r2) implies
                not(r1.getName() = r2.getName()))))
```

which requires the names of readers registered in a library to be different (assuming strings are equal when their corresponding characters are equal).

A method specification is a pair of conditions, called *pre* and *post*, such that if the pre-condition holds before the method is invoked then the post-condition holds after the method has completed. Specifications are given in the form *pre implies post*, for example:

```
Book::addCopy() true implies copies' = copies + 1;
Book::remCopy() true implies copies' = copies - 1
```

which requires that the addCopy and remCopy methods of Book respectively increase and decrease the value of copies when they are invoked.

The formal parameters of a method may be referenced in both the pre- and post-conditions. The change of state in an object is represented using the Z technique of adding ' to state variables in the post-condition. If a method has a return value then it is referenced in a post-condition as reply.

2.2 Collection Types

A number of programming languages have advocated the use of parametric classes including Eiffel [12] and recently Java extensions [1] and [16]. OCL provides a number of builtin collection types including Collection and Set. Each collection class may be supplied with a single parameter which is the type of

```
Library::addBook(book)
  not(self.bookExists(book)) implies
    self'.books->forAll(b | self.books.includes(b) or book = b.name) and
    self.books->forAll(b | self'.books.includes(b))
Library::member(name,named)
  named->exists(n | n.getName() = name) implies reply
Library::copyAvailable(name)
  self.books->exists(b | b.name = name and b.copies > 0)  implies reply
Library::bookExists(name)
  self.member(name,self.books) implies reply
Library::registered(name)
  self.member(name,self.readers) implies reply
```

Fig. 2. Specification of class Library

the elements in the collection. The collection types form a hierarchy and provide collection access and modification methods. The collection types are not the same as classes defined by the user. This is reflected in the special purpose OCL syntax which is used to access the collection methods.

We have taken the view that it is useful to treat collection types as *parameterised classes* that are defined at the same level as user classes in a design. Parameterised types have proved to be very useful in other areas of program development [11] [10]. The specification of the class Reader provides an example of the use of collection methods including and excluding:

```
Reader::addCopy(name)
  true implies self'.copies = self.copies.including(name)
Reader::remCopy(name)
  true implies self'.copies = self.copies.excluding(name)
```

Figure 2 specifies methods of class Library. The associations books and readers are of type Set(Book) and Set(Reader) respectively. In general, associations have multiplicity constraints. Associations with multiplicity other than one-to-one correspond to one or more set-valued attributes. Invariants are used to express multiplicity constraints other than 1 and *. For example if there is an upper limit of 10 library books per reader then the following invariant can be used: Reader: self.copies.size <= 10

The method member is used by both bookExists and registered and provides an example of type conformance (or *inclusion polymorphism* see [3]). The type of self.books is Set(Book) and the type of self.readers is Set(Reader). The domain type of the method member is Collection(Named). OCL permits a method to be supplied with a value whose type is different to the method domain providing the argument type conforms to the domain type. In this case since Set conforms to Collection and both Reader and Book conform to Named the use of member is type correct in bookExists and registered.

$$\llbracket e_1 \rightarrow \texttt{forAll}(v_1 \mid e_2) \rrbracket =$$
$$\llbracket e_1 \rrbracket \rightarrow \texttt{iterate}(v_1 v_2 = \texttt{true} \mid \texttt{if } \llbracket e_2 \rrbracket \texttt{ then } v_2 \texttt{ else false})$$
$$\llbracket e_1 \rightarrow \texttt{exists}(v_1 \mid e_2) \rrbracket =$$
$$\llbracket e_1 \rrbracket \rightarrow \texttt{iterate}(v_1 v_2 = \texttt{false} \mid \texttt{if } \llbracket e_2 \rrbracket \texttt{ then true else } v_2)$$
$$\llbracket e_1 \rightarrow \texttt{select}(v_1 \mid e_2) \rrbracket =$$
$$\llbracket e_1 \rrbracket \rightarrow \texttt{iterate}(v_1 v_2 = \llbracket e_1 \rrbracket \mid \texttt{if } \llbracket e_2 \rrbracket \texttt{ then } v_2 \texttt{ else } v_2.\texttt{excluding}(v_1))$$
$$\llbracket e_1 \rightarrow \texttt{reject}(v_1 \mid e_2) \rrbracket =$$
$$\llbracket e_1 \rrbracket \rightarrow \texttt{iterate}(v_1 v_2 = \llbracket e_1 \rrbracket \mid \texttt{if } \llbracket e_2 \rrbracket \texttt{ then } v_2.\texttt{excluding}(v_1) \texttt{ else } v_2)$$
$$\llbracket e_1 \rightarrow \texttt{collect}(v_1 \mid e_2) \rrbracket =$$
$$\llbracket e_1 \rrbracket \rightarrow \texttt{iterate}(v_1 v_2 = \texttt{Bag}\{\} \mid v_2.\texttt{including}(\llbracket e_2 \rrbracket))$$

Fig. 3. Translation of Iteration Expressions

2.3 Iteration Expressions

OCL provides *iteration expressions* to manipulate collections. An iteration expression selects elements from a collection one at a time and performs an operation. The result of the iteration is defined in terms of the results of all the component operations. For example, let c be a collection, v be a variable and e be a condition then c->forAll(v|e) is true when e is true for all the elements of c.

OCL provides the following types of iteration expression: forAll for universal quantification; exists for existential quantification; select and reject for filtering and collect for transformation. All types of iteration expression can be expressed using a basic form: c->iterate(v1 v2 = i | e) where c is a collection, v1 is a variable bound to each element in the collection in turn, i is a value, e is any OCL expression and v2 is a variable set by the iteration. Before evaluation v2 is set to i. After each iteration, v2 is set to the value of e. The result of the iteration expression is the final value of v2.

Figure 3 defines a semantics preserving translation $\llbracket . \rrbracket$ from *general* OCL iteration expressions to *basic* OCL iteration expressions. In each case the new variable v_2 is not free in e_2. The total translation is the unique extension to a homomorphism over OCL syntax. Consider the specification of member in figure 2. After translation, the result is:

```
Library::member(name,named)
  named->iterate(n v = false |
    if n.getName() = name
    then true
    else v) implies reply
```

3 Syntax and Semantics of Static Models

A class diagram denotes a collection of system states and an OCL expression is a condition on states. The condition can be viewed as an expression denoting a value with respect to a system state that contains values for all the free variables in the expression. This section defines a value semantics for OCL expressions.

A *system state* is a collection of object states. An *object state* (α, τ, ϕ) consists of a unique object identifier α, a type τ and a partial function ϕ from variables to values and functions over values. The function ϕ is referred to as a *value context* and can be extended with an association between variable v and value x to produce $\phi[v \mapsto x]$.

A value, $x, y, \ldots$, may be atomic (integer, character, boolean, *etc.*), a collection of values of the same type or an object state. A function maps indexed collections of values to a single value. An object state is *well-formed* when all occurrences of the same object identifier label equivalent object states and when occurrences of types conform to the type signatures given in the UML class diagram. Sets, bags and sequences are represented as objects with appropriate states. For notational convenience a set object will be written $\{x_1, \ldots, x_n\}$ and a bag object as $<x_1, \ldots, x_n>$ (sequences are handled the same way and will not be included).

An OCL expression denotes a boolean value (or undefined). Component expressions may denote values. We conflate these two categories for the sake of simplicity. An OCL expression e is defined:

$$e ::= v \mid k \mid v(e, \ldots, e) \mid e.v \mid e.v(e, \ldots, e) \mid \text{Set}\{e, \ldots, e\} \mid$$
$$e \rightarrow \text{iterate}(v\ v = e \mid e) \mid \text{if } e \text{ then } e \text{ else } e \mid$$
$$e \rightarrow \text{forAll}(v \mid e) \mid e \rightarrow \text{exists}(v \mid e) \mid e \rightarrow \text{select}(v \mid e) \mid$$
$$e \rightarrow \text{reject}(v \mid e) \mid e \rightarrow \text{collect}(v \mid e)$$

where v is a variable and k is an atomic constant. Prefix and infix operators are not differentiated from unqualified method calls.

Attributes and methods of collections are accessed the same way as attributes and methods of objects, for example `c.size` and `c1.includesAll(c2)`. Methods of collections that bind identifiers *always* include exactly one identifier: multiple identifiers are provided by *currying* the expression, for example: `e1->forAll(i1,i2|e2)` is curried as `e1->forAll(i1|e1->forAll(i2|e2))`.

Let e be an OCL expression, ϕ be a value context and x be a value then x is a value of e with respect to ϕ when $\phi \vdash e \Rightarrow x$ as defined in figure 4. The semantics of class invariants is defined with respect to a value context containing a binding for the variable **self** its state variables and its associations. For a method specification the value context contains additional bindings for **self'** etc.

Theorem 1. *Let ϕ be a value context, e be an OCL expression then there exists at most one value x such that $\phi \vdash e \Rightarrow x$.*

The proof is by induction on the structure of e with respect to the rules in figure 4. In each case we assume by induction at most one value for each of the sub-expressions of e; when defined, the uniqueness of x follows from its definition in each rule. $\square$

$$\phi \vdash v \Rightarrow \phi(v) \qquad (1)$$

$$\phi \vdash k \Rightarrow k \qquad (2)$$

$$\frac{\phi \vdash e_i \Rightarrow x_i \; i = 1, \ldots, n}{\phi \vdash v(e_1, \ldots, e_n) \Rightarrow \phi(v)(\tilde{x})} \qquad (3)$$

$$\frac{\phi \vdash e \Rightarrow (\alpha, \tau, \phi') \quad v \in dom(\phi')}{\phi \vdash e.v \Rightarrow \phi'(v)} \qquad (4)$$

$$\frac{\phi \vdash e_1 \Rightarrow \mathbf{true} \quad \phi \vdash e_2 \Rightarrow x}{\phi \vdash \mathbf{if}\ e_1\ \mathbf{then}\ e_2\ \mathbf{else}\ e_3 \Rightarrow x} \qquad (5)$$

$$\frac{\phi \vdash e_1 \Rightarrow \mathbf{false} \quad \phi \vdash e_3 \Rightarrow x}{\phi \vdash \mathbf{if}\ e_1\ \mathbf{then}\ e_2\ \mathbf{else}\ e_3 \Rightarrow x} \qquad (6)$$

$$\frac{\phi \vdash e \Rightarrow (\alpha, \tau, \phi') \quad \phi \vdash e_i \Rightarrow x_i \; i = 1, \ldots, n \quad v \in dom(\phi')}{\phi \vdash e.v(e_1, \ldots, e_n) \Rightarrow \phi'(v)(\tilde{x})} \qquad (7)$$

$$\frac{\phi \vdash e_i \Rightarrow x_i \; i = 1, \ldots, n}{\phi \vdash Set\{\tilde{e}\} \Rightarrow \{\tilde{x}\}} \qquad (8)$$

$$\frac{\phi \vdash e_1 \rightarrow S \quad \phi[v_1 \mapsto x_i] \vdash e_2 \Rightarrow y_i \text{ for each } x_i \in S}{\phi \vdash e_1 \rightarrow \mathbf{forAll}(v_1 \mid e_2) \Rightarrow \bigwedge_{i=1,n} y_i} \qquad (9)$$

$$\frac{\phi \vdash e_1 \rightarrow S \quad \phi[v_1 \mapsto x_i] \vdash e_2 \Rightarrow y_i \text{ for each } x_i \in S}{\phi \vdash e_1 \rightarrow \mathbf{exists}(v_1 \mid e_2) \Rightarrow \bigvee_{i=1,n} y_i} \qquad (10)$$

$$\frac{\phi \vdash e_1 \Rightarrow S \quad \phi[v_1 \rightarrow x] \vdash e_2 \Rightarrow \mathbf{true} \text{ for each } x \in S_1 \quad \phi[v_1 \rightarrow x] \vdash e_2 \Rightarrow \mathbf{false} \text{ for each } x \in S_2 \quad S_1 \cup S_2 = S}{\phi \vdash e_1 \rightarrow \mathbf{select}(v_1 \mid e_2) \Rightarrow S_1} \qquad (11)$$

$$\frac{\phi \vdash e_1 \Rightarrow S \quad \phi[v_1 \rightarrow x] \vdash e_2 \Rightarrow \mathbf{true} \text{ for each } x \in S_1 \quad \phi[v_1 \rightarrow x] \vdash e_2 \Rightarrow \mathbf{false} \text{ for each } x \in S_2 \quad S_1 \cup S_2 = S}{\phi \vdash e_1 \rightarrow \mathbf{reject}(v_1 \mid e_2) \Rightarrow S_2} \qquad (12)$$

$$\frac{\phi \vdash e_1 \Rightarrow S \quad \phi[v_1 \mapsto x_i] \vdash e_2 \Rightarrow y_i \text{ for each } x_i \in S}{\phi \vdash e_1 \rightarrow \mathbf{collect}(v_1 \mid e_2) \Rightarrow \langle y_1, \ldots, y_n \rangle} \qquad (13)$$

$$\frac{\phi \vdash e_1 \Rightarrow \{x_1, \ldots, x_n\} \quad \phi \vdash e_2 \Rightarrow x \quad \phi[v_1 \mapsto x_i, v_2 \mapsto y_i] \vdash e_3 \Rightarrow y_{i+1} \; i = 1, \ldots, n, \; y_1 = x}{\phi \vdash e_1 \rightarrow \mathbf{iterate}(v_1\ v_2 = e_2 \mid e_3) \Rightarrow y_{n+1}} \qquad (14)$$

Fig. 4. OCL Semantics

510

Theorem 2. *Figure 3 defines a full and faithful transformation for iteration expressions. For any OCL expression e and any value context ϕ, if $\phi \vdash e \Rightarrow x$ then $\phi \vdash \llbracket e \rrbracket \Rightarrow x$ and if $\phi \vdash \llbracket e \rrbracket \Rightarrow x$ then $\phi \vdash e \Rightarrow x$.*

The proof is by induction on the structure of the expression e. For example, consider a `forAll` expression $e = e_1 \rightarrow \text{forAll}(v_1 \mid e_2)$. In order to show that the transformation is *full* there are two cases to consider. Let $\phi \vdash e_1 \Rightarrow S$ for some set S and assume by induction that $\phi \vdash \llbracket e_1 \rrbracket \Rightarrow S$. When $\phi \vdash e \Rightarrow$ true then by definition $\phi[v_1 \mapsto x] \vdash e_2 \Rightarrow$ true for all $x \in S$. Assuming by induction $\phi[v_1 \mapsto x] \vdash \llbracket e_2 \rrbracket \Rightarrow$ true then by definition $\phi[v_1 \mapsto x, v_2 \mapsto \text{true}] \vdash$ if $\llbracket e_2 \rrbracket$ then v_2 else false $\Rightarrow$ true. Therefore, by definition $\phi \vdash \llbracket e_1 \rrbracket \rightarrow$ iterate($v_1 \ v_2 = \text{true} \mid e_2$) $\Rightarrow$ true as required. When $\phi \vdash e \Rightarrow$ false then by definition $\phi[v_1 \mapsto x] \vdash e_2 \Rightarrow$ false for some $x \in S$. Assume by induction that $\phi[v_1 \mapsto x] \vdash \llbracket e_2 \rrbracket \Rightarrow$ false and therefore $\phi[v_1 \mapsto x, v_2 \mapsto y] \vdash \llbracket e_2 \rrbracket \Rightarrow$ false for any y. By definition, $\phi[v_1 \mapsto x, v_2 \mapsto y] \vdash$ if $\llbracket e_2 \rrbracket$ then v_2 else false $\Rightarrow$ false. Therefore, by definition $\phi \vdash \llbracket e_1 \rrbracket \rightarrow$ iterate($v_1 \ v_2 = \text{true} \mid e_2$) $\Rightarrow$ false as required. All other cases in the proof have the same structure. $\square$

4 Type Checking

OCL expressions and class diagrams must be shown to be type consistent. To do this we define a relationship between class diagrams interpreted as types and OCL expressions [2]. An expression is well-typed with respect to a class diagram when it is possible to determine for any system state consistent with the class diagram whether or not the OCL expression holds.

This section defines how class diagrams are interpreted as *types*, and uses a static model type theory to define a well-typed relation between OCL expressions and types.

4.1 Static Models

An OCL type τ is defined $\tau ::= b \mid c(\tau, \ldots, \tau) \mid (\tau, \ldots, \tau) \rightarrow \tau$ where b is an atomic type such as `Boolean` or `Integer`, c is a class constructor and $\rightarrow$ constructs function types. A function type may only occur as part of a class signature.

Each well formed OCL value x has a unique type τ written $x : \tau$. If x is a constant k then $k : typeOf(k)$. If x is a set $\{\tilde{x}\}$ where each of the elements has the same type τ then $\{\tilde{x}\} : Set(\tau)$. If x is a function f such that every elements of the domain of f is of type $\tilde{\tau}$ and the range type if τ then $f : \tilde{\tau} \rightarrow \tau$. If x is an object (α, τ, ϕ) such that $dom(sig(\tau)) = dom(\phi)$ and for all $v \in dom(\phi)$, $\phi(v) : sig(\tau)(v)$ then $x : \tau$. If ϕ is a value context and $\mathcal{A}$ is a signature then $\phi : \mathcal{A}$ when $dom(\phi) = dom(\mathcal{A})$ and for all $v \in dom(\phi)$, $\phi(v) : \mathcal{A}(v)$.

A type scheme ρ is defined $\rho ::= t \mid b \mid c(\rho, \ldots, \rho) \mid (\rho, \ldots, \rho) \rightarrow \rho$ where t is a type variable. Type schemes denote a set of types constructed by substituting types for variables in the type scheme. Type scheme ρ_1 is an instance of type scheme ρ_2 when ρ_1 is the result of substituting type schemes for variables in ρ_2.

A class context C is a set of classes. In the following definitions the class context is assumed to be known. Each class is named $c(\bar{t})$ where t are type parameters. Let $c(\bar{\rho})$ represent the class constructed by substituting each type scheme ρ_i for the corresponding t_i in the definition of the class. Each class has a single super-class $super(c(\bar{\rho})) = c'(\bar{\rho}')$ such that the type variables in $\bar{\rho}'$ are at most those contained in $\bar{\rho}$. Each class has a signature $sig(c(\bar{\rho}))$ that is a partial function from variables to type schemes. Signatures associating variables with types are referred to as *type contexts*. The operator _' is applied to a signature and appends the character ' to each name in the signature. Each class has a set of method names $meth(c(\bar{\rho}))$. Each method m of a class has a set of specifications $specs(c(\bar{\rho}))(m)$. Each specification is $(\bar{v}, e_1, e_2)$ where $\bar{v}$ are formal parameters, e_1 is the pre-condition and e_2 is the post-condition.

An invariant is $(c(), e)$ where $c()$ is a non-parametric class and e is the condition. A static model is a pair $(C, \mathcal{I})$ where $\mathcal{I}$ is a set of invariants.

4.2 Type Conformance

Type checking ensures that no illegal field reference can be made and that all values supplied to basic operations are contained within the operation's domain. Since inheritance cannot remove fields, it is sufficient to require that the actual type of an OCL expression is a sub-type, or conforms to, the expected type.

$$C \vdash b_1 \leq b_2 \quad \text{when basic types } b_1 \text{ and } b_2 \text{ conform.}$$
$$C \vdash c(\bar{t}) \leq c(\bar{t}) \quad \text{when } C \text{ contains a class } c(\bar{t}).$$
$$C \vdash c_1(\bar{t}) \leq c_2(\bar{\rho}) \quad \text{when } C \text{ contains a class } c_1(\bar{t}) \text{ and } super(c_1(\bar{t})) = c_2(\bar{\rho}).$$
$$C \vdash c_1(\bar{\rho}_1) \leq c_2(\bar{\rho}_2) \quad \text{when } C \vdash c_1(\bar{\rho}_1') \leq c_2(\bar{\rho}_2') \text{ and } (\bar{\rho}_1, \bar{\rho}_2) \text{ an instance of } (\bar{\rho}_1', \bar{\rho}_2').$$
$$C \vdash c_1(\bar{\rho}_1) \leq c_2(\bar{\rho}_3) \quad \text{when } C \vdash c_1(\bar{\rho}_1) \leq c_2(\bar{\rho}_2) \text{ and } C \vdash c_2(\bar{\rho}_2) \leq c_3(\bar{\rho}_3).$$
$$C \vdash \bar{\tau} \to \tau \leq \bar{\tau}' \to \tau' \quad \text{when for each } i, \, C \vdash \tau_i' \leq \tau_i \text{ and } C \vdash \tau \leq \tau'.$$
$$C \vdash \mathcal{A} \leq \mathcal{A}' \quad \text{when } \mathcal{A} \text{ and } \mathcal{A}' \text{ are signatures such that } dom(\mathcal{A}') \subseteq dom(\mathcal{A})$$
$$\text{and for each } v \in dom(\mathcal{A}'), \, C \vdash \mathcal{A}(v) \leq \mathcal{A}'(v)$$

Fig. 5. Definition of Type Conformance

UML uses *name equivalence* for types as opposed to *structural equivalence*. A class $c_1(\bar{\rho}_1)$ conforms to another class $c_2(\bar{\rho}_2)$ with respect to a given set of classes C when c_1 inherits from c_2 in C and the actual parameters conform and are consistent with C. Type scheme ρ_1 conforms to ρ_2 when $C \vdash \rho_1 \leq \rho_2$ which is the smallest relation defined by the rules given in figure 5.

Consider the example static model defined in section 2. The type conformance rules give rise to the following: Reader() $\leq$ Named(); Set(Reader()) $\leq$ Set(Named()); Set(Reader()) $\leq$ Collection(Named()).

The following theorem establishes the uniqueness of an upper bound on a class:

Theorem 3. *Let $c_1(\tilde{\rho}_1)$ be a class scheme and let c_2 be a class constructor. For all type schemes $\tilde{\rho}_2$ such that $C \vdash c_1(\tilde{\rho}_1) \leq c_2(\tilde{\rho}_2)$ there exists a unique type scheme $\tilde{\rho}_3$ such that $C \vdash c_2(\tilde{\rho}_3) \leq c_2(\tilde{\rho}_2)$ and $C \vdash c_1(\tilde{\rho}_1) \leq c_2(\tilde{\rho}_3)$*

The proof follows since $\leq$ is a partial order. Suppose that $\tilde{\rho}$ and $\tilde{\rho}'$ are two non-equal candidates then $C \vdash c_2(\tilde{\rho}) \leq c_2(\tilde{\rho}')$ and $C \vdash c_2(\tilde{\rho}') \leq c_2(\tilde{\rho})$. By the definition of type conformance this can only occur if $\tilde{\rho}$ is an instance of $\tilde{\rho}'$ and vice versa, leading to a contradiction $\tilde{\rho} = \tilde{\rho}'$ as required. $\square$

Theorem 4. *There is an algorithm el which, given a type τ returns the most specific element type $el(\tau) = \tau_e$ such that $C \vdash \tau \leq \texttt{Collection}(\tau_e)$.*

Let τ be a type such that $C \vdash \tau \leq \texttt{Collection}(\tau_e)$ for some type τ_e. By definition there must be some unique class $c(\tilde{\tau}) = \tau$ such that after a finite collection of unfoldings $super(\ldots super(c(\tilde{\tau}))) = \texttt{Collection}(\tau_e)$. $\square$

Theorem 5. *Given a class context C and a pair of types τ_1 and τ_2 there exists an algorithm conf such that $conf(\tau_1, \tau_2)$ returns true when τ_1 conforms to τ_2 and false otherwise.*

Consider the form of the types τ_1 and τ_2. If either is an atomic type then *conf* can immediately report success or failure. If both types are instances of the same class then we assume by induction the existence of an algorithm for checking the component types and therefore *conf* reports success when all corresponding component types succeed. Otherwise, the types must be instances of different classes, in which case *conf* reports success when $conf(super(\tau_1), \tau_2)$ succeeds. Since each class has only finitely many super-classes, *conf* will terminate. $\square$

4.3 Type Theory

Let a type context $\mathcal{A}$ be a partial function from variables to types. Given a class context $\mathcal{C}$, an OCL expression e, a type context $\mathcal{A}$ whose domain is the set of free variables of e and a type τ then the relation $\mathcal{C}, \mathcal{A} \vdash e : \tau$ states that τ is a type for the values denoted by e in the given context.

The typing relation is defined as a theory in figure 6. The rules are described in outline as follows. The types of free variables are given by the current context (rule 15). The types of constants are pre-defined (rule 16). Predefined operators and methods of the current object must be supplied with arguments whose types conform to the domain of the operator (rule 17). The type of an attribute is determined with respect to the signature of an object type (rule 18). If a method call is qualified with an object (rule 19) then the type of the method is determined with respect to the signature of the object type; the actual argument types must conform to the formal argument types. The type of the test expression in a conditional (rule 21) must be **Boolean**; notice that we require that both consequent and alternative expressions have the same type. The component expressions of a set (rule 22) must all be the same type.

An iteration expression (rule 20) expects the type of e_1 to conform to a collection type with τ_e as the type of the elements in the collection. As it stands

$$C, A \vdash v : A(v) \qquad (15)$$

$$C, A \vdash k : typeOf(k) \qquad (16)$$

$$\frac{\begin{array}{c} C, A \vdash e_i : \tau_i' \\ A(v) = \tilde{\tau} \to \tau \\ C \vdash \tau_i' \leq \tau_i \end{array}}{C, A \vdash v(\bar{e}) : \tau} \quad (17)$$

$$\frac{\begin{array}{c} C, A \vdash e : \tau \\ sig(\tau)(v) = \tau' \end{array}}{C, A \vdash e.v : \tau'} \quad (18)$$

$$\frac{\begin{array}{c} C, A \vdash e_i : \tau_i' \\ C, A \vdash e : \tau \\ sig(\tau)(v) = \tilde{\tau} \to \tau' \\ C \vdash \tau_i' \leq \tau_i \end{array}}{C, A \vdash e.v(\bar{e}) : \tau'} \quad (19)$$

$$\frac{\begin{array}{c} C, A \vdash e_1 : \tau_1 \\ C, A \vdash e_2 : \tau_2 \\ C \vdash \tau_1 \leq Collection(\tau_e) \\ C, A[v_1 \mapsto \tau_e, v_2 \mapsto \tau_2] \vdash e_3 : \tau_2 \end{array}}{C, A \vdash e \to \texttt{iterate}(v_1 v_2 = e_2 \mid e_3) : \tau_2} \quad (20)$$

$$\frac{\begin{array}{c} C, A \vdash e_1 : Boolean \\ C, A \vdash e_2 : \tau \\ C, A \vdash e_3 : \tau \end{array}}{C, A \vdash \texttt{if } e_1 \texttt{ then } e_2 \texttt{ else } e_3 : \tau} \quad (21)$$

$$\frac{C, A \vdash e_i : \tau}{C, A \vdash Set\{\bar{e}\} : Set(\tau)} \quad (22)$$

Fig. 6. Type Theory for OCL

rule 20 is ambiguous since there may be many different types τ_e that satisfy the requirement; however, by theorem 3 we know that if τ_1 is a sub-class of *Collection* then there is a unique lower bound on the element type τ_e.

Theorem 6. *Let C be a class context, A be a type context and e be an OCL expression. There exists at most one type τ such that $C, A \vdash e : \tau$ providing that e contains no expressions denoting unrestricted empty collections.*

The proof is by induction on the structure of the OCL expression e with respect to the rules defined in figure 6. In each case the uniqueness of the type follows by assuming the uniqueness of the types for all sub-expressions. Theorem 3 ensures that there is a unique type for iteration expressions. The restriction on empty collections is required because an expression $Set\{\}$ has an infinite number of types by rule 22. Note that the translation defined in figure 3 introduces an empty bag, but rule 20 can deduce the type of bag elements locally since they are required to be the same type as that of e_3. $\square$

4.4 Correctness of the Type Theory

The OCL type theory must be shown to be correct. This is done by establishing that it is sound and complete. Soundness is established by showing that when the type theory attributes a type τ to an expression, the semantics associates the expression with a value whose type conforms to τ. Completeness is established by showing that when the semantics associates a value x with an expression e then the type of x is that defined by the type theory.

514

Theorem 7. *The type theory for UML static models is sound. For any class context C, type context A, OCL expression e and type τ such that $C, A \vdash e : \tau$ then for any $\phi : A'$ such that $C \vdash A' \leq A$ and $\phi \vdash e \Rightarrow x$ such that $x : \tau'$ then $C \vdash \tau' \leq \tau$.*

The proof is by induction and proceeds by case analysis on the structure of e. When $e = v$ the theorem holds by definition of $\leq$. When $e = k$ both types are *typeOf(k)*. When $e = v(\tilde{e})$ assume by induction that the theorem holds for the argument expressions. By contravariance, the function $\phi(v)$ is applicable to the argument values when application is defined by the type theory and the type of the result must conform to the type $ran(A(v))$. When $e = e'.v$ assume by induction that the theorem holds for e' and therefore holds for e by the definition of $\leq$. When $e = e'.v(\tilde{e})$ assume by induction that the theorem holds for the argument expressions and the object expression, when a suitable method exists then by definition of $\leq$ there must be a suitable function and by contravariance the range type of the function conforms to the range type declared for the method. When $e = e_1 \rightarrow \texttt{iterate}(v_1 v_2 = e_2 \mid e_3)$ then assume by induction that the theorem holds for e_1 and e_2. Since $C, A[v_1 \mapsto \tau_e, \tau_2] \vdash e_3 : \tau_2$ then assume by induction that $\phi[v_1 \mapsto x_i, v_2 \mapsto y_i] \vdash e_3 \Rightarrow y_{i+1}$ such that $y_i : \tau_i'$ and $C \vdash \tau_i' \leq \tau_2$. When $e = \texttt{if } e_1 \texttt{ then } e_2 \texttt{ else } e_3$ then assume by induction that the condition produces $\texttt{true}$ or $\texttt{false}$ and that the theorem holds for both the consequent and alternative. When $e = \texttt{Set}\{\tilde{e}\}$ assume by induction that the theorem holds for each sub-expression and since e must be well formed (all values must be the same type) the theorem holds by definition of $\leq$. $\square$

Theorem 8. *The type theory is complete. For any value context $\phi : A$, OCL expression e and value x such that $\phi \vdash e \Rightarrow x$ where $x : \tau$ with respect to a class context C then $C, A \vdash e : \tau$.*

The proof is by induction and proceeds by case analysis on the structure of e. When $e = v$ then the theorem holds by definition of $\phi : A$. When $e = k$ then $k : \textit{typeOf}(k)$ for all contexts. When $e = v(\tilde{e})$ then assume by induction that the theorem holds for the argument expressions and therefore it holds by the definition of $\phi : A$. When $e = e'.v$ then assume the theorem holds for the object expression e' and therefore it holds for e by the definition of a well formed object (α, τ, ϕ'). When $e = \texttt{if } e_1 \texttt{ then } e_2 \texttt{ else } e_3$ then the theorem must hold for the condition expression e_1. If e is well formed then both the consequent and alternative expressions produce values of the same type and therefore the theorem holds for e by an inductive assumption on both e_2 and e_3. When $e = e'.v(\tilde{e})$ assume that the theorem holds for the object expression and the argument expressions. If e' denotes a well formed object then the types of v agree and therefore the theorem holds for the result of the function application. When $e = e_1 \rightarrow \texttt{iterate}(v_1 v_2 = e_2 \mid e_3)$ then assuming that the theorem holds for e_1 and e_2 it also holds providing that e_3 produces values of the same type τ_2. When $e = \textit{Set}\{\tilde{e}\}$ then providing the component values are all of the same type then the theorem holds. $\square$

4.5 Type Consistency of Static Models

A static model $(\mathcal{C}, \mathcal{I})$ is *type consistent* when all of the invariants $(c(), e) \in \mathcal{I}$ are well typed with respect to $\mathcal{C}$ and when all ground classes in the context are well typed. Let $\mathcal{A}_0$ be a type context containing types for the builtin variables. An invariant is well typed when:

$$\mathcal{C}, \mathcal{A}_0(sig(c()))[\texttt{self} \mapsto c()] \vdash e : \texttt{Boolean}$$

Let $c()$ be a class defined in $\mathcal{C}$, $c()$ is well typed if for every method $m \in meths(c())$ with type $sig(c())(m) = \tilde{\tau} \to \tau$, for every specification $(\tilde{v}, e_1, e_2) \in specs(c())(m)$ the following holds: let $\mathcal{A} = \mathcal{A}_0(sig(c()))[\texttt{self} \mapsto c(), v_i \mapsto \tau_i]$ in

$$\mathcal{C}, \mathcal{A} \vdash e_1 : \texttt{Boolean}$$

$$\mathcal{C}, \mathcal{A}(sig'(c()))[\texttt{self'} \mapsto c(), \texttt{reply} \mapsto \tau] \vdash e_2 : \texttt{Boolean}$$

5 Implementation

Given a static model $\mathcal{M}$ section 4.5 defines the conditions under which $\mathcal{M}$ is type consistent. In order to provide a tool to perform consistency checking for UML static models we must prove that there exists a suitable algorithm.

Theorem 9. *Given a static model $\mathcal{M}$ there is an algorithm con such that con($\mathcal{M}$) returns true when $\mathcal{M}$ is type consistent and false otherwise.*

Let $\mathcal{M} = (\mathcal{C}, \mathcal{I})$. The algorithm *con* independently steps through the classes $\mathcal{C}$ and the invariants $\mathcal{I}$. In both cases the proof relies on the existence of an algorithm *type* such that given a type context $\mathcal{A}$ and an OCL expression e, if there exists a type τ such that $\mathcal{C}, \mathcal{A} \vdash e : \tau$ then $type(\mathcal{C}, \mathcal{A}, e) = \tau$ otherwise *type* reports failure. By theorem 6 the type τ is unique if it exists and by theorem 2 we need only find an algorithm for $[\![e]\!]$. The existence proof is by induction on the structure of e using the rules defined in figure 6. In each case the type of the expression, if it exists, is a function of the types of its sub-expressions. The algorithms *conf* and *el* are used to determine $\mathcal{C} \vdash \tau_1 \leq \tau_2$ and the type of elements τ_e in an iteration expression respectively. $\square$

A consistency checker for UML static models is currently being implemented in Java. The input to the checker is a textual representation of static models although the tool does not rely on this representation and in principle could accept input from a graphical tool given a suitable intermediate representation. The textual language is parsed using a program generated automatically by JavaCC from a static model grammar. The parser produces a static model represented as a tree of Java objects. The tool then performs consistency checking and reports success or failure.

6 Conclusion

The aim of this work is to define rules for checking the type consistency of UML static models. This has been achieved by defining an abstract representation for UML static models, defining a type theory for OCL expressions, showing the type theory is correct with respect to a value semantics for OCL, defining the conditions under which a static model is type consistent and finally establishing the existence of a consistency checking algorithm.

This work contributes to the ongoing efforts to formalise the UML and thereby establish the UML as a sound basis for the development of software systems and more generally addressing methods and techniques which contribute to rigorous object-oriented development including [5], [4], [6] and [7]. In addition, this work will be of use to tool developers wishing to provide computer support for UML.

The following points address the limitations of consistency checking described in this paper and possible areas for future work:

- *Intersection types* have not been used. These would allow heterogeneous sets and expressions of different types in the consequent and alternative parts of conditional expressions. Their addition would appear to be straightforward given a definition for least upper bounds with respect to $\leq$.
- Expressions denoting empty collections are not allowed. Their admission complicates type checking since $C, \mathcal{A} \vdash Set\{\} : Set(\tau)$ for any type τ. One approach to this problem is to use conditional type schemes [13] such that $C, \mathcal{A} \vdash e : \rho$ such that all possible typings τ are instances of ρ.
- UML models contain *dynamic* information expressed via sequence, collaboration, statechart and activity diagrams which must be shown to be consistent with static information. In each case the diagram can be given a semantics which is linked to the value semantics for OCL defined in this paper.
- The OCL may be used as part of a UML based software development method. A semantics for OCL is a prerequisite for the development of such methods which are likely to rely on activities including checking type consistency, checking the logical consistency of OCL (*i.e.* establishing the existence of at least one UML model satisfying the conjunction of OCL constraints) and checking OCL refinements. The OCL semantics defined in this paper is a first step towards providing a basis for consistency checking and refinement.

References

1. Brache G., Odersky M., Stoutamire D. & Wadler P. (1998) Making the future safe for the past: Adding Genericity to the Java Programming Language. *in proceedings of the 13th Annual ACM SIGPLAN Conference on Object-Oriented Programming Systems, Languages and Applications*, (OOPSLA 98).
2. Cardelli L. (1984) Basic Polymorphic Type Checking. *Science of Computer Programming*, 8(2), 147 – 72.
3. Cardelli L. & Wegner P. (1985) On understanding types, data abstraction and polymorphism. *ACM Computing Surveys.* 17(4).

517

4. Clark, A. N. & Evans, A. S.: Foundations of the Unified Modeling Language. Presented at the Second Northern Formal Methods Workshop, June 14, 1997. Published in the proceedings of the 2nd BCS-FACS Nothern Methods Workshop, Ilkley, 1997. BCS FACS Electronic Workshops in Computing.
5. Clark, A. N. & Evans, A. S.: Semantic Foundations of the Unified Modelling Language. In the proceedings of the First Workshop on Rigorous Object-Oriented Methods: ROOM 1, Imperial College of Science Technology and Medicine, London, June, 1997.
6. Clark, A. N.: A Semantics for Object-Oriented Systems. Proceedings of the Third Northern Formal Methods Workshop. September 1998. BCS FACS Electronic Workshops in Computing, 1999.
7. Clark, A. N.: A Semantic Framework for Object-Oriented Development. Technical Report, Department of Computing, University of Bradford, 1999.
8. Cook, D. & Daniels J.: *Designing Object Systems: Object-Oriented Modelling with Syntropy* , Prentice Hall, 1994.
9. D'Souza, D. & Wills, A.: Components and Frameworks with UML: The Catalysis Approach. Addison-Wesley, 1999.
10. Goguen, J.: Principles of parameterized programming. In Ted Biggerstaff and Alan Perlis (Eds.), Software Reusability, Volume I: Concepts and Models, pp. 159 – 225. Addison Wesley, 1989.
11. Meyer, B.: Genericity Vs. Inheritance. In Meyrowitz N. (Ed.) ACM Symposium on Object-Oriented Programming: Systems, Languages and Applications (OOP-SLA), ACM, 1986, pp. 391 – 405.
12. Meyer, B: *Eiffel the Language*. Prentice-Hall Object-Oriented Series, 1992.
13. Ohori, A., & Buneman, P.: Static type Inference for Parametric Classes. Proceedings of the 1989 ACM Conference on Object-Oriented Programming Systems Languages and Applications, pp 445 – 456.
14. Booch G., Jacobson I. & Rumbaugh J.: The Unified Modeling Language User Guide. Addison Wesley Longman, 1998.
15. UML Consortium. Object Constraint Language Specification Version 1.1. Available from http://www.rational.com 1997.
16. Odersky M. & Wadler P. (1997) Pizza into Java: Translating theory into practice. *Symposium on Principles of Programming Languages*, pp 146 – 159.

Analysing UML Use Cases as Contracts

Ralph-Johan Back, Luigia Petre and Iván Porres Paltor

Turku Centre for Computer Science (TUCS), FIN-20520 Turku, Finland.
{Ralph.Back, Luigia.Petre, Ivan.Porres}@abo.fi

Abstract. The Unified Modeling Language (UML) consists in a set of
diagrams that describe a system under development. A *use case* diagram
specifies the required functionality of the system, showing the collaboration among a set of actors that are to perform certain tasks. We
complement the use case diagrams by providing formal documents (like
specifications or programs), called *contracts* that regulate the behaviour
of the actors involved. The contract is written in a language with a precise semantics and logic for reasoning - the refinement calculus - and thus
it can be *analysed*. To express contracts we need to specify the problem
domain of the system; we describe classes and UML class diagrams using
also the refinement calculus. Thereby, we integrate the functional view
of a system, described by the use case diagram with the object-oriented
view for the same system, described by the class diagram.

1 Introduction

The Unified Modeling Language (UML) [5] is a powerful and expressive diagrammatic notation for describing object-oriented software systems. The UML
standard [16, 18–20] defines several kinds of diagrams that are used to describe
different aspects or views of a system. Class diagrams, e.g., show the object-
oriented structure of the system while statecharts show the dynamic behaviour
of objects. In this paper, we concentrate on how to describe and analyse use case
diagrams.

Use cases were introduced for the first time by Jacobson in [12] and further
developed since [8, 13, 14]. Use case diagrams show the interaction of the system
with external entities, the so-called actors and describe the functionality of the
system as a black box, without revealing its internal structure. At the time of
their introduction they were considered too informal and simple. Yet, it is this
simplicity that makes them so popular: they are abstract and also accessible.

We can also consider a use case as a contract between the system and the
actors. If the actors act as described in the use case, the system promises to
deliver the functionality also described in the use case. The notion of *contract*
has been already introduced by several authors. Within the component-oriented
field, a contract specifies the obligations component providers must meet and
the expectations component clients may have [21]. Reuse contracts [11] consist of
interface descriptions for sets of collaborating participant components. Di Marzo
et. al [15] use contracts for stepwise refining the models of a system during its

development. Each model has a contract attached, that expresses the essential properties that should be provided by each refinement step. We take here the view of contracts as proposed by Back and von Wright in [3]. A computation involves a number of actors who carry out actions according to a formalised contract that has been laid out in advance.

A contract is defined by a contract statement. A contract statement describes in what ways the actors can modify the state of the system. Both the properties of contract statements and the state itself are formalised within the refinement calculus [1]. The refinement calculus is an extension of Dijkstra's weakest pre-condition calculus [9]. The refinement calculus does not only provide a rigorous mathematical foundation for contracts but also the tools to manipulate them. A contract can be analysed to check whether an actor can achieve a specific goal or not by using the contract. The state of the contract models the problem domain of the system while the contract itself embeds the functional requirements for the same system. Thus, this formal model allows for the integration of two different views for the same system (a functional and an object-oriented one).

UML use cases and contracts are created during two different phases of a software engineering process. While use cases are usually developed within the modelling phase, contracts are used as an intermediate step between the informal specification of the system and its detailed design and implementation. As a contract allows the analysis of the system it describes, the development of contracts is especially useful for ensuring the correctness of the final system and its usage, i.e. for the development of mission critical systems.

By using the refinement calculus notion of contracts we give in this paper a formal counterpart for the UML use case diagrams and use case models. Contracts are not meant as a replacement of the use case diagrams but as a formal complement of them, more usefull in the latter stages of the software engineering process. We proceed as follows. In section 2 we review the use cases using an example that we then develop throughout the paper, and discuss certain advantages and disadvantages related to use cases. Section 3 consists in a short presentation of contracts within the refinement calculus. The formal specification of the case study using contracts is described in section 4. Section 5 presents the analysis method we propose and also an application of it on the example. We summarise the main achievements of the paper in section 6.

2 Use Cases

In their initial form, use cases were defined as follows [12]: "A use case is a sequence of transactions in a system, whose task is to yield a measurable value to an individual actor of the system". In the same reference it is said "the set of use case descriptions specifies the complete functionality of the system".

Currently, most software engineering processes start with an initial specification from the customer that describes informally what the future system is supposed to do. This information is used to identify the actors interacting with the system, their goals and to build the use cases. The collection of use cases logically decomposes the functionality of the system; each use case specifies a certain

520

requirement to be performed by the system. Use cases are usually described using UML diagrams. The behaviour of a use case can be described by interaction diagrams, prose or other layouts [4]. Fig. 1 shows the use case diagram for the example we will develop throughout the paper.

The example is called 'Private Library'. The members of the library share together their collections of books. They need a system that allows them to borrow books from each other, renew a loan, and return the books by deadlines specified when the books are borrowed. There is a fee that must be paid unless the book is returned by the specified date. The fee is 1 unit/day within the first 10 days after the deadline, 2 units/day in the following 10 days, and so on, n units/day within the period of days $[10(n-1)+1, 10n]$. A member can not borrow another book or renew an existing loan if s/he has old debts to pay. That is, if the member owes the library, but the late books are still in her/his possession, then new loans or renewals of existing loans are still permitted provided that the old debts are paid. A member of the library is not allowed to borrow a book that s/he owns. For simplicity, the set of members and the set of books are assumed to be constant in the system. These requirements are illustrated in Fig. 1. The actors of the system are the members of the library and the librarian, who is in charge of the system. A member can borrow a book, return a book, renew a loan and these tasks can include the payment of a fee.

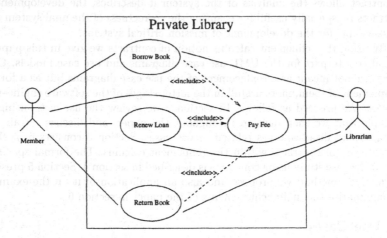

Fig. 1. Use Case Diagram

The main advantages one can get by creating use cases are:

- Capturing the externally-required functionality of the system.
- Identifying the different goals for individual actors.
- Identifying candidate objects for the problem domain.
- Gaining an understanding of the problem domain.
- Gaining an understanding of the proposed solution.

Another benefit of use cases comes from the fact that they are *accountable*, i.e. they can act as a contract between the users and the developers. Still, use cases also have a number of shortcomings:

- They are informal. This is an advantage at an earlier stage in the development process, but later on, informal requirements can be easily misinterpreted.
- It is difficult, if not impossible, to check whether the system provides the functionality expected by the actors. To put it in another way, it is difficult to ensure that the actors can achieve their goals by using the system.
- They are essentially functional in character, even though in UML, they are used to develop object-oriented systems. There is a missing link between functional use case diagrams and object-oriented class diagrams.

These features constitute the starting point for our work: the fact that we can express the information captured by use cases more precisely, using a consistent syntax, with well defined and precise semantics, for regulating the allowed activities between the actors and the system. We propose a solution to these problems by complementing the use cases using formal *contracts* within the refinement calculus:

- Contracts are based on a rigorous formalism. The interpretation of a contract is unambiguous.
- Contracts can be analysed mathematically. We can prove achievability and correctness of contracts.
- The state space of a contract, its universe, can be described in terms of a UML class diagram. Contracts link the functional requirements of the system with its object-oriented decomposition.

As the main contribution of this paper, we will show how an UML use case diagram, as the one in Fig. 1, can be completely described using *contracts*. Moreover, due to the precise form of the contract, we show how to analyse the contract and its correspondent use cases for the achievability of the goals of the involved actors. The following section describes the contracts in more detail.

3 Contracts

A use case diagram is, in fact, an informal model for the computation required from the system under development. This computation is carried out by a collection of actors and the system. We identify the notions of 'actor' and 'system' in UML with the notion of an *agent* in our formalism. An agent is an entity identified by a given name and, most important, it has free will. A computation can now be seen as a number of agents (programs, modules, systems, users, etc...) who carry out actions according to a document (specification, program) given in advance. When reasoning about a computation, we can view this document as a *contract* between the agents involved. Below we describe the notion of a contract as put forward by Back and von Wright in [2, 3].

522

The world that a contract talks about is described as a state σ. The state space Σ is the set of all possible states σ. The state is observed as a collection of *attributes* $x_1, ..., x_n$, each of which can be observed and changed independently of the others. An attribute x of type Γ is really a pair of two functions, the *value function valx* : $\Sigma \to \Gamma$ and the *update function setx* : $\Gamma \to \Sigma \to \Sigma$. The function *valx* returns the value of the attribute x in a given state, while the function *setx* returns a new state where x has a specific value, while the values of all other attributes are unchanged. Attributes are partitioned into *objects*. We simplify our presentation in this paper by assuming a constant collection of objects for the systems we model. This restriction excludes the dynamic creation and destruction of objects. Our formalism handles also these issues, but since they are not the focus of the paper, they were omitted here. E.g., object identifiers can be seen as indexes of arrays that stand for the attributes. In this way, the usual rules for arrays can be used to reason about collections of objects [3]. Assume we have the following attributes: $o1.name$, $o2.name$, $o1.age$, $o2.age$, $o3.author$, $o3.title$. We group these attributes into three objects: object o_1 has attributes $o_1.name$, $o_1.age$, object o_2 has attributes $o_2.name$, $o_2.age$, and object o_3 has attributes $o_3.title$, $o_3.author$. There may also be attributes that are not part of any object: they represent local information for the contract. As the grouping of attributes into objects is just a logical partition of the set of attributes, we can reason in the sequel as having the state modelled by a flat collection of attributes.

An agent changes the state by applying a function f to the present state, yielding a new state $f(\sigma)$. The update of some attribute, part of an object or not, does not influence the values of the other attributes.

An *expression* like $x + y$ is a function on states. Evaluating an expression in a state gives a value: $(x + y)(\sigma) = valx(\sigma) + valy(\sigma)$. We use expressions to observe properties of the state. They are also used in assignments like $x := x + y$. This assignment denotes a state changing function that updates the value of x to the value of the expression $x + y$, i.e.

$$(x := x + y)(\sigma) = setx(valx(\sigma) + valy(\sigma))(\sigma)$$

A function $f : \Sigma \to \Sigma$ that maps states to states is called a *state transformer*. We also make use of predicates and relations over states. A *state predicate* is a boolean function $p : \Sigma \to Bool$ on the state. A *boolean expression* is an expression that ranges over truth values, and is used to describe predicates on the state. For instance, $x \leq y$ is a boolean expression stating that $valx(\sigma) \leq valy(\sigma)$ in a given state σ. A *state relation* $R : \Sigma \to \Sigma \to Bool$ relates the state σ to a state σ' whenever $R(\sigma)(\sigma')$ holds. We permit a generalised assignment notation for relations. For instance, $(x := x'|x' > x + y)$ relates state σ to state σ' if the value of x in σ' is greater than the sum of the values of x and y in σ and all other attributes are unchanged. This notation generalises ordinary assignment: we have that $\sigma' = (x := e)(\sigma)$ iff $(x := x'|x' = e)(\sigma)(\sigma')$.

A contract is described with *contract statements*, i.e. atomic changes of the state the contract talks about. The syntax of these statements is as follows:

$$S ::= \langle f \rangle \mid \mathsf{assert}_a\ p \mid \mathsf{update}_a\ R \mid S_1\ ;\ S_2 \mid \mathsf{choice}_a\ S_1 \sqcup S_2 \mid (\mathsf{rec}_a\ X \cdot S_1)\quad (1)$$

Here f stands for a state transformer, p for a state predicate, R for a state relation, '$\mathsf{rec}_a\ X$' denotes the recursion over the variable statement X, and a is an agent name. The agents carry out a contract statement as follows:

The *update* $\langle f \rangle$ changes the state according to the state transformer f. If the initial state is σ_0 then the next state is $f(\sigma_0)$. An *assignment statement* is a special kind of update where the state transformer is an assignment. For example, the assignment statement $\langle x := x + y \rangle$ or just $x := x + y$ (from now on we will drop the angle brackets for the assignment statements) sets the value of attribute x to the sum of the values of attributes x and y.

The *assertion* '$\mathsf{assert}_a\ p$' is a requirement that agent a must satisfy in a given state. E.g., '$\mathsf{assert}_a\ x + y = 0$' expresses that the sum of (the values of attributes) x and y in the state must be zero and that this should be guaranteed by agent a. If the assertion holds at the indicated state when the contract is carried out, then the state is unchanged, and the rest of the contract is executed. If, on the other hand, the assertion does not hold, then agent a has *breached* the contract.

The *relational update* '$\mathsf{update}_a\ R$' is a contract statement that permits the agent a to choose between all final states related by state relation R to the initial state. If no such final state exists, then agent a has breached the contract. For instance, in the contract statement '$\mathsf{update}_a\ x := x' \mid x < x'$' agent a chooses for x a value x' that is larger than the current value of x, without changing the values of any other attributes.

In the *sequential composition* '$S_1\ ;\ S_2$' the statement S_1 is first carried out, followed by S_2. A choice '$\mathsf{choice}_a\ S_1 \sqcup S_2$' obliges agent a to choose between carrying out S_1 or S_2. The agent a is free to choose either alternative. The *binary* choice extends naturally to the choice among a finite set of alternatives.

The *recursive contract* '$(\mathsf{rec}_a\ X \cdot S)$' is interpreted as the contract statement S, but with each occurrence of the variable statement X in S treated as a recursive invocation of the whole contract '$(\mathsf{rec}_a\ X \cdot S)$'. We assume that the recursion is done on behalf of the agent a, who is responsible for the recursion to terminate. Nontermination (infinite unfolding) means that agent a has breached the contract. At any recursive call, a can choose whether to stop (in which case a has breached the contract) or to continue. So a has never a choice between different courses of action, but only between continuing or terminating the contract.

Programs can be seen as special cases of contracts where two agents are involved: the *user* and the *computer system*. In simple batch-oriented programs, choices are only made by the computer system, which resolves any internal choices in a manner that is unknown to the user of the system (nondeterminism).

We can easily extend the simple language of contracts to include other program constructs, such as 'abort' or 'skip' statements, conditionals and iteration. An abort statement can be explained as $\mathsf{abort} = \mathsf{assert}_{user}\ false$. If executed, it signifies that the user has breached the contract, releasing the computing system from any obligations to carry out the rest of the contract. We can also introduce the contract skip which leaves the state unchanged. We define skip $= \langle id \rangle$ where 'id' is the identity function. The *conditional contract* if $x \geq 0$ then $x := x + 1$ else $x := x + 2$ fi stands for the contract

$$\text{choice}_{system} \ (\text{assert}_{system} \ x \geq 0 \ ; \ x := x + 1) \sqcup (\text{assert}_{system} \ x < 0 \ ; \ x := x + 2)$$

The computer system can here choose between two options. We will assume that an agent does not want to breach a contract. The agent will therefore always choose the alternative for which the guarding assertion is true; choosing the other alternative would breach the contract. *Iteration* is defined in terms of recursion:

$$\text{while } g \text{ do } S \text{ od } = (\text{rec}_{user} \ X \cdot \text{if } g \text{ then } S \ ; \ X \text{ else skip fi })$$

This interpretation means that we consider nontermination of the loop as an error which the user should try to avoid.

As a final extension of the language for contracts, we introduce procedures:

$$S ::= \ldots \mid q_a.$$

Here q_a stands for a procedure call by agent a. A procedure declaration

$$\text{proc } q : \text{pre } b \ ; \ S$$

has a name q, a precondition b and a body S. When a procedure is called, there is usually an agent a responsible for the call. The procedure call is then interpreted as '$\text{assert}_a \ b \ ; \ S$', i.e. the agent is responsible for verifying the precondition of the procedure. If agent a has called the procedure in a state that does not satisfy its precondition, then a has breached the contract. A procedure without precondition, '$\text{proc } q : S$' is interpreted as 'S', i.e. as if the precondition would be 'true'. Value and variable parameters for procedure calls are specified using the keywords val and var respectively; when a procedure with parameters is called, then the formal parameters are substituted for the actual parameters. Preconditions can be used to verify the parameters passed to a procedure body. E.g., if a procedure requires that two of its parameters p_1 and p_2 refer to two different objects (no aliasing), we can add the precondition pre $p_1 \neq p_2$.

Summarising, there is a state σ, representing the domain of a system. This state is observable via a set of attributes (program variables), that are grouped into objects. Contracts act over (possibly) different states and modify certain attributes using the agents in order to change the state.

When several objects have the same structure, they belong to a single *class*. For instance, objects o_1 and o_2 belong to a class *Person* with attributes *name* and *age*. An agent can modify the attributes belonging to an object by invoking a method of the object. Fig. 2(a) shows the specification of a class where x denotes a list of public attributes, declared with their types, x_0 a list of initial values and q a list of methods. The initial values can be specified for each attribute in the init section; if the initialisation is missing, then the attribute has an arbitrary value in its type set. The methods of the class are public procedures. They act only over the attributes of the class and over their own parameters, if any.

```
             class   ClassName              contract  ContractName
             var     x                      agent     a
             init    x : = x₀               var       l
             proc    q                       proc      q
             end                            begin
                                                S
                                            end

           (a) Syntax of Class             (b) Syntax of Contract
```

Fig. 2. Formal specification forms

Fig. 2(b) shows the specification of a contract where a stands for the list of agents involved, l for the local variables, q for the list of procedures used in the contract, and S for the body of the contract. We use the formal specification forms in Fig. 2 in the next section, for expressing the *Private Library* example.

4 Case Study: Private Library

The use case model is a UML model used to specify the functionality of a system. We want to express the *Private Library* use case model using a contract. For this, we first need to specify the problem domain of the system. The class diagram for the *Private Library* example is shown in Fig. 3. We essentially need to model two notions within our system: a *person* and a *book*, therefore we have a class for each of them. Classes and their relationships can also be expressed using the format shown in Fig. 2(a). These two representation of the problem domain are equivalent, except that the representation based on the refinement calculus allows the behaviour of the operations of the classes to be defined in a precise way.

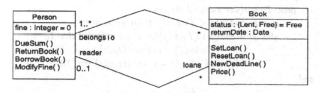

Fig. 3. Class Diagram

The complete description of class *Book* and its behaviour is shown in Fig. 4. The attribute *belongsTo* contains references to the owners of the book. This attribute is used to navigate the *belongsTo* association. When a book is borrowed, the attribute *returnDate* stores the deadline for returning it. The other two attributes are *status* and *reader*. *Status* stores whether the book is *Free* or *Lent*, while *reader* denotes the instance of *Person* that has borrowed the book. Initially, the book is not lent (*status = Free*) and the *reader* attribute has the value Nobody, as stated in the init section of the class. The methods *SetLoan* and *ResetLoan* are used to properly set the attributes of the object when a loan is performed or cancelled. When an existing loan is prolonged, then the attribute *returnDate* is modified by the method *NewDeadLine*. Finally, the fee on the book is computed by the method *Price*: the fee is zero only if the deadline is still in the future or in the present day. Otherwise, the fee is computed as described before using a formula obtained after simple arithmetical deductions.

The class *Person* is shown in Fig. 5. The attribute *loans* stores the references to the books that are borrowed by the person. The attribute *fine* represents the static due sum the person has to pay to the library. Initially there is no loan and the person has no debts to pay. The debts are of two kinds. First, there is a static (old) debt, due to late books, returned but not paid; this debt is modelled by attribute *fine*. Second, there is a dynamic debt, due to late books still in possession of the person; this debt is returned by method *DueSum*. There are two methods for properly setting the attributes of the object when a loan is performed or cancelled: *BorrowBook* and *ReturnBook*. The method *ModifyFine* updates the *fine* attribute, while the method *DueSum* computes the dynamically due sum to the system. The construction 'for all i in I do $p(i)$' stands for the sequential composition of $p(i)$, for all $i \in I$.

```
class Book
    var belongsTo : set of Person ; reader : Person ∪ {Nobody};
        status : {Lent, Free} ; returnDate : Date
    init status, reader : = Free, Nobody
    proc SetLoan(val person : Person) :
        pre status = Free;
        reader, status, returnDate : = person, Lent, Today() + 4 weeks
    proc ResetLoan :
        pre status = Lent;
        reader, status : = Nobody, Free
    proc NewDeadLine(val person : Person) :
        pre person = reader;
        returnDate : = Today() + 4 weeks
    proc Price(var sum : Integer) :
        if returnDate < Today() then sum := 0
        else sum : = (5 * (Today() − returnDate) div 10 +
                (Today() − returnDate) mod 10) *
                ((Today() − returnDate) div 10 + 1) fi
end
```

Fig. 4. Class Book

Using these class descriptions as the domain of our case study, we specify the entire functionality of the system in the contract in Fig. 6. It embeds all the use cases as pictured by the diagram in Fig. 1 and also captures the requirements described in section 2.

There are two agents involved in this contract: a *Member* and a *Librarian*, denoted in Fig. 6 as M and L respectively. The former represents the actions of a member of the library, while the latter models the reaction of the library system. The use cases involve two objects, a person and a book, referred in the

```
class Person
    var fine : Integer ; loans : set of Book
    init loans, fine : = ∅, 0
    proc BorrowBook(val book : Book) :
        pre self ∉ book.belongsTo ∧ book ∉ loans;
        loans : = loans ∪ {book}
    proc ReturnBook(val book : Book) :
        pre book.reader = self;
        loans : = loans \ {book}
    proc ModifyFine(val diff : Integer) :
        pre 0 ≤ fine + diff;
        fine : = fine + diff
    proc DueSum(var sum : Integer) :
        sum : = 0;
        for all book in loans do sum : = sum + book.Price()
end
```

Fig. 5. Class Person

```
contract Private Library
    agent M, L
    var book : Book, person : Person, amount : Natural;
    proc BorrowBook :            //Borrow a book use case
        if person.fine ≠ 0 then PayFee_M fi ;
        assert_M person.fine = 0 ∧ book.status = Free;
        book.SetLoan(person)_L;
        person.BorrowBook(book)_M
    proc RenewLoan :             //Renew a loan use case
        if person.fine ≠ 0 then PayFee_M fi ;
        assert_M person.fine = 0 ∧ book.reader = person;
        book.NewDeadLine(person)_L
    proc ReturnBook :            //Return a book use case
        assert_M book.reader = person;
        if person.DueSum() + person.fine ≠ 0 then
            person.ModifyFine(book.Price)_L;
            PayFee_M fi ;
        person.ReturnBook(book)_M;
        book.ResetLoan()_L
    proc PayFee :                //Pay the fee use case
        update_M amount : = pay | 0 ≤ pay ≤ person.fine;
        person.ModifyFine(−amount)_L
begin
    update_M person : = p' | p' ∈ Person ;
    update_M book : = b' | b' ∈ Book ;
    choice_M BorrowBook ⊔ RenewLoan ⊔ ReturnBook ⊔ PayFee
end
```

Fig. 6. Private Library Contract

contract with two local variables: *person* and *book*. The *Member* is responsible for initialising these variables, and then it is free to choose either of the four use cases, as shown in the begin...end clause of the contract. The use cases-procedures *Borrow Book, Renew Loan,* and *Return Book* include a call to the use case *PayFee*, as required by the use case diagram. In order for the former two use cases to respect the requirements, the person should not have old debts, i.e. *person.fine* = 0. Therefore, the call of *PayFee* use case is performed iff this condition does not hold. The *Member* is responsible for paying its old debt in full. Otherwise, it breaches the contract, as shown by the assertion following the if...fi statement in use cases *BorrowBook* and *RenewLoan*. If the contract is not breached so far, then the main action can take place, i.e. establishing a new loan or respectively a new deadline for an existing loan. Within the use case *ReturnBook*, the whole debt (*person.DueSum()* + *person.fine*) can be paid, if not null. First the fine is increased by the price of the book (which is zero if the book is not late), then the use case *PayFee* is called and finally the main action of returning the book is performed. The *Member* can also choose to pay directly its debts, by choosing *Pay Fee* use case. The variable *amount* is used only within this use case. We declare it for the whole contract as we did not want to introduce local variables for procedures in section 3; however, they can be modelled in our formalism [3].

The formal model of contract we use for describing the functionality of the system suits the object-oriented methodologies for developing systems. Thus, the use case diagrams and the class diagrams can form the first level of the specification (the *modelling* level), while the contract embeds both of them within its precise form, allowing for analyses on the system to be performed. The latter aspect is treated by the following section.

5 Analysis

In order to analyse a contract we need to express the precise meaning of each statement we use, i.e. we need the semantics of contract statements. For our basic language (1), the semantics is given within the refinement calculus using the weakest precondition predicate transformer [2,3]. For each statement S and postcondition q, the predicate $wp_a(S,q)$ is defined, denoting the weakest precondition for agent a to achieve the goal q by executing S. Note that different agents may have different wp predicates in order to achieve the same goal.

$$wp_a(\langle f \rangle, q)(\sigma) = q(f(\sigma))$$
$$wp_a(x := e, q) = q[x/e]$$
$$wp_a(\text{assert}_b\ p, q) = \begin{cases} \neg p \lor q, & a \neq b, \\ p \land q, & a = b. \end{cases}$$
$$wp_a(\text{update}_b\ R, q) = \begin{cases} R(\sigma) \subseteq q, & a \neq b, \\ R(\sigma) \cap q \neq \emptyset, & a = b. \end{cases}$$
$$wp_a(S_1\ ;\ S_2, q) = wp_a(S_1, wp_a(S_2, q))$$
$$wp_a((\text{choice}_b\ S_1 \sqcup S_2), q) = \begin{cases} wp_a(S_1, q) \land wp_a(S_2, q), & a \neq b, \\ wp_a(S_1, q) \lor wp_a(S_2, q), & a = b. \end{cases}$$

From this basic set of preconditions we can deduce formulas for other statements. E.g., we have:

$$wp_a(\text{abort}_b, q) = \begin{cases} \text{true}, & a \neq b \\ \text{false}, & a = b. \end{cases}$$

$$wp_a(\text{skip}, q) = q$$

$$wp_a(\text{if } p \text{ then } S_1 \text{ else } S_2 \text{ fi }, q) = \text{if } p \text{ then } wp_a(S_1, q) \text{ else } wp_a(S_2, q) \text{ fi}$$

$$wp_a(p_b, q) = \begin{cases} \neg r \vee wp_a(P, q), & a \neq b, \\ r \wedge wp_a(P, q), & a = b. \end{cases}$$

Here 'proc p : pre r ; P' is a procedure. If the procedure has any parameters, the formal values are substituted with the actual ones in the body P. A formula for the weakest precondition for the recursive statement can be expressed using a fixpoint notation. The details of the definitions above are studied in [3].

Using the wp predicate transformer we can verify whether an agent can achieve its goal or not. In our running example, we can analyse whether a member of the library can borrow a book or not by means of the contract. In order to borrow a book, the agent M chooses one identity (*person*) and a book to borrow (*book*) and invokes one use case. The agent will successfully borrow the book if after invoking the use case the book is in the *loans* list of the person: $book \in person.loans$. We can determine the weakest precondition to achieve this goal by computing:

$wp_M(\text{choice}_M \ BorrowBook \sqcup RenewLoan \sqcup ReturnBook \sqcup PayFee,$
$book \in person.loans)$.

As given in the weakest precondition rule, we have to compute the disjunction of the weakest precondition of each procedure. However, in order to successfully borrow a book, the *Member* is expected to choose the *Borrow Book* use case. We compute therefore the precondition $wp_M(BorrowBook, book \in person.loans)$. If we obtain a predicate different from false , then we proved that with our contract the *Member* can borrow a new book for a specified person under a certain condition. Figure 7 shows the calculation. It omits some of the steps of the proof, but a complete calculation can be found in [4]. We obtain the following result:

$wp_M(BorrowBook, book \in person.loans) =$
$book.status = Free \ \wedge person \notin book.belongsTo \ \wedge book \notin person.loans$

We can interpret this result in the following way: in order to borrow a book, the book must not be borrowed by the same person or by others ($book.status = Free \wedge book \notin person.loans$), and the book must not be owned by the person ($person \notin book.belongsTo$). The computation of the weakest precondition for the *PayFee* use case shows that the book can be borrowed provided that the *Member* pays the required fees (Fig. 8).

Besides the achievability analysis, other properties can also be stated and proved in a similar manner. For instance safety properties may be stated as

$wp_M(BorrowBook_M, book \in person.loans)$

$= \{wp$ rule for a procedure call, the precondition is true $\}$
$wp_M($if $person.fine \neq 0$ then $PayFee_M$ fi ;
    assert$_M$ $person.fine = 0 \wedge book.status = Free;$
    $book.SetLoan(person)_L$; $person.BorrowBook(book)_M,$
    $book \in person.loans)$

$= \{$sequential composition rule for $wp \}$
$wp_M($if $person.fine \neq 0$ then $PayFee_M$ fi ;
    assert$_M$ $person.fine = 0 \wedge book.status = Free;$
    $book.SetLoan(person)_L, wp_M(person.BorrowBook(book)_M,$
      $book \in person.loans))$

$= \{wp$ rule for a procedure call $\}$
$wp_M($if $person.fine \neq 0$ then $PayFee_M$ fi ;
    assert$_M$ $person.fine = 0 \wedge book.status = Free;$
    $book.SetLoan(person)_L, person \notin book.belongsTo \wedge book \notin person.loans \wedge$
    $wp_M(person.loans := person.loans \cup \{book\}, book \in person.loans))$

$= \{wp$ rule for assignment$\}$
$wp_M($if $person.fine \neq 0$ then $PayFee_M$ fi ;
    assert$_M$ $person.fine = 0 \wedge book.status = Free;$
    $book.SetLoan(person)_L,$
    $person \notin book.belongsTo \wedge book \notin person.loans \wedge book \in person.loans \cup \{book\})$

$= \{$set theory, logic$\}$
$wp_M($if $person.fine \neq 0$ then $PayFee_M$ fi ;
    assert$_M$ $person.fine = 0 \wedge book.status = Free;$
    $book.SetLoan(person)_L, person \notin book.belongsTo \wedge book \notin person.loans)$

$= \{$sequential composition, procedure call and assignment rules for $wp \}$
$wp_M($if $person.fine \neq 0$ then $PayFee_M$ fi ;
    assert$_M$ $person.fine = 0 \wedge book.status = Free,$
    $book.status \neq Free \vee person \notin book.belongsTo \wedge book \notin person.loans)$

$= \{$sequential composition rule for wp and wp rule for assert, $logic\}$
$wp_M($if $person.fine \neq 0$ then $PayFee_M$ fi ,
    $person.fine = 0 \wedge book.status = Free \wedge person \notin book.belongsTo \wedge$
    $book \notin person.loans)$

$= \{wp$ rule for if $\}$
if $person.fine \neq 0$ then $wp_M(PayFee_M, person.fine = 0 \wedge$
    $book.status = Free \wedge person \notin book.belongsTo \wedge book \notin person.loans)$
else $person.fine = 0 \wedge book.status = Free \wedge person \notin book.belongsTo \wedge$
    $book \notin person.loans$ fi

$= \{wp$ for $PayFee$ procedure (see Fig. 8)$\}$
if $person.fine \neq 0$ then
    $book.status = Free \wedge person \notin book.belongsTo \wedge book \notin person.loans$
else $person.fine = 0 \wedge book.status = Free \wedge person \notin book.belongsTo \wedge$
    $book \notin person.loans$ fi

$= \{$logic$\}$
    $book.status = Free \wedge person \notin book.belongsTo \wedge book \notin person.loans$

Fig. 7. Weakest precondition of the *Borrow Book* use case

$wp_M(PayFee_M, person.fine = 0 \wedge$
$\quad book.status = Free \wedge person \notin book.belongsTo \wedge book \notin person.loans)$

$= \{wp$ rule for a procedure call, the precondition is true $\}$
$wp_M(\text{update}_M \ amount := pay \mid 0 \le pay \le person.fine;$
$\quad person.ModifyFine(-amount)_L, person.fine = 0 \wedge$
$\quad book.status = Free \wedge person \notin book.belongsTo \wedge book \notin person.loans)$

$= \{$sequential composition rule for $wp \}$
$wp_M(\text{update}_M \ amount := pay \mid 0 \le pay \le person.fine,$
$\quad wp_M(person.ModifyFine(-amount)_L, person.fine = 0 \wedge$
$\quad book.status = Free \ \wedge person \notin book.belongsTo \wedge book \notin person.loans))$

$= \{wp$ rule for update $\}$
$(\exists \ pay \cdot 0 \le pay \le person.fine \wedge wp_M(person.ModifyFine(-pay)_L,$
$\quad person.fine = 0 \wedge book.status = Free \wedge person \notin book.belongsTo \wedge$
$\quad book \notin person.loans))$

$= \{wp$ rule for $person.ModifyFine\}$
$(\exists \ pay \cdot 0 \le pay \le person.fine \wedge (0 > person.fine - pay \vee (person.fine - pay = 0 \wedge$
$\quad book.status = Free \wedge person \notin book.belongsTo \wedge book \notin person.loans)))$

$= \{$choosing $pay = person.fine$, logic$\}$
$\quad book.status = Free \wedge person \notin book.belongsTo \wedge book \notin person.loans$

Fig. 8. Weakest precondition of the *Pay Fee* use case

a constraint for the overall contract and it can be checked whether they are preserved during all the steps of the interpretation of a contract. Thus, the semantics of contract statements proves to be a powerful tool for analysing the properties of a system under development.

6 Conclusions and Related Work

In this paper we have presented an analysis method for the UML use cases. The analysis method is based on the notion of a *contract*, understood as a formal specification of the behaviour of the UML use cases. The interpretation of a contract is given in terms of the weakest precondition operator and it is unambiguous. Our intention is to enhance the (informal) use case diagrams by providing the contracts as a formal counterpart. Thus, a contract is a rigorous and analysable description of a use case diagram while a use case diagram is a diagrammatic representation of a contract. We have also shown that using our approach, the functional and the object-oriented views of the system can be integrated and analysed in a uniform way. We can analyse if the system can provide the functionality described in the use cases but the goals of the agents and their preconditions are described in terms of objects from the use-case model.

The formal approach we proposed is especially indicated for the development of mission critical systems. The contract is an intermediate document between more informal specifications (use case diagrams, class diagrams) and the final program. The novelty of the approach is the fact that a contract allows both the analysis of the future program and also the analysis of the ways in which that program will be used by potential users (the agents). The fact that a contract *looks like* a program provides a more intuitive grasp for programmers, while the well-definedness of the contract offers accurate premises for analysers.

The importance of formalising UML diagrams has already been proved in the literature, e.g. [7,10]. In [6] UML is given a mathematical foundation that integrates its multiple views over a system under development. The formalisation of the structure of the UML use cases is discussed in [22], but the authors do not show how to analyse the use cases. Our contract language has a textual notation, extended with special constructs that reflect the choices and decisions of the actors. OCL [17] is a textual language that is part of the UML standard and it is used to constraint the behaviour of UML elements. Unfortunately, OCL specifications can not capture the interaction with the actors and, therefore, they are not as expressive as contracts.

The activity carried out by the agents and the objects during the interpretation of a contract can also be represented using interaction diagrams, e.g. sequence diagrams. In this way, we can also use a contract to generate sequence diagrams that describe the behaviour of the use cases using the UML notation. Since these diagrams are directly derived from the contract, they conform to the requirements of the users.

We are currently extending the contract language to describe other features of UML, like construction and destruction of objects, visibility of model elements, ≪include≫ and ≪extend≫ associations of use cases and generalisation of use cases. All these features will be part of a UML editor extended with additional features for formal specification and mathematical analysis. Regarding the requirements specification, the tool will support animation and mathematical analysis of the use cases, determining a behaviour as function of the specific choices the actors made and also the initial conditions under which some actors can achieve some given goals. Thus, our concepts will be graphically illustrated and thereby easier to parse and use.

Acknowledgement We would like to thank the anonymous referees for their useful comments on the first version of this paper.

References

1. R. J. R. Back. *Correctness Preserving Program Refinements: Proof Theory and Applications.* Vol. 131 of *Mathematical Centre Tracts*, Mathematical Centre, Amsterdam, 1980.
2. R. J. R. Back and J. von Wright. Contracts, Games and Refinement. In *C. Palamidessi and J. Parrow (eds), 4th Workshop on Expressiveness in Concurrency.* Electronic Notes of Theoretical Computer Science, Elsevier, 1997.
3. R. J. R. Back and J. von Wright. *Refinement Calculus: A Systematic Introduction.* Graduate Texts in Computer Science, Springer-Verlag, 1998.
4. R. J. R. Back, L. Petre and I. Porres. *Formalising UML Use Cases in the Refinement Calculus.* TUCS Technical Report, No. 279. Turku Centre for Computer Science, 1999.

5. G. Booch, J. Rumbaugh and I. Jacobson. *The Unified Modeling Language User Guide*. Addison-Wesley, 1998.
6. R. Breu, R. Grosu, F. Huber, B. Rumpe and W. Schwerin. Systems, Views and Models of UML. In *M. Schader and A. Korthaus (Eds), The Unified Modeling Language, Technical Aspects and Applications*, pp. 93-109, Physica-Verlag, Heidelberg, 1998.
7. R. Breu, U. Hinkel, C. Hofmann, C. Klein, B. Paech, B. Rumpe and V. Thurner. Towards a Formalization of the Unified Modeling Language. In *Proceedings of ECOOP'97 - 11th European Conference on Object-Oriented Programming*. Lecture Notes in Computer Science 1241, pp. 344-366, Springer-Verlag, 1997.
8. M. Christerson and P. Jonsson. Tutorial 1: Object-Oriented Software Engineering. In *Tutorial Notes, Presented at OOPSLA'95*, ACM, 1995.
9. E. W. Dijkstra. *A Discipline of Programming*. Prentice–Hall International, 1976.
10. R. France, A. Evans, K. Lano and B. Rumpe. The UML as a formal modeling notation. In *Computer Standards & Interfaces.*, Vol. 19, pp. 325-334, 1998.
11. K. De Hondt, C. Lucas and P. Steyaert. Reuse Contracts as Component Interface Descriptions. In *Proceedings of WCOOP'97 - Second International Workshop on Component-Oriented Programming*, Turku Centre for Computer Science, TUCS General Publication No. 5, pp. 43-49, 1997.
12. I. Jacobson. Object-Oriented Development in an Industrial Environment. In *Proceedings of OOPSLA'87, special issue of SIGPLAN Notices*. Vol. 22, No. 12, pp. 183-191, 1987.
13. I. Jacobson, M. Christerson, P. Jonsson and G. Övergaard. *Object-Oriented Software Engineering: A Use Case Driven Approach*. Addison-Wesley, 1992.
14. I. Jacobson, M. Ericsson and A. Jacobson. *The Object Advantage: Business Process Reengineering With Object Technology*. Addison-Wesley, 1995.
15. G. Di Marzo Serugendo, N. Guelfi, A. Romanovsky and A. F. Zorzo. Formal Development and Validation of Java Dependable Distributed Systems. In *Proceedings of ICECCS'99 - Fifth IEEE International Conference on Engineering of Complex Computer Systems*, 1999.
16. Rational Software et.al. *UML Notation Guide*. OMG document ad/97-08-05, 1997.
17. Rational Software et.al. *Object Constraint Language Specification*. OMG document ad/97-08-08, 1997.
18. Rational Software et.al. *UML Proposal Summary*. OMG document ad/97-08-02, 1997.
19. Rational Software et.al. *UML Semantics*. OMG document ad/97-08-04, 1997.
20. Rational Software et.al. *UML Summary*. OMG document ad/97-08-03, 1997.
21. W. Weck. Inheritance Using Contracts & Object Composition. In *Proceedings of WCOOP'97 - Second International Workshop on Component-Oriented Programming*, Turku Centre for Computer Science, TUCS General Publication No. 5, pp. 105-112, 1997.
22. G.Övergaard and K. Palmkvist. A Formal Approach to Use Cases and Their Relationshops. In *Proceedings of UML'98 International Workshop - Beyond the Notation*, Lecture Notes in Computer Science, pp. 309-317. Springer-Verlag, 1998.

Closing the Gap Between Object-Oriented Modeling of Structure and Behavior

Holger Giese, Jörg Graf and Guido Wirtz

Computer Science, Westfälische Wilhelms-Universität Münster
Einsteinstraße 62, 48149 Münster, GERMANY

Abstract. The UML as standardized language for visual object-oriented modeling allows to capture the requirements as well as the structure and behavior of complex software systems. With the increasing demands of todays systems, behavior aspects like concurrency, distribution and reactivity become more important. But the language concepts of the UML for describing behavioral aspects are weak compared to its concepts for describing structures. Besides a lack of visual expressiveness, a deeper integration with the structure specification is missing. In order to close this gap, an expressive language for modeling object-oriented behavior is proposed with the OCoN approach. It describes contracts, object scheduling as well as control and data flow of services in a Petri-net-like form. A seamless visual embedding of contract specifications into service and object scheduling specifications is provided by different net types.

1 Introduction and Background

Todays software systems for business, telecommunication and industry often obtain a high inherent complexity concerning their structure and behavior. Their development demands construction techniques like multi-layer architectures, fine grain class structures, distribution, concurrency, reactivity, etc. to meet their requirements and change over time. The resulting software architecture [28] has to support maintenance and configuration aspects. The object-oriented modeling principle allows to abstract and (de)compose system properties systematically. It offers tools to transform these properties into appropriate object-oriented structures and behavior.

With the UML [23] exists an object-oriented modeling language which is widely accepted as de-facto standard. Its rich set of notations allows to express system requirements, system structures and behavior independent from any specific software development processes. Moreover, *visual* or diagrammatic specifications play a key role in the UML. This is essential for describing complex relationships which rarely obtain a simple linear structure and, hence, make linear *textual* descriptions a bad choice. The (possibly concurrent) control flows through a system when performing different tasks, the different roles and responsibilities taken, as well as the change of states for resources, are critical to understand a systems behavior and should be visible. Moreover, to distinguish between entities which are to be processed and those which are carriers of activity, is helpful. Unfortunately, the UML constructs for modeling such behavior are not as suitable as those for structure modeling. The object-oriented approach has adopted a single model – *entity relationship* [6] – for describing structural aspects and the main approaches [4, 25, 18, 7, 30] differ only concerning the drawing and annotation syntax. In contrast, the behavioral formalisms support fundamental different behavior models. *Cooperative* modeling which puts its focus on roles and the interaction needed when performing a task is supported by *collaboration* and *sequence diagrams* (derived from MSCs [17]); *activity* and *statechart diagrams* [14] are used to model systems in a *reactive* way. Moreover, the adoption of these notations to object-orientation often lacks a precise semantics and the UML does not provide any consistent behavior model. Although the behavior elements are embedded by the UML meta-model, the visibility model and the accessible associated instances are often not represented consistent in the behavioral diagrams. A deep integration with the structural language concept of classifiers (i.e. subsystems, classes, interfaces) and relationships is missing. Thus, a suitable object-oriented behavior language concept with a precise semantics, which covers structural aspects like distribution as well as behavioral aspects like concurrency, reactivity and fault tolerance, is not provided by the UML. A solution has to incorporate both aspects in order to support an integrative way of object-oriented system modeling. The Object Coordination Net (OCoN) extension to the UML is an attempt to close this gap. Its precise semantics, its focus on the *coordination* aspect and the support for a *seamless* way of structure and behavior modeling with the UML makes it a suitable aspirant.

The rest of the paper is organized as follows. In section 2, a running example to further discuss behavior modeling questions is introduced. The requirements for behavior modeling and the concepts provided by the UML are discussed in section 3. The OCoN approach and its basic concepts are introduced in section 4. A solution for the behavior modeling problems of the UML using OCoNs is presented. The article is closed with a comparison and final remarks.

2 Example Overview

A media store software system is used as a running example to explain the different examined aspects of structure and behavior modeling in this paper. The example is inspired from the nutshell case study *"video store system"* found in CATALYSIS [8]. The considered *media store system* (MSS) should support the management of renting medias like books, video tapes and so on.

The four layer architecture of the MSS (figure 1) where each layer provides a set of interfaces, emphasizes *hierarchical design* and *separation of concern* [32], by separating system wide concerns vertically into layers. Each layer is further decomposed horizontally into different subsystems. The subsystem separation is strengthened using the *design by contract* idea of [21]. It extends each interface to a contract by adding a protocol. These protocols specify the components possible external interaction. To reduce complexity, all following considerations

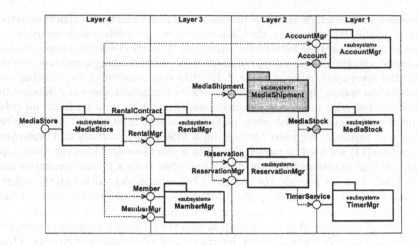

Fig. 1. Coarse grain structure of the media store example

are restricted to the subsystem MediaShipment. (figure 1, highlighted, and figure
2) which is responsible for the shipment of ordered medias. It organizes Ship-
mentOrders which are created from transfered ShipmentData. A ShipmentOrder
represents a single shipment of a set of ordered medias for a MSS member. It
carries all relevant data for a shipment, e.g., address and payment information of
a member, media codes as well as the processing status of the shipment. Besides
ShipmentOrder, the implementation of the subsystem MediaShipment uses one or
more transport services TransService, a single media stock MediaStock and an Account as collaborators. The single steps of this task can be described as follows:
extract the ShipmentData from the ShipmentOrder, check out the ordered medias
from the MediaStock, confirm check out in the ShipmentOrder with the current
date, deliver checked out medias to the member address, confirm the delivery in
the ShipmentOrder and handle all needed payment, i.e., pay for TransService and
debit the rent from the member.

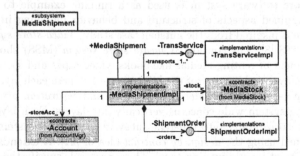

Fig. 2. Fine grained structure of the MediaShipment subsystem

3 UML Behavior Modeling Concepts

The simplest notation to describe the behavior for a specific task like shipping the media for an order are *sequence diagrams*. In figure 3 (upper part) the first part of a processShiping operation is described. The behavior is assigned to several classifiers taking part in the collaboration. Interactions are described by arrows between the *object life lines* applying the *"idealized time"* abstraction.

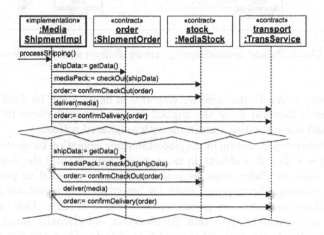

Fig. 3. processShipping behavior with a *sequence diagram*

This kind of abstraction works for the strict sequential version (upper part), but fails when parallel processing or non-deterministic processing orders should be specified (lower part). Then, no single idealized run-time order can be found. For example, a non-deterministic order concerning the checkOut and confirmDelivery call to stock_ can not be described. Thus, this specification technique often results in some kind of over-specification, by choosing one arbitrary strict sequential order for each life line.

The *Activity diagram* formalism provides parallel or non-deterministic behavior defining independent substate spaces operating in parallel with the statechart *and*-state-decomposition mechanism. As demonstrated in figure 4, the modeling

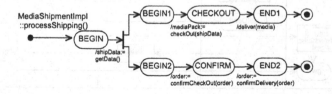

Fig. 4. Parallel processShipping with an *activity diagram*

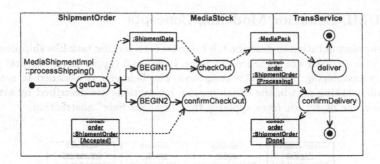

Fig. 5. Parallel processShipping *activity diagram* with *flow objects*

problem only tackled with *sequence diagrams* in figure 3 can be described. The parallelism is modeled using the implicit *and*-decomposed states BEGIN1 and BEGIN2 and corresponding transitions with annotated actions.

An extended specification of the processShipping operation integrates the *data flow* using so called *flow objects*. In this version (figure 5) still the steps for payment are omitted. *Swim lanes*, as syntactical sugar, are used to partition the graph to organize the responsibilities for activities, but a real assignment of responsibilities to context or parameter instances is missing. This time, *activity states* are used to describe calls like checkOut. Additionally, square brackets are used to denote the protocol state of flow objects. Thus, the flow of a ShipmentOrder instance through a confirmCheckOut and confirmDelivery call can be visualized.

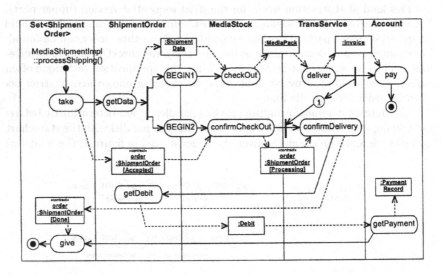

Fig. 6. Attempt to model resource state with *activity diagrams*

539

For *activity diagrams*, the informally introduced *flow objects* do not describe any synchronization. Instead, the object flow has to be encoded explicitly into the underlying *state machine*. When even resources (associated instances) and their protocol states are relevant for the synchronization, the mechanisms presented in the UML version 1.1 [23] are not suitable. For example, a protocol for the used TransService instance, which restricts a confirmDeliver in a manner that it is only available after a corresponding deliver call has taken place, can not be modeled in a compositional way. To overcome this limitation, the UML *state machine* formalism has been extended in version 1.3 [22] with a *sync state* construct, that allows to describe partial states just like places in Petri nets do.

In figure 6, a possible description of the complete *processShipping* operation is shown. Remarkable is that resources can not directly be used as synchronization source. Instead, an additional *sync state* (circle with 1), which corresponds to the resource, has to be used. Thus, the resulting behavior is described in terms of technical elements with only informal correspondence to classifiers. The OCoN approach, as outlined later, is build on top of the Petri net formalism. Thus, no informal ad hoc extensions like *sync states* and *flow objects*, which are not covered by the original Harel statechart notion [14,15], are necessary. Instead, the influence of flowing object, resources and their protocol states can be described in a natural way, as, e.g., described in figure 13.

The UML also provides the *collaboration diagram* notation, which extends the *class diagram* view by adding textual annotations with sequence numbers to describe the behavior. But the behavior aspect, e.g., flow of processed objects or data, is not supported in a visual fashion. Instead, an extended sequence

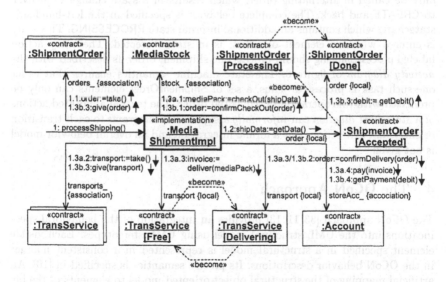

Fig. 7. *Collaboration diagram* with resource states

numbering scheme is used, which results in a complex set of textual annotations spread all over the *class structure*. A version that integrates resource flow is presented in figure 7.

Besides task based notations, the external or complete *reactive* behavior of classifiers can be described in the UML with *statechart diagrams*. The external protocol or life cycle of an object might be specified as well as the whole internal state space.

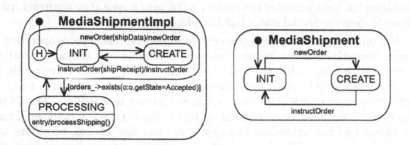

Fig. 8. *Statechart* for life cycle behavior of MediaShipment

As presented in figure 8, the MediaShipmentImpl class of the MediaShipment subsystem and its external representation, the MediaShipment contract, are of similar structure. The MediaShipment interface protocol is described using the statechart on the right-hand-side. The operations newOrder and intructOrder can only be called in alternating order, which results in a state change from INIT to CREATE and back. The complete behavior is specified in the left-hand-side statechart, which contains an additional internal state PROCESSING. This state is entered when an element of orders_ is in state Accepted. The entry action labeled processShipping should process one shipment order as described with the *activity diagram* of figure 6. The state machine semantics ensures that exact one such task is processed. Thus, a set of ShipmentOrder elements can only be processed with this operation in sequential order. The textual annotated actions of a *statechart diagram* can *informally* assign *action diagrams* to each transition or entry and exit action. But a formal integration and an overall behavior model is not given.

4 The OCoN Approach

The OCoN approach [33, 11, 13] provides an integration of the behavioral specifications into the UML structural model using UML stereotypes. Each visible element specified in a structural model is represented in a consistent manner in the OCoN behavior descriptions. Its precise semantics is specified in [10]. An artificial mapping of the structural object-oriented model to elements of the behavior formalism as done for the UML behavior notations and, e.g., proposed

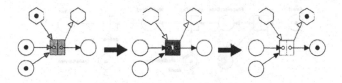

Fig. 9. Simple step semantic of an *action*

for the object-oriented SDL version in [9] can thus be avoided and a complete behavioral model covering even dynamic evolving groups of objects and their interaction is provided.

The behavior is described using events, resource states and transitions for the event processing, resource usage and resource state changes. This net based modeling provides an object-oriented behavior model with orthogonal event and resource processing. By supporting the resources metaphor and using events and service-calls as abstraction from message passing, a higher level of abstraction is achieved. It is possible to explicitly handle resources using mechanisms like selection, locking and consumption. Thus, the needed scheduling for requests or inner activity may be specified as detailed as appropriate. Three different net types are used: *Protocol nets* are used to restrict the supported processing orders. They describe an obligation for the providing instance. *Resource allocation nets* describe the reactive object behavior. A behavior is initiated either as reaction to an external request or to an observed event. *Service nets* describe the behavior associated with a task.

The whole spectrum from untyped abstract behavior to typed operations with concrete data flow and resource allocation is available (see figures 10 - 13). Basic events (Event is the supertype of all events) can be used to express basic synchronization without any specific data flow. Event pools (circles) are distinguished from resource pools (hexagons). For resource pools further the *exclusive* and *imported* case (double border lines) are distinguished to indicate whether the pool is shared with other entities or not. Action symbols (squares) may consume and produce elements of the pools without any identified operation or carrier of activity. Each edge from a pool to an action specifies a pre-condition that must be fulfilled to enable that action. Enabled actions may occur and produce the resulting elements specified by the post-condition edges. An action first consumes its pre-conditions and becomes active. When the execution terminates the post-conditions are produced (see figure 9). This abstract behavior specifications can be refined step by step through typing event and resource pools, adding the resources as carriers of activity (white arrow head) and determining the corresponding operation and its signature for each action.

The upper part of figure 10 presents a sequential order of abstract actions and operation calls, while the whole net including the lower part describes a possibly parallel behavior. It describes a first approximation of the ShipmentOrder task behavior of the MediaShipment class as described in detail in section 3.

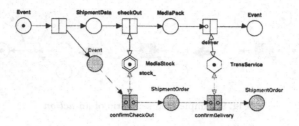

Fig. 10. A parallel processShipping *scenario*

The data flow of associated resources can be embedded into a usage scenario. Thus, after refining the action on the left to a getData operation, the corresponding object flow for the ShipmentOrder *protocol net* can be described as presented in figure 11. First, the contained data of a ShipmentOrder in state [Accepted] are determined using the getData operation. Then the ShipmentOrder instance is further processed as a parameter in confirmCheckOut and confirmDelivery operation calls with resulting states [Processing] and [Done], respectively. The embedded object flow reflects parts of the protocol of the ShipmentOrder contract presented in figure 12. Besides the getData operation call, all protocol usages and the resulting protocol state changes are hidden in the operation calls confirmCheckOut and confirmDelivery.

The OCoN behavior descriptions are integrated into an UML structure model by three distinct classifiers which are build using the UML stereotype mechanism. The ≪contract≫ stereotype represents a contract combining an interface with a behavior protocol as described in section 2. It consists of the elements of an UML interface and a *protocol net* describing the provided interaction orders. A class is modeled using the stereotype ≪implementation≫ which consists of a *resource allocation net* and several associated *service nets* for each task. The

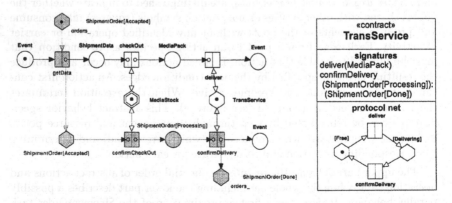

Fig. 11. A processShipping *net* with *data flow* and the TransService contract

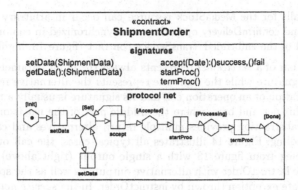

Fig. 12. ShipmentOrder contract

service nets are represented using the ≪service≫ classifier and describe a task specific behavior just like methods in programming languages. If such a service contains relevant coordination aspects, a behavior description with a *service net* is intended. Otherwise, the description can be delayed until the implementation level and a programming language might be used (see figures 11 and 15).

The protocol for the TransService resource presented in figure 11 (right-hand-side) can be further integrated by adding corresponding resource pools for every relevant state and describing the processed resource state changes in the net. This way the *service net* of the ≪service≫ processShipping (figure 13) can describe the final intended behavior, including parallel control flow, object flow and data flow as well as the resource handling. The checkOut and confirmCheckOut

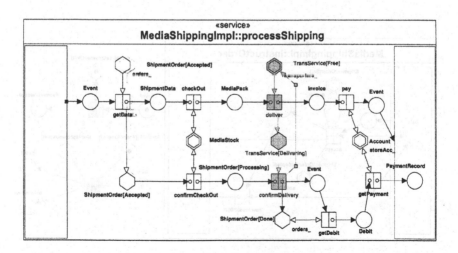

Fig. 13. The processShipping *service net*

544

operation calls for the MediaStock resource can occur in arbitrary order while the deliver and confirmDelivery operations are synchronized in conformance with the protocol of the embedded TransService contract (figure 13, highlighted).

The left bar of a *service net* represents always the input parameter list of the operation signature while the right one represents the alternative return values. The visualization of an operation call and its signature is usually a miniaturized version of this net and bars, where the middle part of the net is omitted. This service call abstraction provides multiple return alternatives and can cover exception handling. Figure 14 illustrates all typical cases: the call of the service processShipping from figure 13 with a single output (right above), the call of accept inside instructOrder with alternative outputs as well as the special case of the InvalidDate exception thrown by instructOrder. In the *service net*, an alternative output in the right bar is used for the exception, too. But it is omitted in the signature visualization of the call and a special exception edge is used instead (figure 14, right below). Hence, the exceptional behavior is visually distinguished from an usual alternative return result.

For the implementation MediaShipmentImpl and its abstract external contract MediaShipment, see figure 15. Much like the complete and external statecharts from figure 8, the contract abstracts from the inner activity. The public operations newOrder and instructOrder are specified in the *resource allocation net* using *forward actions* drawn with a shadow. The inner activity is described using a usual *call action*. It is initiated when at least one ShipmentOrder element of the orders_ resource is in state [Accepted] and the instance itself is in state [Init]. Then, one resource is locked and assigned to the new created instance of a processShipping *service net*. The resource is released in state [Done] when the *service net* terminates. The *protocol net* of the contract MediaShipment shown in figure 15 specifies the guaranteed external protocol. It contains only the offered

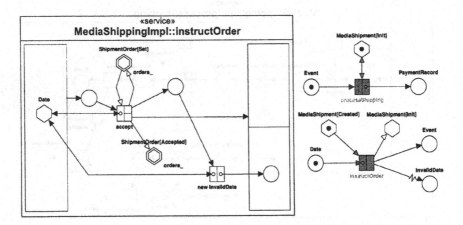

Fig. 14. The instructOrder *service net*

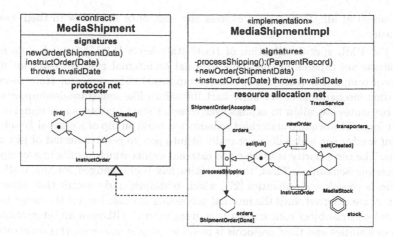

Fig. 15. MediaShipment ≪contract≫ and ≪implementation≫

operations newOrder and instructOrder, as well as the information that they may be accessed only in alternating order.

5 Comparison

As demonstrated in section 3, the behavior modeling abilities of the UML are limited w.r.t. their visual expressiveness. When the behavior is complex or aspects like concurrency are considered, the provided techniques are not appropriate. The inherent object-oriented questions of identifying *responsibilities* or the *resource allocation* can not be modeled in a suitable fashion. The described problems to tackle object-oriented behavior modeling with one of the presented UML notations are caused to a great extent by the origins of these notations. The applied abstraction concepts like idealized life lines, sequence numbers or explicit state modeling do not fulfill the requirements of visual behavior modeling for object-oriented designs.

Sequence diagrams allow to describe external or internal traces. MSCs [17], which are initially a visualization notation for SDL [16] observations, are their origin which indicates already that they are not dedicated to specification purposes. The idealized object life line abstraction enforces that protocol state changes, non-determinism or concurrency can not be expressed conveniently and resembles more an observation than a specification of the behavior. Even worse, the specification of concurrent interactions in traces introduces additional synchronization, which might be not intended by the designer or is hard to implement. The *collaboration diagrams* (formerly *object diagrams* [26, 5, 4]) are just a popular textual encoding of the trace information of a *sequence diagram* via sequence numbers (see BOOCH [4], FUSION [7] and BON [30]). The graphical view of the *class diagram* can be reused, but the behavior specification is not visual at all. They do not express any control or data flow in their visual notation.

The UML specific adoption of Harel statecharts [14,15] named *statechart diagrams* are capable of specifying external or internal reactive behavior of a system component (subsystem, class, component). The powerful state decomposition mechanisms of the statechart formalism like *and*-state-decomposition, history states etc. allow to express even complex state machines in a visual compact form. The original statechart semantics is build on top of a general *broadcast event mechanism* which does not really fit into *peer to peer* connected object systems. The top priority is assigned to external events which results in a complex processing semantics. Thus, the semantics has been changed for the UML to a *run-to-completion* semantics [27], which is simpler and ensures that external stimuli are deferred until the internal processing is done, but on the other hand excludes intra-object concurrency to a great extent. Although an integration of responsibilities and their protocols is possible using sync states, this construct is more of an ad hoc extension towards Petri nets than really integrated into the formalism. Thus, the already complex statechart semantic becomes even more incomprehensible. Even worse, sync states are only a technical tool of specifying state spaces in a compact form and there is no clear guideline how to use them in modeling resource usage and object flow effects. In general, the modeling of a behavioral problem leads to complex explicit state encodings using *statechart diagrams* and does not reflect the underlying resource coordination as explored when the OCoN approach is applied.

Activity diagrams, introduced in [1], are derived from the statechart formalism, too. They can describe a task specific behavior using marked start and end (pseudo) states like, e.g., in OORAM [24]. The explicit *and*-decomposition of states can be done in a graphically less restricted fashion, e.g., a surrounding superstate is implicitly assumed. Also an extension to cover flow objects has been defined informally. The combination of *state machines* and *activity diagrams* to integrate the reactive behavior with task specifications like provided for an OCoN implementation by integrating *service nets* via *forward actions* into a *resource allocation net* is not supported. When the statechart transition annotations are used to emulate this integration, the resulting complete behavior is restricted to the explicitly modeled parallelism. There can be only one active activity diagram for each transition edge at the same time and thus object internal concurrency can not be expressed very well.

Such additional state space encoding efforts and problems with concurrency are avoided when Petri nets are used. They are "a pure model of *coordination*" as Wegner stated in [31]. The OCoN approach provides a direct integration with the object-oriented structural model of the UML. Most object-oriented Petri net approaches like [2, 3, 19, 29, 20] instead chose Petri net specific solutions for synchronization and structure modeling.

The option of complex state decomposition mechanisms for the external protocol specifications is abandoned for the OCoN approach and instead the *seamless* embedding of protocols is of top priority. All aspects of this *seamless embedding*

547

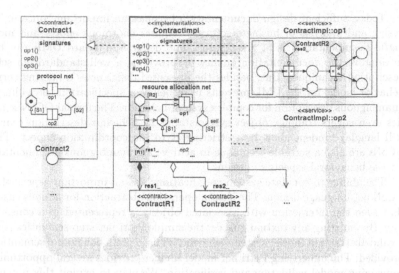

Fig. 16. Seamless integration of behavior specifications

are demonstrated in figure 16. The used parts of a *protocol net* can be embedded into *resource allocation nets* or *service nets* as needed (see [13]). The behavior for a task described by a *service net* is represented in the calling net by an usual *call action* or in the *resource allocation net* by a *forward action*. Contract protocols, which are too complex, are not only a matter of expressive hyper graph foldings as provided by the statechart formalism. Such complex protocols indicate a bad design and the interaction and structure should be better decomposed into multiple independent contracts (see *separation of concern* [32]), as, e.g., provided in role-based design methods like [24]. Thus, fortunately the *scalability* of the approach is preserved in practice. The inherent *true concurrency* of Petri nets allows further to express intra object concurrency by concurrent actions in the *resource allocation net* and/or multiple called *service net* instances that are processed concurrently. The step semantic presented in figure 9 provides an intuitive and simple behavioral semantic. The underlying behavioral model can even be used to simulate a design in order to evaluate it.

6 Conclusion

The object-oriented analysis and design principle provides appropriate techniques concerning structural aspects of complex systems. But, as demonstrated in section 3 and 5, the ability to model the system dynamics is limited. The presented OCoN approach provides a seamless integration of the structural context elements and puts it's focus on their *coordination*. It allows to describe even complex and highly concurrent systems in a way supporting essential object-oriented concepts like responsibilities and contracts.

Today, sufficient design documentation becomes an important concern. The design and analysis of business software requires enormous efforts and meaningful documentation is of great value. The de-facto standard UML can help to improve the documentation quality by providing a well standardized set of description techniques. But, besides the presented deficiencies a missing correct behavior model for the UML prevents from behavior specifications providing an unambiguous description for complex system behavior. The formal OCoN semantics even covers complex object systems and the behavior description remains still language independent by restricting it to the coordination aspect. Thus, OCoNs are also a suitable technique to achieve an unambiguous documentation for the behavioral aspects of designs.

The ability to simulate system specifications was an important aspect of the OCoN net language design. The chosen operation abstraction for actions ensures that even the interaction with associated objects is represented in a consistent way. By omitting any textual guards, the simple Petri net step semantics is still a valid abstraction for the system behavior. Thus, a plain behavioral semantics is provided. The underlying Petri net model further provides several opportunities concerning model validation and verification. We plan to exploit this potential and improve the tool support beyond the existing modeling and simulation tool. The OCoN approach has its origins in the distributed system design and visual languages area. Further evaluation case studies are in process for *embedded systems, coordination* [12] and *component design.*

References

1. Software Standard Integration Definition for Function Modeling. Techreport, 1993. Federal Information Processing Standards Publication 183, Obtainable from National Technical Information Service, US Department of Commerce, Springfield, VA 22161.
2. E. Battiston, F. De Cindio, and G. Mauri. Objsa Nets: A Class of High-level Nets Having Objects as Domains. In G. Rozenberg, editor, *Advances in Petri Nets*, LNCS 424, pages 20–43. Springer, 1988.
3. Olivier Biberstein and Didier Buchs. An Object Oriented Specification Language based on Hierarchical Petri Nets. In *IS-CORE Workshop (ESPRIT)*, Amsterdam, September 27-30 1994.
4. Grady Booch. *Object-Oriented Analysis and Design with Applications*. Addison-Wesley, Menlo Park CA, 1993. (Second Edition).
5. R. Buhr. *System Design with Machine Charts, A CAD Approach with Ada Examples*. Enegelwood Cliffs, NJ, Prentice Hall, 1989.
6. Peter Pin-Shan Chen. The Entity-relationship Model - Towards a Unified View of Data. *ACM Transactions on Database Systems*, 1(1):9–36, March 1976.
7. Derek Coleman, Patrick Arnold, Stephanie Bodoff, Chris Dollin, Helena Gilchrist, Fiona Hayes, and Paul Jeremaes. *Object-Oriented Development: The Fusion Method*. Prentice-Hall, 1994.
8. Desmond Francis D'Souza and Alan Cameron Wills. *Objects, Components, and Frameworks with UML: The Catalysis Approach*. Prentice-Hall, 1999.
9. Jan Ellsberger, Dieter Hogrefe, and Amardeo Sarma. *SDL, Formal Object-oriented Language for Communicating Systems*. Prentice Hall, 1997.

10. Holger Giese. Object Coordination Nets 2.0 – Semantics Specification. Techreport 15/99-I, University Münster, Computer Science, May 1999.
11. Holger Giese, Jörg Graf, and Guido Wirtz. Modeling Distributed Software Systems with Object Coordination Nets. pages 107–116, July 1998. PDSE'98, Kyoto, Japan.
12. Holger Giese, Jörg Graf, and Guido Wirtz. Contract-based Coordination of Distributed Object Systems. PDPTA'99, Las Vegas, Nevada, July 1999.
13. Holger Giese, Jörg Graf, and Guido Wirtz. Seamless Visual Object-Oriented Behavior Modeling for Distributed Software Systems. In *IEEE Symposium On Visual Languages, Tokyo, Japan*, September 1999.
14. D. Harel. Statecharts: A Visual Formalism for complex systems. *Science of Computer Programming*, 3(8):231–274, 1987.
15. D. Harel. On Visual Formalisms. *Science of Computer Programming*, 31:514–530, 1988.
16. International Telecomunication Union (ITU). *Specification and Description Language (SDL) Recommendation Z.100*, 1992.
17. International Telecomunication Union (ITU). *Message Sequence Chart (MSC) Recommendation Z.120*, 1996.
18. I. Jacobson, M. Christerson, P. Jonsson, and G. Övergaard. *Object-Oriented Software Engineering: a use case driven approach*. Addison-Wesley, 1992.
19. C.A. Lakos. Object Petri Nets - Definition and Relationship to Coloured Nets. Techreport, Network Research Group, Department of Computer Science, University of Tasmania, April 1994.
20. Christoph Maier and Daniel Moldt. Object Colored Petri Nets - a Formal Technique for Object Oriented Modelling. Workshop PNSE'97, Hamburg, Germany, September 1997.
21. Bertrand Meyer. *Object-Oriented Software Construction*. Prentice Hall, 1997. 2nd edition.
22. Object Management Group. *OMG Unified Modelling Language 1.3 (alpha R5 draft)*, March 1999.
23. Rational Software Corporation. *Unified Modelling Language 1.1*, September 1997.
24. Trygve Reenskaug. *Working with Objects: The OOram Software Engineering Method*. Addison-Wesley/Manning, 1996.
25. J. Rumbaugh, M. Blaha, W. Premerlani, F. Eddy, and W. Lorensen. *Object-Oriented Modeling and Design*. Prentice Hall, 1991.
26. E. Seidewitz. *Object Diagrams*, NASA Goddard Space Flight Center, May 1985.
27. Bran Selic, Garth Gullekson, and Paul Ward. *Real-Time Object-Oriented Modeling*. Wlley, 1994.
28. Mary Shaw and Davis Garlan. *Software Architecture: Perspectives on an emerging Discipline*. Prentice Hall, 1996.
29. C. Sibertin-Blanc. Cooperative NETs. In R. Valette, editor, *Applications and Theory of Petri Nets 1994, 15th International Conference, Zaragoza, Spain*, LNCS 815, pages 471–490. Springer, June 1994.
30. K. Waldén and J.-M. Nerson. *Seamless Object-Oriented Software Architecture: analysis and design of reliable systems*. Prentice Hall, 1995.
31. Peter Wegner. Coordination as Constrained Interaction. COORDINATION '96, Cesena, Italy, LNCS 1061, pages 28–33. Springer, April 1996.
32. R. Wirfs-Brock, B. Wilkerson, and L. Wiener. *Designing Object-Oriented Software*. Prentice Hall, 1990.
33. Guido Wirtz, Jörg Graf, and Holger Giese. Ruling the Behavior of Distributed Software Components. PDPTA'97, Las Vegas, Nevada, July 1997.

Black and White Diamonds

Brian Henderson-Sellers[1] and Franck Barbier[2]

[1] University of Technology, Sydney, PO Box 123, Broadway,
NSW 2007, Australia
brian@socs.uts.edu.au,
WWW home page: http://www-staff.uts.edu.au/~brian
[2] Université de Nantes, IRIN, 2 rue la Houssinère,
BP 92208, F44322 Nantes Cedex 3, France
Franck.Barbier@irin.univ-nantes.fr

Abstract. This study of the semantics of UML's shared aggregation and composition (black and white diamonds) is based on previous detailed analyses of the semantics of aggregation in object modelling in which primary axioms were identified. All forms of aggregation must comply with these primary axioms. We conclude that both kinds of UML Aggregation do not possess the full complement of primary characteristics and that their secondary characteristics, which define various "flavours" of aggregation, are overlapping and incomplete. We recommend revisions to UML's two kinds of aggregation: completion of the primary set of axiomatic characteristics and then careful selection of secondary characteristics for defining black and white diamond aggregation.

1 A Very Brief History of Aggregation in Object Modelling

Aggregation in object modelling means various things to various people. In this paper, we examine the historical influences on UML's "aggregation" (both shared aggregation and composition), analyze its current status (Version 1.3), identify anomalies and inconsistencies and suggest necessary improvements.

One major influence on UML's concept of "aggregation" is OMT. OMT's aggregation stressed the transitivity of the "is-a-part-of" association relationship — yet transitivity is shown in [1] and more thoroughly in [2] to apply in only a few special cases of aggregation. Rumbaugh et al. [3] (p58) also suggest that if objects can exist independently of each other (parts are separable from the whole), then aggregation is not appropriate. This is contrary to e.g. [4] and [5] who discuss overlapping lifetimes in which aggregates are apposite and yet in which the parts may exist independently of the whole.

A second obvious influence on UML Aggregation is [6] in which aggregation represents whole/part and may be "by-value" or "by-reference" (kinds of physical containment): more recently promulgated in [7]. In addition, it is seen as unidirectional — navigation from aggregate to its parts [6] (p190). Aggregation by-value is said to imply coincident lifetimes and propagation of the destruction operation whereas aggregation by-reference is said to permit parts to exist independently of the whole and for parts to be shared.

Another major influence on the meaning of aggregation was contributed by Odell [8] in 1994 in which he reiterated results from the cognitive science literature [1] in describing six types of aggregation or meronymic relationship, focussing on three secondary characteristics: configurational, homeomerous and inseparable. Odell [8] described aggregation (a.k.a. composition) simply as "a mechanism for forming an object *whole* using other objects as its *parts*". It remained for Kilov and Ross [9] to attempt a more substantial definition, a definition largely ignored in the object community, although echoed in [5,10] and developed further in [2].

The terminology used in aggregation modelling is diverse: composition, containment, aggregation and shared aggregation are all used. In order to avoid these potential ambiguities in the following discussion, we call the supertype of all kinds of "aggregation" the *WP relationship*.

Primary characteristics

		+ consequent properties
P1. whole-part		C1. propagation of one or
P2. emergent property		more operations
P3. resultant property		C2. ownership
P4. irreflexivity at instance level		C3. abstraction
P5. antisymmetry at instance and type level; and		
therefore asymmetry at instance level		

Secondary characteristics

			+ consequent properties
S1. encapsulation (visibility)	Y	N	(also needing secondary characteristics)
S2. overlapping lifetimes	9 possible cases		
S3. transitivity (if same kind)	Y	N	C4. existential dependency
S4. shareability	Y	N	C5. propagation of
S5. configurational	Y	N	destruction operation
S6. separability	Y	N	
S7. mutability	Y	N	

Detailed design/code level

S8. by-value/by-reference	by-value	by-reference
S9. used/not used	used	not used
S10. mandatory/optional	mandatory	optional

Fig. 1. Primary, secondary and derived characteristics of the WP relationship. The values of the secondary characteristics are generally binary (the exception being lifetimes with 9 values). This gives a total of $9 \times 2^9 = 4608$ combinations

There is a basic set of axioms which describe the WP relationship. Some of these have been identified in various articles e.g. [5] but often intermixed with non-critical characteristics. In a careful analysis, [2] have proposed a *necessary and sufficient* set of axioms (Fig. 1) within a framework, initially catalyzed by that of [5], of primary, secondary and derived characteristics. The primary characteristics (labelled P1–P5 in Fig. 1) are those which are axiomatic. In other words, *all* WP relationships must possess these primary characteristics, defined briefly in Table 1.

Table 1 Simplified definitions relevant to the primary characteristics of Figure 1 (P1–P5)

P1 whole–part: qualitatively, a whole consists of (several) parts. While the existence of the whole is contingent on at least some of its parts, the whole *does* have an independent (ontological) existence which transcends its parts — in other words, the whole can be regarded as an abstraction by which means its constituent parts are either not relevant or not visible.

P2 emergent property: a property of the whole not evident in the parts

P3 resultant property: a property of the whole which can be deduced from the part(s)

P4 irreflexivity: a given object cannot be both whole and part at the same time

P5 antisymmetry: if *a* has a WP relationship to *b* and *b* has a WP relationship to *a*, then objects *a* and *b* must be the same object

asymmetry: if an object is part of a whole, that whole cannot be a part of its own part

[2] thus define the WP relationship as a relationship which represents a whole joined to its parts (P1) in which the whole possesses at least one emergent property (i.e. consequent on but not derivable from its parts) (P2) and at least one resultant property (i.e. derivable from its parts) (P3). Furthermore, the relationship is asymmetric at the instance level (antisymmetric, P5, plus irreflexivity, P4) yet antisymmetric at the type level (P5). This defines the WP relationship — nothing more is needed until we wish to discriminate between different "flavours" of the WP relationship. This is done by invoking specific values of the secondary characteristics (labelled S1–S10 in Fig. 1 — see also Table 2). In addition, the derived or consequent characteristics (labelled C1–C5 in Fig. 1) should *not* be used in any *definition* of the WP relationship since they are consequential on adoption of the primary (and in some cases secondary (C4 and C5)) characteristics. A "chart of concepts" (as introduced in [12]) is shown in Fig. 2. It clearly shows which characteristics are dependent on prior knowledge of other characteristics.

As can be seen in Fig. 1, the secondary characteristics identified in [2] appear to delineate a total of 4096 possible styles of aggregation. It is clearly unrealistic to propose this number of kinds of aggregation as being of use or value in OO modelling. The best way to rationalize this choice is to identify the one or two most important (most useful) secondary characteristics to which can be allocated a unique name and notation and then denote the remaining secondary characteristics like any other partition — probably by the use of stereotypes.

553

Table 2 Simplified definitions relevant to the OOAD secondary characteristics of Fig. 1 (S1–S7)

S1 encapsulation: when the internals are not visible from outside

S2 lifetimes: the lifetime of the whole and the part must overlap. Nine cases can be identified [5][11]

S3 transitivity: when aRb and bRc, transitivity implies aRc where R is a relationship, a, b & c objects

S4 shareability: the ability of the part to belong to two or more wholes at the same time

S5 configurational: in which there are functional or structural relationships between the parts

S6 separability: piece(s) can be removed from the whole without destroying either

S7 mutability: individual pieces can be exchanged for an equivalent object without destroying the integrity of the whole. Requires part–whole separability

S8–S10 are implementation-related and not relevant to the discussion here

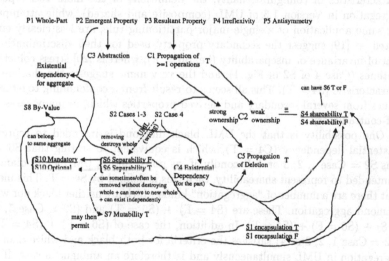

Fig. 2. "Chart of concepts" for major secondary and consequent characteristics of the WP relationship

2 The Current "Definitions" of UML's Two Aggregation Types

There is confusion in the current documentation regarding UML's two aggregation types: called composite aggregation (or simply composition) and shared aggregation (sometimes just called aggregation). The terminology used sometimes tends to be confusing. While aggregation is often used as a shortened form of shared aggregation ("white diamond") it is also sometimes used as the generic supertype of both e.g. [13] (pp2-54 and 2-55). In [3], the generic supertype is called "aggregation association" (p226) although the subtype of "aggregation" is equated, on page 146, to a whole–part relationship. The following statement (also on page 146) that "There is a stronger form of aggregation, called composition" suggests a metamodel in which Composition is a subtype of Aggregation which itself is a subtype of Association.

In this section, we evaluate the definitions of the WP relationship in the UML literature, placing emphasis on the OMG documents [13–15] and the books written by members of Rational Software Corporation e.g. [3,7,20]. We also include, but with less emphasis, third party documents such as [16] [although this describes Version 1.0] as being a frequently quoted document based directly on the original source material.

While OML (e.g. [17,18] partitions WP relationships based on the secondary characteristics of configurationality, the definitions of the main two types of aggregation in Version 1.3 of UML (composite and shared), while attempting the same application of a single major partitioning rule, are less clearly enunciated — [19] suggest the secondary property used for their discrimination is that of invariance or inseparability (i.e. $S6 = F$); whereas [20] stress coincident lifetimes (Case 4 of S2 in Fig. 1); and the very name suggests the shareability characteristic ($S4 = T$). This all seems to result from a confounding together of values from several secondary and derived properties which, in many cases, are self-contradictory.

One possibility is that the UML black diamond is intended to represent existential dependency ($C4 = T$), which is equivalent to stating that $S6 = F$ plus $S2 = $ Case 1, 2, 3 or 4. Secondly, if, as noted above, UML white diamond is intended to represent shareability, then this is defined by $S4 = T$. This means that there are a number of "aggregation" kinds which are neither black nor white diamond aggregation. These are ($S4 = F$) + ($S6 = T$) and ($S2 = $ Case 5, 6, 7 or 8) + ($S6 = F$) + ($S4 = F$). In addition, the case of ($S6 = F$) + ($S4 = T$) + ($S2 = $ Case 1, 2, 3 or 4) satisfies the criteria for both black and white diamond aggregation in UML simultaneously and is therefore an ambiguous case. If this is what was intended, then it is clearly an inadequate definition of black and white diamond WP relationships. In this paper, we explore these contradictions in UML's aggregation types and make some suggestions for the future evolution of UML in this respect.

In [14] and [13], UML (Versions 1.1 and 1.3 respectively) defines both styles of "aggregation" as a WP relationship [14] (p150) and [13] (pB-5) ($P1 = T$) which is both transitive ($S3 = T$) and antisymmetric ($P5 = T$). Secondly, in both shared aggregation and composition, parts may move between owners. In [2] (section 3.3), it was shown that this means that the parts are separable (S6 = T). Thus both black and white diamonds represent cases of separability. This is at odds with [19] (p81) who state that composition, at least, implies inseparability. Inseparability would also seem to be correct for the interpretation of composition as an existential dependency as discussed above. Finally, both WP relationships are inherently bidirectional since in UML they are semantically equivalent to and derived directly from associations — almost all other authors consider aggregations to be essentially unidirectional.

Nowhere in the UML documentation or books can we find discussion (or even passing mention) of the other key attributes of aggregation: emergent property (P2), resultant property (P3) and instance irreflexivity (P4). This leads [21] to state that, since both kinds of UML "aggregation" (WP relationship) are defined to be antisymmetric and transitive, it would include, for example, a greater than relation suggesting the number 1 is part of number 2. Furthermore, [16] (p80) note that, while UML's composition, with strong ownership and coincident lifetimes of parts and wholes leads to the destruction of the whole cascading to the parts, exactly the same occurs for an association with a 1..1 multiplicity. In other words, the omission in [14,15] of any statement regarding resultant and emergent properties leads to an incorrect/inadequate definition for the WP relationship.

In summary so far, we have (a) UML composition (black diamond): P1 = T, P5 = T, S3 = T, S6 = T, (S8 = BY_VALUE)[1]; and (b) UML shared aggregation (white diamond): P1 = T, P5 = T, S3 = T, S6 = T, (S8 = BY_REFERENCE).

Having reviewed some general features of both black and white diamonds in UML Version 1.3, we now explore their semantics individually and in more detail.

2.1 Black Diamonds

UML's composition is defined as being a WP relationship (P1 = T) in which each part belongs to at most one whole at any one time (S4 – F) [14] (p38) and [13] (p2-54); and that there is strong ownership (it is not clear whether this is meant to add anything or merely redefine the fact that shareability = FALSE). This is shown in the notation by multiplicity at the aggregate end which may not exceed 1 [15] (p62) and [13] (p3-69). The WP relationship is described as one in which there are coincident lifetimes [15] (p62), [13] (p3-69), [5] i.e. all parts are created at the same time as the whole and destroyed at the same time (S2 = Case 4 of Fig. 1 plus C5 = T) e.g. [3] (p150). Thus parts cannot have an independent existence [3] (p149) and thus may be said to be inseparable from the whole (S6 = F) – an interpretation supported strongly in [19] (p80). This is a contradiction to the separable nature identified above in which the owner may be changed over time (S6 = T). When ownership is allowed to change (as in

---

[1] A partition stated in [7] (p164) but *not* upheld in [13].

UML Version 1.3 composition), this (surely) contradicts the notion that lifetimes are coincident since the part could now outlive the whole.

In [20], in explaining composition, we are advised that "Parts with non-fixed multiplicity may be created after the composite itself, but once created they live and die with it" (see also [14] (p150) and [13] (pB-5). In other words these parts do *not* necessarily have coincident lifetimes (S2 = Case 2 or 4). This is a contradiction (to S2 = Case 4). Booch *et al.* [20] (p147) then go on to say parts with non-fixed multiplicity may "be explicitly removed before the death of the composite" (see also [14] (p148), [13] (pp2-21, 2-54) and [22] (p91). So this confirms separability (S6 = T) and is a contradiction to the sentence preceding it (that parts live and die with the whole) [20] (p147). We might, however, reinterpret this in the context of an existential dependency (C4 = T) in which parts cannot exist outside of the whole (S6 = F), they may be added and/or deleted (deleted, not removed which implies a continued existence elsewhere) at any time *within the lifetime of the whole* (S2 = Case 1, 2, 3 or 4). Coincident lifetimes (S2 = Case 4) then becomes just *one* of the possibilities. However, if the part(s) still exists at the time of destruction of the whole, then the part must also be destroyed (C5 = T). This suggests that what [20] had in mind was

black diamond = existential dependency

On the other hand, [3] (p148) state that "Other constraints, such as existence dependency, are specified by the multiplicity, not the aggregation marker." This clearly rules out any interpretation of black diamond as an existence (existential) dependency.

UML docs [14] (p38) and [13] (p2-54) state that composition implies propagation semantics, especially deletion (C1 = T, C5 = T). Propagation of the deletion operation requires lifetime ends to be coincident (S2 = Case 2 or Case 4), shareability to be untrue (S4 = F) and for there to be propagation of operations (C1 = T). Since shareability = FALSE for UML black diamonds, a constraint is needed to ensure this. Sharing is forbidden formally in [22] (p91/92) for the two cases where the parts are from the same or from different classes.

For composition (black diamond), [22] (p91) suggest that there exists an existential dependency from aggregate to parts or from parts to aggregate — a result of the combination of S6 = F and S2 = Case 4 (actually Case 1, 2, 3 or 4 would suffice). Existential dependency refers to the case where the parts are dependent on the whole for their very existence. The opposite, where the aggregate is dependent on its parts, called "Existential Dependency for the Aggregate" in [22] (p91), is, in fact, merely the case of an emergent property (P2 = T) coupled with one (or more) mandatory parts (S10 = MANDATORY) — which are of necessity also inseparable (S6 = F). Consequently, destroying one of these parts *must* destroy the aggregate.

In other words, the values of S2 and S6 are confused or self-contradictory. Clearly, a more consistent definition is urgently required.

2.2 White Diamonds

Shared aggregation in UML, on the other hand, is said to have no semantics other than that of association. Booch *et al.* [20] (p68) notes that the white (or open) diamond distinguishes the whole from the part "no more, no less". It therefore appears to add nothing beyond an association. It is stated to represent shared objects. But unless an object is encapsulated within another, *all* objects can be shared and, as pointed out in [18], this is, in any case, a property of the object not a property of the relationship (here aggregation). Shared aggregation is described as being denoted by weak ownership i.e. the part may be included in several aggregates [14] (pp18, 38) and [13] (p2-54) (S4 = T) and its owner may change with time (S6 = T) [14] (p38) and [13] (p2-54). Perhaps more importantly, what we believe is meant is the notion of *shareability* rather than the notion of shared (as is used currently in the definition of UML) which is the realization of that shareability. Shareability then becomes a derived property consequent upon the application of encapsulation (S1 = F) to the aggregation. This ties in with the implication in [13] that "black diamond" aggregation represents encapsulation (S1 = T).

UML docs [14] (p38) and [13] (p2-54) states that in an aggregation deleting the/either aggregate does NOT imply deletion of parts (C5 = F). Therefore, lifetimes are independent (S2 $\neq$ Case 4; S6 = T). This is contradicted by [7] (p79), who states that (white diamond) aggregation in UML *does* require propagation of the deletion operation to the parts (C5 = T) — this is presumably an error in that book.

Martin and Odell [19] (p80) says white diamond represents *optionality*.

Henderson-Sellers [23] suggests OML's membership (defined by S5 = F) may be an association in UML but [5] implies membership is in fact the white diamond of UML — a suggestion reinforced by [24].

Suggested notation changes for (shareable) aggregation (white diamond) are given in [25], who suggest {separable} and {inseparable} at the part end; indicating that they believe a white diamond in UML may be either separable or inseparable — although there is no indication in the UML source documents that suggests anything other than separability for the UML white diamond.

The ambiguities here are S6 and C5 (which is in fact consequent upon S6). However, S6 = F comes from only one, derivative source and C5 = T [7] appears to be erroneous.

2.3 Summary of Current State

In summary, documents on UML suggest the main characteristics of Composition (black diamond), summed across all definitive sources, to be:

P1 = T, P5 = T, S1 = T, S2 = Case 4 yet S2 = Case 2 or 4 and S2 = Case 1, 2, 3 or 4, S3 = T, S4 = F, S6 = T yet S6 = F, S10 = MANDATORY, C1 = T, C4 = T, C5 = T

UML's white diamond represents an association relationship which can be summarized in terms of primary and secondary characteristics as: P1 = T, P5 =

T, S1 = F, S2 ≠ Case 4, S3 = T, S4 = T, S6 = T, (S6 = F), C5 = F, (C5 = T). [The terms in brackets here were shown in Sect. 2.2 to be of dubious validity.]

In comparing black and white diamonds, S1 and S4 are clearly used as major discriminating characteristics between shared and composite aggregation and, if S6 = F and/or S2 = Case 4 turn out to be appropriate for composite aggregation, then S6 and/or S2 also become key discriminators (giving a total of 4) between UML's two existing kinds of WP relationship. In addition, while S8 is used as the only discriminating characteristic in one source, since this occurs in none of the other UML documents, it would appear to be erroneous and we thus exclude it from further consideration as a major partition.

The similarities and differences can be roughly summarized as follows:

UML black and white — owner may change over time, transitive, antisymmetric

UML black only — propagation semantics; coincident lifetimes (probably) — or at least part lifetimes included within lifetime of whole; probably inseparable

UML white — belongs to one or more aggregate; lifetimes independent — components can have independent existence (i.e. separable)

We should also note here what kinds of WP relationship *cannot* be represented by either black or white diamonds in UML Version 1.3. These are

- all configurational relationships (this includes Odell's Component–integral object relationship which he notes as being the most commonly occurring WP relationship and OML's aggregation: [17]
- all non-transitive relationships. This excludes many is-a-member-of relationships
- other combinations such as (i) non-encapsulated, non-shared; (ii) encapsulated, shared WP relationships — although whether all these other combinations are valuable is a moot point.

3 Proposal to Create a Useful Definition for UML WP Relationship

There are two distinct parts to the proposal. First, the supertype of WP Relationship needs to be described in terms of *primary properties* (Sect. 3.1). This WP Relationship could then, at the metalevel, inherit from the Relationship metaclass[2] in UML Version 1.3 (see [26] for an example of how this has already been applied). Secondly, two (possibly 4) partitions need to be identified (Sect. 3.2). If two partitions are used (the likely preferable choice for most modellers), then (and only then) *one* specific *secondary* characteristic needs to be selected. If four kinds of aggregation are to be modelled, then *two* appropriate secondary characteristics need to be identified. For example, one might choose the inseparable characteristic to discriminate between black and white diamonds. *All other characteristics* are then left undefined. The addition of such definitions/values should

---

[2] or possibly a revised form of the Association metaclass

then be represented in terms of additional stereotypes, annotations, multiplicities etc. In the second case, we might choose those WP Relationships which are both inseparable and have coincident lifetimes. These would be represented by black diamonds. The white diamond must then represent three styles of WP Relationship:

- inseparable but without the imposition of coincident lifetimes (an example might be an invoice with its invoice lines)
- coincident lifetimes but separable (an example might be section/paragraph in which a "cut and paste" operation related to a paragraph is a separation while the deletion of the section is by definition the deletion of all its paragraphs)
- separable but without coincident lifetimes (an example here would be membership relationships between, say, a sports club and its membership)

It is of course possible that a combination of characteristic values are chosen to represent, say, black diamond aggregation in UML such that all the other occurrences of WP relationship are represented by a white diamond. For example, black diamond (composition) might be defined by S2 = Case 4 (coincident lifetimes) plus S4 = F (non-shareable) plus S6 = F (inseparable). White diamond would then include a wide variety of kinds of aggregation including all the shareable WP relationships but also those which are non-shareable and separable, non-coincident lifetimes and inseparable and so on. Defining white diamond as only shareable (namely S4 = T) would give essentially three partitions: black diamond ((S2 = Case 4) + (S4 = F) + (S6 = F)); white diamond (S4 = T); and the rest — which, although WP relationships, could only be modelled by an association and not by any WP relationship. The resultant metamodel would be somewhat clumsy!

It is therefore important to identify which characteristics are consequent upon others. If A is consequent on B, then if B is true, A is automatically true and need not be included in the definition. In that context, some of the characteristics which often get confused are:

1. Lifetimes and separability: Lifetimes DO NOT IMPLY anything about separability
2. Separability and independence: Separability (Martin and Odell say black diamond = invariant) implies part can exist independently
3. Propagation of destruction operation and separability: Propagation of destruction semantics implies inseparability
4. (possibly) Mandatory versus optional: Mandatory implies inseparable; Optional implies separable.
5. and possibly even Mutability and separability: in [27], it is demonstrated that immutability ⇒ inseparability which is the same as separability ⇒ mutability.

Some other points to note:

- If lifetimes are linked such that they live and die together
  $\Rightarrow$ propagation of construction and destruction operator *and* Case 4 and inseparable
- If inseparable + contained lifetimes, we have existential dependency (a.k.a. weak or dependent entity in ER modelling).

3.1 Primary Characteristics

The current UML documentation identifies P1 and P5. Clearly, all that remains in the OMG/UML standards documentation is to add the definitions according to P2, P3 and P4.

Once the primary characteristics, applicable to *all* WP Relationships, have been clearly stated, we can proceed to identify specific kinds of "aggregation" in order to define UML black and white diamonds unambiguously.

3.2 Secondary Characteristics

Secondary characteristics need to be chosen in order to create the two (possibly four) flavours of WP Relationship. The choices we investigate in this section aim to respect the initial spirit of UML and its progenitors of OMT and Booch (as discussed in Sect. 1).

First, let us consider the case of choosing just *one* as the main partitioning discriminant. (We will discuss possible combinations of secondary characteristics later.)

Choosing a Single Secondary Characteristic One could choose configurational (S5 = T as in OML) or invariant/inseparable (S6 = F as [19] suggest the UML was intended to do). Then all the other characteristics are represented by stereotypes and other annotations. Either would be a good choice. However, in an attempt to retain the general spirit of the (currently ill-defined) black and white diamonds in UML [3,13,20], we note that the main partitions sought appear to be

1. coincident lifetimes for at least some of the parts belonging to an aggregate; or
2. shareability (incorrectly labelled sharing in UML docs); or
3. separability.

Coincidence of lifetimes is probably the weakest choice of the three. Very few WP relationships have all their parts with a lifetime exactly equal to that of the whole. In addition, as noted above, lifetimes do not prescribe whether parts are or are not inseparable (a misleading assertion in [5]. A much stronger and useful statement would be made by focussing directly on separability (choice 3).

Using shareability as the main partitioning characteristic has its appeal. It is simple and straightforward and already embodied in the UML Version 1.3 white diamond. However, defining black diamond simply as non-shareable components can make no statement about whether parts are or are not separable; what their lifetimes are; whether the destruction operation is propagated; and so on.

As the best choice for a *single* secondary characteristic, we would recommend separability; a choice which is also compatible with the current UML documentation and books. As well as defining separability *per se*, defining black diamond as having inseparable parts permits other consequences. For example, if the lifetime of the parts is contained within that of the whole (and the parts are inseparable), then we have the case of existential dependency or weak entity of ER modelling — perhaps this was what was intended by UML's black diamond all along? The advantage of this is that an existential dependency is one in which parts can live and die with the whole *or* may be added and deleted during the lifetime of the whole. This means that if we have parts, p_i and a single whole w defined as $p_i(i = 1$ to $n) : PART$ and $w : WHOLE$ then

$$\exists p_i(i \neq 1) \text{ s.t. } p_i.t_B \geq w.t_B \text{ and } p_i.t_D \geq t_D$$
$$\text{given that } \exists p_1 \text{ s.t. } p_1.t_B = w.t_B \text{ and } p_1.t_D = w.t_D$$

where t_B is the time of birth and t_D is the time of death. This definition describes, formally, Cases 1–4 of S2. In addition, we need to specify $\forall p_i : PART$, $\exists w R p_i$ where R is the whole–part relationship. This states that (1) we have a whole–part relationship and that (2) all parts *must* participate in this relationship (i.e. they cannot exist outside of it).

The key here is that the parts cannot be separated from the whole and thus can have no independent existence outside of the whole. We have already seen that this is fully in sympathy with the ideas expressed in [20]. It is also easy to use as a base for a compound aggregation style in which several parts (inseparable and embedded lifetimes) are mandatory[3] and others optional — optional parts may still satisfy black diamond definitions but many will be separable and thus use a white diamond. The first "problem" here is that, for most examples, there are more separable parts than non-separable. In this commonly used car example, it is really only the chassis that is inseparable [28] — yet most OO modellers use the car as a prime example of composition. Actually it is best classified as a Composite–integral object flavour of aggregation [8,23] in which there is a strong configurational nature to the parts (the engine, wheels, driveshaft etc.) which must all be accurately configured together.

A second "problem", or at least a point to be noted, is that this definition makes no statement about shareability, which must therefore be dropped from the white diamond definition. However, since shareability is a common characteristics of objects (unless they are fully encapsulated) as noted by [18], then this should not cause any serious problems. Finally, inseparable parts overlain with encapsulation (S1 = T) leads naturally (Fig. 2) to a value of S8 = BY_VALUE — another oft-favoured interpretation.

Another clean, but different, solution would be to adopt S5 as the main secondary characteristics used for partitioning. This has the advantage of a long

---

[3] A mandatory part is one which, if it is destroyed, will automatically lead to the destruction of the whole.

562

history based in linguistic and cognitive science research as well as in software engineering e.g. [1, 8, 23]. Here we would choose black diamond to represent all configurational WP relationships (with the advantage that now black diamond composition includes the commonest kind: Component–integral object) and white diamond models all the remaining, non-configurational WP relationships; as is done in OML [18]. This is also nice since these latter kinds of WP relationship have some family resemblance and can be called "Membership" [23].

Formally, we may state that $\forall p_i (i = 1..n) : PART, wRp_i$ where $p_j C p_k$ $(\forall j, k \in [1..n] \mid j \neq k)$ where C is some relationship between the parts which enforces the details of how they should be configured.

The disadvantage is that this choice says nothing directly about existential dependency, separability or shareability — these must all be modelled by use of appropriate stereotypes, multiplicities or other annotations (in the UML notation).

Choosing Two Secondary Characteristics Application of two partitioning characteristics creates four subsets, which of course requires four symbols for their discrimination (this would necessitate adding two new symbols to the UML notation set).

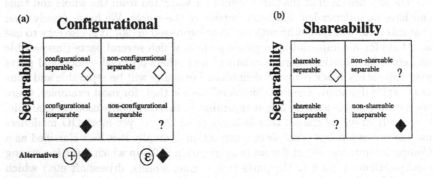

Fig. 3. Four subsets of the WP relationship creating by multiple (2) partitioning (a) using configurational and separability as the example (also shown are alternative symbols) and (b) using shareability and separability

In Sect. 3.2, the most likely candidates for a single partitioning characteristic were S5 (configurational) and S6 (separability). It seems logical that we should investigate the possible value in a four-partition solution to first consider the possibility of using S5 and S6 as the pair of determining characteristics. The four subsets these define are shown in Fig. 3. The nearest in nature to the spirit of the black diamond appears to be configurational and inseparable (although it should be noted that this excludes the commonest form of aggregation as

563

noted earlier). Since the spirit of the white diamond seems to revolve around separability, one might propose that two of the remaining three subsets are allocated the white diamond: essentially those which are separable. But this leaves the non-configurational/inseparable partition without a symbol.

An alternative notation would be simply use a double symbol. In other words, use black and white diamond to indicate separability or not (as in Option 1) and then, in addition, use the configurational/membership symbols first used in OML [26] — an example is shown in Fig. 3(a).

A second obvious alternative warranting evaluation is that of S4 together with S6 (Fig. 3(b)). This choice reflects the ideas in UML Version 1.3 which stress the shareability of the white diamond and the inseparability of the black. Again the problem is allocation of symbols in the UML notation. It seems reasonable to allocate the black diamond symbol to the non-shareable and inseparable partition and the white diamond symbol to those which are shareable and separable. But it is not clear how we should notate the two partitions of non-shareable/separable and shareable/inseparable; nor indeed whether these are useful partitions.

Finally, it should be noted that topological inclusion or "containment" is not[4] a type of WP relationship [1,8,23] and could be given its own notation, as is done in OML.

3.3 Implementation in the UML Metamodel

Aggregation (of all kinds) in the UML Version 1.3 metamodel is controlled by a number of meta-attributes on the metaclass AssociationEnd. The relevant ones are: aggregation: AggregationKind = {none, aggregate, composite} and changeability: ChangeableKind = {changeable, frozen, addOnly}. Any modifications to these metaattributes cannot be decided until a choice has been made from the options discussed above. For example, if chosen, separability might deserve a new metaattribute to facilitate CASE tool support.

4 Discussion and Recommendations

Some of the ambiguities and contradictions in UML may be the result of its informality — in the (mis)use of the English language in the definitions and the lack of any formal underpinning [30]. For example, the contradiction in [20], highlighted earlier, regarding coincidence of lifetimes may be the result of a misunderstanding of the word "coincident": "exactly contemporaneous" [31]. If two things coincide, then they "occupy the same space of time". This is Case 4 of S2. What was probably intended was Cases 1–4 which might better be described as cases in which the lifetimes of the parts are *contained within* the lifetime of the whole.

Another term of confusion turns out to be "may change ownership" which leads us to understand that the parts need to be (a) separated from their current owner and then (b) rejoined to a new whole. Finally, the word "remove"

---

[4] although it is often given, incorrectly, as an exemplar of aggregation e.g. [29] (p210).

(as in the UML definition stating "parts may be removed before the death of the composite": [3] (p226) implies "to move, shift or convey from one place to another" [32] whereas the meaning intended [30] is to (remove and) destroy.

Our specific recommendations for clarifying black and white diamond aggregation in UML are:

1. Add to primary definition of WP relationship (a) emergent property, (b) resultant property and (c) irreflexivity at instance level.
2. Remove transitivity from basic definition — its retention is very limiting.
3. Remove "change of ownership" from basic definition — making everything separable goes against the spirit of composition. Change of ownership should only apply to white diamond.
4. Remove coincident lifetimes from definition of composition and focus instead on separability which leads easily to existential dependency (C4 = T).
5. Add the discriminating definition for composition and aggregation (black and white diamonds) from the chosen option.
6. "Shareability" should be the terminology not "shared".
7. Ensure configurational and separable WP relationships — the most common according to Odell — are included in the definition of either black or white diamond.
8. Ensure that it is clear that in the UML documents topological inclusion (a.k.a. containment) can *in no way* be considered to be implicated in the definition of the WP relationship or any flavours of "aggregation". Topological inclusion is a distinct relationship.
9. Finally, check attributes on AssociationEnd metaclass — modifications may be needed to ensure that it is not possible to accidentally create an invalid or inconsistent type/"flavour" of the WP relationship.

In addition, we would recommend some new additions (rather than modifications to existing relationships):

10. the addition of the containment relationship and associated notation into UML to represent topological inclusion.

References

1. Winston, M.E., Chaffin, R., Herrmann, D.: A taxonomy of part–whole relations. Cognitive Science **11** (1987) 417–444
2. Henderson-Sellers, B., Barbier, F.: What is this thing called aggregation? Procs. TOOLS29 (eds. R. Mitchell, A.C. Wills, J. Bosch and B. Meyer), IEEE Computer Society Press (1999) 216–230
3. Rumbaugh, J., Jacobson, I., Booch, G.: *The Unified Modeling Language Reference Manual*, Addison-Wesley, Reading, MA, 550pp (1999)
4. Civello, F.: Roles for composite objects in object-oriented analysis and design. Procs. OOPSLA (1993) 376–393
5. Saksena, M., France, R.B., Larrondo-Petrie, M.M.: A characterization of aggregation. OOIS'98 (eds. C. Rolland and G. Grosz) Springer (1998) 11–19
6. Booch, G.: *Object-Oriented Analysis and Design with Applications* (2nd edition), Benjamin/Cummings Publishing Co., Inc., Redwood City, CA, USA, 589pp (1994)

7. Quatrani, T.: *Visual Modeling with Rational ROSE and UML*, Addison-Wesley, Reading, MA, USA, 222pp (1998)
8. Odell, J.J.: Six different kinds of composition. J. Obj.-Oriented Prog. **6(8)** (1994) 10–15
9. Kilov, H., Ross, J.: *Information Modeling. An Object-Oriented Approach*, Prentice Hall, Englewood Cliffs, New Jersey, USA, 268pp (1994)
10. Parsons, J., Wand, Y.: Using objects for systems analysis. Comms. ACM, **40(12)** (1997) 104–110
11. Younessi, H.,: personal communication to lead author, 4 August 1999
12. Castellani, X.: An overview of the Version 1.1 of the UML defined with charts of concepts. Procs. ≪UML≫'98. Beyond the Notation (1998) 13–24
13. OMG:, OMG Unified Modeling Language Specification (draft), Version 1.3 alphaR2, January 1999 (unpubl.) (1999)
14. OMG: UML Semantics. Version 1.1, 15 September 1997, OMG document ad/97–08-04 (unpubl.) (1997)
15. OMG: UML Notation. Version 1.1, 15 September 1997, OMG document ad/97-08-05 (unpubl.) (1997)
16. Fowler, M., Scott, K.: *UML Distilled. Applying the standard object modeling language*, Addison-Wesley, Reading, MA, 179pp (1997)
17. Firesmith, D.G., Henderson-Sellers, B.: Clarifying specialized forms of association in UML and OML. JOOP/ROAD **11(2)** (1998) 47–50
18. Firesmith, D.G., Henderson-Sellers, B.: Upgrading OML to Version 1.1: Part 1. Referential relationships. JOOP/ROAD **11(3)** (1998) 48–57
19. Martin, J., Odell, J.J.: *Object-Oriented Methods. A Foundation (UML edition)*, Prentice-Hall, Upper Saddle River, NJ, USA, 408pp (1998)
20. Booch, G., Rumbaugh, J., Jacobson, I.: *The Unified Modeling Language User Guide*, Addison-Wesley, Reading, MA, USA, 482pp (1999)
21. Bock, C., Odell, J.J.: A more complete model of relations and their implementation: roles. JOOP **11(2)** (1998) 51–54
22. Gogolla, M., Richters, M.: Equivalence rules for UML class diagrams. Procs. ≪UML≫'98. Beyond the Notation (1998) 87–96
23. Henderson-Sellers, B.: OPEN relationships — compositions and containments, JOOP/ROAD **10(7)** (1997) 51–55, 72
24. Booch, E.G.: personal communication to first author on the meaning of black and white diamond in UML (1998)
25. Saksena, M., Larrondo-Petrie, M., France, R.B., Evett, M.P.: Extending aggregation constructs in UML. Procs. ≪UML≫'98 (1998) 273–280
26. Firesmith, D., Henderson-Sellers, B., Graham, I.: *OPEN Modeling Language (OML) Reference Manual*, SIGS Books, New York, 276pp (1997); Cambridge University Press, New York (1998)
27. Barbier, F., Henderson-Sellers, B.: The whole–part relationship in object modelling: a definition in cOlOr. submitted to *Inf. Soft. Technol.* (1999)
28. Bock, C., Odell, J.J.: A user-level model of composition. Report on Object Analysis and Design **2(7)** (1996) 5–8
29. Jacobson, I., Booch, G., Rumbaugh, J.: *The Unified Software Development Process*, Addison Wesley Longman Inc., Reading, MA, USA, 463pp (1999)
30. Selic, B.: personal communication to authors (1999)
31. *The Shorter Oxford English Dictionary*, Volume I, Clarendon Press, Oxford (1973)
32. *The Shorter Oxford English Dictionary*, Volume II, Clarendon Press, Oxford (1973)

Interconnecting Objects via Contracts

Luís Filipe Andrade[1] and José Luiz Fiadeiro[2]

[1] OBLOG Software S.A.
Alameda António Sérgio 7 – 1 A,
2795 Linda-a-Velha, Portugal
landrade@oblog.pt

[2] LabMAC & Dept. of Informatics
Faculty of Sciences, University of Lisbon
Campo Grande, 1700 Lisboa, Portugal
llf@di.fc.ul.pt

Abstract. The evolution of today's markets and the high volatility of business requirements put an increasing emphasis on the flexibility of systems, i.e. on the ability for systems to accommodate the changes required by new or different organisational needs with a minimum impact on the implemented services. In this paper, we put forward an extension of UML with a semantic primitive – contract – for representing explicitly the rules that determine the way object interaction needs to be coordinated to satisfy business requirements, as well as the mechanisms that make it possible to reflect changes of the business requirements without having to modify the basic objects that compose the system. Contracts are proposed as extended forms of association classes whose semantics rely on principles that have been used in Software Architectures and Distributed System Design for supporting dynamic reconfiguration.

1 Introduction

Market evolution, and the consequent evolution of business requirements, make the corresponding problem domains become larger and more complex. As a result, the information systems that are required to support business activities are becoming more sophisticated and complex too. Such complex systems can be seen as large collections of coarse-grained parts with a high degree of interaction. From an object-oriented point of view, they are best seen as collections of interacting, concurrent and distributed objects that are responsible for keeping information and performing the required business activities.

Market evolution and volatility of business requirements have a very deep influence on organisations and their information systems. More and more, large organisations face two important problems in this respect:

- How should information systems be conceived and developed in order to support the continuous evolution of the core business and the evolution of information system technology?
- How to make information system development and maintenance scalable in the context of highly volatile business application domains?

When systems are conceived as collections of interacting objects, the problems that we have just identified require that we be able to express, and make available as first-class citizens, the constraints and the rules that capture the business requirements of the application domain. Because business rules determine the way object behaviour and interaction needs to be coordinated, it is necessary that these coordination aspects be available explicitly in the system models so that they can be changed, as a result of modifications that occur at the level of the business requirements, without having to modify the basic objects that compose the system. That is, we should make the evolution of the system *compositional* with respect to the evolution of the application domain in the sense that changes occurring in the domain should be mapped directly to the architecture of the information system.

Several authors already made similar observations in the past, namely in [9], which became the subject matter of the ISO General Relationship Model (ISO/IEC 10165-7). They have been amply confirmed by the experience that *Oblog Software* has accumulated in developing banking applications, an area in which fierce competition has dictated the need for services to be easily reconfigured. For instance, new account packages are defined almost every other day to attract more customers and prevent old customers from being lured by the new services offered by competitors. Time-to-market is a business decision that can be severely conditioned by the capacity of the bank's information system to support and accommodate the definition of new types of package with minimal impact on the applications that manage the individual accounts.

To a certain extent, the modelling of interactions between objects can be made explicit through what in UML is known as *association classes*. Indeed, association classes allow us to represent, in conceptual models, interactions as first-class citizens, i.e. at the same level of importance as the objects themselves. An example can be given, using the banking example mentioned in the previous paragraph, in terms of the interaction that is required between customers and accounts. For that purpose, an association class *ownership* can be used that, as a class, has attributes and methods that model the activities associated with the ownership. The fact that account packages come in all colours and flavours can be captured by the existence of more than one ownership class between classes *account* and *customer*. For instance, besides the standard package, we can well envisage a *VIP-ownership* association class with an attribute *credit* that establishes the amount by which the account can be overdrawn.

Association classes are somewhat too generous in the way they promote relationships to the status of objects, namely in what concerns interfacing with classes external to the association, which may hinder the kind of flexibility that we have motivated above. We shall discuss such issues in the body of the paper. The main problem, however, lies in the mechanisms that are usually made available for implementing associations because they do not provide the degree of flexibility that is required by the need to create, delete or modify such classes without having to change the implementation of the roles. For instance, a solution that requires the code of *account* to be redefined and/or recompiled each time a new account package is put on the market is not acceptable. Hence, the usual implementation of associations in terms of attributes does not provide this degree of flexibility.

Our purpose in this paper is to propose a new primitive for object-oriented modelling – *contracts* – which is an extension of association classes at the representation level as discussed above, but relies on implementation mechanisms that ensure the degree of flexibility required by the need to reflect changes in the business rules. Our solution capitalises on previous work on Software Architectures [15] and Configurable Distributed Systems [11], and relies on a modularisation technique – *superposition* – developed for Parallel Program Design [8] that we show to be implementable in current component-based techniques like CORBA and Java Beans.

In section 2, we further motivate the concept of contract through the analysis of different models of the bank account package. In section 3, we propose a textual notation for and give examples of the application of contracts. In section 4, we outline a semantics for contracts that is based on categorical mechanisms previously used for software architectures. Finally, in section 5, we outline an implementation for contracts as a design pattern that ensures the levels of flexibility motivated above.

2 Motivation

Let us resume the discussion on account packages so that more motivation for the envisaged notion of contract can be provided before we start formalising the concept. In order to achieve the degree of flexibility that one wants from contracts, namely the ability to define new interactions or modify existing interactions without changing the partners in the contract, it is essential that interactions be promoted to first-class citizens. That is to say, it is necessary that interactions be provided with class properties and features like their own attributes and methods. An example for this need can be given in terms of the method withdrawal (amount) of account.

A customer that owns an account has, typically, an attribute owns:account so that transactions performed by that customer can be modelled through calls of the form owns.withdrawal(a). The condition balance≥amount is typically declared on account as a precondition on withdrawal(amount) to ensure that there are enough funds for the withdrawal to be made. This precondition establishes a contractual relationship between account and customer, both in the technical sense developed by B.Meyer [12] and as one of the business rules that establish the way customers are required to interact with accounts.

In order to understand the problems raised by the need to reconfigure systems as a reaction to the constant evolution of market rules, assume that it becomes necessary to make certain instances of the interaction between customers and accounts more flexible by creating "VIP-packages" according to which accounts can be overdrawn up to an amount agreed in advance between the customer and the bank. A naïve solution to the problem of adapting the existing system to the new business rules would be to enrich account with a new operation – VIP-withdrawal – for the more flexible withdrawals. A disadvantage of this solution is in the additional burden that needs to be placed on the supplier – the account – to decide when the new operation is to be used.

The typical OO-solution to this new situation is different: it consists in defining a subclass VIP-account of account with a new attribute credit and a weaker precondition on withdrawal(amount): balance+credit≥amount. In this way, the more flexible contractual relationships can be established directly between the client (customer) and the specialisation of the supplier (the VIP-accounts). Nevertheless, there are two main drawbacks in this solution. On the one hand, it introduces, in the conceptual model, classes that have no counterpart in the real problem domain. It is the customers who are VIPs, not the accounts. However, having placed the contractual relationship between customers and accounts in account, one is forced to model the revised contract through a specialisation of the previous one, which implies the definition of the artificial subclass of account. The second disadvantage is not methodological but a technical one. The new solution still requires changes to be performed on the rest of the system because the other classes need to be made aware of the existence of the new specialised class so that links between instances can be established through the new class.

The disadvantages in both solutions result from the fact that the contract between customers and accounts is being placed on the side of the supplier (the account), which is what is favoured by Meyer's notion of contract [12]. This makes it difficult to accommodate new forms of the contract that depend on the client side of it. Hence, it makes more sense to place the contract at the level of the relationship that exists between customers and accounts by promoting it to what in UML is known as an *Association Class*, i.e. an association that can also have class properties and features. In the account package example, the desired promotion could be achieved by introducing an association class ownership in which the details that pertain to the envisaged coordination of the joint behaviour of clients and accounts, including the preconditions that apply to the interactions, would be placed.

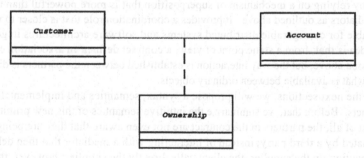

Changes to the ownership contract, such as the addition of the attribute credit and the weaker precondition on the interaction, can now be put on specialisations of the new class without having to change the role classes. Indeed, resorting to association classes keeps the model faithful to the application domain by representing explicitly the business rules that coordinate the interaction between the entities involved.

As a consequence, changes in the business rules should be more easily accommodated in the model without having to modify the classes that model the entities involved and, hence, leave a vast core of the information system unchanged.

From the discussion held above about the disadvantages of the attribute-based representation of the relationship between accounts and customers, it seems clear that the best way of implementing the contract through the association class is for a new operation to be declared for ownership that can act as a mediator between the roles. Upon a call from the client, the mediator would have the responsibility of determining whether the contractual relationship between the partners is valid and, in case it is, delegate on the supplier to proceed. In this way, it is, indeed, possible to achieve the required flexibility for accommodating changes in the business rules simply by modifying the contracts as required, e.g. at the level of the preconditions of the mediators, without having to modify the partners in the contract.

Although the advantages of making relationships first-class citizens in conceptual modelling has been recognised by many authors (e.g. [9]), which lead to the ISO General Relationship Model (ISO/IEC 10165-7), things are not as clean with this solution as they may seem. On the one hand, the fact that a mediator is used for coordinating the interaction between two given objects does not prevent direct relationships from being established that may violate the contract. In the case of the account package, nothing prevents a designer from connecting directly a customer to an account, possibly breaching the contract because the precondition has now been moved from the account to the mediator. On the other hand, the solution is not incremental (or additive) in the sense that the addition of new business rules cannot be achieved by simply introducing new contracts. Indeed, different contracts may interact with each other thus requiring an additional level of coordination among the mediators themselves, i.e. the introduction of contracts between contracts, etc. This leads to models that are not as abstract as they ought to be due to the need to make explicit (even program) the relationships that may exist between different contracts.

The notion of contract that we put forward in this paper circumvents these problems by relying on a mechanism of superposition that is more powerful than the use of mediators as outlined above. It provides a coordination role that is closer to what is available for configurable distributed systems and software architectures in general. The idea is that, from a static point of view, a contract defines an association class as discussed above, but the way interaction is established between the partners is different from what is available between ordinary objects.

In the next sections, we will propose a syntax, semantics and implementation for contracts. Before that, we summarize the intuitive semantics of this new primitive:

- first of all, the partners in the contract are not even aware that they are being coordinated by a third party; instead of interacting with a mediator that then delegates execution on the supplier, the client calls directly the supplier; however, the contract "intercepts" the call and superposes whatever forms of behavior are prescribed; this means that it is not possible to bypass the coordination being imposed through the contract;
- the same transparency applies to all other clients of the same supplier: no changes are required on the other interactions that involve either partner in the contract;

hence, contracts may be added, modified or deleted without any need for the partners, or their clients, to be modified as a consequence;

- the effect of superposing a contract is cumulative; because the superposition of the contract consists, essentially, of synchronous interactions, and synchronization is transitive, different contracts will superpose their coordinating behavior, achieving a cumulative effect; in fact, an algebra of contracts can be defined much in the same way as the algebraic operations on architectural connectors defined in [16].

3 Notation and Examples

The notation that we adopt for contracts is the one made available in OBLOG, a UML-compliant language that is being defined, together with associated tools, by *Oblog Software* [www.oblog.pt]. Constructs in OBLOG have a textual and a graphical notation. The graphical notation that has been chosen for contracts is reminiscent of UML association classes: the basic difference is that the rectangle of the association class has been replaced by a scroll.

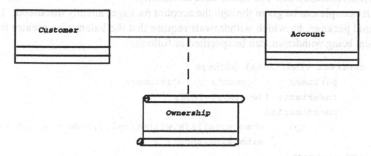

Given an application of a contract to relevant classes of a system (the partners in the contract), the instances of the partners that can actually become coordinated by instances of the body can be subject to conditions. The typical case is that the instances are required to be related through some association between the partners. Such conditions are specified as invariants in the contract body.

In OBLOG, the textual format of contracts is as follows:

```
contract <name>
partners <list-of-partners>
invariant <the relation between the partners>
constants
attributes
operations
coordination <interactions-with-partners>
behaviour        // the contract's own behaviour
    <additional behaviour being superposed>
end contract
```

Each interaction under "coordination" is of the form

```
<name> : when <condition> do <set of actions> with <condition>
```

The name of the interaction is necessary for establishing an overall coordination among the various interactions and the contract's own actions. This is similar to what happens in parallel program design languages like Interacting Processes [5]. The condition under "when" establishes the trigger of the interaction. Typical triggers are the occurrence of actions in the partners. The "do" clause identifies the reactions to be performed, usually in terms of actions of the partners and some of the contract's own actions. Together with the trigger, the reactions of the partners constitute what we call the synchronisation set associated with the interaction. Finally, the "with" clause puts further constraints on the actions involved in the interaction, typically further preconditions.

The intuitive semantics (to be formalised in the following sections) is that, through the "when" clause, the contract intercepts calls to the partners or detects events in the partners to which it has to react. It then checks the "with" clause to determine whether the interaction can proceed and, if so, coordinates the execution of the synchronisation set. All this is done atomically.

An example can be given through the account packages already discussed. The traditional package, by which withdrawals require that the balance be greater than the amount being withdrawn, can be specified as follows:

```
contract Traditional package
        partners x : Account; y : Customer;
        invariants ?owns(x,y)=TRUE;
        coordination
             tp:     when y.calls(x.withdrawal(z)) do x.withdrawal(z)
                     with x.Balance() > z;
        end contract
```

Notice that, as specified by the invariant, this contract is based on the ownership association previously discussed. This contract involves only one interaction. It relates calls placed by the customer for withdrawals with the actual withdrawal operation of the corresponding account. The customer is the trigger of the interaction: the interaction requires every call of the customer to synchronise with the withdrawal operation of the account but enables other withdrawals to occur outside the interactions, e.g. by other joint owners of the same account. The constraint is the additional precondition already discussed. Notice that the constraint applies only to the identified pair of customer and account, meaning that other owners of the same account may subscribe to different contracts (e.g. the VIP-contract already discussed).

The notation involving the interaction in this example is somewhat redundant because the fact that the trigger is a call from the customer to an operation of the account immediately identifies the reaction to be performed. In situations like this, OBLOG allows for abbreviated syntactical forms of interaction. However, in the paper, we will consistently present the full syntax to make explicit the various aspects involved in an interaction. In particular, the full syntax makes it explicit that the call put by the client is intercepted by the contract, and the reaction, which includes the call to the supplier, is coordinated by the contract. Again, we stress that such interactions are atomic, implying that the client will not know what kind of coordination is being superposed. From his point of view, it is the supplier that is being called.

In general, we allow for contracts to have features of their own. The examples below illustrate some of these situations. The first example addresses the VIP-package already discussed. This contract requires a constant and an attribute for establishing the intended coordination.

```
contract VIP package
        partners x : Account; y : Customer;
        constants CONST_VIP_BALANCE: Integer;
        attributes Credit : Integer;
        invariants
                ?owns(x,y)=TRUE;
                x.AverageBalance() >= CONST_VIP_BALANCE;
        coordination
                vp:     when y.calls(x.withdrawal(z)) do x.withdrawal(z)
                        with x.Balance() + Credit() > z;
    end contract
```

It is important to stress that features declared for the body are all private to the contract: they cannot be made available for interaction with objects other than the partners. Indeed, the contract does not define a public class.

We end this section with an example that illustrates a more involved form of coordination. Consider a package of two accounts of the same customer, a savings account and a checking account, that allows for a more efficient management of the customer's funds. The customer defines the minimum balance he or she wants in the checking account; by doing so, the package automatically transfers funds between the two accounts in order to maintain that amount always available in the checking account and the maximum possible amount in the savings account to earn interest.

```
contract Flexible package
        partners c : CheckingAccount; s : SavingsAccount;
        attributes m_minimum : Integer;
        invariants c.owner=s.owner;
        coordination
                stoc:   when (c.?GetBalance() < m_minimum)
                        do s.TransferFrom(m_minimum-c.?GetBalance(),c);
                ctos:   when (c.?GetBalance() > m_minimum)
                        do c.TransferFrom(c.?GetBalance()-m_minimum,s);
    end contract
```

In this case, the triggers of the interactions are not calls of the partners. It is the contract itself that has the initiative of monitoring the state of the partners and trigger transactions between them.

For simplicity, the examples given in the paper are based on binary relationships. However, contracts may involve more than two partners. In this case the invariant and coordination clauses may refer to all partners. Among other uses, such contracts may be seen to correspond to synchronisation agents, as presented in the Actors model [1], that coordinate the rules of engagement of various objects participating simultaneously in the same task.

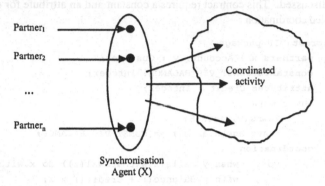

Partner₁

Partner₂

...

Partnerₙ

Coordinated activity

Synchronisation
Agent (X)

4 Semantics of Contracts

The semantics that we propose for contracts is based on a program design language and model – CommUnity [3] – that is similar to Unity [2] and Interacting Processes [5]. A basic difference between CommUnity and these other languages is that it replaces the shared-variables model of interaction by an object-based coordination model centred around private state and shared actions.

CommUnity is not a full-fledged object-oriented language but its primitives are expressive enough to capture the aspects that are involved in contracts. In particular, the mechanism that it provides for interconnecting simple programs into complex systems is the one that explains how contracts can be used to coordinate the behaviour of given objects in a system – superposition or superimposition [8].

Templates of object behaviour can be expressed in CommUnity according to the following structure:

```
object P is
var output   out(V)
       input    inp(V)
init   I
do    []    g: [B(g) →    ||    a:=F(g,a)]
      g∈Γ              a∈D(g)
```

where
- V is the set of *variables*. Variables can be declared as *input* or *output* and they correspond to communication channels that are used by the object. Input variables are used for receiving data from the environment of the object. They provide a

simplified model for data transmission usually associated with input parameters of methods. Output variables provide for both queries in the sense of UML, i.e. observations of the state of the object, and the output parameters usually associated with methods. They are not used as representations of the state of the object but, rather, as abstractions of the state that are convenient for the specification of interactions with the rest of the system.

- Γ is the set of *action names*; each action name has an associated guarded command. Actions represent interactions between the program and the environment. Hence, their execution is also under the control of the environment in the sense that their occurrence may require synchronisation with actions of other objects in the system. Each action g is typed with a set $D(g)$ consisting of the output variables whose values may change as a result of the occurrence of action g. For every variable v, we also denote by $D(v)$ the set of actions whose occurrence can change v.
- I is a proposition over the set of output variables – the initialisation condition.
- For every action $g \in \Gamma$, $B(g)$ is a proposition over the set of variables – its *guard*. When $B(g)$ is false, g cannot be executed.
- For every action $g \in \Gamma$ and local variable $v \in D(g)$, $F(g,v)$ denotes the value that v has after the execution of g.

As an example, consider the template corresponding to a bank account:

```
object account is
var output balance
      input   amount
init
do     withdraw: [→ balance := balance–amount]
    [] deposit: [→ balance := balance+amount]
```

As explained, the input variable *amount* accounts for the input parameters that are necessary for both actions. On the other hand, the output variable *balance* can be used as a query of the state of the account (no commitment is made about the representation of the state) as used, for instance, in the specification of the flexible package given in the previous section.

A model-theoretic semantics of CommUnity is presented in [10] based on labelled transition systems.

Consider now the specification of interactions between objects. In previous papers [e.g. 3,4] we have argued in favour of the use of Category Theory as a mathematical framework for expressing system configurations following Goguen's work on General Systems Theory [7]. In such a framework, interconnections are expressed via morphisms. The notion of morphism in CommUnity captures what in the literature on parallel program design is known as superposition. The original idea of superposition is of a mechanism meant to support a layered approach to systems design by which we are allowed to build on already developed components (drawing on the services they provide) by "augmenting" them (by, say, extending their state space and/or their actions/control activity) while *preserving their properties*. Most of the conditions expressed in the definition below are standard when defining superposition and are more thoroughly discussed in [3].

A morphism $\sigma: P_1 \rightarrow P_2$ consists of a (total) function $\sigma_{var}: V_1 \rightarrow V_2$ and a partial mapping $\sigma_{ac}: \Gamma_2 \rightarrow \Gamma_1$ s.t.:

1. For every $v \in V_1$, $o \in out(V_1)$, $i \in inp(V_1)$, we have $Sort_2(\sigma_{var}(v)) = Sort_1(v)$, $\sigma_{var}(o) \in out(V_2)$, and $\sigma_{var}(i) \in out(V_2) \cup inp(V_2)$;

2. For every $g \in \Gamma_2$ s.t. $\sigma_{ac}(g)$ is defined and $v \in V_1$: $\sigma_{var}(D_1(\sigma_{ac}(g)) \subseteq D_2(g)$ and $\sigma_{ac}(D_2(\sigma_{var}(v)) \subseteq D_1(v)$;

3. For every $g \in \Gamma_2$ s.t. $\sigma_{ac}(g)$ is defined and $v \in D_1(\sigma_{ac}(g))$: $\vdash (F_2(g, \sigma_{var}(v)) \models \sigma(F_1(\sigma_{ac}(g), v)))$;

4. $\vdash (I_2 \supset \sigma(I_1))$;

5. For every $g \in \Gamma_2$ s.t. $\sigma_{ac}(g)$ is defined, $\vdash (B_2(g) \supset \sigma(B_1(\sigma_{ac}(g))))$.

A morphism $\sigma: P_1 \rightarrow P_2$ identifies a way in which P_1 is "augmented" to become P_2 so that P_2 can be considered as having been obtained from P_1 through the superposition of additional behaviour. A typical situation, and the one that interests us for contracts, is when this additional behaviour is given itself as an object Q, e.g. a regulator for constraining the behaviour of the original object or an observer that monitors given properties of the underlying object. In such situations, superposition is achieved through the interconnection of the two objects so that P_2 is a form of parallel composition $P_1 \| Q$. Because of this, when describing superposition, we usually call P_1 the component and P_2 the system. In fact, superimposition in the sense of IP [5] is explained, precisely, as a parallel composition operator. The morphism view is the one that corresponds to Unity [2]. As we shall see below, both views are related through a categorical construction called *colimit*.

The map σ_{var} identifies, for every variable of the component, the corresponding variable in the system, and σ_{ac} identifies the action of the component that is involved in each action of the system, if ever. Condition 1 states that sorts, visibility and locality of variables are preserved. Notice, however, that input variables of P_1 may become output variables of P_2. This is because the result of interconnecting an input variable of P_1 with an output variable of another component Q of P_2 results in an output variable of P_2 (the input/output communication is internalised but the values are still made available for communication to the outside). Condition 2 means that the domains of variables are preserved and that an action of the system that does not involve an action of the component cannot change any variable of the component. Conditions 3 and 4 correspond to the preservation of the functionality of the component program: (3) the effects of the actions have to be preserved or made more deterministic and (4) initialisation conditions are preserved. Condition 5 allows guards to be strengthened but not weakened. Strengthening the guard is typical of superposition and reflects the fact that all the components that participate in the execution of a joint action have to give their permission.

System configuration in the categorical framework is expressed via diagrams. Morphisms as in the diagram below can be used to establish synchronisation between actions of programs P_1 and P_2 as well as the interconnection of input variables of one component with output variables of the other component for data transmission during interaction.

This kind of interaction can be established in a configuration diagram through interconnections of the form depicted above, where *channel* is, essentially, a set of input variables and a set of actions. Each action of *channel* acts as a *rendez-vous* point where actions from the components can meet (synchronise): any action a of *channel* establishes the synchronisation set that consists of all pairs $<a_1,a_2>$ of actions of P_1 and P_2, respectively, that are mapped through the morphisms to a. Hence, action names act as interaction points as in IP [5]. On the other hand, each variable of the channel provides for an input/output communication to be established between the components.

This semantics of the configuration diagram is the one that is obtained through the colimit of the diagram. Taking the colimit of a diagram is a categorical construction that collapses the configuration into an object by internalising all the interconnections. In the case of actions, it represents every synchronisation pair $<a_1,a_2>$ of actions of P_1 and P_2, as identified above, with a single action $a_1\|a_2$ whose occurrence captures the joint execution of the synchronised actions. The assignments performed by the joint action are exactly the assignments performed locally by each of the synchronised actions; its guards are the conjunctions of the guards of the components, i.e. $B(a_1\|a_2)=B(a_1)\wedge B(a_2)$. This is how a contract superposes new forms of behaviour. See [3] for more details on the colimit semantics of configuration diagrams.

More complex configurations can be described by diagrams of the form

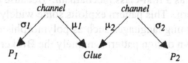

where *Glue* is a program that describes how the activities of both components are coordinated. Such configurations typically arise from the superposition of architectural connectors to given components of the system [4] with the aim of establishing, and coordinating, interactions between the components. This is also the way we formalise contracts.

Indeed, consider the general form of contracts as depicted in section 3. Each instance of a binary contract, i.e. a contract with two partners, gives rise to a configuration diagram as above where the P_i are the partners and the *Glue* is the contract body. The channels identify the interactions specified in the contract under "coordination". More precisely, the name of each interaction gives rise to an action name in *Glue*. For each interaction a, $\sigma_i^{-1}(\mu_i(a))$ is the action of P_i involved in that interaction, i.e. morphisms establish the interactions as claimed above. The different modes in which synchronisation can be programmed in CommUnity for achieving the required interaction are explained in [16]. The conditions of the "with" clause of contracts become guards of the actions that correspond to the interactions. Notice that, as explained through the colimit semantics, the effect of the interactions is to superpose the guards of the actions involved, which includes the "with" conditions. This is how contracts achieve the overall coordination effect.

Finally, it is important to stress that the incremental nature of contract superposition is captured in this formal framework through the properties of categorical diagrams. Again, see [16] on this aspect as applied to architectural connectors.

5   A Design Pattern for Contracts

As already explained in the previous sections, a contract works as an active agent that coordinates the contract partners. In this section, we are concerned with the way these coordination mechanisms can be implemented. When defining an implementation, we need to have in mind that, as motivated in the introduction, we should be able to superpose a contract to given objects in a system and coordinate their behaviour as intended *without having to modify the way these objects are implemented*. This degree of flexibility is absolutely necessary when the implementation of these objects is not available or cannot be modified, as in legacy systems. It is also a distinguishing factor of contracts when compared with existing mechanisms for modelling object interaction, and one that makes contracts particularly suited in business domains where the ability to support the definition and dynamic application of new forms of coordination is a significant market advantage.

Different standards for component-based software development have emerged in the last few years, among which CORBA, JavaBeans and COM are the current trend in industry. However, none of these standards provide a convenient and abstract way of supporting superposition as a first-class mechanism. Because of this, we propose our solution as a design pattern. This pattern exploits some widely available properties of object-oriented programming languages such as polymorphism and subtyping, and is based on other well known design patterns, namely the Broker, and the Proxy or Surrogate [6].

The class diagram below depicts the proposed pattern. In what follows, we explain, in some detail, its basic features, starting with the participating classes.

SubjectInterface-i – as the name indicates, it is an abstract class (type) that defines the common interface of services provided by AbstractSubject-i and Subject-i.

Subject-i – This is a concrete class that implements a broker maintaining a reference that lets the subject delegate received requests to the abstract subject (AbstractSubject-i) using the polymorphic entity proxy. At run-time, this entity may point to a RealSubject-i if no contract is involved, or point to a PartnerConnector-i that links the real subject to the contracts that coordinate it.

AbstractSubject-i – is an abstract class that defines the common interface of RealSubject-i and PartnerConnector-i. The interface is inherited form SubjectInterface-i to guarantee that all these classes offer the same interface as Subject-i (the broker) with which real subject clients have to interact.

RealSubject-i – is the concrete domain class with the business logic that defines the real object that the broker represents. The concrete implementation of provided services is in this class.

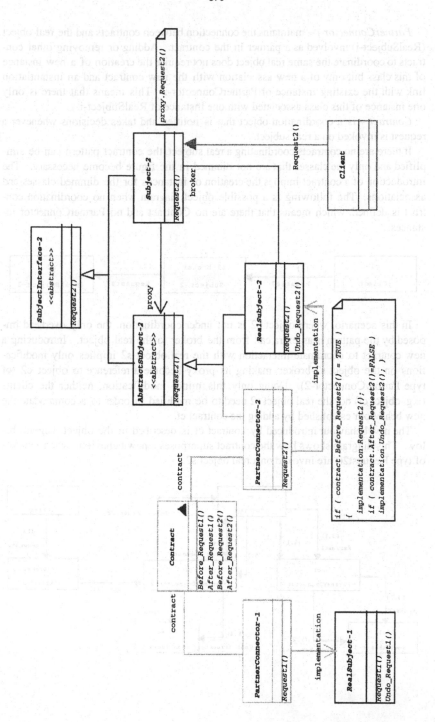

580

PartnerConnector-i – maintains the connection between contracts and the real object (RealSubject-i) involved as a partner in the contract. Adding or removing other contracts to coordinate the same real object does not require the creation of a new instance of this class but only of a new association with the new contract and an instantiation link with the existing instance of PartnerConnector-i. This means that there is only one instance of this class associated with one instance of RealSubject-i.

Contract – is a coordination object that is notified and takes decisions whenever a request is invoked on a real subject.

If there are no contracts coordinating a real subject, the contract pattern can be simplified and only the classes that are not dimmed in the figure become necessary. The introduction of a contract implies the creation of instances for the dimmed classes and associations. The following is a possible object diagram when no coordination contract is defined, which means that there are no Contract and no PartnerConnector instances.

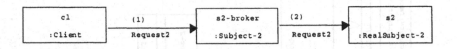

In this scenario, where object s2 is not under coordination, the only overhead imposed by the pattern is an extra call from the broker to the real object. Introducing a new contract to coordinate interaction with the real object s2 implies only modifications on the object s2-broker, making its proxy become a reference to object c2 (of type PartnerConnector-2). Doing only this minor modification, neither the clients (e.g. object c1) nor the real object s2 need to be modified in order to accommodate the new behaviour established by adding the contract ct.

The new behaviour introduced by contract ct is described in the object diagram below. This diagram shows how the contract superposes a new behaviour when requests of type Request2() are invoked on a real object s2.

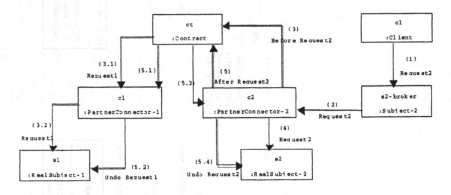

In this diagram we can see that once Request2() is sent to the s2-broker, it delegates its execution on the proxy reference (in this case on c2 instead of s2, as seen on the previous figure).

In c2 the implementation of subject services has the following format

```
(2): Request2() is {
        (3): If (contract.Before_Request2() = TRUE ) Then {
        (4):    implementation.Request2();
        (5):    If (contract.After_Request2() = FALSE )
                Then implementation.Undo_Request2();
    }
}
```

That is to say, before the partner connector c2 gives rights to the real object s2 to execute the request, it intercepts the request and gives right to the contract ct to decide if the request is valid and perform other actions. This interception allows us to impose other contractual obligations on the interaction between the caller and the callee. This is the situation of the first model discussed in section 2 where new pre-conditions were established between Account Withdrawals and their Customers. On the other hand, it allows the contract to perform other actions before or after the real object executes the request. Only if the contract authorises can the connector ask s2 to execute and commit, or undo execution because of violation of post-conditions established by the contract.

As stated at the beginning of this section, current component-based technology does not provide a convenient way for coordinating components. The benefit of having this form of coordination available as a primitive construction when specifying components and their interactions is that it avoids the burden of having to code such a pattern. In the meanwhile, tools which, like OBLOG, provide automatic code generation from high level specifications, must hide the implementation complexity of coordination, allowing the developer just to specify the contract itself.

6   Concluding Remarks

In this paper, we proposed an extension of UML with a new semantic primitive – contracts – for making explicit, in conceptual models, the business rules that coordinate the behaviour of given objects in a system, and allowing for this business rules to be added, or modified, without having to modify the way those objects are implemented. This proposal was made based on the experience that we have had in conceiving and implementing combined packages of banking products. In those packages, contracts emerged as enablers for more flexible domain models, and for the effective reuse of architectural design patterns to implement the system and accommodate changes with minimal impact.

In this respect, it seems important to make clear the difference between the approach that we advocated in the paper and proposals which, like [13,14], have been concerned with the representation and implementation of general laws that apply to the system as a whole, for instance to the way messages are exchanged or inheritance hierarchies are established, and, therefore, cannot be localised in any given component. Whereas such general laws need to be enforced by the environment in which the system is developed, for instance by regulating the creation, destruction and manipulation of objects, our interest is in more localised contractual laws that directly coordinate interaction between objects of given classes and, thus, can be represented, at an architectural level, as connectors. We feel that the two approaches are complementary, and we intend to study ways in which this complementary can be brought about into languages and methods.

A semantics was also proposed for contracts that is based on a categorical formalisation of the notion of superposition that has been used in parallel program design for achieving the kind of incrementality in design that motivated the use of contracts. This semantics was previously used for formalising the notion of connector as used in Software Architectures, which further highlights the similarities between the proposed primitive and the mechanisms that have been developed in Software Engineering for Re-configurable Distributed Systems.

We also showed how contracts can be implemented with Object-Oriented Programming Languages like C++ or Java by proposing a design pattern that makes it possible to introduce or remove contracts without affecting the objects under coordination. With this solution, only the contract compilation and program linkage is required whenever new contracts or changes to existing ones occur. On the other hand, this contract pattern allows us to coordinate third-party components (changing the behaviour of black boxes) without having to redefine or even recompile them. This makes contracts particularly suited for component-based programming.

This new primitive is also intended as an extension of B.Meyer's notion of contract [12], awarding it the role of coordinator of object interaction. Indeed, we showed that there are advantages in moving contracts from supplier classes to the association classes (contracts) that capture interconnections, namely as a way of enabling modifications to contracts that reflect specialisations of the client classes. It is our belief that, even if the design solution that we offered is not adopted, there are many advantages in using the proposed primitive at the more abstract modelling levels as we hope was evident from the example that we were able to fit in the paper.

Acknowledgements

We would like to thank the referees for their encouragement and suggestions. This work was partially supported by Fundação para a Ciência e a Tecnologia under contract PRAXIS XXI 2/2.1/TIT/1662/95.

583

References

1. G.A.Agha, *ACTORS: A model of Concurrent Computation in Distributed Systems*, MIT Press 1986.
2. K.Chandy and J.Misra, *Parallel Program Design - A Foundation*, Addison-Wesley 1988.
3. J.L.Fiadeiro and T.Maibaum, "Categorical Semantics of Parallel Program Design", *Science of Computer Programming* 28, 1997, 111-138.
4. J.L.Fiadeiro and A.Lopes, "Semantics of Architectural Connectors", in *TAPSOFT'97*, LNCS 1214, Springer-Verlag 1997, 505-519.
5. N.Francez and I.Forman, *Interacting Processes*, Addison-Wesley 1996.
6. E.Gamma, R.Helm, R.Johnson and J.Vlissides, *Design Patterns: Elements of Reusable Object Oriented Software*, Addison-Wesley 1995.
7. J.Goguen, "Categorical Foundations for General Systems Theory", in F.Pichler and R.Trappl (eds) *Advances in Cybernetics and Systems Research*, Transcripta Books 1973, 121-130.
8. S.Katz, "A Superimposition Control Construct for Distributed Systems", *ACM TOPLAS* 15(2), 1993, 337-356.
9. H.Kilov and J.Ross, *Information Modeling: an Object-oriented Approach*, Prentice-Hall 1994.
10. A.Lopes and J.L.Fiadeiro, "Using Explicit State to Describe Architectures", in *Proc. International Conference on Fundamental Aspects of Software Engineering (FASE'99)*, J-P.Finance (ed), LNCS 1577, Springer-Verlag 1999, 144-160.
11. J.Magee and J.Kramer, "Dynamic Structure in Software Architecures", in *4th Symp. on Foundations of Software Engineering*, ACM Press 1996, 3-14.
12. B.Meyer, "Applying Design by Contract", *IEEE Computer*, Oct.1992, 40-51.
13. N.Minsky and D.Rozenshtein, "A Law-based Approach to Object-oriented Programming", in Proc. OOPSLA'87, ACM Sigplan Notices 22(12), 1987, 482-493.
14. N.Minsky, "Law-governed Regularities in Object Systems; Part1: an Abstract Model", in *TAPOS* II(4), 1996.
15. M.Shaw and D.Garlan, *Software Architecture: Perspectives on an Emerging Discipline*, Prentice-Hall 1996.
16. M.Wermelinger and J.L.Fiadeiro, "Towards an Algebra of Architectural Connectors: a Case Study on Synchronisation for Mobility", in *Proc. 9th International Workshop on Software Specification and Design*, IEEE Computer Society Press 1998, 135-142.

How Can a Subsystem Be Both a Package and a Classifier?

Joaquin Miller [1] and Rebecca Wirfs-Brock [2]

[1] EDS, 10 South 5th Street Suite 1100
Minneapolis MN 55402 USA
joaquin@acm.org

[2] Wirfs-Brock Associates, 24003 SW Baker Road
Sherwood OR 97140 USA
rebecca@wirfs-brock.com

Abstract. The UML specifies that a subsystem is both a package and a classifier. This paper explores what that could possibly mean and explains why that was the right choice. It points out a key to the use of the concept in CASE tools, mentions the historical precedent for that key, and challenges CASE tools to support the flexibility that architects and designers need. Along the way, the paper reviews a method for discovering a good partition of a system into subsystems, describes a scheme for using UML to build a model of a system, and suggests some changes to the UML.

System has two clusters of meaning in English:

I: An organized or connected group of objects.

II: A set of principles, a scheme, method. [1]

We describe a system for modeling systems. That is: We review a method for discovering a good partition of a system into subsystems. We offer a set of principles for use of the concept, system, in specifying an organized or connected group of objects. And we describe a scheme for building a specification of a system. Then we pose some challenges for CASE tool builders.

We will use 'subsystem,' as shorthand for 'instance of a UML Subsystem.'

The purpose of this contribution is to clarify (one interpretation of) the meaning of 'subsystem', and to challenge tool builders to provide excellent support for using subsystems in models.

The first section briefly states what we intend by 'subsystem.' This will narrow the subject of this paper to the meaning we have in mind.

The second section reviews one method for using subsystems to design a system. This method discussion, though not the point of this paper, is relevant, because it is the process of using subsystems in design that creates the need for the tool support we challenge tool builders to offer.

In the third section we assert some principles.

The body of the paper, in the fourth section: explains our interpretation of the UML concept, subsystem; answers the question in the title of this paper; takes the reader through some model changes that might take place using subsystems to design; and makes a brief comment on UML notation.

Finally, we present our challenges.

1 What is a subsystem?

A system is an organized or connected set of parts. A subsystem is a kind of system. Because it is a system, a subsystem has parts. It is a subsystem because it happens to be a part of a larger system.

The concept, subsystem, serves as a tool for both organization and abstraction. Used for organization, a subsystem is a tool for partitioning a system: a subsystem represents one of the parts of the system. Used for abstraction, a subsystem is a tool for hiding complexity: a subsystem hides its own parts, that is, the details of its internal structure.

The UML Subsystem represents a combination of and a compromise between different needs felt by several of the UML partners.[1] We discuss only the needs for organization and abstraction in designing and presenting a specification. We use 'subsystem' only in the sense described in this section. (We will briefly discuss other uses of UML Subsystem near the end.)

2 Review of a method

Partitioning a system is one of the ways of reducing complexity by separating concerns. Dividing a system into subsystems not only reduces complexity for architects and designers and simplifies the work of programmers; it also provides project managers a way to organize the work of development.

Principles for discovering a good partition have been well known for decades. Myers identified the central concepts as module, coupling, and strength (now called cohesion). [3] These concepts have since been restated in terms of objects. [4], [5] Many other authors have made contributions to understanding subsystems.

For two reasons we briefly review a method for partitioning a system. This will provide a background for the principles presented in Section 3. It will also motivate the need for the tool support we challenge tool builders to offer.

We choose a particular method, [5], but another would serve as well for the purposes of this paper.

---

1 Hewlett-Packard, IBM, ICON Computing, i-Logix, IntelliCorp, EDS, ObjecTime, Oracle, Platinum Technology, Ptech, Rational Software, Reich Technologies, Softeam, Sterling Software, Taskon and Unisys.

2.1 Things to do

Divide responsibilities. Identify a set of subsystems and assign responsibilities to each of them. Evenly distribute system intelligence. Specify responsibilities as generally as possible. Keep behavior with related information. Keep information about one thing in one place. Share responsibilities among related objects.

Model Interactions. In the context of the collaborating subsystems, identify all the interactions between subsystems.

Simplify. Minimize the number of interactions a subsystem has with other subsystems. Minimize the number of other subsystems to which a subsystem delegates. Minimize the number of interfaces that a subsystem presents.

Evaluate the result. The goal is to maximize the cohesion of each subsystem and minimize the coupling between subsystems. Ask if there is opportunity to improve the design.

Repeat. Repeat the process, making changes to the design at each step. Continue design iterations while significant improvements are found.

Of course, it is good to build and execute architecture prototypes as a part of thi process; this is very likely to uncover reasons to change the design. And, as parts of the design appear to be stable, development may start. Certainly, this process will, in any case, continue while development is underway, as the surprises come.

2.2 Direction of work

The previous section discusses the work of designing subsystems. But that work can start from different places and proceed in different directions.

Bottom up. In [5], designers were advised to find classes, then to build and refactor class hierarchies, and finally to identify subsystems. This approach gives the subsystem designer a well thought out and well specified set of parts to use in specifying subsystems.

Top down. Critics of [5] said that subsystems should be identified first before any objects or classes are modeled. Architecture should precede design at the object level, they said, or the design method will not scale to large systems. Many prefer an approach that starts with the responsibilities of the system, then designs a set of subsystems, and specifies objects (and classes) only later in the process.

Middle out. In practice, a lot of work is actually done starting with a small number of potential or candidate subsystems and working in both directions. To better understand or further specify our model we may partition one of the parts of a subsystem, turning an object into a subsystem. And to gain a more abstract view, we may pick several subsystems and make them the parts of a larger subsystem. Or we may find a generalization that permits one subsystem to do the work that was done by two. Furthermore, subsystems or objects that are part of one subsystem may be moved to another.

This is the most practical approach for many problems.

3 Principles

The argument of the paper is based on certain principles, which the authors hold and present here simply as assertions.

A system is "something of interest as a whole or as comprised of parts. ... A component of a system may itself be a system, in which case it may be called a subsystem." [2]

"The basic elements of architectural description are components,[2] connectors, and configurations." [6] In object modeling, a component of a system is a subsystem or an object. (A connector, if it is of interest as consisting of parts, is a subsystem, too!)

During the process of designing a system, many changes will be made:
— Any object may become a subsystem as the design develops.
— Likewise, any subsystem may become an object. That is, the parts of the subsystem may be combined into a single object.
— We may replace a subsystem with some already specified object, or with a complex part that we treat as a simple object in our design.
— We may add an object to a model, but not know whether it will later become a subsystem. (That is, not know whether it will later come to be of interest as composed of parts.)
— After making an object or subsystem a part of one subsystem, we may move it to another subsystem.

Often, we will start to do something, but not be ready to complete it:
— We may fasten a connector to an object, but not know where we will connect the other end.
— Or feel an urge to specify a connector without connecting it to anything yet.
— We may add objects to models, without (yet) specifying that they are a part of a particular subsystem.

Always, we will use the power of abstraction. In particular:
— At a given level of abstraction, we may wish to hide the fact that something is a subsystem. (That is, hide the fact that it has parts.)

All these changes will be intermingled with shifts in viewpoint or level of abstraction. Neither our modeling language nor or methods, nor our tools have any reason for limiting these changes in any way. Rather, they must enable them, and make them easy for us.

---

2 'Component' in the usual English sense: a constituent element or part. The authors of [2] and [6] do not mean UML components.

588

4 A Scheme

In the UML, a subsystem "represents a behavioral unit."[3] It is defined to be both a classifier and a package. The UML classifier is an abstraction used to specify the features common to classes, interfaces and other UML model elements. "A classifier ... describes behavioral and structural features." "The purpose of the [UML package] is to provide a general grouping mechanism." "In fact, its only meaning is to define a namespace for its contents."

At first glance, it is a bit of a puzzle that something could be only a way to group things and define a namespace for them, and at the same time be a way to describe behavior. But we will show that the UML partners made the right choice.

The model elements grouped by a UML Subsystem into three categories:
— Operations
— Specification element
— Realization elements
We will mention each of these in the next section.

4.1 Levels of description

Sometimes, we want to represent a subsystem as a single object, hiding the parts; at other times, we want to show the parts.

Outside. From the outside, a subsystem is treated as a single model element. It appears as a whole, collaborating with other parts of the system to fulfill its responsibilities. Its collaborators treat the subsystem as a black box. Thought of in this way, subsystems are yet another encapsulation mechanism. The services provided by a subsystem are represented by interfaces and the corresponding operations. Other model elements further specify the behavior. (For example, a state machine, perhaps with action specifications.) These other model elements are what UML calls the specification elements of the subsystem. The operations and the specification elements of the subsystem (as well as any interfaces defined for the subsystem) are what specify the system without reference to its parts.

An instance of a UML classifier is an appropriate representation of a subsyste when it is being treated as a whole and as a black box.

Inside. From the inside, a subsystem reveals itself to have a complex structure. It is a system of objects collaborating with each other to fulfill distinct responsibilities that contribute to the purpose of the subsystem: the fulfillment of its responsibilities. These model elements specify the subsystem in terms of its parts; they are what UML calls the realization elements of the subsystem.

---

3 Quotations for which no reference is given are from the OMG UML specification current at this writing, [7]. The latest specification is available at: http://uml.shl.com .

A UML package is an appropriate container for the objects, relationships and other model elements that are the parts of a subsystem. The package will also hold other model elements that help specify the way these parts work together, for example, collaborations and interactions.

Outside	Inside
Interfaces Operations Specification elements	Realization elements

4.2 Solution to the puzzle

Abstraction is "the process of suppressing irrelevant detail to establish a simplified model." [2]

We see the escape from our puzzle: how can anything be both a package and a classifier? For us, as architects or designers, a UML Subsystem is*not* both a package and a classifier. A UML Subsystem is *either* a package or a classifier, depending on our level of abstraction.

[As mentioned above, we use 'subsystem,' as shorthand for 'instance of a UML Subsystem.' So we rephrase: A subsystem (an instance of a UML Subsystem) is *either* a package or an object, depending on our level of abstraction.]

The relationship of the subsystem considered as a single model element to the objects and relationships that are the parts of that subsystem is that of abstraction. The subsystem is an abstraction of its parts.

We can think of a subsystem as simply an object, if we view it from a particular level of abstraction and suppress the details of its realization. Those details are irrelevant in the black box view. Or we can think of a subsystem as only a package, if we view it from a more detailed level, but suppress the detail that it is considered an object in another view

An object (an instance of a UML class) is an appropriate representation of a subsystem when it is being treated as a black box.

Of course, we also need to specify the correspondences between the specification of the subsystem and the specification of its parts and their collaborations.

4.3 Drawing the pictures

This section presents what amounts to a simple modeling language. It will be clear that what we propose can be expressed using the variety of relationship kinds, keywords and stereotypes available in UML.

In the spirit of abstraction, we use a very simplified model. Because of our choice of words, this will appear to be a model about pictures, but actually it is a simplified model of the UML. (The pictures are not UML notation, nor a proposed alternative.)

Our model includes:

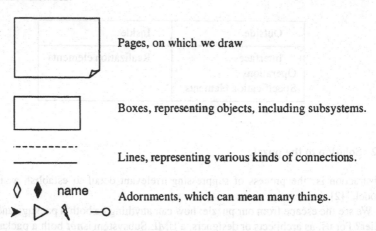

Pages, on which we draw

Boxes, representing objects, including subsystems.

Lines, representing various kinds of connections.

Adornments, which can mean many things.

In this simplified model, building a specification is represented as drawing boxes on pages, connecting the boxes with lines, and placing adornments on the lines. [The drawings of boxes, lines, and adornments in this simplified model are not diagrams in UML notation. But the reader (in particular, CASE tool makers) will see what we are getting at.]

Let us now illustrate how we might move between levels of abstraction and view of our model as we build one such specification. To start, we draw a box on a page and give it a name. This box represents an object, s0. At some point, we decide that s0 is a subsystem. It still appears as a box, because we are viewing it from outside, as a black box.

A box on a page.

When we wish to specify the parts of s0, we zoom in, and s0, which was a box in the view above, and becomes a page, as shown below.

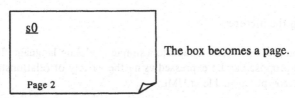

The box becomes a page.

We will use this page to specify s0 when seen from the viewpoint that considers it as being composed of parts. Since no parts yet exist, we specify them and add them to our model.

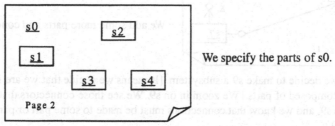

We specify the parts of s0.

If we zoom back out, we will see s0 as a box, unchanged. If we decide that s3 is a subsystem, we can zoom in on it, and we will see another page, labeled s3, where we can specify the parts of s3. But let's zoom back out and specify the connections of the parts of s0.

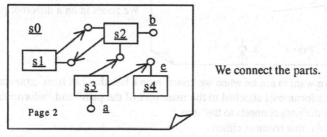

We connect the parts.

[In this diagram, the lines are connectors, and mean: is connected to. The UML provides a rich selection of relationships and relationship stereotypes.]

When we zoom back out again, what do we want to see? Not the parts of s0, and not the connections of the parts. But we do want to be able to specify that the two unconnected interfaces, a and b, are to appear as interfaces of s0.

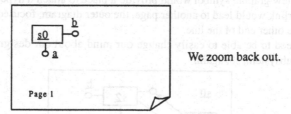

We zoom back out.

[Reminder: These are drawings in our simplified model, not diagrams using UML notation.]

Now we will work on our design at this level of abstraction. We add some objects to this page. We specify interfaces on the new objects, and make some connections from the objects to subsystem s0.

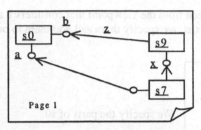

We add some more parts and connections.

Now we decide to make s9 a subsystem. [It means we decide that we are interested in s9 as composed of parts.] We zoom in on s9. We see those connectors that are connected to s9, and we know that connections must be made to some part or parts of subsystem s9.

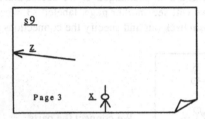

We zoom in on a different part.

What we want to see on when we zoom in on a box that has lines connected to it are the lines, "adornments attached to the main part of the path" and "adornments at [the] end where the path connects to the" box.

Obviously, this requires either:

— relaxing the current UML requirement that "paths are always attached to other graphic symbols at both ends (no dangling lines)," or

— adding a new graphic symbol to represent the model element at the other end of the line (perhaps a small open pentagon, which is widely used as a off page connector symbol).

Adding a new graphic symbol would provide a place to attach a hyperlink. Following that hyperlink would lead to another page, the outer diagram, focused on the model element at the other end of the line.

We also need to be able to easily change our mind about our design. Consider the parts of the subsystem, s0, again.

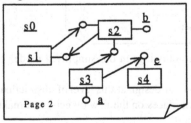

We must be free to decide that we want s4 to be, not a part of s0, but another subsystem appearing as a peer of s0. When s4 is moved, the model will be:

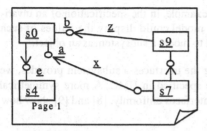 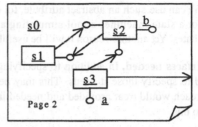

This should be a single step for the modeler, with the modeling tool making the several necessary changes.

As we said at the start of this section, this model is about what we want to specify with UML. Obviously, we would like our drawing tool to work this way also.

4.4 UML Notation

We believe it will be a clarifying improvement to consider a subsystem to be a kind of object (or, when reduced to code, an instance of a component) rather than an instance of some other kind of classifier. (When not reduced to code, the subsystem is an abstraction, since some details are hidden, since it is a simplified model.)

UML does not consider a subsystem to be an object. However, the ideas above can be implemented while adhering to the adopted UML specification.

The three-compartment notation allows operations, specification elements and realization elements to be shown at once in separate compartments. But any of the compartments can be suppressed.

4.5 Desirable subsystem properties

A modelling language needs to provide rich capabilities for specifying subsystems. UML already deals with some aspects quite nicely, others it left out.

Attributes. UML specifies that a subsystem may not have attributes. But it is often very useful to attribute attributes to a subsystem. This should present no problem, especially if we are able to see an attribute as an abstraction, rather than as a data member of a programming language object. We mean to say that an attribute can and should be expressed at the same level as its object. So, if we think of a subsystem as an abstraction, then attributes should not be understood as code-specific declarations. Instead, for example, as responsibilities of the subsystem.

The specifier of such an abstract attribute will then need to show how it is related to the realization elements of the subsystem. (For example, a designer might specify that the value of an abstract attribute is derivable from the values of some attributes of some objects among the realization elements. Or an attribute of a subsystem might be refined as an operation on an object among the realization elements.)

594

We can use such an abstract attribute, for example, in the specification of an invariant or a state machine; a tool simulating a model could display the values of such attributes. Yet another reason it will be useful to model a subsystem as an object.

Interfaces needed. In addition to specifying the interfaces a subsystem provides, we need to specify those it needs. This may be specified in UML. A more symmetrical approach would treat provided and needed interfaces uniformly. [8] and [9] show how to do this.

Component. When construction of a system reaches the point where we have code, UML adds another concept, component. A subsystem in the specification may appear as a component in the code. The parts of a subsystem may appear as components in the code.[4] If a subsystem is implemented as a component, it is proper and correct to call this component a subsystem.

Generalization. A UML Subsystem is a generalizable element. That's good. Being a generalizable element, a subsystem may be abstract or not. That's good, too.

Instantiability. A UML Subsystem may be instantiable or not. This is not the place to discuss what the UML specification might mean by 'is instantiable.' Nor whether there might be a difference between having an instance in a model, and having an instance in the universe of discourse of the model (for example, at run time). (Whether there might be an instance in a model that did not correspond to a concrete thing in the universe of discourse. Whether a model might include an instance of an abstract class.)

However all that may be, allow us to mention some possibilities:

In an implementation using the Façade pattern, a subsystem will appear at run time as a single programming language object that corresponds to the outside view of the subsystem. (This façade object will be a realization element of the subsystem.)

Following CORBA, a subsystem might appear at run time as a CORBA object.

If a component platform is used, a subsystem might appear at run time as an Enterprise Java Bean or CORBA component.

If whatever implements a subsystem at run time maintains an IP address and port, or corresponds to a CORBA object reference, it has a unique identifier, even if there is no single object that is the subsystem.

In these cases, in many other cases, and in all the cases that fall under our use of the concept, 'subsystem,' a subsystem at run time will have the property that distinguishes it from all other things: the quality that it is itself, and not something else. It will have identity. [1], [10]

That does not mean that at run time a subsystem will always comprise a single object or component or even that it will have an identifier. It will often be the case that only the objects from the inside of the subsystem (the objects that are realization ele-

---

4 Unless these two possibilities are prohibited by the UML well-formedness rule that:
"A Component may only implement DataTypes, Interfaces, Classes, Associations, Dependencies, Constraints, Signals, DataValues and Objects."

ments of the subsystem) will appear as objects at run time. In that case, the run time subsystem comprises this collection of objects and their links.

Abstraction. We ask the reader to agree with us that abstraction is useful even when talking about run time, about "physical," about the real world. Whether there is an object at run time that corresponds to (is an instance of) a given subsystem in a model is both a question of levels of abstraction and an implementation decision. Someone might ask: "Is there is an object at run time that corresponds to this subsystem in the model?" As to levels of abstraction, a quite reasonable answer is: "That depends on what you mean by object." As to implementation decisions, a quite reasonable answer is: "We haven't decided yet."

There is no reason for UML to dictate the set of possibilities. The builders of each specification that uses UML will say what they mean.

4.6 Just what is a subsystem, anyway

We have presented our understanding of (one use of) the UML concept, subsystem. It is not easy to tell from the UML specification if this is included in what it wants to mean by subsystem. The structure ("abstract syntax") is clear. And the discussion of the specification and realization elements is as we have described. But the meaning ("semantics") of the concept is murky.

How it happened. To testify to history:[5] Among the UML partners different needs were felt, all of which needs could be met by something called 'subsystem.' Some felt a need for "a grouping mechanism for specifying a behavioral unit of a physical system." Others felt a need for what we have presented as our meaning of 'subsystem.' There is adequate historical precedent for these different uses of the term, 'subsystem.' And these different uses have much in common. The UML Subsystem represents a combination of and a compromise between ideas to meet these different felt needs.

What resulted. We feel that this combination and compromise has resulted in a portmanteau concept and unclear text in the specification. For example: A subsystem has operations, but "... has no behavior of its own." A subsystem may be instantiable, but "... there are no explicit instances of a subsystem; instead, the instances of the model elements within the subsystem form an implicit composition to an implicit subsystem instance, whether or not it is actually implemented."

It appears to us that this language wants to describe the subsystem when used "merely as a specification unit for the behavior of its contained model elements."

What to do. We suggest the UML define 'subsystem' along the lines of what RM-ODP calls a composite object: an object expressed as "a combination of two or more objects yielding a new object, at a different level of abstraction." [2] From the outside

5 Personal experience of one of the authors (Miller).

viewpoint a subsystem will be an object. From the inside viewpoint it will be a set objects and links. (Those who prefer the style will use UML classes and associations.)

When the object seen from the outside viewpoint is intended "merely as a specification unit for the behavior of" the objects seen from the inside viewpoint, and not to be thought of as existing at run time, the model will specify that. A more down home concept than is not instantiable would serve that purpose.

When we decide an object is a subsystem (has parts), let the tool create a package for the parts.When we decide we want to compose a subsystem from a set of objects we have already specified, let the tool put them in a package we can view as an object.

5 Challenges

Architects and designers need CASE tools that handle the UML Subsystem in a way which reinforces the principles we have offered and which provide strong support for designing subsystems in a middle-out, top-down or bottom-up fashion.

As illustrated by the review of a method in Section 2, the typical process of designing subsystems is iterative and involves restructuring the design. During the design process, objects (or the classes they are instances of) will be moved from one subsystem to another; subsystems will be removed from the design and others added; what was a subsystem will become an simple object, and what was a simple object will become a subsystem.

We challenge the builders of CASE tools to make the restructuring of a design a easy as possible. Build your tools so it is easy for architects and designers to:
— Shift levels of abstraction (with connections mapped and preserved by the tool).
— Move objects and subsystems from one subsystem to another (as easily as cut and paste or drag and drop).
— Change an object into a subsystem (as easily as selecting a menu item).
— Change a subsystem into an object (and have something helpful done with what were the parts of the subsystem).
— Combine a set of objects into a subsystem.

For the day that those challenges are met, here are some easy ones. Architects and designers will be well served by a UML and CASE tools that allow them to:
— Treat a subsystem as a full-fledged object, when viewed from outside.
— Specify attributes for a subsystem, to represent what [5] calls responsibilities for knowing, to be used in specifying invariants or state machines, or to be displayed by model simulators.
— Specify the interfaces that a subsystem requires, using a graphical notation as simple and compact as the lollipop, which indicates that an interface is provided.
— Having specified that a subsystem requires a particular interface, connect the graphical representation of that requirement to the graphical representation of an interface on another subsystem that satisfies the requirement.

6 Conclusions

Architects and designers think at multiple levels of abstraction. This means they see subsystems both as objects and as containers of parts. They find it natural to shift focus between these two viewpoints. The ideas we have presented in this paper are not new. Years ago, when functional decomposition was popular, data flow diagrams were used. In those old, slow days, this zooming in and out, with connections preserved and mapped, is exactly how good data flow modeling tools worked.

With modest extensions to UML, and with serious effort by toolmakers, UML and CASE tools can better support working at multiple levels of abstraction.

7 Acknowledgements

Thanks to all the reviewers for their comments. Special thanks to Jim Rumbaugh for his extensive criticisms, helpful suggestions and encouragement. And many thanks to all the authors whose ideas have contributed to our thinking.

References

1. Oxford English Dictionary (Second Edition), Oxford University Press, Oxford (1994)
2. Information Processing–Open Distributed Processing–Reference Model–Foundations, X.902 | IS 10746-2. International Organization for Standardization, Geneva (1995) http://enterprise.shl.com/RM-ODP/
3. Myers, G. J.: Reliable Software through Composite Design. Van Nostrand Reinhold, New York (1975)
4. Cox, B. J., Novobilski, A. J.: Object-oriented Programming–An Evolutionary Approach. Addison-Wesley, Menlo Park, California (1986)
5. Wirfs-Brock, R., Wilkerson, B., Wiener, L.: Designing Object-Oriented Software. Prentice Hall, Englewood Cliffs (1990)
6. Shaw, M., Garlan, D.: Software Architecture–Perspectives on an Emerging Discipline. Prentice-Hall, Englewood Cliffs (1996)
7. Object Management Group, OMG Unified Modeling Language Specification—Version 1.3. Object Management Group, Framingham, Massachusetts (1999) http://www.omg.org/cgi-bin/doc?ad/99-06-08
8. Reenskaug, W., Reenskaug, T., Lehne, O. A.: Working With Objects: The OOram Software Engineering Method. Prentice Hall, Englewood Cliffs (1995)
9. Selic, B., Gullekson, G., Ward, P.T.: Real-Time Object-Oriented Modeling. John Wiley & Sons (1994)
10.Khoshafian, S. N., Copeland,G.P.: Object Identity, OOPSLA '86 Proceedings, in SIGPLAN Notices 21, 11 (1986) 406-416.

Using UML/OCL Constraints for Relational Database Design

Birgit Demuth and Heinrich Hussmann

Dresden University of Technology, Department of Computer Science

Abstract. Integrating relational databases into object-oriented applications is state of the art in software development practice. In database applications, it is beneficial if constraints like business rules are encoded as part of the database schema and not in the application programs. The Object Constraint Language (OCL) as part of the Unified Modeling Language (UML) provides the posssibility to express constraints in a conceptual model unambiguously. We show how OCL, UML and SQL can be used in database constraint modeling, and discuss their advantages and limitations. Furthermore, we present patterns for mapping OCL expressions to SQL code.

1 Introduction

Integrating relational databases into object-oriented applications is state of the art in software development practice. Many object-oriented projects choose to use a relational database management system (RDBMS) for a variety of reasons, e.g. compatibility with existing systems and databases. In the last years, some authors have described transformations from object-oriented class diagrams to database schemas, in order to support the systematic object-oriented development of database applications [1], [3]. However, these approaches are restricted to a simple class-to-table mapping, mainly based on attributes and associations. They do not at all exploit the full power of relational database technology. Advanced static database concepts like automatic check of assertions or maintenance of database integrity by automatically triggered operations are not covered by these mappings. On the other hand, it is well known that the development of database applications benefits from business rules being encoded as part of the database schema, using assertions and triggers. This reduces the amount of required coding and ensures that all applications working on the same database adhere to the same business rules, in particular the same integrity constraints. This paper discusses the use of UML and its constraint language OCL for the design of integrity constraints for relational databases. There are several advantages of such an approach:

- The UML as a widely accepted standard language for object-oriented models ensures support by many CASE tools.
- The OCL provides an abstract and precise language for specifying integrity constraints as invariants for object models. The navigation-oriented approach

of OCL fits somehow with the concept of databases (although it does not completely correspond to relational query languages).
- Current RDBMS implementations often use product-specific dialects of SQL to denote assertions and triggers. The usage of OCL makes the specification of integrity constraints independent of the choice of the RDBMS.
- To write down constraints in a constraint language like OCL during object-oriented design is a methodically recommended but practically time-consuming step. An automatic generation of SQL code from OCL preserves the effort that is invested into writing the OCL specification.
- Finally, specifications of operations by pre- and postconditions in OCL form are a good starting point for the automatic generation of SQL code for database queries, updates or triggers corresponding to these operations.

This paper is structured as follows: Section 2 introduces constraints from a more generalized point of view and provides an overview about relational database constraints expressed in SQL. The essential part of this paper, Section 3, presents a family of patterns for mapping OCL constraints to relational database integrity constraints by means of a UML model example. We illustrate how OCL, UML and SQL-92 can be used in database constraint modeling, and discuss their advantages and limitations. Section 4 gives conclusions and future directions.

2   Constraints from a Database Perspective

2.1   Classification and Comparison of Constraint Languages

A **constraint** is defined as a restriction on one or more values of (part of) an object-oriented model or system [19]. This generic definition can be interpreted in different ways. But, by definition, a constraint is always coupled with a model.

From a methodological viewpoint, constraints are required to specify information, in particular business rules, that otherwise cannot be expressed in the model. In current UML, there is no clear borderline between the object-oriented modeling language and the constraint language. An example is the restriction of the multiplicity of an association, which can be expressed either directly by using a syntactical construct from UML class diagrams or by an OCL constraint (see also [2]). These UML constructs can be understood as convenient shorthand notation for OCL constraints.

In general, constraints can be classified according to various criteria. For a first comparison of OCL and database constraints, the following criteria are useful:

- Classification according to the system view: Constraints can be defined in various views of the system, as the static, dynamic and functional view. These three views correspond to three different types of constraints: In a static view, a constraint usually is an invariant, i.e. a condition which has to hold for any "snapshot" of the state of the whole system or a part of it. In a dynamic view, constraints are used mainly to express the condition under which a transition from one state into another is allowed (guards). In a functional view, finally, the output values and the induced state transformation of an operation are described with respect to the input values. This is done in OCL by pre- and postconditions, but there are also other possibilities (e.g. axiomatic specification).

– Classification according to the policy for dealing with constraint violations: There are various interpretations for the situation where a constraint, in particular a static constraint (invariant), is not fulfilled. The implementation can be considered as faulty (as is the view in software verification), the recent modification to the state can be made undone (as is the view in database systems), or actions can be taken to automatically correct the state (a view also used in database systems). If a constraint is intended to automatically re-establish a correct state, it is called an *operational* constraint. A *declarative* constraint just states the condition that has to be fulfilled without specifying any consecutive action.

There are other classifications for constraints that are not relevant here. For example, in the most general meaning of the term, a constraint may also be a logical property which is not necessarily decidable (i.e. stating that an operation always has to terminate). Here we assume that all constraints are executable algorithms on a system state.

The following table uses the criteria from above to compare OCL with the integrity constraint checking mechanisms implemented in modern RDBMS:

	OCL	RDBMS
System view	static, dynamic, functional	static
Violation policy	declarative	declarative, operational

The table shows clearly that there is a common intersection between the constraint mechanisms of OCL and RDBMS. Constraints in this intersection are static (i.e. invariants) and declarative which we will consider in this paper. The difference between the two constraint mechanisms is that OCL is defined on a very abstract conceptual (UML) model of the system, whereas database constraints are formulated in rather low-level terms. Therefore, it is an interesting question how and to which extent OCL can be used for the high-level specification of relational database constraints.

2.2 Relational Database Constraints

What UML and OCL calls a constraint has been known in database technology for a long time under various names, for example integrity constraints [8], [18] [15], integrity rule [4], and consistency constraints [10]. We will use the term *integrity constraint* to refer to constraints in relational databases. In [8], types of integrity constraints are identified that are encountered frequently in database schemas:

- An *implicit* constraint represents an integrity rule which is part of the data model and must be specified on individual relations.
- An *explicit* constraint may occur in an application. It is often called a *business rule*.
- An *inherent* constraint does not have to be specified in a schema but are assumed to hold by the definition of the relational model. It can be compared with a UML meta model constraint.

The following table summarizes the semantics of implicit, explicit and inherent constraints in the relational model and their specification in SQL-92 [11]. Only the very well-known trigger concept has been standardized later (SQL-99 [7]). Every SQL constraint is checked either at the end of every SQL insert, update or delete statement (IMMEDIATE mode) or at the end of the transaction (DEFERRED mode).

type of integrity constraint	relational model	specification in SQL
implicit	primary key/ entity integrity rule [4]	PRIMARY KEY (NOT NULL)
	foreign key/ referential integrity rule [4]	FOREIGN KEY REFERENCES
explicit	column constraint and table constraint	CHECK, NOT NULL, UNIQUE
	assertion	CONSTRAINT ... CHECK, CREATE ASSERTION
	domain	CREATE DOMAIN
	trigger	CREATE TRIGGER
inherent	atomic attribute values	implicit

Only trigger and referential constraints with actions (SET DEFAULT, SET NULL, CASCADE) are operational constraints. The other ones (including FOREIGN KEY REFERENCES ... NO ACTION) represent declarative constraints.

3 Mapping of UML and OCL Constraints to Relational Database Integrity Constraints

3.1 A Relational Database Schema for a UML model

Considering the usage of UML and OCL constraints for relational database design we have to deal at first with the impedance mismatch between the object-oriented and the relational model. However, if we bridge the gap between the two models we are able to exploit powerful integrity maintenance mechanisms of advanced RDBMS in object-oriented applications. The figure below contains a similiar UML model example of the OCL specification [14] we will use as an example. In the translation to a database schema, we assume the simple and

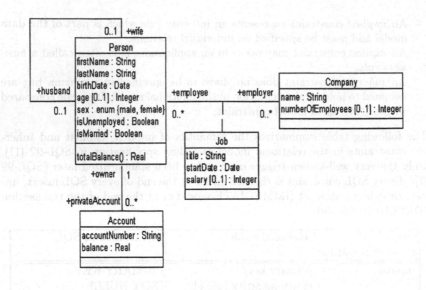

commonly used class-to-table mapping [1]. Every class is represented as one table, called class table, on the relational side. Relationships in class diagrams are implemented by foreign key references. According to this mapping, the following tables are generated for our example:

```
create table PERSON  ( PID integer primary key,
                       FIRSTNAME varchar not null,
                       LASTNAME varchar not null,
                       BIRTHDATE date not null,
                       AGE integer,
                       SEX SEXTYPE not null,
                       ISUNEMPLOYED boolean not null,
                       ISMARRIED boolean not null,
                       WIFE_HUSBAND integer references PERSON )

create table COMPANY ( CID integer primary key,
                       NAME varchar not null,
                       NUMBEROFEMPLOYEES integer )

create table ACCOUNT ( AID integer primary key,
                       ACCOUNTNUMBER varchar not null,
                       BALANCE float not null,
                       OWNER integer not null references PERSON )

create table JOB     ( PID integer references PERSON,
```

```
CID integer references COMPANY,
TITLE string not null,
STARTDATE date not null,
SALARY float,
primary key (PID,CID) )
```

create domain SEXTYPE character check (value in ('m','f'))

For dealing with object identifiers, we have chosen the existence-based identity approach [1]. An object identifier attribute is added to each class table and is made the primary key. Therefore, identifiers such as PID in Person are defined as integers. Some RDBMS provide semantic support for such identifiers. It should be noticed that booleans are supported not before SQL-99 [7]. A many-to-many relationship including an association class such as the association between Person and Company is mapped to a distinct table (JOB). The primary key of this table is the combination of the primary keys from each class table. A one-to-many association such as the relationship between Person and Account is implemented with a buried foreign key (OWNER) in the table corresponding to the class on the *many* side (ACCOUNT). Because the multiplicity of Person is exactly one in the UML model, we do not allow the OWNER foreign key to be null. A one-to-one relationship is generally implemented by a buried foreign key in either class. Because our one-to-one relationship is a symmetric association that models the marriage between persons, we did not introduce an asymmetric WIFE or HUSBAND attribute in PERSON, but, as can be seen above, a symmetric WIFE_HUSBAND attribute. Enumerations such as the sex datatype in the class Person can be mapped to SQL domains (SEXTYPE). It should be noticed that some of the important UML model elements and constraints like multiplicity are already encoded by SQL integrity constraints. Besides the above explained integrity constraints, the multiplicity zero of attributes is modeled by null values in relational databases. The default UML value of exactly one is specified as a *not null* constraint in the attribute definition. Note that all SQL integrity constraints are checked automatically at any write transaction.

The translation described here is rather straightforward and based on standard literature. However it should be emphasized that most current commercial CASE tools only generate a significantly simpler database schema.

3.2 OCL Mapping Patterns

We describe the mapping of OCL to SQL constraints by a family of patterns, where each pattern encapsulates the idea for translation of an OCL language concept. Each pattern is given by its name, a short description and an example in OCL and SQL notation followed by a discussion. Whereas the example illustrates a typical use case of the pattern, the discussion especially points out mapping problems. We use the OCL syntax of the OMG draft 1.3 [14]. The patterns are founded on a comprehensive study of possible transformations of OCL expressions to SQL [16].

OCL Invariants. As stated above, we only consider OCL invariants to transform them to SQL integrity constraints. A naive approach would be the mapping of OCL invariants to SQL table constraints in the following manner:

OCL:

```
context Company inv enoughEmployees:
self.numberOfEmployees > 50
```

SQL:

```
create table COMPANY( ... ,
constraint ENOUGHEMPLOYEES check (NUMBEROFEMPLOYEES > 50))
```

The OCL context is expressed by the create table statement. If we refer to any column of that table (e.g., NUMBEROFEMPLOYEES) just by its name, the context is always the current row of that table. The current row could be seen as the contextual instance `self` in OCL. The problem, however, is to refer to the current row in nested subqueries when we map more complex OCL expressions (see the NAVIGATION pattern below). Though SQL's correlation names could solve this problem, they can not be used for the *context table* of the constraint (in our example COMPANY). Therefore, we must find a mapping where the equivalent SQL constraint includes the context table. This leads to the definition of our first transformation pattern.

Name: OCL INVARIANT
Description: An OCL invariant of the form

```
context <class name> inv <constraint name>:
<OCL expression(self)>
```

is transformed to

```
<class name>.allInstances -> forAll (<OCL expression(self)>).
```

Then, the OCL forAll operation can be mapped to a SQL predicate, and the whole constraint is written as:

```
create assertion <constraint name>
check (not exists
(select * from <context table> SELF
 where not (<OCL expression(self)>)))
```

The evaluation of the subquery nested in the exist predicate searches for objects which violate the OCL constraint. If there are no such objects, the constraint is satisfied. Since integrity constraints often involve multiple tables, they should be specified as standalone constraints (assertions). Alternatively to assertions, table constraints or column constraints can be used.

Example :
  OCL:

```
context Company inv enoughEmployees:
self.numberOfEmployees > 50
```

  SQL:

```
create  assertion ENOUGHEMPLOYEES
check (not exists (select * from COMPANY SELF where
                      SELF.NUMBEROFEMPLOYEES > 50)))
```

Discussion: A SQL constraint can be evaluated based on the null values and a three-valued logic to *unknown*. Although there are some uncertainties in the OCL definition [9], the same three-valued logic is specified for OCL (in OCL, an unknown value is called *undefined*). OCL says nothing about how undefined is handled when true or false is expected. For example, is a constraint broken if it is evaluated to undefined? Because of this uncertainty we assume the same treatment of OCL constraints like in SQL. That is, if the expression is evaluated to undefined, then the constraint is satisfied.

An OCL invariant is specified as an expression over different OCL types. The grouping of OCL types into Basic Types and Collection-Related Types [14] is not very helpful when considering the usage and mapping of OCL expressions in the context of relational databases. OCL Basic Types include such very basic predefined types like Integer as well as types defined in a UML model. In contrary to the OCL specification, Warmer and Kleppe [19] distinguish basic types and model types to separate user-defined types from real basic types. Although the type hierarchy of OCL is somehow confusing [9], it is clear that OCLAny is the supertype of all types in the UML model (*model types*). Therefore, we distinguish model types from basic and collection types in the following. The access to the features of UML classes (attributes, query operations, enumerations defined as attribute types and navigations that are derived from relationships between instances) and its transformation to SQL depends on the class-to-table mapping. All other OCL operations can be transformed in a model independent and therewith database schema independent way if we evaluate them on the basis of the results for each subexpression over model types (see [16]). Collection-related types comprise the abstract supertype Collection as well as the concrete Set, Bag and Sequence types.

Another issue is the result of an OCL expression. The expression that represents the OCL constraint is always of type Boolean and can be specified by a SQL predicate resp. a SQL search condition (see OCL INVARIANT). Each OCL subexpression results either in a Boolean value or in an object of any other OCL type. All expressions that are not of type Boolean should be encoded in SQL by a basic value expression resp. a column reference, or a query expression. If this descriptive specification is not possible, one may always encode the constraint in SQL extended by procedural statements in a computationally complete language and store it as a *stored procedure* in the database [12].

If we consider the OCL types and the results of OCL expressions from the SQL perspective we can identify further OCL mapping patterns. The table below indicates which fields are covered by the different OCL mapping patterns. Metamodel properties of OCL types such as **Person.attributes** are not taken into account because it mostly makes no sense to specify database integrity constraints on the database schema itself.

OCL expression over	Result type of an OCL expression		
	Boolean	Basic value excluding Boolean	Collection
Basic Types	BASIC TYPE		
Model types: - Class - Association class - Attribute	CLASS AND ATTRIBUTE		CLASS AND ATTRIBUTE
- Association end	NAVIGATION		
- Operation	OPERATION		
Collection types	COMPLEX PREDICATE	BASIC VALUE	QUERY

In the following patterns, we assume the above described OCL INVARIANT pattern. We consider the transformation of <OCL expression(self)> resp. of its subexpressions and use the correlation name SELF in SQL code to map the contextual instance **self**.

Basic Types. There is only one simple pattern. Expressions over basic objects can not result in collections.

Name: BASIC TYPE

Description: The basic predefined OCL types (Real, Integer, String, Boolean), their values as well as their unary, multiplicative, additive, relational and logical operations mostly have a direct SQL counterpart. If the OCL subexpression is of type Boolean like in the example we specify a SQL search condition. Other primitive subexpression, such as the integer value 18, are part of a SQL predicate.

Example :

OCL:

```
[context Person]
self.age > 18 and self.isUnemployed = true
```

SQL:

```
[from PERSON SELF]
self.AGE > 18 and SELF.ISUNEMPLOYED = true
```

Discussion: Operators that have no direct SQL counterpart can be transformed into equivalent expressions, for example <b1> implies <b2> into not <b1> or <b2>.

Model Types. If we want to map the access to model types, we have to deal with objects of all model types and their properties, which means in database context classes, association classes, attributes, associationEnds and operations.

Name: CLASS AND ATTRIBUTE

Description: Basically, each OCL subexpression which refers to a class, association class or an attribute can be mapped to a SQL query. In the above explained class-to-table-mapping, the access to an attribute is simplified by the SQL notation `<class_table>.<attribute>` resp. `SELF.<attribute>` like we have already used it in the patterns above.

Example : see the OCL INVARIANT pattern

OCL:

```
Person.allInstances
```

SQL:

```
select PID from PERSON
```

Discussion: Here two serious problems arise which are based on the impedance mismatch of the relational and the object-oriented paradigm. The first one is the identification of objects. We have chosen the existence-based approach. Therefore the set of all instances of Person is evaluated by selecting the PID attribute. However, the semantics of an artificial primary key of a table is not exactly the same as that of an object identifier. A primary key only identifies tuples representing objects within the table scope. Furthermore, a primary key can basically be altered and reused. The second problem is the equality of objects. What does

```
object = (object2: OCLAny): Boolean
```

exactly mean? From a database point of view it is useful to distinguish between values (in OCL only basic values such as integers) and objects (for example instances of Person) [5]. Any object is uniquely identified by an object identifier (OID). In our relational database schema mapping this means that the equality of persons has to be checked by the primary key PID of PERSON.

Name: NAVIGATION

Description: Basically, each OCL subexpression that refers to an association end can be mapped to a SQL query. Starting from an specific object, we often have to navigate associations to other objects and their properties to specify a constraint. Mapping such navigations is done by queries with nested subqueries and depends on the kind of association (one-to-one, many-to-one or many-to-many relationship) and its mapping to the database schema.

Example :

OCL:

```
[context Company]
self.employee
```

608

SQL:

```
[from COMPANY SELF]
select PID from PERSON where PID in
         (select PID from JOB where CID in
                  (select CID from COMPANY where CID = SELF.CID))
```

Discussion: The value of a navigation expression is one object (in our relational context one tuple) if the multiplicity of the association end has a maximum of one, otherwise it is a set of objects. After evaluation of any expression, one can always apply another property to the result to get a new result value.

Name: OPERATION

Description: Only operations (or methods) with *isQuery* being true can be used in OCL expressions. In advanced RDBMS, operations can be modeled as user-defined functions and implemented in different ways. Currently, RDBMS vendors integrate Java into their systems to implement user-defined functions as static methods [13]. These methods can be called in a SQL statement.

Example :
   OCL:

```
[context Person]
self.totalBalance() > 0
```

   Java/SQL:

```java
public class Example {
  //the method will be called in a SQL query
  //pid represents the contextual instance self
  public static double totalBalance(int pid)
    throws SQLException {
    Connection c =
    DriverManager.getConnection("JDBC:DEFAULT:CONNECTION");
    PreparedStatement stmt = c.prepareStatement(
     "select sum(BALANCE)from ACCOUNT where OWNER =?");
    stmt.setInt(1, pid);
    ResultSet sum = stmt.executeQuery();
    sum.next();
    return sum.getDouble(1);
    }
}
```

Discussion: The example shows the implementation of an operation by a Java method that uses JDBC (Java Database Connectivity) to perform database operations. These techniques are currently under standardization. The first RDBMS prototypes with integrated Java are available [13].

OCL Collection Types. Apart from the specification as a literal, the only way to get a Collection is by navigation or by the allInstances operation. These collections are represented as tables resp. relations in the relational model. In opposite to the pure relational model where all relations are sets, the SQL approach provides the possibility to construct bags and sequences beside sets. For example,

```
select distinct TITLE from JOB
```

evaluates a set of titles of all job instances whereas the deletion of the keyword distinct creates a bag.

Name: COMPLEX PREDICATE

Description: The pattern covers all collection operations that result in a Boolean value (includes, excludes, includesAll, excludesAll, isEmpty, notEmpty, exists, forAll, isUnique, sortedby, equality of collections (=)). Such OCL (sub)expressions can be basically mapped to a SQL predicate with or without nested subqueries.

Example :

OCL:

```
[context Company]
self.employee->forAll(isUnemployed = false)
```

SQL:

```
[from COMPANY SELF]
not exists
(select PID from
            (select PID from PERSON where PID in
            (select PID from JOB where CID in
            (select CID from COMPANY where CID = SELF.CID)))
        where
            PID in (select PID from PERSON
                        where  not (ISUNEMPLOYEED = false)))
```

Discussion: Notice that the evaluation of `self.employee` as a set of objects (see NAVIGATION pattern) is given as derived table in the from clause of the exists predicate query. Such complex queries are only supported by a few RDBMS.

Name: BASIC VALUE

Description: The pattern covers all collection operations that result in a Real, Integer or String value (size, count, sum). Such OCL (sub)expressions can be basically mapped to a scalar subquery using SQL aggregate functions (degree and cardinality of the result = 1).

Example :
OCL:

```
[context Person]
self.privateAccount.balance -> sum
```

SQL:

```
[from PERSON SELF]
select case when sum(BALANCE) is null then 0 else sum(BALANCE)
from ACCOUNT where OWNER = SELF.PID)
```

Discussion: Notice that Account objects could not exist. Then, sum(BALANCE) results in unknown. According to the semantics of the OCL property sum, it must be transformed to the Real value zero. Such a SQL query is generated using the NAVIGATION pattern specialized for many-to-one relationships.

Name: QUERY
Description: The pattern covers all collection operations that result in a collection again:

- construction of collections (Set, Bag, Sequence) including transformation of different collections (asSet, asBag, asSequence) and flattening of collections
- specification of a subset of a collection (select, reject)
- specification of a collection which is derived from some other collections (collect)
- typical set operations such as union, intersect and difference adopted to sets, bags and sequences.

Such OCL (sub)expressions can be basically mapped to SQL queries with or without nested subqueries.

Example :
OCL:

```
[context Company]
self.employee-> select(p: Person | p.sex = female)
```

SQL:

```
[from COMPANY SELF]
select PID from (select PID from PERSON where PID in
                (select PID from JOB where CID in
                (select CID from COMPANY
                             where CID = SELF.CID)))
         where PID in
               (select P.PID from PERSON P
                             where P.SEX = 'f')
```

Discussion: A detailed representation of possible transformations is given in [16]. Such OCL (sub)expressions can be basically mapped to a SQL predicate with or without nested subqueries.

A special role is played by the not yet mentioned iterate operation. The iterate operation is very generic and therefore more complicated than the other collection operations which can be partially described in terms of iterate. In its most general form it can only be transformed to SQL by a so-called *procedural mapping pattern*. In this case we need a computationally complete language. That means, an iterate operation must be encoded by a stored procedure. In [16] it is shown how this can be done by Sybase' stored procedures using Transact SQL.

3.3 Problems and Limitations

We have shown that it is possible to generate equivalent SQL code for OCL constraints. Our goal to avoid procedural SQL code has been achieved for the full OCL language, with the only exception of the iterate operation. It must be noticed that our model type mapping patterns are dependent of the assumed class-to-table mapping. We have used the whole SQL-92 language (full level) to encode complex OCL expressions. Especially, the specification of a derived table in the from clause is a fundamental feature to formulate any OCL expressions in SQL in a declarative manner. We know only one of the major commercial RDBMS that provides this expressiveness. Using other (current versions of) RDBMS makes it necessary to map many further OCL operations besides iterate to stored procedures. This has been studied for a less expressive commercial SQL dialect in detail in [16]. A stored procedure for database integrity checking must be called at the end of every transaction. But, there is the hope that more RDBMS vendors provide full SQL-92 conformant database servers very soon.

In consideration of OCL we have referred to the most recent OMG draft [14]. Existing unclarities such as the type hierarchy and the undefined values problem have to be adopted to our patterns after their clarification.

Although some authors emphasize that a UML class diagram and a data model are different things [17], UML is a powerful tool to model persistent objects and therewith databases [[2], [1]]. But for UML to achieve a wide acceptance in database modeling some refinements of UML may be helpful. For example, it is useful to define a standard UML constraint *unique* to model key attributes. A unique attribute constraint could be mapped very simple to a (predefined) unique SQL constraint:

```
create table ACCOUNT (..,ACCOUNTNUMBER varchar not null unique,..)
```

In OCL one can formulate this very often needed constraint by an iterate based expression such as

```
[context Account]
Account.allInstances -> forAll ( a1, a2 |
    a1<>a2 implies a1.accountNumber<>a2.accountNumber )
```

This OCL expression would be mapped to a complex SQL assertion. The following shorthand notation introduced by the most recent OCL specification makes the above OCL constraint more readable and easier to transform to SQL.

```
[context Account]
Account.allInstances -> isUnique (accountNumber)
```

There are many equivalent UML/OCL expressions and notations, within one notation (OCL resp. UML) as well as between both notations [19]. If a code generator will be developed we have to consider these equivalences. A first step should be a normalization of UML and OCL constraints.

4 Conclusion

In this paper, we reported on a systematic study of the use of OCL for the specification of database integrity constraints. It has turned out that OCL is essentially adequate for this purpose. We have proposed a minor addition to UML/OCL which would make specification and transformation into SQL easier. When assuming the most advanced SQL standard available, there is only one language construct of OCL (iterate) which cannot be mapped to declarative SQL in every case. However, this seems not to be a serious problem, since in practical OCL specification the iterate operator is rarely used, and all OCL constructs derived from iterate (like forAll, select) can be mapped properly. The practical value of UML-to-SQL mappings as described in this paper can only be judged with the help of tools, in order to create sufficiently large examples. So we plan as a next step the realization of a code generation tool for SQL code from OCL, based on our Dresden UML Toolset [6]. Dresden UML Toolset provides a platform for experimental research on the basis of UML (code generation, reverse engineering from legacy applications and extensions of UML for formal specification such as OCL). It integrates commercial UML tools with experimental tools developed at the university and combines experimental UML tools. One of our first steps in development of Dresden UML Toolset was the implementation of an OCL basic library. Currently, we are developing a modular OCL compiler which generates Java code to evaluate OCL expressions using the OCL basic library. The OCL compiler has an interface for the code generation of SQL constraints. Such planned OCL-to-SQL tool will enable also experiments with different RDBMS systems and evaluation of the performance achieved by the generated integrity constraints. Further theoretical investigations are required in several directions. It would be useful to make the OCL-to-SQL mapping more independent of the code generation algorithm for the database schema. Moreover, the generation of SQL code can be extended from integrity constraints to the specification of query and update operations. There should be a sublanguage of OCL pre- and post-conditions from which executable SQL code can be generated. This approach can form a basic technology for the automatic generation of the core part of a database application from a high-level specification. The fact that UML and OCL are standardized languages gives this approach a potential for much higher impact than the proprietary application generation tools available today.

Acknowledgment: The authors thank Alexander Schmidt for his extensive study of OCL-to-SQL transformations and valuable contributions during his diploma project in 1998.

References

1. Blaha, M., Premerlani, W.: Object-Oriented Modeling and Design for Database Applications. Prentice Hall, 1998
2. Booch, G., Rumbaugh, J., Jacobson, I.: The Unified Modeling Language User Guide. Addison-Wesley, 1999
3. Bruce, K., Whitenack, B.: Crossing Chasms - A Pattern Language for Object-RDBMS Integration. Knowledge Systems Corp., ftp://members.aol.com/kgb1001001/Chasms/chasms.pdf
4. Date, C.: An Introduction to Database Systems. Volume I, Fifth Edition, Addison-Wesley, 1990
5. Demuth, B., Geppert, A., Gorchs, T.: Algebraic Query Optimization in the CoOMS Structurally Object-oriented Database System. in: Freytag, Ch., Maier, D., Vossen, G. (Ed.): Query Processing For Advanced Database Systems. Morgan Kaufmann, 1994
6. Dresden UML Toolset, http://www-st.inf.tu-dresden.de/UMLToolset
7. Eisenberg, A., Melton, J.: SQL: 1999, formerly known as SQL-3. ACM SIGMOD Record, 22(1999)1, 131-138
8. Elmasri, R., Navathe, S.: Fundamentals of Database Systems. Benjamin/Cummings, 1989
9. Gogolla, M., Richters, M.: On Constraints and Queries in UML. in: Schader, M., Korthaus, A., (Ed.): The Unified Modeling Language. Technical Aspects and Applications. Physica-Verlag, 1998
10. Korth, H., Silberschatz, A.: Database System Concepts. Second Edition. McGraw-Hill, 1991
11. Melton, J., Simon, A.: Understanding the New SQL: A Complete Guide. Morgan Kaufmann, 1993
12. Melton, J.: SQL's Stored Procedures. A Complete Guide to SQL/PSM. Morgan Kaufmann, 1998
13. Olson, St. et. al.: The Sybase Architecture for Extensible Data Management. Bulletin of the Technical Committee on Data Engineering, IEEE Computer Society, 21(1998)3, 12-24
14. OMG UML Specification v. 1.3 draft
15. Ricardo, C.: Database Systems. Principles, Design, and Implementation. Macmillan, 1990
16. Schmidt, A.: Untersuchungen zur Abbildung von OCL-Ausdruecken auf SQL. Technische Universitaet Dresden, Diplomarbeit, 1998
17. Simons, A., Graham, I.: 37 Things that Don't Work in Object-Oriented Modeling with UML. in: Kilov, H., Rumpe, B. (Ed.): Second ECOOP Workshop on Precise Behavioral Semantics. Technical Report, Technische Universitaet Muenchen, TUM-19813, Juni 1998
18. Teorey, T.: Database Modeling & Design. The Fundamental Principles. Second Edition, Morgan Kaufmann, 1990
19. Warmer, J., Kleppe, A.: The Object Constraint Language. Precise Modeling with UML. Addison-Wesley, 1999

Towards a UML Extension for Hypermedia Design*

Hubert Baumeister[1], Nora Koch[1,2], and Luis Mandel[2]

[1] Institut für Informatik
Ludwig-Maximilans-Universität München
Oettingenstr. 67
D-80538 München, Germany
baumeist@informatik.uni-muenchen.de

[2] Forschungsinstitut für Angewandte Software Technologie (FAST e.V.)
Arabellastr. 17
D-81925 München, Germany
{koch,mandel}@fast.de

Abstract. The acceptance of UML as a de facto standard for the design of object-oriented systems, together with the explosive growth of the World Wide Web has raised the need for UML extensions to model hypermedia applications running on the Internet. In this paper we propose such an extension for modeling the navigation and the user interfaces of hypermedia systems. Similar to other design methods for hypermedia systems we view the design of hypermedia systems as consisting of three models: the conceptual, navigational and presentational model. The conceptual model consists of a class diagram identifying the objects of the problem domain and their relations. The navigational model describes the navigation structure of the hypermedia application by a class diagram specifying *which* navigational nodes are defined and an object diagram showing *how* these navigational nodes are visited. Finally, the presentational model describes the abstract user interface by composite objects and its dynamic behavior by state diagrams. Each model is built using the notations provided by the UML, applying the extension mechanism of the UML, i.e. stereotypes and OCL constraints, when necessary.

Keywords: Unified Modeling Language, Object-Oriented Design, Multimedia, Hypermedia, WWW

1 Introduction

The development of Web applications for Intranets and the Internet in the case of personal homepages or tiny information systems could be seen as an easy job. Such applications are normally written once and probably never modified; directly implemented without taking into account aspects like navigation design. However, when designing applications like Web sites of companies or large Web-based information systems, having hundreds of nodes to be visited, interacting

---

* This work was partially supported by the Bayerische Forschungsstiftung.

with many different programs and running as distributed systems, a methodology together with a modeling language is needed in order to document the system, to communicate the desired structure and behavior, to visualize and control the systems architecture, showing places where simplification, factorization and reuse can be applied.

Different modeling languages can be used, but the importance of using a standard is clear: it provides a common language which facilitates the communication among project partners as well as to the external world and future readers of the system documentation. Further, tools based on this standard can be used to develop, test and validate the models. UML [BRJ99] has been accepted as the de facto industrial standard for modeling object-oriented systems. It can be extended by using the stereotype mechanism, tagged values and the Object Constraint Language (OCL) [WK99].

Different methods have been proposed for the design of hypermedia systems. From among these we have chosen the OOHDM [SRB96] as the basis for our design method presented in [KM99], because it is an object-oriented method. Further, it allows a concise specification of the navigation by introducing the concept of navigational contexts and it includes a step for modeling the user interface. However, the OOHDM notation is not UML-based.

In this paper we present a UML extension – fully compliant with the UML standard – for modeling hypermedia systems supporting our method. The classical diagrams provided by the UML, like class diagrams or statecharts are not sufficient to model all aspects of hypermedia systems, for instance, to model the navigational space and to represent this model graphically.

Hypermedia systems are complex software systems, therefore most of the methodologies for development complex software systems can be applied to their development process. In particular, the development of a hypermedia system can be divided into five tasks: requirements capture, analysis, design, implementation and test.

The goal of requirements capture is to find out the functional and non-functional requirements of the system. The functional requirements can be captured using use cases and scenarios. An example of a non-functional requirement for Web systems is the decision about the users of an application. If a broad audience is intended, one has to restrict to techniques that all Web browsers understand. Usually this implies the restriction to plain HTML. The use of Javascript, Java Applets, Active-X and plug-ins drastically reduces the reachable audience.

The result of the analysis task is the conceptual model, which is a UML class diagram consisting of classes and associations found in the problem domain. This class diagram may be supplemented by other UML diagrams, like sequence diagrams specifying the behavior of the system.

The conceptual model is the starting point for the design task. This is one aspect where the method for the development of hypermedia applications differs from the development of other software systems. Based on the conceptual model the navigational structure of hypermedia applications is defined which consists of the navigational class model and the navigational structure model. The navigational class model specifies *which* classes and associations from the conceptual class diagram are available for navigation and the navigational structure diagram specifies *how* the navigation is performed.

The last step in the design task is the presentational design, where a rough version of the user interface is produced. This is done by first defining the user interface objects which are composed of primitive user interface objects, like text, anchor and image as well as other composite user interface objects, giving hints to the final appearance of the user interface objects on the screen. The final decisions about the layout are made in the implementation task. In a second step the behavior of these objects are defined.

The implementation task maps the objects of the presentational model to Web pages, server side scripts and client side scripts, style sheets, etc. in the case the target platform is the Web. Existing software tools for the user interface design can be used to implement the presentational model, e.g. the NeXT Interface Builder, MacroMind Director, VisualWave and JBuilder.

A test model describes the test cases with the purpose of verifying the use cases and checking the navigability of the structure, e.g. dangling links and unreachable pages.

This paper focuses on the notation and techniques used in the design phase and is structured as follows: Section 2 outlines some related work. Sections 3, 4 and 5 describe the conceptual, navigational and presentational models. Finally, we present concluding remarks and an overview of future work in Section 6.

2 Related Work

Some of the works in the hypermedia modeling field only focus on the notation, like the UML extension proposed by Conallen, or on the design process, such as RMM, OOHDM, EORM and WSDM. The latter use standard notation, like E-R notation, OMT or UML, merely for the conceptual design and define their own notation and graphical techniques for the other steps.

Conallen [Con98] defines a set of UML stereotypes for the Web, but does not present a method for the design of Web applications. It includes stereotypes for components, classes, methods and associations, such as server component, client component, server page, client page, form, frameset, link, redirect and submit. These stereotypes are appropriate to model layout and implementation aspects, but not to define the navigation structure of the Web applications.

The Relationship Management Methodology (RMM) [ISB95] addresses the design and construction of hypermedia applications by a process of seven steps. RMM is at the same time a top down and a bottom up approach. During the E-R design step, entities and relationships are identified, which will become nodes and links in the resulting hypermedia application. The second step, slice design, involves grouping entity attributes for presentation. Slices are "presentation units" which appear as pages of hypermedia applications. Separation of contents and presentational aspects is not fulfilled in this step. RMM specifies navigation by access primitives, such as link, grouping (menus), index and guided tour. The techniques proposed for the user interface design in [BBI96] are elaboration of mock-ups and prototyping.

The Enhanced Object-Relationship Model (EORM) is defined as an iterative process concentrating on the enrichment of the object-oriented model by the representation of relations between objects (links) as objects. According to [Lan96], this has the following advantages: relations become semantically rich as they are extensible constructs, they can participate in other relations and they can be part of reusable libraries. The method is based on three frameworks of reusable libraries: for class definition, composition (link class definition) and GUIs.

The Object-Oriented Design Method (OOHDM) comprises four activities; they are conceptual modeling, navigational design, abstract interface design and implementation [Ros96]. These activities are performed in a mix of incremental, iterative and prototype-based development style. This method sees an application as a view over the conceptual model. The concept of navigational context is introduced to describe the navigational structures. It is a powerful concept that allows different groupings of navigational objects with the purpose to navigate them in different contexts. A special notation is used for the representation of the navigation structure. In earlier papers OMT was proposed [SRB96] as the notation of the conceptual schema; later paper use UML instead [SR98]. However, OOHDM diagrams are not UML compliant as they use own notation for perspectives of attributes in the class diagrams and proposes other kind of diagrams for the navigational and abstract user interface design.

The Web Site Design Method (WSDM) is a user-centered approach defining the navigation objects based on the information requirements of the users of a Web application [TL97]. WSDM consists of three main phases: user modeling, conceptual design and implementation design. In the user modeling phase the potential users of the Web site are identified and classified. Different perspectives are defined for the user classes; these are different ways user classes look at the same information. The navigational model consists of a number of navigation tracks expressing how users of a particular perspective can navigate through the available information. This method defines its own graphical notation for the objects of the navigational model. The navigational design achieves Web applications that have a very hierarchical structure.

3 Conceptual Model

The result of the analysis task is the conceptual model of the problem domain defined by classes relevant to the domain and associations between these classes. The goal is to capture the domain semantics with as little concern as possible of the navigational and presentational aspects. Activities of the conceptual design step are to find classes, to specify attributes and operations, to define hierarchical structures and to determine sub-systems. Well-known object-oriented modeling techniques, such as aggregation, composition, generalization and specialization are used to achieve this purpose. Classes and associations are described by attributes and operations; they can be organized into UML packages. The conceptual model is the starting point for one or more navigational designs , i.e. more than one hypermedia application. It is represented by a UML class diagram.

The UML extensions presented in this paper are illustrated by the design of a Web site for a service company. The objective of this Web site is to offer information about the company itself, the employees and their relationship to projects and departments. Projects are performed for customers and may establish relationships to partners. For demonstration purposes, we restrict ourself in the example to the classes and associations shown in Fig. 1. Many other aspects may be added, e.g. information about products, publications, events, press releases and job offers.

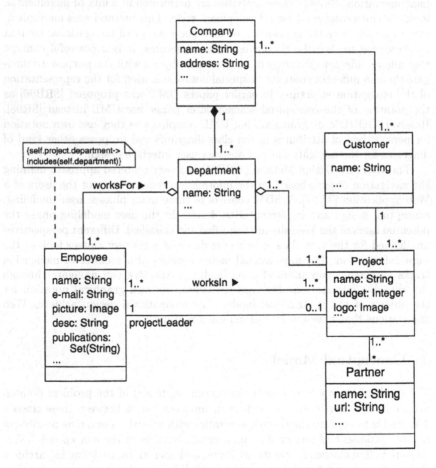

Fig. 1. Conceptual Model of a Company's Web Site.

4 Navigational Model

A hypermedia application is organized into nodes and links that establish relationships of type navigation between the nodes. Nodes that are obtained from objects of the conceptual model are called navigational objects. The navigational design is a critical step in the development of every hypermedia application. Even applications with a non-deep hierarchical structure and a small number of nodes may have a complex navigation structure. Links improve navigability on the one hand, but imply on the other hand, higher risk of loosing orientation. Building a navigational model is not only helpful for the documentation of the application structure, it also allows for a more structured navigability. The navigational design defines the structure of the hypermedia application by building two models: the navigational class model and the navigational structure model.

4.1 Navigational Class Model

The navigational class model defines a view on the conceptual model showing which classes of the conceptual model can be visited through navigation in the application. This model is built with a set of navigational classes and associations, which are obtained from the conceptual model, i.e. each navigational class and each association of the navigational class model is mapped to a class, respectively to an association in the conceptual model. In the navigational class model navigability is specified for associations, i.e. direction of the navigation along the association is indicated by an arrow attached to the end of the association's line. It is possible to attach an arrow to both ends of an association. In this case the user can navigate in both directions along the association. We distinguish for each link a navigational source object and a navigational target object; the latter one may be determined dynamically.

A navigational class is defined as a stereotyped class «navigational class» (cf. Fig. 2) with the same name as the corresponding class of the conceptual model. We use a small box as the icon for the stereotype «navigational class». Navigational objects are instances of these navigational classes connected by links (in UML terms) that are instances of the associations of the navigational model. The navigational class model is usually a sub-graph of the conceptual model where classes and associations which are not needed for navigation are eliminated or reduced to attributes of other classes. The values of these attributes can be computed from some conceptual objects. The formula to compute the derived attribute is given by an OCL expression. A derived attribute is denoted in UML by a slash (/) before its name. Navigational classes and associations with navigability are graphically represented in a UML class diagram.

The navigational class model for our example is shown in Fig. 2. The first step for the construction of this model consists of determining which classes and associations of the domain model are relevant for the company's Web site. All classes and associations of the conceptual model with the exception of Customer and the associations linking Customer to Company and Customer to Project, respectively, are included in the navigational model. The customers of a company

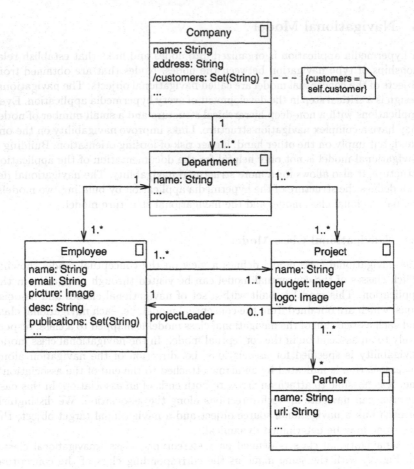

Fig. 2. Navigational Class Model of a Company's Web Site.

are only included as a derived attribute of class **Company**. This is to conceal the information which customer participates in which project from the visitor of the Web site.

4.2 Navigational Structure Model

The navigational structure model is based on the navigational class model. It defines the navigation structure of the application, i.e. how navigational objects are visited. Additional model elements are required to perform the navigation between navigational objects: menus, indexes, external nodes and navigational contexts. We introduce the concept navigational nodes to denote navigational objects as well as the above mentioned model elements.

Navigational Context A navigational context (context for short) consists of a sequence of navigational nodes. It includes the definition of links that connect each navigational node belonging to a navigational context to the previous and to the next navigational node within the context. In addition, links to the first and to the last, as well as circular navigation may be defined. This concept was introduced by OOHDM to permit different groupings of the navigational objects. This way a navigational object can be navigated in different contexts. For example, the information about the "Forsoft" project is shown as one of the "projects of the department R&D" and one of the "projects of the employee Baumeister".

A navigational context is depicted as an object with stereotype «navigational context» and has associated an OCL-expression defining its sequence of navigational nodes. Navigation is performed within a navigational context, but navigational context changes are possible. Continuing with the company's Web site example, if the "Forsoft" project is visited in the context of "projects of R&D", it is possible to continue navigation with other projects the same employee has been working at.

We distinguish between navigational contexts (simple contexts), grouped contexts and filtered contexts (cf. Fig. 3). An example of a simple navigational context is "all projects" (**projects by name**). A grouped context is a sequence of sequences of navigational nodes, such as **projects by department** denoting a sequence of contexts, each of them are the projects of one department. It represents a partition of a sequence of navigational objects into sub-sequences given by a common value of an attribute or a common object related by an association. As the notation for grouped context we use the stereotype «grouped context» and write **Class by attribute/rolename/classname** inside the box.

A filtered context («filtered context») allows a dynamic selection of a collection of elements from a navigational context that satisfies a property. This property is supplied usually by the user in a query which is part of the filtered context. An example of a filtered context is the result of the query: "employees that have been working at the Forsoft project since 1997".

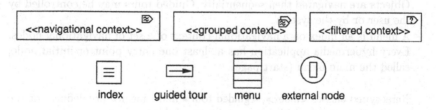

Fig. 3. Stereotypes for the Navigational Structure Model.

We group navigational contexts for a same navigational class in a UML package (cf. Fig. 4). Navigational contexts are related by special associations which allow for context changes. This is possible since the same object is part of different sequences (navigational contexts). A stereotyped association is defined; we call it «change». It permits navigation into another context and to return to the starting point before the navigational context was changed. The possibility to change from one context to another within a package is the default semantics for a package of contexts when the changes are not explicitly drawn. In case only certain context changes are planned, a diagram of the package must show which context changes are allowed. As shown in Fig. 4, only a context change from project by department to project by employee is permitted.

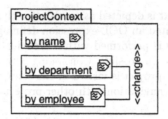

Fig. 4. Package of Navigational Contexts.

Access Primitives To define a navigational model it is necessary to specify how navigational contexts are accessed. The model elements defined for this access (called access primitives) are: guided tours, indexes and menus.

- An index allows direct access to each element within a navigational context. Usually an index comprises a list of descriptions of the nodes of the navigational context, from which the user selects one. These descriptions are the starting points of the navigation.
- A guided tour gives access to the first object of a navigational context. Objects are navigated then sequentially. Guided tours may be controlled by the user or by the system.
- A menu is an index on a navigational context of a set of navigational nodes. Every hypermedia application has at least one entry point or initial node, called the main menu (start page).

Stereotyped classes «index», «guided tour» and «menu» are defined for the access primitives index, guided tour and menu (cf. Fig. 3).

External Node An external node («external node») is a navigational node belonging to another hypermedia application, i.e. this node is not part of the application that is being modeled. Partners are modeled as external nodes in our example.

Navigational Structure Diagram To show how these navigational contexts, access primitives and external nodes collaborate we use a UML object diagram. The navigational structure model is illustrated by the company's Web site example (cf. Fig. 5). The following navigational contexts are needed to support the navigation specified in the navigational class model: department by name, employee by name and by department as well as project by name, by department and by employee.

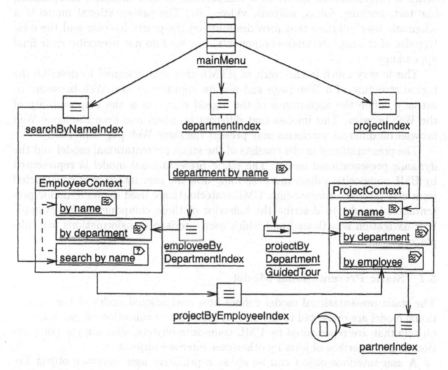

Fig. 5. Navigational Structure Model of the Company's Web Site

The main menu is the starting point to access different navigational contexts related to the defined navigational classes. Indexes are added to permit direct access to the objects of a navigational context. That is the case of departmentIndex, employeeIndex and projectIndex. A guided tour through all the projects of a department has been incorporated to show users the activities of one department.

5 Presentational Model

The next step in the process of designing a hypermedia application is to design its user interface. While the navigational design defines *what* is the navigational structure of an application, the task of the presentational design is to define *how* this navigational structure is presented to the user. Note that the same navigational structure may yield different presentational designs depending, among other things, on the restrictions of the intended target platform and the used technology.

Most of the methods for hypermedia design suggest the development of prototypical pages for the design of the user interface. Instead, we propose first to define a presentational model as a composition of user interface components, like text, anchors, forms, buttons, videos, etc. The presentational model is a schematic user interface that provides hints on the position, color and the relative size of the user interface components, but does do not prescribe their final appearance.

This is very much in the spirit of HTML that was designed to describe the logical structure of a Web page and not its appearance in a Web browser. To actually define the appearance of the logical elements is the responsibility of the Web browser. This implies that different browsers and even the same Web browser on different platforms may present the same Web page differently.

The presentational model consists of the static presentational model and the dynamic presentational model. The static presentational model is represented by UML composition diagrams describing how the user interface is constructed from user interface components. UML statecharts are used in the dynamic presentational model to describe the behavior of these components, for example how navigation is activated and which user interface transformations will take place.

5.1 Static Presentational Model

The static presentational model defines how navigational nodes of the navigational model are presented to the user. It consists of a collection of user interface objects that are represented by UML composite objects, showing the composition of user interface objects by other user interface objects.

A user interface object can be either a primitive user interface object like text and button, or a composition of user interface objects. A special kind of composite user interface object is a presentational object that depends on the state of a navigational object. User interface objects can have their own state, like a button which can be either in state up or down, and they react to user events. Their behavior is defined in the dynamic presentational model. In Fig. 6 we present the stereotypes for the most frequently used interface objects: anchor, text, image, audio, button, form, image, video, collection and anchored collection. The user can define additional user interface objects by composition and by subclassing a user interface class. The user interface objects have the following semantics:

- An anchor is a clickable area and is the starting point of a navigation. They are presented in the literature mostly as part of links, seldom as independent objects. A similar distinction between the concept of link and anchor can be

625

found in the Dexter Hypertext Reference Model [HS94]. An anchor consists of a presentation, which can be a text, an image, a video, another interactive object, etc. together with a link.

- A text is a sequence of characters together with formatting information.
- A button is a clickable area that has some action associated to it, which is defined by the dynamic presentational model. Example of actions are playVideo, displayImage and stopAudio.
- A form is used to request information from the user, for example his name or a search string. A form consists of input fields, menus, checkboxes etc.
- Images, audio and video are multimedia objects. Images can be displayed; audio and video can be started, stopped, rewinded and forwarded. To provide these functionality interactive user interface objects, such as buttons or anchors may be associated to these multimedia objects.
- An external application is included in the current application, but is not related to the navigational objects of the application. An external application can be implemented for example by applets, Active-X components or embedded objects.
- The user interface objects collection and anchored collection are introduced to provide a convenient representation of a composite consisting of a set of user interface objects and a set of anchors respectively.

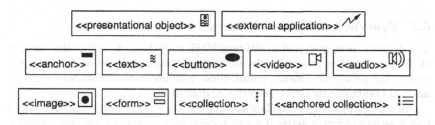

Fig. 6. Often Used User Interface Objects.

A presentational object («presentational object» cf. Fig. 6) is a user interface object that depends on the state of a navigational node and is the composition of other user interface objects including other presentational objects.

On the left side of Fig. 7 we show the presentational object for an Employee and on the right side the presentational object of the navigational context employee by name as UML composite object diagrams. The attributes of an Employee are represented by user interface objects in the presentational object employee, e.g. the picture of an employee is represented by an image. This object contains anchors linking to the projects of an employee and to the department the employee is working in.

The presentational object employee is embedded in the presentational object for the navigational context employee by name. The buttons prev and next allow to jump to the previous and to the next element in the context, respectively, while the buttons prev by dept and next by dept permit the change to the navigational context employee by department.

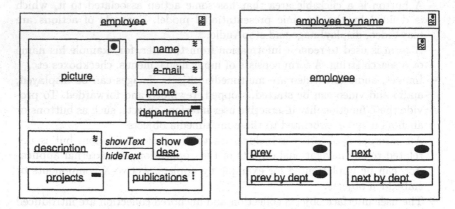

Fig. 7. Presentational Object for Navigational Class Employee and Presentational Object for Context employee by name.

5.2 Dynamic Presentational Model

The dynamic presentational model uses UML statecharts to define the reaction of the user interface objects in the static presentational model on external user events, like mouse movements, mouse clicks and keyboard presses, as well as on internal events like timeouts, activations, deactivations.

To control which user interface objects are perceptible to the user and which user interface object is active, i.e. can receive user events, we use two variables: perception and activation. The variable perception contains the list of currently perceptible user interface objects which are not part of other user interface objects. Perceptible means audible in the case of audio and visible in case of all other user interface objects. Each user interface object in the list held by perception can be thought of as being presented in its own window on the screen. Whenever an element is put into the perception the internal event show is generated and whenever an element is removed the event hide is generated.

The variable activation contains the user interface object from the list of perceptible user interface objects which is currently active, which means that it can react to user input and will receive external user interface events. Whenever a user interface object is assigned to the variable activation it will receive the internal event activate and the previous object assigned to activation, if any, will receive the internal event deactivate. For example, a video user interface object may choose to start playing when it receives activate and to stop playing when it receives deactivate.

When a user interface object receives an event it may change the variables perception and activation, generate new events and send messages to other user interface objects or navigational objects. A composite user interface object delegates the events it receives to its components. Most user interface objects have a default behavior which the designer of a hypermedia application need not specify. An exception are buttons. The dynamic presentational model has to define the actions that are performed when a button is pressed. In Fig. 8 an example is given how statecharts are used to define the behavior of the button showDescr of the presentational object employee (cf. Fig. 7) to toggle the display of the description text user interface object whenever the button is pressed.

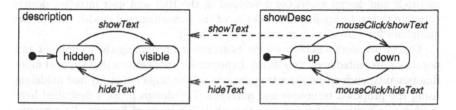

Fig. 8. UML Statechart Specifying the Behavior of the showDescr Button.

The variable perception can be changed in several ways. First, a user interface object can be added to the list of user interface objects. This implies that a new window displaying this user interface object is generated. Second, a user interface object may be removed from the list. Thus, removing the window displaying that user interface object. Third, a user interface object in the list may be replaced by another user interface object in that list. In this case, the window displaying the user interface object that was removed now displays the added user interface object. These three options are called *macro navigation*. In contrast, we call *micro navigation* the replacements of components of a user interface object in the list of perceptible user interface objects. In the context of Web design, macro navigation can be implemented by fetching a new HTML page from the server, while micro navigation can be implemented using frames, which allow to replace only part of the display.

6 Conclusions and Future Work

The UML extension of hypermedia design presented here is a model-based approach whose modeling techniques are UML diagrams and whose graphical representation only uses UML notation. Limiting to the notation proposed by the UML instead of introducing a new notation has the advantage of using a well-known standard and that UML is supported by many case tools. UML is extended to model the navigation and the presentation according to the UML extension mechanisms. These mechanisms are based on the definition of stereotypes and the use of OCL. As far as we know this is the only hypermedia design

628

method that is full UML compliant in every step. The central point of the design are the navigational models that aid hypermedia designer in the specification of a clear structure of hypermedia systems and improve orientation of the users in the application navigation space. The concept of navigational context and the graphical notation of the navigational class and navigational structure diagram achieve to represent complex hypermedia structures in a summarized and clear way.

In contrast to our approach, which deals with the design of a complete hypermedia application, user interface design methods focus only on the design of the user interface part of an application (cf. [Pre94]). In our approach the user interface design is captured mainly in the presentational model. However, methods and design guidelines developed in the HCI and user interface design community can be applied to the design of the presentational model, such as the guidelines for data display mentioned in [Shn98].

Our future work will concentrate its attention on refining the techniques and notations presented here. Specifically hypermedia applications that support more functionality, such as database transactions will be object of study and modeling using the proposed technique and notation. The design process described here is part of a methodology covering the whole life cycle of hypermedia systems. The description of the tasks requirement capture, analysis, implementation and test is on work. Case tools that support UML can be used for the conceptual, navigational and presentational design. We are developing an editor for the UML-based hypermedia design that not only adds the defined stereotypes as model elements, but also includes generation rules for the navigational model based on the conceptual model as well as for the presentational model based on the navigational model.

References

[BBI96] V. Balasubramanian, M. Bieber, and T. Isakowitz. Systematic hypermedia design. Technical report, CRIS Working Papers series. Stern School of Business, New York University, 1996.

[BRJ99] G. Booch, J. Rumbaugh, and I. Jacobson. *Unified Modeling Language: User Guide.* Addison Wesley, 1999.

[Con98] J. Conallen. Modeling Web application design with UML, 1998. Available at http://www.conallen.com/ModelingWebApplications.html.

[HS94] F. Halasz and M. Schwartz. The Dexter Hypertext Reference Model. *Communications of the ACM*, 37(2):30–39, February 1994.

[ISB95] T. Isakowitz, E. Stohr, and P. Balasubramanian. A methodology for the design of structured hypermedia applications. *Communications of the ACM*, 8(38), 1995.

[KM99] N. Koch and L. Mandel. Using UML to design hypermedia applications. Technical Report 9901, Institut für Informatik, Ludwig-Maximilians-Universität, München, March 1999.

[Lan96] D. Lange. An object-oriented design approach for developing hypermedia information systems. *Journal of Organizational Computing and Electronic Commerce*, 6(3):269–293, 1996.

629

[Pre94] Jenny Preece. *Human-Computer Interaction.* Addison-Wesley, 1994.

[Ros96] G. Rossi. *OOHDM: Object-Oriented Hypermedia Design Method (in portuguese).* PhD thesis, PUC-Rio, Brasil, 1996.

[Shn98] Ben Shneiderman. *Designing the User Interface.* Addison-Wesley, 1998.

[SR98] D. Schwabe and G. Rossi. Developing hypermedia applications using OOHDM. In *Workshop on Hypermedia Development Process, Methods and Models, Hypertext'98*, 1998.

[SRB96] D. Schwabe, G. Rossi, and S. Barbosa. Systematic hypermedia design with OOHDM. In *Proceedings of the ACM International Conference on Hypertext (Hypertext'96)*, 1996.

[TL97] O. De Troyer and C. Leune. WSDM: a user-centered design method for Web sites. In *Proceedings of the 7th Internatinal World Wide Web Conference*, 1997.

[WK99] J. Warmer and A. Kleppe. *The Object Constraint Language.* Addison Wesley, 1999.

Why Unified is Not Universal
UML Shortcomings for Coping with Round-Trip Engineering

Serge Demeyer, Stéphane Ducasse, and Sander Tichelaar

Software Composition Group, University of Berne
Neubrückstrasse 10, CH-3012 BERNE
{demeyer,ducasse,tichel}@iam.unibe.ch
http://www.iam.unibe.ch/~scg/

Abstract. UML is currently embraced as "the" standard in object-oriented modeling languages, the recent work of OMG on the Meta Object Facility (MOF) being the most noteworthy example. We welcome these standardisation efforts, yet warn against the tendency to use UML as the panacea for all exchange standards. In particular, we argue that UML is not sufficient to serve as a tool-interoperability standard for integrating round-trip engineering tools, because one is forced to rely on UML's built-in extension mechanisms to adequately model the reality in source-code. Consequently, we propose an alternative meta-model (named FAMIX), which serves as the tool interoperability standard within the FAMOOS project and which includes a number of constructive suggestions that we hope will influence future releases of the UML and MOF standards.

Keywords: meta model, unified modeling language (UML), meta-object facility (MOF), interoperability standard, famoos information exchange (FAMIX)

1 Introduction

With the advent of UML, the progress in CASE technology has reached a next stage of maturity. Indeed, the consensus on a common notation helps both tool vendors and program designers to concentrate on more relevant issues than the direction in which arrows should be drawn, or the question whether to represent classes as rectangles or clouds.

One of these more relevant issues is the notion of *round-trip engineering*: the seamless integration between design diagrams and source code, between modeling and implementation. With round-trip engineering a programmer generates code from a design diagram, changes that code in a separate development environment and recreates the adapted design diagram back from the source code. The object-oriented development processes with their emphasis on iterative development (see [1], [5], [14], [9], [8]) undoubtedly make round-trip engineering a relevant issue.

A second related issue that has become quite relevant is the one of *tool interoperability*. While many of the early CASE tools tried to cover the whole development process, practice has shown that such a generic approach has trouble competing with a series of individual specialised tools. Consequently, CASE tools are becoming more and more open, permitting developers to assemble their favourite development environment from different tools purchased from different vendors yet co-operating via a single interoperability standard.

The OMG has anticipated this tool interoperability evolution by encouraging and adopting the Meta Object Facility (MOF) as a standard. The goal of the MOF is *"to provide the specification of a rich semantics to enable two systems or applications to meaningfully share information. This goal is achievable by providing domain specific metamodels (such as the OOAD metamodel - UML) that conform to the MOF metamodeling architecture."* ([6] - section 1.1.1.1 - Goals and Objectives).

In its current form, the MOF is primarily intended to serve as an exchange standard between OOAD tools. Consequently, the first concrete exchange standard that has been specified using the MOF concerns exchange of UML models. Yet, the increasing demand for round-trip engineering features in OOAD tools will cause tool vendors to use the built-in extension mechanism of UML to cope with more implementation oriented data exchange.

This paper argues that UML *in stricto sensu* is not sufficient as a tool-interoperability standard for integrating round-trip engineering tools. Indeed, since UML is specifically targeted towards OOAD, it lacks some concepts that are necessary in order to adequately model source-code, in particular the concept of a *"method invocation"* and an *"attribute access"*. Of course it is possible to extend UML to incorporate these concepts, but then the protection of the standard is abandoned and with that the reliability necessary to achieve true interoperability.

We start this paper with the requirement extraction concerning round-trip engineering tools using a well-known technique of a scenario (section 2). Afterwards we proceed with an investigation of how to satisfy these requirements using extensions of the UML meta model (section 3). In the section thereafter, we argue that UML extensions cannot achieve tool interoperability (section 4) and consequently propose an alternative meta-model named FAMIX and relate that to the MOF (section 5). Finally, we summarise the results of this work in the conclusions (section 6).

2 A Round-trip Engineering Scenario

Since we claim that UML is not sufficient to serve as a tool-interoperability standard for integrating round-trip engineering tools, it is necessary to be precise about what exactly is a round-trip engineering tool and what kind of requirements it imposes on an interoperability standard. As commonly accepted in today's analysis practices, we define a round-trip engineering tool and its requirements by means of a *scenario*.

The driving force underlying the scenario is the observation that round-trip engineering tools should at least support a smooth transition between implementation and design. Thus, it is not that we neglect analysis; it is just that most tools on the market cover design and implementation. Note as well that at first glance the scenario may seem a bit naive to serve in practice. Yet we have successfully applied the described tool prototypes on a number of industrial case studies.[1] As such we assure that the scenario is both *characteristic* for what practitioners expect from round-trip engineering tools and *realistic* in the sense that it is applicable in the context of industrial development processes.

2.1 Scene 1: Detecting Design Anomalies via Metrics

Carmen is part of a team developing a Geographical Information System called GEOS. The kind of functionality required in GEOS is quite domain specific, so the project adopted an iterative development style with C++ as the implementation language. The development of GEOS started some eight years ago and the system is currently in its 3.7.1 release. Lately, developers have been complaining that it becomes difficult to add functionality.

Carmen is asked to do some code reviewing to see if it is possible to improve the GEOS class structure. Unfortunately, the source code has grown quite large (1 million lines of code - 2837 classes) and Carmen would like some tool support to help her identifying potential design anomalies. Therefore, she selects a metrics tool that allows her to measure various aspects of classes (size, inheritance, cohesion, ,....) and focus her attention on those classes where the measurements exceed certain threshold values.

Character. Carmen is the code reviewer of the team. Like all good code reviewers she relies mainly on reading the code to form her opinion. Yet, she appreciates all tools that help her filtering out potential problems.

Goal(s) of this Scene. This scene introduces a metric tool as one possible element of a round-trip engineering environment. The tool depicted in this scene fits the definition of a round-trip engineering tool because the metrics are interpreted on the design level yet are collected from the implementation. Such metric tools are very important in an iterative development process, because they help to control and steer this process. (See [11] for a practical treatment on how to incorporate object-oriented metrics in a development process and [7] for an overview of the state-of-the-art in object-oriented metrics.)

Data Model Requirements. Metric tools need to access the *complete* source code model, as they must collect data about the whole system. As such, the schema of the model must take special precautions concerning the memory footprint of the entities and especially the associations in the source code model, as the sheer number of them may be very large.[2] To measure object-oriented

---

[1] Both the tool prototypes and the case studies stem from the FAMOOS project, an ESPRIT project whose goal it is to produce a set of reengineering techniques and tools to support the development of object-oriented frameworks (see http://www.iam.unibe.ch/~famoos/).

[2] We have made the following observation concerning the growth of the number of entities and associations in a source code model: (a) the number of methods and

source code, a metrics tool requires knowledge about inheritance associations between classes and the containment associations between classes, methods and attributes. Moreover, for some of the size metrics and all of the cohesion metrics, a tool requires knowledge about method invocations and attribute accesses.

2.2   Scene 2: Assessing disjoint Classes via Program Visualisation

Carmen has identified one suspicious class which appears quite big, yet has very low cohesion and a large number of subclasses. Moreover, the class is a core part of GEOS as it is part of a bridge pattern that is used to make the GEOS objects persistent. She believes that this class has too many responsibilities and she wants to check whether it is possible to split the class in two separate classes. She assumes that distributing the responsibilities over two smaller classes will make subclassing easier and thus improve the inheritance hierarchy of the GEOS system.

To check whether the class can indeed be split, Carmen applies a special visualisation tool. The tool displays a graph containing attributes and methods as nodes plus attribute accesses and method invocations as edges. The tool has the special feature to incorporate a graph layout heuristic that minimises the number of crossings between edges (see Figure 1). With such a visualisation, Carmen observes two clusters in the methods and attributes and concludes that the class may indeed be split.

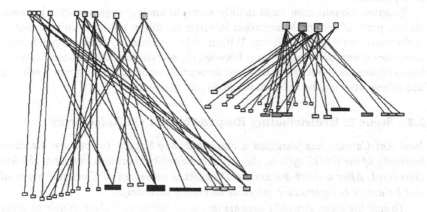

Fig. 1. Visualisation of a class, i.e. the way the methods (in the bottom) access the attributes (on the top). The two clusters indicate that the class may be split.

---

attributes is an order of magnitude larges than the number of classes; (b) the number of invocations and accesses grows almost quadratically with the number of methods. Thus, a project with 150 classes has about 1200 methods and 4000 invocations while for 700 classes this increases untill 7000 methods and 30,000 invocations.

Character. Being a good code reviewer, Carmen never relies on a single tool to help her assess the quality of the code. Rather, she has a whole suite of complementary tools that she applies when the situation calls for it.

Goal(s) of this Scene. This scene has two main purposes. First, it introduces another round-trip engineering tool, namely program visualisation. Visualising a program is often interesting because it allows the human brain to study multiple aspects of a complex structure in parallel and as such can be of great help in program understanding. Consequently, a program visualisation tool like depicted in this scene is a round-trip engineering tool because its output is interpreted at design level, yet it takes its input from the implementation (see [13], [10], [3] for examples of program visualisation in reverse engineering in an object-oriented context).

Second, and more importantly, this scene emphasises the need for various highly specialised tools within round-trip engineering. Because of this variety and specialisation, it is unlikely that all these tools will be purchased from the same tool vendor, hence the need for tool interoperability.

Data Model Requirements. Program visualisation tools are examples of the need for designated access to portions of the source code model. Flexible integration with the data model is crucial, as such a tool wants to visualise any kind of dependency that is present in the source code and needs to customise the lay-out depending on the type of dependency. Memory footprint is less an issue as only slices of the source code must be visualised. Therefore, program visualisation tools want to stay relatively close to a standard model, yet require minor extensions to represent additional information concerning the type of dependency.

Program visualisation tools mainly serve to analyse dependencies between various parts of the implementation in order to obtain a better understanding of the inner workings of a system. Within object-oriented systems, dependencies stem from inheritance associations between classes; containment associations between classes, methods and attributes; invocation associations between methods; and access associations between methods and attributes.

2.3   Scene 3: Redistributing Responsibilities via Refactoring

Now that Carmen has identified a class that may be split to improve the class hierarchy of the GEOS system, she contacts Benedikt to explain him what she has discovered. After a short discussion, Benedikt is convinced of Carmen's proposal and he agrees to restructure the class hierarchy accordingly.

To split the class, Benedikt uses his favourite coding tool which is able to apply a series of low-level refactorings (such as create new class, move attribute, move method) to accomplish the desired redistribution of responsibilities. Afterwards, Benedikt runs a series of regression tests to see whether the split of the class did not affect the working system.

Character. Benedikt enters the scenario, playing the role of the code warrior. Since Benedikt is working with code daily, he uses sophisticated development tools (testing, refactoring, browsers) to make him highly productive.

Goal(s) of this Scene. In this scene, we illustrate the notion of refactoring. A single refactoring corresponds to a low-level semantic preserving restructuring, for instance moving an attribute inside the class hierarchy after checking all its references, or for instance renaming a method and patching all places where it is invoked. The idea is to combine several low-level refactorings to improve the design of a class hierarchy, thus refactorings fit our definition of a round-trip engineering. (See the Ph.D. work of Opdyke for the early definitions of refactoring [12] and [16] for a description of a full-fledged refactoring tool. Refactorings are applicable in practice, as is illustrated by [4]).

Data Model Requirements. A refactoring tool is an example of a tool that modifies the source code, thus should know about the exact source code location of data model entities. Since such a tool must update the corresponding data model items without breaking any other tool relying on it, refactoring tools want to stay as close as possible to a standard model of the source code.

To apply refactorings —be it manually or with a tool— it is necessary to check given preconditions and often patch existing references. For instance, before moving an attribute, one must check where this attribute is accessed. Or while renaming a method, one must patch all places where it is invoked. Thus any tool that supports refactorings must at least know about which methods invoke which other ones and which methods access which attributes.

2.4 Consequences

Accepting the above scenario (sections 2.1 to 2.3) has some important consequences concerning what to expect from a round-trip engineering tool.

- Round-trip engineering is more than a mere succession of reverse and forward engineering steps. Rather, it is a succession of activities, where each activity involves some reverse and some forward engineering aspects. *Consequently, round-trip engineering tools should support a tight integration between reverse and forward engineering.*
- A round-trip engineering tool is never a monolithic application. Rather, it consists of a wide variety of specialised utilities that are applied when the situation calls for it. Because of this variety and specialisation, *it is unlikely that all tools will be purchased from the same tool vendor, hence the need for tool interoperability standards.*
- Because round-trip engineering demands for a tight interaction between reverse and forward engineering, *the supporting tools need an adequate representation of source code.* At minimum they should incorporate the core object-oriented implementation model depicted in Figure 2. Thus they should know about (i) classes, methods and attributes; (ii) the belongs-to relation between classes, methods and attributes; (iii) the invocation relation between methods;[3] (iv) the access relation between methods and attributes.

---

[3] Note that the (method) invocation association should take polymorphism into account. This implies that one invocation has several candidate target methods. The actual target can only be resolved at run-time.

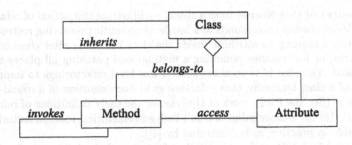

Fig. 2. The Core Object-Oriented Implementation Model

3 Embedding Implementation Concepts into UML

Given the core object-oriented implementation model depicted in Figure 2, the question is whether this can be embedded into UML. Comparing the core implementation model with the UML meta model [17], [18] we make the following observations (summarised in Figure 3).

(a) The UML meta model defines a large number of concepts that do not appear in the implementation model. "Aggregation" and "Constraint" are two examples but there are many more.

(b) There is a substantial overlap between the core object-oriented implementation model and the UML meta model. With some flexibility it is possible to map "Inheritance" onto "Generalisation" and "Class", "Method" and "Attribute" on their respective counterparts bearing the same name.

(c) The implementation model includes two concepts that do not map directly onto UML equivalents. These two are "(method) Invocation" and "(attribute) Access".

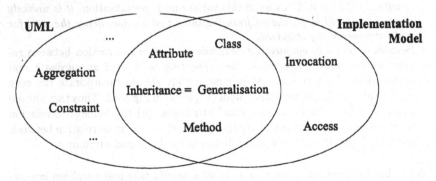

Fig. 3. Comparing the UML meta model with the Core Object-Oriented Implementation Model

Given observations (a) and (b), it should be possible to embed the core implementation model within the UML meta model, if only we find a solution for observation (c). That is, we must find a way to represent the concepts of a method invocation and an attribute access in UML. In the following subsections we will analyse some possibilities to extend the UML meta model to incorporate these two concepts.

3.1 The Behavioural Elements Approach

There are a number of UML concepts in the Behavioural Elements package (see [18], part 3) that come close to the "Invocation" and "Access" concepts of the implementation model (see Figure 2). It is out of the scope of this paper to describe all possibilities in detail, so we restrict ourselves to an analysis of the most plausible concept, namely "Action". This concept is characteristic for other suitable behavioural elements like "Message", "MessageInstance", and "AttributeLink".

Most notably, the class "Action" has one subclass that is attractive for our purposes, namely "CallAction". Checking the UML semantics, we read that "an action is a specification of an executable statement (...), realised by sending a message to an object or modifying a value of an attribute" ([18], p. 68). And then "a call action is an action resulting in an invocation of an operation on an instance" ([18], p. 68). Knowing that an attribute access can always be mimicked by an invocation of some special purpose accessor method, a "CallAction" seems a suitable candidate for modeling the implementation concepts "Invocation" and "Access". So lets examine these a bit further to see whether it fulfils our requirements, in particular how to retrieve both origin and target of an invocation and an access.

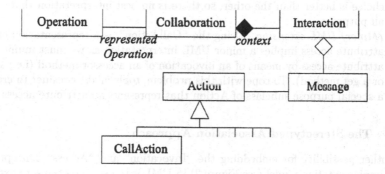

Fig. 4. CallAction and how to navigate back to its origin

First, to retrieve the possible targets of an instance of "CallAction", we must interpret the "target" attribute of the action which, citing the UML semantics "resolves into zero or more specific Instances which are the intended recipients of the dispatched Request" ([18], 8.2 Abstract Syntax / Action - p. 68). Second, to retrieve the origin of an instance of "CallAction", we must —as is depicted in Figure 4— navigate from an "Action" over a "Message", over an "Interaction", over a "Collaboration" to finally arrive at the originating "Operation" ([18], Figure 15: Collaborations). Note that, "in a collaboration it is specified what properties instances should have" ([18], 9.4 Semantics / Collaboration - p. 86), thus the operation retrieved that way is associated with an "Instance" as well. Therefore, we infer that the "CallAction" is representing a dynamic association between instances, not a static association between methods.

Consequently, choosing "CallAction" as a representation for the implementation concepts "Invocation" and "Access" has the following implications.

- *Verbose Construct.* To express a single method invocation we need to build a quite complicated construct (a chain of instances from the classes "Operation", "Collaboration", "Interaction", "Message" and "Action") which consumes a considerable amount of memory and is slow in processing. Remember our metric tools (see scenario 2.1) which must store and analyse thousands of these invocation dependencies and it is clear that this construct is not always optimal.
- *Interpretation Issues.* Using "CallAction" as a representation of an "Invocation" leaves room for interpretation. Indeed, since the "CallAction" is a dynamic association between operations invoked on instances and "Invocation" is supposed to be a static association between methods, we must supply an instance and interpret it as a representative of its class. There are several possibilities to do such: one is choosing a distinct object for each originating and target method, a second is optimising the former by sharing objects representing the most common superclass for a given method, the third is to have one object for each class. Depending on the tool requirements, one choice is better than the other, so there is no best interpretation that suits all purposes.
- *(Minor) UML extension.* Using the "CallAction" as a representation of an attribute access implies a minor UML interpretation as we must mimic the attribute access by means of an invocation of an accessor method (i.e., a set or a get method). To cope with this problem, tools might consider to create a special purpose subclass of Action that represents an attribute access.

3.2 The Stereotyped Association Approach

Another possibility for embedding the "Invocation" and "Access" concepts of the implementation model (see Figure 2) in UML is to use stereotypes to extend an existing UML concept. The concept of "Association" is particularly interesting for our purpose, because it declares the presence of a relationship between classes and as such may serve to represent the static relationships between items contained in that class.

Stereotyping an "Association" to represent an invocation or access mainly involves the specification of a number of tags that are automatically attached to instances of the stereotyped association. These tags will then maintain (a) a reference to the location in source code, (b) the name of the actual method that is initiating the invocation or access (c) the name of the actual attribute being accessed or (d) the name of the actual method being invoked plus the arguments that are passed, plus all that is necessary to deal with polymorphism.

Consequently, using a stereotyped association to represent associations and accesses has the following implications.

- *Large Amount of Associations.* A stereotyped association remains an association and will be stored as such in the underlying UML model. The sheer amount of invocations and accesses is likely to cause problems when other tools try to visualise such a model. Consequently, this approach should only be used when a small number of associations will be created, thus when we represent small slices of an implementation.
- *(Minor) UML extension.* Stereotypes represent one of the built-in extensibility mechanisms in UML. Thus, by stereotyping we abandon the protection of the UML standard, but then stereotypes are so common that this should not cause major difficulties.
- *(Minor) Interpretation Issues.* There is a minor interpretation issue because an association is supposed to connect classes, while our stereotyped associations connect items contained within classes. Using the appropriate tagged values it is possible to deal with this issue, however it requires careful naming conventions especially when dealing with polymorphism.

3.3 The Special Purpose Extension Approach

The final possibility presented in this paper is the usage of the meta meta model underneath UML to add the special purpose "Invocation" and "Access" concepts to the meta model (see [18] Table 1: Four Layer Metamodeling Architecture).

As mentioned in the UML standard "This capability depends on unique features of certain UML-compatible modeling tools, or direct use of a meta-metamodel facility, such as the CORBA Meta Object Facility." ([18], p. 51). Although the operation in itself is not so difficult —it boils down to the definition of two classes with all the necessary attributes to hold whatever is required, much in the same way as with stereotypes— the fact that not all tools will be able to deal with the extension is a major obstacle.

Thus, using the special purpose extensions has the following implications.

- *(Major) UML extension.* Using the four layer meta modeling architecture to extend the UML meta model is a major UML extension. Consequently, it should be used sparingly, as not all tools will be able to benefit from the extensions.
- *No overhead.* Special purpose extensions are not shared with other tools, hence do not involve any memory overhead for representing information required by other tools.
- *Displaying the Extensions.* One possible drawback of this approach might be that such special purpose extensions of the meta model cannot be displayed

in a tool. However, by adding two special compartments in the graphical representation of a class one can deal with this problem. One compartment would then list all invocations per method, the other compartment would show all accesses per method.

3.4 Consequences

The solutions described in sections 3.1 to 3.3 are probably not the only possibilities to embed the concepts of an "Invocation" or an "Access" into the UML meta model. However, they are characteristic for the kind of solutions we may expect when round-trip engineering tools choose UML for their underlying representation. Hence, from these three solutions we can derive some important implications.

- There are several solutions for embedding implementation concepts into the UML meta model. Thus, *it is feasible to use UML as an underlying representation* for the round-trip engineering tools described in section 2.
- Between the several possibilities, there is no optimal solution for all purposes. Thus, without formal agreement, *tools will adopt the solution that is best suited to their needs.*
- Each solution involves either non-standard interpretations of the UML meta model, or some extensions of the UML standard or both. Thus, *none of the solutions is based on the strict UML standard.*

4 Tool Interoperability

Let us now revise our round-trip engineering scenario (sections 2.1 to 2.3) from a tool oriented perspective, assuming that each of the three tools are supplied by different vendors. Moreover, let us assume that each tool retrieves and stores its knowledge about the software system using a common repository. Finally, let us assume that this repository has an API that is specified according to the MOF standard, thus uses the CORBA/IDL description of the UML meta model.

As you remember, Carmen first applies the metrics tool to measure various aspects of the classes in the GEOS system and identify suspicious classes. This metrics tool makes use of the repository's API to enumerate all classes, methods and attributes and calculate the corresponding measurements. However, to compute the coupling and cohesion metrics the tool needs to know about the (method) invocations and (attribute) accesses, and these cannot be supplied by the repository. Therefore, the tool instructs a special purpose propriety utility to parse the method bodies and return the required invocations and accesses. Of course the tool wants to save this information for later use, hence uses the "The Special Purpose Extension Approach" (section 3.3) to store this into the repository. The choice of the extension mechanism is best suited for a metrics tool because it needs to construct all invocation and access associations that occur in the system, hence need a representation that has very little memory overhead.

Next, Carmen visualises the access patterns between methods and attributes to check whether the class can be split. The visualisation tool retrieves the necessary methods and attributes from the repository, but the attribute accesses are obtained via a propriety utility that parses method bodies on the fly. Again, the tool wants to save this extra information into the repository. However, this time the "The Stereotyped Association Approach" (section 3.2) is best suited, because it stays quite close to the standard yet allows to represent additional information concerning the type of dependency.

Then, Benedikt splits the class with his refactoring tool. Retrieving the methods and attributes of the class to be split is of course done via the repository. But then a third special purpose propriety utility is necessary to collect the invocations and accesses from the source code. Here as well, the refactoring tool will save this extra information in the repository. However, since the refactorings modify the internal UML model, a refactoring tool favours a model that is as close as possible to the UML standard, hence uses the "The Behavioural Elements Approach" (section 3.1). Unfortunately, after the class has been split, the refactoring tool cannot instruct the repository how to update the extra information that is stored there by the metrics and visualisation tool, due to the fact that these are non-standard UML extensions. Thus, the repository is now in an inconsistent state and Carmen cannot trust that her tools will function properly.

4.1 Consequences

The revision of the scenario from a tool-oriented perspective reveals why the UML meta model is not sufficient to serve as an interoperability standard between round-trip engineering tools. To summarise, the fact that UML lacks the concepts of an "(method) Invocation" and an "(attribute) Access", has the following drawbacks.

- Each tool is forced to have its propriety parsing utility that extracts the lacking data from source code. Anyone who has tried to build a reliable C++ parser will confirm that this is a highly specialised and difficult task that should be done once and then reused by others.
- Each tool is forced to extend UML to express the insufficiencies. Worse, each tool will define its own extensions that are not understood by others.
- Once a tool modifies the implementation, the repository risks to be out of synch causing malfunctioning of tools.

From these drawbacks we conclude that, to achieve true interoperability between round-trip engineering tools, it is necessary to build a special purpose meta model that closely reflects the reality in source code, yet is independent of the implementation language. The following section briefly introduces such a meta model.

5 An Alternative: FAMIX

Within the FAMOOS project, a number of geographically dispersed programming teams experimented with various tool prototypes to support reengineering activities. Almost immediately, we encountered the kind of problems described in section "4 Tool Interoperability" and have been looking for a satisfactory solution ever since. Because UML had clear shortcomings, we defined a language independent meta model named FAMIX (http://www.iam.unibe.ch/~famoos/FAMIX/), which we present here as one possible alternative to UML.

The core of the FAMIX model corresponds to the one in Figure 2. However, this is too simplistic to serve in practice and we include concepts to represent crucial source code items like functions, global and local variables, formal parameters, packages, etc. The complete FAMIX model is depicted in Figure 5.

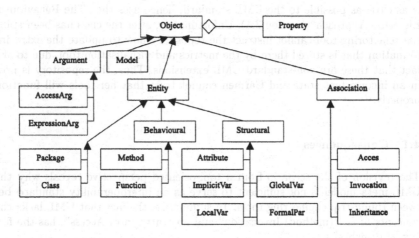

Fig. 5. The Complete FAMIX Meta Model

At the time of writing, parsing technology exists to generate FAMIX models from C++, Java, Smalltalk and Ada. The generated information has been successfully employed in metric and program visualisation experiments [2], [3], [15] and we are currently investigating how well it could support refactorings. Thus, we are fairly confident that the FAMIX model may support the scenario presented in Section 2.

Yet, the scenario is only there to show that there is more to round-trip engineering than obtaining UML models from source code. Of course, tools for UML extraction are considered quite important in industry and this position has been reflected by the FAMOOS partners. Therefore, we conducted an experiment to implement a mapping from FAMIX to UML [19]. Based on this experience, we are convinced that it is advantageous to have a separate source code meta-model instead of embedding source-code information into the UML meta model (see [6],

Table 1: Four Layer Metamodeling Architecture). The reason is that with two clearly separated meta-models, it is feasible to explore different mappings from the one into the other, which is relevant when generating code as well as when extracting UML from source code. We consider this insight important and hope that it will influence the development of other MOF meta models and standards.

6 Conclusions

In this paper, we have supplied a proof-of-concept of the feasibility to use UML as the underlying representation for a round-trip engineering tool. This proof-of-concept follows from the three solutions we have presented for embedding the concepts of a "*method invocation*" and an "*attribute access*" into the UML meta model. While these three solutions are probably not the only possible ones, they are characteristic for the kind of solutions we may expect when representing implementation constructs in UML.

However, the fact that one must extend UML to represent implementation concepts, together with the fact that there are several possibilities to do so has dire consequences on tool interoperability. First of all, it implies that different tool vendors can and will choose different extensions for modeling the same implementation construct. But more importantly, it implies that UML *in stricto sensu* cannot serve as an interoperability standard between round-trip engineering tools.

Does this imply that OMG's standardisation work on UML and MOF is wasted? On the contrary, it implies that the really interesting work is just about to begin. Indeed, the fact that UML is not the most adequate representation for implementation models suggests that we need a meta model besides UML. And since the MOF is actually a meta meta model, it can be used to explore and express various mappings between meta models to achieve what has been called a "Universal Design Language"[6]. We hope that the constructive suggestions in this paper will contribute and influence future work on such a "Universal Design Language".

Acknowledgements

This work has been funded by the Swiss Government under Project no. NFS-2000-46947.96 and BBW-96.0015 as well as by the European Union under the ESPRIT programme Project no. 21975. We thank the partners in the FAMOOS consortium for supporting our work on the FAMIX model and incorporating it in their tools. We also want to thank the members of the SSEL group in the University of Brussels for a fruitful discussion on the pros and cons of UML.

References

[1] Grady Booch. *Object Oriented Analysis and Design with Applications*. The Benjamin Cummings Publishing Co. Inc., 2nd edition, 1994.

[2] Serge Demeyer and Stéphane Ducasse. Metrics, do they really help ? In Jacques Malenfant, editor, *Proceedings LMO'99 (Languages et Modèles à Objets)*, pages 69–82. HERMES Science Publications, Paris, 1999.

[3] Serge Demeyer, Stéphane Ducasse, and Michele Lanza. A hybrid reverse engineering platform combining metrics and program visualization. In Françoise Balmas, Mike Blaha, and Spencer Rugaber, editors, *WCRE'99 Proceedings (6th Working Conference on Reverse Engineering)*. IEEE, October 1999.

[4] Martin Fowler, Kent Beck, John Brant, William Opdyke, and Don Roberts. *Refactoring: Improving the Design of Existing Code*. Addison-Wesley, 1999.

[5] Adele Goldberg and Kenneth S. Rubin. *Succeeding With Objects: Decision Frameworks for Project Management*. Addison-Wesley, Reading, Mass., 1995.

[6] Object Management Group. *Meta Object Facility (MOF) Specification*. OMG Document ad/97-08-14. Object Management Group, September 1997.

[7] Brian Henderson-Sellers. *Object-Oriented Metrics: Measures of Complexity*. Prentice Hall, 1996.

[8] Ivar Jacobson, Grady Booch, and James Rumbaugh. *The Unified Software Development Process*. Addison-Wesley, 1999.

[9] Ivar Jacobson, Martin Griss, and Patrik Jonsson. *Software Reuse*. Addison-Wesley/ACM Press, 1997.

[10] Danny B. Lange and Yuichi Nakamura. Interactive visualization of design patterns can help in framework understanding. In *Proceedings of OOPSLA'95*, pages 342–357. ACM Press, 1995.

[11] Mark Lorenz and Jeff Kidd. *Object-Oriented Software Metrics: A Practical Approach*. Prentice-Hall, 1994.

[12] William F. Opdyke. *Refactoring Object-Oriented Frameworks*. Ph.D. thesis, University of Illinois, 1992.

[13] Wim De Pauw, Richard Helm, Doug Kimelman, and John Vlissides. Visualizing the behavior of object-oriented systems. In *Proceedings OOPSLA '93*, pages 326–337. ACM Press, 1993.

[14] Trygve Reenskaug. *Working with Objects: The OOram Software Engineering Method*. Manning Publications, 1996.

[15] Tamar Richner and Stéphane Ducasse. Recovering high-level views of object-oriented applications from static and dynamic information. In Hongji Yang and Lee White, editors, *Proceedings ICSM'99 (International Conference on Software Maintenance)*. IEEE, September 1999.

[16] Don Roberts, John Brant, and Ralph E. Johnson. A refactoring tool for smalltalk. *Journal of Theory and Practice of Object Systems (TAPOS)*, 3(4):253–263, 1997.

[17] Rational Software, Microsoft, Hewlett-Packard, Oracle, Sterling Software, MCI Systemhouse, Unisys, ICON Computing, IntelliCorp, i Logix, IBM, ObjecTime, Platinum Technology, Ptech, Taskon, Reich Technologies, and Softeam. *Unified Modeling Language (version 1.1)*. Rational Software Corporation, September 1997.

[18] Rational Software, Microsoft, Hewlett-Packard, Oracle, Sterling Software, MCI Systemhouse, Unisys, ICON Computing, IntelliCorp, i Logix, IBM, ObjecTime, Platinum Technology, Ptech, Taskon, Reich Technologies, and Softeam. *Unified Modeling Language - UML Semantics (version 1.1)*. Rational Software Corporation, September 1997.

[19] Sander Tichelaar and Serge Demeyer. SNiFF+ talks to Rational Rose – interoperability using a common exchange model. In *SNiFF+ User's Conference*, January 1999.

Timed Sequence Diagrams and Tool-Based Analysis – A Case Study

Thomas Firley, Michaela Huhn, Karsten Diethers, Thomas Gehrke*, and
Ursula Goltz

Institut für Software, Abteilung Programmierung,
Technische Universität Braunschweig
{firley,huhn,diethers,gehrke,goltz}@ips.cs.tu-bs.de

Abstract. We use UML timed Sequence Diagrams to specify the real-time behaviour of a communication protocol of audio/video components. The Sequence Diagrams build the requirements specification against which an implementation of the protocol developed by the Bang & Olufsen company is proven correct.

To obtain a complete requirements specification, we have to mark the UML Sequence Diagrams as *optional* or *mandatory* behaviour. Then the Sequence Diagram interactions with their timing constraints and periods are transferred to a setting of timed automata. We use the UPPAAL tool for verification. In particular, we show that the implementation of the protocol conforms to the Sequence Diagram specification concerning the correct data transfer on the bus.

1 Introduction

Approaches like Realtime UML [Dou98], the ROOM method [SGW94], or the ACCORD approach [LGT98] aim to introduce the advantages of object-orientation, in particular reusability and evolutivity of the designs, into the development of real-time systems. Several of these real-time extensions are already supported by commercial tools like Rhapsody, ObjecTime (ROOM), and Object-Geode (UML-RT) [Enc96].

However, the development of real-time systems remains a complex and error-prone task even if object-oriented concepts are used. Therefore one is interested in validation techniques which allow to analyse the behaviours of a design already in early phases. As a first step, nearly all tools provide some basic consistency checks for object-oriented designs: For instance, warnings indicate the states which cannot be entered or left.

Recently, also formal verification techniques have been considered for the validation of more complex requirements on the designs. In [LH99], a graphical interface for (a variant of) the ROOM model to the widely accepted SPIN model checker [HP96] is presented. In [AHP96] the timing consistency of Message Sequence Charts is investigated. A similar analysis for Sequence Diagrams is mentioned in [SvG98].

* This author was supported by the DFG project EREAS

Here we consider formal analysis on the basis of timed Sequence Diagrams, but take the timing consistency within *one* Sequence Diagram for granted. Our starting point is the common use of Sequence Diagrams as attachments to use cases, where they describe the dynamic aspects of certain scenarios [BjRJ99]. If in later design phases the behaviour of the system is modelled in more detail, one may ask whether the design is able to perform a sequence of interactions according to the Sequence Diagram specification. In that sense our approach can be regarded as a check on the timing consistency between *different* dynamic models used in the object-oriented design of real-time systems. From a slightly different viewpoint, we consider timed Sequence Diagrams as a requirements specification which should be satisfied by other dynamic models of the system that are closer to implementation.

To check if scenarios described in timed Sequence Diagrams can be performed by a system, we have to indicate which Sequence Diagrams describe *mandatory* or *optional* behaviour, and in which order the Sequence Diagrams shall occur. Harel and Damm extended Message Sequence Charts by a similar classification in [DH99].

To perform the formal analysis we transfer timed Sequence Diagrams as defined in [SvG98] to a timed automata setting [AD94]. Timed automata are well-suited for our purposes since the real-time constraints of the diagrams can be translated directly to clock conditions of timed automata. The transformation has to be directed by the designer who has to provide the connection between the different dynamic models: The designer has to associate the messages from the diagrams to transitions in the state-based models of the real-time design. Then the transformation to the timed automata setting can be done automatically.

We show the feasibility of our approach by investigating a medium size case study taken from a real-time protocol of the Bang & Olufsen company [HSLL97]. The interactions of the audio/video components with the single bus are modelled as timed Sequence Diagrams. Using the UPPAAL tool [LPW95,LPW97] we prove that an implementation of the protocol performs the data transfers correctly.

The paper is structured as follows: In Section 2 we briefly describe timed Sequence Diagrams. Section 3 is concerned with the transformation of Sequence Diagrams to timed automata. In Section 4 the verification of the case study is described. Section 5 concludes.

2 Real-Time extensions to Sequence Diagrams

2.1 Sequence Diagrams

UML contains several diagram types to model dynamic behaviour. The UML Interaction Diagrams can be attached to use cases to show the interactions of the system and some actors, or different subsystems or classes by indicating a sequence of messages which is exchanged between the participating instances. There are two different forms of Interaction Diagrams: The focus of Collaboration Diagrams is on the relationship between the instances participating in a communication. To accentuate the temporal aspects of the communication process, Sequence Diagrams are used. They can either express one possible run (*instance form*) or they define all possible sequences by using loops and branches (*generic form*). In a Sequence Diagram the instances are horizontally arranged and represented by named rectangles. Below the rectangle of an instance a lifeline is attached. The flow of time is displayed in vertical direction. Arrows between the lifelines of two instances represent messages which are sent in the direction of the arrow. The stick arrow is deployed to specify a flat flow of control and concurrent objects. In the case study, messages are synchronous. The position of the instances in horizontal direction has no semantic relevance.

2.2 Syntax of Real-Time Extensions

To define timing constraints in standard UML, labels can be attached at the beginning and the end of a message arrow. The labels are interpreted as time stamps and can be used in timing constraints, e.g., to specify the minimum or maximum time gap between two marked points in the diagram or to define the duration of a periodic sequence. We will only deal with specific constraints (see Section 3.1). We adopt the notation of [SvG98] which extends the Sequence Diagrams of UML to express loops. In this notation, a sequence of messages which is repeated several times is surrounded by a rectangle. The loop condition can be placed at the top or at the bottom of the rectangle. We use the notation LOOP N TIMES $\{expr\}$. The constraint at the right side of LOOP defines the set of possible values for N. If the constraint is missing, N is an arbitrary natural number. To deal with different occurrences of a labelled event in loops, we introduce the following convention: Before a loop, a_{first} can be used in constraints to refer to the first occurrence time of an event with time stamp a in the loop. After a loop, a_{last} refers to the last occurrence of the tagged event in the loop. Within a loop, a_{next} denotes the time of the event occurrence in the following iteration. Figure 1 (a) shows examples of this notation.

3 Tool-based Verification

We aim to formally check whether a real-time design behaves according to a collection of timed Sequence Diagram scenarios. Therefore, we have to transfer the problem to an appropriate formalism which is able to handle real-time constraints. A further restriction is imposed by the fact that verification of practical examples is only feasible if tool support is available. Thus we selected timed automata as our design representation [AD94] and the UPPAAL tool [LPW97].

3.1 Timed Automata

Timed automata were introduced by Alur and Dill [AD94] for formal reasoning on real-time systems. They extend finite automata by real-time clocks and acceptance conditions. The transitions of a timed automaton may be labelled with actions, reset operations for clocks and guards containing timing constraints

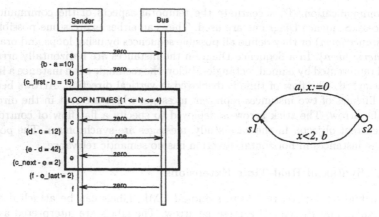

Fig. 1. (a) Loops and timing constraints   (b) Example of a timed automaton

on clocks. However, to keep the verification problems decidable, the timing constraints and acceptance conditions are restricted to simple arithmetic expressions and comparisons: Only constraints of the form $c_1 \approx x \approx c_2$ and $c_1 \approx x - y \approx c_4$ are permitted where the c_is are non-negative constants or ∞ (infinity), x, y denote clocks, and $\approx$ denotes a comparison, i.e., $\approx \in \{=, \leq, <\}$.

Figure 1 (b) shows an example of a timed automaton. The automaton has two states s_1 and s_2 and it may perform the actions a and b toggling between the states. Additionally, the automaton has a clock x. A run of the automaton is an infinite sequence of timed transitions which obey to the timing constraints. Every time a is performed the clock is reset. The automaton may stay in both of the states and let time elapse. But it must leave s_2 within 2 time units after entering, because the transition back to s_1 may only be taken while the value of the clock is less than 2. If staying longer in s_2 the execution gets stuck, which is not a valid behaviour.

In this paper, we work with networks of the extended version of timed automata which are used in the UPPAAL-tool [LPW95,LPW97]. In addition to clocks, integer variables are available[1]. Communication is possible on shared memory (i.e., via global variables) or by output and input actions via channels. Input and output actions are synchronized by a synchronization function. Furthermore, a special kind of states is introduced. While a process may stay in a regular state while time is elapsing, in so-called *committed* states time must not elapse. Moreover, a process must leave a committed state in the next step of the system.

---

[1] The admissible constraints on integer variables are restricted in an analogous manner as those on clocks.

3.2 The Verification Process

In this paper, we consider timed Sequence Diagrams as requirements specification. We assume that the real-time design under consideration is already represented as a collection of UPPAAL timed automata.

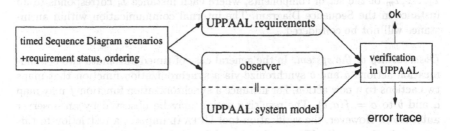

Fig. 2. The verification process

To be able to verify the requirements specified within the timed Sequence Diagrams the following steps are necessary.

- The instances addressed in the Sequence Diagrams have to be associated with components of the real-time design under development. In our setting, the instances will be mapped to sets of automata.
- The specification language offered by the UPPAAL-tool for the requirements only allows to express reachability properties like *in all reachable states some predicate is satisfied* or dually *a state satisfying some predicate is reachable*. Thus, to express complex behaviour like communication sequences as they occur in timed Sequence Diagrams, we have to construct an observer automaton. The observer monitors the system and is changing its state according to the Sequence Diagram scenarios. It will eventually reach a final state if the system conforms to the Sequence Diagram scenario. In case that the system violates the requirements, the observer will enter an error state.
  For the construction of the observer, the designer has to map the messages occurring in the Sequence Diagrams to actions the observer shall perform. Additionally, the designer has to declare the requirement status of a Sequence Diagram. Thus, it has to be indicated whether a family of Sequence Diagrams describes *mandatory* or *optional* behaviour. Also the order in which Sequence Diagrams shall be observed has to be given. Afterwards the observer and a proof obligation can be constructed automatically from the collection of timed Sequence Diagrams. The proof obligation will be that no error state can be reached and that it is possible to reach a final state.
- The system under development has to be slightly prepared to become observable by an observer. First the designer has to select a set of actions which shall correspond to messages which occur in the Sequence Diagrams, i.e., these actions are candidates for observation. Then the automata modelling the real-time design can be semi-automatically extended by our techniques to become observable.

3.3 Translation to UPPAAL timed automata

Instances. The real-time system consists of different components communicating among each other. These components do not necessarily reflect the instances whose communication we want to consider. Therefore we choose a partition $\mathcal{I}_1, \ldots, \mathcal{I}_m$ of the set of components, where each instance $\mathcal{I}_i$ corresponds to an instance in the Sequence Diagram. The internal communication within an instance will not be considered.

Observability of the system. In the general case of timed automata two communication actions a and b synchronize via a synchronization function that maps two actions to a new action. For instance a synchronization function f may map a and b to $c = f(a, b)$. The resulting action may be observed by an observer automaton. However, the verification tool UPPAAL imposes a restriction to this communication model. The restricted model uses *input* and *output actions* over certain *channels* to realize synchronous communication. For instance, $a!$ is an output action over the channel a and $a?$ is the complementary input action. Only complementary actions may communicate and the result is not visible for other automata. Hence we have to introduce additional observable actions which must not change the behaviour of the modelled system. The observer has to know which communication actions take place in the system and which are the participating instances $\mathcal{I}_i$.

We use the following approach to connect the observer to the system: After a reception on channel a the receiving instance B sends a new action $a\_receiver\_B!$. The observer is synchronized to this action by $a\_receiver\_B?$. Moreover, the sending instance A sets a fresh global variable $a\_sender := A$ to enable the observer to identify A as the source of the communication by the condition $a\_sender == A$. Figure 3 shows the modification of instance B to make the reception of a communication action a in the system model observable. The

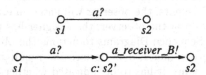

Fig. 3. Original and observable transition

new intermediate state $s2'$ is marked as committed. These committed states are an UPPAAL-specific extension which ensures that neither time elapses nor any other automaton performs actions while control is in such a state. Note that our construction only works if the target state of a communication action is not a committed state since otherwise the observer cannot do some necessary observation steps. However, in practical examples this does not pose severe problems. In case this situation occurs, more sophisticated preparation techniques have to be applied, like pushing communications a transition further or doubling a state. We did not investigate automatization for these steps, because these cases occur rarely and cannot be handled uniformly. One has to make sure that the communication order and timings are not changed.

Formally we introduce the notion ACT for the set of communication actions of the system model which have to be observed and we derive the set ACT$'$ of actions and assignments which are added to the system to make the communication observable and the set $\overline{\text{ACT}}'$ containing the complementary actions and assignments which are used in the observer to identify a communication action of ACT in the system model and the corresponding sending and receiving instances.

$$\text{ACT}' := \{\, a\_receiver\_\mathcal{I}! \,|\, \text{for all instances } \mathcal{I} \text{ and all actions } a? \in \text{ACT} \,\} \quad (1)$$
$$\cup \{\, a\_sender := \mathcal{I} \,|\, \text{for all instances } \mathcal{I} \text{ and all actions } a! \in \text{ACT} \,\}$$

$$\overline{\text{ACT}}' := \{\, (a\_sender == \mathcal{I}_1, a\_receiver\_\mathcal{I}_2?) \quad (2)$$
$$|\, a\_sender := \mathcal{I}_1 \in \text{ACT}' \text{ and } a\_receiver\_\mathcal{I}_2! \in \text{ACT}' \,\}$$

Now we have to define a connection between the observable communication pairs and the messages of the Sequence Diagram. Let M be the set of considered messages of the Sequence Diagram, consisting of triples $(n, \mathcal{I}_1, \mathcal{I}_2)$, where n is the name of a message, $\mathcal{I}_1$ is the sending instance and $\mathcal{I}_2$ is the receiving instance. Then the injective function m from M to $\overline{\text{ACT}}'$ is defined as

$$m(n, \mathcal{I}_1, \mathcal{I}_2) := (a\_sender == \mathcal{I}_1, a\_receiver\_\mathcal{I}_2?) \quad \text{and} \quad n \sim a \quad (3)$$

The designer has to associate message names n from the Sequence Diagrams to communication channels a of the system via the relation $\sim$.

An example of the relationship between messages and observer actions is shown in Figure 4. In the left part of the figure we see the message (a, A, B). The

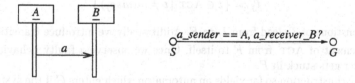

Fig. 4. Sequence Diagram and simple observer

message name a corresponds to the communication on channel a. Therefore ACT equals $\{a!, a?\}$. If we consider the two instances A and B, we get the set $\text{ACT}' = \{a\_sender := A, a\_sender := B, a\_receiver\_A!, a\_receiver\_B!\}$. The system is modified accordingly. The set $\overline{\text{ACT}}'$ which is important for the observer construction is defined as

$$\{(a\_sender == A, a\_receiver\_A?), \quad (a\_sender == A, a\_receiver\_B?),$$
$$(a\_sender == B, a\_receiver\_A?), \quad (a\_sender == B, a\_receiver\_B?)\}$$

As the set M contains only (a, A, B) the basic part of the observer only reads the communication $(a\_sender == A, a\_receiver\_B?)$. The basic automaton is shown as the right part of Figure 4.

Construction of the Observer. The observer automaton is constructed according to the visual order of messages in the Sequence Diagram [AHP96]. For every instance $\mathcal{I}$ there is a local total order $\leq_\mathcal{I}$ over the input and output events which corresponds to the order in which the events are displayed. The relation

$$\leq_v := (\bigcup \leq_\mathcal{I}) \cup \tag{4}$$
$$\{(e_o, e_i)| \ e_o \text{ and } e_i \text{ are output and input events of the same message}\}$$

is called the *visual order*, which is a partial order. For simplicity we will consider only Sequence Diagrams in which the messages are totally ordered. The general case can be treated by adding all interleavings accordingly.

First we construct an automaton which observes the desired behaviour but ignores the timing contraints. Suppose that there are N_0 messages $msg_i \in M, (1 \leq i \leq N_0)$ in the Sequence Diagram. Then the observer consists of a starting state $S = I_0$ and a successful final state $G = I_{N_0}$, the goal, and intermediate states I_i with $i \in \{1, \ldots, N_0 - 1\}$ between them. Transitions labelled with $m(msg_j)$ lead from I_{j-1} to I_j.

Then we have to add transitions to the automaton, which are used when observing a behaviour not according to the Sequence Diagram.

We introduce an error state F, which is entered if a behaviour is recognized that is not according to the Sequence Diagram specification. For each state I_j with $j \geq 1$ we calculate a set of transitions

$$T_j := \{\, t \in \overline{\mathrm{ACT}}' \,|\, t \neq m(msg_j)\,\} \tag{5}$$

All transtions of T_j lead from I_j to F. Additionally we introduce transitions for any element of $\overline{\mathrm{ACT}}'$ from F to itself. Once we observe a faulty behaviour the observer gets stuck in F.

The construction so far yields an automaton which enters G if the system has the same behaviour as described in the Sequence Diagram. The state F is entered if a wrong behaviour is observed. If the observer is in one of the intermediate states $I_1, \ldots, I_{n-1}$, an incomplete behaviour has been observed, which may be extended to a correct run.

Timing Constraints. We now add the timing constraints to the basic observer automaton. For each timing constraint we introduce a realtime clock $x_i, 1 \leq i \leq k$, where k is the number of timing constraints. Each timing constraint compares

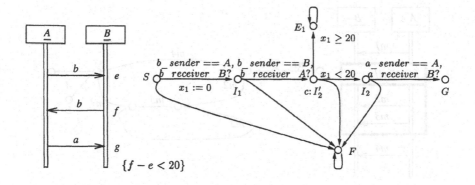

Fig. 5. Observer with timing constraints

the duration of a set of actions with a constant. In the construction we reset the clock to 0 if the first of these actions occurs. Directly after the final action concerned with the timing constraint, we introduce a new state in the observer. If the original transition of the basic observer, which should observe the final action, leads to I_j, then the inserted state is called $c: I'_j$. Thus it is a committed state where no time delay is allowed. We add two transitions with source $c: I'_j$. The first is labelled with the timing constraint and leads to the original state I_j. The second is labelled with the negation of the timing constraint and leads to the new error state E_i. This construction allows to state, that whenever the state E_i is entered, the timing constraint i is violated. If E_i cannot be entered the constraint is always fulfilled.

Figure 5 shows a simple example with three messages and the constraint that the delay between the first two messages must not exceed 20 time units. In the timed automaton we introduce the clock x_1, which is reset when the first message is observed. Immediately after the second message, we decide if the constraint is fulfilled and change the state accordingly. Note that we omitted all labels of transitions to F in Figure 5 for clarity reasons.

Loops. Observing loops is slightly more difficult, because we want to allow constraints concerning more than one cycle. The technique is exemplified in Figure 6. Suppose that four messages are in the Sequence Diagram. The message m_1 precedes the loop, the messages m_2 and m_3 are cycled within the loop and m_4 follows the loop.

In Figure 6, action s may set a counter for the loop. The loop condition l is checked after leaving the loop. Note that this is only possible if the actions m_2 and m_4 are different. If they are not different, the automaton is nondeterministic which poses a restriction to the requirements which may be checked, because the observer might choose the wrong transition and end up in an error state.

In order to distinguish between the action m_2 in the first interation and m_2 in any of the following iterations, we double this action. Constraints that refer to the

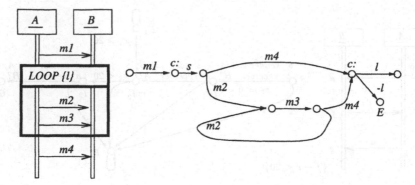

Fig. 6. Observer for a loop

first iteration cycle may be added to the upper m_2-transition, while constraints referring to one of the following iterations and the counter incrementation has to be added to the lower m_2-transition. Constraints that only refer to actions within one iteration cycle do not pose any problem. If a time stamp in a loop is indexed with *first* or *next* which is not the first message in the loop, the automaton has to split the first cycle and the following ones up to this transition as it is done here for the message m_2.

3.4 Verification

In addition to the observer we need proof obligations, expressed in a simple branching time temporal logic, which are verified automatically by the tool.

Besides the regular logical operators the UPPAAL temporal logic has got temporal operators and path quantifiers. The temporal operators, □ *(always)* and ◇ *(eventually)*, allow to refer to particular moments in an execution. The path quantifiers ∃ *(there exists a run)* and ∀ *(for all runs)* allow to refer to a specific run or to regard all possible executions at once. Only two combinations of these operators are allowed. ∀□P means *in all reachable states predicate P holds* and ∃◇P means *a state satisfying predicate P is reachable.*

Sequence Diagrams show either an instancious view of a system or a generic one. We formulate three different kinds of interpretations for Sequence Diagrams. For each of them requirements and in two cases additional transitions in the observer are necessary.

Mandatory Behaviour. When the behaviour of a Sequence Diagram is regarded as *mandatory*, no other behaviour of the system is allowed than the behaviour given in the diagram. The following obligations describe this condition:

$$\forall\Box\neg F \tag{6}$$

$$\forall\Box\neg E_i \text{ for every timing constraint } i \tag{7}$$

$$\exists\Diamond G \tag{8}$$

655

These requirements consider the state F as an error state. Besides the timing constraints an exact correspondence to the communication structure is demanded.

Optional Behaviour. The behaviour described in a Sequence Diagram is *optional*, if it may happen at an arbitrary point in the run of the system. In this case, we have to ignore the behaviour not described in the Sequence Diagram. Therefore we add transitions from the starting state S to itself for all actions of $\overline{\text{ACT}}'$. Thus we get a nondeterministic automaton. We cannot guarantee that we will not enter an error state. The only requirement that we can formulate is:

$$\exists \Diamond G \qquad (9)$$

If-Then Behaviour. If we require that whenever the behaviour of a first Sequence Diagram α has been observed, the system will immediately behave as described in a second diagram β, then the observers for both have to be connected by introducing a new transition from G_α to the start of β, S_β. The state G_α has to be marked as committed. The state S_α needs transitions to itself for all elements of $\overline{\text{ACT}}'$. The requirement α is optional, but whenever its behaviour has been observed, the requirement β gets mandatory. The obligations for this interpretation are

$$\forall \Box \neg F_\beta \qquad (10)$$
$$\forall \Box \neg E_{i\beta} \text{ for every timing constraint } i \qquad (11)$$

For the requirement α *is eventually observed followed by* β, we add the proof obligation

$$\exists \Diamond G_\beta. \qquad (12)$$

4 Case Study: Protocol for Audio/Video Components

To illustrate our approach, we specify the audio/video protocol described in [HSLL97] by a set of Sequence Diagrams. The company Bang & Olufsen uses this protocol to define in which way the components of a stereo system may access a common bus to communicate. In particular, the protocol distinguishes the phases of initialization of the communication process, the transmitting of a data frame and the detection and handling of collisions. Due to the lack of space we only consider the transmission of messages here.

4.1 The Protocol

Frames. Communication between the components of a stereo system takes place by the transmission of data frames. Each frame consists of at least 17 so called T-messages. The protocol considers 5 different T-messages: A T_5-message marks the beginning of a frame and a T_4-message signals the end of a frame. Inbetween, a sequence of at least 15 messages (T_1-T_3) occurs which encode the information to transfer. After sending a frame, a component waits $50000\mu s$ before initializing the next communication.

T-messages. T-messages are subdivided in protocol periods. Two T-messages are separated by a *zero*-signal of one period. A message T_i, $1 \leq i \leq 5$, is encoded by a *one*-signal of $2 * i$ periods. Figure 7 (a) shows the Sequence Diagram of the transmission of a data frame. The diagram shows only the messages which change the state of the bus. All constraints refer to microseconds.

Periods. A protocol period has a duration of $1562\mu s$. Additional sampling of the bus is required at the beginning (S_1) and at the middle of a period (S_2) to detect collisions. For physical reasons, the sender has to put its message on the bus at W ($0\mu s < (W - S_1) < 600\mu s$) (see Figure 7 (b)). In the implementation, W is arbitrarily set to $40\mu s$ after S_1. The message $m2$ has the value *zero* in the first and the last iteration, otherwise the value *one*. If no collision occurs the values of $m1$ and $m3$ result from the previous setting of the bus at W. The Sequence Diagram in Figure 8 contains all messages which are exchanged during the transmission of a frame.

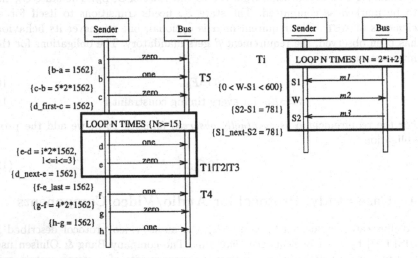

Fig. 7. (a) Transmission of a data frame   (b) Bus sampling for collision detection

4.2   The Implementation

Our case study was already modelled as a set of UPPAAL timed automata in [HSLL97]. The system consists of several automata, three for each sending unit. In addition one automaton, modelling the global bus, is present. Only the sending is verified since the listening involves the same techniques. As the sending units are symmetric, it is sufficient to consider one of them. For a complete verification, however, a second sender for the modelling of collisions would be necessary.

Each sender consists of three automata, which communicate internally. The interesting communication is that with the bus. Reading is done through the

657

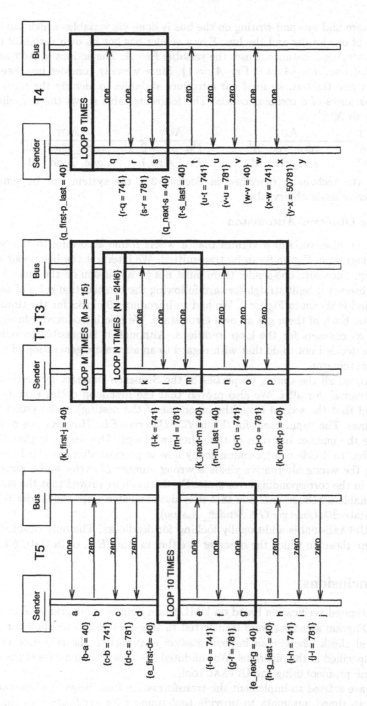

Fig. 8. Communication within one frame

channels *zero* and *one* and writing on the bus is done via variables which are in the scope of one sender and the bus. Every sender has got its own variables to communicate, e.g.,. instance A uses the variable Pn_A. The actions in ACT are thus $\{zero?, one?, Pn\_A := 0, Pn\_A := 1\}$. Since we only consider instances, one sender and the bus, we need not introduce variables to identify the participating instances of a communication. The following table shows the mapping from ACT to ACT':

ACT	ACT'	ACT	ACT'
zero?	zero_rec_A!	$Pn\_A := 0$	zero_send_A!
one?	one_rec_A!	$Pn\_A := 1$	one_send_A!

Using the techniques described in Section 3.3, the system can be semi-automatically made observable.

4.3 The Observer Automaton

We build an observer which verifies that a whole frame according to the Sequence Diagram in Figure 8 can be transmitted. We interpret the behaviour as mandatory, since with one sender there must not be a collision on the bus.

The observer is built straightforward following the construction rules of section 3.3 and is shown in Figure 9. We had to introduce 29 clocks for the timing constraints. Each of them got an own error state. We also introduced 4 integer variables as counters for the loop iterations. Although it is possible to reuse clocks, we decided not to do that with regard to an automatic generation of the observer automaton.

We proved all the timing properties of the Sequence Diagram by verifying $\forall\Box\neg Observer.E_i$ for all i. We also proved that the first loop cycles exactly 8 times, and that the second loop (the inner one of the nesting) cycles either 2, 4, or 6 times. The requirement for this is $\forall\Box\neg Observer.EL_i$. However, we could not prove the number of cycles of the other two loops. The reason is that the observer has to decide nondeterministically how to proceed after the third loop. Choosing the wrong alternative yields a wrong number of cycles and a thus a deadlock in the corresponding error state. Furthermore we proved that the state G is reachable $\exists\Diamond Observer.G$ and that it is also reachable when the sender is in its final state $\exists\Diamond(Observer.G \wedge Sender\_A.stop)$.

The UPPAAL-tool is additionally looking for deadlocks. The only deadlocks found were those, in which the observer is either in state EL_3 or in state EL_4.

5 Conclusions

We investigated verification based on UML timed Sequence Diagrams. The Sequence Diagram scenarios were transferred to timed automata to allow for an automated check whether an implementation conforms to the real-time constraints specified in the diagrams. We validated the approach in a case study on a real-time protocol using the UPPAAL tool.

We have started to implement the transformation from Sequence Diagrams to UPPAAL timed automata to provide tool support for verification of timed

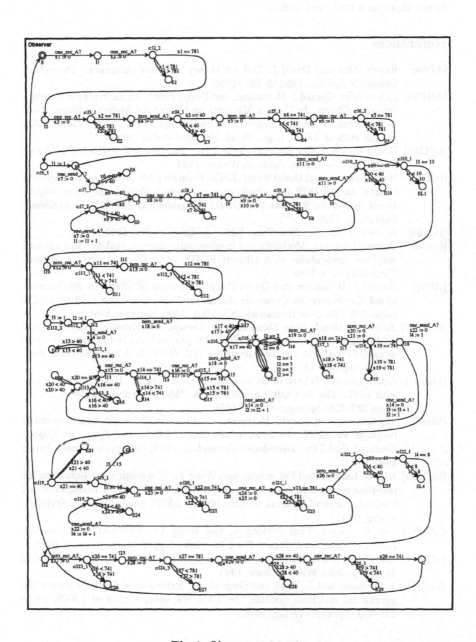

Fig. 9. Observer automaton

Sequence Diagrams requirements. In future work, we want to investigate the transformation of UML behavioural system models, in particular Statechart Diagrams containing timing constraints to timed automata. The aim is a formal analysis of the timing consistency between *different* dynamic models giving different views on a real-time design.

References

[AD94]   Rajeev Alur and David L. Dill. A theory of timed automata. *Theoretical Computer Science*, 126:183–235, 1994.

[AHP96]   Rajeev Alur, Gerard J. Holzmann, and Doron Peled. An analyzer for Message Sequence Charts. In Tiziana Margaria and Bernhard Steffen, editors, *Tools and Algorithms for the Construction and Analysis of Systems (TACAS'96)*, volume 1055 of *LNCS*, pages 35–48. Springer-Verlag, 1996.

[BjRJ99] Grady Booch, james Rumbaugh, and Ivar Jacobson. *The Unified Modeling Language User Guide*. Addison-Wesley, 1999.

[DH99]   Werner Damm and David Harel. LSCs: Breathing life into Message Sequence Charts. In *3rd IFIP Int. Conference on Formal Methods for Open Object-Based Distributed Systems, (FMOODS'99)*, pages 293–312. Kluwer Academic Publishers, 1999.

[Dou98]   Bruce P. Douglass. *Real-Time UML*. Addison-Wesley, 1998.

[Enc96]   Vincent Encontre. Modeling and implementing correct, scalable and efficient real-time applications with ObjectGEODE. *1rst Quarter Edition of Real-Time Magazine*, 1996.

[HP96]   Gerard J. Holzmann and Doron Peled. The state of SPIN. In *8th International Conference on Computer Aided Verification*, volume 1102 of *LNCS*, pages 385–389, New Brunswick, NJ, USA, 1996. Springer Verlag.

[HSLL97] Klaus Havelund, Arne Skou, Kim G. Larsen, and Kristian Lund. Formal modelling and analysis of an audio/video protocol: An industrial case study using UPPAAL. In *Proceedings of the 18th IEEE Real-Time Systems Symposium*, pages 2–13, 1997.

[LGT98]   Agnès Lanusse, Sébastian Gérard, and Francois Terrier. Real-time modelling with UML: The ACCORD approach. In *UML '98*, volume 1618 of *LNCS*, pages 287–296. Springer-Verlag, 1998.

[LH99]   Stefan Leue and Gerard Holzmann. v-Promela: A visual, object-oriented language for SPIN. In *Proc. of the 2nd IEEE Intern. Symp. on Object-Oriented Real-Time Distributed Computing*. IEEE Computer Society Press, 1999.

[LPW95]   Kim G. Larsen, Paul Pettersson, and Wang Yi. Diagnostic model-checking for real-time systems. In *Proc. of the 4th DIMACS Workshop on Verification and Control of Hybrid Systems*, volume 1066 of *LNCS*, pages 575–586. Springer-Verlag, 1995.

[LPW97]   Kim G. Larsen, Paul Pettersson, and Wang Yi. UPPAAL in a nutshell. *Intern. Journal on Software Tools for Technology Transfer*, 1(1+2), 1997.

[SGW94]   Bran Selic, Garth Gullekson, and Paul T. Ward. *Real-Time Object-Oriented Modeling*. John Wiley & Sons, 1994.

[SvG98]   J. Seemann and J. Wolff von Gudenberg. Extension of UML Sequence Diagrams for real-time systems. In *UML '98*, volume 1618 of *LNCS*, pages 225–233. Springer-Verlag, 1998.

Timing Analysis of UML Sequence Diagrams

Xuandong Li[1,2]* and Johan Lilius[1]

[1] Turku Centre for Computer Science (TUCS)
Department of Computer Science, Åbo Akademi University
Lemminkäisenkatu 14, FIN-20520 Turku, Finland
{xuandong.li, Johan.Lilius}@abo.fi
[2] State Key Laboratory of Novel Software Technology
Department of Computer Science and Technology
Nanjing University, Nanjing
Jiangsu, P.R.China 210093
lxd@nju.edu.cn

Abstract. For real-time systems, UML sequence diagrams describe interaction among objects, which show the scenarios of system behaviour. In this paper, we give the solution for timing analysis of simple UML sequence diagrams which describe exactly one scenario without any alternatives and loops, and develop an algorithm for checking the compositions of UML sequence diagrams, which describe multiple scenarios, for timing consistency.

1 Introduction

The Unified Modeling Language (UML) is a general-purpose visual modeling language that is designed to specify, visualize, construct and document the artifacts of software systems [4,9]. UML provides a number of diagrams to describe particular aspects of software artifacts. These diagrams can be classified depending on whether they are intended to describe structural or behavioral aspects of systems. UML sequence diagrams describe behavioral aspects of systems. They describe a collaboration of interacting objects, where the interactions are exchanges of messages, and provide a scenario-based specification which offers an intuitive and visual way of describing design requirements. Such specifications focus on message exchanges among communicating entities in real-time and distributed systems. Like any other aspect of the specification and design process, the specifications in UML sequence diagrams are susceptible to errors, and their analysis is necessary.

France et al. [5] already discuss the need of formal semantics for UML. In this paper we are specifically interested in formalizing and analyzing UML sequence diagrams with timing constraints. In previous work [1-3,7], timing analysis of UML sequence diagrams and of message sequence charts has been restricted to

---

* Partly supported by National Natural Science Foundation of China and 863 High Technology Research Development Project.

timing constraints that consist of linear inequalities on the occurrence times of events.

In this paper we make two extensions to this work:

- first we consider more general and expressive timing constraints and show that we can reduce the problem of timing consistency checking to linear programming, and
- secondly we propose an extension to UML sequence diagrams for describing multiple scenarios, and we give an algorithm that checks timing consistency of such diagram compositions..

The paper is organised as follows. In next section, we give a simple introduction on UML sequence diagrams. Section 3 gives a solution for timing analysis of simple UML sequence diagrams by linear programming. An algorithm is developed in section 4 for checking the compositions of UML sequence diagrams for timing consistency. The last section discusses the related work and contains some conclusion.

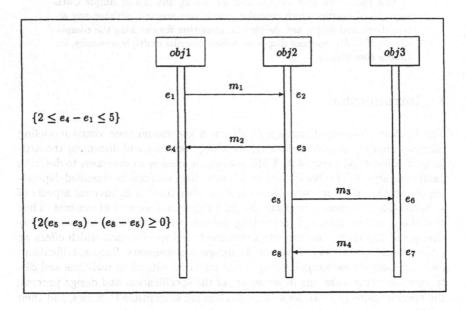

Fig. 1. A simple UML sequence diagram

2 UML Sequence Diagrams

A UML sequence diagram describes an interaction, which is a set of messages exchanged among objects within a collaboration to effect a desired operation or result. Its focus is on the temporal order of the message flow. A UML sequence diagram has two dimensions: the vertical dimension represents time, and the horizontal dimension represents different objects. Each object is assigned a column, the messages are shown as horizontal, labeled arrows. Here we just consider simple UML sequence diagrams which describe exactly one scenario without any alternatives and loops. In section 4, we will consider the compositions of UML sequence diagrams, which include loops and alternatives and describe multiple scenarios. For example, a simple UML diagram is depicted in Figure 1.

By events we mean message sending, message receiving, creating an object, or deleting an object. Each event is given a name which represents its occurrence time. So, timing constraints can be described by boolean expressions on event names. Here we let any timing constraint be of the form

$$a \le c_0(e_0 - e_0') + c_1(e_1 - e_1') + \ldots + c_n(e_n - e_n') \le b,$$

where $e_0, e_0', e_1, e_1' \ldots, e_n, e_n'$ are event names, a, b and $c_0, c_1, \ldots, c_n$ are real numbers (b may be ∞). For example, in the simple UML diagram depicted in Figure 1, the boolean expression $2 \le e_4 - e_1 \le 5$ represents the timing constraint that the separation in time between receiving message m_2 and sending message m_1 is between 2 and 5 time units. Furthermore, if we require that the separation in time between receiving the message m_4 and sending the message m_3 is not greater than two times the one between sending the message m_3 and sending the message m_2, we can describe the requirement by the timing constraint $2(e_5 - e_3) - (e_8 - e_5) \ge 0$.

3 Timing Analysis of UML Sequence Diagrams

For analyzing UML sequence diagrams, we formalize UML sequence diagrams as follows.

Definition 1. A UML sequence diagram is a tuple $D = (O, E, V, C)$ where

- O is a finite set of objects;
- E is a finite set of events corresponding to sending a message, receiving a message, creating an object, and deleting an object;
- V is a finite set whose elements are of the form (e, e') where e and e' are in E and $e' \ne e$, which represents a visual order displayed in D;
- C is a set of boolean expressions, which represents the timing constraints enforced on D. □

For example, the simple UML diagram depicted in Figure 1 can be represented by the tuple (O, E, V, C) where

$O = \{obj1, obj2, obj3\}$,
$E = \{e_1, e_2, e_3, e_4, e_5, e_6, e_7, e_8\}$,
$V = \{(e_1, e_2), (e_2, e_3), (e_3, e_4), (e_3, e_5), (e_5, e_6), (e_6, e_7), (e_7, e_8)\}$, and
$C = \{2 \le e_4 - e_1 \le 5,\ 2(e_5 - e_3) - (e_8 - e_5) \ge 0\}$.

We use *event sequences* to represent the untimed behaviour of UML sequence diagrams. Any event sequence is of the form $e_0 \hat{\ } e_1 \hat{\ } \ldots \hat{\ } e_m$, which represents that e_{i+1} takes place after e_i for any i $(0 \leq i \leq m-1)$.

Definition 2. For any UML sequence diagram $D = (O, E, V, C)$, a event sequence $e_0 \hat{\ } e_1 \hat{\ } \ldots \hat{\ } e_m$ is a untimed behaviour of D if and only if the following condition holds:

- all events in E occur in the sequence, and each event occurs only once, i.e. $\{e_0, e_1, \ldots, e_m\} = E$ and $e_i \neq e_j$ for any i, j $(i \neq j, 0 \leq i, j \leq m)$; and
- $e_1, e_2, \ldots, e_m$ satisfy the visual order defined by V, i.e. for any e_i and e_j, if $(e_i, e_j) \in V$, then $0 \leq i < j \leq m$. □

For example, for the simple UML sequence diagram depicted in Figure 1, the event sequence $e_1 \hat{\ } e_2 \hat{\ } e_3 \hat{\ } e_4 \hat{\ } e_5 \hat{\ } e_6 \hat{\ } e_7 \hat{\ } e_8$ represents a untimed behaviour. It is not difficult to give an algorithm to check if there is a untimed behaviour for a given UML sequence diagram. Since we focus on timing analysis of UML sequence diagrams in this paper, we assume that for any UML sequence diagram, there is at least one event sequence expressing its untimed behaviour.

We use *timed event sequences* to represent the behaviour of UML sequence diagrams. A timed event sequence is of the form $(e_0, t_0) \hat{\ } (e_1, t_1) \hat{\ } \ldots \hat{\ } (e_m, t_m)$ where e_i is a event and t_i is a nonnegative real numbers for any i $(0 \leq i \leq m)$, which describes that e_0 takes place t_0 time units after the system starts, then e_1 takes place t_1 time units after e_0 takes place, so on and so forth, at last e_m takes place t_m time units after e_{m-1} takes place. It follows that for any i $(0 \leq i \leq m)$, the occurrence time of e_i is $\sum_{j=0}^{i} t_j$. Let $\tau(\sigma) = \sum_{i=0}^{m} t_i$ for any timed event sequence $\sigma = (e_0, t_0) \hat{\ } (e_1, t_1) \hat{\ } \ldots \hat{\ } (e_m, t_m)$.

Definition 3. A timed event sequence $\sigma = (e_0, t_0) \hat{\ } (e_1, t_1) \hat{\ } \ldots \hat{\ } (e_m, t_m)$ is a behaviour of a UML sequence diagram $D = (O, E, V, C)$ if and only if the following condition holds:

- $e_0 \hat{\ } e_1 \hat{\ } \ldots \hat{\ } e_m$ represents a untimed behaviour of D, and
- $t_0, t_1, \ldots, t_m$ satisfy the timing constraints described by C, i.e. for any boolean expression $a \leq \sum_{i=0}^{n} c_i(f_i - f_i') \leq b$ in C, $a \leq c_0\delta_0 + c_1\delta_1 + \ldots + c_n\delta_n \leq b$ where for each i $(0 \leq i \leq n)$, if $f_i = e_j$ and $f_i' = e_k$, then

$$\delta_i = \begin{cases} t_{k+1} + t_{k+2} + \ldots + t_j & \text{if } j > k \\ -(t_{j+1} + t_{j+2} + \ldots + t_k) & \text{if } j < k \end{cases} .$$

Let $\mathcal{L}(D)$ denote the set of the timed event sequences representing the behaviour of D. □

For example, for the simple UML sequence diagram depicted in Figure 1, the timed event sequence

$$(e_1, 1) \hat{\ } (e_2, 2) \hat{\ } (e_3, 1) \hat{\ } (e_4, 2) \hat{\ } (e_5, 4) \hat{\ } (e_6, 3) \hat{\ } (e_7, 5) \hat{\ } (e_8, 4)$$

represents a behaviour, but the timed event sequence

$$(e_1, 1)\hat{\ }(e_2, 2)\hat{\ }(e_3, 1)\hat{\ }(e_4, 2)\hat{\ }(e_5, 4)\hat{\ }(e_6, 3.5)\hat{\ }(e_7, 5)\hat{\ }(e_8, 4)$$

does not represent any behaviour because it does not satisfy the timing constraint $2(e_5 - e_3) - (e_8 - e_5) \geq 0$.

For describing timing constraints on the occurrence time of events, we introduce a special event ε which represents the start of system. For any UML sequence diagram $D = (O, E, V, C)$ such that $\varepsilon \in E$, any timed event sequence in $\mathcal{L}(D)$ is of the form

$$(\varepsilon, 0)\hat{\ }(e_1, t_1)\hat{\ }(e_2, t_2)\hat{\ }\ldots\hat{\ }(e_m, t_m).$$

Now we consider timing analysis of UML sequence diagrams. A UML sequence diagram D is *timing consistent* if and only if $\mathcal{L}(D) \neq \emptyset$. In [3,2,1], the problem of checking message sequence charts for timing consistency is reduced to computing negative cost cycles and shortest distances in a weighted directed graph by using temporal constraint network techniques. In [7], the same techniques is used for timing consistency analysis of a class of UML sequence diagrams in which any timing constraint is of the form $a \leq e_1 - e_2 \leq b$. In the worst case, the time efficiency of these algorithms for solving the problem is $O(|E|^3)$. In the following, we consider timing analysis of UML sequence diagrams with more general and expressive timing constraints, and show that the problem can be solved by linear programming.

Let $D = (O, E, V, C)$ be a UML sequence diagram, $E = \{e_0, e_1, \ldots, e_m\}$, and t_i represent the occurrence time of e_i for any i ($0 \leq i \leq m$). By Definition 3, for any $(e, e') \in V$, $t_i - t_j \leq 0$ where $e = e_i$ and $e' = e_j$, and for any timing constraint in C

$$a \leq c_0(f_0 - f_0') + c_1(f_1 - f_1') + \ldots + c_n(f_n - f_n') \leq b,$$

$t_0, t_1, \ldots, t_m$ must satisfy

$$a \leq c_0\delta_0 + c_1\delta_1 + \ldots + c_n\delta_n \leq b,$$

where for any i ($0 \leq i \leq n$), if $f_i = e_j$ and $f_i' = e_k$ then $\delta_i = t_j - t_k$, which form a group of linear inequalities on $t_0, t_1, \ldots, t_m$, denoted by $lp(D)$. Hence, the problem of checking D for timing consistency can be solved by checking if the group $lp(D)$ of linear inequalities has no solution, which can be solved by linear programming.

Furthermore, for a UML sequence diagram $D = (O, E, V, C)$, by linear programming we can solve the problem of checking if every timed event sequence representing the behaviour of D satisfies a given timing constraint

$$a \leq c_0(f_0 - f_0') + c_1(f_1 - f_1') + \ldots + c_n(f_n - f_n') \leq b.$$

Let t_i represent the occurrence time of e_i for any i ($0 \leq i \leq m$). The problem can be solved by finding the maximum value of the linear function $\sum_{i=0}^{n} c_i\delta_i$ subject to the linear constraint $lp(D)$ where for any i ($0 \leq i \leq n$), if $f_i = e_j$ and $f_i' = e_k$ then $\delta_i = t_j - t_k$, and checking whether it is not greater than b and smaller than a, which can be solved by linear programming.

4 Checking Compositions of UML Sequence Diagrams for Timing Consistency

A simple UML sequence diagram describes exactly one scenario. For describing multiple scenarios and specifying real-time systems, we need to consider the compositions of UML sequence diagrams. At the moment the UML standard does not define compositions of UML sequence diagrams. Here we suggest to introduce high-level graphs, which are similar to *High-Level* MSCs defined in the message sequence chart standard [6], to describe compositions of UML sequence diagrams, and develop an algorithm for checking them for timing consistency.

4.1 Compositions of UML Sequence Diagrams

The compositions of UML sequence diagrams can be described by a hierarchical graph. For example, Figure 2 shows a composition of UML sequence diagrams, which describes a simple connection establishment protocol in a telecommunication system exampled in [2]. In Figure 2, there are three simple UML diagrams (D_1, D_2, D_3) and a high-level graph (D_4) which describes the composition of these simple UML sequence diagrams. $D1$ describes a connection request, $D2$ describes the successful establishment of the connection, and $D3$ describes an unsuccessful establishment of the connection. $obj1$ is a service provider, $obj2$ is a local protocol machine, and $obj3$ is a remote protocol machine. D_4 describes the composition of D_1, D_2, and D_3: the iterating branch describes a repeated request to establish the connection, while the non-iterating branch describes a successful connection establishment. For describing the timing constraints enforced on between two events in different UML sequence diagrams, a set of boolean expressions of the form $a \le e - e' \le b$ can be used as a complement.

Definition 4. A composition of UML sequence diagrams (CUD) is a tuple $S = (U, N, succ, ref, M)$ where

- U is a finite set of simple UML sequence diagrams satisfying that for any $D = (O, E, V, C) \in U$ and $D' = (O', E', V', C') \in U$, if $D \ne D'$, then $E \cap E' = \emptyset$;
- $N = \{\top\} \cup I \cup \{\bot\}$ is a finite set of nodes partitioned into the three sets: singleton-set of *start* node, *intermediate* nodes, and singleton-set of *end* node, respectively;
- $succ \subset N \times N$ is the relation which reflects the connectivity of the nodes in N such that any node in N is reachable from the start node and that the end node is reachable from any node in N;
- $ref : I \mapsto B$ is a function that maps each intermediate node to UML sequence diagram in U; and
- M is a finite set of timing constraints of the form $a \le e - e' \le b$ where e and e' occur in different simple UML sequence diagrams and $0 \le a \le b$ (b may be ∞). □

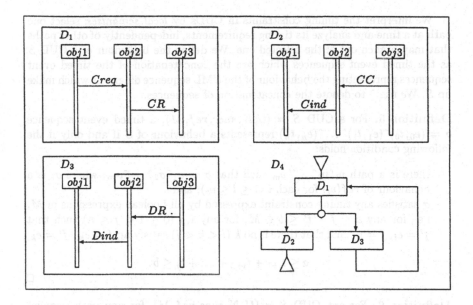

Fig. 2. A composition of UML sequence diagrams

For a CUD $S = (U, N, succ, ref, M)$, a *path segment* is a sequence of intermediate nodes $v_1{}^\smallfrown v_2{}^\smallfrown \ldots {}^\smallfrown v_n$ satisfying that $(v_{i-1}, v_i) \in succ$ for any i $(2 \le i \le n)$. A path segment is called *simple* if all its nodes are distinct. A *path* is a path segment $v_1{}^\smallfrown v_2{}^\smallfrown \ldots {}^\smallfrown v_n$ such that $(\top, v_1) \in succ$ and $(v_n, \bot) \in succ$. A *simple* path is a path which is a simple path segment. For a simple path segment $v_1{}^\smallfrown v_2{}^\smallfrown \ldots {}^\smallfrown v_n$ such that $(\top, v_1) \in succ$, if there is v_i $(1 \le i \le n)$ such that $(v_n, v_i) \in succ$, then the sequence $v_i{}^\smallfrown v_{i+1}{}^\smallfrown \ldots {}^\smallfrown v_n$ is a *loop* and v_i is a *loop-start node*.

To avoid the ambiguity in interpreting timing constraints related to loops, timing constraints are not allowed to combine event occurrences inside and outside of a loop, i.e., for a CUD $S = (U, N, succ, ref, M)$, all timing constraints in M of the form $a \le e \cdot e' \le b$ must satisfy the following condition:

- for any loop $v_1{}^\smallfrown v_2{}^\smallfrown \ldots {}^\smallfrown v_m$, if e occurs in $ref(v_i)$ $(1 \le i \le m)$ and e' does not occur in any $ref(v_j)$ $(1 \le j < i)$, then there is no simple path segment $v_1'{}^\smallfrown v_2'{}^\smallfrown \ldots {}^\smallfrown v_n'$ such that e' occurs in $ref(v_1')$, e does not occur in any $ref(v_k')$ $(1 \le k \le n)$, and that $v_n' = v_1$;
- for any loop $v_1{}^\smallfrown v_2{}^\smallfrown \ldots {}^\smallfrown v_m$, if e' occurs in $ref(v_i)$ $(1 \le i \le m)$ and e does not occur in any $ref(v_j)$ $(i < j \le m)$, then there is no simple path segment $v_1'{}^\smallfrown v_2'{}^\smallfrown \ldots {}^\smallfrown v_n'$ such that $v_1 = v_1'$, e occurs in $ref(v_n')$, and that e' does not occur in any $ref(v_k')$ $(1 \le k \le n)$; and
- for any loop $v_1{}^\smallfrown v_2{}^\smallfrown \ldots {}^\smallfrown v_m$, if e occurs in $ref(v_i)$ and e' occurs in $ref(v_j)$, then $1 \le j < i \le m$.

We interpret the timing constraints in CUDs by *local semantics*: select one path at a time and analyze its timing requirements, independently of other paths that may branch out of the selected one. We define the behaviour of a CUD S as the timed event sequences which are the concatenation of the timed event sequences representing the behaviour of the UML sequence diagrams which make up S. We use ^ to denote the concatenation of sequences.

Definition 5. For a CUD $S = (U, N, succ, ref, M)$, a timed event sequence $\sigma = (e_0, t_0)^\wedge(e_1, t_1)^\wedge \ldots ^\wedge(e_n, t_n)$ represents a behaviour of S if and only if the following condition holds:

- there is a path $v_1{}^\wedge v_2{}^\wedge \ldots {}^\wedge v_m$ such that $\sigma = \sigma_1{}^\wedge\sigma_2{}^\wedge \ldots {}^\wedge\sigma_m$, where σ_i is a behaviour of $ref(v_i)$ for each i $(1 \le i \le m)$; and
- σ satisfies any timing constraint expressed by all boolean expressions in M, i.e., for any $a \le f - f' \le b \in M$, for any i,j $(0 \le i < j \le n)$ such that $f' = e_i$, $f = e_j$, and that there is no k $(i < k < j)$ satisfying $f = e_k \vee f' = e_k$,

$$a \le t_{i+1} + t_{i+2} + \ldots + t_j \le b.$$

□

Definition 6. For any CUD $S = (U, N, succ, ref, M)$, for any path segment $\rho = v_1{}^\wedge v_2{}^\wedge \ldots {}^\wedge v_m$, let $\mathcal{L}(\rho)$ be the set of all timed event sequences which are of the form $\sigma = (e_0, t_0)^\wedge(e_1, t_1)^\wedge \ldots ^\wedge(e_n, t_n)$ and satisfy that

- $\sigma = \sigma_1{}^\wedge\sigma_2{}^\wedge \ldots {}^\wedge\sigma_m$, where σ_i is a behaviour of $ref(v_i)$ for each i $(1 \le i \le m)$; and
- σ satisfies all timing constraints expressed by all boolean expressions in M, i.e., for any $a \le f - f' \le b \in M$, for any i,j $(0 \le i < j \le n)$ such that $f' = e_i$, $f = e_j$, and that there is no k $(i < k < j)$ satisfying $f = e_k \vee f' = e_k$,

$$a \le t_{i+1} + t_{i+2} + \ldots + t_j \le b.$$

□

4.2 An algorithm for checking timing consistency

For a CUD S, a path ρ is timing consistent if and only if $\mathcal{L}(\rho) \ne \emptyset$. A CUD S is timing consistent if and only if all paths are timing consistent. In the following, we develop an algorithm to check CUDs for timing consistency.

Let $S = (U, N, succ, ref, M)$ be a CUD, and $\rho = v_1{}^\wedge v_2{}^\wedge \ldots {}^\wedge v_m$ be a finite path. Suppose that $ref(v_i) = (O_i, E_i, V_i, C_i)$ for any i $(1 \le i \le m)$. From Definition 3 and 6, it follows that all timed event sequence $\sigma \in \mathcal{L}(\rho)$ are of the form $\sigma_1{}^\wedge\sigma_2{}^\wedge \ldots {}^\wedge\sigma_m$ where $\sigma_i = (e_{i1}, t_{i1})^\wedge(e_{i2}, t_{i2})^\wedge \ldots ^\wedge(e_{in_i}, t_{in_i})$ for all i $(1 \le i \le m)$. Furthermore, all t_{ij} $(1 \le i \le m, 1 \le j \le n_i)$ must satisfy all timing constraints in M and C_i $(1 \le i \le m)$. This set of timing constraints forms a group of linear inequalities. Thus, we can check if $\mathcal{L}(\rho) = \emptyset$ by checking if the group of linear inequalities has no solution, which can be solved by linear programming. So, for a finite path, we can reduce the problem into a linear programming problem. However we know that for a CUD, there could be infinite paths and the number of paths could be infinite. So we need a way to reduce the problem to finite set of finite paths.

Let $S = (U, N, succ, ref, M)$ be a CUD, and $\rho = v_1 \hat{\ } v_2 \hat{\ } \ldots \hat{\ } v_m$ be a loop. If the following condition holds:

- for any $ref(v_i) = (O_i, E_i, V_i, C_i)$ $(1 \leq i \leq m)$, any timing constraint in C_i of the form $a \leq \sum_{j=0}^{n} c_j(e_j - e'_j) \leq b$ is such that $a \leq 0$ and $b \geq 0$; and
- for any timing constraint $a \leq e - e' \leq b \in M$, if e occurs in $ref(v_j)$ and e' occurs in $ref(v_k)$ $(1 \leq k < j \leq m)$, then $a = 0$ and $b \geq 0$;

then ρ is a *unbounded* loop; otherwise ρ is a *bounded* loop. Notice that for any loop ρ of the form $v_1 \hat{\ } v_2 \hat{\ } \ldots \hat{\ } v_m$,

- if ρ is a unbounded loop, then there is $\sigma \in \mathcal{L}(\rho)$ such that

$$\sigma = (e_1, 0) \hat{\ } (e_2, 0) \hat{\ } \ldots \hat{\ } (e_n, 0),$$

i.e. there is $\sigma \in \mathcal{L}(\rho)$ such that $\tau(\sigma) = 0$;
- if ρ is a bounded loop, then there is no $\sigma \in \mathcal{L}(\rho)$ such that

$$\sigma = (e_1, 0) \hat{\ } (e_2, 0) \hat{\ } \ldots \hat{\ } (e_n, 0),$$

i.e. any $\sigma \in \mathcal{L}(\rho)$ is such that $\tau(\sigma) > 0$.

Let $S = (U, N, succ, ref, M)$ be a CUD, ρ be a bounded loop. For any timing constraint $a \leq e - e' \leq b$ in M where $b \neq \infty$, we say that it *constrains* ρ if and only if there is a path segment of the form

$$v_1 \hat{\ } v_2 \hat{\ } \ldots \hat{\ } v_{i-1} \hat{\ } v_i \hat{\ } v_{i+1} \hat{\ } \ldots \hat{\ } v_m$$

such that

- ρ starts from v_i,
- e' occurs in $ref(v_1)$ and e occurs in $ref(v_m)$,
- e and e' do not occur in ρ and in any v_l $(2 \leq l \leq m-1)$, and
- all v_j $(1 \leq j \leq i)$ are distinct and all v_k $(i \leq k \leq m)$ are distinct.

Theorem 1. A CUD $S = (U, N, succ, ref, M)$ is timing consistent if and only if the following condition holds:

- any loop ρ is such that $\mathcal{L}(\rho) \neq \emptyset$,
- any simple path is timing consistent, and
- any timing constraint $a \leq e - e' \leq b$ in M such that $b \neq \infty$ does not constrain any bounded loop.

Proof. The details of the proof is presented in the appendix. □

```
currentpath := ⟨T⟩; loopset := ∅;
repeat
    node := the last node of currentpath;
    if node has no new successive node
    then delete the last node of currentpath
    else begin
            node := a new successive node of node;
            if node is in currentpath then
              begin
                if the loop ρ is such that L(ρ) = ∅ return false
                else put the loop into loopset;
              end;
            if node ≠ ⊥ and node is not in currentpath
            then append node to currentpath;
         end
until currentpath = ⟨⟩;
```

```
currentpath := ⟨T⟩;
repeat
    node := the last node of currentpath;
    if node has no new successive node
    then delete the last node of currentpath
    else begin
            node := a new successive node of node;
            if (node, ⊥) ∈ succ then
              begin
                if the path corresponding to currentpath
                   is not timing consistent
                then return false;
              end;
            if node ≠ ⊥ and node is not in currentpath
            then append node to currentpath;
         end
until currentpath = ⟨⟩;
```

check if any timing constraint $a \leq e - e' \leq b$ in M such that $b \neq \infty$ does not constrain any bounded loop. The algorithm is shown in Figure 4.

return true.

Fig. 3. Algorithm for checking CUDs for timing consistency

671

```
for each timing constraint a ≤ e − e′ ≤ b (b ≠ ∞) in M do
 begin
   node := the node including e′; currentpath1 := ⟨node⟩;
   repeat
     node := the last node of currentpath1;
     if node has no new successive node
     then delete the last node of currentpath1
     else begin
         node := a new successive node of node;
         if there is a bounded loop in loopset starting from node
             and e and e′ do not occur in the loop
         then
               currentpath2 := ⟨node⟩;
               repeat
                 node := the last node of currentpath2;
                 if node has no new successive node
                 then delete the last node of currentpath2
                 else begin
                     node := a new successive node of node;
                     if e is in node then return false
                     if node ≠ ⊥, e′ is not in node, and
                         node is not in currentpath2
                     then append node to currentpath2;
                     end
               until currentpath2 = ⟨⟩;

         if node ≠ ⊥, e is not in node, and node is not in currentpath
         then append node to currentpath;
         end
   until currentpath1 = ⟨⟩;
 end
```

Fig. 4. Algorithm checking if any timing constraint does not constrain any bound loop

Based on Theorem 1, for a given CUD $S = (U, N, succ, ref, M)$, we can give an algorithm to check it for timing consistency (c.f. Figures 3 and 4). The main data structures in the algorithm include three lists of nodes *currentpath*, *currentpath1*, and *currentpath2* which are used to record the current paths, and a set *loopset* of loops which is used to record all loops. The algorithm consists of three steps. First, we find all loops and check if any loop ρ is such that $\mathcal{L}(\rho) \neq \emptyset$. Then, we traverse all simple paths and check if they are timing consistent. Last, we check if there are timing constraints $a \leq e - e' \leq b$ in M with $b \neq \infty$ that do not constrain any bounded loops. This step is done with a nested depth-first search. We start from the node that includes e' and look for a simple path segment ending at a loop-start node *node* from which a bounded loop not including both e and e' starts. If we find one, we start a new depth-first search looking for a simple path segment from *node* to a node that includes e.

The algorithm is based on depth-first search. The space consumption is proportional to the size of the longest path in the CUD. In the algorithm, we need to solve linear programs. Linear programming is well studied, and can be solved with a polynomial-time algorithm in general. Indeed many softwares packages have been developed to efficiently find solutions for linear programs. In the algorithm we need to solve a linear program when we check if a loop ρ is such that $\mathcal{L}(\rho) \neq \emptyset$, and when we check if a path corresponding to *currentpath* is timing consistent. So the number of the linear programs we need to solve equals the number of all loops and simple paths in the CUD.

5   Conclusion

In this paper we study timing consistency of UML sequence diagrams. First we consider more general and expressive timing constraints and show that the problem of timing consistency checking can be reduced to linear programming. Secondly we propose an extension to UML sequence diagrams for describing multiple scenarios, and give an algorithm that checks timing consistency.

In [7], UML sequence diagrams are extended for real-time systems with loops, and timing consistency analysis is considered: timing constraints are of simple form $a \leq e - e' \leq b$, and are not over any loop. Instead in this paper more general and expressive timing constraints are considered, and timing constraints are allowed to be over loops so that the compositions of UML sequence diagrams are much more complicated to check. In [1], some algorithms for analyzing message sequence charts with interval delays are presented, and a corresponding tool is described. In [3,2], this timing analysis is extended to MSC specifications, which are compositions of message sequence charts. However, the problem of analyzing MSC specifications for timing consistency is not solved completely there because just a sufficient condition for timing consistency is given, which is not enough to develop an algorithm to analyze MSC specifications for timing consistency. The problem of checking MSC specifications for timing consistency has been solved completely by us in [8]. Compared with that work, this paper makes two new contributions: First, in this paper the approach is algorithmic and leads itself to direct implementation, and secondly the compositions of UML sequence diagrams that we consider here are more complex to check because boolean expressions on event names are used to describe the timing constraints among UML sequence diagrams instead of just between split time events like in MSC specifications.

We are implementing the algorithms into vUML, which is a tool for verifying UML designs being developed at TUCS.

References

1. R. Alur, G.J. Holzmann, D. Peled. An Analyzer for Message Sequence Charts. In *Software-Concepts and Tools* (1996) 17: 70-77.
2. H. Ben-Abdallah and S. Leue. Expressing and analyzing timing constraints in message sequence chart specifications. Technical Report 97-04, Department of Electrical & Computer Engineering, University of Waterloo.
3. H. Ben-Abdallah and S. Leue. Timing Constraints in Message Sequence Chart Specifications. In *Formal Description Techniques X*, Proceedings of the Tenth International Conference on Formal Description Techniques FORTE/PSTV'97, Osaka, Japan, November 1997, Chapman & Hall.
4. Grady Booch and James Rumbaugh and Ivar Jacobson. *The Unified Modeling Language User Guide*, Addison-Wesley, 1998.
5. R. France and A. Evans and K. Lano and B. Rumpe. The UML as a formal modeling notation. In *Computer Standards & Interfaces*, 19:325-334,1998.
6. ITU-T. Recommendation Z.120. ITU - Telecommunication Standardization Sector, Geneva, Switzerland, May 1996.
7. J. Seemann, J. WvG. Extension of UML Sequence Diagrams for Real-Time Systems. In *Proc. International UML Workshop*, Lecture Notes in Computer Science, Springer, 1998.
8. Xuandong Li, Johan Lilius. Timing Aanlysis of Message Sequence Charts. TUCS Technical Report 255, Turku Centre for Computer Science, Finland, March 1999.
9. J. Rumbaugh and I. Jacobson and G. Booch. *The Unified Modeling Language Reference Guide*, Addison-Wesley, 1998.

A   Proof of Theorem

Theorem 1. A CUD $S = (U, N, succ, ref, M)$ is timing consistent if and only if the following condition holds:

- any loop ρ is such that $\mathcal{L}(\rho) \neq \emptyset$,
- any simple path is timing consistent, and
- any timing constraint $a \leq e-e' \leq b$ in M such that $b \neq \infty$ does not constrain any bounded loop.

Proof. One half of the claim, if S is timing consistent then the condition given in the theorem holds, can be proved as follows. Since S is timing consistent, any simple path is timing consistent. Suppose that there is a loop $\rho = v_1\hat{\ }v_2\hat{\ }\ldots\hat{\ }v_m$ such that $\mathcal{L}(\rho) = \emptyset$. Since v_1 is reachable from the start node and the end node is reachable from v_1, there is a path ρ' of the form $\rho_1\,\hat{}\,\rho\,\hat{}\,\rho_2$. Since $\mathcal{L}(\rho) = \emptyset$, $\mathcal{L}(\rho') = \emptyset$, which contradicts that S is timing consistent. Thus, any loop ρ is such that $\mathcal{L}(\rho) \neq \emptyset$. Suppose that there is a timing constraint $a \leq e - e' \leq b$ $(b \neq \infty)$ in M which constrains a bounded loop ρ. It follows that there is a path segment of the form

$$v_1\,\hat{}\,v_2\,\hat{}\ldots\hat{}\,v_{i-1}\,\hat{}\,v_i\,\hat{}\,v_{i+1}\,\hat{}\ldots\hat{}\,v_m$$

where ρ starts from v_i, e' occurs in $ref(v_1)$ and e occurs in $ref(v_m)$, e and e' do not occur in ρ and in any v_l $(2 \leq l \leq m - 1)$, and all v_j $(1 \leq j \leq i)$ are distinct and all v_k $(i \leq k \leq m)$ are distinct. By repeating ρ many times, we can construct a path segment ρ' of the form

$$v_1\,\hat{}\,v_2\,\hat{}\ldots\hat{}\,v_{i-1}\,\hat{}\,\underbrace{\rho\,\hat{}\,\rho\,\hat{}\ldots\hat{}\,\rho}_{k}\,\hat{}\,v_{i+1}\,\hat{}\ldots\hat{}\,v_m.$$

Since ρ is a bounded loop, any $\sigma \in \tau(\sigma) > 0$. It follows that for any given real number c, there is k such that for any

$$\sigma \in \mathcal{L}(\underbrace{\rho\,\hat{}\,\rho\,\hat{}\ldots\hat{}\,\rho}_{k}),$$

$\tau(\sigma) > c$. Since $b \neq \infty$, $a \leq e - e' \leq b$ is not satisfied by ρ', i.e. $\mathcal{L}(\rho') = \emptyset$. Since v_1 is reachable from the start node and the end node is reachable from v_m, there is a path ρ'' of the form $\rho_1\,\hat{}\,\rho'\,\hat{}\,\rho_2$. Since $\mathcal{L}(\rho') = \emptyset$, $\mathcal{L}(\rho'') = \emptyset$, which is contradicts that S is timing consistent. Thus, any timing constraint $a \leq e - e' \leq b$ $(b \neq \infty)$ does not constrain any bounded loop.

The other half of the claim, if the condition given in the theorem holds then S is timing consistent, can be proved as follows. The claim can follow from that if the condition in the theorem holds, then any path ρ is such that $\mathcal{L}(\rho) \neq \emptyset$. Let ρ be a path. If ρ is a simple path, then by the condition given in the theorem, $\mathcal{L}(\rho) \neq \emptyset$. If ρ is not a simple path, then there are some repetitions of loops in ρ. Suppose that ρ is of the form

$$v_1\,\hat{}\,v_2\,\hat{}\ldots\hat{}\,v_m\,\hat{}\,\rho_1\,\hat{}\,v_1'\,\hat{}\,v_2'\,\hat{}\ldots\hat{}\,v_n'$$

where $\rho_1 = v_1''\,\hat{}\,v_2''\,\hat{}\ldots\hat{}\,v_k''$ is a loop and $v_1' = v_1''$. By removing ρ_1 in ρ, we get $\rho' = v_1\,\hat{}\,v_2\,\hat{}\ldots\hat{}\,v_m\,\hat{}\,v_1'\,\hat{}\,v_2'\,\hat{}\ldots\hat{}\,v_n'$, which is still a path. By the condition given in the theorem, $\mathcal{L}(\rho_1) \neq \emptyset$. Suppose that $\mathcal{L}(\rho') \neq \emptyset$. Since $\mathcal{L}(\rho_1) \neq \emptyset$ and any timing constraint $a \leq e - e' \leq b$ $(b \neq \infty)$ in M does not constrain any bounded loop, and since it is not allowed for timing constraints to combine event occurrences inside and outside of a loop, $\mathcal{L}(\rho) \neq \emptyset$. It follows that if $\mathcal{L}(\rho) = \emptyset$, then $\mathcal{L}(\rho') = \emptyset$. Suppose that $\mathcal{L}(\rho) = \emptyset$. It follows that $\mathcal{L}(\rho') = \emptyset$. By removing loop from ρ repeatedly, we can get a simple path which is not timing consistent, which contradicts the condition given in the theorem. Thus, $\mathcal{L}(\rho) \neq \emptyset$. Therefore, we have proved that any path ρ is such that $\mathcal{L}(\rho) \neq \emptyset$, i.e. the claim holds. $\square$

The Normal Object Form:
Bridging the Gap from Models to Code

Christian Bunse and Colin Atkinson

Fraunhofer Institute for Experimental Software Engineering
Technopark II, Sauerwiesen 6
D-67661 Kaiserslautern, Germany
{Christian.Bunse, Colin.Atkinson}@iese.fhg.de

Abstract. The value of graphical modeling within the analysis and design activities of object-oriented development is predicated on the assumption that the resulting models can be mapped correctly, optimally and efficiently into executable (normally textual) code. In practice, however, because of the large potential mismatch in abstraction levels, the mapping of graphical models into code is often one of the weakest and most error prone links in the chain of development steps. This paper describes a practical approach for addressing this problem based upon the definition of a restricted extension of the UML known as the Normal Object Form (NOF). The basic purpose of the NOF is to provide a set of UML modeling concepts which are "semantically close" to those found in object-oriented programming languages. Highly abstract UML models can then be mapped into corresponding executable code by means of a series of semantically small refinement (intra-UML) and translation (extra-UML) translation steps, rather than in one large (often ad hoc) step. This not only increases the chances of a correct and optimal mapping, but also significantly improves the traceability of UML constructs to and from code constructs, with all the associated advantages for maintenance and reuse.

1 Introduction

The success of the Unified Modelling Language reflects a growing consensus in the software industry that "modelling is a good thing." However, models created in the earlier phases of development (e.g. analysis and design) are of little value unless they can be readily mapped into correct and efficient executable forms, which in today's technology means code in high-level object-oriented programming languages. Any problems in the transformation path from models to code not only have a negative impact on the quality of the delivered software system, but also hinder its future maintenance and/or reuse. Furthermore, they reinforce the widely held suspicion that modelling is "just paperwork" without any serious connection to the "real" business of code generation.

Ironically, the very richness and generality of the UML is something of an "achilles heel" in this regard. This is because power without control is dangerous. Developers attempting to use the UML have such a wide selection of modelling concepts at their disposal, ranging from low-level implementation oriented concepts to very high-level abstract concepts, that they can easily lose their way and end up facing a daunting gap to span in order to translate their models into code. Not surprisingly, the wider the semantic gap to be bridged, the greater the chance of mappings which are inadequate or incorrect.

It is not the UML per se which is really responsible for addressing this problem, but rather the methods which are intended to support it. However, few if any, of the current UML-oriented methods pay much attention to this issue. The books which define the leading methods at best include a chapter discussing "implementation issues", usually in a general and ad hoc way [10, 3, 6]. Similarly, books on computer languages rarely spend more than a chapter discussing how features of the language relate to modelling concepts such as those in the UML [14, 15, 9]. As a result, the mapping of graphical models into code is one of the most neglected links in modern software development processes.

Perhaps one reason for this is the widely held view that case tools have already solved this problem. Widely advertised capabilities such as "round-trip engineering" give the impression that at the press of a button a case tool can translate a rich UML model into a complete, optimal, executable program. However, although the code creation capabilities of modern case tools can be helpful when used appropriately, no tool is yet capable of handling all the nuances and trade-offs involved in creating an optimal and efficient implementation of UML models. The "one-size-fits-all" mapping schemes found in most case tool usually end up being incomplete and/or suboptimal for most situations.

Metamodelling approaches such as 'Design by Translation' [12, 13] or metamodelling tools such as Platinum Technology's 'Paradigm Plus' represent a step forward over simple case tools since they enable the mapping scheme for each metamodel concept to be defined independently by the user. However, this technology still assumes that the user "knows" what mapping scheme to use. Without an underlying theory or methodology for mapping UML models into code, a user can just as easily define an inappropriate mapping in such a tool as when applying the mapping by hand.

These problems all point to the need for a well-defined and flexible methodology for supporting the translation of UML models into executable code in a way that takes into account prevailing non-functional requirements. This paper describes an attempt to provide such methodological support. After first outlining the principles underlying the approach in section 2, the bulk of the paper in section 3 describes the Normal Object Form, a restricted extension of the UML which aims to encapsulate and enhance the implementation-oriented elements of the notation. This is followed in section 4 by a description of an accompanying methodology to support the flexible and optimal implementation of NOF elements based on a modified form of design pattern.

2 Refinement versus Translation

There are a large number of different object-oriented methods to choose from when considering the use of the UML in a software development project, with an equally large number of different processes and modelling approaches. Despite their prima-facie differences, however, they all share the same underlying assumption that high-level "analysis" models will be developed during the early phases of development, and lower level executable code (i.e. the implementation) will result from the later phases.

The terms "earlier" and "later" may no longer apply in a strict waterfall sense due to the prevalence of incremental processes, but the basic idea of progress depending on abstract models being transformed to concrete code is more or less universal:

Two basic transformations take place in turning high-level abstract models into concrete executable code - *refinement* and *translation*. Refinement is a relationship between two descriptions of the same thing, with one, the *abstraction*, containing less information than the other, the *realization*. In the context of software development, refinement can be viewed as a relationship between two descriptions of a software entity, the abstraction or high-level description, and the realization or low-level description closer to implementation. Translation, in contrast, is the description of a given phenomenon in two different ways, but at the same level of abstraction. A classic example from every day life is the translation of a piece of text from one natural language (say English) to another (say German). If done correctly, the information content (i.e. the meaning) of both versions should be identical. In the context of software development, translation results in the description of a given software entity in two different ways (e.g. graphical and textual), but with the same information content.

2.1 Separating Concerns in Implementation

Although from a conceptual point of view these two ideas are clearly distinct, they are rarely distinguished in practical software development methods. On the contrary, most methods bundle them together into a series of "shopping list" style implementation guidelines which attempt to describe an "implementation" for each distinct modelling feature on a case-by-case basis. However, this approach leads to various problems -

1. *large semantic gap* - for high level modelling constructs with no direct programming language counterpart, trying to bridge the semantic gap to code in one large jump significantly increases the chances of errors or poor mappings,
2. *undocumented decisions* - even if a reasonable mapping is obtained when performing the mapping in one step, the decisions that it embodies are undocumented and thus unavailable for future maintainers of the system,
3. *loss of reuse opportunities* - although the refinement concepts applied within such a "single step" mapping are often applicable to various languages, bundling them up with language specific translations makes them unavailable for reuse in mappings to other languages,
4. *replication of information* - the previous problem (3) implies that implementation guidelines targeted to different languages often replicate language independent refinement concepts.

A concrete example of the problem in a UML context is shown in Fig. 1, which depicts a UML association being mapped into C++. Such an association in the UML conveys a limited amount of information. It indicates only that instances of the two classes may be linked in some way at run-time, and says nothing about the precise nature of the links, nor how they should be implemented in a programming language such as C++. As a result there are usually many ways to implement the association, depending on

678

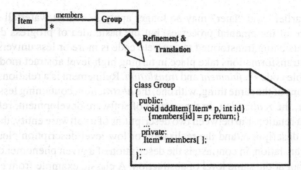

Fig. 1. Example 1: Refinement & Translation

the exact properties required. In the case of a high-level association such as this, some of the major implementation considerations are which class should be the client and which the server, whether the server should be a data member of the client or just a local method variable, whether the server should be embedded with the client or just referenced by it and so on. By making all these decisions in one step, and bundling them all together within the resulting code, they remain implicit and undocumented. This not only makes the code hard to check for correctness (i.e., is this the correct implementation of the association giving the prevailing non-functional requirements), but also difficult to understand. This, in turn, causes problems during maintenance when one cannot easily identify why the association was implemented in a particular way and thus what changes are acceptable.

Following the time honored principle of "separation of concerns" we believe the only way to seriously address this problem is to cleanly decouple refinement and translation within the software development process. Instead of bundling both transformations together into a single step they should be performed independently. According to this scheme analysis and design models would be developed in the usual way as before, but before translation they would first be refined within the UML to a lower level of detail. Only when models have been obtained at the appropriate level would they be translated directly and straightforwardly into code. Applying this approach to the previous example, we obtain the additional description of the association illustrated in Fig. 2. This is still in the UML but at a lower level of detail. The two most important benefits of this approach are that the refinement relationships are clearly visible as explicit UML constructs, and the size of the individual mappings steps (and thus the likelihood of error) is significantly decreased.

2.2 When to Refine and When to Translate?

The idea of separating refinement from translation in the implementation of high-level modelling concepts would seem to offer numerous advantages, but its practical realization begs one major question - to what level should "high-level" models be refined before they are ready for translation, and how can this level be identified. In other words, what exactly is "high -level". The goal of the Normal Object Form (or NOF) is to provide an answer to this question.

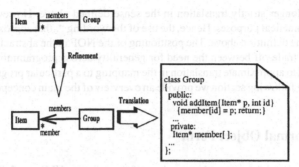

Fig. 2. Example 2: Refinement then Translation

As illustrated in Fig. 3 (a generalization of Fig. 2) the purpose of the NOF is to define a set of UML modelling constructs which are "semantically close" to object-oriented programming features, and which can therefore be mapped into elements of a program in a manner that approximates translation (i.e without a significant change in abstraction level). We call this set the Normal Object Form, or NOF, because in a sense it represents a "normal" form, akin to that used in relational databases, to which UML models must be 'reduced' before translation can begin. On might think of a UML model as being "normalized" by the application of refinement rules to prepare it for translation into code.

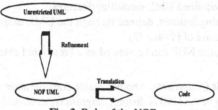

Fig. 3. Role of the NOF

Notice that the definition given above does not mention a particular programming language, but instead refers to "object-oriented programming features". This reveals a second goal of the NOF which is to capture the concepts which are common to the majority of mainstream object-oriented languages, not those specific to a particular language. Only then will the true benefits of separating refinement from translation be available. As experienced programmers are well aware, the underlying concepts of object-oriented programming are basically the same whatever language or implementation vehicle is actually used to apply them. For example, the various implementation options discussed above in the implementation of an association apply to most mainstream object-oriented programming languages. In fact, this core set of concepts lies at the heart of the universal applicability and success of object-oriented design patterns [7, 5]. In a sense, the NOF can be thought of as providing a UML embodiment of the common, core features of object-oriented programming.

This generality has a price, however. By positioning the NOF at a level to capture the common concepts of the main object-oriented programming languages (e.g. Java, C++, Eiffel, Smalltalk, Ada), the mapping of NOF elements to any one particular languages is often no longer strictly translation in the sense defined above. However it is close enough for practical purposes. Hence, the use of the wording "..that approximates translation.." in the definition above. The positioning of the NOF in the abstraction hierarchy represents a trade-off between the need for generality across programming languages, and the need to approximate translation in the mapping to a particular programming language. In the following section we provide an overview of the main concepts in the NOF.

3 The Normal Object Form

At first sight it might appear that a *subset* of the existing UML modelling concepts would be sufficient to satisfy the goal identified previously. Certainly the UML already contains many low-level features that have a very close correspondence to object-oriented language concepts, such as classes, methods, packages, etc. However, it turns out that there are also numerous other fundamental object-oriented programming features which are not represented directly within the UML. For example, the various choices for the implementation of an association, identified in the previous section, are not well supported by current UML modelling constructs. Therefore, in defining the NOF we also found it necessary to add some additional concepts using the UML's in-built extension mechanisms (i.e. stereotypes, tagged values and constraints). In short, the definition of the NOF consists of three distinct parts:
- a subset of the predefined UML modelling features,
- additional modelling features, defined through the UML extension mechanism,
- constraints on the use of (1) and (2).
In a sense therefore, the NOF can be viewed as a "restricted extension" of the UML.

Explicit vs. Implicit Constructs. Another major consideration in the definition of the NOF was how many of the UML's distinct diagram types should be affected? In principle, all of them could be, because they all have a bearing on the properties of the software system, and thus ultimately on the way it should be implemented. In practice, however, it turns out that information in many of the diagram types is usually "folded into" other types as part of the refinement process.

As result, it is useful to distinguish between *explicit* and *implicit* modeling elements. Explicit modelling elements have a directly visible counterpart in object-oriented programming languages, whereas implicit constructs do not. Consider use case diagrams, for example. These have a strong influence on the methods of the classes from which the system will be constructed, but no object-oriented programming language enables use cases to be explicitly represented. The information provided by these diagrams is therefore *implicitly* contained within the source-code but is not explicitly visible. The same is also true for the UML's statechart diagrams. These describe characteristics of a system which, although important, are not usually directly visible in object-oriented programs. The constraints they define are instead indirectly manifest within the instance variables and methods of the class concerned.

Generally speaking, object-oriented programming languages tend to focus on the explicit description of the structural aspects of a system, with the high-level behavioral aspects being more implicit. The information in the UML's dynamic models is therefore usually "folded into" the static structure models through refinement steps, and is not directly "translated" into programming language constructs. In practice, therefore, we find that only two of the main diagram families have a significant role in describing the "as is" implementation of an object-oriented program:

- Static Structure Diagrams (i.e., Class Diagrams and Object Diagrams)
- Implementation Diagrams (i.e., Component Diagrams and Deployment Diagrams)

Eliding Diagrams. Most of the UML modelling elements can be used at different levels of abstraction. For example in the analysis phase an operation can be specified by a simple name whereas during detailed design it can be specified in more detail (e.g., types, parameters, etc.). In general, the UML allows the modeler to decide when and where certain pieces of information are displayed. While this flexibility is valuable during high-level modelling phases, it becomes a problem in the implementation phase when the "as is" implementation needs to be described. From an individual diagram it is not possible to determine whether the absence of specific information is due to the fact that it has been omitted (from the diagram in hand) or because the corresponding decisions remain to be made. A general goal of the NOF is to rule out such ambiguities by making it as clear as possible how included information should be displayed, and by providing well defined mechanisms for indicating when certain information has been omitted. This achieved in the following two ways:

1. *Constraints* - For each diagram type the NOF provides a set of constraints on the usage of its modelling elements. These constraints define the level of detail required for an implementation level description of a software system (e.g., a parameter always needs a type). Thus, for example, when a class appears in a NOF diagram, its precise NOF implementation stereotype must be defined - that is, whether it is an abstract class (<>), a persistent class (<<persistent>>), a template class (<<template>>) or a utility class (<<utility>>). If a class is not marked as belonging to any of these categories in a NOF diagram, this indicates that a decision has been made to implement it as a normal class.
2. *Elided Diagrams* - In practical modelling work, diagrams often become huge or information becomes dispersed over several different diagrams. To avoid diagram clutter, the NOF allows information to be selectively and explicitly omitted. Diagrams in which information is explicitly omitted are called *elided diagrams*. The explicit omission of information is achieved by means of relationship and annotation ellipses [1]. Relationship ellipses indicate that a class or object has additional relationships (e.g., clientship, inheritance, import; each denoted by its own symbol) in which it participates, but which are not shown on the diagram. Annotation ellipses are used to indicate when an annotation only contains partial information.

3.1 Class Diagrams

UML Subset. At its core, the NOF contains those elements of UML class diagrams which embody the fundamental elements of object-oriented programs. In other words, the NOF contains a subset of the UML class diagram constructs. As always, there are two ways to define a subset of an existing set: identifying all the elements of the set which are not in the subset, or identifying all the elements that are in the subset. Unfortunately, in a paper of this size it is not possible to do either completely for the NOF. Therefore, to provide an intuitive understanding of the composition of the NOF Class diagram subset we briefly identify some of the key elements of UML class diagrams which are in the NOF, and some of the key elements which are not.

Naturally the fundamental constructs of object-orientation are part of the NOF. The most fundamental of all is the concept of a class as a descriptor for a set of objects. The NOF supports not only classes in the general sense but also specific class types including abstract, template, and metaclasses because these are explicitly visible in most object-oriented programming languages. Naturally other concrete elements of classes such as attributes and operations are part of the NOF. In addition most languages provide some mechanism by which one class can incorporate structure and behavior from another, more general, classes (i.e. inheritance). This requires the inclusion of inheritance into the NOF, although with certain restrictions on the use of the predefined inheritance stereotypes. Important class diagram constructs which are not in the NOF include:

- *Specialized Compartments*. These compartments are used to show specialized abstract properties of a class (e.g., responsibilities, business rules, etc.) By definition therefore, such compartments play their major role in the analysis phase to help developers understand the domain, but do not play an important role in implementation. Consequently they are not included in the NOF. The only compartments which are acceptable in the NOF are those for attributes, operations and exceptions.
- *Association Class*. Association classes describe an association that is also a class. Although it is stated [11] that an association class is not the same as a class connecting two other classes, no existing object-oriented language supports any other implementation [8] (i.e., a dictionary class is used). Consequently association classes are not part of the NOF.
- *Class-in-state*. Classes with a state machine may have many states. The class-in-state modelling element describes a state that objects of that class can hold. Due to its close relation to activity diagrams, it is another way to accomplish the same goal as dynamic classification. Therefore it is not necessary as a part of the NOF.
- *Trace Dependency*. All dependencies which describe historical connections between elements (e.g., <<trace>>) do not influence the implementation of a system and are therefore not part of the NOF.
- *Derived Elements*. These are not part of the NOF because they are used for the purpose of clarity and do not provide additional semantic information.
- *N-Ary Associations*. N-ary associations are associations between three or more classes. However, just as for association classes no currently existing object-oriented language provides direct support for such associations; they have to be "simulated" using multiple binary associations.

- *Qualifiers.* These are used to partition a set of objects connected with an object via an association. Qualifiers are not part of the NOF for two reasons. First, due to their definition as attributes of an association (see also association class). Second, they are clearly analysis elements, which model an important semantic situation, but do not influence the general strategy for implementing an association.

Additional Diagram Elements. Although the UML is a powerful tool for describing object-oriented software systems it is not possible to describe all properties of programs entirely in the UML subset within the NOF. Certain extensions (i.e., new elements) are needed. The NOF includes legal extensions to the UML defined using the built-in extension mechanism. Most of the UML extensions in the NOF occur in connection with associations. This is because the fundamental implementation variations for associations are not fully supported in the present version of the UML. Although associations can vary in many ways at the analysis and design level, such as in their arity (e.g. binary, ternary, etc.) and their multiplicity (e.g. one-to-one, one-to-many, many-to-many), at the implementation level there are far fewer variations. All inter-object relationships are essentially implemented by the same basic mechanism: one object holding a pointer to, or the value of, another object. Even the implementation of associations by 'Relation Tables' makes use of these mechanisms, by implementing the table as a class in its own rights which routes the communication.

Following ION [1], we call this basic relationship between classes *"clientship"*. Clientship is an asymmetric relationship; the client needs to be aware of the identity of the server class, but the server requires no knowledge of the client class. All clientship relationships are therefore represented with a UML navigation arrow indicating the direction of the client/server relationship. As with all program-level relationships, clientship implies a compilation (and thus static) dependency. In total there are four different, orthogonal properties of clientship relationships, each with two possible values. Each possible value for each property has a corresponding UML association stereotype:

1. *Attached vs. Detached:* One of the most important characteristics of a clientship relationship is whether the client holds a reference to the server or whether it holds the actual state (i.e., the value) of the server. When the client holds a reference to the server the clientship is said to be *detached*, whereas when the client actually holds the state of the server the clientship is said to be *attached*.
2. *Permanent vs. Transient:* Another important characteristic of a clientship relationship is how long the class has visibility of a particular instance of the server class. If the client holds a reference to, or the value of, the server in its main data structure, the clientship is said to be *permanent*. If, on the other hand, the client has visibility of a server object only for the duration of a single method, the clientship is said to be *transient*.
3. *Proper vs. Intimate:* Normally, a client class only has access to the "official", publicly visible methods of the server. This is termed *proper* clientship. Most object-oriented languages also allow client classes to be given privileged access to the server (e.g., the friend-construct in C++). This is termed *intimate* clientship.
4. *Direct vs. Indirect:* The final property of a clientship relationship is whether the client holds visibility of the server, or whether the client relies on a second server for visibility of the first. The first situation is known as *direct* clientship, and the second *indirect* clientship.

Individual clientship relationship between two objects, or their corresponding classes, must make a choice between all four binary attributes. However, showing all four stereotypes, in full, on a clientship arrow would unduly clutter a NOF static structure diagram. Therefore, as well as defining default properties (detached, permanent, proper, direct) the NOF defines stereotypes corresponding to meaningful permutations of the four orthogonal characteristics (see Table 1).

In general there are 16 different possible combinations of the four binary clientship properties, but some of them may not necessarily fit well together from a programming point of view. A typical example is a clientship relationship characterized by 'Attached, Permanent, Proper, Indirect'. Such a combination defines a relationship where the client contains the server and all intervening objects which, in the worst case, may lead to really large objects.

Combination of Characteristics	stereotype	Description
Attached, Permanent, Proper, Direct	<<embedded>>	Client holds value of public parts of the server in main data-structure directly.
Attached, Permanent, Intimitate, Direct	<<private>>	Client holds value of public/private parts of the server in main data-structure directly.
Attached, Transient, Proper, Direct	<<public local>>	Client holds value of public parts of the server for method execution directly.
Attached, Transient, Intimitate, Direct	<<local>>	Client holds value of public/private parts of the server for method execution directly.
Detached, Permanent, Proper, Direct	<<standard>>	Client holds direct reference to public parts of the server in main data-structure.
Detached, Permanent, Proper, Intimitate	<<Dictionary>>	Client holds indirect reference to public parts of the server in main data-structure.
Detached, Permanent, Intimitate, Direct	<<Friendship>>	Client holds direct reference to public/private parts of the server in main data-structure.
Detached, Transient, Proper, Direct	<<Parametric>>	Client holds direct reference to public parts of the server for method execution.
Detached, Transient, Intimitate, Indirect	<<Parametric Dictionary>>	Client holds indirect reference to public parts of the server for method execution.
Detached, Transient, Intimitate, Direct	<<Parametric Friend>>	Client holds direct reference to public/private parts of the server for method execution.

Table 1. Possible Clientship Relationships

Diagram Constraints. The modeling elements within the NOF (i.e., UML subset and additional elements) can be used at different levels of abstraction. A typical example for different levels of abstraction is that of a class. In the analysis phase it is often sufficient to specify a class only by its name whereas in detailed design it has to be specified in more detail (e.g., attributes, operations, etc.). The NOF supports modelling at the implementation level by defining a set of constraints which clearly describe the level of detail needed for an individual element. Table 2 presents a short selection of constraints to give an overview of their nature. These examples clearly indicate how constraints can enforce modelling to the desired level of detail. The first example in Table 2 is typical. In analysis it is sufficient to know what attributes exist for a given class whereas in programming it must be clear if these attributes are public, protected, or private members of that class.

1.	All attributes of a class must have visibility markings. In other words, it must be clear whether they are to be implemented as public, private or protected members (in the case of C++).
2.	A method must have a visibility marking, a list of parameters (if existing), and a return type
3.	A parameter is specified in the following form: name:type=default-value. The default value may be suppressed.
4.	Each class must have at least a constructor and destructor method.
5.	The methods of a class have to be grouped by using one of the following stereotypes <<constructor>>, <<destructor>>, <<update>>, etc.
6.	A clientship relationship has to be augmented with multiplicity markings.
7.	A clientship relationship has to be augmented with roles.

Table 2. Class Diagram Constraints

3.2 Object Diagrams

The basic purpose of an object-oriented program is to create a set of objects which, at run-time will interact in such a way as to satisfy the needs of the users. The UML provides the basic mechanisms needed to describe this aspect of a system's implementation in the form of an object diagram, but not entirely in an appropriate form. This is because the way in which object diagrams are used in analysis is not appropriate for describing the "as is" implementation of a program. In analysis, object diagrams are generally used to depict a *typical* set of named instances and a *typical* set of links in order to provide an illustration of the kind of object structures that are meaningful for the application under development. There is no real sense of precisely when the object structure exists, and from whose perspective it is defined.

In order to describe structural aspects of the "as is" implementation of an object-oriented system it is necessary to show actual object structures at various points in a programs execution, not just *typical* structures which *may* exist. Therefore, the primary difference between object diagrams in the NOF and those in regular UML is the way they are used.

UML subset. As a consequence, virtually all UML features for creating object diagrams are available in the NOF. According to [2] object diagrams typically contain objects as concrete instances of classes and links which represent the instantiation of a relationship between classes. Furthermore they may contain comments and constraints just like other UML diagrams. Since these elements can be used at a level of abstraction near to the implementation of a software system they are included as part of the NOF.

Additional Diagram Elements. In a running object-oriented program the set of objects in existence at any one point in time is constantly changing. Moreover, the absolute

686

names of objects, if they have any meaning, are known only to the run-time system. Individual objects only know about (and are able to communicate with) other objects through the names of their instance variables. Therefore, in order to provide meaningful views of the "as is" implementation of an object-oriented program, mechanisms are needed to describe precisely what set of objects an object diagram is depicting and when.

The predefined UML object diagram features neglect another important aspect of a running system known as the *creation tree*. Apart from the outermost object (or main program) all other objects in a running program have to be created by some other object, and failing to ensure this takes places properly is one of the major sources of errors in object-oriented programs.

To rectify these problems, the NOF makes two main enhancements to the basic object diagram concepts in the UML:
- the first is to identify two distinct kinds of object diagrams - *snapshot* object diagrams which describe a group of objects and their links at a particular instance in time, and *history* object diagrams, which describe the same information accumulated over a period of time.
- the second is to make every object diagram relative to one specific object in the diagram known as the *root*.

These two enhancements are closely related, because the information contained in each kind of object diagram is defined relative to the root object. In the case of a snapshot diagram, the instance of time represented by the diagram corresponds to a particular state of the root object, and in the case of a history diagram, the presented information is accumulated with respect to the root object. More specifically, a history object diagram shows an accumulation of all objects which the root object is linked to during its lifetime.

The NOF defines various stereotypes and constraints to support and enforce this usage of object diagrams as illustrated in Fig. 4. This examples illustrates the set of links which are meaningful for a user of a library who is currently borrowing a book. A user who has no books on loan would have a different set of links.

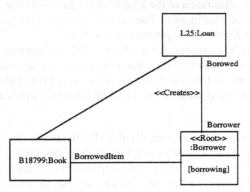

Fig. 4. Snapshot Object Diagram

Figure 4 illustrates most of the important object diagram stereotypes and extensions defined in the NOF. First, it shows the use of the <<*Root*>> stereotype to show that the anonymous object: *Borrower* is the root of this particular object diagram. Second, it illustrates the use of the stereotype <<*Creates*>> to indicate that this object is responsible for generating the *Loan* object *L25*. Although this object has a name, the *Borrower* does not know this name. It is only able to access the *Loan* object through the role name *Borrowed*, which corresponds to the name of the instance variable in the class defining borrowers. Thirdly, the presence of the state definition *[borrowing]* in the body of the object indicates that this diagram is a snapshot diagram depicting the links which *Borrower* has when it is in the *borrowing* state.

Diagram Constraints. The object diagram elements within the NOF (i.e., UML sub-set and additional elements) can be used at different levels of abstraction. Therefore, in order to support the creation of object diagrams which clearly depict the "as is" structure of a software system at the implementation level the NOF contains a set of constraints on the use of these features. Table 3 presents a short selection of constraints to give an overview of their nature. The selection illustrates that the constraint are broadly defined (i.e., cover various aspects). They can range from general constraints such as an object diagram has one root object, to very specific constraints such as the format of attribute definitions.

1.	Each object diagram has one and only one root object
2.	A root object is indicated with the stereotype <<root>>
3.	The attributes of an object are described in the following format: attributename:type = value
4.	An object in a specific state is described as: objectname::classname [statename]
5.	An object diagram containing the object-in-state element is called snapshot diagram
6.	The creation relationship between two object is marked by the stereotype <<creates>>
7.	Role names correspond to the name of an instance variable

Table 3. Object Diagram Constraints

3.3 Implementation Diagrams

As their name implies, implementation diagrams (i.e., component- and deployment diagrams) are designed for the purpose of capturing "as is" implementation details of an object-oriented system. Therefore, they are suitable for inclusion in the NOF unchanged. Component diagrams can be used to describe the individual modules of software capable of independent execution in a distributed environment, such as that provide by CORBA, and deployment diagrams show how these modules are actually deployed in a physical network. The only way in which the NOF affects implementation diagrams is through the definition of a set of constraints describing where and how they should be used.

4 SORT

In order to be of practical value the NOF needs appropriate methodological support. This is the goal of *SORT* (Systematic Object-Oriented Refinement and Translation). *SORT* provides a practical technique for leveraging the NOF, and the concept of refinement/translation separation, by packaging useful refinements and translations [4].

In view of the success of the pattern cataloguing approaches pioneered by Gamma et. al. [7] and Buschmann et. al. [5], SORT refinement and translation guidelines are packaged in a similar style. However, there is a subtle differences between the patterns defined in *SORT* and those of Gamma and Buschmann. Whereas the latter essentially capture good (i.e. useful) object-oriented structures/behaviors, *SORT* patterns capture good (i.e. useful) mappings <u>between</u> object-oriented structures/behaviors.

Two forms of patterns are recognized in SORT: *refinement patterns*, which describe "good" refinements within the UML for reaching structures at the implementation level specified by the NOF, and *translation patterns*, which describe the "good" mapping of UML-NOF models to a specific object-oriented programming language (e.g., C++). The latter are similar to "idioms" [5] in that they are language specific, however, as mentioned above they represent more of a mapping guideline than a useful programming practice. An example of each of the two pattern forms can be found in Fig. 5.

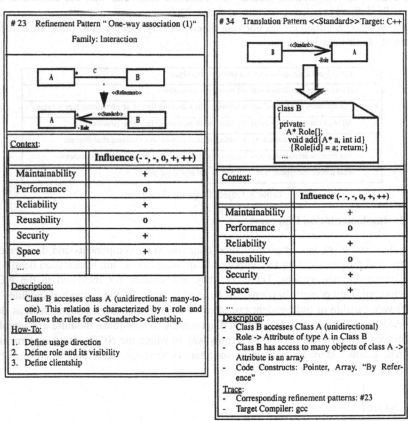

Fig. 5. Pattern Examples

Of course, there is rarely a single pattern which provides the best mapping (refinement or translation) of a given structure under all circumstances. Generally, there are several potential mappings, and the one which is most appropriate in a particular context depends on the associated non-functional requirements (e.g. performance needs, space limitations, reliability etc.). Therefore patterns have to provide a context description which allows a developer to choose the one most suitable for his/her particular needs. Providing such information allows *SORT* to offer a level of context sensitivity which is impossible in automated mapping tools, while at the same time still being reasonably systematic. Developers are told precisely what to refine or translate, how to perform these refinements/translations, and when to perform the relevant activity. At the same time, however, they can tailor the particular refinements and translations in different ways that depend on the relevant context factors. More detailed information on SORT and its patterns can be found in [4].

5 Conclusion

The phenomenal interest in the Unified Modelling Language provides a unique opportunity to increase the amount of modelling work performed in the software development industry, and through this to increase quality standards. However, this chance will be lost if developers are given no effective and practical link between UML diagrams and the executable implementations of their systems.

This paper has described a strategy for addressing this need which is based on three time-honored and fundamental engineering principles. The first is the principle of "separation of concerns". The basic tenet of the approach is to cleanly distinguish and separate two independent issues in transforming high level models into executable code - namely, refinement and translation. The second is the principle of "exploiting commonality". The approach is explicitly aimed at exploiting the common, core concepts of object-oriented languages, and through this to define a set of general refinement patterns which are independent of language idiosyncrasies. The third is the principle of "divide and conquer". By clearly separating and capturing individual refinement and translation steps, the approach divides a single, "semantically large" mapping step into a series of intellectually smaller steps. This not only serves to document the steps, but significantly improves the likelihood that the overall mapping will be correct.

These ideas are embodied within a restricted extension of the UML known as the Normal Object Form, and are supported by a methodology for their practical application known as SORT. One of the main advantages of the SORT approach is that it is independent of the original source of the UML diagrams. In other words, the SORT approach is independent of, and usable with, any of the mainstream UML development methodologies. The ultimate goal of this work is to define sufficient refinement patterns to enable any UML compliant diagram to be normalized into NOF form ready for translation into code.

Another important property of this approach is its "tool friendliness". As discussed in the introduction, by trying to support the single-step implementation of all UML modelling concepts, no matter what their level of abstraction, case tools may actually be doing developers a disservice. This is because they typically have to apply a "one-size-fits-all" mapping strategy which fails to document the rationale for the mapping, and often fails to provide the best mapping for the circumstances in hand. The SORT approach, based on the NOF, promises to improve this situation by enabling tools to concentrate what they do best, namely context independent translation between different representations of the same concept. The subtle, context sensitive aspects of the implementation process can then be left to humans.

In the long term, formally defined languages and notations may become available which will enable the translation of graphical models to executable code to be performed with formal rigor and 100% accuracy. Until that time however, we believe the approach outlined in this paper, which is currently under development at the Fraunhofer Institute for Experimental Software Engineering, provides a practical way of introducing some rigor into the implementation phase of object-oriented development, and in providing a certain degree of traceability and verifiability between graphical models and the object-oriented programs which implement them.

References

[1] Colin Atkinson and Michel Izygon. ION a notation for the graphical depiction of object-oriented programs. Technical report, University of Houston-Clear Lake, 1995.

[2] G. Booch, J. Rumbaugh, and I. Jacobson. *The Unified Modeling Language User Guide*. Addison-Wesley, 1999.

[3] Grady Booch. *Object Oriented Analysis and Design with Applications*. Benjamin/Cummings, Redwood City, California, 2nd edition, 1994.

[4] Christian Bunse and Colin Atkinson. Improving quality in object-oriented software: Systematic refinement and translation of models to code. Technical Report IESE-Report No. 036.99/E, Fraunhofer Institute for Experimental Software Engineering, Kaiserslautern, Germany, 1999.

[5] F. Buschmann, R. Meunier, H. Rohnert, P. Sommerlad, and M. Stal. *Pattern-Oriented Software Architecture – A System of Patterns*. John Wiley and Sons, 1996.

[6] D. Coleman, P. Arnold, S. Bodoff, C. Dollin, H. Gilchrist, F. Hayes, and P. Jeremaes. *Object-Oriented Development: The Fusion Method*. Prentice Hall, 1993.

[7] E. Gamma, R. Helm, R. Johnson, and J. Vlissides. *Design Patterns – Elements of Reusable Object-Oriented Software*. Addison-Wesley, 1995.

[8] Richard C. Lee and William M. Tepfenhart. *UML and C++. A Practical Guide To Object-Oriented Development*. Prentice Hall, 1997.

[9] Bertrand Meyer. *EIFFEL - The Language*. Prentice Hall, 1992.

[10] James Rumbaugh, Michael Blaha, William Premerlani, Frederick Eddy, and William Lorensen. *Object-Oriented Modeling and Design*. Prentice Hall, 1991.

[11] James Rumbaugh, Ivar Jacobson, and Grady Booch. *The Unified Modeling Language Reference Manual*. Object Technology Series. Addison-Wesley, 1999.

[12] Sally Shlaer and Stephen J. Mellor. The shlaer-mellor method. Pages on the WWW which can be found at: http://www.projtech.com/, 1998.

[13] Sally Shlaer and Stephen J. Mellor. Recursive design of an application-independent architecture. *IEEE Software*, 14(1):61–72, January 97.

[14] Bjarne Stroustrup, editor. *The C++ Programming Language*. Addison Wesley, 1993.

[15] US Department of Defense. *Reference Manual for the Ada Programming Language*, 1983.

Modeling Exceptional Behavior

Neelam Soundarajan and Stephen Fridella

Computer and Information Science
The Ohio State University
Columbus, OH 43210
e-mail: {neelam,fridella}@cis.ohio-state.edu

Abstract. While exception handling mechanisms are very useful for implementing systems, they are equally useful when building business models. The standard method of modeling exceptions in UML is to include them in class diagrams as a kind of *signal*, as stereotyped classes. However this alone is insufficient, since by the very nature of exceptions, the circumstances under which particular exceptions may be raised, as well as the details of what actions the class handling the exception will perform in doing so, tend to be rather complex and in general difficult to express pictorially. In this paper, we consider how this information may be specified using the *Object Constraint Language (OCL)*. We illustrate our approach by applying it to a simple example.

1 Introduction

Exception handling mechanisms are useful for *implementing* systems, and indeed that is why several OO languages such as *Java, C++*, and *Eiffel* provide such mechanisms. Exceptions are useful also when building business models, in other words at, what Fowler [3] calls, the *conceptual* level. At the implementation level exceptions are useful because they allow us to structure systems so that the normal code is cleanly separated from the exception handling code, thereby making the system easier to understand and to maintain. It is for exactly the same reason that exceptions are useful at the conceptual level too; the OO designer can first design the business model taking account of only the normal cases; and once this is done, he[1] can extend the model, by adding appropriate exceptions, to deal with the exceptional cases.

How do we include information about these exceptions in the UML diagram for the application? The standard approach [1] is to model exceptions as a special kind of *signal*, as stereotyped classes. But exceptions, whether in the implementation or in the business model, can be difficult to understand. For this reason, representing exceptions as signals in UML diagrams, although it conveys much useful intuition about which classes may raise which exceptions, by itself is not

---

[1] Following standard practice, we use he, his etc. as abbreviations for 'he or she', 'his or hers' etc.

always sufficient. It is important to characterize more precisely, the conditions under which specific exceptions may be raised, as well as what information will be available when particular exceptions are raised. In this paper we investigate ways of doing this using the *Object Constraint Language (OCL)* [6].

A standard OCL specification of a class C consists of precise specifications in the form of *pre-* and *post-* conditions for each of the operations that C provides, and that may be used by its clients. The pre-condition of each operation specifies the condition that must be satisfied at the time that the operation is invoked; this condition may involve the values of any arguments that the operation receives, as well as the values of any of the attributes of C. The post-condition of the operation similarly tells us what conditions will be satisfied by the values of the attributes of C when the operation finishes. And, in the case of operations that return a result, it also provides information about the result. The pre- and post-condition specifications embody the principle of *Design by Contract (DBC)* [4]: If the client (the caller) satisfies his part of the contract, that of ensuring the pre-condition of the operation at the time of the call to the operation, then (and only then) the service provider guarantees his part of the contract, that of ensuring the post-condition at the time the operation finishes. *Nothing* is guaranteed if at the time of the call to the operation, its pre-condition is not satisfied.

What do we do if this operation raises an exception? One approach would be to think of exceptions as 'catastrophic' events. In other words, an exception essentially tells the client class that the invoked operation *failed* for some reason. In effect this is what treating an exception as a particular kind of signal without providing any additional information amounts to; indeed, the icon that the three amigos suggest [1] for these signals is the classic 'danger' icon, consisting of a big exclamation point inside a triangle. Again, in *Eiffel* [4], exceptions are treated as failures (although by using some standard library facilities it is possible for the operation of an *Eiffel* class to provide some modest indication to the client, of the reason for its failure). While exceptions corresponding to such events as, say, power failures during the execution of the actual implementation model may qualify as catastrophes, languages like *Java* and *C++* allow for a more flexible treatment of exceptions. In these languages, exceptions are full-fledged objects, and can be used in very flexible ways to pass information back to the client including, in particular, information about the reason for the exception being raised. Similarly, considering exceptions as full-fledged objects (rather than merely signals with little information content), allows us to exploit them in interesting ways in the business model. In the next section we will see a simple banking example in which we use exception objects in this manner.

We also need to see how to specify exactly the conditions under which the exceptions will be raised, as well as precisely what information the exception object will contain if it is raised. In section 3 we will propose a simple and natural extension to OCL specifications to include such information. In section 4 we will return to the banking example and see how it can be specified using our notation, and how the information contained in the specification can be used by the designer of the client class. In the final section we reiterate the importance of exceptions as full-fledged objects in OO design, summarize our approach to extending OCL specifications to include information about operations that raise exceptions, and briefly consider some other aspects of UML that need to be enriched to deal with exceptions.

2 Banking Example

Consider the simple banking example shown in Figure 1. As the figure shows, a Bank has one or more Accounts; each Account is held by a single Customer (in other words, no joint accounts); and a Customer has zero or one Account. The Account class provides standard operations for checking the current balance in the account, and for depositing and withdrawing money from the account. The Account class is simple enough that one might be tempted not to include any further information about the model than is provided in Figure 1. But even such a simple example presents potential problems. For example, does the bank impose any service charges on the transactions? If yes, do the charges apply to both withdrawals and deposits? Do the charges depend in any way on the balance in the account?

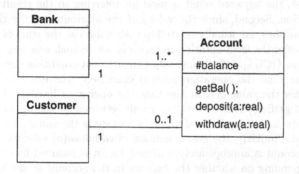

Fig. 1. Bank and Account classes

Suppose that the bank does impose a service charge of, say, 1.00 (one dollar) on each transaction but only if the current balance in the account is below 1000.00. In Figure 2, we use the OCL notation to specify this behavior in the form of *pre-* and *post*-conditions of the particular operations. The pre-condition of each operation specifies the condition that must be satisfied at the time that the operation is invoked; this condition may involve the values of any arguments that the operation receives, as well as the the values of any attributes of the Account class. Thus the pre-condition of the withdraw(a) operation, for example, tells us that we can only withdraw positive amounts; this pre-condition does not

694

```
Account :: getBal() : real
pre : true
post : (result = balance) ∧ (balance = balance@pre)

Account :: deposit(a:real) : void
pre : a > 0.0
post : (balance@pre ≥ 1000.00  ⇒  balance = balance@pre + a)
        ∧ (balance@pre < 1000.00  ⇒  balance = balance@pre + a − 1.0)

Account :: withdraw(a:real) : void
pre : a > 0.0
post : (balance@pre ≥ 1000.00  ⇒  balance = balance@pre − a)
        ∧ (balance@pre < 1000.00  ⇒  balance = balance@pre − a − 1.0)
```

Fig. 2. Specification of Account class

impose any requirements on the values of the attributes of the account at the time the operation is applied, but we will change that shortly.

The post-condition of the operation similarly tells us what conditions will be satisfied by the values of the attributes of Account when the operation finishes. And, in the case of operations that return a result, it also provides information about the result. OCL provides two important notations for use in post-conditions. First, the keyword result is used for referring to the result returned by an operation. Second, since the values of the attributes of the class when the operation finishes, are usually related to their values at the start of the operation, OCL provides the @pre notation in terms of which such relations can be easily expressed: in OCL specifications, x@pre in the post-condition denotes the value of x at the time the operation started execution, whereas x in the post-condition denotes the value of x at the time the operation finishes. Thus the post-condition of getBal() tells us that the result returned by the operation is the same as balance, and that the final value of balance is the same as when the operation started. Similarly, the post-condition of withdraw(a) tells us that the balance in the account is appropriately updated, i.e., it is reduced by either a or by (a + 1.0) depending on whether the balance in the account at the time the operation was invoked (i.e., balance@pre) was not or was less than 1000.0.[2]

In practice, it is frequently the case that the value of one or more attributes of the class when an operation finishes, are the same as at the start of the operation. To make it easy to specify such behavior, we use one other notation

---

[2] OCL allows us to write the two clauses of the post-condition as two separate post-conditions for the operation, one corresponding to the case when balance@pre was less than 1000.0, the other corresponding to when it was not. As a matter of style, we prefer to combine all the cases into a single post-condition to ensure that we don't overlook any particular case. But this is purely a matter of taste. Similarly when writing what we will call exception-post-conditions, we will combine all the cases into a single post-condition, although it would be perfectly possible to write them as distinct post-conditions.

in post-conditions, although it is not part of OCL: in post-conditions, !{x} will mean that the value of x when the operation finishes is the same as when it started; thus this is equivalent to (x = x@pre). Similarly, !{x,y} means that the final values of x and y are the same as their corresponding initial values; in strict OCL this would be written as: (x = x@pre) ∧ (y = y@pre) Using the '!' notation, the post-condition of getBal() could have been written as:

(result = balance) ∧ !{balance}

Consider again the withdraw(a) function. What happens if it is invoked with an argument value that is greater than the current balance? According to the specification in Figure 2, such an operation is indeed allowed and the result will be that the value of balance will be reduced by the amount a (or (a + 1.0), depending on balance@pre) and hence will become negative. Real banks don't, of course, allow all their customers arbitrary overdrafts, so let us change the model to make it a bit more realistic. One simple approach would be make the pre-condition of withdraw() more demanding:

Account :: withdraw(a:real) : void
pre : (a > 0.0) ∧ (balance ≥ a + 1.0)
post : (balance@pre ≥ 1000.00 ⇒ balance = balance@pre − a)
∧ (balance@pre < 1000.00 ⇒ balance = balance@pre − a − 1.0)

Given such a specification, we do not, in the withdraw(a) operation, have to worry about the possibility that there might be insufficient funds in the account. While this approach is not incorrect, it is rather inflexible. Rather than strengthening the pre-condition of Account::withdraw(), a more flexible design, would be for the Account class operation to raise an *exception* if there are insufficient funds in the account, and let the client class, which in this example will be the Bank class, decide how proceed in this case.

But what does it mean to raise an exception? What can the client class do if an exception is raised? Suppose, for example, that the Bank class contains an operation that is carrying out a whole series of transactions on different Account objects by invoking deposit() and withdraw() operations on the corresponding Account objects. And suppose one of these withdraw() operations raises an exception because the particular account has insufficient funds. If the exception were just a signal of some kind indicating to the Bank class that one of the transactions failed, that does not allow the Bank class to take appropriate action; indeed, since we would expect the exception-handling part of the Bank class operation in question to be separate from the main body of that operation, this part may not even know which account object had insufficient funds that resulted in the exception being raised. Hence, a more useful design would be for the exception to include additional information such as, for example, a reference to the particular account that was the source of the problem; that would allow the exception handling portion of the Bank operation to take suitable corrective action – such as freezing the account in question, or allowing the overdraft if the past history of the account warrants it, or simply not performing that particular transaction but continuing with the rest, etc. Of course some of these courses of action (such as freezing the account) will require the Account class to provide suitable operations and we will assume that such operations have been added to the Account class.

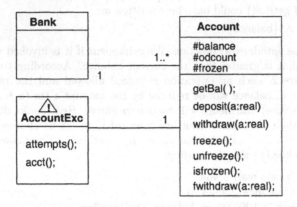

Fig. 3. Bank, Account, and AccountExc classes

In Figure 3, we have introduced the exception class AccountExc (and, in the interest of reducing clutter, we have omitted the Customer class). We have also made some changes in the Account class; let us consider these changes first. We have added two attributes #odcount and #frozen to the Account class. These will keep track respectively of the number of attempts made to overdraw funds from this particular account, and whether this account is currently frozen or not. We have also introduced four new operations: freeze(), unfreeze(), isFrozen(), and fwithdraw(); the first allows the client (the Bank class) to freeze an account, the second unfreezes the account, the third tells us if the account is frozen, and the fourth allows a *forced* withdrawal, essentially an overdraft.

The AccountExc is a fairly simple class. First note that we have flagged this as an exception class by including the 'caution' icon. The class provides two operations; the first, attempts(), returns the number of attempts that have been made to overdraw funds from the account that caused this exception to be raised. The second, and more interesting is the operation acct() which returns (a reference to) the account object that caused the exception. In the figure we did not indicate the attributes in which this information is maintained; we will introduce them when we specify these classes more precisely.

If the withdraw() operation is invoked with the balance in the account being insufficient, or if the account is currently frozen, the operation will create a new instance, which we will often refer to as ε, of the AccountExc class and 'throw' it. This exception object ε will include a reference to this particular account, and it is this reference that the acct() operation will return when it is applied to the object ε. This will allow the exception handling code in the client class to know which particular account was responsible for the exception being raised, and can then proceed to freeze the account or perhaps, if the the number of overdraw attempts is not too high, allow an overdraft (using fwithdraw()), etc.

But how do we specify all this information? Since it is the withdraw() operation that creates and throws the exception, it would be reasonable to except the specification of that operation to provide this information. In the next section, we will extend the pre- post-condition OCL specification of operations to provide information about exceptions that the operation may raise. Essentially, we will allow the specification of an operation to include a set of *exception-* or *e-post-conditions*, one corresponding to each exception that the operation may raise. The e-post-conditions will give us information about the conditions that will be satisfied if the operation throws that particular exception. As we will see, this post-condition can also, by judiciously using the @pre mechanism, allows us to specify the conditions under which this exception will be thrown. In section 4 we will return to the banking example and precisely specify the behaviors of the classes of Figure 3.

3 Specifying Exceptions

Suppose C is a class and f is one of its operations. The standard OCL specification of f specifies its pre- and post-conditions. Such a specification is appropriate only if f is guaranteed to terminate normally, i.e., without raising an exception (provided, of course, that its pre-condition was satisfied when it began execution). Suppose instead that f may either terminate normally or it may terminate by raising an exception of type $E1$, or of type $E2$, or of type $\dots$, En. We will specify such an f as follows:

```
C :: f
pre   : ...
post  : ...
E1.post : ...
E2.post : ...
    :
En.post : ...
```

The normal post-condition (which we will refer to as the n-post-condition) post, will, as usual, be a condition on the values of the attributes of C at the time that f terminates normally and the result, if any, returned by f (as well as, by using the @pre notation, the values of the C's attributes at the time f started execution). The e-post-condition $E1$.post will be a condition on the values of the attributes of C at the time that f terminates by creating a new exception ε of type $E1$ and raising it, and the values of the attributes of ε at the same time; $E1$.post may also refer, using the @pre notation, to the values of the attributes of C at the time that f started execution, but it may not refer to the result returned by the operation. This is because if f terminates by raising an exception, it does not return a result. It *does* return (or, rather, 'raises') the exception object ε, so we include information about the values of the attributes of this object in $E1$.post. The other e-post-conditions are similar.

This specification should be interpreted as follows: If the execution of f begins with the condition specified in pre being satisfied, then it is guaranteed that one of the following will happen: Either the execution of f will terminate normally with the values of the attributes of C at the time of termination (and the value returned by f if any) satisfying the condition specified in post; Or the execution of f terminates by creating and throwing an exception of type $E1$ or $E2, \dots$, En with the values of the attributes of C at the time of termination and the state of the exception object ε satisfying the condition specified in $E1$.post or $E2$.post or ... or En.post , respectively.

A number of points are worth noting. First, at most one exception may be raised by an operation. Thus, either the operation terminates normally *or* it terminates by raising a single exception. There are no other possibilities.

Second, the raising of an exception does *not* mean that the design-by-contract principle is being violated in any sense. Indeed, the kind of specification we have introduced allows us to characterize the exact conditions under which various types of exceptions may be raised. Consider a simple (but completely artificial) example:

$C :: f : \text{int}$
$\overline{\text{pre} : (0 \leq x \leq 2)}$
$\text{post} : (x@\text{pre} = 0) \land (x = 1) \land (\text{result} = 1)$
$E1.\text{post} : (x@\text{pre} = 1) \land (x = 2) \land (\varepsilon.y = -1)$
$E2.\text{post} : (x@\text{pre} = 2) \land (x = 2) \land (\varepsilon.z = -2)$

This specification tells us that $C.f$ maybe invoked only if the value of the x attribute of the C object is between 0 and 2. If this condition is satisfied when f is invoked, this specification assures us that this invocation of f may terminate normally, or it may terminate by raising an exception of type $E1$, or by raising an exception of type $E2$. f will terminate normally provided the value of the x attribute when f started execution was 0; and in this case, f will return 1 as the result, and the value of x when f finishes will be 1. If, on the hand, the value of x at the time that f started execution, was 1, then f will terminate by throwing an exception of type $E1$, the value of x at that time will be 2, and the y attribute of the thrown exception object will be -1. And, finally, if the the value of x at the time that f started execution, was 2, then f will terminate by throwing an exception of type $E2$, the value of x at that time will be 2, and the z attribute of the thrown exception object will be -2. Note also that nothing is asserted about what will happen if f is invoked with the pre-condition not being satisfied. That is outside the contract.

Third, in this particular example, the three possible outcomes are disjoint.

That is, for any particular value (within the range allowed by the pre-condition) for x when f started execution, exactly one outcome is possible. But this is a feature of this example; our specification notation does not require it. Suppose, for example, that we replaced the clause (x@pre = 2) in $E2$.post by [(x@pre = 2) ∨ (x@pre = 1)], and made no other changes, that would mean that if the initial value of x was 1, the operation would terminate by raising an exception of type $E1$ or of type $E2$. It is important to be able to allow such freedom in the specification because this allows us to build flexible business models. But note that 'flexibility' does not mean 'imprecise'; the specification, with this substitution, is as precise as the original specification. It is just that it leaves the freedom, to the implementor of the class, of raising either the exception of type $E1$ or of type $E2$ if f were to start with x being 1. D'Souza [2] also points out the importance of such flexibility.

Clearly, our approach to specifying and modeling exceptions is quite different from that of treating exceptions as 'failures' of some kind. Meyer [4], for example, essentially treats an exception as a violation of the contract by the operation raising the exception. We believe that if we are to exploit the power of using exceptions as full-fledged objects that pass useful information back to the client class (beyond the minimal fact that the operation 'failed'), then rather than treating it as a violation of the contract, it is more useful to extend the contract to cover this case; in other words, precisely specify under what conditions which exceptions may be thrown, and precisely specify information about the state of the specific exception objects that may be thrown.

4 Specifying Account and AccountExc classes

In this section we will specify the behavior of the Account class and AccountExc class of Figure 3 in terms of n- and e-post-conditions. To make the presentation simple, we will omit the account-balance based transaction fee that we considered in our earlier discussion. Before looking at the formal specifications, let us consider the behavior of each of the Account class operations informally.[3]

The attribute balance represents the current balance in the account; odcount gives us the number of attempts made so far to overdraw funds from this account; the value of frozen indicates whether the account is currently frozen. The operation getBal() returns the current balance. deposit(a) adds the specified amount

---

[3] There is nothing wrong with doing this. Just because we have a notation for formal specifications does not mean we should not also work with informal explanations. Indeed, the whole idea is to use several different ways of looking at the the classes and the operations – the UML class diagrams, informal explanations of what the operations do, and the formal specifications. And other techniques such as use-case diagrams provide additional handles to help us understand the behavior of the entire system; indeed, that is the point of the *unified* modeling language. We will return to this point in the final section.

to the balance. The operations freeze() and unfreeze() allow the client to change the value of frozen. The operation isFrozen() returns the current value of that attribute. These operations are all specified in Figure 5.

Account :: getBal() : real
pre : true
post : (result = balance) ∧ !{balance, odcount, frozen}

Account :: deposit(a) : void
pre : (a > 0)
post : (balance = balance@pre + a) ∧ !{odcount, frozen}

Account :: isFrozen() : bool
pre : true
post : (result = frozen) ∧ !{balance, odcount, frozen}

Account :: freeze() : void
pre : true
post : (frozen = true) ∧ !{balance, odcount}

Account :: unfreeze() : void
pre : true
post : (frozen = false) ∧ !{balance, odcount}

Fig. 4. Specification of Account class (part 1)

Note that in the n–post-condition of each operation we have used the '!' notation to specify which variables are unmodified from their original values. Note also that there are no e-post-conditions in any of these operations; that means that none of these operations may raise an exception (at least not if their respective pre-conditions are satisfied when the operations are invoked).

The interesting operations are of course, withdraw(a) and fwithdraw(a). Since these might throw exceptions of type AccountExc, let us first consider the attributes of that class. This class has two attributes, noattempts, and account. The second attribute contains a reference to the particular account object on which an operation was being attempted that resulted in this exception being thrown. The value of the first attribute is the number of attempts that had so far been made to overdraw funds from that particular account at the time that the exception was thrown.

The operation withdraw(a) works as follows: if the account is currently frozen, it raises an exception of type AccountExc; this exception object has its account attribute set equal to the account in question, and the noattempts attribute set equal to the current value of the odcount attribute of this account. If the account is not currently frozen, and if the current balance in the account is greater than or equal to the amount a, then balance is appropriately updated, and the operation returns normally. If the balance is insufficient, the value of odcount is incremented by 1, and an exception of type AccountExc is raised. The operation fwithdraw(a) is similar, except that it does not raise an exception if the balance is insufficient.

Account :: withdraw(a:real) : void

pre : $a > 0$

n-post : (frozen@pre = false) $\wedge$ (balance@pre $\geq$ a)
  $\wedge$ (balance = balance@pre $-$ a) $\wedge$!{frozen, odcount}

AccountExc.post :
  ((frozen@pre = true) $\vee$ (balance@pre $<$ a)) $\wedge$!{balance, frozen}
  $\wedge$ ((frozen = true) $\Rightarrow$ (!{odcount})
  $\wedge$ (frozen = false) $\Rightarrow$ (odcount = odcount@pre $+$ 1))
  $\wedge$ ((ε.account = self) $\wedge$ (ε.noattempts = odcount))

Account :: fwithdraw(a:real) : void

pre : $a > 0$

n-post : (frozen@pre = false) $\wedge$ (balance = balance@pre $-$ a) $\wedge$!{frozen, odcount}

AccountExc.post :
  (frozen@pre = true) $\wedge$!{balance, frozen, odcount}
  $\wedge$ ((ε.account = self) $\wedge$ (ε.noattempts = odcount))

Fig. 5. Specification of Account class (part 2)

The n-post-condition withdraw() requires the account not to be frozen at the time of the call to the operation, and the balance in the account at that time to be sufficient. The AccountExc-post-condition on the other hand asserts that this exception will be thrown if the account was already frozen or if the funds in the account are insufficient; this post-condition further tells us that in the former case odcount is not updated when the exception is thrown, but in the latter case it is incremented by 1 (matching our informal understanding of the operation). More interesting is the last line of this post-condition; it tells us that the account attribute of the exception being thrown is (a reference to) the self object, i.e., the account object on which the withdraw() operation was applied; the final clause tells us that the attribute noattempts of the exception is appropriately set. The specification of fwithdraw() is similar but simpler since this operation does not raise an exception even if the balance is insufficient.

We also need to specify the AccountExc class. This is a fairly simple class and its operations just return the corresponding attribute values. We capture this precisely in the OCL specifications for this class in Figure 6.

Ultimately, of course, the purpose of such precise specifications is to help the designer of the client class, Bank in our particular example. Earlier we considered the possibility that this class might be carrying out a series of deposit() and withdraw() transactions on a collection of Accounts. Suppose one of the withdraw() operations applied to one of these accounts raises an exception. The exception

```
AccountExc :: attempts() : Integer
pre : true
post : !{noattempts, account} ∧ (result = noattempts)

AccountExc :: acct() : Account
pre : true
post : !{noattempts, account} ∧ (result = account)
```

Fig. 6. Specification of AccountExc class

handling-code in the Bank class could be written, in *Java*-like syntax, as follows:

```
catch (AccountExc e) {
    Account a = e.acct();
    if (e.attempts() ≥ 2) { a.freeze(); }
    else ...
}
```

This catches any exception of type AccountExc; it then checks the number of attempts already made to overdraw funds in this account, and if it is two or more, freezes the account. Note how we are using the information provided by the exception object. According to the specification in Figure 6, the acct() method gives us (a reference to) the account that the exception refers to which, by the specifications Figure 5, is in fact the account that the overdraft was attempted on. The specifications in Figures 5 and 6 also tell us that the attempts() operation will return the number of overdraft attempts made so far on this account. Thus by referring to the information provided in the e-post-conditions of the operations that raise the exceptions, and by referring to the information in the specification of the operations of the exception class, we can, in the client class, make effective use of the information contained in the exceptions.

We conclude this section with a final observation about our example. Consider the 'else' clause in the exception handling code above. What would be an appropriate action to perform in this situation? The natural answer would be to use the fwithdraw() operation to allow the overdraft. We do know the account on which the fwithdraw() operation should be applied; it is the one we got as e.acct(). But what should be the argument to fwithdraw(), in other words, what amount should we force-withdraw? We do not currently have that information. This is in fact, an incompletely designed AccountExc class. A more useful design would have included not just the attributes account, noattempts, but also withdrawAmount, being the amount of money that we were trying to withdraw in this transaction when the exception was thrown. It would be easy to add this attribute, and the corresponding operation wAmount() that returns this value. With this change and the corresponding changes in the specifications if Figures 5 and 6, we will be able to write an appropriate else clause.

5 Discussion

Exceptions are useful not only when implementing OO systems, but also when building business models. The question we have focused on in this paper is, 'What information should we include in our UML models about the exceptional behavior of classes?' Clearly the answer to this question depends on why exceptions occur. A common view [5, 4] of exceptions has been that they represent catastrophic failures such as, for example, power failures, or network crashes, etc. Under this view, the client class, upon receiving the signal that an exception occurred is required to recover as best as it can (or if it cannot, itself fail in turn, and signal its client). A more useful view, both when implementing systems as well as when designing business models, is to think of exceptions as unusual but not disastrous situations. And that exceptions are in fact, full-fledged objects that can be used to convey a range of useful information to the client class. This is the view that languages like *Java* support, and is the view we have tried to model in this paper. The power of this approach is clear even in the relatively simple banking example that we considered. It was only because the AccountExc objects contained information, in particular about the identity of the Account object that was responsible for the exception, that the client class could proceed appropriately once the exception was thrown.

Note that we are *not* advocating that exceptions be used as if they were just another control structure; indeed, greater the number of exceptions a class may throw, greater will be the complexity of its OCL specification in our approach. What situations are best handled with exceptions rather than regular code is really a *design* question. A reasonable case could be made that the bank account example we considered did not really warrant the use of exceptions. We used this example mainly as a simple setting in which to explain what specifications of exceptions look like in our approach, not as a paradigmatic example of where exceptions are appropriate.

Since exceptions are simply objects, it is easy to represent them in UML class diagrams. We can simply treat them as instances of a stereotype ≪exception≫ with the additional decoration of a caution icon as recommended by [1]. While modeling exceptions in this manner (as in our Figure 3) provides useful information, this is not sufficient. The point is that, by their very nature, exceptions are difficult to understand, and it is important to specify precisely when they might be thrown, and exactly what information will be contained in them when they are thrown. The OCL component of UML is designed to enable precise specification of the behavior of operations of a class. However, standard OCL does not deal with operations that may raise exceptions. In Section 3 of this paper, we proposed a simple extension to OCL to deal with such operations. The main idea underlying our extension was to specify an operation by means of a pre-condition, a normal-post-condition (or n-post-condition), *and* a set of exception-post-conditions (or e-post-conditions), rather than in terms of just a pre-condition and a (normal) post-condition. Each e-post-condition allows us to

704

specify the conditions that will hold if the operation were to terminate by throwing the corresponding exception, just as the normal post-condition allows us to specify the conditions that will hold if the operation were to terminate normally. In Section 4 we saw how this approach could be used to model precisely the behavior of the banking example.

Thus the key contributions of this paper have been to:

- Present a case for why treating exceptions as full-fledged objects rather than just signals is extremely useful not just at the implementation level, but also at the conceptual level.
- Present a simple extension to OCL to allow us to specify precisely the behavior of applications that make use of exceptions in this manner.
- Demonstrate the approach by applying it to a simple illustrative example.

We conclude with a couple of observations. First, UML has other components than just class diagrams and OCL specifications. For example, dynamic views of the system via use-cases and interaction diagrams are extremely important. Suppose an operation is invoked; the invocation is of course represented on the interaction diagram, as is the return. What if the operation were to throw an exception? For a full picture of the system, it is clearly essential to represent this event, as well as the subsequent catching and handling of the exception by the caller on the interaction diagram. It would seem natural to introduce corresponding stereotypical actions to represent these events and use them in the interaction diagram.

Second, we believe that an AccountExc object containing a reference to the Account object that threw that exception is not incidental to this particular example, but rather an instance of a common pattern applicable to exceptions in general. In other words, we believe that it would be useful in general for exception objects to include a reference to the object that was responsible for creating and throwing that exception. Hence it would be useful to include this as part of the definition of the exception-stereotype so this information does not have to be repeatedly included in each class diagram. And in OCL specifications of exception classes, it may be appropriate to use a special notation, such as @source, to refer to the source object. This would eliminate the need for every exception class providing an operation similar to the acct() operation of AccountExc to give us this information.

References

1. G. Booch, J. Rumbaugh, and I. Jacobson. *The Unified Modeling Language User Guide.* Addison-Wesley, 1999.
2. D. D'Souza and A. Willis. *Objects, Components, and Frameworks with UML.* Addison Wesley, 1999.
3. M. Fowler. *UML Distilled.* Addison-Wesley, 1997.

4. B. Meyer. *Object-Oriented Software Construction*. Prentice Hall, 1997.
5. J. Rumbaugh, M. Blaha, W. Premerlani, F. Eddy, and W. Lorensen. *Object-Oriented Modeling and Design*. Prentice-Hall, 1991.
6. J. Warmer and A. Kleppe. *The Object Constraint Langauge*. Addison-Wesley, 1999.

Advanced Methods and Tools for a Precise UML
Panel

Moderator:
Andy Evans[1],

Panelists:
Steve Cook[2], Steve Mellor[3], Jos Warmer[4], and Alan Wills[5]

[1] Department of Computer Science, University of York, UK.
[2] IBM EMEA Object Technology Practice, UK.
[3] Project Technology, Inc. US.
[4] Klasse Objecten, Netherlands.
[5] TriReme International, UK.

1   Introduction

Imagine for a moment you are a software 'architect' in the year 2003. You're working at home as usual, and decide to use your quantum computer to do some system modelling. Imagine also that the UML is 'still' the de-facto language for software engineering. As a language it has made some big advances of the last few years. The last three versions (3.0-5.0) have all had a precise semantics - even the mysteries of aggregation have been resolved - and its applicability has been widened to every kind of system imaginable. Standardisation has also been good for the software profession. CASE vendors and methodologists, no longer able to invent new notations, have devoted their energies to building increasingly sophisticated tools and processes. Thus, the tool and methods you are about to use incorporates a maturity of software technology that has never been realised before...

The question for the panel is, what would you expect of the tool you imagine using, and the methods it supports? How will a precise UML (assuming it *is* or *can* be made precise) benefit software development tools and processes over the next five years? What features will be desirable in future tools and methods, and what type of semantics will they require to support these features? How might tool vendors go about building such tools? How might methodologists go about using the semantics to compliment or improve their work?

To answer these questions, four leading methodologists and tool developers will give their vision of the future. A snapshot of their thoughts, and views on the way forward, are given below.

2   Steve Cook

Let's look at the question "Is rigorous proof achievable in UML" This is an interesting question from a theoretician's point of view. But from the perspective of today's typical software developer, such a question would be met with blank incomprehension. Just to understand and formulate the need for a proof is outside the experience of most developers, let alone actually to do one. Twelve years on from the publication of Eiffel, we have just about reached the point where a reasonable proportion of developers have heard of the idea of a pre-condition, although the number who can actually formulate one correctly is a small subset of these.

So I want to distinguish strongly between precision in the definition of UML and precision in its use. Long experience tells us that normal software developers will not, in the foreseeable future, be willing to use abstract formal languages and notations to design software, regardless of how theoretically desirable it might be to do so. Even such simple semi-formal languages as OCL meet great resistance from the typical developer or student of programming. We might surmise that the reason for this is that the mental interpretation of declarative, logical statements is intrinsically rather hard and requires a skill in abstract thinking which is uncommon. Whether the fault is in the education system or in our genes, programmers prefer to write code, because they can test it to find out whether it is correct; and analysts prefer to remain vague, because they know that the programmers will have to sort it out. The only way to get end users to be precise is to provide tools that make it very simple and obvious to do so.

On the other hand, the designers of UML itself have a strict obligation to ensure that the structure and semantics are consistent and well defined as UML evolves. Without this, the language will fail to provide the promised interchangeability of designs, UML skills will be a matter of local interpretation, and the evident advantages of standardization will be only very partially achieved. Furthermore, the ability to construct tools that can help modelers be more precise will be fatally compromised.

It's critically important to realize that UML is a family of languages, each with its own semantics. For example, one can easily envision two versions of UML identical except for some detailed specifics about the amount of concurrency that is permitted within statecharts. It would be quite wrong, in my view, to decide which of these specific semantics applies at the level of the kernel UML specification. Instead there needs to be a set of mechanisms which allow these semantics to be applied in the precise definition of standardized UML family members. It follows from this that the kernel UML should have rather sparse semantics, consisting of simple definitions of well-formedness and simple axioms about dynamic semantics. The profile mechanism must be composable and provide rich and robust mechanisms for extending and refining the structural and dynamic semantics.

Assuming that UML becomes well-defined in this way, the main added value for the next generation of tools would be model interchangeability and the ability to recognize and interpret profiles. Building on this, tools could be constructed to assist the end user to produce consistent and expressive models, including tools for automated consistency checking.

Biography

Steve Cook is an IBM Distinguished Engineer. He is currently Chief Architect of IBM's Object Technology Practice. He was the lead author of IBM's submission (with ObjecTime Limited) to the OMG's Object-Oriented Analysis and Design Facility standard, which introduced several elements including OCL (Object Constraint Language) into the UML definition. He was the co-author (with John Daniels) of the Syntropy method, a formalized variant of OMT which was the main precursor to OCL. He has been working with object-oriented methods for 20 years.

3 Stephen Mellor

A View of the Present

Assumptions: I assume that a precisely defined, verifiable, executable, and translatable UML is a Good Thing and leave it to others to make that case.

Where We Are. In the summer of 1999, the UML has definitions for the semantics of its components. These definitions address the static structure of UML, but they do not define an execution semantics. They also address (none too precisely) the meaning of each component, but there are "semantic variation points" which allow a component to have several different meanings. Multiple views are defined, but there is no definition of how the views fit together to form a complete model. When alternate views conflict, there is no definition of how to resolve them. There are no defined semantics for actions.

Work In Progress. In November 1998, the Object Management Group (OMG) made a request for proposal (RFP) for a definition of the semantics of actions. The schedule calls for submissions in the autumn of 1999 and an accepted standard by 2000. The RFP encourages proposals to address the question of how the semantics of actions fit in with the (undefined) execution semantics of the remainder of UML.

What Are The Issues? Even with defined semantics for actions and a model of execution, model interchange-the goal, surely, of the standard-is still not possible because views conflict and they do not fit together to form a complete model. Further, the plethora of semantic variation points make semantic model interchange impractical, even if it is possible to import and export diagrams.

The bottom line is that models today are expenses. They need to be assets.

A Path to the Future

Primitives and Composition. A precise UML should be constructed from as small a set as possible of primitive, well-defined, building blocks. This approach will lead to a smaller and simpler UML with no loss of power. The rationale for this assertion is that a compositional approach permits higher-order constructs to be built from smaller, more primitive, pieces. Consequently, if only one component can be defined in terms of primitives, the UML has a smaller, simpler, definition than at present.

This still leaves open the issue of whether some constructs should be removed from the UML to make it easier to learn, use, and communicate with. This argument certainly deserves to be made, but not here.

Multiple Views. To determine what requires formalization, the UML must distinguish clearly between essential, derived, auxiliary, and deployment views. An essential view models precisely and completely some portion of the behavior of a subject matter, while a derived view shows some projection of an essential view. For example, if the state chart is an essential view, then a collaboration diagram that shows communications between state charts, but not the details, is a projection. On the other hand, were a collaboration diagram defined to be essential, then all the communications on a state chart would perforce be derived. Because the state chart would then comprise both essential and derived elements, the former view, in which the state chart is essential, is to be preferred.

The distinction between essential and derived views does not establish a sequence for the construction of the various views. A developer may begin with a collaboration diagram, using it to come up with the state charts, then revise-preferably automatically-the collaboration diagrams to conform to the essential view.

An auxiliary view, such as a use case diagram, supports the essential and derived views. An auxiliary view models informal aspects of the system and does not require formalization.

A deployment view models the packaging (or repackaging) of existing components. The existing deployment diagrams certainly model deployment views. One can also make the argument that any model which shows processors or tasks also shows a deployment view. Taking this one more step, if a writer intends a class diagram to be interpreted as a decision about which classes should exist, then a class diagram also represents a deployment view. This writer intends a 'class' to be a logical grouping of the definition of data and behavior, but not necessarily the structure for the implementation.

In summary, formalization of the execution semantics is required for the essential views only. The derived models are, well, derived, and auxiliary views don't require formalization. We take deployment again later.

A Coherent Set of Essential Views. As an example, consider the following coherent set of essential views. First, a class diagram has essential compartments for the class name and attributes only. The operation compartment shows derived operations only. Second, exactly one state chart describes the behavior for each class. No other (essential) state charts exist. The primitives for the state chart comprise only transitions, events, and the data that accompany them, (non-hierarchical) states and an action set associated with each state (i.e. a Moore state model). Finally, each action set comprises synchronous actions of only four types: data access, transformations of input data into output data, transformations of input data into mutually exclusive control outputs, and asynchronous signal generators. All four types may operate on single values or collections. In this formulation, data access from one class to another is a derived operation on the destination class, which could be shown-preferably automatically-on the class diagram.

The execution model defines run-to-completion semantics for the action set; preserves order of signal events only between sender and receiver object instance pairs; requires the developer foreswear data access conflict (or be oblivious to it), and treats all events as having the same priority.

Pace state chart fans. We can now define the elements of the UML state chart in terms of the set of primitives that comprise the basic execution model. We may formulate concurrent states in terms of two state charts that act on two sets of instances that mirror each other; action sets on transitions (a Mealy machine) as action sets on destination sets, adding states if actions conflict; deferred events as an additional (hidden) class to queue the deferred event and a state model for the associated deferred behavior, and so on.

Should this prove impossible in any particular case, we require the invention of a new semantic unit, which may not appear on any diagram. Alternatively, perhaps we have a contradiction. Good luck finding that contradiction in today's 808-page specification!

Semantic Variation. A semantic variation point is a Bad Thing. For one, it reduces the interchangeability and understandability of UML models because the same element has multiple meanings. To the extent that semantic variation points interact, so the complexity of the UML as a whole increases dramatically.

The perspicacious reader will have observed that the execution model above de-emphasizes operations. This perspective conflicts with a synchronous, transaction oriented, perspective assumed, inter alia, by the Meta-Object Facility (MOF). One could define a completely separate execution model that emphasizes operations and transactions to the detriment of states and events. Some components may have the same definition in both execution models. This, however, is a rather different matter than the indiscriminate use of semantic variations.

The perspective above constitutes a coherent set of essential views defined by composition. There are no "semantic variation points" One can make the argument that this perspective is too sparse to be useful. What about, say, actions on transitions, priorities on events, and arbitrary methods defined for classes?

Deployment. The UML Summary describes the UML as "a language for specifying, constructing, visualizing, and documenting the artifacts of a software-intensive system". As such, the UML incorporates many elements intended for implementation and deployment. For example, consider a deadline-driven real-time system that controls an manufacturing plant, say. At first blush, this system requires priorities on events and tasks. However, the underlying behavior can be stated without priorities. Events merely synchronize the actions of various state models. And the very concept of a task is a deployment issue. A core UML, therefore, is a very sparse beast-all the better to define its semantics.

How then may we define the organization of the software target? The target may comprise tasks, processors, classes of certain kinds, operations on those classes that may do more that access data. We could redraw the application model to show this deployment, and the various deployment diagrams would necessarily use the elements as currently defined in the UML. While this set of elements is extremely large, it is nonetheless limited.

As an alternative, we may define the elements of the target using the core UML. The resulting model is a model of the components that comprise the target. Hence, we may define classes Periodic Task, Event- Driven Task, Function, Parameter, Application Class and so on. The target model is just another UML model. It models the organization of the software components and it is independent of the application. In fact, it is a model of an application-independent software architecture.

The object instances of this target model can be primitive, so the decision to have three periodic tasks and seven event-driven will manifest itself as three instances of the class Periodic Task and seven instances of the class Event-Driven Task. Other object instances in the target are elements-even down to the individual actions-copied from the original application model. The populated target model has exactly the same semantics as the original application model. It's just organized differently.

It is then a simple matter to spit out the content of this model in code. With supporting tools, the target model-the application-independent software architecture-acts as a model compiler for the application.

A Vision for the Future

With the components identified above in hand, the nature of methods and tools change significantly. First, let's summarize what we (can soon) have:

- a small set of primitives that can be composed
- a clear distinction between essential and other views of a model
- a coherent (small) set of components that comprise the essential view (with syntactic sugar)
- an execution model for the components that comprise the essential view without semantic variations

There are three main features to consider: separation of subject matter, execution, and translation.

Separation of Subject Matter. The separation of the application model from the application-independent software architecture makes each model an asset.

There may be multiple implementations of that same application model, each of which has different performance properties depending on the pattern of usage for the system. An application model can be redeployed using a different model compiler, so the application model does not change repeatedly as the business undergoes technology change. Similarly, the model compilers are themselves assets that can be reused across multiple applications that share the pattern of usage.

The model compiler is general. Any application that has the same pattern of usage for the system can make use of that same model to produce code. Consequently, this target model can be sold to a broad variety of customers. There is a need for only a relatively limited number of these target models.

The models-both the application models and the target models-are now assets. The application model can be re-deployed by acquiring a new target model. And vice-versa.

Execution, Simulation, and Verification. Because the essential view has a well-defined execution model, we can build a model of a problem and simulate it. This will shorten the analysis cycle because a model can be executed without making detailed decisions about the design of the system, only the content of the underlying problem.

Translation. In reality, systems comprise multiple layers. Each layer can be constructed as a UML model. The object instances derive from the layers above. The bottom-most layer is the one that, when all is said and done, the entire system resides. We may then generate code from that final layer by serializing the model into text that can be compiled.

Methods and Tools: What Do We Need?

The fundamental issue in model building, and therefore the methods required to build such models, is abstraction. How does the developer come to a set of abstractions that properly captures the domain?

This task is eased by the simplification of the UML to contain only essential elements. Now we can obviate all packaging elements, there is no need to choose to package a model in a particular way for the sake of efficiency. The next step, then, is to define a "normal form" for model structure. This normal form is implicit in the coherent set of models defined above. The abstractions must represent the invariant in a problem domain, and all variants must be expressed in data. Methods will focus on how to find those invariants. Believe me, it won't be use cases.

As for tools, the fundamental issue is the logical structure of the repository.

- First, model builder tools will act as front ends for that repository. Separate files per model will not work. Goodbye, Rational Rose
- Second, tools that execute the model in the repository are required. Specifically, there is a need for offline model verification tools, online interpretive

simulation tools, and online animation tools. The last is frankly not very useful, but it demonstrates well. The other tools will be used for regression testing and model debugging respectively.
- Third, we need a wide variety of different model compilers, each targeting different architectures.
- Fourth, a set of bridging tools that populate one model from the content of another is needed. The issue here is generality so that the bridging tools can populate any arbitrary model.
- Fifth, a standard "language" for turning a populated model into code is required.

This set of tools enables a market for complete models of domains, expressed in UML, that can be deployed onto other models, including model compilers.

Methods and Tools: What Do We Have?

We have:

- a method
- repository-based tools
- verification and simulation tools
- the beginnings of a market in model compilers
- little, so far, in the way of general purpose bridging tools
- languages (but not a standard) for generating code
- the stirrings of a market for domain models

All we need now is to make the market aware that all this is possible, build tools around the standards defined by the core, executable UML, and make it so, Mr. Crusher.

3.1 Biography

Stephen J. Mellor is best known in the real-time community for his contribution to the development of Object-Oriented Analysis, Recursive Design, and the Shlaer-Mellor Method. He is Vice President of Project Technology, Inc., where consults, teaches, and researches applications of the method. Mr. Mellor is currently working with the Object Management Group (OMG) to define semantics for actions. Mr. Mellor is also a member of the editorial board for IEEE Software.

4 Jos Warmer

With the emergence of the UML as a standard for modeling new opportunities arrive and they will change the future of tools. UML has several characteristics that allow for this development. UML includes an explicit meta-model and an explicit interchange format: XMI. Any tool that supports UML will be able to read and write models using XMI. If the tools implement this correctly, UML models can readily be interchanged between different tools from different vendors. When this becomes reality, the tool market can change fundamentally.

How tools cooperate today

Until today, object oriented case tools from different vendors are mostly incompatible. Once you start using a specific tool to develop your models, you are locked into this tool. Changing to another case tool is a very costly decision, because almost all information needs to be entered into the new case tool by hand. Also, the choice of add-on tools is restricted to those available for the chosen case tool. For vendors that develop add-ons or plug-ins for case tools, this situation is bad. Let us assume that we are a company developing a code generator for e.g. the Java language. We need to decide beforehand to which case tool this will become an add-on. Let us assume that we choose case tool C1 to be our target. The market potential of our code generator is now limited to the market share of the targeted case tool C1. Being good software engineers, we design a good architecture of the code generator: we develop it as two components. One component will interface to the case tool C1; the other component generates the actual code. When we want to target a second case tool C2, only the interface component needs to be rewritten. At first sight this looks like a solution, but the task is more complex than expected. To start with, we need to use different methods to extract information from C2. The real surprise follows shortly: the structure and content of the information we get from C2 is completely different from what we got from C1. Both C1 and C2 use a different meta-model (sometimes explicit, most often implicit) and there is no well-defined way to transform the meta-model of C2 into that of C1. Our code generator, however, was based on the information we got from C1, and therefore based on the meta-model of C1. In the end we need to rewrite a large part of our code generation component specifically for C2.

How tools cooperate tomorrow

With UML the scenario above will change drastically. First of all, the XMI interchange format allows us to write the interchange component of our code generator once. Because all UML tools export the UML model in XMI format, our new interchange component can read the output of all case tools. Secondly, the structure and contents of the information that we get will be identical for each case tool. It is based on the UML meta- model. The fundamental change that we see is that we do not need to make a choice to support one or more specific case tools. The only thing we decide is to develop a code generator for UML. We will be able to combine our code generator with any existing UML tool. Even new tools, which didn't exist at the time we started, will be able to be combined with our code generator. Consequences for tools For tool vendors UML will open up a much broader market, because they can target their tools to the whole UML market without restrictions. This will generate stronger competition for widely used tools. A specific group of more specialized tools, like advanced model validators or proofing systems, are barely available on the marketplace today, because their market share is too small. With UML, these tools will now have a wider market, and they might become economically feasible.

Limitations of tool cooperation

Although UML will enable much more tool support, there are several areas where UML won't help and the old situation won't change much. The UML meta-model and interchange format only defines the contents of a UML model, not the visual diagrams that a modeler draws. The visual information cannot be interchanged yet. This means that the choice of the visual modeling tool will still result in a vendor lock-in. This limitation might even affect the possibility to combine different non-visual tools. It is therefore important that the OMG will start work on a standard for interchanging visual UML information.

New types of tools

UML defines an explicit meta-model and includes a complete description of the semantics of all modeling constructs. Having this semantic description of the meaning of all constructs, might allow tools do a more intelligent job. For example, until now information found in sequence diagrams (messages sent and received) and state charts (events being accepted, states being defined) has been used mostly as analysis and design documentation, but it doesn't necessarily have a well-defined and checkable relationship with the actual code. Using the semantics defined in the UML tools might be able to validate the actual code against the sequence diagrams and state charts. It remains unclear for now whether the UML semantics is defined rigorously enough to allow for these kinds of validations. If not, the UML semantics specification should be enhanced and made more precise.

Conclusion

Although the subject is about tools for UML, it is important to realize that it is not UML alone that makes a difference. Without the XMI interchange standard none of this would ever be possible. The inclusion of XMI is the enabling factor for this process. It is therefore understandable that none of the above has taken place with the UML 1.1 and 1.2, because XMI was still lacking. With UML 1.3, the industry can finally start the real work on generic UML tools.

Concluding, I expect that we will see a much broader variety of specialized tools that use the UML meta-model and XMI as their basis. More tools will become widely available. It will probably take several years before we will see a real change, because most vendors are still struggling to get their tools up to date with the new UML 1.3 and XMI standard. Also a new type of tools might emerge, although this will certainly take much longer.

Bibliography

Jos Warmer is working as a trainer and consultant object technology with Klasse Objecten. Before starting to work with Klasse Objecten Jos worked within IBM's Object Technology Practice. During the development of the UML 1.1 until now he has been IBM's representative as a member of the UML core team. He was responsible for the development of the Object Constraint Language (OCL). He is still a member of the UML Revision Task Force, which prepares new versions of the UML, under the responsibility of the OMG. Jos Warmer is one of the authors of 'The Object Constraint Language: Precise Modeling with UML' and of Dutch books on OMT and UML. You can reach Jos by email: J.Warmer@klasse.nl.

5 Alan Wills

Tools support methods. Here is a selection of methodinos that my clients find very useful. To make them work, we have to assume certain clear relationships between the different parts of the UML notation, and between the UML and the program code. Currently, no such relationships are generally agreed for UML; and the techniques are under-supported by tools.

I'll explain each technique, show how precise UML semantics are necessary, and show what a supporting tool could do. The ideas are part of Catalysis, developed by Desmond D'Souza and me.

Business modeling with consistency and completeness checks

We start most projects from a business domain model, separate from any particular piece of software, defining the vocabulary common to all components. Scenarios, stories about typical sequences and flows in the business, form a good start; but we need to identify exactly what types of 'actions' (use-cases, tasks, interactions, operations) and objects are involved: they will ultimately be reflected in the software.

We go round this cycle, asking questions of the domain experts & source documents. You can enter at any step:

- Method – Identify actions by asking "what happens?", looking for verbs, etc. Document each as an action: draw it and its participants using the UML use-case ellipse:

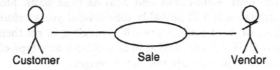

Customer   Sale   Vendor

Write a postcondition (and precondition) defining its effects on the participants. (We prefer to start with pre/post instead of giving a use-case sequence, because there may be several different ways in which the same effect could be achieved: we want to say what's common to all of them, defering detail.)

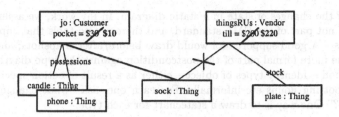

The postconditions are expressed both informally and in terms of changes in the associations and attributes of the participants. We can illustrate this with a before/after instance diagram ('snapshot') – the bolder lines represent 'after'. The postcondition can be written:

```
action sale (cust: Customer, vendor: Vendor, thing: Thing,
price:Price)
```

```
post        cust.possessions == cust.possessions@pre + thing
    and vendor.stock == vendor.stock@pre - thing
- - thing is transferred from vendor's stock to customer's
possessions
    and     cust.pocket == cust.pocket@pre - price
    and vendor.till == vendor.till + price
- - price is transferred from customer's pocket to vendor's till
```

This gives us a type diagram of the vocabulary we've used in the action spec:

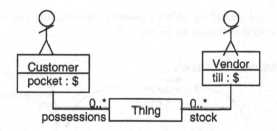

So by being precise about the specification of the action (use-case) we've made a clear relationship between the static or logical part of the model, and the dynamic part. As far as our modeling cycle goes, we have a way of discovering new objects, attributes and associations: they come up when you try to describe the postcondition of an action.

More informally, we can ask about any action, 'who or what is affected, or affects the outcome?'

Precision – Notice that we can't really do this unless we assign a semantics to UML in which the ellipses we draw in a dynamic diagram can be described in terms of the changes of state in a static diagram. In my work, we assume this; but it is not part of the UML standard, and there are no tools that support it.

Tools – A good support tool would draw before/after snapshots; and would check the more formal part of the postcondition against the type diagram. ?

Method – Identify types of object – either as a result of action specification, or by spotting nouns etc informally. For each one, ask "what changes can it undergo?" One way is to draw a statechart for each type:

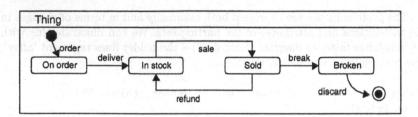

Notice that 'sale' is one of the actions we can see here. In the Catalysis interpretation, the transitions on statecharts are actions: this gives us a useful way to discover actions: we can now write postconditions of 'refund', 'deliver', etc, and possibly discover new types of object, associations and attributes in the process of describing them.

Precision – This technique relies on a clear relationship between the transitions on statecharts and the actions in the use-case model. We also use the idea that states are boolean attributes, usually derived from other attributes or associations: for example, 'sold' means that the Thing is part of the 'possessions' of a Customer.

Tools – A tool should be able to generate the actions and attributes, and highlight those that still need to be defined.

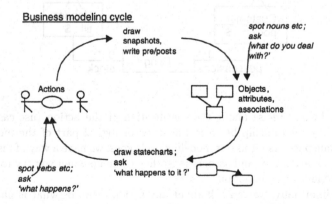

We continue around this cycle, finding actions and objects, until there are no new ones, and we don't have to modify any definitions. We stick to a chosen level of abstraction, avoiding going into very detailed actions – the beauty of a postcondition is that it can state the relationship between the states before and after a transaction that might take a long time and involve all kinds of smaller tasks that we don't need to concern ourselves with initially.

Tools – A good support tool for domain modeling should help us to create new actions and associations, and highlight those that need further definition.

Imprecision

Although I've described the modeling cycle in terms of precisely-defined postconditions etc, it is also possible apply the technique less formally. Postconditions written in informal language still use a vocabulary of terms, about the relationships between things: those words can be spotted and should turn up as associations and attributes.

Tools – an informal noun-spotter and verb-spotter would help to take us round the modeling cycle.

Model construction from views and frameworks

The same modeling patterns tend to crop up repeatedly. This is often about the relationship between objects; not just the static relationships, but the dynamic protocols describing how they interact. The Observer pattern is a good example. To use the pattern, you don't usually create a class called Observer and another called Subject: you take some existing classes of your own and add in the associations and operations required by the pattern. Nor do you necessarily copy code from the patterns book: what's important is not how each operation is coded, but what it achieves – its postcondition. What we need is a way of defining 'macros' that can be combined with other chunks of model in a well defined way. So for example, if our Customer- Vendor definitions are a framework (in its own package), we could apply them to specific types:

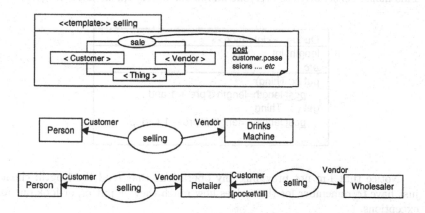

The application of the template establishes a relationship by adding the appropriate attributes and associations and actions to the target type definitions. A tool should be able to 'unfold' the resulting model on demand, but preferably normally show the model as drawn.

Precision – UML lacks a semantics of model template composition. There is one described in the Catalysis book ([1] chapter 9). Particular issues include how to combine postconditions – not the same as subtyping.

Tools – While some of the tools on the market have a way of applying a pattern, it's a one-off operation: you lose any trace of the original intention of the designer. A tool should preserve and present the original framework application, showing the 'unfolded' resulting model only if the designer asks for it.

The same mechanism is useful when we have several views of a domain or a set of requirements: different interested parties have their own ideas about what it should do, and each applies their own set of constraints on the final design. Combining views together is therefore a process of adding constraints.

Refinement and Conformance

Object and component design are all about encapsulation. We therefore like to separate the specification of an interface from the definition of the class that implements it – Java, IDL, and Eiffel provide explicitly for this. Then there may be many implementors of an interface.

Interface definitions as compilers understand them aren't sufficient for us as designers: a list of operation signatures doesn't say anything about what the operations are supposed to achieve. We can use postconditions to define what's expected of an operation; if we don't like formal language, we can write them in English or whatever, or draw illustrative snapshots as above.

The postconditions use a vocabulary of attributes and associations, which are not all directly visible in the implementation. For example, if I want to say of a Queue, that the put operation increases the length (more formally: put post: length==length@pre + 1 and .etc), then I have to have an attribute 'length'. This makes sense, because everyone knows that every queue has a length:

```
Queue   <<interface>>
length: int
etc
put (x: Thing)
     post length=length@pre + 1 and ....
get ( ) : Thing
     post length==length@pre -- 1 and ....
```

Notice that I'm not providing any operations that read the length: clients just have to remember for themselves how much they've put in, or look out for exceptions.

Now suppose someone comes to me with a class that implements Queue as a linked list. I want to check to see if their put increases the length as specified. The first obstacle is that they don't have a variable called 'length': it's just a chain of nodes. But they can write a function that 'retrieves' my length by counting up the nodes. Then we can test to see what happens to the length. Such a function is just for the purposes of verification against my spec, and it doesn't need to perform well, and might be omitted in production versions.

Precision – UML doesn't currently have a clear notion of what an attribute means. In this case, we take it to mean something that doesn't have to appear in the code, but can be written as a read-only function for QA purposes. (More in [1], chapter 213.)

The retrievals are part of the refinement relation – again, this is not clear in UML as it stands.

Tools – Refinement is about traceability. Any change to a postcondition mentioning 'length' affects the parts of an implementation that implement it – which are traceable through being mentioned in the retrieve function. I want tools to help me find out how changes propagate through my design, and refinements are part of that.

Testing

In case postconditions seem a little too abstract for everyday use, let me point out that they are equivalent to test harnesses: if your design meets the postcondition, you have done it properly. The advantages of pre/postconditions and invariants are: that they can be much more succinct than the program code; and they are general across all the different implementations. The programming language Eiffel includes executable invariants, pre and postconditions used for testing. All modern RAD approaches advocate writing the test material before completing the design.

Pre/postconditions are generally written in OCL within a UML model; but they can just as well be written in Java. The benefit of OCL is that it is more succinct: for example, you can write conditions across all members of a set, instead of having to code a loop.

Precision – we want a clear definition of what program code does and does not conform to a given UML specification.

Model refinement and systems integration

The model written for a high-level specification is usually simpler than the class structure in the program code: the former is intended to explain the externally-visible behaviour of a system or component; the latter is supposed to make it efficient, flexible, and built from reusable parts. While we would like to keep the two models similar for traceability, we also need to be able to map from one to the other.

One reason for the model/implementation difference is that the system has been constructed from disparate parts. While our sales business has quite a simple model, the sales support system may be made up of an accounts subsystem, a deliveries system, and so on, each of which has its own partial model.

Precision – UML needs a clear definition of what it means for a model to be implemented by several partial ones, especially where they overlap. (See [1] ch.10.11).

Biography

Alan Cameron Wills is a consultant in object and component based development, working with a wide variety of clients in Europe and America since 1990, in fields as diverse as finance, telecoms, and manufacturing control. He gained a PhD in the application of formal methods to object oriented programming in 1991. With Desmond D'Souza, he is author of the Catalysis method, detailed in the book "Objects, Components and Frameworks in UML" (Addison-Wesley, 1999). Alan is Principal Consultant with TriReme International Ltd, based in the UK. Contact details: alan@trireme.com. Tel: +44 161 225 3240. http://www.trireme.com/

References

1. Desmond D'Souza and Alan Cameron Wills. Objects, Components and Frameworks in UML: the Catalysis approach , Addison Wesley, 1998.

Author Index

Abdurazik, Aynur, 416
Abi-Antoun, Marwan, 17
Andrade, Luís Filipe, 566
Atkinson, Colin, 49, 675

Back, Ralph-Johan, 518
Barbier, Franck, 550
Baumeister, Hubert, 614
Belaunde, Mariano, 188
Berner, Stefan, 249
Bidoit, Michel, 399
Birchenough, Alan, 265
Boger, Marko, 204
Booch, Grady, 1
Boulestin, Michel, 446
Bunse, Christian, 675

Carrington, David, 83
Champeau, Joel, 446
Chonoles, Michael Jesse, 131
Clark, Tony, 503
Cook, Steve, 131, 372, 706

Demeyer, Serge, 630
Demuth, Birgit, 598
Dhaussy, Philippe, 446
Diethers, Karsten, 645
D'Souza, Desmond, 131, 265
Ducasse, Stéphane, 630
Dykman, Nathan, 236

Egyed, Alexander, 2
Ek, Anders, 446
Engels, Gregor, 473
Evans, Andy, 140, 706

Fiadeiro, José Luiz, 566
Firesmith, Don, 49
Firley, Thomas, 645
Franch, Xavier, 292
Fridella, Stephen, 691

Gehrke, Thomas, 645
Giese, Holger, 534
Glinz, Martin, 249
Gogolla, Martin, 156, 489

Goltz, Ursula, 645
Graf, Jörg, 534
Griss, Martin, 236

Harel, David, 324
Haugen, Øystein, 446
Henderson-Sellers, Brian, 49, 550
Hennicker, Rolf, 399
Herzberg, Dominikus, 330
Hilliard, Rich, 32
Howse, John, 384
Hruby, Pavel, 308
Hücking, Roland, 473
Huhn, Michaela, 645
Hussmann, Heinrich, 598

Iyengar, Sridhar, 131

Joos, Stefan, 249

Kabous, Laila, 339
Kähkipuro, Pekka, 356
Kent, Stuart, 140, 384
Kessler, Robert, 236
Kim, Soon-Kyeong, 83
Kleppe, Anneke, 372
Knapp, Alexander, 116
Kobryn, Chris, 131
Koch, Nora, 614
Koskimies, Kai, 172

Leblanc, Philippe, 446
Li, Xuandong, 661
Lilius, Johan, 430, 661

Mandel, Luis, 614
Matthes, Florian, 204
Medvidovic, Nenad, 2, 17
Mellor, Steve, 706
Miller, Joaquin, 584
Mitchell, Richard, 372
Møller-Pedersen, Birger, 446

Nebel, Wolfgang, 339

Offutt, Jeff, 416

Ostroff, Jonathan S., 67
Övergaard, Gunnar, 99

Paige, Richard F., 67
Petre, Luigia, 518
Porres Paltor, Iván, 430, 518

Rácz, Ferenc Dósa, 172
Radfelder, Oliver, 489
Ramackers, Guus, 131
Ribó, Josep M., 292
Richters, Mark, 156, 489

Sane, Aamod, 265
Sauer, Stefan, 473
Selic, Bran, 446
Sendall, Shane, 278
Soundarajan, Neelam, 691

Sourrouille, Jean Louis, 457
Strohmeier, Alfred, 278
Suzuki, Junichi, 220

Tichelaar, Sander, 630
Tort, Françoise, 399

Wagner, Annika, 473
Warmer, Jos, 372, 706
Weigert, Thomas, 446
Wienberg, Axel, 204
Wills, Alan, 372, 706
Wirfs-Brock, Rebecca, 584
Wirsing, Martin, 399
Wirtz, Guido, 534

Yamamoto, Yoshikazu, 220

Lecture Notes in Artificial Intelligence (LNAI)

Vol. 1574: N. Zhong, L. Zhou (Eds.), Methodologies for Knowledge Discovery and Data Mining. Proceedings, 1999. XV, 533 pages. 1999.

Vol. 1582: A. Lecomte, F. Lamarche, G. Perrier (Eds.), Logical Aspects of Computational Linguistics. Proceedings, 1997. XI, 251 pages. 1999.

Vol. 1585: B. McKay, X. Yao, C.S. Newton, J.-H. Kim, T. Furuhashi (Eds.), Simulated Evolution and Learning. Proceedings, 1998. XIII, 472 pages. 1999.

Vol. 1599: T. Ishida (Ed.), Multiagent Platforms. Proceedings, 1998. VIII, 187 pages. 1999.

Vol. 1600: M. J. Wooldridge, M. Veloso (Eds.), Artificial Intelligence Today. VIII, 489 pages. 1999.

Vol. 1604: M. Asada, H. Kitano (Eds.), RoboCup-98: Robot Soccer World Cup II. XI, 509 pages. 1999.

Vol. 1609: Z. W. Ras, A. Skowron (Eds.), Foundations of Intelligent Systems. Proceedings, 1999. XII, 676 pages. 1999.

Vol. 1611: I. Imam, Y. Kodratoff, A. El-Dessouki, M. Ali (Eds.), Multiple Approaches to Intelligent Systems. Proceedings, 1999. XIX, 899 pages. 1999.

Vol. 1612: R. Bergmann, S. Breen, M. Göker, M. Manago, S. Wess, Developing Industrial Case-Based Reasoning Applications. XX, 188 pages. 1999.

Vol. 1617: N.V. Murray (Ed.), Automated Reasoning with Analytic Tableaux and Related Methods. Proceedings, 1999. X, 325 pages. 1999.

Vol. 1620: W. Horn, Y. Shahar, G. Lindberg, S. Andreassen, J. Wyatt (Eds.), Artificial Intelligence in Medicine. Proceedings, 1999. XIII, 454 pages. 1999.

Vol. 1621: D. Fensel, R. Studer (Eds.), Knowledge Acquisition Modeling and Management. Proceedings, 1999. XI, 404 pages. 1999.

Vol. 1623: T. Reinartz, Focusing Solutions for Data Mining. XV, 309 pages. 1999.

Vol. 1630: M. M. Huntbach, G. A. Ringwood, Agent-Oriented Programming. XIV, 386 pages. 1999.

Vol. 1632: H. Ganzinger (Ed.), Automated Deduction – CADE-16. Proceedings, 1999. XIV, 429 pages. 1999.

Vol. 1634: S. Džeroski, P. Flach (Eds.), Inductive Logic Programming. Proceedings, 1999. VIII, 303 pages. 1999.

Vol. 1637: J.P. Walser, Integer Optimization by Local Search. XIX, 137 pages. 1999.

Vol. 1638: A. Hunter, S. Parsons (Eds.), Symbolic and Quantitative Approaches to Reasoning and Uncertainty. Proceedings, 1999. IX, 397 pages. 1999.

Vol. 1640: W. Tepfenhart, W. Cyre (Eds.), Conceptual Structures: Standards and Practices. Proceedings, 1999. XII, 515 pages. 1999.

Vol. 1647: F.J. Garijo, M. Boman (Eds.), Multi-Agent System Engineering. Proceedings, 1999. X, 233 pages. 1999.

Vol. 1650: K.-D. Althoff, R. Bergmann, L.K. Branting (Eds.), Case-Based Reasoning Research and Development. Proceedings, 1999. XII, 598 pages. 1999.

Vol. 1652: M. Klusch, O.M. Shehory, G. Weiss (Eds.), Cooperative Information Agents III. Proceedings, 1999. XI, 404 pages. 1999.

Vol. 1669: X.-S. Gao, D. Wang, L. Yang (Eds.), Automated Deduction in Geometry. Proceedings, 1998. VII, 287 pages. 1999.

Vol. 1674: D. Floreano, J.-D. Nicoud, F. Mondada (Eds.), Advances in Artificial Life. Proceedings, 1999. XVI, 737 pages. 1999.

Vol. 1688: P. Bouquet, L. Serafini, P. Brézillon, M. Benerecetti, F. Castellani (Eds.), Modeling and Using Context. Proceedings, 1999. XII, 528 pages. 1999.

Vol. 1692: V. Matoušek, P. Mautner, J. Ocelíková, P. Sojka (Eds.), Text, Speech, and Dialogue. Proceedings, 1999. XI, 396 pages. 1999.

Vol. 1695: P. Barahona, J.J. Alferes (Eds.), Progress in Artificial Intelligence. Proceedings, 1999. XI, 385 pages. 1999.

Vol. 1699: S. Albayrak (Ed.), Intelligent Agents for Telecommunication Applications. Proceedings, 1999. IX, 191 pages. 1999.

Vol. 1701: W. Burgard, T. Christaller, A.B. Cremers (Eds.), KI-99: Advances in Artificial Intelligence. Proceedings, 1999. XI, 311 pages. 1999.

Vol. 1704: Jan M. Żytkow, J. Rauch (Eds.), Principles of Data Mining and Knowledge Discovery. Proceedings, 1999. XIV, 593 pages. 1999.

Vol. 1705: H. Ganzinger, D. McAllester, A. Voronkov (Eds.), Logic for Programming and Automated Reasoning. Proceedings, 1999. XII, 397 pages. 1999.

Vol. 1706: J. Hatcliff, T. Æ. Mogensen, P. Thiemann (Eds.), Lectures on Partial Evaluation. Proceedings, 1998. IX, 433 pages. 1999.

Vol. 1711: N. Zhong, A. Skowron, S. Ohsuga (Eds.), New Directions in Rough Sets, Data Mining, and Granular-Soft Computing. Proceedings, 1999. XIV, 558 pages. 1999.

Vol. 1712: H. Boley, A Tight, Practical Integration of Relations and Functions. XI, 169 pages. 1999.

Vol. 1714: M.T. Pazienza (Eds.), Information Extraction. IX, 165 pages. 1999.

Vol. 1715: P. Perner, M. Petrou (Eds.), Machine Learning and Data Mining in Pattern Recognition. Proceedings, 1999. VIII, 217 pages. 1999.

Vol. 1721: S. Arikawa, K. Furukawa (Eds.), Discovery Science. Proceedings, 1999. XI, 374 pages. 1999.

Lecture Notes in Computer Science

Vol. 1692: V. Matoušek, P. Mautner, J. Ocelíková, P. Sojka (Eds.), Text, Speech and Dialogue. Proceedings, 1999. XI, 396 pages. 1999. (Subseries LNAI).

Vol. 1693: P. Jayanti (Ed.), Distributed Computing. Proceedings, 1999. X, 357 pages. 1999.

Vol. 1694: A. Cortesi, G. Filé (Eds.), Static Analysis. Proceedings, 1999. VIII, 357 pages. 1999.

Vol. 1695: P. Barahona, J.J. Alferes (Eds.), Progress in Artificial Intelligence. Proceedings, 1999. XI, 385 pages. 1999. (Subseries LNAI).

Vol. 1696: S. Abiteboul, A.-M. Vercoustre (Eds.), Research and Advanced Technology for Digital Libraries. Proceedings, 1999. XII, 497 pages. 1999.

Vol. 1697: J. Dongarra, E. Luque, T. Margalef (Eds.), Recent Advances in Parallel Virtual Machine and Message Passing Interface. Proceedings, 1999. XVII, 551 pages. 1999.

Vol. 1698: M. Felici, K. Kanoun, A. Pasquini (Eds.), Computer Safety, Reliability and Security. Proceedings, 1999. XVIII, 482 pages. 1999.

Vol. 1699: S. Albayrak (Ed.), Intelligent Agents for Telecommunication Applications. Proceedings, 1999. IX, 191 pages. 1999. (Subseries LNAI).

Vol. 1700: R. Stadler, B. Stiller (Eds.), Active Technologies for Network and Service Management. Proceedings, 1999. XII, 299 pages. 1999.

Vol. 1701: W. Burgard, T. Christaller, A.B. Cremers (Eds.), KI-99: Advances in Artificial Intelligence. Proceedings, 1999. XI, 311 pages. 1999. (Subseries LNAI).

Vol. 1702: G. Nadathur (Ed.), Principles and Practice of Declarative Programming. Proceedings, 1999. X, 434 pages. 1999.

Vol. 1703: L. Pierre, T. Kropf (Eds.), Correct Hardware Design and Verification Methods. Proceedings, 1999. XI, 366 pages. 1999.

Vol. 1704: Jan M. Żytkow, J. Rauch (Eds.), Principles of Data Mining and Knowledge Discovery. Proceedings, 1999. XIV, 593 pages. 1999. (Subseries LNAI).

Vol. 1705: H. Ganzinger, D. McAllester, A. Voronkov (Eds.), Logic for Programming and Automated Reasoning. Proceedings, 1999. XII, 397 pages. 1999. (Subseries LNAI).

Vol. 1706: J. Hatcliff, T. Æ. Mogensen, P. Thiemann (Eds.), Lectures on Partial Evaluation. Proceedings, 1998. IX, 433 pages. 1999. (Subseries LNAI).

Vol. 1707: H.-W. Gellersen (Ed.), Handheld and Ubiquitous Computing. Proceedings, 1999. XII, 390 pages. 1999.

Vol. 1708: J.M. Wing, J. Woodcock, J. Davies (Eds.), FM'99 – Formal Methods. Proceedings Vol. I, 1999. XVIII, 937 pages. 1999.

Vol. 1709: J.M. Wing, J. Woodcock, J. Davies (Eds.), FM'99 – Formal Methods. Proceedings Vol. II, 1999. XVIII, 937 pages. 1999.

Vol. 1710: E.-R. Olderog, B. Steffen (Eds.), Correct System Design. XIV, 417 pages. 1999.

Vol. 1711: N. Zhong, A. Skowron, S. Ohsuga (Eds.), New Directions in Rough Sets, Data Mining, and Granular-Soft Computing. Proceedings, 1999. XIV, 558 pages. 1999. (Subseries LNAI).

Vol. 1712: H. Boley, A Tight, Practical Integration of Relations and Functions. XI, 169 pages. 1999. (Subseries LNAI).

Vol. 1713: J. Jaffar (Ed.), Principles and Practice of Constraint Programming – CP'99. Proceedings, 1999. XII, 493 pages. 1999.

Vol. 1714: M.T. Pazienza (Eds.), Information Extraction. IX, 165 pages. 1999. (Subseries LNAI).

Vol. 1715: P. Perner, M. Petrou (Eds.), Machine Learning and Data Mining in Pattern Recognition. Proceedings, 1999. VIII, 217 pages. 1999. (Subseries LNAI).

Vol. 1716: K.Y. Lam, E. Okamoto, C. Xing (Eds.), Advances in Cryptology – ASIACRYPT'99. Proceedings, 1999. XI, 414 pages. 1999.

Vol. 1717: Ç. K. Koç, C. Paar (Eds.), Cryptographic Hardware and Embedded Systems. Proceedings, 1999. XI, 353 pages. 1999.

Vol. 1718: M. Diaz, P. Owezarski, P. Sénac (Eds.), Interactive Distributed Multimedia Systems and Telecommunication Services. Proceedings, 1999. XI, 386 pages. 1999.

Vol. 1721: S. Arikawa, K. Furukawa (Eds.), Discovery Science. Proceedings, 1999. XI, 374 pages. 1999. (Subseries LNAI).

Vol. 1722: A. Middeldorp, T. Sato (Eds.), Functional and Logic Programming. Proceedings, 1999. X, 369 pages. 1999.

Vol. 1723: R. France, B. Rumpe (Eds.), «UML»'99 – The Unified Modeling Language. XVII, 724 pages. 1999.

Vol. 1726: V. Varadharajan, Y. Mu (Eds.), Information and Communication Security. Proceedings, 1999. XI, 325 pages. 1999.

Vol. 1727: P.P. Chen, D.W. Embley, J. Kouloumdjian, S.W. Liddle, J.F. Roddick (Eds.), Advances in Conceptual Modeling. Proceedings, 1999. XI, 389 pages. 1999.

Vol. 1728: J. Akoka, M. Bouzeghoub, I. Comyn-Wattiau, E. Métais (Eds.), Conceptual Modeling – ER '99. Proceedings, 1999. XIV, 540 pages. 1999.

Vol. 1729: M. Mambo, Y. Zheng (Eds.), Information Security. Proceedings, 1999. IX, 277 pages. 1999.

Vol. 1734: H. Hellwagner, A. Reinefeld (Eds.), Scalable Coherent Interface – SCI. XXI, 490 pages. 1999.